Study Guide to Accompany

Buckwalter/Riban

# College Physics

Study Guide to Accompany
Buckwalter/Riban

# College Physics

MARLLIN L. SIMON

Auburn University
Auburn, Alabama

**McGraw-Hill Book Company**
New York St. Louis San Francisco Auckland Bogotá Hamburg
London Madrid Mexico Milan Montreal New Delhi
Panama Paris São Paulo Singapore Sydney Tokyo Toronto

STUDY GUIDE TO ACCOMPANY BUCKWALTER-RIBAN:
**COLLEGE PHYSICS**



3 4 5 6 7 8 9 0 EDWEDW 8 9 4 3 2 1 0 9

ISBN 0-07-057532-0

The editors were Irene Nunes, Karen Misler, and Steven Tenney; the production supervisor was Denise L. Puryear. Edwards Brothers, Inc. was printer and binder.

To Jonathan T. Simon, who has not read this study guide but who understands in his own twelve-year-old way why it is important.

I am most grateful to Sandra Simmons and Cheryl Jackson of Auburn University, for the word processing in the manuscript; to John Chen, architecture student at Auburn University, for drawing the figures; to Jeffery Burdett, physics student at Auburn University, for layout work and assistance with the appendix; to Dana Klinck, Hillsborough Community College, Tampa, Florida for reviewing the entire manuscript and making numerous helpful suggestions; and to the editoral staff at McGraw-Hill Book Company, Doug Burke and especially Irene Nunes, for their assistance in bringing the manuscript into this final form.

Any errors which remain in the study guide are, of course, fully my responsibility, and I welcome corrections and comments to improve future editions.

## CONTENTS

Study Guide to Accompany
Buckwalter/Riban

# College Physics

# 1 Preliminaries and Definitions

NEW IDEAS IN THIS CHAPTER

| Concepts and equations introduced | Text Section | Study Guide Page |
|---|---|---|
| Physical quantities and their units | 1.3,4,5 | 1 |
| Unit conversion | 1.6 | 2 |
| Significant figures | 1.7 | 6 |
| Scientific notation and powers of 10 | 1.8 | 7 |

PRINCIPAL CONCEPTS AND EQUATIONS

1. Physical Quantities and Their Units (Sections 1.3, 1.4, 1.5)

Review: The international system of units, abbreviated SI, has seven fundamental quantities, which are defined in terms of basic phenomena rather than by equation. In the study of mechanics, the fundamental quantities needed are length, mass, and time (Table 1.1).

| Fundamental Physical Quantity | SI Basic Unit | SI Unit Abbreviation |
|---|---|---|
| Length | Meter | m |
| Mass | Kilogram | kg |
| Time | Second | s |

Table 1.1

Most of the quantities used in physics have units that are a combination of fundamental quantities. Such quantities are called derived quantities and are defined by an equation. For example, the derived quantity area has units of $m^2$ since the equation for area is $A = (\ell)(w)$ and both $\ell$ and $w$ have units of m.

Practice: Obtain the SI units for each of the following derived physical quantities.

| | |
|---|---|
| 1. Area: $A = (\ell)(w)$ | $(m)(m) = m^2$ |

| | |
|---|---|
| 2. Volume: $V = (\ell)(w)(h)$ | $(m)(m)(m) = m^3$ |
| 3. Mass density: $\rho = m/V$ | $kg/m^3$ |
| 4. Mass flow rate: $m/t$ | kg/s |
| 5. Volume flow rate: $V/t$ | $m^3/s$ |
| 6. Speed: $v = \Delta s/\Delta t$ | m/s |
| 7. Acceleration: $a = \Delta v/\Delta t$ | $(m/s)/s = m/s^2$ |
| 8. Force: $F = ma$ | $kg \cdot m/s^2$ |
| 9. Work: $W = Fd\cos\theta$ | $(kg \cdot m/s^2)(m) = kg \cdot m^2/s^2$ |
| 10. Pressure: $P = F/A$ | $(kg \cdot m/s^2)/m^2 = kg/s^2 \cdot m$ |

Related Text Problems: 1-1, 1-2, 1-3.

## 2. Unit Conversion (Section 1.6)

Review: The two most common systems of units are the international system and the British system. When dealing with units, you need to be able to convert not only from one system to the other but also from one unit to another within a system. In order to convert a quantity from one unit to another, you need the relationship between the units (called the conversion factor). In SI, the the conversion factor is easily obtained from the frequently used prefixes shown in Table 1-2.

Table 1.2

| Prefix | Abbreviation | Power of ten |
|---|---|---|
| micro | μ | $10^{-6}$ |
| milli | m | $10^{-3}$ |
| centi | c | $10^{-2}$ |
| kilo | k | $10^{3}$ |
| mega | M | $10^{6}$ |

Table 1.3 lists a number of typical conversions, the relationship between the units of interest and the conversion factor.

| Desired Conversion | Relationship | Conversion Factor |
|---|---|---|
| Kilometers to meters | 1 km = $10^3$ m | $\frac{10^3\ \text{m}}{1\ \text{km}} = 1$ |
| Centimeters to meters | 1 cm = $10^{-2}$ m | $\frac{10^{-2}\ \text{m}}{1\ \text{cm}} = 1$ |
| Millimeters to meters | 1 mm = $10^{-3}$ m | $\frac{10^{-3}\ \text{m}}{1\ \text{mm}} = 1$ |
| Kilograms to grams | 1 kg = $10^3$ g | $\frac{10^3\ \text{g}}{1\ \text{kg}} = 1$ |
| Microseconds to seconds | 1 μs = $10^{-6}$ s | $\frac{10^{-6}\ \text{s}}{1\ \mu\text{s}} = 1$ |

Table 1.3

Note: The principle involved in converting units is that you can multiply or divide any quantity by 1 without changing its value. In converting from kilometers to centimeters, for example, we simply multiply by 1 as many times as needed:

$$(50\ \text{km})\left(\frac{10^3\ \text{m}}{1\ \text{km}}\right)\left(\frac{1\ \text{cm}}{10^{-2}\ \text{m}}\right) = 50 \times 10^5\ \text{cm}$$

When dealing with units that are not based on powers of 10, you have to either memorize the conversion factors or obtain them from refernce material. Table 1.4 gives some of the conversion factors you will use frequently in solving the problems at the end of this chapter.

| Desired Conversion | Relationship | Conversion Factor |
|---|---|---|
| Hours to seconds | 1 h = 3600 | $\frac{1\ \text{h}}{3600\ \text{s}}$ |
| Inches to feet | 12 in = 1 ft | $\frac{12\ \text{in}}{1\ \text{ft}}$ |
| Feet to yards | 3 ft = 1 yd | $\frac{3\ \text{ft}}{1\ \text{yd}}$ |
| Feet to miles | 5280 ft = 1 mi | $\frac{5280\ \text{ft}}{1\ \text{mi}}$ |

Table 1.4

In order to convert from the British system to SI and vice versa, you will need to obtain conversion factors from reference material. Those needed for the problems in Chapter 1 are given in Table 1.5.

| Desired Conversion | Relationship | Conversion Factor |
|---|---|---|
| Meters to feet | 1 m = 3.28 ft | $\frac{3.28 \text{ ft}}{1 \text{ m}}$ |
| Meters to yards | 1 m = 1.09 yd | $\frac{1.09 \text{ yd}}{1 \text{ m}}$ |
| Meters to miles | 1610 m = 1 mi | $\frac{1 \text{ mi}}{1610 \text{ m}}$ |

Table 1.5

Conversion factors may be determined for derived physical quantities by combining the information found in tables 1.3, 1.4 and 1.5. For example we may convert from area in square meters to area in square inches as follows:

$$1 \text{ m}^2 = (1 \text{ m}^2)\left(\frac{3.28 \text{ ft}}{1 \text{ m}}\right)^2\left(\frac{12 \text{ in}}{1 \text{ ft}}\right)^2 = 1 \text{ m}^2\left(\frac{10.8 \text{ ft}^2}{1 \text{ m}^2}\right)\left(\frac{144 \text{ in}^2}{1 \text{ ft}^2}\right) = 1560 \text{ in}^2$$

From this relationship, we obtain the conversion factor 1560 $\text{in}^2/\text{m}^2$ = 1.

Knowing this, we may convert an area in square meters (say 5.00 $\text{m}^2$) to an area in square inches.

$$(5.00 \text{ m}^2)\left(\frac{1560 \text{ in}^2}{\text{m}^2}\right) = 7800 \text{ in}^2$$

Note: When converting units, be careful to insert the conversion factors so that all unwanted units cancel.

Practice: Using only the information in Tables 1.3, 1.4, and 1.5, determine the conversion factor for the following conversions:

| | |
|---|---|
| 1. Kilometers to inches | The relationship is<br>$1 \text{ km} = (1 \text{ km})\left(\frac{10^3 \text{ m}}{1 \text{ km}}\right)\left(\frac{3.28 \text{ ft}}{1 \text{ m}}\right)\left(\frac{12 \text{ in}}{1 \text{ ft}}\right)$<br>$= 39{,}400 \text{ in}$<br>The conversion factor is 39,400 in/1 km. |

| | |
|---|---|
| 2. Cubic inches to cubic meters | The relationship is<br>$1\ in^3 = (1\ in^3)(\frac{1\ ft}{12\ in})^3(\frac{1\ m}{3.28\ ft})^3$<br>$= 0.0000164\ in^3$<br>The conversion factor is $0.0000164\ m^3/1\ in^3$. |
| 3. Miles per hour to meters per second | The relationship is<br>$1\ \frac{mi}{h} = (1\ \frac{mi}{h})(\frac{1\ h}{3600\ s})(\frac{1610\ m}{1\ mi}) = 0.444\ m/s$<br>The conversion factor is (0.444 m/s)/(mi/h). |

Using the conversion factors just obtained, make the following conversions:

| | |
|---|---|
| 4. $10^6$ in to kilometers | $(10^6\ in)(\frac{1\ km}{39{,}400\ in}) = 25.4\ km$ |
| 5. 500 $in^3$ to cubic meters | $(500\ in^3)(\frac{0.0000164\ m^3}{1\ in^3}) = 0.00820\ m^3$ |
| 6. 55 mi/h to meters per second | $(55\ mi/h)(\frac{0.444\ m/s}{1\ mi/h}) = 24.4\ m/s$ |

If 1 zit = 2 zots; 3 zubs = 4 zuds, and 1 zud = 1/5 zot, determine the following:

| | |
|---|---|
| 7. 1 zot = ? zit | Given: 2 zots = 1 zit<br>Then: 1 zot = 1/2 zit |
| 8. 2 zuds = ? zubs | Given: 4 zuds = 3 zubs or 1 zud = 3/4 zubs<br>Then: 2 zuds = $(2\ zuds)(\frac{3/4\ zub}{zud}) = \frac{3}{2}$ zub |

| | |
|---|---|
| 9. 1 zit = ? zuds | Given: 1 zit = 2 zots<br>1 zud = 1/5 zot or 1 zot = 5 zuds<br>Then: $1 \text{ zit} = (1 \text{ zit})(\frac{2 \text{ zots}}{1 \text{ zit}})(\frac{5 \text{ zuds}}{1 \text{ zot}}) = 10 \text{ zuds}$ |
| 10. 1 zit = ? zus | Given: 1 zit = 2 zots<br>1 zud = 1/5 zot or 1 zot = 5 zud<br>3 zus = 4 zud or 1 zud = 3/4 zubs<br>Then<br>$1 \text{ zit} = (1 \text{ zit})(\frac{2 \text{ zots}}{1 \text{ zit}})(\frac{5 \text{ zuds}}{1 \text{ zot}})(\frac{3/4 \text{ zubs}}{1 \text{ zud}}) = 7.5 \text{ zubs}$ |

Related Text Problems: 1-4 through 1-17.

## 3. Significant Figures (Section 1.7)

Review: In order to work physics problems efficiently, you need to know the following five things about significant figures.

1. How to determine which digits in a number are significant. This is done by scanning the number from left to right. The first nonzero digit and all following digits (including zeros) are significant.

2. How to round off to the correct number of significant figures. If a series of values are used in a calculation, the value with the least number of significant figures determines the number of significant figures in the answer. To round off a number, look at the digit to the right of the last significant figure (i.e., the first unwanted digit). If it is less than 5, retain the previous digit without change and drop all unwanted digits. If it is 5 or greater, increase the previous digit by one and drop all unwanted digits.

3. How to add and subtract using significant figures. In addition and subtraction, drop every digit in the result which falls in a column containing a nonsignificant figure. It is usually convenient to round off the numbers to the correct number of significant figures before adding or subtracting.

4. How to multiply and divide using significant figures. The result of multiplying or dividing should have no more significant figures than the least significant of the original numbers. When doing a series of calculations keep one extra digit so as not to accumulate errors in rounding off intermediate results. Then round off to the final result to the proper number of signficant figures.

5. How to treat a pure number. Pure numbers are considered to have an unlimited number of significant figures.

Throughout your text and this study guide, it is generally assumed that the given data are precise enough to yield an answer having three significant figures. Thus, if we state that an object travels a distance of 5 m with an average speed of 2.5 m/s, it is to be understood that the distance covered is 5.00 m and the average speed is 2.50 m/s.

<u>Practice</u>:

| | |
|---|---|
| 1. Determine the number of significant figures in (a) 205; (b) 452.0; (c) 116.72; (d) 0.03; (e) 0.043; (f) 1.030 | (a) 3; (b) 4; (c) 5; (d) 1; (e) 2; (f) 4 |
| 2. Round off 147.6082 to (a) 6, (b) 5, (c) 4, and (d) 3 significant figures | (a) 147.608; (b) 147.61; (c) 147.6; (d) 148 |
| 3. Perform the indicated operation to the correct number of significant figures | |
| (a) $\begin{array}{r} 4.64 \\ +\ 7.261 \\ \hline \end{array}$ (b) $\begin{array}{r} 5.20 \\ +\ 0.327 \\ \hline \end{array}$ | (a) $\begin{array}{r} 4.64 \\ +\ 7.26 \\ \hline 11.90 \end{array}$ (b) $\begin{array}{r} 5.20 \\ +\ 0.33 \\ \hline 5.53 \end{array}$ |
| (c) $\begin{array}{r} 19.2 \\ +\ 0.03 \\ \hline \end{array}$ (d) $\begin{array}{r} 176.4 \\ +\ 15 \\ \hline \end{array}$ | (c) $\begin{array}{r} 19.2 \\ +\ 0.0 \\ \hline 19.2 \end{array}$ (d) $\begin{array}{r} 176 \\ 15 \\ \hline 191 \end{array}$ |
| (e) $\begin{array}{r} 27.43 \\ -\ 19.027 \\ \hline \end{array}$ (f) $\begin{array}{r} 152.3 \\ -\ 140. \\ \hline \end{array}$ | (e) $\begin{array}{r} 27.43 \\ -\ 19.03 \\ \hline 8.40 \end{array}$ (f) $\begin{array}{r} 152 \\ -\ 140 \\ \hline 12 \end{array}$ |
| (g) 12.3 x 4.0 | (g) 12.3 x 4.0 = 49 |
| (h) 17.36 x 1.27 | (h) 17.36 x 1.27 = 22.0 |
| (i) 42.73 ÷ 0.250 | (i) 42.74/0.250 = 171 |

<u>Related Text Problems</u>: 1-18 through 1-28.

## 4. Scientific Notation and Powers of 10 (Section 1.8)

Review: When working with very large or very small numbers it is convenient to express them in scientific notation, that is, a number between 1 and 9.999....multiplied by a power of 10:

(a) 10000 ..... 00. = $1.00 \times 10^{+b}$ and 0.000....01 = $1.00 \times 10^{-a}$

b digits a digits

The rules for performing mathematical operations using powers of 10 are as follows:

$$10^a \times 10^b = 10^{a+b} \; ; \; 10^a \div 10^b = 10^{a-b} \; ; \; (10^a)^P = 10^{(a)(P)}$$

Practice:

| | |
|---|---|
| 1. Express in scientific notation to the correct number of significant figures:<br><br>(a) 0.00720; (b) 2703; (c) 0.083 | (a) $7.20 \times 10^{-3}$<br>(b) $2.703 \times 10^{3}$<br>(c) $8.3 \times 10^{-2}$ |
| 2. Express in scientific notation to three significant figures:<br><br>(a) 186,400; (b) 0.003146<br>(c) 5272; (d) 342.67 | (a) $1.86 \times 10^{5}$<br>(b) $3.15 \times 10^{-3}$<br>(c) $5.27 \times 10^{3}$<br>(d) $3.43 \times 10^{2}$ |
| 3. Perform the indicated operation and express all answers in scientific notation to three significant figures:<br><br>(a) $10^4 \div 10^{-2}$<br>(b) $3746 \div 0.02501$<br>(c) (0.00732)(5802.6) | (a) $1.00 \times 10^{6}$<br>(b) $\frac{3.75 \times 10^{3}}{2.50 \times 10^{-2}} = 1.50 \times 10^{5}$<br>(c) $(7.32 \times 10^{-3})(5.80 \times 10^{3})$<br>$= 4.25 \times 10^{1}$ |

Related Text Problems: 1-24 through 1-28.

## PRACTICE TEST

Take and grade this practice test. Doing so will allow you to determine any weak spots in your understanding of the concepts taught in this chapter. The following section prescribes what you should study further to strengthen your understanding.

Table 1.6 gives information about several physical quantities.

| Quantity | Distance | Mass | Time | Speed | Acceleration |
|---|---|---|---|---|---|
| Symbol | S | m | t | v | a |
| SI Abbreviation | m | kg | s | m/s | $m/s^2$ |

Table 1.6

Obtain the derived SI units for the quantities defined by the following expressions:

_______ 1. Momentum: $p = mv$
_______ 2. Kineic energy: $KE = mv^2/2$
_______ 3. Distance: $S = at^2/2$
_______ 4. Force: $F = ma$

A car travels 900 km in 15 h and consumes 75.0 liter of gasoline that costs $24.00.

_______ 5. How far can this car travel on 5 liters of gas?
_______ 6. What would gasoline cost for a 2000 km trip?
_______ 7. What was the rate of consumption of gasoline?
_______ 8. What was the rate of change of position in meters per second?

A solid sphere has a mass of 2.107 kg and a radius of 0.500 m. Determine the following to the correct number of significant figures:

_______ 9. Surface area, $A = 4\pi r^2$
_______ 10. Volume, $V = 4\pi r^3/3$
_______ 11. Mass density, $\rho = m/V$

Make the following conversions:

_______ 12. 40 mi/h to meters per second
_______ 13. 200 $in^2$ to square centimeters
_______ 14. 10 m/s to miles per hour
_______ 15. 100 km to inches

Perform the indicated operations and express your answer in scientific notation to the correct number of significant figures:

_____ 16. 12,000 ÷ 0.0040
_____ 17. 110.02 x 0.025
_____ 18. 102.56 + 0.0382

(See Appendix I for answers.)

## PRINCIPAL CONCEPTS AND EQUATIONS PRESCRIPTION

Your score on the practice test is an excellent measure of your understanding of this chapter. You should now use the following chart to write your own prescription for curing any of your physics ills. Look down the leftmost column to the number of the question(s) you answered incorrectly, read across that row to see which section(s) of the study guide you should return to for further study, and then do the suggested text problems to gain additional experience in working with the particular concept.

| Practice Test Question | Concepts and Equations | Prescription | |
|---|---|---|---|
| | | Principal Concept | Text Problems |
| 1 | Fundamental and derived quantities | 1 | 1-1,2 |
| 2 | Fundamental and derived quantities | 1 | 1-2,3 |
| 3 | Fundamental and derived quantities | 1 | 1-1,3 |
| 4 | Fundamental and derived quantities | 1 | 1-1,2 |
| 5 | Units and conversions | 1,2 | 1-10,11 |
| 6 | Units and conversions | 1,2 | 1-11,14 |
| 7 | Units and conversions | 1,2 | 1-11,14 |
| 8 | Units and conversions | 1,2 | 1-4,5 |
| 9 | Significant figures | 3 | 1-18,23 |
| 10 | Significant figures | 3 | 1-17,19 |
| 11 | Significant figures | 3 | 1-15,22 |
| 12 | Unit conversion | 2 | 1-5,8 |
| 13 | Unit conversion | 2 | 1-4,7 |
| 14 | unit conversion | 2 | 1-8,10 |
| 15 | Unit conversion | 2 | 1-6,13 |
| 16 | Scientific notation and significant figures | 3,4 | 1-24,27 |
| 17 | Scientific notation and significant figures | 3,4 | 1-25,28 |
| 18 | Scientific notation and significant figures | 3,4 | 1-20,26 |

# 2 Linear Motion

## RECALL FROM PREVIOUS CHAPTERS

| Previously learned concepts and equations frequently used in this chapter | Text Section | Study Guide Page |
|---|---|---|
| Significant figures | 1.7 | 6 |
| Scientific notation and powers of 10 | 1.8 | 8 |

## NEW IDEAS IN THIS CHAPTER

| Concepts and equations introduced | Text Section | Study Guide Page |
|---|---|---|
| Distance and displacement | 2.2 | 11 |
| Average speed and average velocity | 2.3 | 15 |
| Average velocity and instantaneous velocity | 2.3 | 19 |
| Average acceleration | 2.4 | 21 |
| Equations for translational motion: | 2.4 | 22 |
| $v_{av} = \Delta s/\Delta t$  $v = v_o + at$ | | |
| $v_{av} = (v_o + v)/2$  $\Delta s = v_o t + at^2/2$ | | |
| $a_{av} = \Delta v/\Delta t$  $2a\Delta s = v^2 - v_o^2$ | | |

## PRINCIPAL CONCEPTS AND EQUATIONS

### 1. Difference Between Distance and Displacement (Section 2.2)

Review: Distance is the total length of travel without regard to direction or changes in direction. It is expressed as the number of units of length (for example, meters) you travel to get from one point to another. Displacement is the straight-line distance between two points. It must tell the number of units of length in the straight-line path as well as the direction of this path.

Figure 2.1 shows the path of a jogger. Notice that she has jogged $2.20 \times 10^3$ m but her displacement is $2.00 \times 10^2$ m north of her starting point.

Figure 1.

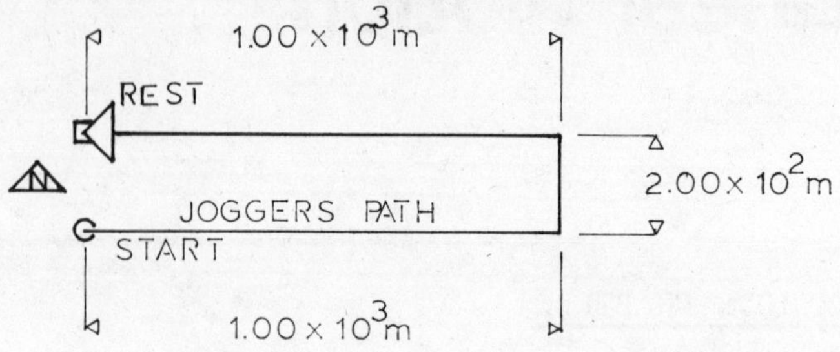


In order to determine distance and displacement, you must first establish a coordinate system. The location of the origin of this system is arbitrary, but the axes must be marked off in standard units of length. In fig. 2.1, the origin is the jogger's starting point and the axes run east-west and north-south and are marked off in meters. In your text and this study guide, the following notation is used:

- $s$ represents a displacement
- $\Delta s$ represents a change in displacement
- $|\Delta s|$ represents the magnitude of a change in displacement; distances are determined from the magnitude of the changes in displacement

Practice: An object is traveling along the linear track shown in Fig. 2.2. It travels from A to B to C, reverses direction, and travels back through B to A. An origin has been arbitrarily chosen and the x axis marked off in meters. In order to illustrate several ideas, two cases will be considered. The difference between the cases is the location of the origin. The time the object is at each position is as follows:

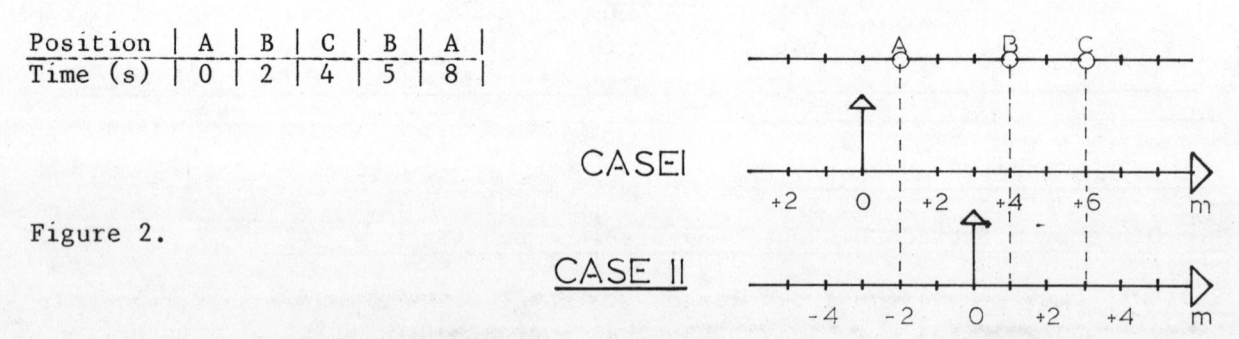

| Position | A | B | C | B | A |
|---|---|---|---|---|---|
| Time (s) | 0 | 2 | 4 | 5 | 8 |

Figure 2.

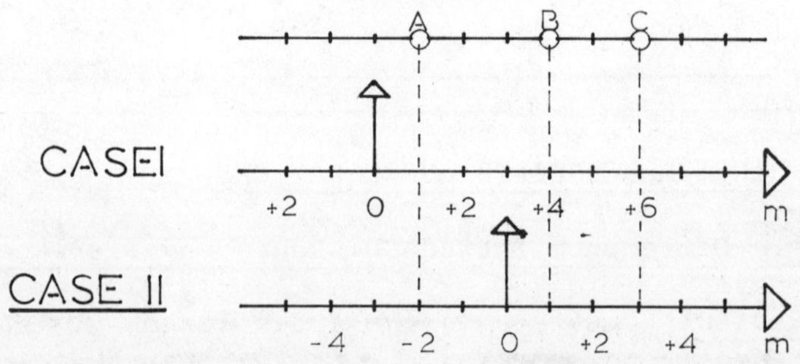


Determine the following (remember to cover the right-hand column while you are obtaining your answers).

| | |
|---|---|
| 1. The displacement of A, B, and C relative to the origin | I. $s_A = +1$ m, $s_B = +4$ m, $s_C = +6$ m<br>II. $s_A = -2$ m, $s_B = +1$ m, $s_C = +3$ m |

| | |
|---|---|
| 2. The change in displacement of the object during the first 2 s, as it travels from A to B | At $t = 0$ s, the displacement is $s_A$. At $t = 2$ s, the displacement is $s_B$. During this 2 s interval, the change in displacement is $\Delta s_{A \to B} = s_B - s_A$<br><br>I. $\Delta s_{A \to B} = +4 \text{ m} - (+1 \text{ m}) = +3 \text{ m}$<br>II. $\Delta s_{A \to B} = +1 \text{ m} - (-2 \text{ m}) = +3 \text{ m}$ |
| 3. The distance traveled by the object during the first 2 s | The distance traveled during the first 2 s is the magnitude of the change in displacement<br><br>I. $\lvert \Delta s_{A \to B} \rvert = 3 \text{ m}$<br>II. $\lvert \Delta s_{A \to B} \rvert = 3 \text{ m}$ |
| 4. The displacement of the object (relative to the origin) at $t = 5$ s | At $t = 5$ s, the object has returned to B, hence its displacement relative to the origin is<br><br>I. $s_B = +4 \text{ m}$<br>II. $s_B = +1 \text{ m}$ |
| 5. The change in displacement of the object during the first 5 s, as it travels from A to C and then back to B | At $t = 0$ s the displacement is $s_A$. After 5 s, the displacement is $s_B$. Hence the change in displacement during the first 5 s of travel is $\Delta s_{A \to B} = s_B - s_A$<br><br>I. $\Delta s_{A \to B} = +4 \text{ m} - (+1 \text{ m}) = +3 \text{ m}$<br>II. $\Delta s_{A \to B} = +1 \text{ m} - (-2 \text{ m}) = +3 \text{ m}$ |
| 6. The distance traveled by the object during the first 5 s | This distance is the sum of the magnitude of the change in displacement $\Delta s_{A \to C}$ and $\Delta s_{C \to B}$; hence<br><br>$\lvert \Delta s_{A \to C} \rvert + \lvert \Delta s_{C \to B} \rvert = \lvert s_C - s_A \rvert + \lvert s_B - s_C \rvert$<br><br>I. $\lvert +6 \text{ m} - (+1 \text{ m}) \rvert + \lvert +4 \text{ m} - (+6 \text{ m}) \rvert$<br>$5 \text{ m} + 2 \text{ m} = 7 \text{ m}$<br><br>II. $\lvert +3 \text{ m} - (-2 \text{ m}) \rvert + \lvert +1 \text{ m} - (+3 \text{ m}) \rvert$<br>$5 \text{ m} + 2 \text{ m} = 7 \text{ m}$ |

| | |
|---|---|
| 7. The change in displacement during the last 3 s | At the start of the time interval under consideration, the displacement is $s_B$. At the end of this time interval, the displacement is $s_A$. The change in displacement during this time interval is<br>$\Delta s_{B\to A} = s_A - s_B$<br><br>I. $\Delta s_{B\to A} = +1\ m - (+4\ m) = -3\ m$<br>II. $\Delta s_{B\to A} = -2\ m - (+1\ m) = -3\ m$<br><br>The object's displacement is changed 3 m to the left (note the minus sign) during the last 3 s of travel. |
| 8. The distance traveled during the last 3 s | The distance traveled is the magnitude of the change in displacement $\lvert\Delta s_{B\to A}\rvert$; hence<br><br>I. $\lvert\Delta s_{B\to A}\rvert = 3\ m$<br>II. $\lvert\Delta s_{B\to A}\rvert = 3\ m$ |

Note: Numerous other questions regarding displacement, change in displacement, and distance may be asked for this situation. Using this practice as a guide, you should determine some of these questions and their answers.

Note: All displacements include a direction (+ for displacement to the right and - for displacement to the left) and a magnitude (expressed in length units) that depends on the location of the origin. See steps 1 and 4 of the preceding.

Note: All changes in displacement include a direction (+ for a change to the right and -for a change to the left) and a magnitude (expressed in length units) that does not depend on the location of the origin. See steps 2, 5, and 7 of the preceding.

Note: The magnitude of the change in displacement is equal to the distance traveled if and only if the object does not change direction. All distances are positive. See steps 2, and 3, and steps 7, and 8 of the preceding.

Note: The magnitude of the change in displacement is not equal to the distance traveled if the object changes direction. See steps 5 and 6 of the preceding.

Note: Don't try to memorize all these notes. Understand the idea being expressed in the note and then practice the physics until you are comfortable with the idea.

## 2. Difference Between Average Speed and Average Velocity (Section 2.3)

Review: Average speed is the rate at which a body is changing its distance traveled:

$$\text{Average speed} = v_{av} = \frac{\text{total distance traveled}}{\text{total time of travel}} = \frac{d}{\Delta t}$$

where d represents the total distance traveled and is obtained from the magnitude of changes in displacement.

Average speed is always a positive number, and it gives no information about the moving object's direction. For example, we might say that the object is moving with an average speed of 10 m/s. This number gives no information about direction.

Average velocity is the rate at which a body is changing its displacement:

$$\text{Average velocity} = v_{av} = \frac{\text{change in displacement}}{\text{total time of travel}} = \frac{\Delta s}{\Delta t}$$

Since average velocity is defined in terms of a change in displacement, it must have a sign to indicate direction. This sign will always be the same as the sign of the change in displacement. If an object undergoes a change in displacement $\Delta s = +10.0$ m in a time of 2.00 s, the object's average velocity is

$$v_{av} = \frac{\Delta s}{\Delta t} = \frac{+10.0 \text{ m}}{2.00 \text{ s}} = +5.00 \text{ m/s}$$

Practice: We will determine average speeds and velocities for the situation shown in Fig. 2.2, using only the coordinate system for case I.

| Position | A | B | C | B | A |
|---|---|---|---|---|---|
| Time (s) | 0 | 2 | 4 | 5 | 8 |

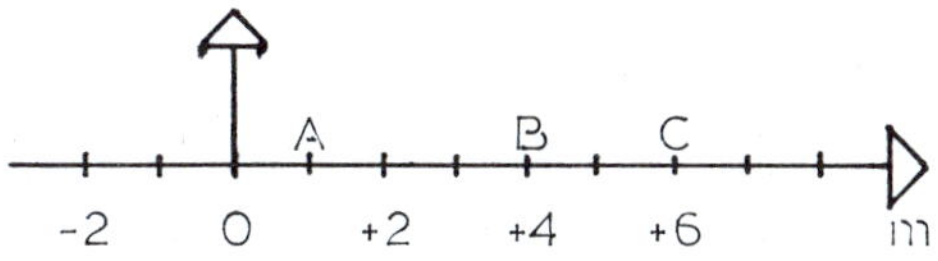


Figure 2.3

An object travels from A to B to C, reverses direction, and travels back through B to A. The table gives the time that the object is at each position.

Determine the following:

| | |
|---|---|
| 1. The average velocity of the object as it travels from A to B | The change in displacement is $\Delta s_{A\to B} = s_B - s_A = +4 \text{ m} - (+1 \text{ m}) = +3$ m. The time of travel is $\Delta t = 2$ s. The average velocity is $v_{avg} = \Delta s_{A\to B}/\Delta t = +1.50$ m/s. |

| | |
|---|---|
| 2. The average speed of the object as it travels from A to B | The total distance traveled is $d = \lvert\Delta s_{A\to B}\rvert = 3$ m. The time of travel is $\Delta t = 2$ s. The average speed is $v_{av} = d/\Delta t = 1.50$ m/s. |
| 3. The average velocity of the object during the first 5 s of travel | During the first 5 s, the object travels from A to C and then back to B. The change in displacement is $\Delta s_{A\to B} = s_B - s_A = +4\text{ m} - (+1\text{ m}) = +3$ m. The time of travel is $\Delta t = 5$ s. The average velocity is $v_{av} = \Delta s_{A\to B}/\Delta t = +0.600$ m/s. |
| 4. The average speed of the object during the first 5 s | The total distance traveled is $d = \lvert\Delta s_{A\to C}\rvert + \lvert\Delta s_{C\to B}\rvert = 5\text{ m} + 2\text{ m} = 7$ m. The time of travel is $\Delta t = 5$ s. The average speed is $v_{av} = d/\Delta t = 1.40$ m/s. |
| 5. The average velocity of the object for the entire trip | The change in displacement is $\Delta s_{A\to A} = 0.0$ m. The time of travel is $\Delta t = 8$ s. The average velocity is $v_{av} = \Delta s_{A\to A}/\Delta t = 0.0$ m/s. |
| 6. The average speed of the object for the entire trip | The total distance traveled is $d = \lvert\Delta s_{A\to C}\rvert + \lvert\Delta s_{C\to A}\rvert = 5\text{ m} + 5\text{ m} = 10$ m. The time of travel is $\Delta t = 8$ s. The average speed is $v_{av} = d/\Delta t = 1.25$ m/s. |

Example 2.1 A car travels around a circular track of circumference $6.00 \times 10^4$ m once every 90.0 s. What is (a) its average speed and (b) its average velocity for each lap?

Given: $d = 6.00 \times 10^4$ m = total distance traveled for each lap
$\Delta s = 0.0$ m = change in displacement for each lap
$\Delta t = 90.0$ s = time of travel for each lap

Determine: $v_{av\ speed}$ and $v_{av\ velocity}$ for each lap

Strategy: Knowing the total distance traveled and the time of travel, we can determine the average speed. Knowing the change in displacement and the time for that change, we can determine the average velocity.

Solution:

(a) $$v_{av\ speed} = \frac{d}{\Delta t} = \frac{6.00 \times 10^4 \text{ m}}{90.0 \text{ s}} = 6.67 \times 10^2 \text{ m/s}$$

(b) $$v_{av\ vel} = \frac{\Delta s}{\Delta t} = \frac{0.0 \text{ m}}{90 \text{ s}} = 0.0 \text{ m/s}$$

Example 2.2 A car travels at a constant speed of 20.0 m/s for $2.00 \times 10^3$ s and then suddenly changes to a speed of 30.0 m/s for $1.00 \times 10^3$ s. Determine the average speed for the entire trip.

Given: $v_1 = 20.0$ m/s, $t_1 = 2.00 \times 10^3$ s; $v_2 = 30.0$ m/s, $t_2 = 1.00 \times 10^3$ s

Determine: $v_{av\ speed}$ for the entire trip

Strategy: We can use $v_1$ and $t_1$ to determine $d_1$, the distance traveled in time $t_1$. In like manner, we can use $v_2$ and $t_2$ to determine $d_2$. We can then determine the total distance traveled, the total time of travel, and the average speed.

Solution: $d_1 = v_1 t_1 = (20.0 \text{ m/s})(2.00 \times 10^3 \text{ s}) = 4.00 \times 10^4 \text{ m}$

$d_2 = v_2 t_2 = (30.0 \text{ m/s})(1.00 \times 10^3 \text{ s}) = 3.00 \times 10^4 \text{ m}$

$d_{total} = d_1 + d_2 = 7.00 \times 10^4 \text{ m}$

$t_{total} = t_1 + t_2 = 3.00 \times 10^3 \text{ s}$

$$v_{av\ speed} = \frac{d_{total}}{t_{total}} = \frac{7.00 \times 10^4 \text{ m}}{3.00 \times 10^3 \text{ s}} = 23.3 \text{ m/s}$$

Note: You don't just average the speeds to get the average speed.

Example 2.3 During a 150-km race, the team for car A establishes the following:

Car A is averaging 190 km/h.
Car A can average 260 km/h.
Car B is averaging 200 km/h.
Car B is traveling at its top speed.
The race has been in progress 0.600 h.
Car A is 6.00 km behind car B.

What is the minimum amount team A should advise their driver to increase her speed in order to win the race by at least 2.00 km?

Given: $v_{av\ A} = 190$ km/h; $v_{av\ A\ max} = 260$ km/h; $v_{av\ B} = 200$ km/h; $t = 0.600$ h; $d_{total} = 150$ km; $d_B - d_A = 6.00$ km at $t = 0.600$ h; $d_A - d_B = 2$ km at end of race

Determine: The amount the driver of car A must increase her speed in order to win the race by 2.00 km.

Strategy: Using $v_{av\ B}$, determine how much longer ($t_{left}$) it will take car B to complete 148 km. This will also be the time left for car A to complete 150 km in order to win by 2.00 km. Using $d_{total}$, $v_{av\ A}$, and t, determine how far car A has to go to complete lthe race ($d_{A\ left}$). Knowing the distance and time left, we can determine what car A's average speed must be in order to win by 2.00 km and hence the increase in average speed.

Solution: Time for car B to complete 148 km is

$$t_B = \frac{148\ \text{km}}{200\ \text{km/h}} = 0.740\ \text{h}$$

Time left for car A to complete 150 km in order to win by 2.00 km is

$$t_{left} = 0.740\ \text{h} - 0.600\ \text{h} = 0.140\ \text{h}$$

The total distance traveled by car A at the time the decision about $\Delta v_{av}$ is being made is

$$d_A = 190\ \frac{\text{km}}{\text{h}}\ (0.600\ \text{h}) = 114\ \text{km}$$

The distance left for car A to travel at the time the decision about $\Delta v_{av}$ is being made is

$$d_{A\ left} = 150\ \text{km} - 114\ \text{km} = 36.0\ \text{km}$$

The average speed car A must maintain for the last 36.0 km of the race is

$$v_{av} = \frac{d_{A\ left}}{t_{left}} = \frac{36.0\ \text{km}}{0.140\ \text{h}} = 257\ \text{km/h}$$

The amount by which the driver of car A must increase her speed in order to win the race by 2.00 km is

$$\Delta v_{av} = v_{av\ final} - v_{av\ initial} = 257\ \frac{\text{km}}{\text{h}} - 190\ \frac{\text{km}}{\text{h}} = 67.0\ \text{km/h}$$

Note: Although while example problems 2.1, 2.2, and 2.3 deal with different situations, the principle involved is that of average speed.

Related Text Problems: 2-1 through 2-9.

## 3. Difference Between Average and Instantaneous Velocity (Section 2.3)

Review:

$$v_{av} = \frac{\text{change in displacement}}{\text{total time of travel}} = \frac{\Delta s}{\Delta t} \qquad (2.2)$$

The average velocity of an object over some time interval may be obtained from a plot of displacement vs time (Fig. 2.4). The slope of the straight line connecting two points on the displacement vs time plot is the average velocity over the associated time interval.

Figure 2.4

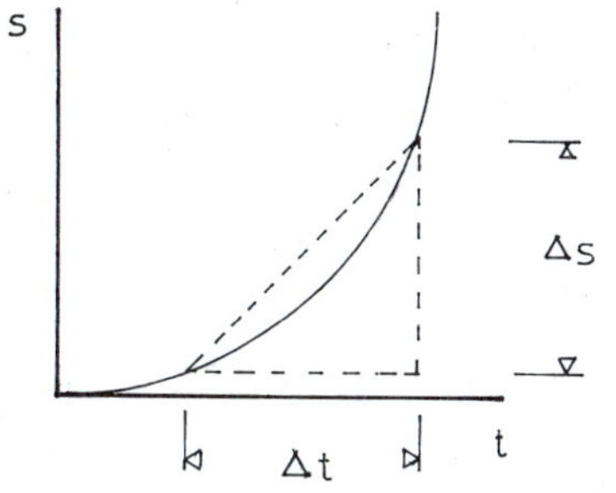


$$v_{avg} = \frac{\Delta s}{\Delta t}$$

$$\text{slope} = \frac{\text{rise}}{\text{run}} = \frac{\Delta s}{\Delta t}$$

$$v_{avg} = \text{slope}$$

The instantaneous velocity is equal to the average velocity over a time interval that is so small we are willing to call it an instant. The instantaneous velocity of an object may be obtained from a plot of displacement vs time (Fig. 2.5). It is the slope of the line tangent to the curve at the instant (time) of interest.

Note: It is crucial that you understand why the instantaneous velocity at some instant is equal to the average velocity over a very small time interval. It is also essential that you understand why the average velocity over a small time interval is the same as the slope of the tangent line to the curve at that time.

Figure 2.5

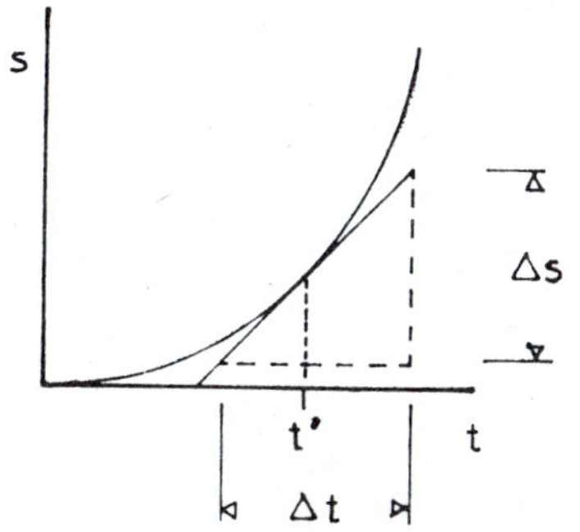


tangent line at t'
$v_{inst}$ at t' is the slope of the tangent line at t'

Practice: Consider an object moving along a linear track. As the object moves, a measuring devise collects displacement and time information. The resultant data and a plot of displacement versus time are shown in Fig. 2.6 for the first 5 s of motion.

Figure 2.6

| time(s) | displacement (m) | |
|---|---|---|
| | label | value |
| 0 | $s_0$ | 1 |
| 1 | $s_1$ | 3 |
| 2 | $s_2$ | 7 |
| 3 | $s_3$ | 13 |
| 4 | $s_4$ | 21 |
| 5 | $s_5$ | 31 |

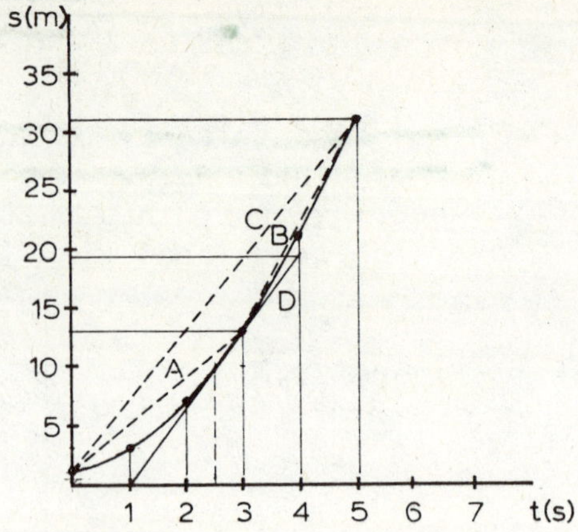


Using the displacement vs time plot, determine the following:

1. The average velocity during the first 3.00 s

The average velocity during the first 3.00 s is the slope of the dashed line labeled A in Fig. 2.6.

$$v_{av} = \frac{\Delta s}{\Delta t} = \frac{s_3 - s_0}{t_3 - t_0}$$

$$= \frac{+13.0 \text{ m} - (+1.00 \text{ m})}{3.00 \text{ s} - 0.0 \text{ s}} = +4.00 \text{ m/s}$$

2. The average velocity during the last 2.00 s

The average velocity during the last 2.00 s is the slope of the dashed line labeled B in Fig. 2.6.

$$v_{av} = \frac{\Delta s}{\Delta t} = \frac{s_5 - s_3}{t_5 - t_3}$$

$$= \frac{+31.0 \text{ m} - (+13.0 \text{ m})}{5.00 \text{ s} - 3.00 \text{ s}} = +9.00 \text{ m/s}$$

3. The average velocity during the first 5.00 s

The average velocity during the first 5.00 s is the slope of the dashed line labeled C in Fig. 26.

$$v_{av} = \frac{\Delta s}{\Delta t} = \frac{s_5 - s_0}{t_5 - t_0}$$

$$= \frac{+31.0 \text{ m} - (+1.00 \text{ m})}{5.00 \text{ s} - 0.0 \text{ s}} = +6.00 \text{ m/s}$$

| | |
|---|---|
| 4. The instantaneous velocity at t = 2.50 s | The instantaneous velocity at t = 2.50 s is equal to the slope of the line tangent to the s vs t curve at t = 2.50 s. The solid line D is tangent to the s vs t curve at t = 2.50 s $v_{inst} = \frac{(19.0 - 0.00)\ m}{(4.00 - 1.00)\ s} = +6.00\ m/s$ |

## 4. Acceleration (Section 2.4)

Review: If an object is changing its velocity, it is accelerating. Acceleration is a description of how an object is changing its velocity. Average acceleration is the time rate of change of velocity:

$$a_{av} = \frac{\Delta v}{\Delta t} = \frac{v - v_o}{t}$$

If the acceleration is constant, the average acceleration is equal to the instantaneous acceleration at any instant. The average acceleration of an object over some time interval may be obtained from a plot of the instantaneous velocity vs time. It is the slope of the straight line connecting two points on the v vs t curve.

Practice: Three objects are traveling along a linear track. A record of velocity vs time is shown in Fig. 2.7.

Figure 2.7

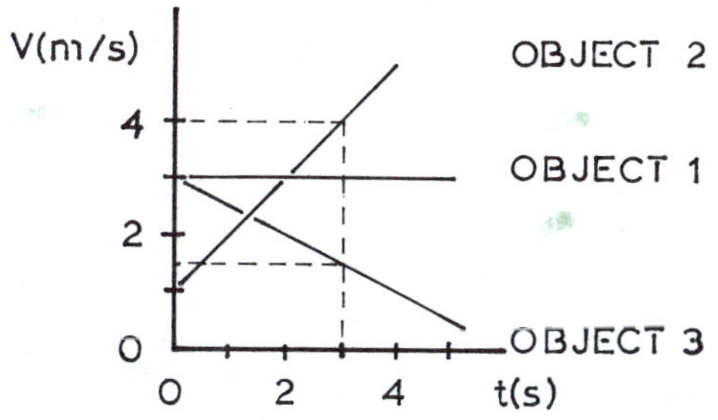


Using Fig. 2.7, determine the following:

| | |
|---|---|
| 1. The acceleration of object 1 | $a = \frac{\Delta v}{\Delta t} = 0\ m/s^2$ Object 1 does not change its velocity; consequently its acceleration is zero. |

| | |
|---|---|
| 2. The acceleration of object 2 | $a = \frac{\Delta v}{\Delta t} = \frac{(4.00 - 1.00)\ m/s}{(3.00 - 0.00)\ s}$<br>$= +1.00\ m/s^2$<br>Object 2 is increasing its velocity by 1.00 m/s every second it is motion. Consequently, its acceleration is $+1.00\ m/s^2$. |
| 3. The acceleration of object 3 | $a = \frac{\Delta v}{\Delta t} = \frac{[(+1.50 - (+3.00)]\ m/s}{(3.00 - 0.00)\ s}$<br>$= -0.500\ m/s^2$<br>Object 3 is decreasing its velocity by 0.500 m/s every second it is motion. Consequently, its acceleration is $-0.500\ m/s^2$. |

Related Text Problems: 2-10 through 2-15.

## 5. You Can Answer Any Question About the Linear Motion of an Object with Equations 2-9 through 2-13 (Section 2.4)

Review: The equations that allow us to answer questions about linear motion are

(2-9) $v = v_o + at$ (2-11) $\Delta s = v_{av}t$ (2-13) $2a\Delta s = v^2 - v_o^2$

(2-10) $v_{av} = \frac{v + v_o}{2}$ (2-12) $\Delta s = v_o t + \frac{1}{2} at^2$

Note: All of these equations except Eq. 2-11 are valid only for constant acceleration.

Note: When working motion problems, students frequently have trouble with the difference between distance and displacement and the difference between speed and velocity. The following statements should help you make the proper distinction.

1. Distances are obtained from the magnitude of changes in displacement.
2. The speed at any instant is always the magnitude of the velocity at that instant.
3. For an object undergoing linear motion and not changing direction,
   (a) the magnitude of the change in displacement is equal to the distance traveled
   (b) the magnitude of the average velocity is equal to the average speed

Note: You should not regard Eq. 2-9 through 2-13 as totally new. Equation 2-9 is obtained from the definition of average acceleration. Equation 2-11 is just our definition of average velocity solved for $\Delta s$. Equation 2-13 is obtained by combining Eqs. 2-9 and 2-12 in such a way as to eliminate t.

Note: In solving problems in physics, it is important to read the problem and decide the following:

(a) which quantities are known
(b) which quantities are unknown
(c) which equations to use to in order to express the unknown quantities in terms of the known quantities. After this step and perhaps a little algebra, you will be able to insert numbers and obtain a value for the desired unknown quantity.

Practice: Case I. An object moving in a straight line with an initial velocity $v_o$ undergoes a constant acceleration a for a time t. What is the object's final velocity, change in displacement while accelerating, and average velocity?

Determine the following:

| | |
|---|---|
| 1. The known quantities | $v_o$, a, t |
| 2. The unknown quantities | v, $\Delta s$, $v_{av}$ |
| 3. The equation to determine final velocity | $v = v_o + at$ |
| 4. The equation to determine change in displacement | $\Delta s = v_o t + \frac{1}{2} at^2$ |
| 5. The equation to determine average velocity | $v_{av} = \frac{\Delta s}{\Delta t}$ |
| 6. A method to check your work | Determine $\Delta s$ with the equation $v^2 - v_o^2 = 2 a\Delta s$ Determine $v_{av}$ with the equation $v_{avg} = (v + v_o)/2$ |

Case II. An object moving with an initial velocity $v_o$ undergoes a constant acceleration a while changing its displacement $\Delta s$. What is the velocity after acceleration, the time elapsed during acceleration and the average velocity during acceleration?

Determine the following:

| | |
|---|---|
| 1. The known quantities | $v_o$, $a$, $\Delta s$ |
| 2. The unknown quantities | $v$, $t$, $v_{av}$ |
| 3. The equation to determine velocity after the acceleration | $v^2 - v_o^2 = 2a\Delta s$ |
| 4. The equation to determine time elapsed during acceleration | $\Delta s = v_o t + \frac{1}{2} at^2$ (This method uses only given information but requires solving a quadratic equation.)<br><br>$v = v_o + at$ (This method uses a previously calculated quantity but avoids solving a quadratic equation.) |
| 5. The equation to determine average velocity | $v_{av} = \Delta s/\Delta t$ |
| 6. A method to check your work | Use the equation in step 4 not previously used to determine t.<br><br>Determine $v_{av}$ with the equation $v_{av} = \frac{v + v_o}{2}$ |

Example 2.4. A train traveling at a speed of 100 m/s is braked uniformly to rest in 20.0 s. Determine the distance the train travels while coming to rest.

Given: $v_o = 100$ m/s; $v = 0.00$ m/s (since the train stops); $t = 20.0$ s

Determine: d, the distance the train travels while coming to rest

Strategy: Use the given information and Eq. 2-9 to determine the acceleration of the train. Once the acceleration is known, use either Eq. 2-12 or 2-13 to determine the change in displacement while slowing down. Assuming the track is linear and knowing the train does not change direction while slowing down, we can use the fact that the distance traveled is the magnitude of the change in displacement.

Solution: Using the given information and Eq. 2-9, we can determine the acceleration of the train:

$$a = \frac{v - v_o}{t} = \frac{(0.00 - 100)\ m/s}{20.0\ s} = -\ 5.00\ m/s^2$$

Now that the acceleration is known, we can use Eq. 2-12 to determine $\Delta s$, the change in displacement while the train is slowing down:

$$\Delta s = v_o t + at^2/2 = (+100\ m/s)(20.0\ s) + (-5.00\ m/s^2)(20.0\ s)^2/2 = +1.00 \times 10^3\ m$$

Since the track is linear and the train does not change direction we can write

$$d = |\Delta s| = 1.00 \times 10^3\ m$$

Example 2.5. A student holds a water balloon outside a fifth-story window (15.0 m off the ground), tosses it vertically upward and then quickly retracts his hand. The balloon hits the ground in 3.00 s. Determine the initial velocity of the balloon and its maximum displacement with respect to the ground.

Given and Diagram:

$Y_o$ = +15.0 m (initial displacement)
$Y$ = 0.0 m (final displacement)
$\Delta Y = Y - Y_o$ = -15.0 m (total change in displacement)
$\Delta Y_{up}$ (upward change in displacement)
$a = -g = -9.80\ m/s^2$ (acceleration due to gravity)
$t$ = 3.00 s (time of flight)

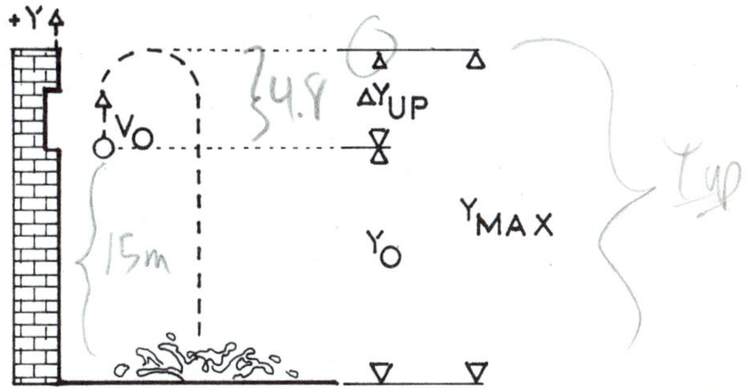


Figure 2.8

Determine: $Y_{max}$, maximum displacement of the balloon with respect to the ground; $v_o$, the initial velocity of the balloon

Strategy: Knowing $\Delta Y$ and t, we can obtain $v_o$ using Eq. 2-12. We know that the balloon will travel upward until v = 0 m/s (at this point it momentarily stops and then starts downward). We can use Eq. 2-13 to determine $\Delta Y_{up}$. Adding $\Delta Y_{up}$ to the initial displacement $Y_o$, we can obtain the maximum displacement of the balloon with respect to the ground.

Solution: Solve Eq. 2-12 for $v_o$: $v_o = (\Delta Y - at^2/2)/t$
Knowing that $\Delta Y$ = -15.0 m when t = 3.00 s, we can obtain $v_o$:

$$v_o = \frac{[-15.0\ m - (-9.80\ m/s^2)(3.00\ s)^2/2]}{3} = +9.70\ m/s$$

We know that the balloon decelerates to v = 0.00 m/s while traveling upward. The displacement of the balloon is $\Delta Y_{up}$ when v = 0.00 m/s. We can obtain $\Delta Y_{up}$ by using Eq. 2-13:

$$\Delta Y_{up} = \frac{v^2 - v_o^2}{2a} = \frac{[0.00\ m/s - (9.70)^2]m^2/s^2}{2(-9.8\ m/s^2)} = +4.80\ m$$

Finally, the maximum displacement of the balloon with respect to the ground is

$$Y_{max} = Y_o + \Delta Y_{up} = +15.0\ m + 4.80\ m = +19.8\ m$$

Example 2.6. A motorist traveling at a constant velocity of +5.00 m/s runs a red light. A police officer sitting at the intersection observes this and starts off after the motorist. If the patrol car can accelerate 1.00 $m/s^2$ until the motorist is caught, determine the time it takes to catch the motorist, the velocity of the patrol car when the motorist is caught, and the displacement of the two cars with respect to the intersection.

Given and Diagram:

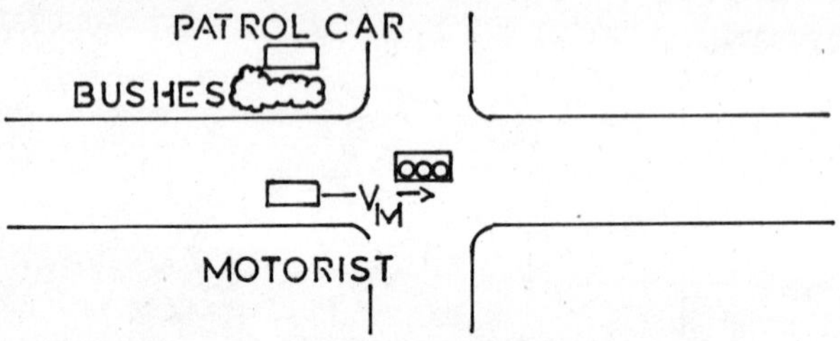


(a)
Motorist runs red light

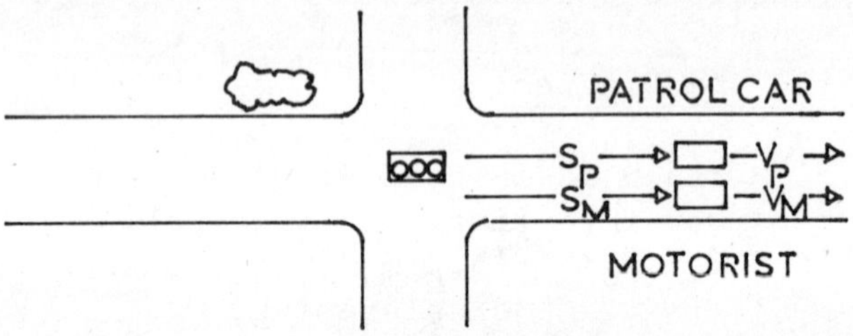


(b)
Patrol car catches motorist ($s_p = s_m$)

$v_m$ = +5.00 m/s (constant velocity of motorist)
$v_p$ = 0.0 m/s (initial velocity of patrol car)
$a_p$ = +1.00 $m/s^2$ (acceleration of the patrol car)

Determine: $t_c$, time for patrol car to catch motorist; $v_p$, velocity of patrol patrol car when motorist is caught; $s_c$ and $s_p$, displacement of both cars when patrol car catches motorist.

Strategy: We can write expressions for the change in displacement of the motorist and patrol car at any time. Since these two changes in displacement are equal when $t = t_c$, we will equate them and solve for $t_c$. The distance from the intersection may be determined by inserting $t_c$ into the change in displacement expression of either the motorist or the patrol car. Knowing the acceleration of the patrol car and $t_c$, we can determine the velocity of the patrol car when the motorist is caught.

Solution: The changes in displacement of motorist and patrol car at any time are:

$$\Delta s_m = v_m t \quad \text{and} \quad \Delta s_p = a_p t^2/2$$

Since $\Delta s_m = \Delta s_p$ when $t = t_c$ we may write

$$v_m t_c = a_p t_c^2/2 \quad \text{or} \quad t_c = 2v_m/a_p = 10.0 \text{ s}$$

The patrol car will catch the motorist in 10.0 s.

Now insert $t_c$ into the equation for either $\Delta s_m$ or $\Delta s_p$ to obtain the change in displacement of either car.

Using the expression for the motorist's change in displacement at $t = t_c$, obtain

$$\Delta s_m = v_m t_c = (5.00 \text{ m/s})(10.0 \text{ s}) = 50.0 \text{ m}$$

Using the expression for the patrol car's change in displacement at $t = t_c$, obtain

$$\Delta s_p = a_p t_c^2/2 = (1.00 \text{ m/s}^2)(10.0 \text{ s})^2/2 = 50.0 \text{ m}$$

The speed of the patrol car at $t = t_c$ is

$$v_p = v_{op} + a_p t_c = (1 \text{ m/s}^2)(10.0 \text{ s}) = 10.0 \text{ m/s}$$

Note: The key to solving this problem is to realize that two things ($\Delta s_p$ and $\Delta s_m$) are equal at a certain time ($t_c$).

Related Text Problems: 2-11 through 2-17, 2-19 through 2-38.

---

## PRACTICE TEST

Take and grade this practice test. Doing so will allow you to determine any weak spots in your understanding of the concepts taught in this chapter. The following section prescribes what you should study further to strengthen your understanding.

An object starts from rest and moves along the x axis as shown in this plot of displacement vs time. Determine:

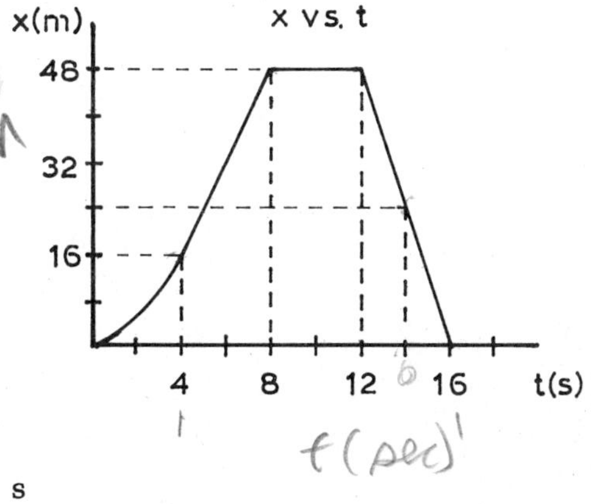


24 1. Displacement at t = 14.0 s
72 2. Distance traveled during the first 14.0 s
-16 3. Change in displacement during the time interval t = 4.00 s to t = 16.0 s
80 4. Distance traveled during the time interval t = 4.00 s to t = 16.0 s
-1.33 5. Average velocity during the last 12.0 s
0 6. Average velocity for the entire trip
6.67 7. Average speed during the last 12.0 s
6.00 8. Average speed for the entire trip
-12 9. Instantaneous velocity at t = 14.0 s

12.0 10. Instantaneous speed at t = 14.0 s
0 11. Acceleration at t = 6.00 s
2.00 12. Acceleration during the first 4.00 s

A hot-air balloon released from the ground at t = 0 s accelerates upward at the rate of 0.100 $m/s^2$. When it is at an altitude of 100 m, a camera is accidentally dropped overboard. Determine the following:

+4.47 13. The velocity of the balloon when the camera is dropped
1.02 14. The upward change in displacement of the camera after it is dropped
0.456 15. The time the camera travels upward after it is dropped
−9.8 16. The acceleration of the camera at the top of its trajectory
5.00 17. The time it takes the camera to hit the ground after it is dropped
−44.5 18. The impact velocity of the camera

(See Appendix I for answers)

---

<u>Principal Concepts and Equations Prescription</u>: Take and grade the practice test. Then use the chart below to determine any weak areas and to write a prescription which will allow you to strengthen your understanding of physics in these areas.

| Practice Test Question | Concepts and Equations | Prescription | |
|---|---|---|---|
| | | Principal Concept | Text Problems |
| 1 | Displacement | 1 | 2-28,35 |
| 2 | Distance | 1 | 2-28,35 |
| 3 | Change in displacement | 1 | 2-28,35 |
| 4 | Distance | 1 | 2-28,35 |
| 5 | Average velocity: $v_{av} = \Delta s/\Delta t$ | 2 | 2-8,28 |
| 6 | Average velocity: $v_{av} = \Delta s/\Delta t$ | 2 | 2-8,28 |
| 7 | Average speed: $v_{av} = d/t$ | 2 | 2-1,3 |
| 8 | Average speed: $v_{av} = d/t$ | 2 | 2-4,5 |
| 9 | Instantaneous velocity | 3 | * |
| 10 | Instantaneous speed | 3 | 2-18 |
| 11 | Acceleration: $a = \Delta v/\Delta t$ | 4 | 2-10,12 |
| 12 | Acceleration: $a = \Delta v/\Delta t$ | 4 | 2-11,15 |
| 13 | $v^2 - v_o^2 = 2a\Delta s$ | 5 | 2-20,21 |
| 14 | $v^2 - v_o^2 = 2a\Delta s$ | 5 | 2-22,24 |
| 15 | $v = v_o + at$ | 5 | 2-25,29 |
| 16 | Acceleration: $a = \Delta v/\Delta t$ | 4 | 2-13,14 |
| 17 | $\Delta s = v_o t + at^2/2$ | 5 | 2-19,26 |
| 18 | $v = v_o + at$ or $v^2 - v_o^2 = 2a\Delta s$ | 5 | 2-20,31 |

*No chapter problems deal specifically with this concept.

---

# 3 Vectors and Multidimensional Motion

## RECALL FROM PREVIOUS CHAPTERS

| Previously learned concepts and equations frequently used in this chapter | Text Section | Study Guide Page |
|---|---|---|
| $v = v_0 + at$ | 2.4 | 22 |
| $\Delta s = v_0 t + at^2/2$ | 2.4 | 22 |
| $2a\Delta s = v^2 - v_0^2$ | 2.4 | 22 |

## NEW IDEAS IN THIS CHAPTER

| Concepts and equations introduced | Text Section | Study Guide Page |
|---|---|---|
| Difference between vector and scalar quantities | 3.1 | 29 |
| Use of vectors to represent physical quantities that have magnitude and direction | 3.1 | 30 |
| Graphical treatment of vectors (representation, components, addition, subtraction) | 3.1,2 | 30,31 |
| Analytical treatment of vectors (resolution into components, addition, subtraction) | 3.2 | 32 |
| Treatment of motion in two dimensions | 3.3,4,5,6 | 37 |

## PRINCIPAL CONCEPTS AND EQUATIONS

In this chapter, we will review the principal concepts and equations related to vectors and then do some practice exercises and consider a few examples.

### 1. Difference Between Vector and Scalar Quantities (Section 3.1)

Review: Scalar quantities have magnitude only. For example, the mass of a bowling ball is 4.50 kg, a class period is 50 min, and a comfortable room temperature is 25°C. Vector quantities have direction as well as magnitude. For example, a quarterback's displacement was 5 yards down field, and the velocity of a car is 90 km/h directed at an angle of 40° E of N.

## 2. Graphical Representation of Vectors (Section 3.1)

Review: A vector may be represented by an arrow. The orientation of the arrow shows the direction of the vector, and the length when drawn to the proper scale represents the magnitude of the vector. Figure 3.1 is the graphical representation of a displacement vector that has a magnitude of 10 m and is directed at an angle of 30° N of E.

Figure 3.1

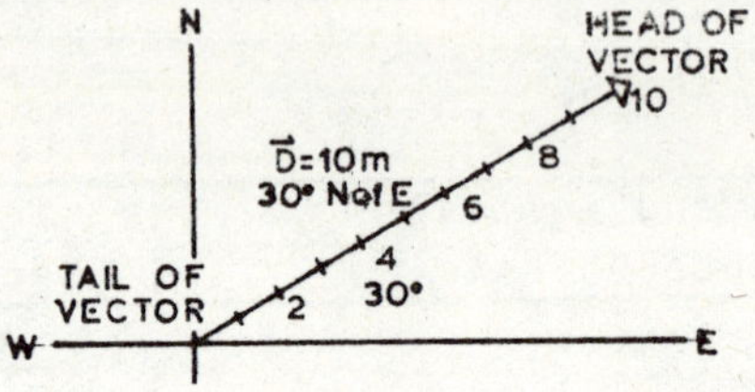


## 3. Graphical Addition and Subtraction of Vectors by the Closed Polygon Method (Section 3.1)

Review: To add vectors graphically by the closed polygon method, we first slide them around on the page, attaching the tail of the second vector to the head of the first, the tail of the third to the head of the second, and so on until all the vectors to be added are connected. Then, we draw the resultant vector from the tail of the first to the head of the last, as shown in Fig. 3.2, which also illustrates that the resultant vector is independent of the order in which we add vectors.

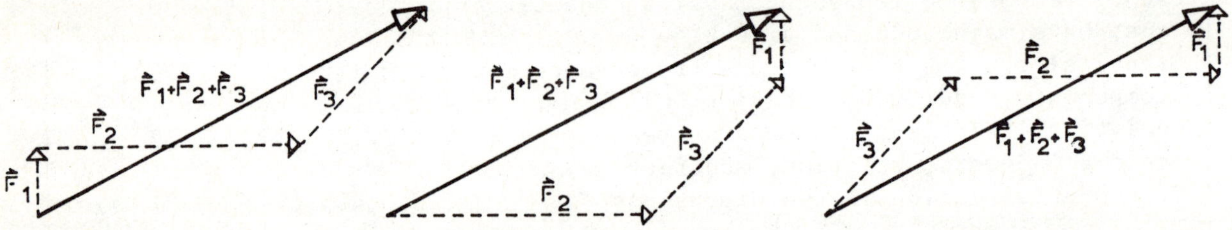

Figure 3.2

Figures 3.3 and 3.4 illustrate that vector subtraction is a special case of vector addition. Vector $-\vec{F}_1$ has the same magnitude as vector $\vec{F}_1$ but is directed in the opposite direction (Fig. 3.3).

Figure 3.3

To subtract $\vec{F}_1$ from $\vec{F}_2$, add $-\vec{F}_1$ to $\vec{F}_2$. This is stated algebracially and shown graphically in Fig. 3.4.

Figure 3.4

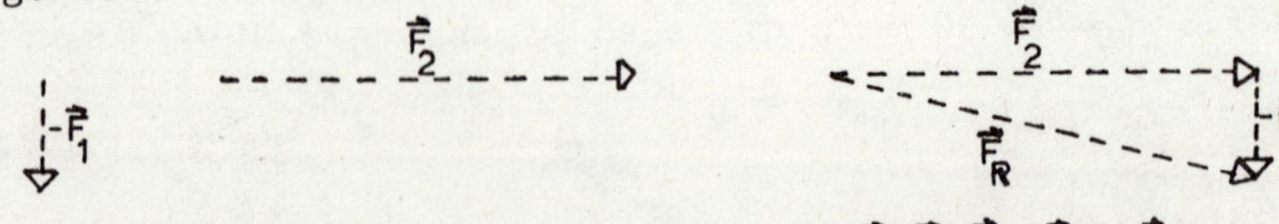

$$\vec{F}_R = \vec{F}_2 + \vec{F}_1 = \vec{F}_2 + (-\vec{F}_1)$$

## 4. Graphical Resolution of Vectors into Components (Section 3.2)

Review: Once a coordinate system is established (regardless of the orientation), we may resolve a vector into components by projecting it onto the axes. As shown in Fig. 3.5, this projection is accomplished by drawing a line from the tip of the arrow perpendicular to the x axis and another perpendicular to the y axis.

Figure 3.5

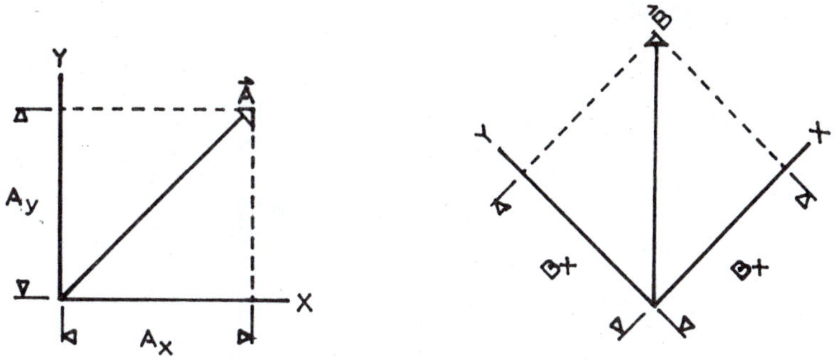


If we are given the components of a vector, we can reverse this procedure and determine the vector graphically.

## 5. Graphical Addition of Vectors by the Component Method (Sections 3.1, 3.2)

Review: From Fig. 3.6, we see that the x and y components of the resultant vector are respectively the sum of the x and y components of $\vec{A}$ and $\vec{B}$. The graphical component method of adding vectors amounts to resolving all vectors (graphically) into their components, adding the components (graphically), and then determining the resultant. This is done for $\vec{C}$ and $\vec{D}$ in Fig. 3.7.

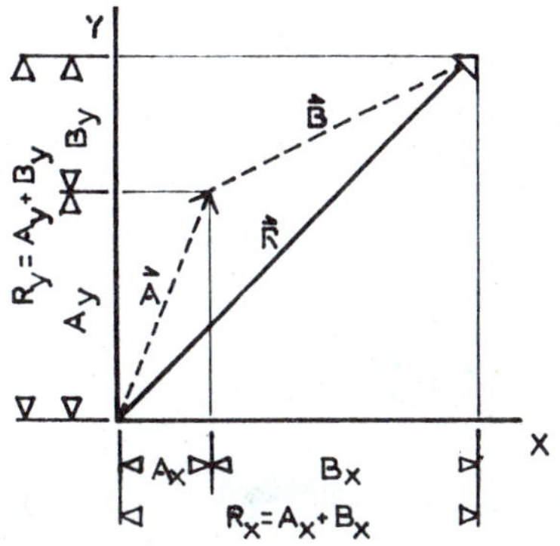


Figure 3.6

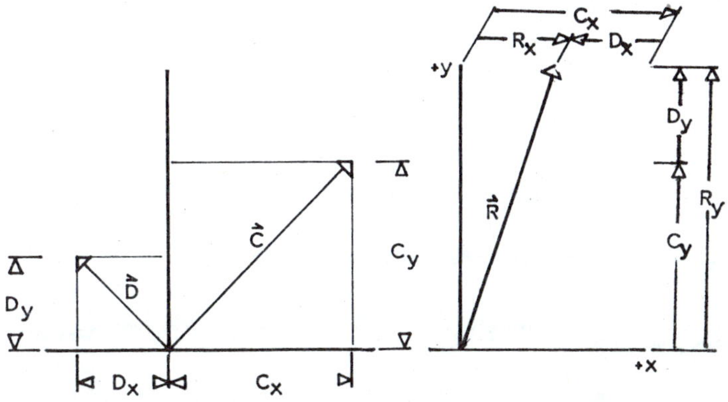


Figure 3.7

Note: It is important that you be able to handle vectors graphically because this method has the advantage of helping you visualize the situation. In the process of solving problems, you will probably never use the graphical component method. However, you will frequently use the analytical component method, which is easily understood and used only by those students who understand and can visualize the graphical component method.

## 6. Analytically Resolving a Vector into Its Components (Section 3.2)

Review: We are now familiar with what is meant by the components of a vector. We can use our knowledge of the trigonometric functions to express components in terms of the magnitude of the vector and an orientation angle. In the following, we analytically resolve the displacement vector $\vec{A}$ (magnitude 10 m and direction 30° N of E) into its components.

Figure 3.8

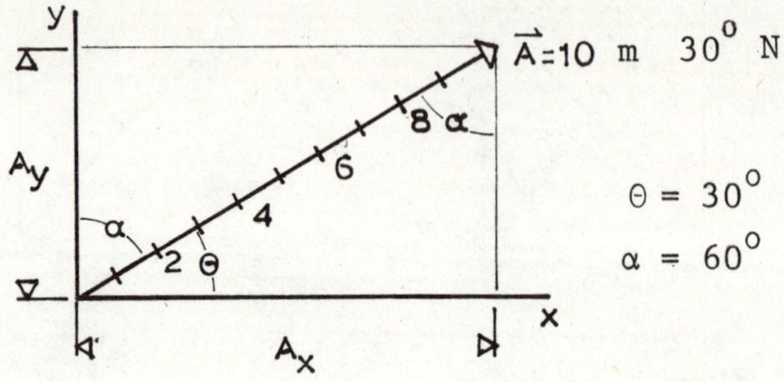


$A_x$ = A cosθ = (10 m)cos30° = 8.66 m or $A_x$ = A sinα = (10 m)sin60° = 8.66 m

$A_y$ = A sinθ = (10 m)sin30° = 5.00 m or $A_y$ = A cosα = (10 m)cos60° = 5.00 m

Note: A common student error is to memorize that the x and y components are obtained by using the cosine and sine functions respectively. If you are given the angle θ, then the cosine function is used to obtain the x component, and the sine function is used for the y component. However, if you are given the angle α, then you must use the sine function for the x component and the cosine function for the y component. To be safe, you should (1) visualize a component and the given angle, (2) decide if the component is opposite or adjacent to the given angle, and (3) decide whether to use the sine or the cosine function.

If we are given the components of a vector, we can analytically obtain its magnitude and direction. To show this, let's start with $A_x$ and $A_y$ and determine the magnitude of $\vec{A}$ and its direction (θ or α).

Given: $A_x$ = 8.66 m and $A_y$ = 5.00 m

Determine the magnitude of $\vec{A}$ by using the Pythagorean theorem:

$$A = (A_x^2 + A_y^2)^{1/2} = [(8.66\ \text{m})^2 + (5.00\ \text{m})^2]^{1/2} = 10.0\ \text{m}$$

Determine the direction of $\vec{A}$ by using the tangent function:

$$\tan\theta = \frac{A_y}{A_x},\ \text{hence}\ \theta = \tan^{-1}\left(\frac{A_y}{A_x}\right) = \tan^{-1}\left(\frac{5.00\ \text{m}}{8.66\ \text{m}}\right) = 30.0°$$

or

$$\tan\alpha = \frac{A_x}{A_y},\ \text{hence}\ \alpha = \tan^{-1}\left(\frac{A_x}{A_y}\right) = \tan^{-1}\left(\frac{8.66\ \text{m}}{5.00\ \text{m}}\right) = 60.0°$$

## 7. Analytically Adding and Subtracting Vectors by the Component Method (Section 3.2)

Review: We know how to add vectors by the component method as a result of our graphical treatment of vectors. We also know how to analytically obtain vector components. If we combine these two bits of information, we can add vectors analytically by the component method. Consider three displacement vectors $\vec{A}$, $\vec{B}$, and $\vec{C}$ added by the analytical component method (Fig. 3.9):

$\vec{A}$ = 10 m N $\qquad$ $\vec{B}$ = 30 m E $\qquad$ $\vec{C}$ = 20 m 30° N of E

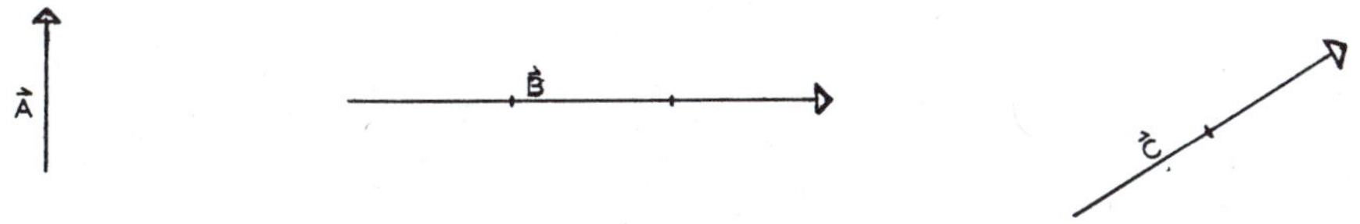

Figure 3.9

$A_x = A\cos90° = (10\text{ m})\cos90° = 0\text{ m}$
$B_x = B\cos0° = (30\text{ m})\cos0° = 30.0\text{ m}$
$C_x = C\cos30° = (20\text{ m})\cos30° = 17.3\text{ m}$
$R_x = A_x + B_x + C_x = 47.3\text{ m}$

$A_y = A\sin90° = (10\text{ m})\sin90° = 10.0\text{ m}$
$B_y = B\sin0° = (30\text{ m})\sin0° = 0\text{ m}$
$C_y = C\sin30° = (20\text{ m})\sin30° = 10.0\text{ m}$
$R_y = A_y + B_y + C_y = 20.0\text{ m}$

Now that the components are known, we can determine the magnitude and direction of the resultant vector (Fig. 3.10).

Figure 3.10

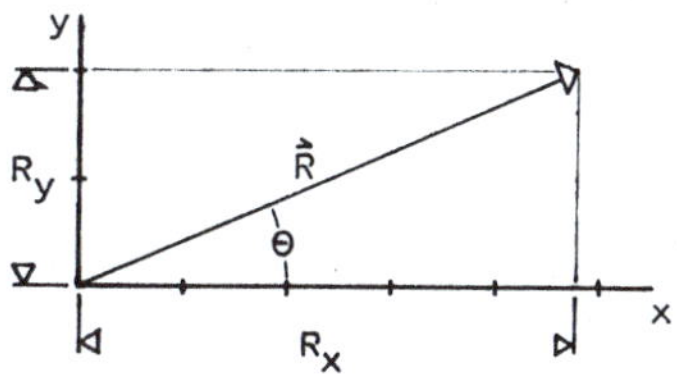

$$R = (R_x^2 + R_y^2)^{1/2} = [(47.3\text{ m})^2 + (20\text{ m})^2]^{1/2} = 51.4\text{ m}$$

$$\tan\theta = \frac{R_y}{R_x} \text{ or } \theta = \tan^{-1}\left(\frac{R_y}{R_x}\right) = \tan^{-1}\left(\frac{20.0\text{ m}}{47.3\text{ m}}\right) = 22.9° \text{ N of E}$$

Practice: Consider the following information about the displacement vectors $\vec{A}$, $\vec{B}$, $\vec{C}$, and $\vec{D}$:

$\vec{A}$

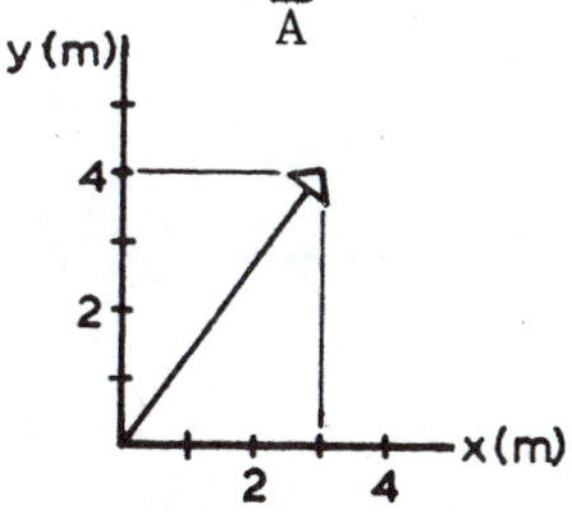

$\vec{B}$

$B_x = -3$ m
$B_y = -2$ m

$\vec{C}$

$\vec{C}$ = 3 m 45° S of E

$\vec{D}$

$\vec{D}$ = 4 m N

Determine the following by graphical methods:

1. Representation of $\vec{D}$ and $\vec{B}$

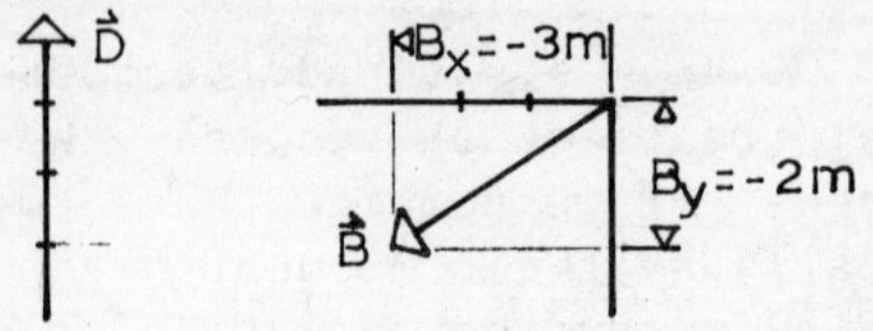


2. The magnitude and direction of $\vec{B}$

Refer to the figure in step 1.
$|\vec{B}| \simeq 3.6$ m measured with a linear scale
$\theta \simeq 34°$ S of E measured with a protractor

3. The x and y components of $\vec{C}$

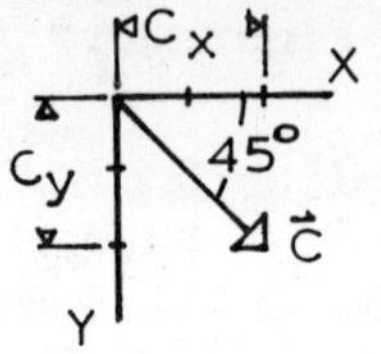


$C_x \simeq 2.1$ m

$C_y \simeq -2.1$ m

$C_x$ and $C_y$ are measured with a linear scale

4. The magnitude and direction of $-\vec{A}$

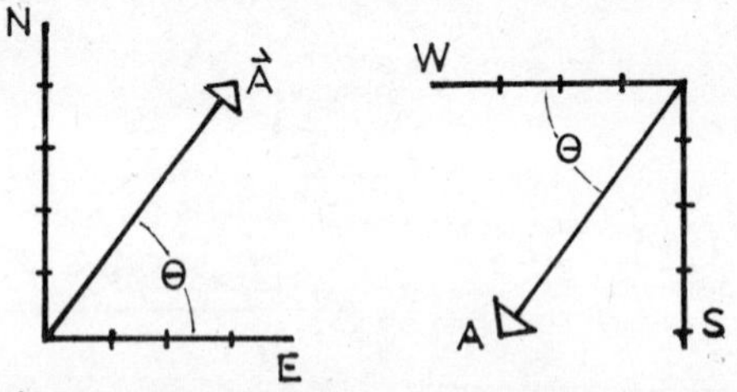


$|-\vec{A}| \simeq 5$ m measured with a linear scale
$\theta \simeq 53°$ S of W measured with a protractor

5. $\vec{A} + \vec{D}$ by the closed polygon method

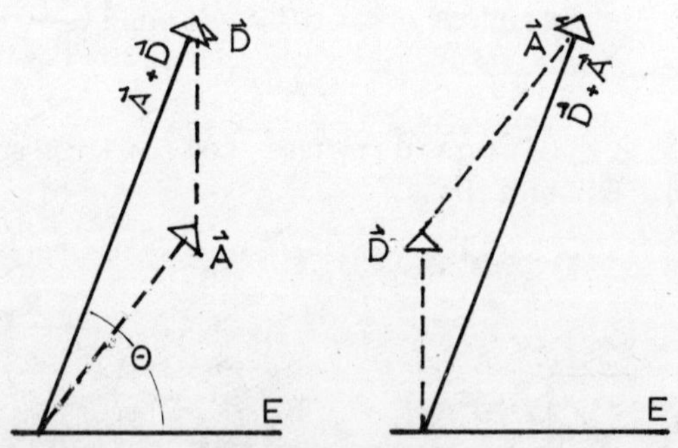


$|\vec{A} + \vec{D}| = |\vec{D} + \vec{A}| \simeq 8.6$ m, $\theta \simeq 70°$

| | |
|---|---|
| 6. $\vec{C} - \vec{B} + \vec{A}$ by the closed polygon method | $\lvert\vec{C} - \vec{B} + \vec{A}\rvert \simeq 9.2$ m, $\theta \simeq 64°$ E of N 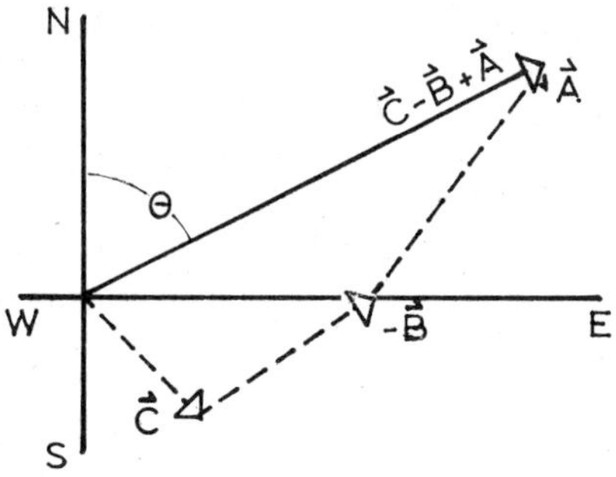  |

Using the same four vectors, determine the following by analytical methods:

| | |
|---|---|
| 7. The x and y components of $\vec{C}$ | 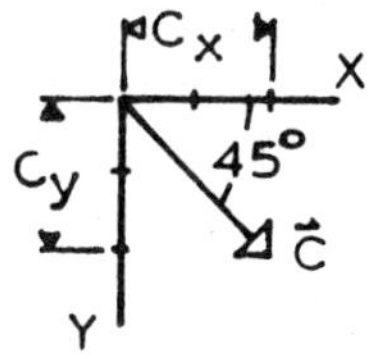  $C_x = C \cos 45° = (3\text{ m})(0.707) = 2.12$ m<br><br>$C_y = -C \sin 45° = -(3\text{ m})(0.707) = -2.12$ m |
| 8. The x and y components of $-2\vec{D}$ |   The components of D are $D_x = 0$, $D_y = +4$ m<br><br>The components of $-2\vec{D}$ are $-2D_x = 0$, $-2D_y = -8$ m |
| 9. The magnitude and direction of $\vec{B}$ | 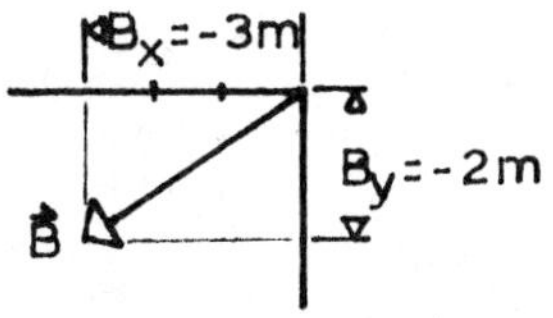  $B = (B_x{}^2 + B_y{}^2)^{1/2} = (9 + 4)^{1/2}$ m $= 3.6$ m<br><br>$\theta = \tan^{-1}(-2/-3) = 33.7°$ S of W |

| | |
|---|---|
| 10. $\vec{C} - \vec{B} + \vec{A}$ by the component method | Draw a rough sketch or refer to step 6.<br><br>$C_x = +2.12$ m  $C_y = -2.12$ m<br>$-B_x = +3.00$ m  $-B_y = +2.00$ m<br>$A_x = +3.00$ m  $A_y = +4.00$ m<br>$R_x = +8.00$ m  $R_y = +3.88$ m<br><br>$\lvert R\rvert = (R_x{}^2 + R_y{}^2)^{1/2} = 8.89$ m<br><br>$\theta = \tan^{-1}(R_y/R_x) = 25.9°$ N of E<br><br>From the sketch in step 6 we see that $\theta = 25.9°$ W of N. |

Example 3.1. An airplane is heading 30° N of E when it takes off at an angle of 20° above the horizontal. Its velocity is 100 m/s. (a) What is the vertical component of its velocity? (b) What is the horizontal component? (c) What is the component of the velocity toward the east?

Given and diagram:

$\theta = 30°$
$\alpha = 20°$
$|\vec{v}| = 100$ m/s

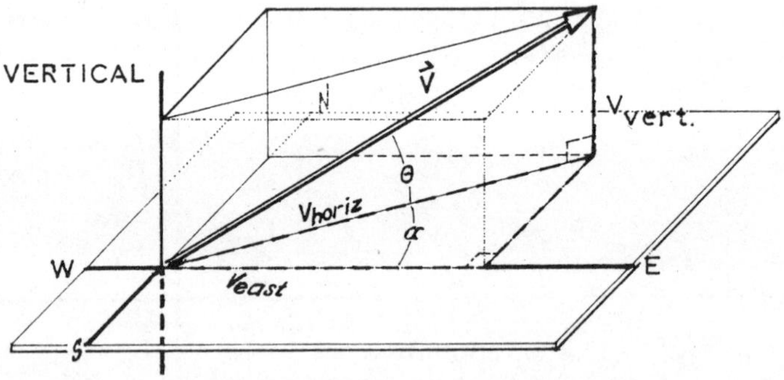


Determine: $v_{vert}$, $v_{hor}$, $v_{east}$

Strategy: We can determine $v_{vert}$ and $v_{hor}$ from $|\vec{v}|$ and θ. We can then use the values of $v_{hor}$ and α to obtain $v_{east}$.

Solution:
$v_{vert} = v \sin\theta = (100 \text{ m/s})\sin 20° = 34.2$ m/s
$v_{hor} = v \cos\theta = (100 \text{ m/s})\cos 20° = 94.0$ m/s
$v_{east} = v_{hor} \cos\alpha = (94.0 \text{ m/s})\cos 30° = 81.4$ m/s

Example 3.2. A reconnaissance plane leaves its home base and flies 100 mi N, 80 mi 60° W of N, and 150 mi 30° N of E. Determine how far the plane is from its home base and the direction it must fly in order to return straight to base.

Given: $\vec{S}_1 = 100$ mi N, $\vec{S}_2 = 80$ mi 60° W of N, $\vec{S}_3 = 150$ mi 30° N of E

Determine: $|\vec{S}_R|$ final distance of plane from base
θ angle at which plane must fly to return to base

Strategy: Establish a coordinate system and resolve the displacement vectors into their components. From these components, determine the components and magnitude of the resultant displacement. This magnitude is the final distance of the plane from base. The direction in which the plane must fly on its return trip is opposite the direction of the resultant displacement.

Solution: First let's do a quick graphical representation to make sure we visualize the problem correctly. Then we can find the components of the various displacements and of the resultant displacement, the magnitude of the resultant displacement, and the direction of the return flight.

$S_{1x} = 0$ mi
$S_{2x} = -S_2 \sin 60° = -69.3$ mi
$S_{3x} = S_3 \cos 30 = 140$ mi
$S_{Rx} = S_{1x} + S_{2x} + S_{3x} = 70.7$ mi

$S_{1y} = 100$ mi
$S_{2y} = S_2 \sin 30° = = 40.0$ mi
$S_{3y} = S_3 \cos 60° = 75.0$ mi
$S_{Ry} = S_{1y} + S_{2y} + S_{3y} = 115$ mi

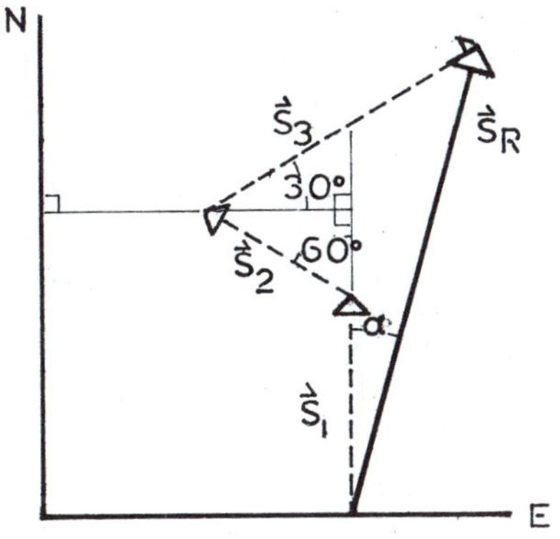


$$|\vec{S}_R| = (S_{Rx}{}^2 + S_{Ry}{}^2)^{1/2} = [(70.7\text{ mi})^2 + (115\text{ mi})^2]^{1/2} = 135\text{ mi}$$

$$\alpha = \tan^{-1}(S_{Rx}/S_{Ry}) = 31.6°\text{ E of N, hence } \theta = 31.6°\text{ W of S}$$

At the end of the flight, the plane is 135 mi directed 31.6° E of N from the base. In order to return to the base it must fly 135 mi directed 31.6° W of S.

Related Text Problems: 3-1 through 3-10.

## 8. Two-Dimensional Projectile Problems (Sections 3.3, through 3.5)

In Chapter 2, we studied one-dimensional (linear) motion and found that we could answer any question about constant-acceleration motion by using Eqs. 2-9, 2-10, 2-11, 2-12, and 2-13 shown below:

$$v = v_o + at \;;\; v_{av} = \frac{v + v_o}{2} \;;\; \Delta s = v_{av} t \;;\; \Delta s = v_o t + \frac{1}{2} at^2 \;;\; 2a\Delta s = v^2 - v_o^2$$

In this chapter, we have found that for projectile motion the x and y components are independent of one another. Consequently, we may write a set of equations (similar to those shown above) for the x and y directions. The most useful of these equations for solving projectile motion problems are:

$$v_x = v_{ox} + a_x t \qquad \Delta x = v_{ox} t + a_x t^2/2 \qquad 2a_x \Delta x = v_x{}^2 - v_{ox}{}^2$$

$$v_y = v_{oy} + a_y t \qquad \Delta y = v_{oy} t + a_y t^2/2 \qquad 2a_y \Delta y = v_y{}^2 - v_{oy}{}^2$$

After the two one-dimensional problems are solved, the results may be combined to answer questions about the two-dimensional motion. For example, once we know $v_x$ and $v_y$, we can combine them to determine v.

In solving two-dimensional motion problems, we will use the following prescription:

1. Make a sketch of the situation under consideration.
2. Pick a convenient coordinate system and use the standard notation that +y is up, −y is down, +x is to the right, and −x is to the left. This means that an acceleration, velocity, or change in displacement directed upward or to the right is positive and one directed downward or to the left is negative.
3. Write down the general x and y component equations.
4. Write down the initial conditions.
5. Insert the initial conditions into the general equations to obtain component equations that are unique for the situation being considered.
6. Read the question and establish the physics of the situation. For example, if asked to find the maximum height of a projectile, you must recognize that it obtains its maximum height when the y component of the velocity is zero. This may be stated algebraically as $y = y_{max}$ when $v_y = 0$.
7. Decide which of the unique component equations to use.
8. Manipulate the unique component equations to obtain expressions for the desired quantities.

<u>Practice:</u> A projectile is fired off a cliff of height H. It is fired with a speed $v_o$ at an angle $\theta$ above the horizontal.

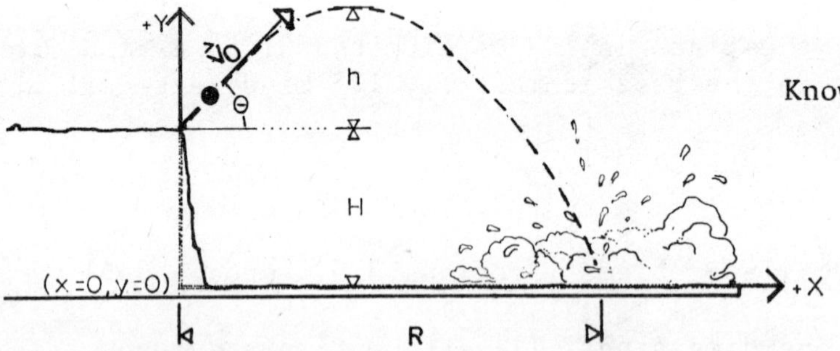


Known quantities $v_o$

$H$

$\theta$

$g$

Determine the following:

| | | |
|---|---|---|
| 1. | The general equations for the x and y components of the motion | $v_x = v_{ox} + a_x t$ <br> $\Delta x = v_{ox}t + a_x t^2/2$ <br> $2a_x\Delta x = v_x^2 - v_{ox}^2$ <br><br> $v_y = v_{oy} + a_y t$ <br> $\Delta y = v_{oy}t + a_y t^2/2$ <br> $2a_y\Delta y = v_y^2 - v_{oy}^2$ |
| 2. | The initial conditions | $a_x = 0$; $v_{ox} = v_o \cos\theta$; $x_o = 0$ <br><br> $a_y = -g$; $v_{oy} = v_o \sin\theta$; $y_o = H$ |
| 3. | An expression for the x component of the velocity at any time | $v_x = v_{ox} + a_x t$; $v_{ox} = v_o \cos\theta$; $a_x = 0$ <br><br> $v_x = v_o \cos\theta$ |

| | |
|---|---|
| 4. An expression for the y component of the velocity at any time | $v_y = v_{oy} + a_y t$; $v_{oy} = vo \sin\theta$; $a_y = -g$<br>$v_y = v_o \sin\theta - gt$ |
| 5. An expression for the x component of the position vector at any time | $\Delta x = v_{ox}t + a_x t^2/2$; $v_{ox} = v_o \cos\theta$<br>$a_x = 0$; $\Delta x = x - x_o$; $x_o = 0$<br>$x = v_o t \cos\theta$ |
| 6. An expression for the y component of the position vector at any time | $\Delta y = v_{oy}t + a_y t^2/2$; $v_{oy} = v_o \sin\theta$<br>$a_y = -g$; $\Delta y = y - y_o$; $y_o = H$<br>$y = H + v_o t \sin\theta - gt^2/2$ |
| 7. Unique equations for the x and y components of the projectile's accelera;tion, velocity and position at any time | $a_x = 0$; $v_x = v_o \cos\theta$; $x = v_o t \cos\theta$<br>$a_y = -g$; $v_y = v_o \sin\theta - gt$<br>$y = H + v_o \sin\theta - gt^2/2$ |

Note: Step 7 is just a summary of steps 3 through 6.

| | |
|---|---|
| 8. The expression obtained when time is eliminated in expressions for $v_y$ and $\Delta y$ | $v_y = v_{oy} \sin\theta + a_y t$; $\Delta y = v_o \sin\theta - gt^2/2$<br>Eliminate t to obtain $2a_y\Delta y = v_y^2 - v_{oy}^2$<br>Insert initial conditions to obtain<br>$-2g\Delta y = v_y^2 - v_o^2 \sin^2\theta$ |

Note: You should be able to answer any question about the projectile's motion by manipulating the equations in steps 7 and 8.

| | |
|---|---|
| 9. An expression for maximum height (h) of projectile above cliff | $\Delta y = h$ when $v_y = 0$, and so use<br>$-2g\Delta y = v_y^2 - v_o^2 \sin^2\theta$, which gives<br>$-2gh = -v_o^2 \sin^2\theta$ or $h = (v_o^2 \sin^2\theta)/2g$ |

| | |
|---|---|
| 10. An expression for the time ($t_{up}$) it takes projectile to reach maximum height | $v_y = 0$ when $t = t_{up}$, and so use<br>$v_y = v_o \sin\theta - gt$, which gives<br>$0 = v_o \sin\theta - gt_{up}$ or $t_{up} = (v_o \sin\theta)/g$ |
| 11. An expression for the time ($t'$) it takes projectile to get back down to height $y = H$ | $\Delta y = 0$ when $t = t'$, and so use<br>$\Delta y = v_o t \sin\theta - gt^2/2$, which gives<br>$0 = v_o t' \sin\theta - g(t')^2/2$ or<br>$t' = (2v_o \sin\theta)/g$<br>Note that this is twice $t_{up}$. |
| 12. An expression for the time of flight ($t_f$) | $\Delta y = -H$ or $y = 0$ when $t = t_f$, and so use<br>$y = H + v_o t \sin\theta - gt^2/2$, which gives<br>$0 = H + v_o t_f \sin\theta - g t_f^2/2$ or<br>$t_f^2 - (2v_o t_f/g)\sin\theta - 2H/g = 0$<br>Solving this quadratic for $t_f$ obtain<br>$t_f = (v_o \sin\theta)/g + \{(v_o \sin\theta/g)^2 + 2H/g\}^{1/2}$ |
| 13. An expression for the range R of the projectile | $x = R$ when $t = t_f$, and so use<br>$x = v_o t \cos\theta$, which gives<br>$R = v_o t_f \cos\theta$<br>$t_f$ has already been determined. |

Note: Do not memorize these equations! If any one detail is changed, these expressions will not work. They are correct only for this particular set of initial conditions and for this specific set of known information.

Example 3.3. A place kicker wishes to kick a field goal straight at the middle of the goalpost from the 40-yard line. He kicks the ball at an angle of 45° and with a speed of 21.2 yards/s. If the horizontal bar of the goalpost is 3 yards off the ground, does he score a field goal?

Given: $\Delta x = 40$ yards, $\Delta y_{min} = 3$ yards, $v_o = 21.2$ yards/s, $g = 10.7$ yards/$s^2$, $\theta = 45°$,

Determine: If a field goal is scored. That is, for a ball kicked at the given angle and speed, we must establish the value of $\Delta y$ for $\Delta x = 40$ yards. If $\Delta y$ is greater than 3 yards, then a field goal will be scored.

Strategy: Obtain expressions for $\Delta y$ and $\Delta x$ of the ball at any time. Eliminate time in these two equations and solve for $\Delta y$ as a function of $\Delta x$. Insert values for $\Delta x$, $v_o$, $g$, and $\theta$ to determine $\Delta y$. If $\Delta y$ is greater than 3 yards, a field goal will be scored.

Solution: The general equations for the components of the motion are:

$$v_x = v_{ox} + a_x t, \quad \Delta x = v_{ox}t + a_x t^2/2, \quad v_y = v_{oy} + a_y t, \quad \Delta y = v_{oy}t + a_y t^2/2$$

For this problem, the initial conditions are:

$$a_x = 0, \quad v_{ox} = v_o \cos\theta, \quad x_o = 0, \quad a_y = -g, \quad v_{oy} = v_o \sin\theta, \quad y_o = 0$$

When the initial conditions are inserted into the general equations, we obtain a set of equations for the components of the motion that are unique for this situation:

$$v_x = v_o \cos\theta, \quad \Delta x = v_o t \cos\theta, \quad v_y = v_o \sin\theta - gt, \quad \Delta y = v_o t \sin\theta - gt^2/2$$

Since we want to know the value of $\Delta y$ for a particular $\Delta x$, we eliminate time from the expressions for $\Delta y$ and $\Delta x$. This leaves us with an expression for $\Delta y$ in terms of $\Delta x$:

$$\Delta x = v_o t \cos\theta, \text{ which leads to } t = \Delta x/(v_o \cos\theta)$$

$$\Delta y = v_o t \sin\theta - gt^2/2 = v_o[\Delta x/(v_o \cos\theta)] \sin\theta - g[\Delta x/(v_o \cos\theta)]^2/2$$

$$\Delta y = \Delta x \tan\theta - g\Delta x^2/(2v_o^2 \cos^2\theta)$$

Now insert values for $\Delta x$, $v_o$, $g$, and $\theta$ to obtain $\Delta y = 1.91$ yards. Since $\Delta y = 1.91$ yards is less than 3 yards, a field goal is not scored.

Related Text Problems: 3-15, 3-17 through 3-31.

Note: After you have worked a number of projectile motion problems, you should notice that they fall into one of two classes:

I. Projected off a cliff or building at some angle
II. Launched from ground level at some angle

These problems may also be modified by varying the landing surface, but they are still essentially the same as the problems we have been working.

Note: Don't memorize the equations for a particular situation. Even if you are successful at thinking up all possible problems and their modifications, your instructor can create a new problem by changing the given information. Learn how to do the problems rather than memorizing a set of equations for a situation you probably won't be asked about.

---

Practice Test:

Take and grade this practice test. Doing so will allow you to determine any weak spots in your understanding of the concepts taught in this chapter. The following section prescribes what you should study further to strengthen your understanding.

Answer questions 1 through 6 for the following three displacement vectors:

$\vec{A}$ = 5 m 30° N of E; $\vec{B}$ = 8 m 60° S of W; $\vec{C}$ = 4 m 40° W of N

_____ 1. What is the x component (including sign) of $\vec{C}$?
_____ 2. What is the y component (including sign) of $\vec{B}$?
_____ 3. What is the x component (including sign) of the resultant displacement vector?
_____ 4. What is the y component (including sign) of the resultant displacement vector?
_____ 5. What is the magnitude of the resultant displacement vector?
_____ 6. What is the direction of the resultant displacement vector?

A baseball is thrown off a dormitory roof. The building is 50 m high and the ball is thrown with a velocity $v_o$ = 30 m/s 60° above the horizontal.

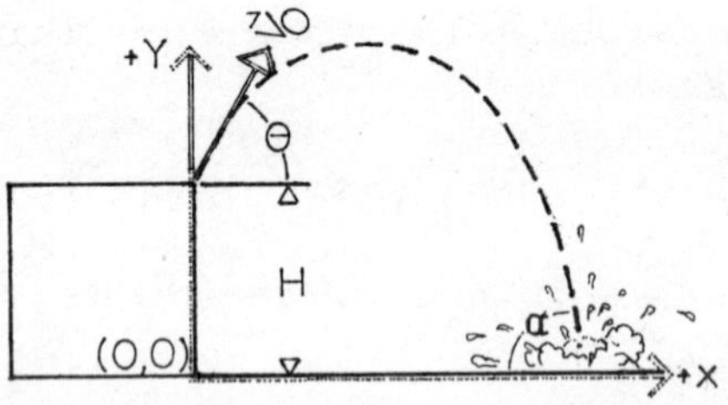


H = 50 m
$\vec{v}_o$ = 30 m/s
θ = 60°
g = 9.8 m/s$^2$

_____ 7. What are the x and y components of the initial velocity?
_____ 8. What are the x and y components of the ball's acceleration the instant it is released?
_____ 9. How long does it take the ball to reach the top of its trajectory?
_____ 10. What is the maximum height (with respect to the ground) of the ball?
_____ 11. What is the ball's velocity at the top of its trajectory?
_____ 12. What is its acceleration at the top of its trajectory?
_____ 13. How long does it take the ball to get back down to its initial height?
_____ 14. What are the x and y components of the ball's velocity when it is back at the initial height?
_____ 15. What is the time of flight?
_____ 16. What is the magnitude of the impact velocity?
_____ 17. What is the direction of the impact velocity?
_____ 18. What is the range?

(See Appendix I for answers.)

---

## PRINCIPAL CONCEPTS AND EQUATIONS PRESCRIPTION

Your score on the practice is an excellent measure of your understanding of this chapter. You should now use the following chart to write you own prescription for curing any of your physics ills. Look down the leftmost column to the number of the question(s) you answered incorrectly, read across that row to see which section(s) of the study guide you should return to for further study, and then do the suggested text problems to gain additional experience in working with the particular concept.

| Practice Test Question | Concepts and Equations | Prescription: Principal Concept | Prescription: Text Problems |
|---|---|---|---|
| 1 | Components of a vector - analytical | 6 | 3-3,8 |
| 2 | Components of a vector - analytical | 6 | 3-3,8 |
| 3 | Adding vectors - analytical | 6 | 3-2,4 |
| 4 | Adding vectors - analytical | 7 | 3-4,5 |
| 5 | Adding vectors - analytical | 7 | 3-5,9 |
| 6 | Adding vectors - analytical | 7 | 3-9,10 |
| 7 | Components of a vector - analytical | 6 | 3-19b,21b |
| 8 | Components of a vector - analytical | 6 | 3-3,8 |
| 9 | $v_y = v_{oy} + a_y t$ | 8 | 3-23c,27a |
| 10 | $2a_y \Delta y = v_y^2 - v_{oy}^2$ | 8 | 3-17,23b |
| 11 | $v_x = v_{ox} + a_x t$ | 8 | 3-21 |
| 12 | Components of a vector - analytical | 6 | 3-3,8 |
| 13 | $\Delta y = v_{oy} t + a_y t^2/2$ | 8 | 3-18,25 |
| 14 | $v_x = v_{ox} + a_x t$ and $v_y = v_{oy} + a_y t$ | 8 | 3-31 |
| 15 | $\Delta y = v_{oy} + a_y t^2$ | 8 | 3-20,23 |
| 16 | $v_x = v_{ox} + a_x t$, $v_y = v_{oy} + a_y t$ | 8,6 | 3-21,26 |
| 17 | Direction of a vector - analytical | 7 | 3-19,22 |
| 18 | $\Delta x = v_{ox} t + a_x t^2/2$ | 8 | 3-22,27b |

# 4 Newton's Laws of Motion

RECALL FROM PREVIOUS CHAPTERS

| Previously learned concepts and equations frequently used in this chapter | Text Section | Study Guide Page |
|---|---|---|
| Resolving vectors into components | 3.2 | 32 |
| Adding vectors analytically | 3.2 | 32 |
| Definition of acceleration: $a = \Delta v/\Delta t$ | 2.4 | 21 |
| Equations for translational motion: $v = v_o + at$, $\Delta s = v_o t + at^2/2$, $2a\Delta s = v^2 - v_o^2$ | 2.4 | 22 |

NEW IDEAS IN THIS CHAPTER

| Concepts and equations introduced | Text Section | Study Guide Page |
|---|---|---|
| Newton's first law of motion: If $\sum \vec{F} = 0$, then $\vec{v}$ = constant | 4.2 | 44 |
| Newton's second law of motion (most common form): $\vec{F} = m\vec{a}$ | 4.3 | 47 |
| Newton's third law of motion: $\vec{F}_{12} = -\vec{F}_{21}$ | 4.5 | 57 |
| Normal force: $\vec{F}_N$ | 4.7 | 58 |
| Static and kinetic friction: $f_s \leq \mu_s F_N$ $\quad f_k = \mu_k F_N$ | 4.7 | 59 |

PRINCIPAL CONCEPTS AND EQUATIONS

## 1. Newton's First Law of Motion (Section 4.2)

Review: Newton's first law of motion states that, unless acted upon by an unbalanced force $\vec{F}_u$, every object at rest will remain at rest and every object in uniform motion will continue its motion in a straight line. The unbalanced force is obtained by vectorially adding all the forces acting on the object: $\vec{F}_u = \sum \vec{F}_i$. We may state the first law in symbols as follows:

If $\vec{F}_u = 0$, then $\vec{v}$ = constant (that is, the object's motion doesn't change)

Note: You should use the following procedure in working any problems involving forces:

1. Construct a diagram showing all forces acting on an object.
2. Establish a convenient coordinate system. If motion is a consideration, you will find it convenient to choose a coordinate system parallel and perpendicular to the possible direction of motion.
3. Replace the object under consideration with a point and draw in all forces roughly to scale. This is called a free-body diagram, and it results in a simpler, less cluttered drawing.
4. Resolve all forces into components along the choosen coordinate system. When you replace one of the original vectors by its components, be sure to indicate this so that you don't use both a vector and its components in a subsequent calculation.

Practice: A 200-N crate is at rest on a horizontal frictionless surface. The forces $\vec{F}_1$ and $\vec{F}_2$ are applied as shown in Fig. 4.1.

Figure 4.1

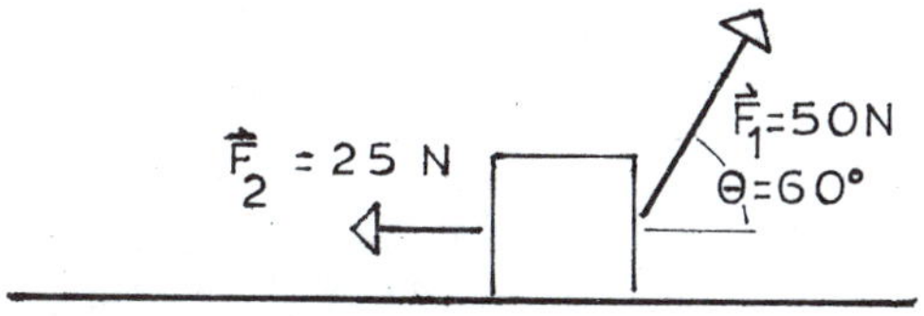


1. Draw a diagram showing all forces acting on the crate and an appropriate coordinate system.

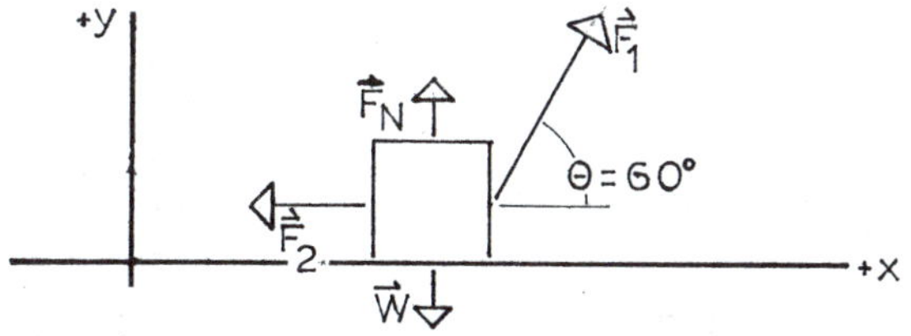


$\vec{W}$ = weight
$\vec{F}_N$ = normal or perpendicular force the surface exerts on the crate
$\vec{F}_1$ and $\vec{F}_2$ = applied forces

2. Draw a free-body diagram with all forces resolved into components along the coordinate system.

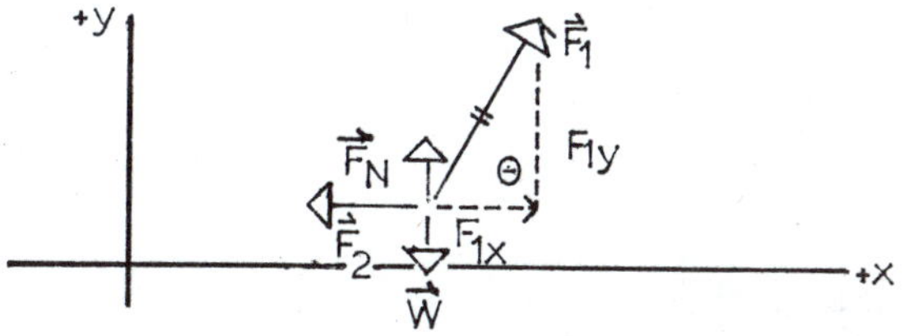


The two short lines through $\vec{F}_1$ indicate that this vector has been replaced by its components.

Determine the following:

3. The x component of $\vec{F}_1$

$F_{1x} = F_1 \cos\theta = (50\text{ N})\cos 60° = 25.0\text{ N}$

| | |
|---|---|
| 4. The x component of $\vec{F}_2$ | $F_{2x} = -F_2 = -25.0$ N |
| 5. The unbalanced force $F_{ux}$ in the x direction | $F_{ux} = F_{1x} + F_{2x} = 25.0 \text{ N} - 25.0 \text{ N} = 0$ |
| 6. Any change in motion in the x direction | Since no unbalanced force exists in the x direction, the crate will not change its motion in this direction. It will remain at rest at the same location. |
| 7. The y component of $\vec{F}_1$ and $\vec{W}$ | $F_{1y} = F_1 \sin\theta = (50 \text{ N})\sin 60 = 43.3$ N<br>$W_y = -W = -200$ N |
| 8. The magnitude of $\vec{F}_N$ | Since $F_{1y} < W$, we expect no change in the crate's motion in the y direction. If there is no change in motion, we expect the unbalanced force in the y direction to be zero, that is:<br>$F_N + F_{1y} + W_y = F_{uy}$, or<br>$F_N = F_{uy} - W_y - F_{1y}$<br>$F_N = 0 \text{ N} + 200 \text{ N} - 43.3 \text{ N} = 157$ N |

Example 4.1. A crate is pulled up a frictionless ramp inclined at an angle of 30° (Fig. 4.2). If the crate weighs 400 N, what is the magnitude of the force $\vec{F}$ needed to pull it up the ramp at a constant speed of 1 m/s?

Given: $W = 400$ N
$\theta = 30°$
$\alpha = 10°$
$v = 1$ m/s

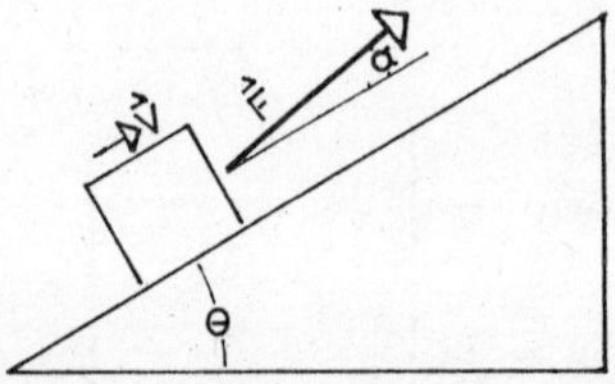


Figure 4.2

Determine: F, the magnitude of the force that will pull the crate up the ramp at a constant speed

Strategy: Construct a diagram showing all forces acting on the crate and choose a convenient coordinate system. Construct a free-body diagram and resolve all vectors into components along the coordinate system. Since the speed is constant, we know that the acceleration parallel to the ramp and hence the unbalanced force parallel to the ramp are zero. If the unbalanced force parallel to the ramp is zero, the component of $\vec{F}$ up the ramp must equal the component of $\vec{W}$ down the ramp. Knowing $\vec{W}$ and $\theta$, we can determine the component of $\vec{W}$ down the ramp and hence the component of $\vec{F}$ up the ramp. Knowing the component of $\vec{F}$ and $\alpha$, we can determine $\vec{F}$.

Solution:

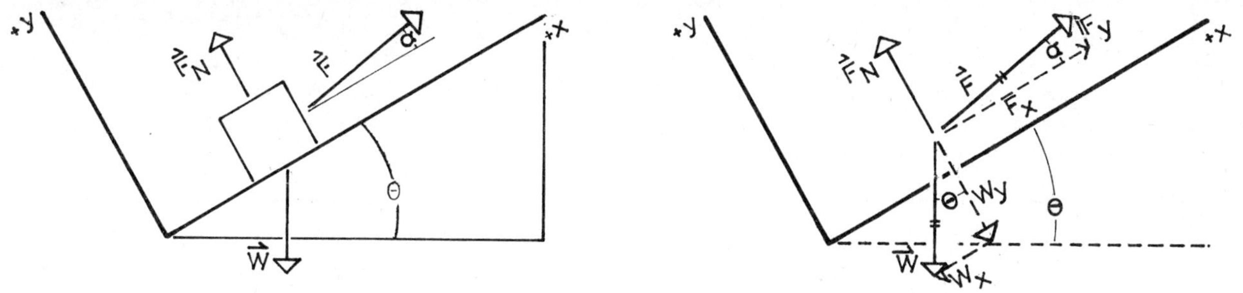


(a) All forces and a convenient coordinate system.

(b) A free-body diagram with all forces resolved into components along the coordinate system

Figure 4.3

A convenient coordinate system is one that has its axes parallel and perpendicular to the motion. The decision to put the positive x axis up or down the ramp is one of personal preference. Most students like to orient the axes so that the distance of the moving object from the origin increases with time. For this reason, we put the positive x axis up the ramp.

The fact that v is constant allows us to write the following:

$$v_x = \text{constant} \rightarrow \Delta v_x = 0 \rightarrow a_x = 0 \rightarrow F_{ux} = 0$$

Since the unbalanced force in the x direction is zero, we know that if we sum all the forces in the x direction, they must add to zero:

$$F_{ux} = F_x + W_x = 0$$

From this, we may write:

$$F_x = -W_x = -(-W \sin\theta) = W \sin\theta = (400 \text{ N}) \sin 30 = 200 \text{ N}$$

Finally, we may determine the magnitude of $\vec{F}$ as follows:

$$F_x = F \cos\alpha \quad \text{or} \quad F = F_x/\cos\alpha = (200 \text{ N})/\cos 10 = 203 \text{ N}$$

Related Text Problems: 4-1, 4-2.

## 2. Newton's Second Law of Motion (Section 4.3)

Review: Newton's second law of motion may be written in symbols as

$$\sum \vec{F} = \Delta(m\vec{v})/\Delta t$$

When the mass is constant and the velocity is changing, this expression becomes

$$\sum \vec{F} = \Delta(m\vec{v})/\Delta t = m\Delta\vec{v}/\Delta t = m\vec{a}$$

When the velocity is constant and the mass is changing, the expression becomes

$$\sum \vec{F} = \Delta(m\vec{v})/\Delta t = \vec{v}\Delta m/\Delta t$$

Practice: A 100-kg box is being pushed across the floor by a force $\vec{F}_1$. This force has a magnitude of 100 N and is directed 60° below the horizontal. The force of friction exerted on the box by the floor is 25 N.

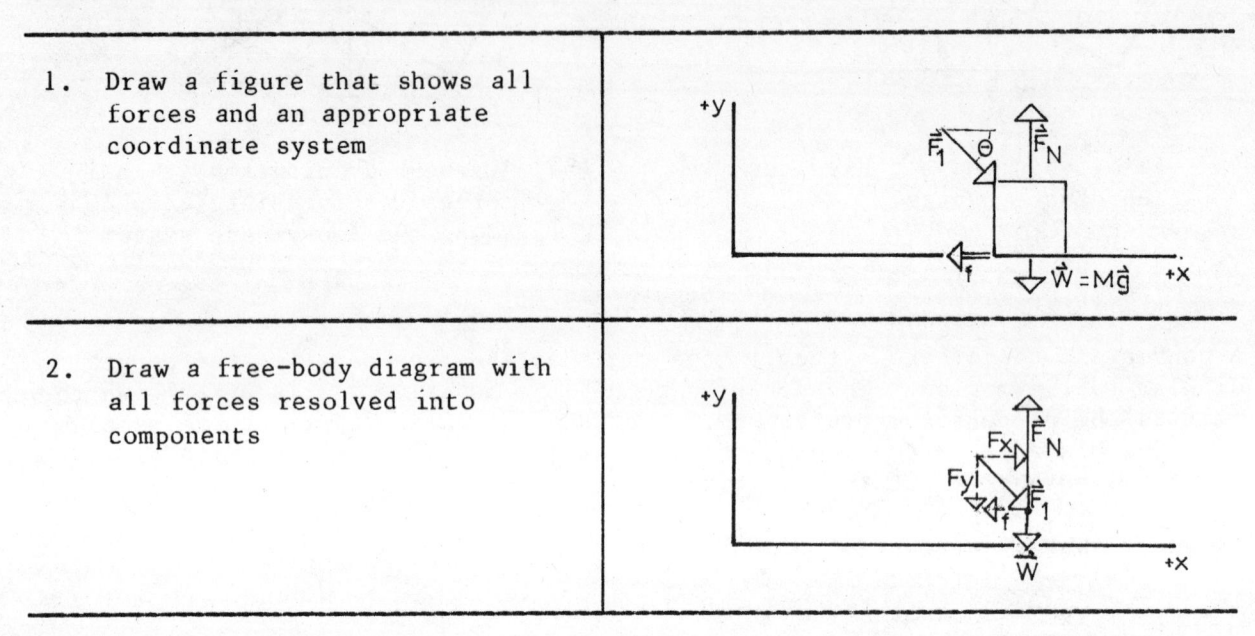

1. Draw a figure that shows all forces and an appropriate coordinate system

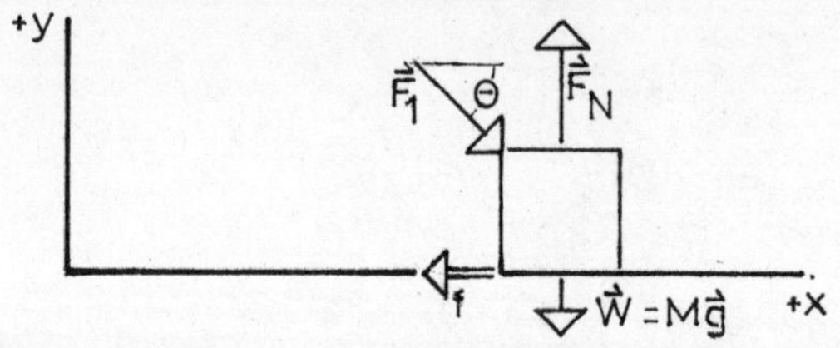


2. Draw a free-body diagram with all forces resolved into components

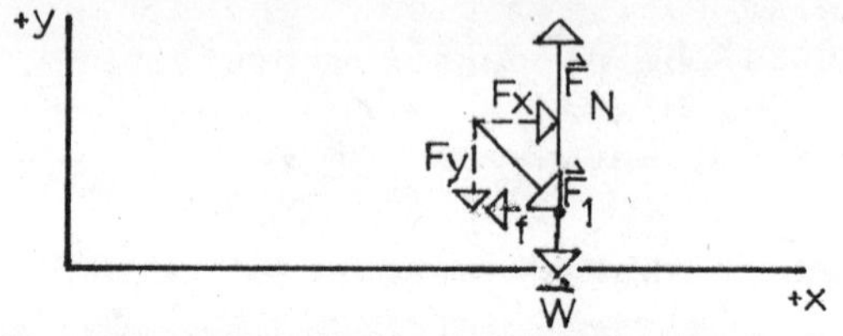


Now determine the following:

| | |
|---|---|
| 3. The component of $\vec{F}_1$ pushing the box in the positive x direction | $F_{1x} = F_1 \cos\theta = (100\ \text{N})\cos 60 = 50.0\ \text{N}$ |
| 4. The x component of the unbalanced force acting on the box | $F_{ux} = F_{1x} - f = 50.0\ \text{N} - 25.0\ \text{N} = 25.0\ \text{N}$<br>This force is in the positive x direction |
| 5. The magnitude of the acceleration of the box | $F_{ux} = ma_x$ or $a_x = F_{ux}/m = 0.250\ \text{m/s}^2$ |

Example 4.2. A hockey puck slides 40 m down the rink before coming to rest. If the frictional drag is one tenth the weight of the puck, what was the puck's initial speed?

Given: $v = 0$, $\Delta s = 40$ m, $f = 0.1W$

Determine: $v_o$, the initial speed of the puck

Strategy: The unbalanced force on the puck is due to friction. Since the unbalanced force is known in terms of the weight of the puck, we can use Newton's second law to determine the acceleration. Knowing $v_f$, a, and $\Delta s$, we can use our knowledge of kinematics ($v^2 - v_o^2 = 2a\Delta s$) to determine $v_o$.

Solution: Since the unbalanced force on the puck is due to friction, we can write

$$F_u = -f = -0.1W$$

The minus sign indicates that this is a retarding force. According to Newton's second law,

$$F_u = ma = Wa/g$$

Equating these two expressions for $F_u$, we obtain

$$Wa/g = -0.1W \quad \text{or} \quad a = -0.1g = -0.980 \text{ m/s}^2$$

Recall from our study of kinematics that $v^2 - v_o^2 = 2a\Delta s$. Since v = 0, a = $-0.980$ m/s$^2$, and $\Delta s$ = 40 m, this reduces to

$$v_o = \pm(-2a\Delta s)^{1/2} = \pm[-2(-0.980 \text{ m/s}^2)(40.0 \text{ m})]^{1/2} = 8.85 \text{ m/s}$$

The positive root is chosen since the initial direction of the motion was taken to be positive. Recall that we gave the retarding force a negative sign.

Example 4.3. The takeoff speed for a particular 50,000-kg jet is 200 km/h, and it can acquire this speed after 50 s. Assuming that the thrust force exerted by the jet is constant, calculate this thrust and the minimum length of runway needed.

Given: $v_o$ = 0 km/h, $v_f$ = 200 km/h, m = $5.00 \times 10^4$ kg, t = 50.0 s

Determine: F, the thrust force of the jet engines
$\Delta s$, the minimum length of runway needed for takeoff

Strategy: Knowing $v_o$ and v determine $\Delta v$. Knowing $\Delta v$ and t, determine a. Determine F from a and m. Knowing $v_o$, v, and a, use your knowledge of kinematics ($v^2 - v_o^2 = 2a\Delta s$) to determine $\Delta s$.

Solution:

$$\Delta v = v - v_o = (2 \times 10^2 \text{ km/h})(10^3 \text{ m/km})(\text{h}/3.60 \times 10^3 \text{ s}) = 55.6 \text{ m/s}$$

$$a = \Delta v/\Delta t = (55.6 \text{ m/s})/\ 50.0 \text{ s} = 1.11 \text{ m/s}^2$$

$$F = ma = (5.00 \times 10^4 \text{ kg})(1.11 \text{ m/s}^2) = 5.55 \times 10^4 \text{ N}$$

$v^2 - v_o^2 = 2a\Delta s$, which may be solved for $\Delta s$ to obtain

$$\Delta s = v^2/2a = (55.6 \text{ m/s})^2/(2)(1.11 \text{ m/s}^2) = 1.39 \times 10^3 \text{ m}$$

Related Text Problems: Since the majority of problems in this chapter deal with Newton's second law, they have been further classified as follows:

Problems involving a straightforward application of Newton's second law: 4-3 through 4-6.

Problems involving Newton's second law and our knowledge of kinematics: 4-7, 4-8, 4-10, 4-12, 4-14 through 4-19, 4-24.

Problems involving Newton's second law in the form $F = v\Delta m/\Delta t$: 4-9, 4-11.

Problems involving Newton's second law to solve elevator problems: 4-13, 4-20, 4-21. (See concept 3 below.)

Problems involving Newton's second law on a system of objects: 4-30, 4-34, 4-36, 4-45. (See concept 4 on page 44.)

## 3. Newton's Second Law Applied to Elevator Problems (Section 4.3)

Note: Newton's second law was introduced and reviewed in Sec. 4.3. However, since its application to elevator problems tends to give students some difficulty, we consider this application in a separate section.

Practice: An elevator is designed to travel at a constant speed of 4 m/s. It stops in a distance of 2 m and can accelerate to its cruising speed in 4 m. The elevator is equipped with an ordinary bathroom scale. A 60-kg woman gets in the elevator and steps on the scale.

1. Sketch this situation and show a concenient coordinate system and all forces acting on the woman.

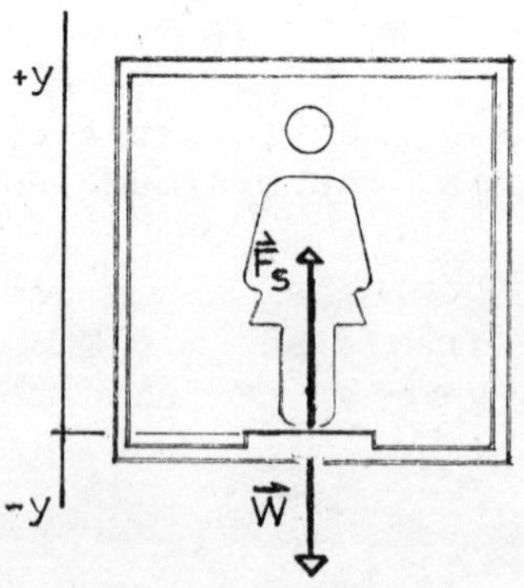


$W = mg$ = woman's weight
$F_s$ = force scale exerts on woman (scale reading)

Now determine the following:

| | |
|---|---|
| 2. The acceleration of the woman when the elevator is sitting still, traveling upward at a constant speed, or traveling downward at a constant speed | When the elevator is sitting still or traveling at a constant speed, its acceleration and hence that of the woman are zero. |
| 3. The acceleration of the woman when the elevator accelerates upward to its cruising speed | $v^2 - v_o^2 = 2a\Delta x \rightarrow a = v^2/2\Delta x$<br><br>$a = (4\ m/s)^2/2(4\ m) = 2.00\ m/s^2$ |
| 4. The acceleration of the woman when the elevator accelerates downward to its cruising speed | This acceleration will have the same magnitude as in step 3 but will be in the opposite direction ($a = -2.00\ m/s^2$) |
| 5. The acceleration of the woman as the elevator slows to a stop while traveling upward | $v^2 - v_o^2 = 2a\Delta x$ or $a = -v_o^2/2\Delta x$<br><br>$a = -(4\ m/s)^2/2(2\ m) = -4.00\ m/s^2$ |
| 6. The acceleration of the woman as the elevator slows to a stop while traveling downward | This acceleration will have the same magnitude as in step 5 but will be in the oposite direction ($a = 4.00\ m/s^2$). |
| 7. An expression for the force the scale exerts on the woman at any time | By Newton's second law, $F_u = ma$, but by the drawing in step 1 we see that the unbalanced force on the woman at any time is $F_u = F_s - mg$. Equate these expressions for $F_u$ and solve for $F_s$ to get $F_s = m(g + a)$. Note that in this expression $g = +9.80\ m/s^2$ since the direction has already been taken into consideration. |
| 8. The scale reading when the elevator is at rest or traveling at a constant speed | For this case, $a = 0$ (step 2).<br><br>$F_s = m(g + a) = (60\ kg)(9.80\ m/s^2) = 588\ N$ |
| 9. The scale reading when the elevator is accelerating upward to cruising speed | For this case, $a = 2.00\ m/s^2$ (step 3).<br><br>$F_s = m(g + a) = (60\ kg)(11.8\ m/s^2) = 708\ N$ |

| | |
|---|---|
| 10. The scale reading when the elevator slows to a stop while traveling upward | For this case, $a = -4.00\ m/s^2$ (step 5). $F_s = m(g + a) = 348\ N$ |
| 11. The scale reading when the elevator is accelerating downward to cruising speed | For this case, $a = -2.00\ m/s^2$ (step 4). $F_s = m(g + a) = 468\ N$ |
| 12. The scale reading when the elevator slows to a stop while traveling downward | For this case, $a = +4.00\ m/s^2$ (step 6). $F_s = m(g + a) = 828\ N$ |

Related Text Problems: 4-13, 4-20, 4-21.

## 4. Newton's Second Law Applied to Systems of Masses (Sections 4.3 and 4.6)

Note: Newton's second law was introduced, reviewed, and applied to individual masses in principal concept 3. However, since the application of the law to a system of masses is somewhat complicated, we will consider this application in a separate section.

Practice: The three masses in Fig. 4.4 are pulled upward by a 120-N force. Let's agree to call all forces that aid the motion positive and all forces that oppose the motion negative. $\vec{T}_A$ and $\vec{T}_B$ represent the tension in strings A and B respectively.

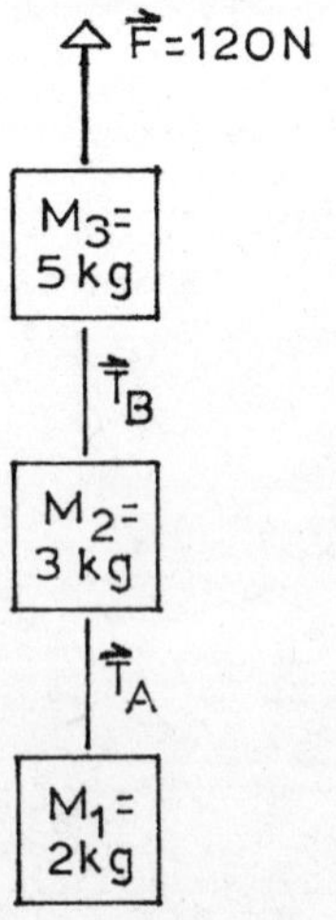


Figure 4.4

Determine the following:

| | |
|---|---|
| 1. The equation of Newton's second law for $M_1$ | 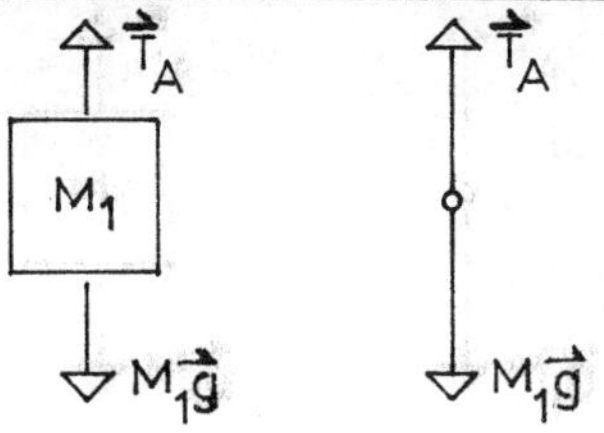  According to Newton's second law, the unbalanced force on $M_1$ is $F_{u1} = M_1a$. Referring to the free-body diagram, we see that $F_{u1} = T_A - M_1g$. Combining these two expressions, we obtain the equation for $M_1$: (a) $M_1a = T_A - M_1g$ |
| 2. The equation of Newton's second law for $M_2$ | 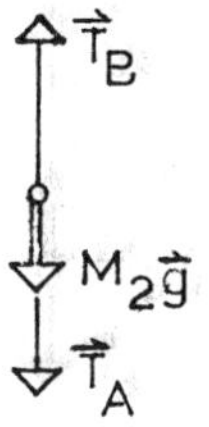  $F_{u2} = M_2a$ <br> $F_{u2} = T_B - T_A - M_2g$ <br> (b) $M_2a = T_B - T_A - M_2g$ |
| 3. The equation of Newton's second law for $M_3$ | 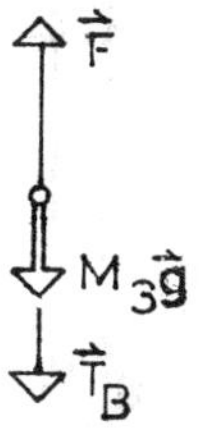  $F_{u3} = M_3a$ <br> $F_{u3} = F - T_B - M_3g$ <br> (c) $M_3a = F - T_B - M_3g$ |
| 4. The equation of Newton's second law for $M_1$ and $M_2$ as a system | 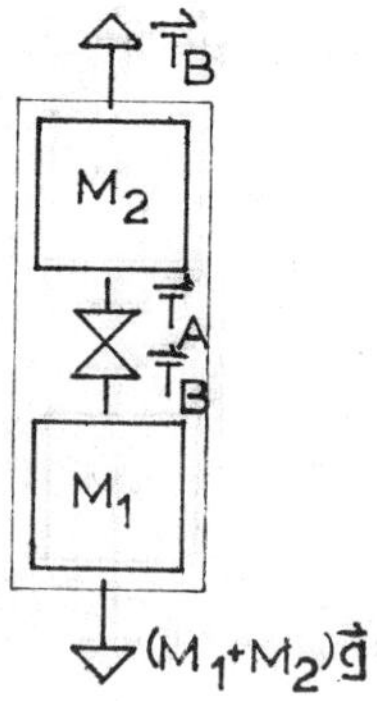  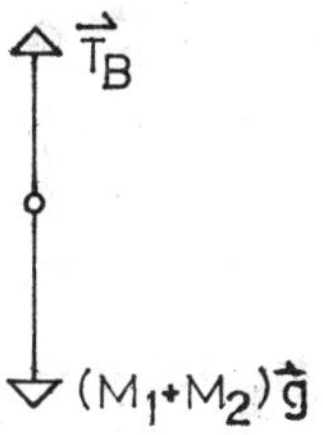  $F_{u12} = (M_1 + M_2)a$ <br> $F_{u12} = T_B - (M_1 + M_2)g$ <br> (d) $(M_1 + M_2)a = T_B - (M_1 + M_2)g$ |

| | |
|---|---|
| 5. The equation of Newton's second law for $M_2$ and $M_3$ as a system | 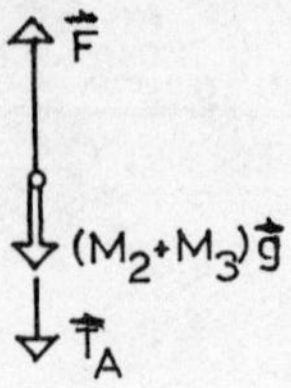<br><br>$F_{u23} = (M_2 + M_3)a$<br>$F_{u23} = F - T_B - (M_2 + M_3)g$<br>(e) $(M_2 + M_3)a = F - T_A - (M_2 + M_3)g$ |
| 6. The equation of Newton's second law for $M_1 + M_2 + M_3$ as a system | 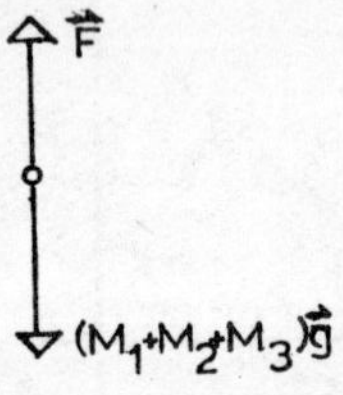<br><br>$F_{u123} = (M_1 + M_2 + M_3)a$<br>$F_{u123} = F - (M_1 + M_2 + M_3)g$<br>(f) $(M_1 + M_2 + M_3)a = F - (M_1 + M_2 + M_3)g$ |

Note: In steps 1 through 6, the mass and the unbalanced force have been matched. We used $F_{u1}$ for $M_1$, $F_{u2}$ for $M_2$, $F_{u3}$ for $M_3$, $F_{u12}$ for $M_1 + M_2$, $F_{u23}$ for $M_2 + M_3$ and $F_{u123}$ for $M_1 + M_2 + M_3$.

| | |
|---|---|
| 7. The acceleration of the system. | Since we have determined the second-law equation for every mass and every combination of masses, we have three methods available for determining a:<br>I. Use Eq. f<br>II. Use Eqs. a and e to eliminate $T_A$<br>III. Use Eqs. c and d to eliminate $T_B$<br><br>Let's use method I.<br><br>(f) $a = [F - (M_1 + M_2 + M_3)g]/(M_1 + M_2 + M_3)$<br>$= 2.20\ m/s^2$ |
| 8. The tension $T_A$ | Now that a is known use either Eq. a or Eq. e<br><br>(a) $T_A = M_1 a + M_1 g = 24.4\ N$ |
| 9. The tension $T_B$ | Now that a is known use either Eq. c or Eq. d.<br><br>(c) $T_B = F - M_3 g - M_3 a = 60.0\ N$ |

Note: In solving problems that involve a system of masses, you will usually write down only enough of Newton's second law equations to allow you to determine the unknown quantities.

Example 4.4. Consider the situation in Fig. 4.5 and then determine the acceleration of the system and the tension in the cord. Ignore friction and assume that the cord does not stretch.

Given: $M_1 = 10$ kg

$M_2 = 20$ kg

$g = 9.80$ m/s$^2$

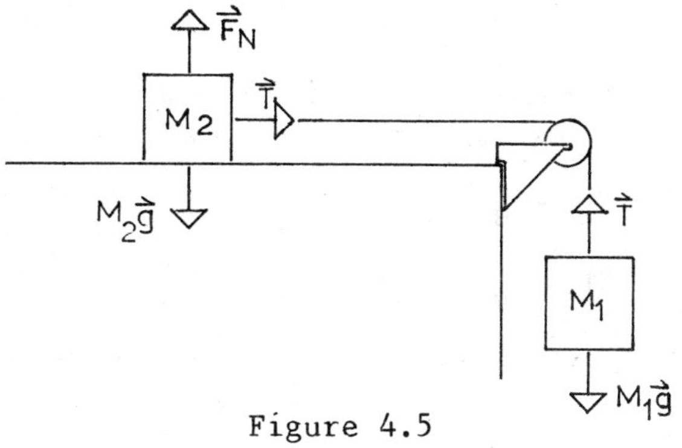


Figure 4.5

Determine: a, acceleration of the system
T, tension in the cord

Strategy: Write down the second-law equation for $M_1$ and $M_2$ individually and for the system $M_1 + M_2$. These three equations involve only the two unknowns a and T. Since we have three equations and two unknowns, we can determine the unknowns and then check our work.

Solution: Let's agree that any force that aids the motion is positive and any that opposes the motion is negative.

Determine the second law equation for $M_1$, $M_2$, and $M_1 + M_2$:

$M_1$: $F_{u1} = M_1a$; $F_{u1} = M_1g - T$; hence $M_1a = M_1g - T$ (a)

$M_2$: $F_{u2} = M_2a$; $F_{u2} = T$; hence $M_2a = T$ (b)

$M_1 + M_2$: $F_{u12} = (M_1 + M_2)a$; $F_{u12} = M_1g$; hence $(M_1 + M_2)a = M_1g$ (c)

Notice that we have three equations and two unknowns; consequently, we may proceed by either of two methods.

Method I. Solve Eq. c for a and then insert this value into either Eq. a or Eq. b to obtain T. Let's do this algebraically:

From (c) $a = M_1g/(M_1 + M_2)$

From (b) $T = M_2a = M_2M_1g/(M_1 + M_2)$

From (a) $T = M_1(g - a) = M_1g\{(1 - [m_1/(m_1 + m_2)\} = M_1M_2g/(M_1 + M_2)$

Method II. Solve Eqs. a and b simultaneously for a and T:

(a) $M_1a = M_1g - T$, but $T = M_2a$ by (b), hence

$M_1a = M_1g - M_2a$ or $a = M_1g/(M_1 + M_2)$

Insert this value for a into either Eq. a or b to determine T.

From (b), $T = M_2a = M_2M_1g/(M_1 + M_2)$

Notice that the expressions from the two methods agree. Now that we have expressions for a and T, let's obtain numerical values.

$$a = M_1g/(M_1 + M_2) = (10\text{ kg})(9.80\text{ m/s}^2)/(10\text{ kg} + 20\text{ kg}) = 3.27\text{ m/s}^2$$

$$T = M_2M_1g/(M_1 + M_2) = (20\text{ kg})(3.27\text{ m/s}^2) = 65.4\text{ N}$$

<u>Note</u>: When working problems dealing with a system of masses, you can check your work by the following methods.

1. Units Check - The expressions for a and T should have units of $m/s^2$ and N, respectively.

2. Algebra Check - Manipulate the algebra two different ways or insert the expression for a back into both expressions for T.

3. Reduction Check - Many complex problems can be reduced to a simpler problem that has previously been solved. If the reduction agrees with the previous solution then you can feel more confident that your work is correct. For example, the expressions for the acceleration and tension in the ramp situation shown in Fig. 4.6 (text problem 4-37) reduce to the problem just solved if we let $\theta = 0$.

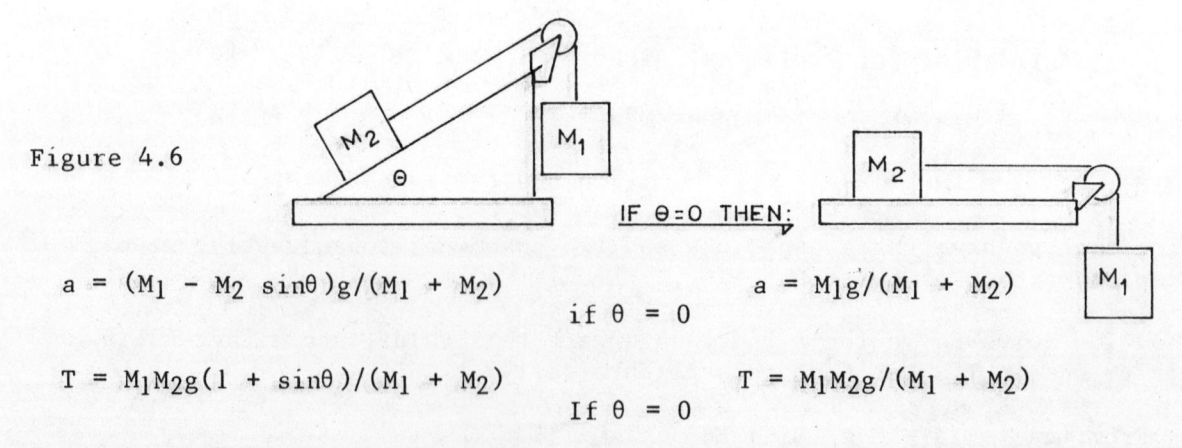

Figure 4.6

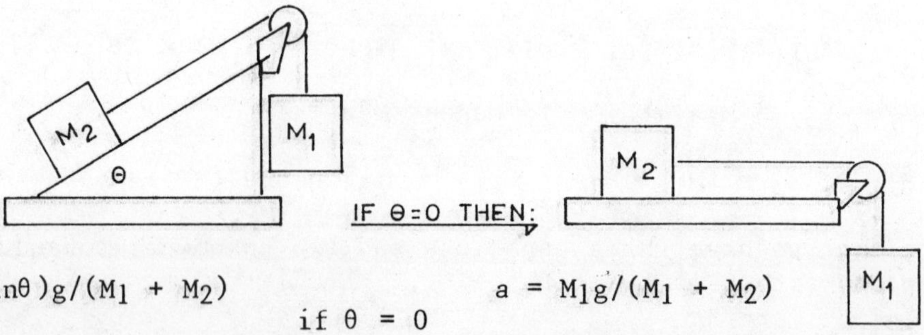


$a = (M_1 - M_2\sin\theta)g/(M_1 + M_2)$ if $\theta = 0$ $a = M_1g/(M_1 + M_2)$

$T = M_1M_2g(1 + \sin\theta)/(M_1 + M_2)$ If $\theta = 0$ $T = M_1M_2g/(M_1 + M_2)$

If $\theta = 0$, then the two figures and the expressions for a and T are identical.

<u>Related Problems</u>: 4-30, 4-34, 4-36, 4-45.

## 5. Newton's Third Law of Motion (Section 4.5)

Review: Newton's third law of motion states that whenever one object exerts a force on a second object, the second object exerts on the first a force equal in magnitude and opposite in direction.

Remember that the two forces to which the third law refers (often called an action-reaction pair) always act on different bodies.

If we let $\vec{F}_{AB}$ represent the force that body A exerts on body B and $\vec{F}_{BA}$ represent the force that body B exerts on body A, then we can write Newton's third law in symbols as

$$\vec{F}_{AB} = -\vec{F}_{BA} \tag{4-9}$$

Practice: A monkey hangs from a branch as shown in Fig. 4.7. All forces on the monkey, tree, and earth have been drawn in and labeled.

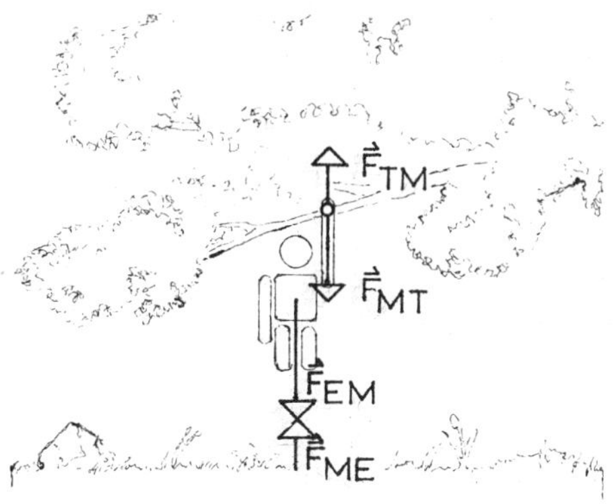

Figure 4.7

$\vec{F}_{EM}$ = force the earth exerts on the monkey (weight of monkey)

$\vec{F}_{ME}$ = force the monkey exerts on the earth

$\vec{F}_{TM}$ = force the tree exerts on the monkey

$\vec{F}_{MT}$ = force the monkey exerts on the tree

Determine the following:

| | |
|---|---|
| 1. Newton's third-law equation involving $\vec{F}_{MT}$ | $\vec{F}_{MT} = -\vec{F}_{TM}$ <br> One force is on the tree ($\vec{F}_{MT}$), and the other is on the monkey ($\vec{F}_{TM}$). |
| 2. Newton's third-law equation involving $\vec{F}_{EM}$ | $\vec{F}_{EM} = -\vec{F}_{ME}$ <br> One force is on the monkey ($\vec{F}_{EM}$), and the other is on the earth ($\vec{F}_{ME}$). |
| 3. Since the monkey's motion is not changing, we may write <br> $\vec{F}_{EM} = -\vec{F}_{TM}$ <br> Is this another third-law equation? | No, because both $\vec{F}_{EM}$ and $\vec{F}_{TM}$ are acting on the monkey. Forces of a third-law pair act on different objects. This particular force statement says that the unbalanced force on the monkey is zero; hence its motion will not change. |

Related Text Problem: 4-35.

## 6. Determinating the Normal Force for Different Situations (Section 4.7)

Review: The normal force $\vec{F}_N$ is the perpendicular force that a surface exerts on an object with which it is in contact.

Practice: Consider the following three situations:

| | | |
|---|---|---|
| The mass M is being pulled across the floor | The mass M is being pushed across the floor | The mass M is being pushed up the ramp |
| 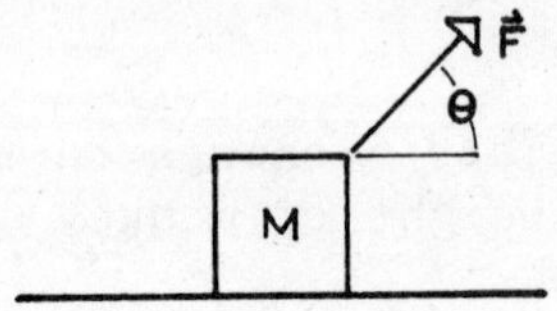 | 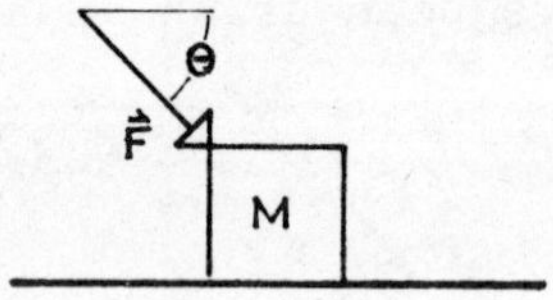 | 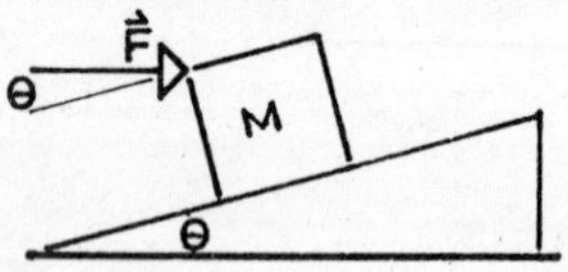 |
| Know: M, F, θ, g | Know: M, F, θ, g | Know: M, F, θ, g |

1. For each situation, draw a figure showing all forces acting on M and a convenient coordinate system.

| | | |
|---|---|---|
| 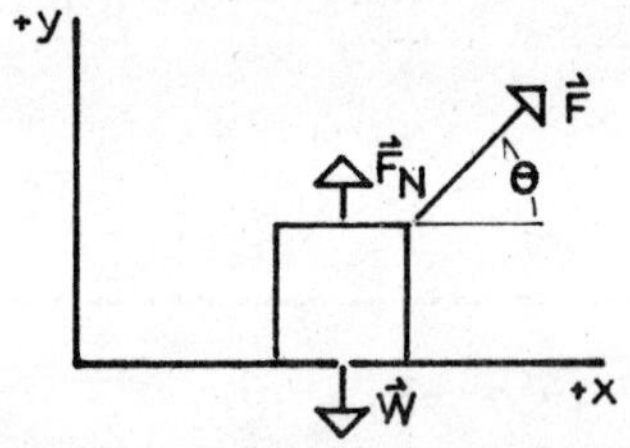 | 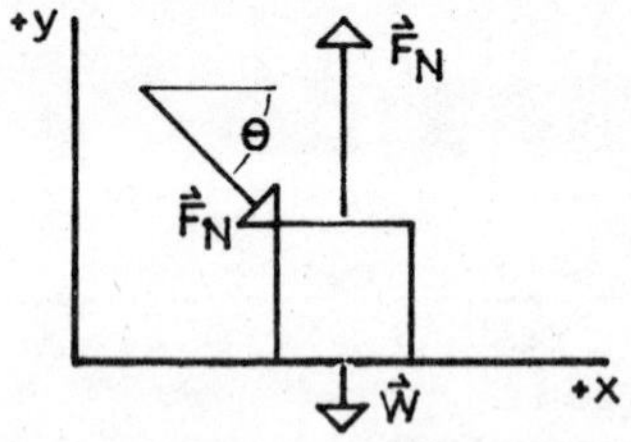 | 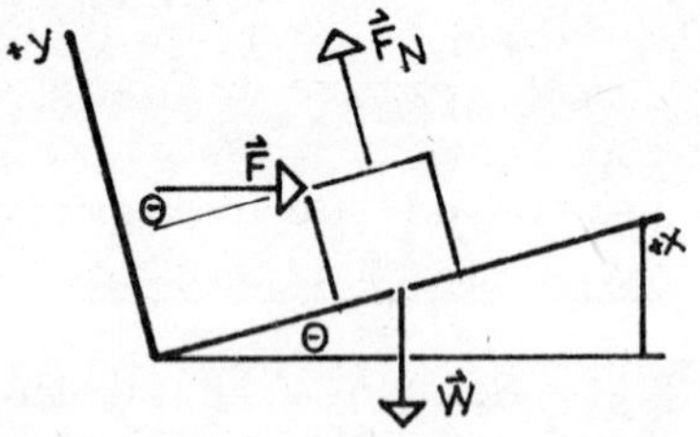 |

2. For each situation, draw a free-body diagram with all the forces resolved into components.

| | | |
|---|---|---|
| 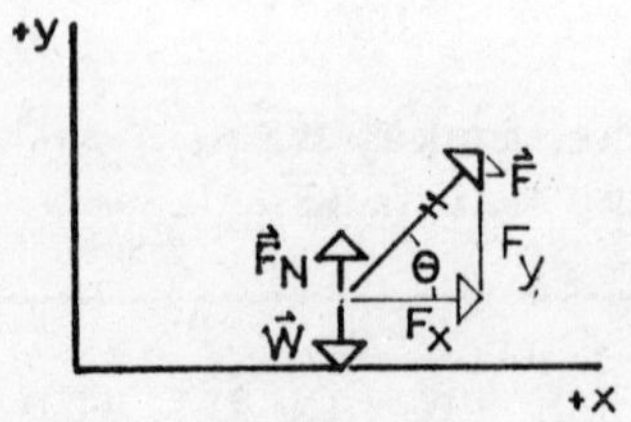 | 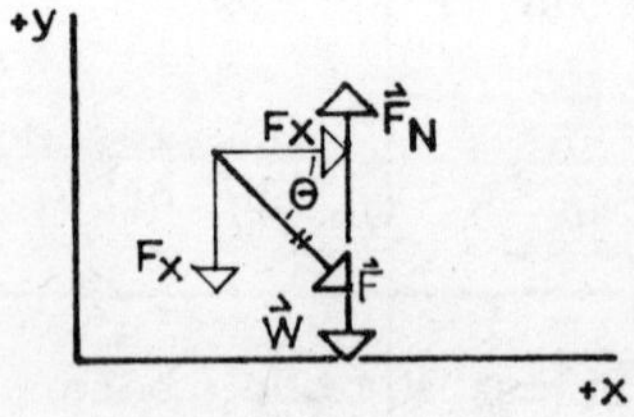 | 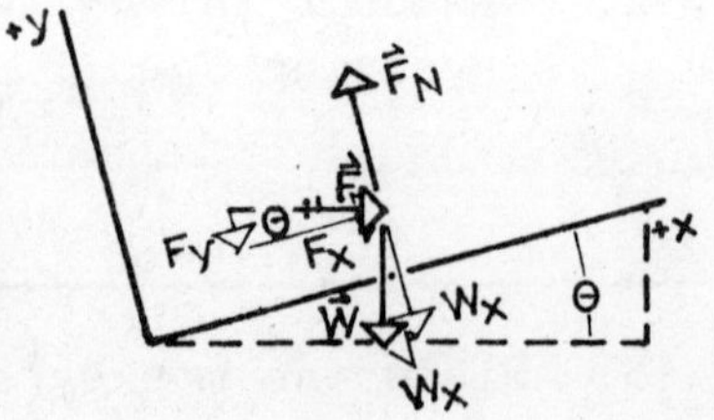 |

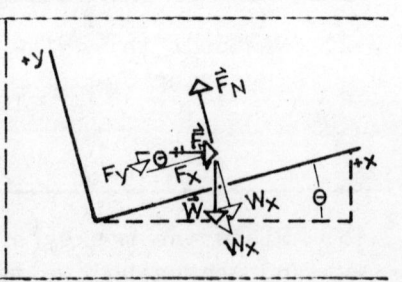

Now determine the following:

3. The total force in the -y direction for each situation

| | | |
|---|---|---|
| W | W + F sinθ | W cosθ + F sinθ |

4. The total force in the y direction for each situation

| | | |
|---|---|---|
| $F_N + F \sin\theta$ | $F_N$ | $F_N$ |

5. The unbalanced force in the y direction for each situation (assuming M does not leave the surface)

| | | |
|---|---|---|
| $F_{uy} = 0$ | $F_{uy} = 0$ | $F_{uy} = 0$ |

6. Any relationship that exists between the total force in the y direction and the total force in the -y direction

| | | |
|---|---|---|
| Since $F_{uy} = 0$, then $F_{+y} = F_{-y}$ or $F_N + F \sin\theta = W$ | Since $F_{uy} = 0$, then $F_{+y} = F_{-y}$ or $F_N = W + F \sin\theta$ | Since $F_{uy} = 0$, then $F_{+y} = F_{-y}$ or $F_N = W \cos\theta + F \sin\theta$ |

7. An expression for the normal force for each situation

| | | |
|---|---|---|
| $F_N = W - F \sin\theta$<br>The surface doesn't have to support the full weight of the object since $\vec{F}$ has an upward component. | $F_N = W + F \sin\theta$<br>The surface must support not only the weight of the object but also the downward component of $\vec{F}$. | $F_N = W \cos\theta + F \sin\theta$<br>The surface must support a component of the weight and a component of the force. |

## 7. Static and Kinetic Friction (Section 4.7)

Review: Friction always acts so as to oppose the start or the continuance of motion. Static friction opposes the start of motion and is given by

$$f_s \leq \mu_s F_N$$

Kinetic friction opposes the continuance of motion and is given by

$$f_k = \mu_k F_N$$

where $\mu_s$ and $\mu_k$ are the coefficients of static and kinetic friction.

<u>Practice</u>: A 100-kg crate sits at rest on the floor, and a horizontal force $\vec{F}$ is applied as shown in figure 4.8.

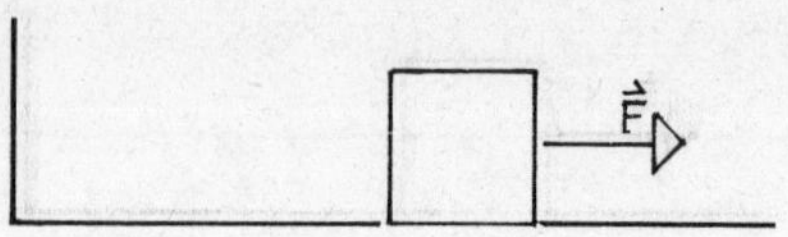


$\mu_s = 0.4$, $\mu_k = 0.2$, $M = 100$ kg

Figure 4.8

| | |
|---|---|
| 1. Draw a figure showing all forces acting on the crate and a convenient coordinate system. | 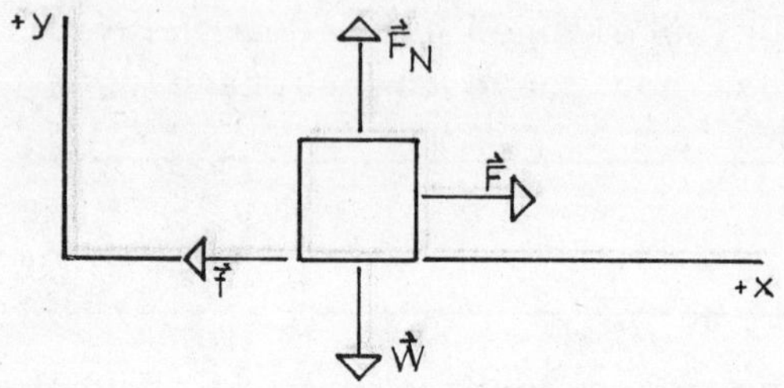  |
| 2. Draw a free-body diagram with all forces resolved into components. | 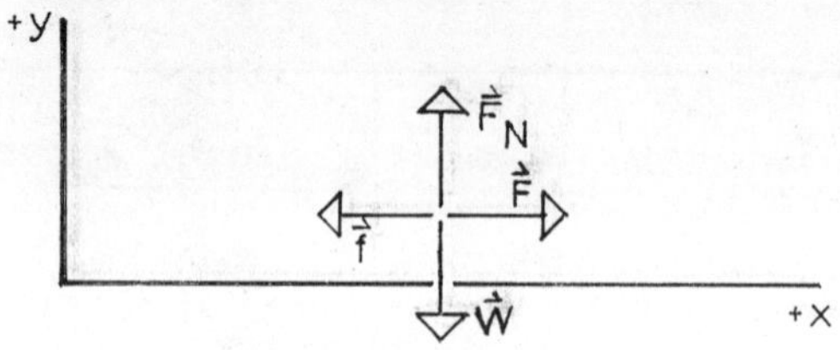  |

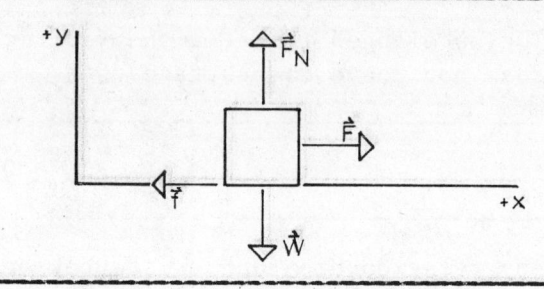

Now determine the following:

| | |
|---|---|
| 3. The normal force the floor exerts on the crate | Since the crate is not changing its motion in the y direction, we know that $F_{uy} = 0$, hence<br>$F_N = W = Mg = (100\ \text{kg})(9.80\ \text{m/s}^2) = 980\ \text{N}$ |
| 4. The maximum force of static friction | $f_{s\text{-}max} = \mu_s F_N = (0.4)(980\ \text{N}) = 392\ \text{N}$ |
| 5. The force of friction when F = 100 N | $f_s \leqslant \mu_s F_N = 392\ \text{N}$<br>That is, $f_s$ can have any value between 0 and 392 N. In this case, its value needs to be only 100 N in order to keep the crate stationary. |
| 6. The acceleration of the crate when F = 100 N | $F_{ux} = F_x - f_s = 100\ \text{N} - 100\ \text{N} = 0\ \text{N}$<br>$Ma_x = F_{ux} = 0$, hence $a_x = 0\ \text{m/s}^2$ |

| | |
|---|---|
| 7. The force of friction and the acceleration when $F = 200$ N | The force $f_s$ can have any value between 0 and 392 N. Since $F = 200$ N, we need $f_s = 200$ N. Since $F = f_s$ and they are in opposite directions, $F_{ux} = 0$ and hence $a_x = 0$. |
| 8. The force of friction and the acceleration when $F = 392$ N if $v_o = 0$ | Since $F = 392$ N, we need $f_s = 392$ N, which gives $F_{ux} = 0$ and $a_x = 0$. |
| 9. The force of friction and the acceleration when $F = 392$ N if $v_o = 0.1$ m/s | Since the object is moving, use $\mu_k$. $f_k = \mu_k F_N = (0.2)(980 \text{ N}) = 196 \text{ N}$ $F_{ux} = F - f_k = 392 \text{ N} - 196 \text{ N} = 196 \text{ N}$ $a_x = \frac{F_{ux}}{m} = \frac{196 \text{ N}}{100 \text{ kg}} = 1.96 \text{ m/s}^2$ |
| 10. The force of friction and the acceleration when $F = 500$ N if $v_o = 0$ m/s | $f_k = \mu_k F_N = (0.2)(980 \text{ N}) = 196 \text{ N}$ $F_{ux} = F - f_k = 500 \text{ N} - 196 \text{ N} = 304 \text{ N}$ $a_x = \frac{F_x}{M} = \frac{304 \text{ N}}{100 \text{ kg}} = 3.04 \text{ m/s}^2$ |

Example 4.5. A 100-kg crate sits on a flatbed truck as shown in Fig. 4.9. The truck is traveling 10 m/s, and the coefficient of static and kinetic friction between the crate and the truckbed are $\mu_s = 0.3$ and $\mu_k = 0.2$. What is the minimum distance in which the truck can stop so that the crate will not slide?

Given: $M = 100$ kg
$g = 9.80$ m/s2
$\mu_s = 0.3$
$\mu_k = 0.2$
$v_o = 10$ m/s
$v_f = 0$

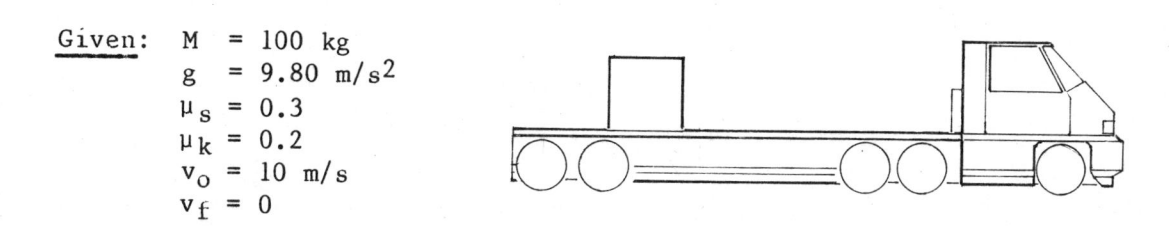

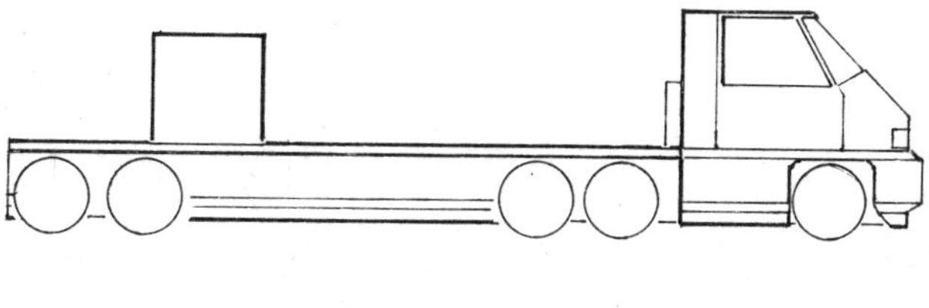

Figure 4.9

Determine: $\Delta s$, the minimum distance in which the truck can stop without the crate sliding.

Strategy: Determine the maximum force ($f_{s-max}$) static friction can exert to keep the crate stationary. If the truck stops so fast that a force bigger than $f_{s-max}$ is needed to keep the crate stationary, it will slide forward; consequenty $f_{s-max}$ dictates the maximum deceleration (-a). Use $f_{s-max}$ in Newton's second law to determine this maximum deceleration. Knowing $v_o$, v, and $a_{max}$, we can use our knowledge of kinematics ($v^2 - v_o^2 = 2a\Delta s$) to determine the minimum distance ($\Delta s$) the truck can stop in without the crate sliding forward.

Solution: (a) The normal force the truckbed exerts on the crate is

$$F_N = W = Mg = (100\ \text{kg})(9.80\ \text{m/s}^2) = 980\ \text{N}$$

The maximum force static friction can exert to hold the crate stationary is

$$f_{s-max} = \mu_s F_N = (0.3)(980\ \text{N}) = 294\ \text{N}$$

The maximum deceleration this force can provide is obtained from Newton's second law:

$$a_f = -f_{s-max}/M = -294\ \text{N}/100\ \text{kg} = -2.94\ \text{m/s}^2$$

If the truck decelerates more than this, the crate will slide. Since we don't want it to slide, we must require that

$$a_{T-max} = a_f = -2.94\ \text{m/s}^2$$

The distance the truck can stop in with this deceleration is

$v^2 - v_o^2 = 2a\Delta s$, where $v = 0$, $v_o = 10$ m/s , and $a = -2.94\ \text{m/s}^2$; hence

$$\Delta s = (v^2 - v_o^2)/2a = -(10\ \text{m/s})^2/(2)(-2.94\ \text{m/s}^2) = 17.0\ \text{m}$$

Related Text Problems: 4-22, 4-36 through 4-46.

---

Practice Test:

Take and grade this practice test. Doing so will allow you to determine any weak spots in your understanding of the concepts taught in this chapter. The following section prescribes what you should study further to strengthen your understanding.

The system shown in Fig. 4.10 is pulled by the force F.

| | | | |
|---|---|---|---|
| $M_A$ | = 20 kg | $\mu_s$ | = 0.20 |
| $M_B$ | = 30 kg | $\mu_k$ | = 0.10 |
| $M_C$ | = 10 kg | $v_o$ | = 0 m/s |
| F | = 100 N | $\theta$ | = 60° |
| g | = 9.80 m/s$^2$ | $\alpha$ | = 30° |

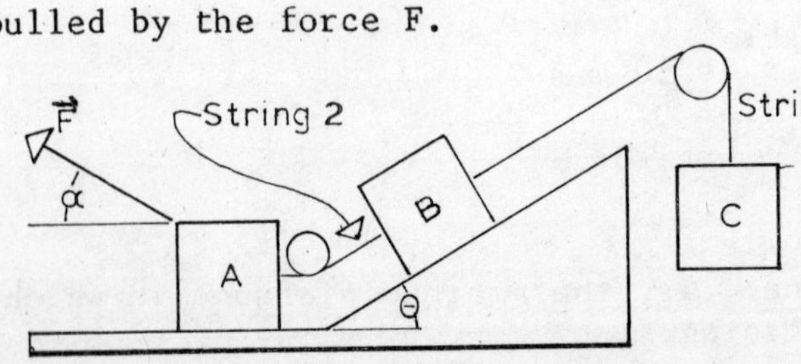


Figure 4.10

Determine the following:

_____ 1. The normal force on B
_____ 2. The normal force on A
_____ 3. The force of friction on A
_____ 4. The force of friction on B
_____ 5. The component of B's weight acting parallel to the ramp
_____ 6. The unbalanced force on the system
_____ 7. The acceleration of the system
_____ 8. The tension in string 1
_____ 9. The tension in string 2
_____ 10. The unbalanced force on A
_____ 11. The unbalanced force on B
_____ 12. The unbalanced force on C
_____ 13. The speed of the system after 1 s
_____ 14. The distance the system travels in 1 s
_____ 15. The speed of the system after it has traveled 1 m

(See Appendix I for answers).

## PRINCIPAL CONCEPTS AND EQUATIONS PRESCRIPTIONS

Your score on the practice test is an excellent measure of your understanding of this chapter. You should now use the following chart to write your own prescription for curing any of your physics ills. Look down the leftmost column to the number of the question(s) you answered incorrectly, read across that row to see which section(s) of the study guide you should return to for further study, and then do the suggested text problems to gain additional experience in working with the particular concept.

| Practice Test Question | Concepts and Equations | Prescription | |
|---|---|---|---|
| | | Principal Concept | Text Problems |
| 1 | Normal force | 6 | 4,37,39 |
| 2 | Normal force | 6 | 4-41,42 |
| 3 | Friction: $f = \mu_k N$ | 7 | 4-43,44 |
| 4 | Friction | 7 | 4-45,46 |
| 5 | Component of a vector | 6 of Ch. 3 | 3-2,3 |
| 6 | Unbalanced force | 1,2 | 4-30,40 |
| 7 | Newton's second law: $\vec{F} = m\vec{a}$ | 2,4 | 4-40,45 |
| 8 | Newton's second law | 2,4 | 4-30,34 |
| 9 | Newton's second law | 2,4 | 4-36,45 |
| 10 | Newton's second law | 2,4 | 4-3,4 |
| 11 | Newton's second law | 2,4 | 4-5,6 |
| 12 | Newton's second law | 2,4 | 4-3,6 |
| 13 | $v = v_0 + at$ | 5 of Ch. 2 | 2-12,19 |
| 14 | $\Delta s = v_0 t + at^2/2$ | 5 of Ch. 2 | 2-22,23 |
| 15 | $v^2 - v_0^2 = 2a\Delta s$ | 5 of Ch. 2 | 2-10,14 |

# 5 Rotational Kinematics and Gravity

## RECALL FROM PREVIOUS CHAPTER

| Previously learned concepts and equations frequently used in this chapter | Text Section | Study Guide Page |
|---|---|---|
| Equations used to analyze translational motion:<br>$v_{av} = \Delta s/\Delta t$ $\quad v = v_o + at$<br>$v_{av} = (v_o + v)/2$ $\quad \Delta s = v_o t + at^2/2$<br>$a = \Delta v/\Delta t$ $\quad 2a\Delta s = v^2 - v_o^2$ | 2.4 | 22 |
| Resolving vectors into components | 3.2 | 32 |
| Newton's second law of motion: $F = Ma$ | 4.3 | 46 |
| Normal force $F_N$ and friction f: $f = \mu F_N$ | 4.7 | 58,59 |

## NEW IDEAS IN THIS CHAPTER

| Concepts and equations introduced | Text Section | Study Guide Page |
|---|---|---|
| Angular measure in radians, degrees, and revolutions | 5.2 | 65 |
| Relationship between linear and angular distance, speed, and acceleration:<br>$\Delta s = r\Delta\theta$ $\quad v = r\omega$ $\quad a_t = r\alpha$ | 5.2,5.3, 5.4 | 68 |
| Equations used to analyze rotational motion:<br>$\omega_{av} = \Delta\theta/\Delta t$ $\quad \omega = \omega_o + \alpha t$<br>$\omega_{av} = (\omega_o + \omega)/2$ $\quad \Delta\theta = \omega_o t + \alpha t^2/2$<br>$\alpha = \Delta\omega/\Delta t$ $\quad 2\alpha\Delta\theta = \omega^2 - \omega_o^2$ | 5.4 | 71 |
| Centripetal acceleration and force:<br>$a_c = v^2/r$ and $F_c = Ma_c = Mv^2/r$ | 5.5 | 73 |
| Newton's law of universal gravitation:<br>$F_G = GM_1M_2/r^2$ | 5.6 | 77 |

## PRINCIPAL CONCEPTS AND EQUATIONS

### 1. Angular Measure in Degrees, Revolutions, or Radians (Section 5.2)

Review: Your past experiences have left you familiar with angular measure in terms of degrees and revolutions. This knowledge is summarized in Fig. 5.1.

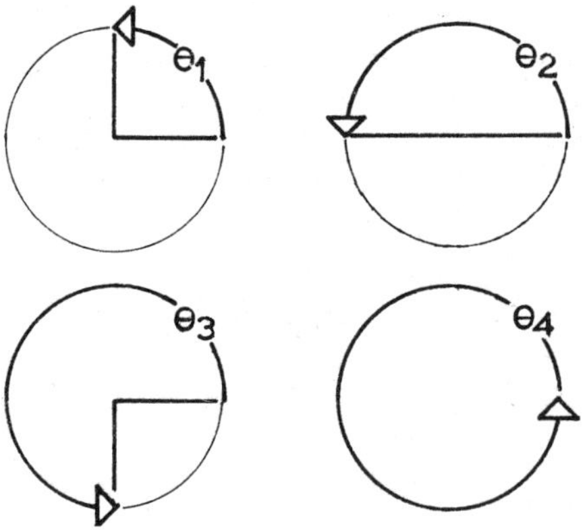


| Angle | Expressed in Degrees | Expressed in Revolutions |
|---|---|---|
| $\theta_1$ | 90° | 1/4 rev |
| $\theta_2$ | 180° | 1/2 rev |
| $\theta_3$ | 270° | 3/4 rev |
| $\theta_4$ | 360° | 1 rev |

Figure 5.1

Once you start working problems, you will learn that a more convenient way to measure angles is in radians. One radian is the angle subtended by an arclength of one radius. As shown in Fig. 5.2, exactly 6.28 (or $2\pi$) radii fit on the circumference of a circle.

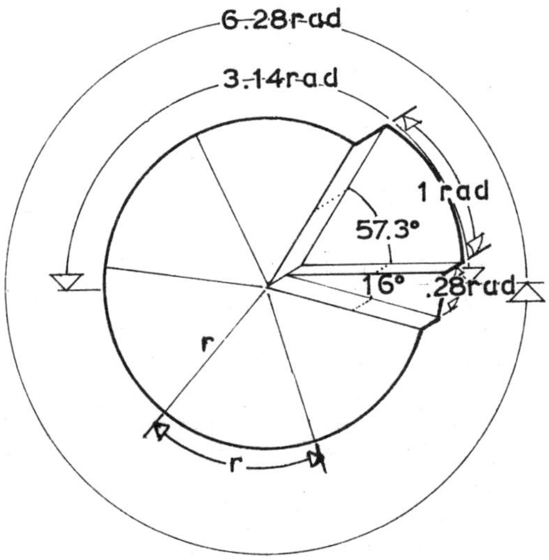


$$1 \text{ rad} = 57.3°$$

$$\pi \text{ rad} = 3.14 \text{ rad} = 180°$$

$$2\pi \text{ rad} = 6.28 \text{ rad} = 360°$$

Figure 5.2.

Since the number of radians in an angular measure is determined by dividing the arclength by the radius, we may write

$$\Delta\theta = \Delta s/r$$

Note: According to this definition, an angle expressed in radians has no units. However, we will carry the word "radian" (abbreviated rad) along in our calculations to confirm that we are using radian measure rather than some other angular measure.

Practice:

1. Express $\theta_1$ and $\theta_2$ in terms of degrees, revolutions, and radians.

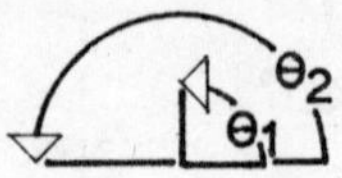


| Angle | Degrees | Rev | Rad |
|---|---|---|---|
| $\theta_1$ | 90 | 1/4 | $\pi/2$ |
| $\theta_2$ | 180 | 1/2 | $\pi$ |

2. Express the $\theta_1$, $\theta_2$, and $\theta_3$ in terms of degrees, revolutions and radians.

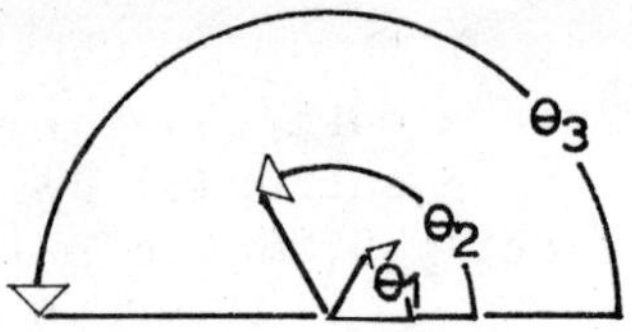


| Angle | Degrees | Rev | Rad |
|---|---|---|---|
| $\theta_1$ | 60 | 1/6 | $\pi/3$ |
| $\theta_2$ | 120 | 1/3 | $2\pi/3$ |
| $\theta_3$ | 180 | 1/2 | $\pi$ |

3. Express the $\theta_1$, $\theta_2$, $\theta_3$ and $\theta_4$ in terms of degrees, revolutions, and radians.

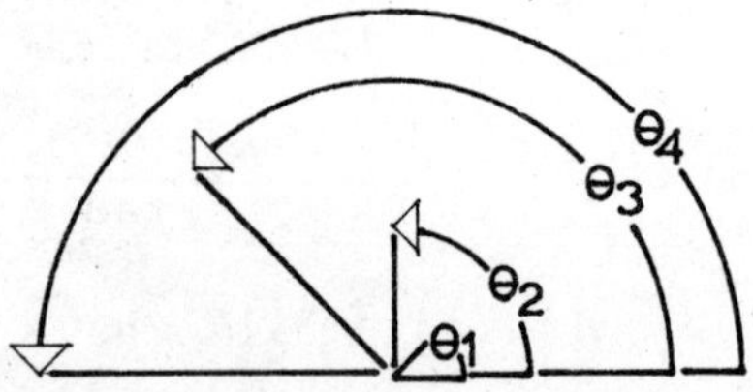


| Angle | Degrees | Rev | Rad |
|---|---|---|---|
| $\theta_1$ | 45 | 1/8 | $\pi/4$ |
| $\theta_2$ | 90 | 1/4 | $\pi/2$ |
| $\theta_3$ | 135 | 3/8 | $3\pi/4$ |
| $\theta_4$ | 180 | 1/2 | $\pi$ |

4. What is an easily remembered relationship that allows you to convert from degrees to revolutions and vice versa?

$1 \text{ rev} = 360°$

5. Express 1/3 rev and 0.8 rev in degrees.

$[(1/3)\text{rev}](360°/\text{rev}) = 120°$
$(0.8 \text{ rev})(360°/\text{rev}) = 288°$

6. Express 120° and 300° in revolutions.

$(120°)(\text{rev}/360°) = (1/3)\text{rev}$
$(300°)(\text{rev}/360°) = 0.833 \text{ rev}$

| | |
|---|---|
| 7. What is an easily remembered relationship that allows you to convert from degrees to radians and vice versa? | $\pi$ rad = 180° |
| 8. Convert 70° and 330° to radians. | 70° ($\pi$ rad/180°) = 1.22 rad<br>330° ($\pi$ rad/180°) = 5.76 rad |
| 9. Convert 0.7$\pi$ rad and 5 rad to degrees. | (0.7$\pi$ rad)(180°/$\pi$ rad) = 126°<br>(5 rad)(180°/$\pi$ rad) = 286° |
| 10. What is an easily remembered relationship that allows you to convert from revolutions to radians and vice versa? | 1 rev = 2$\pi$ rad |
| 11. Convert 1/3 rev and 0.7 rev to radians. | [(1/3)rev](2$\pi$ rad/rev) = 2.09 rad<br>(0.700 rev)(2$\pi$ rad/rev) = 4.40 rad |
| 12. Convert 6 rad and $\pi$/3 rad to revolutions. | (6.00 rad)(rev/2$\pi$ rad) = 0.955 rev<br>[($\pi$/3) rad)](rev/2$\pi$ rad) = 0.167 rev |
| 13. Determine how far a wheel with a radius of 0.500 m advances when it turns through the following angles:<br><br>$\Delta\theta_1$ = 0.8 rev<br>$\Delta\theta_2$ = 200°<br>$\Delta\theta_3$ = 0.3 rad | $\Delta s_1 = r\Delta\theta_1$<br>$= (0.500\text{ m})(0.800\text{ rev})(\frac{2\pi\text{ rad}}{\text{rev}})$<br>= 2.51 m<br><br>$\Delta s_2 = r\Delta\theta_2$<br>= (0.500 m)(200°)($\pi$ rad/180°)<br>= 1.74 m<br><br>$\Delta s_2 = r\Delta\theta_3$ = (0.500 m)(0.300 rad)<br>= 0.150 m |

Note: The calculation in step 13 should convince you that working with angular distances in radians is more convenient than working with revolutions or degrees.

Example 5.1. Two spools of rope are mounted on the same axle. One has a radius of 20 cm, and the other has a radius of 12 cm. As you pull on the rope wound on the larger spool, the smaller spool also unwinds. When you unwind 1 m of the rope from the large spool, through what angle will the smaller spool turn? Express your answer in radians, degrees, and revolutions.

Diagram and Given:

Figure 5.3

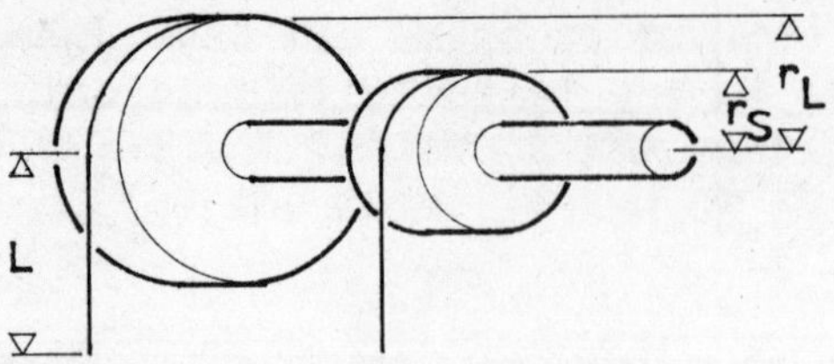


$r_L$ = 20 cm = 0.20 m

$r_S$ = 12 cm = 0.12 m

L = 1.00 m

Determine: The angle through which the smaller spool rotates when you unwind 1 m of rope from the larger spool.

Strategy: Knowing the radius of the larger spool ($r_L$) and the length of rope unwound (L), we can determine the angle of rotation (in radians) of the larger spool. Since the spools are mounted on the same axle, both rotate through the same angle. Now that this angle is known in radians, we can convert to degrees and revolutions.

Solution:

$$\Delta\theta_L = \frac{\Delta s}{r_L} = \frac{1\ \text{m}}{0.2\ \text{m}} = 5\ \text{rad, then} \qquad \Delta\theta_S = \Delta\theta_L = 5\ \text{rad}$$

$$\Delta\theta_S = 5\ \text{rad}\ \left(\frac{180°}{\pi\ \text{rad}}\right) = 286° \qquad \Delta\theta_S = 5\ \text{rad}\ \left(\frac{1\ \text{rev}}{2\pi\ \text{rad}}\right) = 0.796\ \text{rev}$$

Related Text Problems: 5-1, 5-2, 5-4, 5-9a.

## 2. The Relationship Between Linear and Angular Distance, Linear and Angular Speed, and Linear and Angular Acceleration (Section 5.2, 5.3, 5.4)

Review: When working with linear motion, the three physical quantities needed are linear displacement ($\Delta s$), speed (v), and acceleration (a). In this chapter, we learn how to deal with angular motion, and the three physical quantities of interest are angular displacement ($\Delta\theta$), angular speed ($\omega$), and angular acceleration ($\alpha$). The chart below summarizes these linear and angular physical quantities and their relationships:

| Physical Quantity | Linear | Angular | Relation |
|---|---|---|---|
| Displacement | $\Delta s$ | $\Delta\theta$ | $\Delta s = r\Delta\theta$ |
| Speed | $v = \Delta s/\Delta t$ | $\omega = \Delta\theta/\Delta t$ | $v = r\omega$ |
| Acceleration | $a = \Delta v/\Delta t$ | $\alpha = \Delta\omega/\Delta t$ | $a = r\alpha$ |

Practice: Consider the two wheels mounted on the axle shown in Fig. 5.4.

Figure 5.4

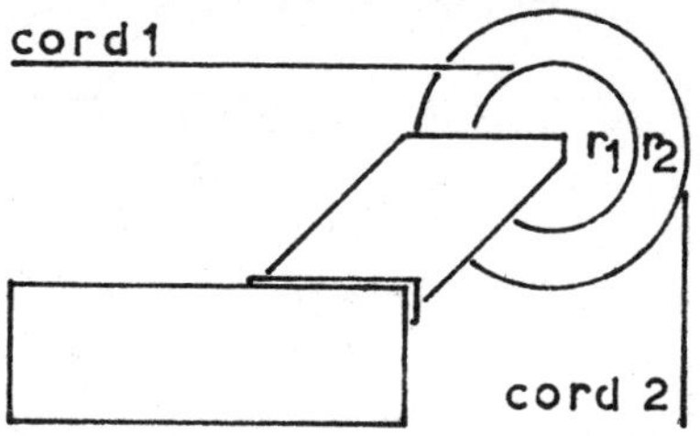


$r_1$ = 15 cm = radius of wheel 1

$r_2$ = 25 cm = radius of wheel 2

The system is initially at rest, and thin cords have been wrapped around both wheels. The cord wrapped around wheel 1 is pulled with a constant force for 20 s, and after that time the cord is unwound at the rate of 10 m/s. As the cord is unwound from wheel 1, another cord is wound onto wheel 2.

Determine the following:

| | |
|---|---|
| 1. The linear acceleration of cord 1 | $a_1 = \Delta v_1/t = (v_1 - v_{o1})/t = 0.500$ m/s$^2$ |
| 2. The length of cord pulled off wheel 1 during the 20.0 s | $s_1 = a_1 t^2/2 = 100$ m |
| 3. The angular distanced traveled by wheel 1 during the 20.0 s | $\theta_1 = s_1/r_1 = 667$ rad |
| 4. The angular distance traveled by wheel 2 during the 20.0 s | Since the wheels are mounted on the same axle, they turn through the same angle§ $\theta_2 = \theta_1 = 667$ rad |
| 5. The amount of cord wrapped onto wheel 2 during the 20.0 s | $s_2 = r_2\theta_2 = 167$ m |
| 6. The linear speed at which cord is wound onto wheel 2 at the end of the 20.0 s | $v_{av} = s_2/t$; $v_{av} = (v_{o2} + v_2)/2$<br>$v_2 = 2v_{avg} = 2s_2/t = 16.7$ m/s |
| 7. The linear acceleration of cord 2 at the end of the 20.0 s | $s_2 = v_{o2}t + a_2t^2/2$ or $v_2 = v_{o2} + a_2t$<br>$a_2 = 2s_2/t^2 = 0.835$ m/s$^2$ or<br>$a_2 = v_2/t = 0.835$ m/s$^2$ |

| | |
|---|---|
| 8. The angular speed of wheel 1 at the end of the 20.0 s | $\omega_1 = v_1/r_1 = 66.7$ rad/s |
| 9. The angular speed of wheel 2 at the end of the 20.0 s | $\omega_2 = \omega_1 = 66.7$ rad/s or<br>$\omega_2 = v_2/r_2 = 66.7$ rad/s |
| 10. The angular acceleration of wheel 1 during the 20.0 s | $\alpha_1 = a_1/r_1 = 3.33$ rad/s$^2$ or<br>$\alpha_1 = \Delta\omega_1/t = 3.33$ rad/s$^2$ |
| 11. The angular acceleration of wheel 2 during the 20.0 s | $\alpha_2 = \alpha_1 = 3.33$ rad/s$^2$<br>We can also use either $\alpha_2 = a_2/r_2$ or $\alpha_2 = \Delta\omega_2/t$ to obtain the same answer. |

Example 5.2. An automobile traveling 20.0 m/s accelerates to 30.0 m/s in a 15.0 s time interval. The outside diameter of the tires is 0.720 m. Determine the following physical quantities:

(a) The initial, final, and average angular speed of the wheels
(b) The tangential acceleration at the outside diameter of the tires and angular acceleration of the wheels
(c) The linear distance traveled and the number of revolutions turned by the wheels during the 15.0 s

Given: $v_o = 20.0$ m/s, $v = 30.0$ m/s, $t = 15.0$ s, $d = 0.720$ m

Determine: $\omega_o$, $\omega_f$, $\omega_{av}$, $a$, $\alpha$, $\Delta s$, N

Strategy: Knowing $v_o$, v, and r, we can determine $\omega_o$ and $\omega$. We can then obtain $\omega_{av}$ from $\omega_o$ and $\omega$. Knowing $v_o$, v, and t, we can determine a and then $\alpha$. Knowing $v_o$, a, and t, we can calculate $\Delta s$. Finally, we can use $\Delta s$ to determine $\Delta\theta$ and then N.

Solution:

(a) $\omega_o = v_o/r = (20.0 \text{ m/s})/0.720 \text{ m} = 27.8$ rad/s

$\omega = v/r = 30.0 \text{ m/s} / 0.720 \text{ m} = 41.7$ rad/s

$\omega_{av} = (\omega_o + \omega)/2 = [(27.8 + 41.7) \text{ rad/s}]/2 = 34.8$ rad/s

(b) $a = \Delta v/\Delta t = (v - v_o)/t = [(30.0 - 20.0) \text{ m/s}/15.0 \text{ s} = 0.667 \text{ m/s}^2$

$\alpha = a/r = (0.667 \text{ m/s}^2)/0.720 \text{ m} = 0.926 \text{ rad/s}^2$

or

$\alpha = \Delta\omega/\Delta t = (\omega - \omega_o)/t = [(41.7 - 27.8)\ r/s]/15.0\ s = 0.927\ rad/s^2$

(c) $\Delta s = v_o t + at^2/2 = (20.0\ m/s)(15.0\ s) + (0.667\ m/s^2)(15.0\ s)^2/2 = 375\ m$

We can obtain $\Delta\theta$ by any one of the following

$$\Delta\theta = \Delta s/r \qquad \Delta\theta = \omega_o t + \alpha t^2/2 \qquad \Delta\theta = (\omega^2 - \omega_o^2)/2\alpha$$

Using the first expression we obtain

$$\Delta\theta = \Delta s/r = (375\ m)/0.720\ m = 521\ rad$$

The number of revolutions turned by the wheel is

$$N = \Delta\theta/(2\pi\ rad/rev) = 521\ rad/(2\pi\ rad/sec) = 82.9\ rev$$

Related Text Problems: 5-1, 5-3, 5-5, 5-6.

## 3. The Equations Used to Analyze Translational Motion Have Rotational Analogs (Section 5.4)

Review: The equations used to analyze translational motion and their rotational analogs are shown in the chart below:

| Translational Motion | Rotational Motion |
|---|---|
| $v_{av} = \Delta s/\Delta t$ | $\omega_{av} = \Delta\theta/\Delta t$ |
| $v_{av} = (v_o + v)/2$ if $a$ = const | $\omega_{avg} = (\omega_o + \omega)/2$ if $\alpha$ = const |
| $a = \Delta v/\Delta t$ | $\alpha = \Delta\omega/\Delta t$ |
| $v = v_o + at$ | $\omega = \omega_o + \alpha t$ |
| $\Delta s = v_o t + at^2/2$ | $\Delta\theta = \omega_o t + \alpha t^2/2$ |
| $2a\Delta s = v^2 - v_o^2$ | $2\alpha\Delta\theta = \omega^2 - \omega_o^2$ |

A little thought will reveal that, when analyzing an object's rotational motion, you are dealing with five physical quantities: $\omega_o$, $\omega$, $\alpha$, $t$, and $\Delta\theta$. In general, you will be given three of these physical quantities and be asked to solve for the other two. If we are dealing with five quantities of which three are known, then ten possible cases exist. In the following practice, we will consider only three of these ten cases. You should consider the others on your own.

| Physical Quantities | $\omega_o$ | $\omega$ | $\alpha$ | $t$ | $\Delta\theta$ |
|---|---|---|---|---|---|
| Given Quantities Case I | ✓ | ✓ | | ✓ | |
| Case II | ✓ | ✓ | | | ✓ |
| Case III | | ✓ | ✓ | ✓ | |

Practice: Case I. Given $\omega_o$, $\omega$, and $t$, determine $\alpha$ and $\Delta\theta$. A mounted wheel is initially rotating with an angular speed of $\omega_o$. It accelerates for a time $t$ and acquires a speed $\omega$.

Determine expressions for the following:

| | |
|---|---|
| 1. The angular acceleration | $\alpha = \Delta\omega/\Delta t = (\omega - \omega_o)/t$ |
| 2. The angle it turns through during this acceleration | $\Delta\theta = \omega_o t + \alpha t^2/2$ or<br>$\Delta\theta = (\omega^2 - \omega_o^2)/2\alpha$ |

Case II. Given $\omega_o$, $\omega$, and $\Delta\theta$, determine $\alpha$ and $t$. A mounted wheel is initially rotating with an angular speed of $\omega_o$. It decelerates with a constant deceleration and comes to rest after rotating through an angle $\Delta\theta$.

Determine expressions for the following:

| | |
|---|---|
| 1. The angular acceleration | $\alpha = \Delta\omega/\Delta t = (\omega - \omega_o)/t$ |
| 2. The time it takes to decelerate to a stop | $\omega = \omega_o + \alpha t$ or $t = (\omega - \omega_o)/\alpha$ |

Case III. Given $\omega$, $\alpha$, and $t$, determine $\omega_o$ and $\Delta\theta$. A mounted wheel is rotating with some unknown angular speed. It decelerates with an angular acceleration $\alpha$ for a time $t$ and has an angular speed $\omega$.

Determine expressions for the following:

| | |
|---|---|
| 1. The initial angular speed | $\omega = \omega_o + \alpha t$ or $\omega_o = \omega - \alpha t$ |

| | |
|---|---|
| 2. The angle it turns through during this deceleration | $\Delta\theta = \omega_0 t + \alpha t^2/2$ or $\Delta\theta = (\omega^2 - \omega_0^2)/2\alpha$ |

Example 5.3. A rotary lawn mower blade requires 10.0 s to reach its maximum speed $\omega_m$ of 300 rpm. Calculate (a) its angular acceleration in rad/s$^2$, (b) the angle through which it turns in the process of coming up to this speed, and (c) the time it takes to achieve a speed of 50.0 rad/s.

Given: $\omega_0 = 0$; $\omega_m = 300 \frac{\text{rev}}{\text{min}} \left(\frac{2\pi \text{ rad}}{\text{rev}}\right)\left(\frac{\text{min}}{60.0 \text{ s}}\right) = 31.4$ rad/s; $t = 10.0$ s

Determine: $\alpha$, $\Delta\theta$, and $t$ to obtain $\omega = 50.0$ rad/s

Strategy: We can determine $\alpha$ using $\omega_0$, $\omega_m$, and $t$ in $\alpha = \Delta\omega/\Delta t$. We can then determine $\Delta\theta$ by either $\Delta\theta = \omega_0 t + \alpha t^2/2$ or $\Delta\theta = (\omega^2 - \omega_0^2)/2\alpha$. Finally, we can determine the time it takes to achieve a certain speed by $\omega = \omega_0 + \alpha t$.

Solution:

a) $\alpha = \Delta\omega/\Delta t = (\omega_m - \omega_0)/(t - 0) = (31.4 \text{ rad/s})/10.0 \text{ s} = 3.14 \text{ rad/s}^2$

b) $\Delta\theta = \omega_0 t + \alpha t^2/2 = (3.14 \text{ rad/s}^2)(10.0 \text{ s})^2/2 = 157 \text{ rad}$

or

$\Delta\theta = (\omega^2_m - \omega_o^2)/2\alpha = (31.4 \text{ rad/s})^2/2(3.14 \text{ rad/s}^2) = 157 \text{ rad}$

c) $\omega = \omega_0 + \alpha t$ or $t = (\omega - \omega_0)/\alpha = (50.0 \text{ rad/s})/(3.14 \text{ rad/s}^2) = 15.9 \text{ s}$

Related Text Problems: 5-6 through 5-17.

## 4. Centripetal Acceleration and Centripetal Force (Section 5.5)

Review: If an object is traveling in a circle of radius r with a speed v, it must be experiencing a central or centripetal acceleration of magnitude

$$a_c = v^2/r$$

This acceleration is the result of the velocity vector continually changing directions. If the object has a mass M, it must be experiencing an unbalanced central force of

$$F_c = Ma_c = Mv^2/r$$

A force smaller than this allows the object to spiral outward, and a larger force causes the object to spiral inward.

Note: Up to this point, we have represented an unbalanced force by $F_u$. In this section, we are introducing an unbalanced force that causes circular motion. For uniform circular motion this unbalanced force is directed toward the center of the circle, subsequently we will call it an unbalanced centripetal force and represent it by $F_c$.

Note: When working circular-motion problems, it will prove convenient to call any force directed toward the center positive and any force directed away from the center negative.

Note: Centripetal-force problems involve four physical quantities: $F_c$, M, v, and r. You will usually be given three of these quantities and be asked to determine the fourth.

Practice: A car is traveling around a flat circular curve, and the following quantities are known: g, μ, M, and r.

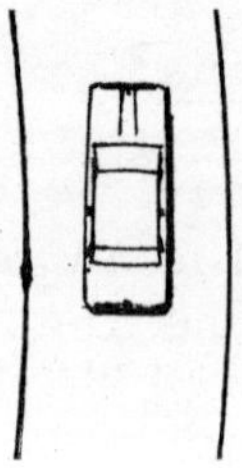

Determine the following:

| | |
|---|---|
| 1. Those forces that provide the centripetal force | Friction 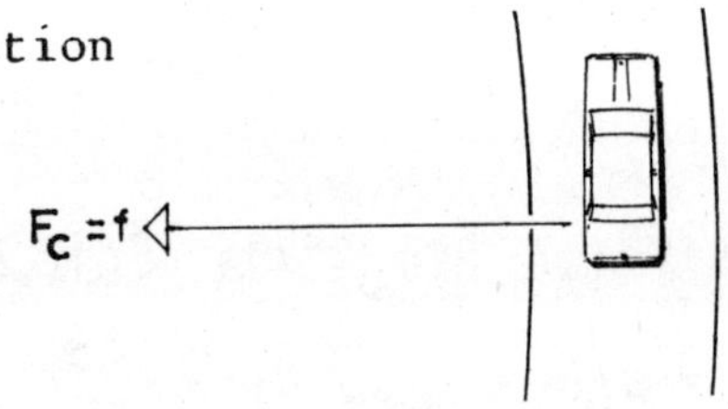  |
| 2. An expression for the centripetal force in terms of known quantities | $F_c = f = \mu F_N = \mu Mg$ |
| 3. An expression for the speed limit of the curve in terms of known quantities | $F_c = \mu Mg$ and $F_c = Mv^2/r$<br>Equate these expressions for $F_c$ and solve for v:<br>$v = (\mu gr)^{1/2}$ |

A conical pendulum is rotating in a horizontal plane, and the following quantities are known: g, L, θ, and M.

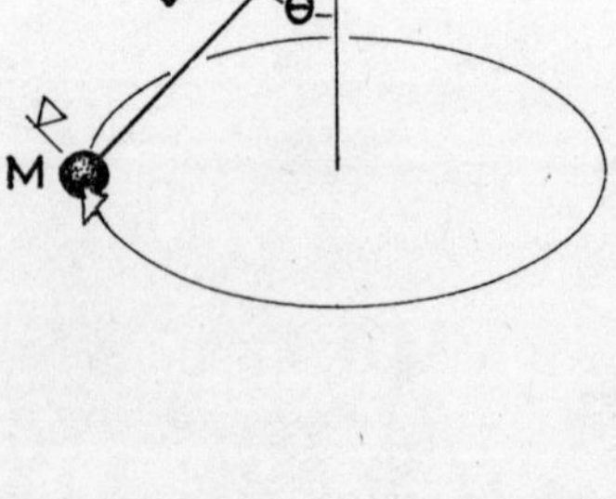

Determine the following:

| | |
|---|---|
| 1. Those forces that provide the centripetal force | The horizontal component of the tension (T) 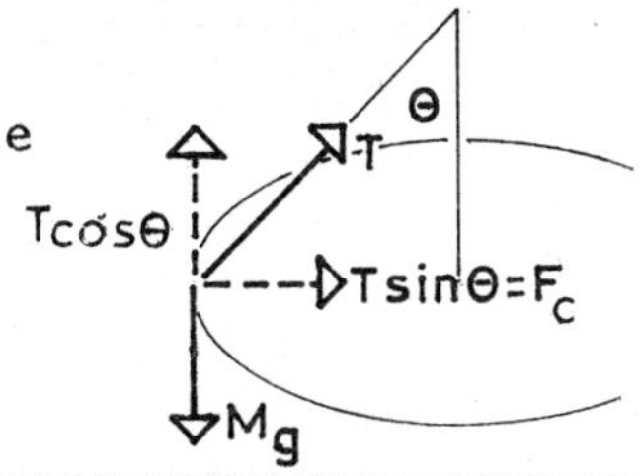  |
| 2. An expression for the centripetal force in terms of known quantities | $F_c = T\sin\theta$ and $Mg = T\cos\theta$<br>Eliminate T to obtain<br>$F_c = Mg\sin\theta/\cos\theta = Mg\tan\theta$ |
| 3. An expression for the speed of the orbiting pendulum bob in terms of known quantities | $F_c = Mg\tan\theta$<br>$F_c = Mv^2/r = Mv^2/L\sin\theta$<br>Equate these expressions for $F_c$ and solve for v:<br>$v = \sin\theta(Lg/\cos\theta)^{1/2}$ |

A satellite is orbiting the earth and the following quantities are known: $M_s$, $R_E$, r, $g_E$.

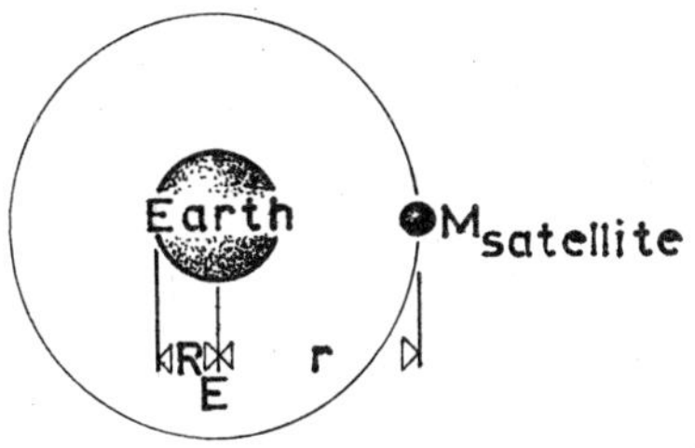


Determine the following:

| | |
|---|---|
| 1. The forces that provide the centripetal force | Gravitational attraction between the two masses |
| 2. An expression for the centripetal force in terms of known quantities | $F_c = F_g = GM_EM_s/r^2$ and $g_E = GM_E/R_E^2$<br>Eliminate $GM_E$ to obtain<br>$F_c = g_ER_E^2M_s/r^2$ |
| 3. An expression for the speed of the satellite in its orbit around the earth in terms of known quantities | $F_c = g_ER_E^2M_s/r^2$ and $F_c = M_sv^2/r$<br>Equate these expressions for $F_c$ and solve for v<br>$v = R_E(g_E/r)^{1/2}$ |

| | |
|---|---|
| 4. An expression for the period of the satellite in terms of known quantities | $v = R_E(g_E/r)^{1/2}$ and $v = 2\pi r/T$<br>Equate these expressions for v and solve for T<br>$T = \dfrac{(2\pi r)^{3/2}}{(R_E g_E)^{1/2}}$ |

A car is traveling around a banked frictionless curve, and the following quantities are known: g, θ, v, m, and $\mu = 0$.

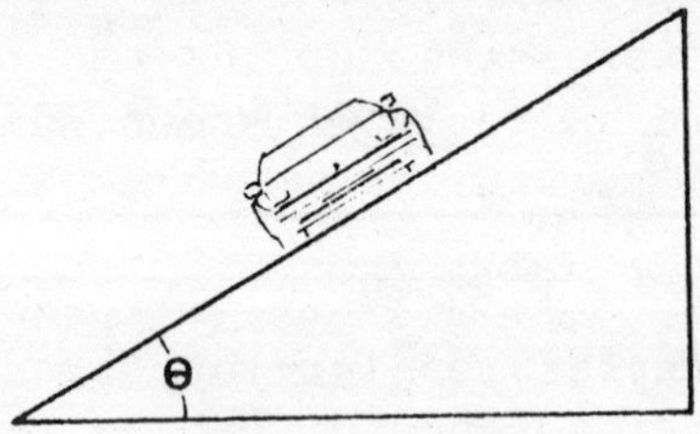


Determine the following:

| | |
|---|---|
| 1. Those forces that provide the centripetal force | 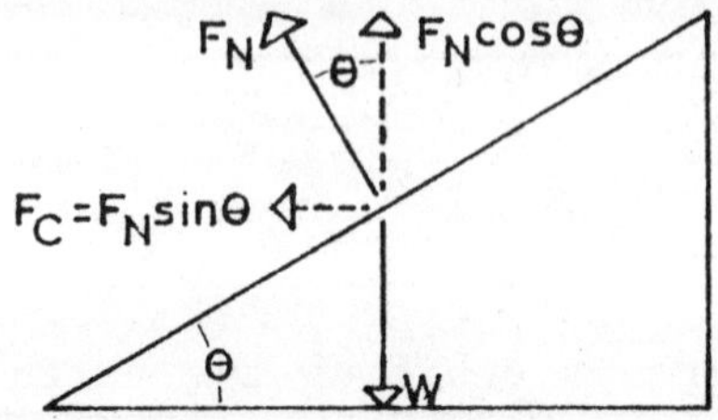 <br>The horizontal component of the normal force provides the centripetal force |
| 2. An expression for the centripetal force in terms of known quantities | $F_c = F_N \sin\theta$ and $F_c = Mv^2/r$<br>Equate these expressions for $F_c$ and then use the fact that<br>$F_N \cos\theta = W = Mg$<br>to obtain<br>$r = v^2/(g \tan\theta)$ |
| 3. An expression for the radius of curvature of the banked curve in terms of known quantities | $F_c = Mg \tan\theta$ and $F_c = Mv^2/r$<br>Equate these expressions for $F_c$ and solve for r:<br>$r = v^2/(g \tan\theta)$ |

Example 5.4. A 400-N girl sits on a bathroom scale while riding a roller coaster. When her car is in a valley that has a radius of curvature of 20.0 m, she notices that the scale reads 600 Nt. When her car tops a hill of 15.0-m radius, she notices that the scale reads 0 N. What is the speed of her car in the valley and at the top of the hill?

Given and Diagram:

$W = 400$ Nt
$r_v = 20.0$ m
$r_h = 15.0$ m
$F_{Nv} = 600$ N
$F_{Nh} = 0$ N
$g = 9.80$ m/s$^2$

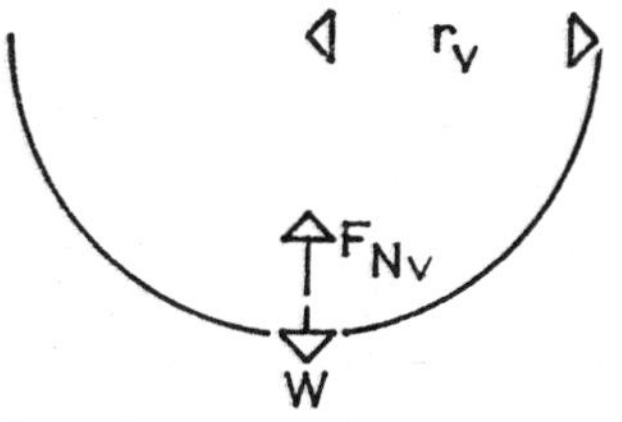

Valley

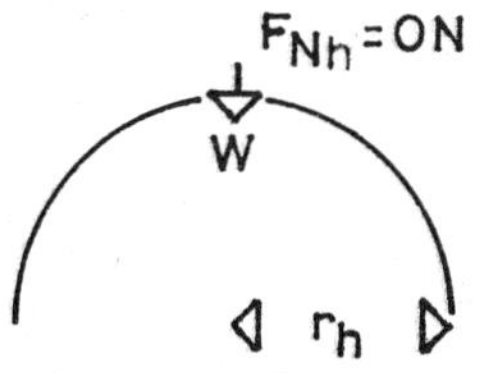

Hill

Determine: $V_v$ and $V_h$

Strategy: Write an expression for the centripetal force in terms of $F_{Nv}$ and W. Then equate this to $Mv^2/r$ and solve for v.

Solution:

Valley $F_c = F_{Nv} - W$, where $F_{Nv}$ = the scale reading = the normal force on the girl when the car is in the valley

$$F_c = Mv_v{}^2/r_v = Wv_v{}^2/gr_v$$

Equate these two expressions for $F_c$ and solve for $v_v$:

$$Wv_v{}^2/gr_v = F_{Nv} - W \quad \text{or} \quad v_v = \{gr_v[(F_{Nv}/W) - 1]\}^{1/2} = 10.0 \text{ s}$$

Hill $F_c = W - F_{Nh}$, where $F_{Nh}$ = the scale reading = the normal force on the girl when the car is on the hill

$$F_c = Mv_h{}^2/r_h = Wv_h{}^2/gr_h$$

Equate these two expressions for $F_c$ and solve for $v_h$:

$$Wv_h{}^2/gr_h = W \quad (\text{since } F_{Nh} = 0) \quad \text{or} \quad V_h = (gr_h)^{1/2} = 12.1 \text{ m/s}$$

Related Text Problems: 5-18 through 5-30, 5-32.

Note: The fact that half the problems in this chapter deal with centripetal acceleration and/or centripetal force is ample evidence that these are important concepts.

## 5. Newton's Law of Universal Gravitation (Section 5.6)

Review: Newton's law of universal gravitation

$$F_G = GM_1M_2/r^2$$

$M_1$ and $M_2$ represent the masses of the two objects, r represents the distance between their centers, and G is a universal constant of proportionality. If we consider the attraction between the mass of an object ($M_o$) placed on the surface of a planet ($M_p$) of radius $R_p$, Newton's law of gravitation looks like

$$F_G = GM_pM_o/R_p^2$$

The quantity $GM_p/R_p^2$, a constant for the planet, is represented by $g_p$ ($= GM_p/R_p^2$). We call this attractive force that the planet exerts on the object the weight of the object on that planet ($W_{op}$). Hence

$$F_G = W_{op} = M_og_p$$

Practice: Consider three planets X, Y, and earth, each with a manufactured satellite of mass $M_S$ orbiting at a radius r. The known information about the planets is as follows:

| Mass | Radius | Gravity |
|---|---|---|
| $M_E$ = known | $R_E$ = known | $g_E$ = known |
| $M_x = 2\ M_E$ | $R_X = \frac{1}{2}\ R_E$ | $g_X$ = unknown |
| $M_Y = \frac{1}{3}\ M_E$ | $R_Y$ = unknown | $g_Y = 2g_E$ |

Determine the following in terms of $M_E$, $R_E$, $g_E$, $M_s$ and r:

| | |
|---|---|
| 1. Weight of the satellite on earth | $W_{SE} = M_Sg_E$ |
| 2. Weight of the satellite on Y | $W_{SY} = M_Sg_Y = M_S(2g_E)$ |
| 3. Acceleration due to gravity on X | $g_X = GM_X/R_X^2 = G(2M_E)/(R_E/2)^2$<br>$= 8GM_E/R_E^2 = 8g_E$ |
| 4. Weight of the satellite on X | $W_{SX} = M_Sg_X = M_S(8g_E)$ |
| 5. Radius of planet Y | $g_Y = GM_Y/R_Y^2 = G(\frac{1}{3}\ M_E)/R_Y^2$ and $g_E = GM_E/R_E^2$<br>Insert these into $g_Y = 2g_E$ to obtain<br>$GM_E/3R_Y^2 = 2GM_E/R_E^2$ or $R_Y = R_E/(6)^{1/2}$ |

| | |
|---|---|
| 6. The orbital speed $V_s$ of the satellite around the earth | $F_c = GM_EM_S/r^2 \qquad F_c = M_SV_s^2/r$<br>Equate expressions for $F_c$ and solve for $V_s$:<br>$V_s = (GM_E/r)^{1/2}$ |
| 7. The period of the satellite around the earth | $V_s = (GM_E/r)^{1/2}$ and $V_s = 2\pi r/T_s$<br>Equate expressions for $V_s$ and solve for $T_s$<br>$T_s = (4\pi^2r^2/GM_E)^{1/2}$ |

Example 5.5. An artificial satellite is put into orbit around the earth. The following information is known: $M_s$ = mass of the satellite
$M_E$ = mass of the earth
$R_E$ = radius of the earth
r = radius of the satellite's orbit
G = the universal gravitational constant

Obtain expressions (in terms of known quantities only) for the speed of the satellite in its orbit, the period of the satellite, the weight of the satellite on the surface of the earth, and the weight of the satellite in its orbit.

Diagram and Given:

$M_s$, $M_E$, $R_E$, r, G

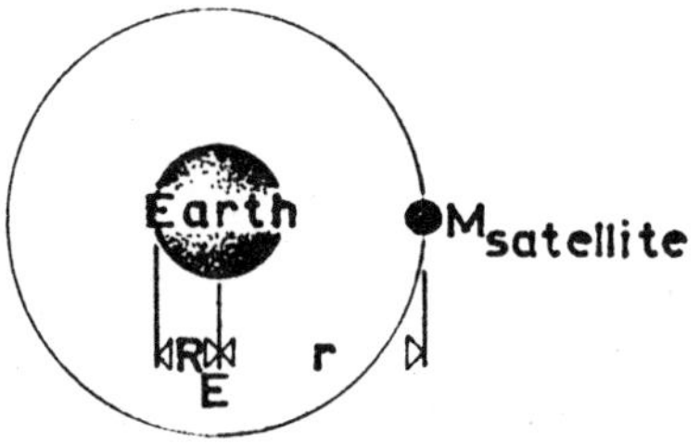


Determine: The speed of the satellite in its orbit ($v_s$), the period of the satellite ($T_s$), the weight of the satellite on earth ($W_{sE}$), and the weight of the satellite in its orbit ($W_{sr}$).

Strategy: The speed of the satellite can be determined by recognizing that the unbalanced centripetal force is supplied by gravitational attraction between $M_E$ and $M_s$. The period of the satellite can be determined from the speed of the satellite and the circumference of its orbit. The weight of the satellite on earth and in its orbit can be obtained from Newton's law of gravitation.

Solution: Since the unbalanced centripetal force is supplied by gravitational attraction, we can set up the following equations:

$$\frac{M_s V_s^2}{r} = G\frac{M_s M_E}{r^2} \quad \text{or} \quad V_s = \left(\frac{GM_E}{r}\right)^{1/2}$$

We can determine the period of the orbit as follows:

$$V_s = \frac{2\pi r}{T_s} \quad \text{or} \quad T_s = \frac{2\pi r}{V_s} = 2\pi(r^3/GM_E)^{1/2}$$

he weight of the satellite on the surface of the earth ($W_{sE}$) and in its orbit ($W_{sr}$) can be obtained from Newton's law of gravitation as follows:

$$W_{sE} = GM_sM_E/R_E^2 \quad \text{and} \quad W_{sr} = GM_sM_E/r^2$$

Related Text Problems: 5-31, 5-33 through 5-36, 5-38 through 5-43.

---

## PRACTICE TEST

Take and grade this practice test. Doing so will allow you to determine any weak spots in your understanding of the concepts taught in this chapter. The following section prescribes what you should study further to strengthen your understanding.

A wheel is mounted on an axle as shown:

$R = 0.200$ m
$r = 1.00$ cm
$h = 2.00$ m
$M_1 = 10.0$ kg
$M_2 = 0.100$ kg
$g = 9.80$ m/s$^2$

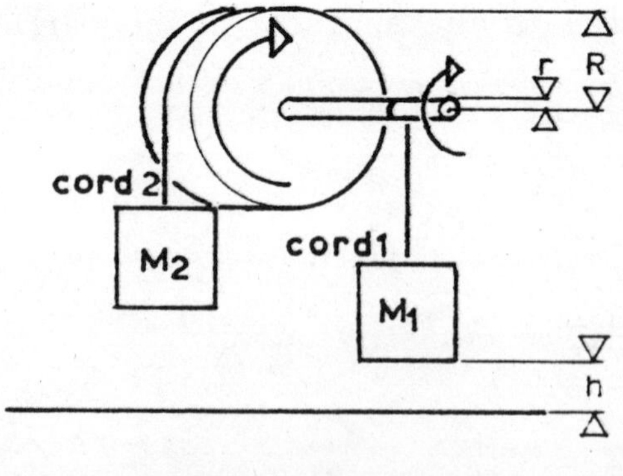


A cord is wrapped around the axle and attached to $M_1$. Another cord is wrapped around the wheel and attached to $M_2$. When the system is released, $M_1$ falls and unwraps cord 1; $M_2$ is lifted as cord 2 is wound on the wheel. $M_1$ falls the distance h in 10.0 s. Based on this information, answer questions 1 through 9.

_______ 1. The linear acceleration of $M_1$
_______ 2. The angular acceleration of the wheel and axle
_______ 3. The linear acceleration of $M_2$
_______ 4. The linear speed of $M_1$ after a time of 5.00 s
_______ 5. The angular speed of the wheel and axle after a time of 5.00 s
_______ 6. The linear speed of $M_2$ after a time of 5.00 s
_______ 7. The distance $M_1$ falls during the first 5.00 s of travel
_______ 8. The number of revolutions made by the wheel and axle during the first 5.00 s of travel
_______ 9. The distance $M_2$ is lifted during the first 5.00 s of travel

An object tied to a cord swings in a vertical circle. In general, the speed of the object will vary, and we know its value at the five labeled positions.

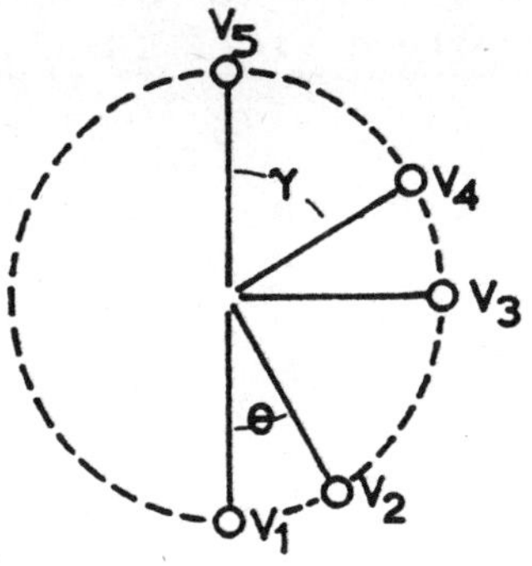


Given: $r = 1.00$ m, $M = 0.50$ kg, $\theta = 30°$, $\gamma = 60°$, $g = 9.80$ m/s$^2$, $v_1 = 6.00$ m/s, $v_2 = 5.50$ m/s, $v_3 = 5.00$ m/s, $v_4 = 4.50$ m/s, $v_5 = 4.00$ m/s

Based on this information, answer questions 10 through 14.

__________10. The tension in the cord for position 1
__________11. The tension in the cord for position 2
__________12. The tension in the cord for position 3
__________13. The tension in the cord for position 4
__________14. The tension in the cord for position 5

A distant planet has a radius one-fourth that of the earth ($R_X = R_E/4$) and a mass one-tenth that of the earth ($M_X = M_E/10$).

__________15. What is the acceleration due to gravity on planet X?
__________16. How much would a 70.0-kg astronaut weigh on planet X?
__________17. If we wish to put an artificial satellite into orbit with a period of 20.0 h around planet X, what is the required size of the satellite's orbit?
__________18. How much would the astronaut of question 16 weigh if placed in the satellite of question 17?

(See Appendix I for answers.)

PRINCIPAL CONCEPTS AND EQUATIONS PRESCRIPTION

Your score on the practice is an excellent measure of your understanding of this chapter. You should now use the following chart to write your own prescription for curing any of your physics ills. Look down the leftmost column to the number of the question(s) you answered incorrectly, read across that row to see which section(s) of the study guide you should return to for further study, and then do the suggested text problems to gain additional experience in working with the particular concept.

| Practice Test Question | Concepts and Equations | Prescription | |
|---|---|---|---|
| | | Principal Concept | Text Problems |
| 1 | $S_1 = S_{o1} + v_{o1}t + a_1t^2/2$ | 5 of Ch. 2 | 2-19,26 |
| 2 | $a_1 = \alpha r$ | 2 | 5-16c,17a |
| 3 | $a_2 = \alpha R$ | 2 | 5-16c,17a |
| 4 | $v_1 = v_{o1} + a_1t$ | 5 of Ch. 2 | 2-20,31 |
| 5 | $\omega = \omega_o + \alpha t$ or $\omega = v_1/r$ | 2,3 | 5-5,11 |
| 6 | $v_2 = \omega R$ or $v_2 = v_{o2} + a_2t$ | 2 | 5-17,18 |
| 7 | $S_1 = S_{o1} + v_{o1} + a_1t^2/2$ | 5 of Ch. 2 | 2-19,26 |
| 8 | $S_1 = r\theta$ | 2 | 5-1 |
| 9 | $S_2 = R\theta$ or $S_2 = S_{o2} + v_{o2}t + a_2t^2/2$ | 2 | 5-1 |
| 10 | Centripetal force | 4 | 5-18,19 |
| 11 | Centripetal force | 4 | 5-20,21 |
| 12 | Centripetal force | 4 | 5-22,23 |
| 13 | Centripetal force | 4 | 5-24,25 |
| 14 | Centripetal force | 4 | 5-18,21 |
| 15 | Newton's law of gravitation | 5 | 5-31,32 |
| 16 | Newton's law of gravitation | 5 | 5-40,35 |
| 17 | Centripetal force and Newton's law of gravitation | 4,5 | 5-34,38 |
| 18 | Newton's law of gravitation | 5 | 5-35,40 |

# 6 Equilibrium and Torques

RECALL FROM PREVIOUS CHAPTERS

| Previously learned concepts and equations frequently used in this chapter | Text Section | Study Guide Page |
|---|---|---|
| Resolving vectors into components | 3.2 | 32 |
| Adding vectors | 3.2 | 32 |
| Normal force $F_N$ and frcition f: $f = \mu F_N$ | 4.7 | 58,59 |

NEW IDEAS IN THIS CHAPTER

| Concepts and equations introduced | Text Section | Study Guide Page |
|---|---|---|
| First condition of equilibrium: $\sum_i \vec{F}_i = 0$ | 6.1 | 83 |
| Torque: $\tau = rF \sin\theta$ | 6.2 | 87 |
| Second condition of equilibrium: $\sum_i \vec{\tau}_i = 0$ | 6.3 | 90 |
| Center of gravity and center of mass | 6.4 | 95 |

PRINCIPAL CONCEPTS AND EQUATIONS

## 1. First Condition of Equilibrium (Section 6.1)

Review: The first condition of equilibrium states that an object is in translational equilibrium if the vector sum of all the forces acting on the object is zero. This may be stated algebraically as follows:

If $\sum_i \vec{F}_i = 0$, then the object is in translational equilibrium

If the vector sum of all the forces acting on an object is zero, then the translational acceleration of the object is zero, and its translational velocity is constant. If the translational velocity is zero, the object is in static translational equilibrium. If the translational velocity is a nonzero constant, the object is in dynamic translational equilibrium. This may be summarized as follows:

If $\sum_i \vec{F}_i = 0$, then $\vec{a} = 0$ and $\vec{v}$ = constant

If $\vec{v} = 0$, then static translational equilibrium exists.

If $\vec{v} \neq 0$ but constant, then dynamic translational equilibrium exists.

When solving problems of this type, you will often find it convenient to write the first condition of equilibrium in component form:

$$\sum_i F_{xi} = 0 \; ; \; \sum_i F_{yi} = 0 \; ; \; \sum_i F_{zi} = 0$$

Practice: Consider the object shown in Fig. 6.1 and the forces acting on it.

$|\vec{F}_1| = 10.0$ N

$|\vec{F}_2| = 15.0$ N

$|\vec{F}_3| = 20.0$ N

$|\vec{F}_4| = 10.0$ N

$|\vec{F}_5| = 5.0$ N

Figure 6.1

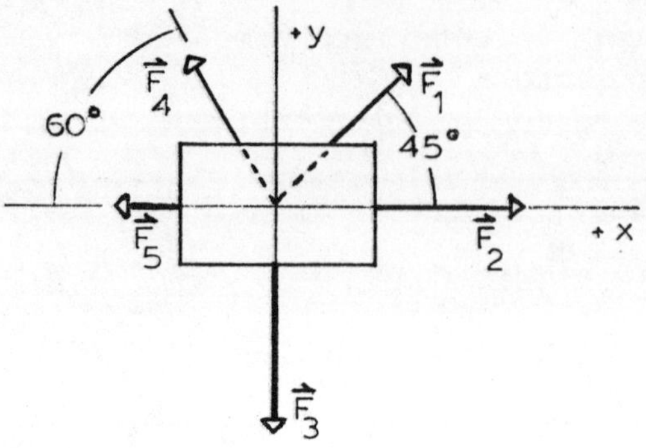


Determine the following:

| | |
|---|---|
| 1. The sum of all forces in the +y direction | $F_{+y} = (F_1 \sin 45°) + (F_4 \sin 60°)$<br>$= [(10.0\text{ N})\sin 45°] + [(10.0\text{ N})\sin 60°]$<br>$= +15.7$ N |
| 2. The sum of all forces in the -y direction | $F_{-y} = -F_3 = -20.0$ N |
| 3. If the object is in vertical equilibrium | Since $\sum_i F_{yi} = F_{+y} + F_{-y} = -4.30\text{ N} \neq 0$,<br>conclude that the object is not in vertical equilibrium. |
| 4. The sum of all forces in the +x direction | $F_{+x} = F_2 + (F_1 \cos 45°)$<br>$= 15.0\text{ N} + [(10.0\text{ N})\cos 45°] = 22.1$ N |
| 5. The sum of all forces in the -x direction | $F_{-x} = -F_5 - (F_4 \cos 60°)$<br>$= -5.0\text{ N} - [(10.0\text{ N})\cos 60°] = -10.0$ N |

| | |
|---|---|
| 6. If the object is in horizontal equilibrium | Since $\sum_i F_{xi} = F_{+x} + F_{-x} = +12.1\ \text{N} \neq 0$, conclude that the object is not in horizontal equilibrium. |
| 7. The magnitude of the force that would put the object in translational equilibrium | $F = (F_x^2 + F_y^2)^{1/2}$ $= [(12.1\ \text{N})^2 + (4.30\ \text{N})^2]^{1/2} = 12.8\ \text{N}$ |
| 8. The direction of the force that would put the object in translational equilibrium | The force must have components $F_x = -12.1$ N and $F_y = +4.30$ N. The direction of the force is given by: $\theta = \tan^{-1} \frac{F_y}{F_x} = \tan^{-1} \frac{4.30\ \text{N}}{12.1\ \text{N}} = 19.6°$ $\theta = 19.6°$ clockwise from the direction of $F_5$ |

Example 6.1. A 40-N object is suspended by a rope, as shown in Fig. 6-2. A girl pushes horizontally on the object so that the rope makes an angle of 15° with respect to the vertical. What is the tension in the rope, and how hard is the girl pushing?

Given and Diagram:

$W = 40.0$ N

$\theta = 15°$

Figure 6.2

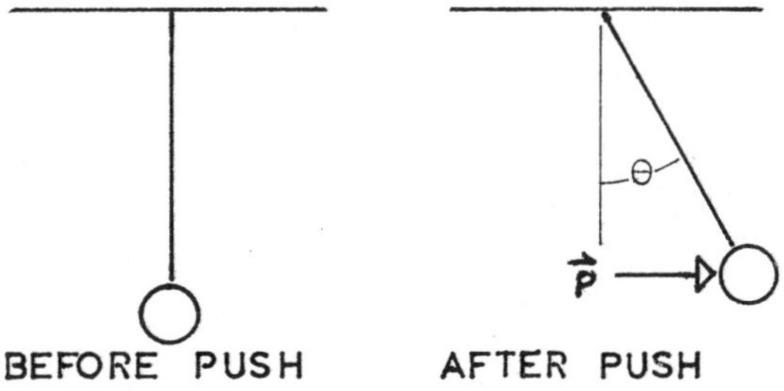


Determine: T, the tension in the rope while the girl is pushing
P, the magnitude of the girl's push

Strategy: Draw a free-body diagram showing all forces acting on the object. Establish a coordinate system, and resolve all forces into components. Use the fact that the object is in vertical equilibrium to determine $T_y$ and, hence, T. Knowing T, we can determine $T_x$. Since the object is in horizontal equilibrium, we can determine P from $T_x$.

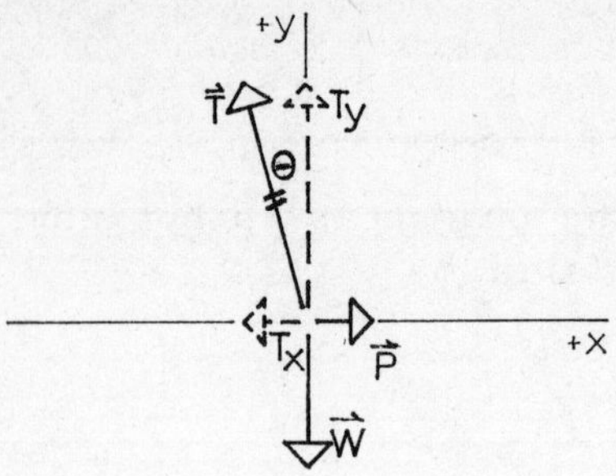


Solution: Since the object is in vertical equilibrium, the sum of the forces in the y direction must be zero.

$$\sum_i F_{yi} = T_y - W = (T \cos\theta) - W = 0 \quad \text{or}$$

$$T = W/\cos\theta = 40.0\ \text{N}/\cos 15° = 41.4\ \text{N}$$

Since the object is in horizontal equilibrium, the sum of the forces in the x direction must be zero.

$$\sum_i F_{xi} = P - T_x = P - (T \sin\theta) = 0 \quad \text{or} \quad P = T \sin\theta = (41.4\ \text{N})\sin 15° = 10.7\ \text{N}$$

Example 6.2. A crate weighing 100 N sits on an incline that makes an angle of 10° with respect to the horizontal (Fig. 6.3). The coefficient of kinetic friction between the crate and the incline is 0.20. What is the magnitude of the push parallel to the incline that will move the crate down the incline at a constant speed?

Given and Diagram:

$W = 100$ N
$\theta = 10°$
$\mu = 0.20$

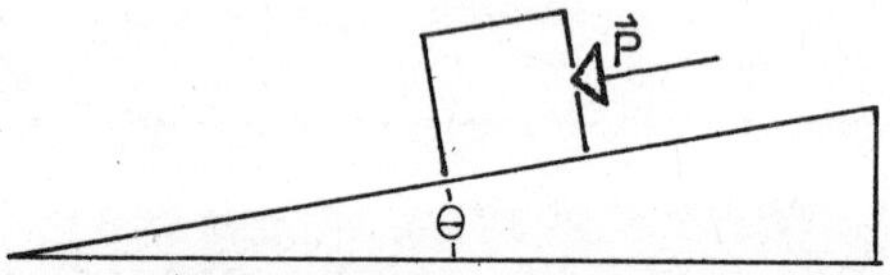


Figure 6.3

Determine: P, the magnitude of the push that will move the crate down the incline at a constant speed

Strategy: Draw a free body diagram showing all forces acting on the object. Establish a coordinate system and resolve all forces into components. Since the crate is to move down the ramp at a constant speed it will be in dynamic equilibrium. Consequently, the sum of the forces in the x direction must be zero. If we write a summation-of-forces statement for the x direction, we can solve for P if we know f. We can obtain f if we know $F_N$. We can obtain $F_N$ by recognizing that the sum of the forces in the y direction must be zero.

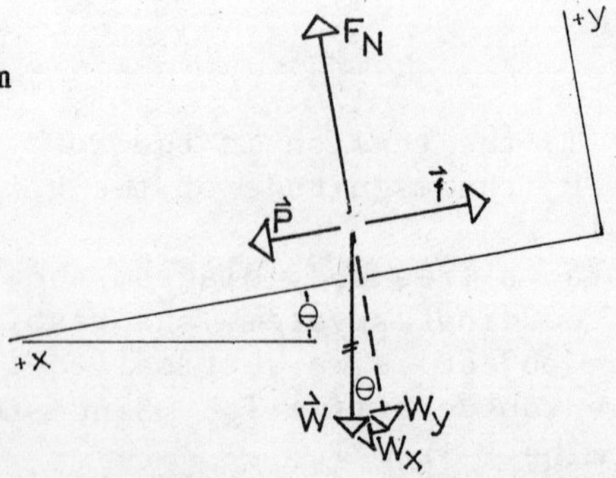


Solution: Since the crate is in equilibrium in the x direction, the sum of force in the x direction is zero.

$$\sum_i F_{xi} = +W_x + P - f = 0$$

or

$$P = f - W_x = f - (W \sin\theta)$$

Since the crate is in equilibrium in the y-direction, the sum of forces in the y direction is zero.

$$\sum_i F_{yi} = +F_N - W_y = 0 \quad \text{or} \quad F_N = W_y = W \cos\theta$$

Now that we know $F_N$, we can determine the friction by $f = \mu F_N = \mu W \cos\theta$.

Then, substituting back into the expression for P, we obtain:

$$P = f - (W \sin\theta) = \mu(W \cos\theta) - (W \sin\theta) = W(\mu \cos\theta - \sin\theta) = 2.33 \text{ N}$$

Related Text Problems: 6-1 through 6-11.

## 2. Torque (Section 6.2)

Review: Torque is a vector quantity that has both magnitude and direction. The magnitude of the torque can be determined by three different methods. Figures 6.4, 6.5, and 6.6 illustrate these methods.

Method I.

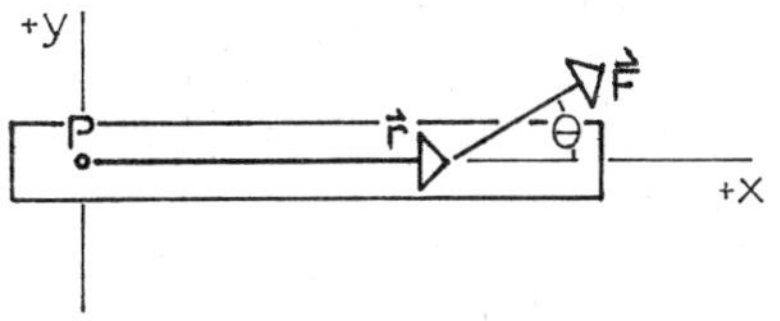


Figure 6.4

P = the point at which we wish to determine the torque
r = the position vector that locates the point of application of the force
F = the force that produces the torque
θ = the angle between the direction of r and F
$\tau = rF \sin\theta$ = the magnitude of the torque

The magnitude of the torque can be determined by multiplying the magnitude of r, the magnitude of F, and $\sin\theta$.

Method II.

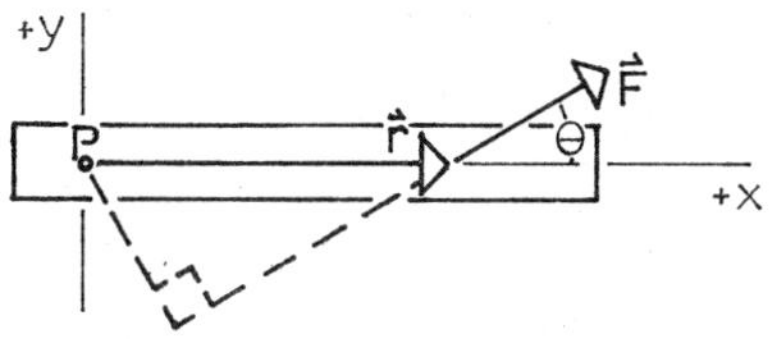


Figure 6.5

The line of action for F is the line along which F is exerted.

The moment arm for F is the perpendicular distance between the line of action for F and the point at which we wish to determine the torque (moment arm = $r \sin\theta$).

$$\tau = F(\text{moment arm}) = F(r \sin\theta) = rF \sin\theta$$

The magnitude of the torque can be determined by multiplying the magnitude of the force and the moment arm.

Method III.

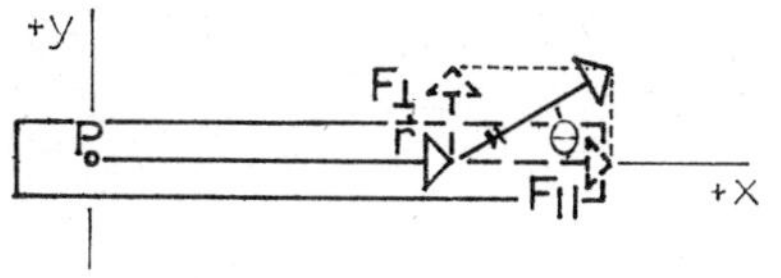


Figure 6.6

$F_\parallel = F \cos\theta$; produces no torque about P

$F_\perp = F \sin\theta$; produces all the torque about P

$$\tau = r F_\perp = r(F \sin\theta) = rF \sin\theta$$

The magnitude of the torque can be determined by multiplying the magnitude of r and the component of F perpendicular to r.

The direction of the torque can be determined by considering the direction of advance of a right-handed screw.

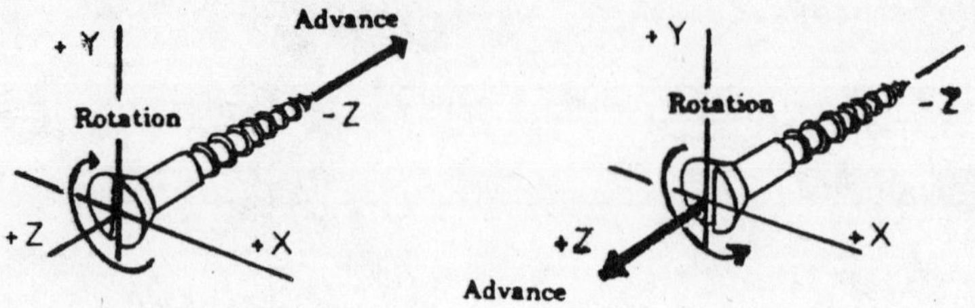


clockwise torque = -

counterclockwise = +

If you rotate the screw clockwise (cw), it advances into the wood (-z direction). Hence, we call the torque that created this rotation a negative torque. A negative torque produces a clockwise rotation.

If you rotate the screw counterclock wise (ccw), it backs out of the wood (+z direction). Hence, we call the torque that created this rotation a positive torque. A positive torque produces a counterclockwise rotation.

Practice: Consider the rectangle of width W and length L shown in Fig. 6.7. It is mounted so that it can rotate freely about axis P.

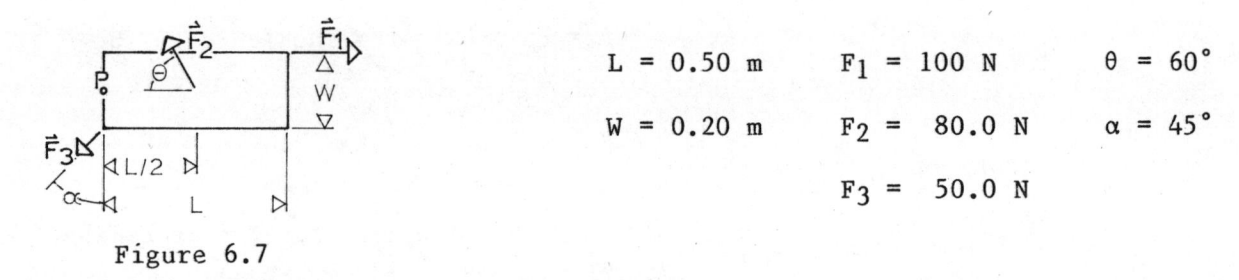

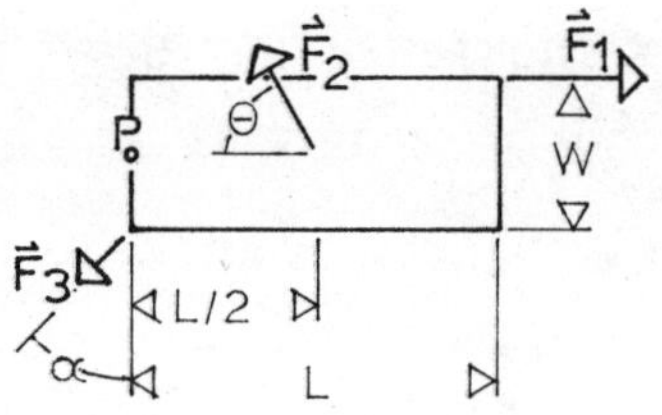


Figure 6.7

| | | |
|---|---|---|
| L = 0.50 m | $F_1$ = 100 N | θ = 60° |
| W = 0.20 m | $F_2$ = 80.0 N | α = 45° |
| | $F_3$ = 50.0 N | |

| | |
|---|---|
| 1. What is the torque that $\vec{F}_1$ exerts about P? | The moment arm for $\vec{F}_1$ is W/2 = 0.100 m<br><br>$\tau_1 = F_1(\text{moment arm})_1 = F_1W/2 = 10.0$ N·m<br><br>Since $\vec{F}_1$ tends to rotate the object in a clockwise manner, it produces a negative torque. Hence $\vec{\tau}_1 = -10.0$ N·m. |

Note: Since $\vec{F}_1$ and its moment arm were easily obtained, we used τ = F(moment arm) to determine the torque (Method II).

| | |
|---|---|
| 2. What is the torque that $\vec{F}_2$ exerts about P? | The component of $\vec{F}_2$ that produces the torque about P is $F_{2\perp} = F_2 \sin\theta = 69.3$ N<br>$\tau_2 = rF_{2\perp} = \frac{L}{2} F_2 \sin\theta = 17.3$ N·m<br><br>Since $\vec{F}_2$ tends to rotate the object in a counterclockwise manner it is a positive torque. Hence $\vec{\tau}_2 = +17.3$ N·m |

Note: Since the magnitude of the position vector locating the point of application of the force and the component of the force perpendicular to the position vector were easily obtained, we used $\tau = rF_\perp$ to determine the torque (Method III).

| | |
|---|---|
| 3. What is the torque that $\vec{F}_3$ exerts about P? | $\tau_3 = r_3 F_3 \sin\alpha = (0.100 \text{ m})(50.0 \text{ N})\sin 45°$<br>$= 3.54$ N·m<br><br>Since $\vec{F}_3$ tends to rotate the object in a clockwise manner, it is a negative torque. Hence $\vec{\tau}_3 = -3.54$ N·m. |

Note: Since we know the magnitude of the position vector locating the point of application of the force, the magnitude of the force, and the angle between the direction of these two vectors, we used $\tau = rF \sin\theta$ to determine the torque (Method I).

| | |
|---|---|
| 4. What is the net or resultant torque about P? | $\vec{\tau}_{Net} = \vec{\tau}_1 + \vec{\tau}_2 + \vec{\tau}_3$<br><br>$= (-10.0 + 17.3 - 3.54)$N·m $= +3.76$ N·m<br><br>The positive value indicates that the object is experiencing a net counterclockwise torque. |

Note: We could have obtained $\tau_1$, $\tau_2$, and $\tau$ by using any one of the three methods. However, by virtue of the given information, Method II was easiest for $\vec{F}_1$, Method III for $\vec{F}_2$, and Method I for $\vec{F}_3$. You need to be familiar with all three methods in order to solve the problems in this and later chapters.

Example 6.3. A disk of 0.200-m radius is mounted on an axle, as shown in Fig. 6.8. A rope is wrapped around the perimeter of the disk and then tied to a 0.500-kg mass. If the 0.500-kg mass if released from rest and travels 2.00 m in a time of 3.00 s, what is the torque tending to rotate the disk?

Given and Diagram:

$R = 0.200$ m = radius of the disk
$M = 0.500$ kg = mass tied to the rope
$h = 2.00$ m = distance M travels in 3.00 s
$t = 3.00$ s = time for M to travel 2.00 m

Figure 6.8

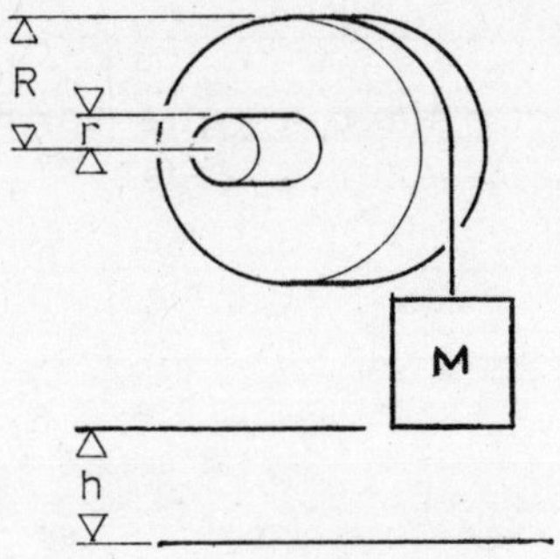


Determime: The torque being exerted on the disk.

Strategy: If we ignore friction, the only torque exerted on the disk is that due to the tension in the cord. If we know the tension in the cord, we can determine the torque by $\tau = RT$. We can write a Newton's Second Law equation for M and determine T if we know the acceleration of M. We can determine the acceleration of M, and hence T and then $\tau$, from the fact that M starts from rest and travels the distance h in a time t.

Solution:

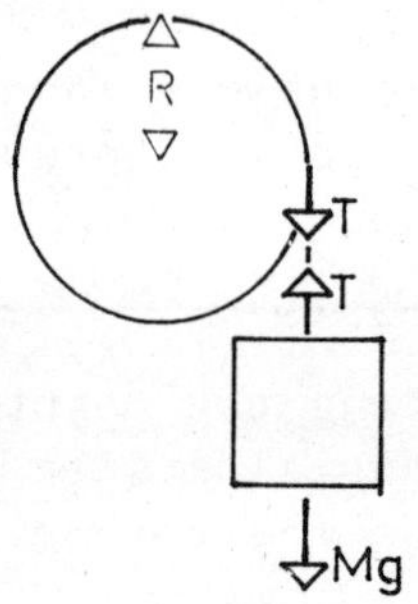


$$\tau = -RT$$

$$Ma = Mg - T \quad \text{or} \quad T = M(g-a)$$

$$\Delta y = v_o t + \frac{1}{2} at^2 \quad \text{or} \quad a = 2h/t^2, \text{ since } \Delta y = h \text{ and } v_o = 0$$

$$\tau = -RT = -RM(g-a) = -RM(g - 2h/t^2) = -0.936 \text{ N·m}$$

Related Text Problems: 6-12, 6-13.

## 3. Second Condition of Equilibrium (Section 6.3)

Review: The second condition of equilibrium states that an object is in rotational equilibrium if the sum of all the torques acting on it is zero. This may be stated algebraically as follows:

If $\sum_i \vec{\tau}_i = 0$, then object is in rotational equilibrium.

If the vector sum of all the torques acting on an object is zero, then the angular acceleration of the object is zero, and its angular velocity is a constant. If the angular velocity is zero, the object is in static rotational equilibrium. If the angular velocity is a nonzero constant, the object is in dynamic rotational equilibrium. This may be summarized as follows:

If $\sum_i \vec{\tau}_i = 0$, then $\alpha = 0$ and $\omega$ = constant
If $\omega = 0$, then static rotational equilibrium exists.
If $\omega \neq 0$ but constant, then dynamic rotational equilibrium exists.

Practice: The plank shown in Fig. 6.9 is in static rotational equilibrium.

Figure 6.9

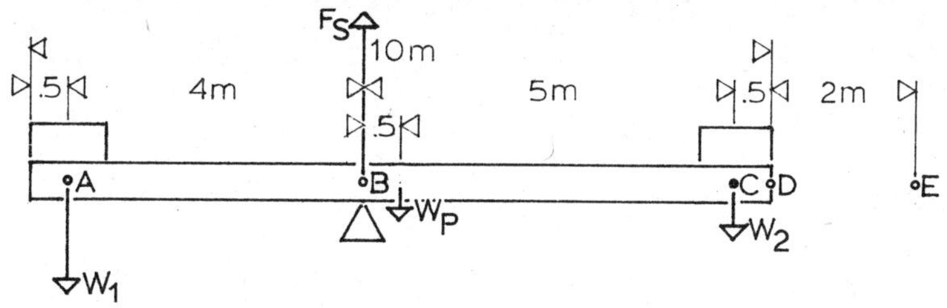


$W_1$ = 100 N

$W_2$ = 75.0 N

$W_P$ = 50.0 N

Determine the following:

| | |
|---|---|
| 1. Total cw torque about B | $\vec{\tau}_{cwB} = -W_P(0.500 \text{ m}) - W_2(5.00 \text{ m}) = -400 \text{ N}\cdot\text{m}$ |
| 2. Total ccw torque about B | $\vec{\tau}_{ccwB} = W_1(4.00 \text{ m}) = (100 \text{ N})(4.00 \text{ m})$ <br> $= 400 \text{ N}\cdot\text{m}$ |
| 3. Total torque about B | $\vec{\tau}_B = \vec{\tau}_{cwB} + \vec{\tau}_{ccwB} = (-400 + 400)\text{N}\cdot\text{m} = 0 \text{ N}\cdot\text{m}$ <br> The plank is in rotational equilibrium about point B. |
| 4. Total cw torque about A | $\vec{\tau}_{cwA} = -W_P(4.5 \text{ m}) - W_2(9.00 \text{ m}) = -900 \text{ N}\cdot\text{m}$ |
| 5. Total ccw torque about A | $\vec{\tau}_{ccwA} = F_s(4 \text{ m})$ <br> $F_s$ = the force that the support must exert upward in order to maintain translational equilibrium. <br> $F_s = W_1 + W_2 + W_P = 225 \text{ N}$ <br> $\vec{\tau}_{ccwA} = (225 \text{ N})(4.00 \text{ m}) = +900 \text{ N}\cdot\text{m}$ |
| 6. Total torque about A | $\vec{\tau}_A = \vec{\tau}_{cwA} + \vec{\tau}_{ccwA} = (-900 + 900)\text{Nm} = 0 \text{ N}\cdot\text{m}$ <br> The plank is in rotational equilibrium about point A. |
| 7. Total cw torque about D | $\vec{\tau}_{cwD} = -F_s(5.5 \text{ m}) = -1.24 \times 10^3 \text{ N}\cdot\text{m}$ |

| | |
|---|---|
| 8. Total ccw torque about D | $\vec{\tau}_{ccwD} = W_1(9.50\text{ m}) + W_p(4.50\text{ m}) + W_2(0.500\text{ m}) = +1.24 \times 10^3\text{ N}\cdot\text{m}$ |
| 9. Total torque about D | $\vec{\tau}_D = \vec{\tau}_{cwD} + \vec{\tau}_{ccwD} = 0\text{ N}\cdot\text{m}$<br>The plank is in rotational equilibrium about point D. |

Note: From steps 3, 6, and 9 it should be evident that an object in rotational equilibrium about any one point is in rotational equilibrium about any other point. This point may be anywhere on or off the object. You should now convince yourself that the total torque about C and E is also zero.

---

Example 6.4. A painter's is 3.00 m long and is resting on two sawhorses that are 0.500 m in from each end, as shown in Fig. 6.10. The plank has a weight of 200 N, and the painter weighs 600 N. How far beyond the sawhorse can the painter stand before the plank starts to tip?

Given and Diagram:

$L = 3.00$ m = length of plank

$W_{pk} = 200$ N = weight of plank

$W_{pt} = 600$ N = weight of painter

Figure 6.10

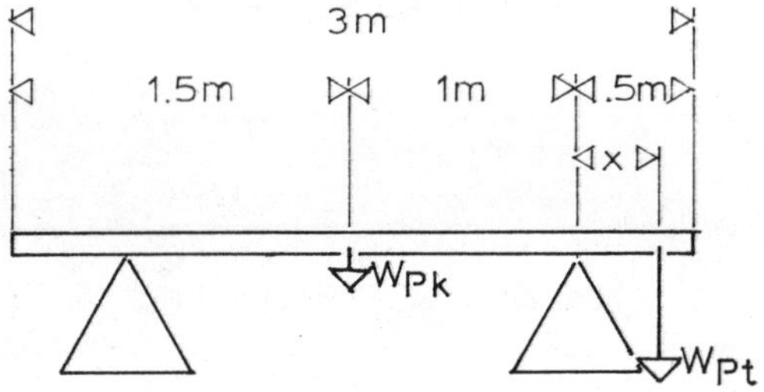


Determine: How far (distance x) beyond the sawhorse the painter can stand before the plank starts to tip.

Strategy: We want to know the largest value for distance x with the system maintaining static rotational equilibrium. When x has its largest value, the clockwise torque due to the weight of the painter and the counterclockwise torque due to the weight of the plank will be equal, and the left sawhorse will be exerting no force on the plank. A summation of torques about the sawhorse on the right allows us to solve for the quantity x.

Solution: Since the system is in rotational equilibrium, we can write the following equations describing the torques about the sawhorse on the right.

$$\sum_i \vec{\tau}_i = \vec{\tau}_{pk} + \vec{\tau}_{pt} = W_{pk}(1.00\text{ m}) - W_{pt}(x) = 0$$

or

$$x = W_{pk}(1.00\text{ m})/W_{pt} = (200\text{ N})(1.00\text{ m})/(600\text{ N}) = 0.330\text{ m}$$

Example 6.5. For a sign suspended as shown in Fig. 6.11., determine the tension in the cable and the force exerted on the beam by the pin.

Given and Diagram:

Figure 6.11

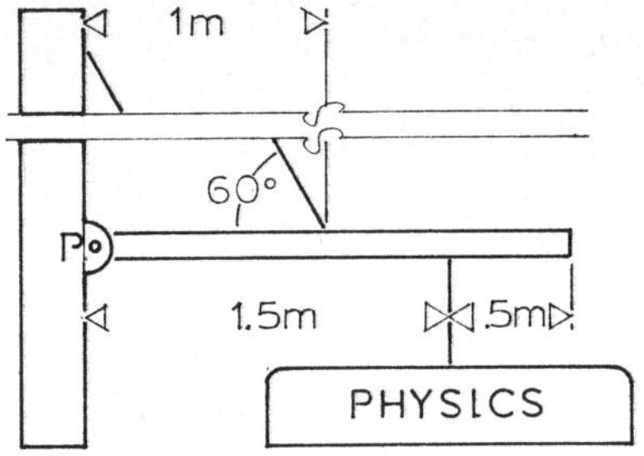


$W_b$ = 100 N = weight of beam

$W_s$ = 200 N = weight of sign

L = 2.00 m = length of beam

Determine: T, the tension in the cable.
$F_p$, the force the pin exerts on the beam.

Strategy: First draw a diagram showing all forces acting on the beam. If we sum the torques about the pin P, we do not have to worry about the force $F_p$. The only unknown in a summation-of-torques statement about P is the tension in the cable. Once T is known, we can sum horizontal and vertical forces, respectively, to obtain the horizontal and vertical components of $F_p$. Finally, we can obtain the magnitude and direction of $F_p$ from its components.

Solution: Since the system is in rotational equilibrium, a summation of torques about any point will be zero. Let's use the point P.

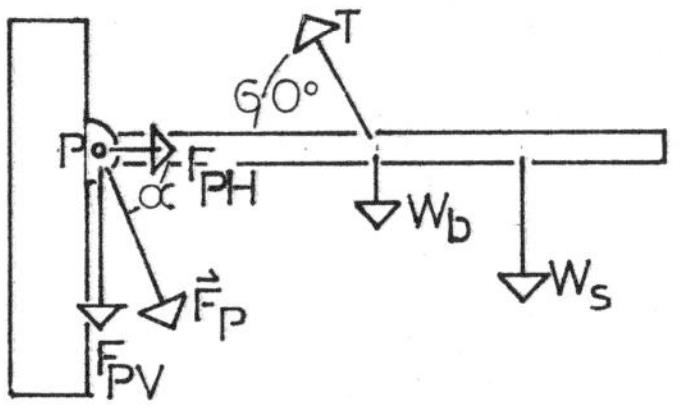


$$\sum_i \vec{\tau}_i = \vec{\tau}_b + \vec{\tau}_s + \vec{\tau}_T = 0$$

Hence

$$-W_b(1.00\text{ m}) - W_s(1.50\text{ m}) + T\sin 60°(1.00\text{ m}) = 0$$

or

$$T = [(100\text{ N})(1.00\text{ m}) + (200\text{ N})(1.50\text{ m})]/(0.866\text{ m}) = 462\text{ N}$$

Since the system is in horizontal equilibrium, a summation of the horizontal forces must be zero.

$$\sum_i F_{hor_i} = F_{ph} - (T\cos 60°) = 0 \quad \text{or} \quad F_{ph} = T\cos 60° = (462\text{ N})\cos 60° = 231\text{ N}$$

Since the system is in vertical equilibrium a summation of vertical forces must be zero.

$$\sum_i F_{vert_i} = F_{pv} + (T\sin 60°) - W_b - W_s = 0 \quad \text{or} \quad F_{pv} = W_b + W_s - T\sin 60° = -100\text{ N}$$

Note: When the above figure was drawn, the actual direction of $\vec{F}_p$ was unknown. Now that we know $F_{ph}$ and $F_{pv}$, we can establish that $\vec{F}_p$ was drawn correctly. If $\vec{F}_p$ was not initially drawn correct, it could be redrawn at this time.

Now that we know the components of $F_p$, we can determine the magnitude and direction of $F_p$.

$$F_p = (F_{pv}^2 + F_{ph}^2)^{1/2} = 252\text{ N} \quad \text{and} \quad \alpha = \tan^{-1}(F_{pv}/F_{ph}) = 23.4°$$

Example 6.6. A ladder 4.00 m long rests against a vertical, frictionless wall with its lower end 1.50 m from the wall, as shown in Fig. 6.12. The ladder weighs 450 N, and its center of gravity is at its geometric center. The static coefficient of friction between the ladder and the ground is 0.300. How far up the ladder can a 650 N man climb before it starts to slip?

Given and Diagram:

$d = 1.50$ m
$L = 4.00$ m
$W_L = 450$ N
$W_M = 650$ N
$\mu = 0.300$

Figure 6.12

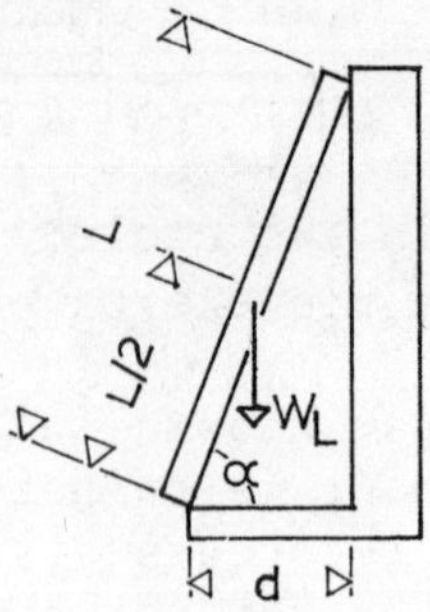


Determine: How far the 650-N man can climb up the ladder before it begins to slip.

Strategy: First draw a diagram showing all forces acting on the ladder. We wish to determine the maximum distance the man can climb up the ladder with the system maintaining static rotational equilibrium.

Since the system is in vertical equilibrium, we can determine the normal force that the ground exerts on the ladder ($F_{NG}$) from $W_L$ and $W_M$. Knowing $F_{NG}$ and $\mu$, we can determine the frictional drag $f_g$ that the ground exerts on the ladder. Since the system is in horizontal equilibrium, we can determine the normal force that the wall exerts on the ladder ($F_{NW}$) from $f_g$. If we use x to represent the distance that the man can go up the ladder without disrupting the rotational equilibrium and sum the torques about the base of the ladder, the only unknown quantity is x.

Solution:

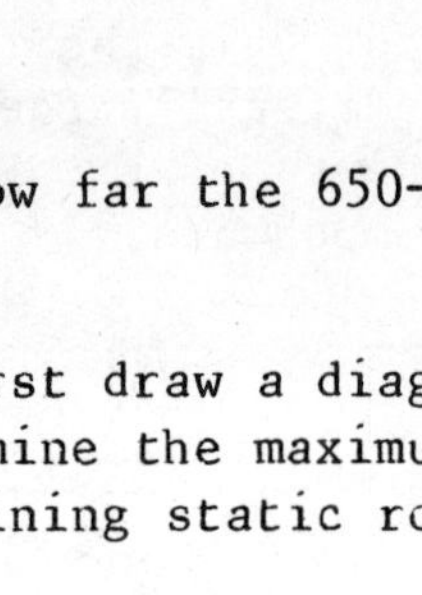

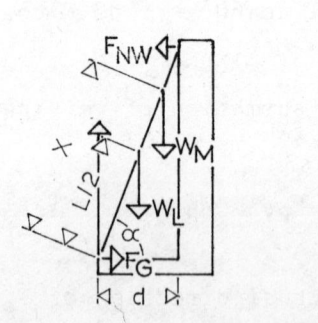

$$\alpha = \cos^{-1}\left(\frac{d}{L}\right) = \cos^{-1}\left(\frac{1.5}{4}\right) = 68.0°$$

Since the system is in vertical equilibrium, we know that the sum of the vertical forces must be zero. Hence,

$$\sum_i F_{vi} = F_{NG} - W_L - W_M = 0 \quad \text{or} \quad F_{NG} = W_L + W_M = 50.0 \text{ N}$$

Now that $F_{NG}$ is known, we can determine the frictional drag $f_g$.

$$f_g = \mu F_{NG} = (0.300)(50.0\ N) = 15.0\ N$$

Since the system is in horizontal equilibrium, we know that the sum of the horizontal forces must be zero. Hence

$$\sum_i F_{hi} = f_g - F_{NW} = 0 \quad \text{or} \quad F_{NW} = f_g = 15.0\ N$$

We wish to determine the largest possible value that x can have with the system maintaining rotational equilibrium. If the system is in equilibrium, the sum of the torques about the base of the ladder must be zero. Hence

$$\sum_i \vec{\tau}_i = \vec{\tau}_{W_L} + \vec{\tau}_{W_M} + \vec{\tau}_{F_{NW}} = -(W_L \frac{L}{2} \cos\alpha) - (W_M x \cos\alpha) + (F_{NW} L \sin\alpha) = 0$$

or

$$x = [(-W_L L \cos\alpha/2) + (F_{NW} L \sin\alpha)]/(W_M \cos\alpha) = 3.62\ m$$

Related Text Problems: 6-14 through 6-31

## 4. Center of Gravity and Center of Mass (Section 6.4)

Review: The center of gravity (CG) of a body or system is defined as that point through which all the weight may be thought to act for the purposes of calculation. If the acceleration due to gravity is the same at all points on an object, the center of gravity and center of mass are the same point.

Practice: A rigid rod 1 m long has a negligible mass and holds three weights as shown in Fig. 6.13.

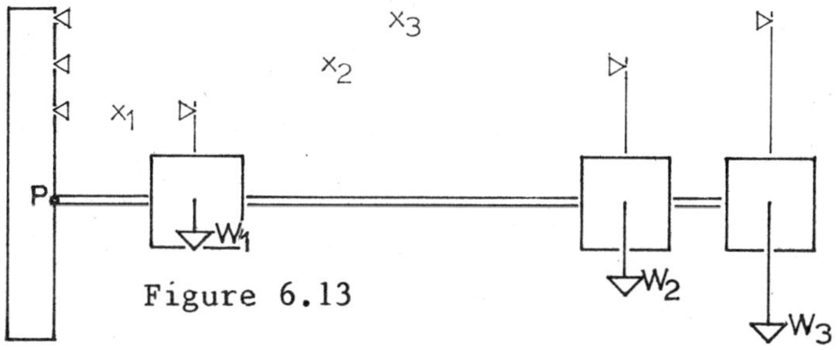

Figure 6.13

$W_1 = 1.00\ N;\quad x_1 = 0.200\ m$

$W_2 = 2.00\ N;\quad x_2 = 0.800\ m$

$W_3 = 3.00\ N;\quad x_3 = 1.00\ m$

Determine the following:

| | |
|---|---|
| 1. The upward force required to support the rod and weights | $F_{up} = W_1 + W_2 + W_3 = 6.00\ N$ |
| 2. The point at which this upward force must be applied in order to support the rod and weights | The CG of the system |

| | |
|---|---|
| 3. Let $X_{CG}$ represent the location of the center of gravity from the point P. Determine the torque that the upward force (Step 1) exerts about P. | $\vec{\tau}_{up} = F_{up}X_{CG} = (6.00\ N)X_{CG}$ |
| 4. The torque that the three weights exert about P | $\vec{\tau}_{wts} = -W_1X_1 - W_2X_2 - W_3X_3 = -4.80\ N{\cdot}m$ |
| 5. The location of the center of mass of the weights | Since the system is in rotational equilibrium, the sum of the torques about P must be zero. $\sum_i \vec{\tau}_{P_i} = -4.80\ N{\cdot}m + (6.00\ N)X_{CG} = 0$ or $X_{CG} = 4.80\ N{\cdot}m/6.00\ N = 0.800\ m$ |

Example 6.7. The device shown in Fig. 6.14 is used to determine the a person's weight and the location of that person's center of gravity. From the given information, determine the girl's weight and the location of her center of gravity.

Given and Diagram:

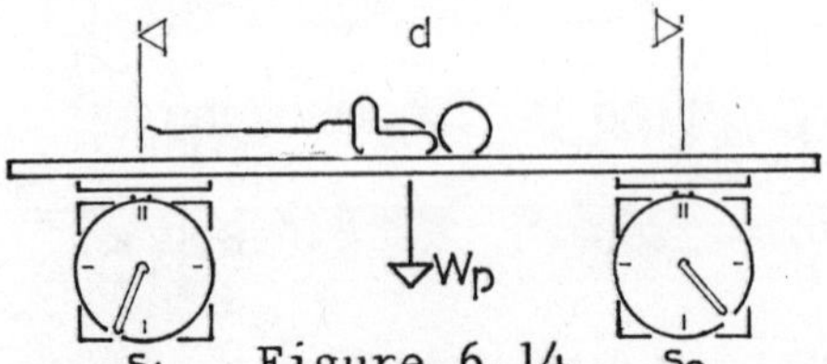

Figure 6.14

$W_P = 40.0\ N$ = weight of the plank
$d = 2.00\ m$ = distance between the supports
$S_1 = 248\ N$ = reading on scale 1
$S_2 = 222\ N$ = reading on scale 2

Determine: $W_g$, the girl's weight
$X_{CG}$, the location of the girl's center of gravity

Strategy: By summing the forces in the vertical direction we can determine the girl's weight. By summing the torques about the point of support for scale 1, we can determine the location of the girl's center of mass.

Solution: Since the system is in translational equilibrium, the summation of the forces in the vertical direction must be zero.

$$\sum_i F_i = +F_{S_1} + F_{S_2} - W_P - W_g = 0 \quad \text{or} \quad W_g = F_{S_1} + F_{S_2} - W_P = 430\ N$$

Since the system is in rotational equilibrium, the summation of torques about the point of support for scale 1 must be zero.

$$\sum_i \vec{\tau}_i = -W_g X_{CG} - W_P \frac{d}{2} + F_{S_2} d = 0 \quad \text{or} \quad X_{CG} = [(F_{S_2} d) - (W_P \frac{d}{2})W_g = 0.940 \text{ m}$$

Related Text Problems: 6-32 through 6-35.

---

PRACTICE TEST

Take and grade this practice test. Doing so will allow you to determine any weak spots in your understanding of the concepts taught in this chapter. The following section prescribes what you should study further to strengthen your understanding.

You push on a box resting on an incline, as shown in the following.

$W = 1.00 \times 10^3$ N = weight of the box
$P = 5.00 \times 10^2$ N = your push
$\mu_s = 0.300$ = static friction
$\mu_k = 0.100$ = kinetic friction

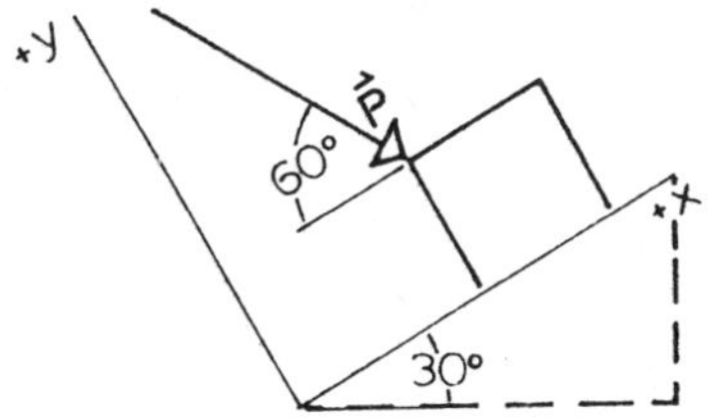


Determine the following:

__________ 1. Component of the weight of the box perpendicular to the incline
__________ 2. Component of your push perpendicular to the incline
__________ 3. Normal force acting on the box
__________ 4. Component of the weight of the box parallel to the incline
__________ 5. Component of your push parallel to the incline
__________ 6. Force of friction acting on the box
__________ 7. Unbalanced force on the box parallel to the incline
__________ 8. Is the box in equilibrium?

A sign is suspended from a uniform pole as shown. The pole is attached to the wall by means of a bracket and pin.

$W_s = 500$ N = weight of the sign
$W_p = 200$ N = weight of the pole
$L = 4.00$ m = length of the pole

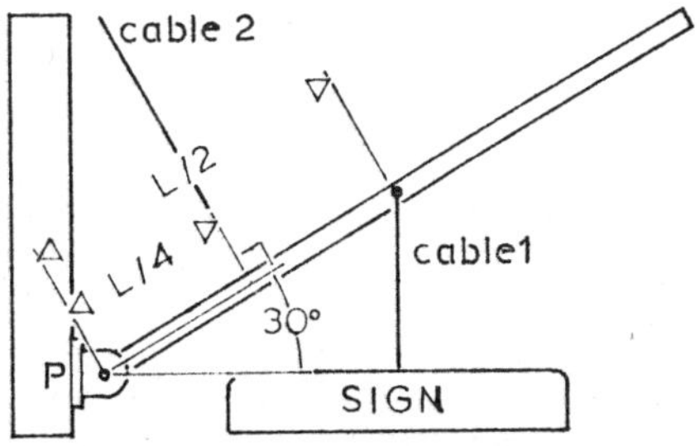


Determine the following:

__________ 9. Torque about pin P due to the weight of the sign
__________ 10. Torque about pin P due to the weight of the pole
__________ 11. Torque about pin P due to cable 2
__________ 12. The tension in cable 2
__________ 13. Horizontal component of the force that pin P exerts on the pole
__________ 14. Vertical component of the force that pin P exerts on the pole

Consider the following figure.

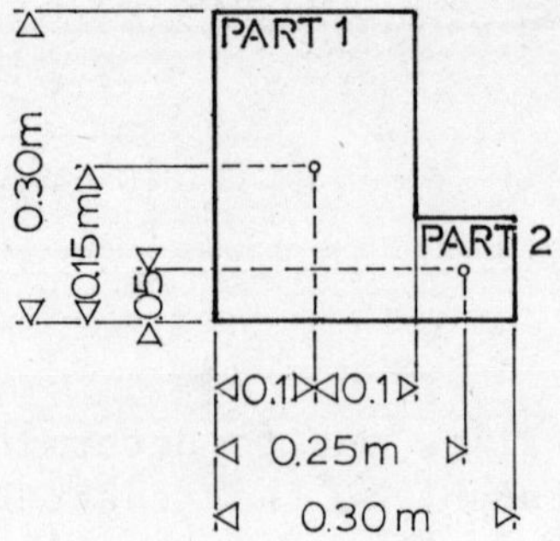


$W_1$ = 12.0 N = is of part 1
$W_2$ = 2.00 N = is of part 2

Determine the following:

__________ 15. The x and y components for the center of gravity of part 1 of the object

__________ 16. The x and y components for the center of gravity of part 2 of the object

__________ 17. The x component of the center of gravity of the object

__________ 18. The y component of the center of gravity of the object

(See Appendix I for answers.)

PRINCIPAL CONCEPTS AND EQUATIONS PRESCRIPTION

Your score on the practice is an excellent measure of your understanding of this chapter. You should now use the following chart to write your own prescription for curing any of your physics ills. Look down the leftmost column to the number of the question(s) you answered incorrectly, read across that row to see which section(s) of the study guide you should return to for further study, and then do the suggested text problems to gain additional experience in working with the particular concept.

| Practice Test Question | Concepts and Equations | Prescription | |
|---|---|---|---|
| | | Principal Concept | Text Problems |
| 1 | Components of a vector - analytical | 6 of Ch. 3 | 3-3,8 |
| 2 | Components of a vector - analytical | 6 of Ch. 3 | 3-3,8 |
| 3 | First condition of equilibrium | 1 | 6-3,6 |
| 4 | Components of a vector - analytical | 6 of Ch. 3 | 3-3,8 |
| 5 | Components of a vector - analytical | 6 of Ch. 3 | 3-3,8 |
| 6 | Force of friction | 7 of Ch. 4 | 4-45,46 |
| 7 | First condition of equilibrium | 1 | 6-7,11 |
| 8 | First condition of equilibrium | 1 | 6-4,5 |
| 9 | Torque | 2 | 6-12,16 |
| 10 | Torque | 2 | 6-13,22 |
| 11 | Second condition of equilibrium | 3 | 6-24,25 |
| 12 | Torque | 2 | 6-26,27 |
| 13 | First condition of equilibrium | 1 | 6-2,4 |
| 14 | First condition of equilibrium | 1 | 6-1,5 |
| 15 | Center of gravity | 4 | 6-32,33 |
| 16 | Center of gravity | 4 | 6-33,34 |
| 17 | Center of gravity | 4 | 6-34,35 |
| 18 | Center of gravity | 4 | 6-32,34 |

# 7 Work and Energy

RECALL FROM PREVIOUS CHAPTERS

| Previously learned concepts and equations frequently used in this chapter | Text Section | Study Guide Page |
|---|---|---|
| Resolving vectors into components | 3.2 | 32 |
| Adding vectors | 3.2 | 33 |
| Normal force $F_N$ and friction f: $f = \mu F_N$ | 4.7 | 58 |

NEW IDEAS IN THIS CHAPTER

| Concepts and equations introduced | Text Section | Study Guide Page |
|---|---|---|
| Work: $W = F_{\parallel} \Delta s$ | 7.3 | 100 |
| Power: $P = W/\Delta t$ | 7.4 | 106 |
| Kinetic energy: $\Delta KE = Mv^2/2$ | 7.5 | 108 |
| Work energy theorem: $W_{Net} = \Delta KE$ | 7.5 | 108 |
| Gravitational potential energy: $\Delta PE = -W_g = Mg\Delta h$ | 7.6 | 110 |
| Work done by an applied force: $W_{app} = \Delta KE + \Delta PE + W_{against\ f}$ | 7.7 | 112 |
| Conservation of mechanical energy: $\Delta KE + \Delta PE = 0$ | 7.8 | 116 |

PRINCIPAL CONCEPTS AND EQUATIONS

1. Work (Section 7.3)

Review: You can use several methods to calculate the work done by a force acting on an object. Three of these methods are reviewed here.

Method A: If a force F acts on an object (Fig. 7.1) and causes a displacement $\Delta s$, the work done by F can be obtained by multiplying the magnitude of the force $|F|$, the magnitude of the displacement $|\Delta s|$, and the cosine of the angle between their directions.

Figure 7.1 $W = F\Delta s \cos\theta$

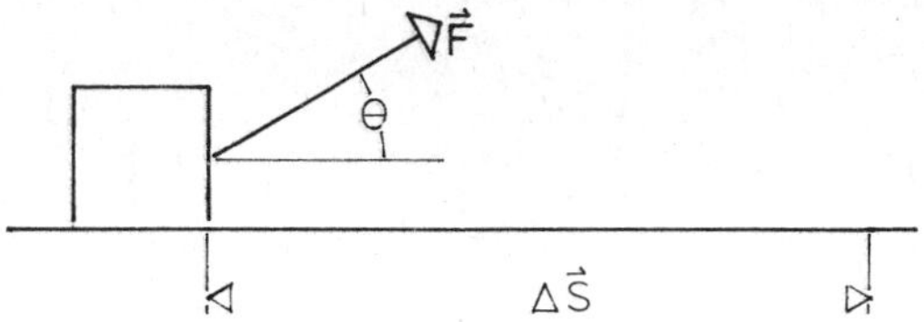


To illustrate this method, consider the object shown in Fig. 7.2 and the forces acting on it.

Figure 7.2

$$F_1 = F_2 = F_3 = 100 \text{ N}$$
$$W = 200 \text{ N}$$
$$\Delta s = 10.0 \text{ m}$$
$$\mu = 0.100$$
$$\theta_2 = 60°$$
$$\theta_3 = 30°$$

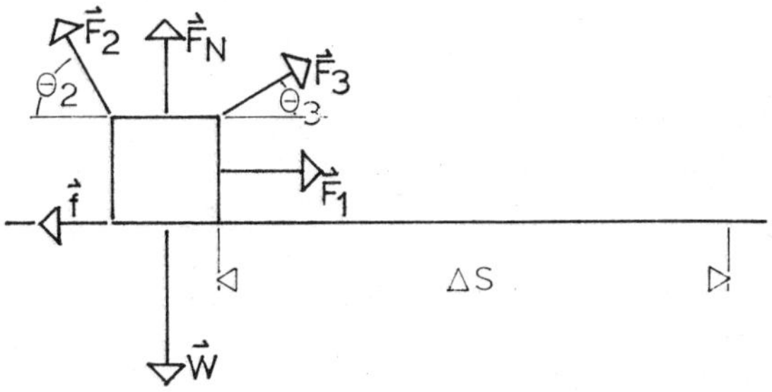


Using method A, we can determine the work done by the forces $\vec{F}_1$, $\vec{F}_2$, $\vec{F}_3$, $\vec{W}$, $\vec{F}_N$, and $\vec{f}$. Before doing the work calculations, we must first determine the magnitude of $\vec{F}_N$ and $\vec{f}$.

Since the object is in vertical equilibrium, we can write:

$$F_N + (F_2 \sin 60°) + (F_3 \sin 30°) = W$$

Solving for $F_N$, we obtain

$$F_N = W - (F_3 \sin 30°) - (F_2 \sin 60°) = 63.4 \text{ N}$$

Knowing $F_N$, we obtain f as follows:

$$f = \mu F_N = 6.34 \text{ N}$$

Now we can calculate the work done by each force.

$$W_1 = F_1 \Delta s \cos 0° = 1000 \text{ J} \qquad W_W = W\Delta s \cos 90° = 0 \text{ J}$$

$$W_2 = F_2 \Delta s \cos 120° = -500 \text{ J} \qquad W_{F_N} = F_N \Delta s \cos 90° = 0 \text{ J}$$

$$W_3 = F_3 \Delta s \cos 30° = 866 \text{ J} \qquad W_f = f\Delta s \cos 180° = -63.4 \text{ J}$$

Knowing the work done by each force,we can calculate the net work done on the object.

$$W_{Net} = W_1 + W_2 + W_3 + W_W + W_F + W_f = 1.30 \times 10^3 \text{ J}$$

Method B: If a force $\vec{F}$ acts on an object (Fig. 7.3) and causes a displacement $\Delta\vec{s}$, the work done by $\vec{F}$ can be obtained by finding the product of the component of the force in the direction of the displacement and the magnitude of the displacement. Note: When finding the component of the force, the positive direction is taken to be the direction of $\Delta\vec{s}$.

Figure 7.3 $W = (F\cos\theta)\Delta s$

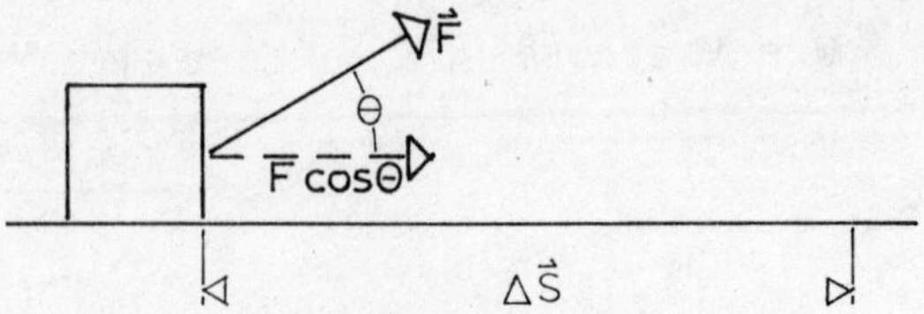


Since $F\cos\theta$ is the component of $\vec{F}$ parallel to $\vec{\Delta s}$, we may write

$$W = F_{\parallel}\Delta s, \quad \text{where } F_{\parallel} = F\cos\theta$$

To visualize this method, consider the object shown in Fig. 7.4 and the forces acting on it.

Figure 7.4

$F_1 = F_2 = F_3 = 100$ N
$W = 200$ N
$\Delta s = 10.0$ m
$\mu = 0.100$
$\theta_1 = 0°$
$\theta_2 = 60°$
$\theta_3 = 30°$

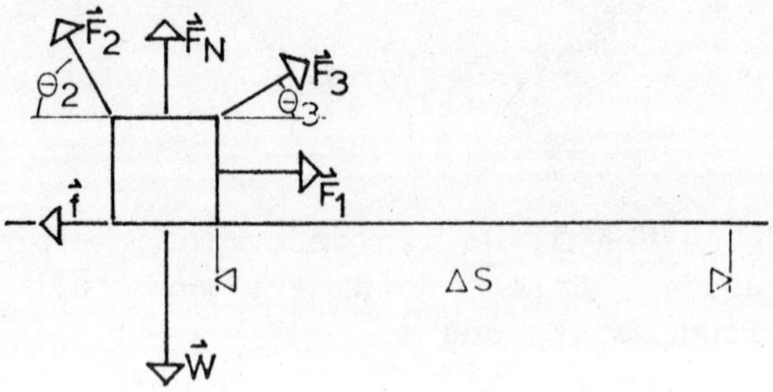


Note: The situation considered here is the same as that considered in method A. Consequently, we know that $F_N = 63.4$ N and $f = 6.34$ N.

To use method B, we must determine the component of the force in the direction of $\Delta s$.

Component of $\vec{F}_1$ in the direction of $\Delta s$: $F_{1\parallel} = F_1\cos\theta_1 = 100$ N
Component of $\vec{F}_2$ in the direction of $\Delta s$: $F_{2\parallel} = -F_2\cos\theta_2 = -50.0$ N
Component of $\vec{F}_3$ in the direction of $\Delta s$: $F_{3\parallel} = F_3\cos\theta_3 = 86.6$ N
Component of $\vec{W}$ in the direction of $\Delta s$: $W_{\parallel} = W\cos 90° = 0$ N
Component of $\vec{f}$ in the direction of $\Delta s$: $f_{\parallel} = -f = -6.34$ N
Component of $\vec{F}_N$ in the direction of $\Delta s$: $F_{N\parallel} = F_N\cos 90° = 0$ N

Knowing the component of each force in the direction of the displacement and the magnitude of the displacement, we can determine the work done on the object by each force.

$W_1 = F_{1\parallel}\Delta S = 1000$ J
$W_2 = F_{2\parallel}\Delta S = -500$ J
$W_3 = F_{3\parallel}\Delta S = 866$ J
$W_W = W_{\parallel}\Delta S = 0$ J
$W_f = f_{\parallel}\Delta S = 63.4$ J
$W_{F_N} = F_{N\parallel}\Delta s = 0$ J

Knowing the work done by each force, we can calculate the net work done on the object.

$$W_{Net} = W_1 + W_2 + W_3 + W_W + W_f + W_{F_N} = 1.30 \times 10^3 \text{ J}$$

Note: Our answers from method A and method B for the work done by each force and the net work are the same.

Method C: If more than one force acts on an object, we can determine the net work done on the object by finding the net or unbalanced force on the object in the direction of $\vec{\Delta s}$ and multiplying by the magnitude of $\vec{\Delta s}$. For example in the case just considered:

$$W_{Net} = W_1 + W_2 + W_3 + W_W + W_F + W_f$$

$$= F_{1\parallel}\Delta s + F_{2\parallel}\Delta s + F_{3\parallel}\Delta s + W_{\parallel}\Delta s + F_{N\parallel}\Delta s + f_{\parallel}\Delta s$$

$$= (F_{1\parallel} + F_{2\parallel} + F_{3\parallel} + f_{\parallel})\Delta s = F_{Net\parallel}\Delta s$$

For this case

$$F_{Net\parallel} = F_{1\parallel} + F_{2\parallel} + F_{3\parallel} + f_{\parallel} = 100\ N - 50.0\ N + 86.6\ N - 6.34\ N = 1.30 \times 10^2\ N$$

$$W_{Net} = F_{Net\parallel}\Delta s = (1.30 \times 10^2\ N)(10.0\ m) = 1.30 \times 10^3\ J$$

Note: Our answers from methods A, B and C for the net work are the same.

Note: Even though methods B and C are based on the same idea, there is a distinction between the two methods.

Method B: we determined the work done by each force and then added these to obtain the net work.
Method C: we determined the net force parallel to $\Delta s$ and then the net work.

Practice: Consider the three situations and the data shown below:

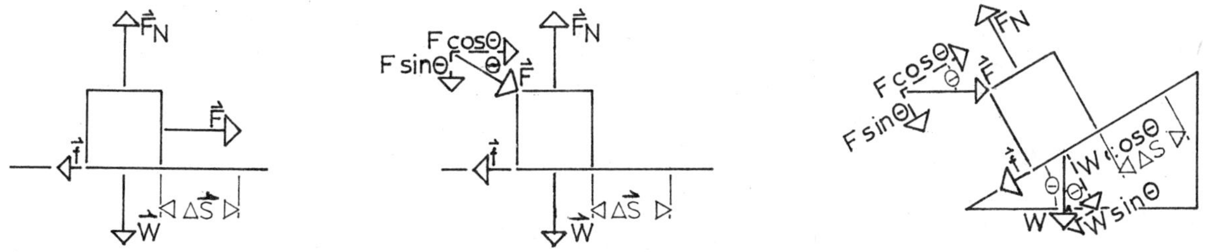


| | | |
|---|---|---|
| $F = 100\ N$ | $F = 100\ N$ | $F = 100\ N$ |
| $W = 100\ N$ | $W = 100\ N$ | $W = 100\ N$ |
| $\Delta s = 10.0\ m$ | $\Delta s = 10.0\ m$ | $\Delta s = 10.0\ m$ |
| $\mu_k = 0.200$ | $\mu_k = 0.200$ | $\mu_k = 0.200$ |
| | $\theta = 60°$ | $\theta = 30°$ |

Determine the following:

1. The normal force $F_N$ acting on each object

| | | |
|---|---|---|
| $F_N = W = 100\ N$ | $F_N = W + (F \sin\theta)$ <br> $= 100\ N + (100\ N) \sin 60°$ <br> $= 186\ N$ | $F_N = (W \cos\theta) + (F \sin\theta)$ <br> $= (100\ N) \cos 30°$ <br> $+ (100\ N) \sin 30°$ <br> $= 136\ N$ |

2. The force of gravity $F_g$ acting on each object

| | | |
|---|---|---|
| $F_g = W = 100\ N$ | $F_g = W = 100\ N$ | $F_g = W = 100\ N$ |

3. The force of friction f acting on each object

| | | |
|---|---|---|
| $f = \mu F_N$<br>$= (0.200)(100\ N)$<br>$= 20.0\ N$ | $f = \mu F_N$<br>$= (0.200)(186\ N)$<br>$= 37.2\ N$ | $f = \mu F_N$<br>$= (0.200)(136\ N)$<br>$= 27.2\ N$ |

4. The work $W_F$ done by the applied force F for each case

| | | |
|---|---|---|
| $W_F = F_\parallel \Delta s$<br>$= (100\ N)(10.0\ m)$<br>$= 1000\ J$ | $W_F = F_\parallel \Delta s$<br>$= F \cos 60° \ \Delta s$<br>$= (100\ N)(0.500)(10.0\ m)$<br>$= 500\ J$ | $W_F = F_\parallel \Delta s$<br>$= F \cos 30° \ \Delta s$<br>$= (100\ N)(0.866)(10.0\ m)$<br>$= 866\ J$ |

5. The work $W_f$ done by the force of friction f for each case

| | | |
|---|---|---|
| $W_f = f_\parallel \Delta s$<br>$= (-20.0\ N)(10.0\ m)$<br>$= -200\ J$ | $W_f = f_\parallel \Delta s$<br>$= (-37.2\ N)(10.0\ m)$<br>$= -372\ J$ | $W_f = f_\parallel \Delta s$<br>$= (-27.2\ N)(10.0\ m)$<br>$= -272\ J$ |

6. The work $W_g$ done by the gravitational force $F_g$ for each case

| | | |
|---|---|---|
| $W_g = F_{g\parallel} \Delta s$<br>$= 0\ J$ | $W_g = F_{g\parallel} \Delta s$<br>$= 0\ J$ | $W_g = F_{g\parallel} \Delta s$<br>$= (-W \sin 30°)\Delta s$<br>$= -500\ J$ |

7. The net work done by all of the forces acting on the object

| | | |
|---|---|---|
| $W_{Net} = W_F + W_f + W_g$<br>$= 800\ J$ | $W_{Net} = W_F + W_f + W_g$<br>$= 128\ J$ | $W_{Net} = W_F + W_f + W_g$<br>$= 94.0\ J$ |

8. The net or unbalanced force $F_u$ acting on the object for each case

| | | |
|---|---|---|
| $F_u = F - f$<br>$= 100\ N - 20.0\ N$<br>$= 80.0\ N$ | $F_u = (F \cos 60°) - f$<br>$= 50.0\ N - 37.2\ N$<br>$= 12.8\ N$ | $F_u = (F \cos 30°) - (W \sin\theta) - F$<br>$= 86.6 - 50.0\ N - 27.2\ N$<br>$= 9.40\ N$ |

9. The net work $W_{Net}$ done by the unbalanced force $F_u$ for each case

| | | |
|---|---|---|
| $W_{Net} = F_{u\parallel}\Delta s$<br>$= (80.0\ N)(10.0\ j)$<br>$= 800\ J$ | $W_{Net} = F_{u\parallel}\Delta s$<br>$= (12.8\ N)(10.0\ m)$<br>$= 128\ J$ | $W_{Net} = F_{u\parallel}\Delta s$<br>$= (9.40\ N)(10.0\ m)$<br>$= 94.0\ J$ |

Example 7.1. A block of mass 20.0 kg is pulled up a rough incline at a constant speed by a rope that makes an angle of 30° with the incline, as shown in Fig. 7.5. If the tension in the rope is 200 N and the coefficient of friction 0.200, determine the net work done on the block in the process of pulling it 10.0 m up the incline.

Given and Diagram:

Figure 7.5

$M = 20.0$ kg
$\mu = 0.200$
$\theta = 30°$
$\Delta s = 10.0$ m
$T = 250$ N

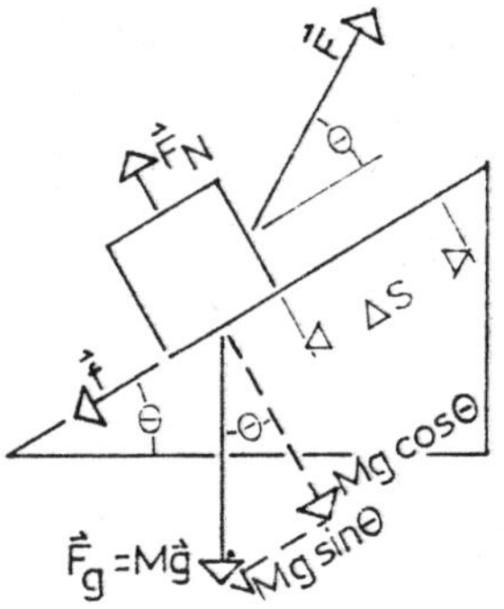


Determine: $W_{Net}$, the net work done on the block in the process of pulling it 10.0 m up the incline.

Strategy: From the given information, we can determine the magnitude of all the forces (T, $F_g$, $F_N$, and f) and the unbalanced force $F_u$ acting on the object. Once the magnitude of each force is known, we can determine its component parallel to $\vec{\Delta s}$. Once the component of each force parallel to $\vec{\Delta s}$ is known, we can proceed by either of the following methods.

Method C (from page 99): Determine $W_{Net}$ from $F_{u\parallel}$ ($W_{Net} = F_{u\parallel}\Delta s$).

Method B (from page 98): Determine the work done by each force and then add these to obtain the net work $W_{Net} = W_g + W_T + W_f + W_{F_N}$.

Solution: First, obtain the magnitude of each force

$T = 250$ N
$F_g = Mg = (20.0\ kg)(9.80\ m/s^2) = 196$ N
$F_N = (Mg\ \cos30°) - (T\ \sin\theta) = 44.7$ N
$f = \mu F_N = (0.200)(44.7\ N) = 8.94$ N

Second, obtain the component of each force parallel to $\vec{\Delta s}$. Recall that the direction of $\vec{\Delta s}$ is regarded as positive.

$T_\parallel = T\ \cos30° = (250\ N)(0.866) = 216.5$ N
$F_{g\parallel} = -Mg\ \sin\theta = -(196\ N)(\sin30°) = -98.0$ N
$F_{N\parallel} = 0$
$f_\parallel = -f = -8.94$ N

From these components, we can obtain the unbalanced force $F_u$ acting on the object.

$$F_{u\parallel} = T_\parallel + f_\parallel + F_{g\parallel} = +110\ N$$

Determine $W_{Net}$ by method C:

$$W_{Net} = F_{u\parallel}\Delta s = (110\ N)(10.0\ m) = 1.10 \times 10^3\ J$$

Determine $W_{Net}$ by method B:

$$W_T = T_{\parallel}\Delta s = 2165\ J \qquad W_{F_N} = F_{N\parallel}\Delta s = 0$$

$$W_{F_g} = F_{g\parallel}\Delta s = -980\ J \qquad W_f = f_{\parallel}\Delta s = -89.4\ J$$

$$W_{Net} = W_T + W_{F_g} + W_{F_N} + W_f = 1.10 \times 10^3\ J$$

Related Text Problems: 7-1 through 7-7, 7-10a, 7-17, 7-19, 7-20 a, b, 7-21a.

## 2. Power (Section 7.4)

Review: Power can be defined as follows:

$P = W/\Delta t$ - Power can be defined as the amount of work done per unit time.

$P = F_{\parallel} v$ - If a force acting on an object has a constant component parallel to the motion ($F_{\parallel}$) and if the velocity is constant, power can be defined as the product of $F_{\parallel}$ and the magnitude of the velocity vector.

$P = \tau\omega$ - If a constant torque acts on an object that has a constant angular speed, power can be defined as the product of this torque and angular speed.

Practice: The crate shown in Fig. 7.6 is pulled 10.0 m up a ramp at a constant speed by means of a cord that is attached to the shaft of a motor.

$v = 0.100$ m/s = speed of the crate up the ramp
$\mu = 0.200$ = coefficient of kinetic friction
$W = 100$ N = weight of the crate
$\theta = 30°$ = angle of incline for the ramp
$\Delta s = 10.0$ m = distance the crate moves
$r = 0.100$ m = radius of the motor shaft

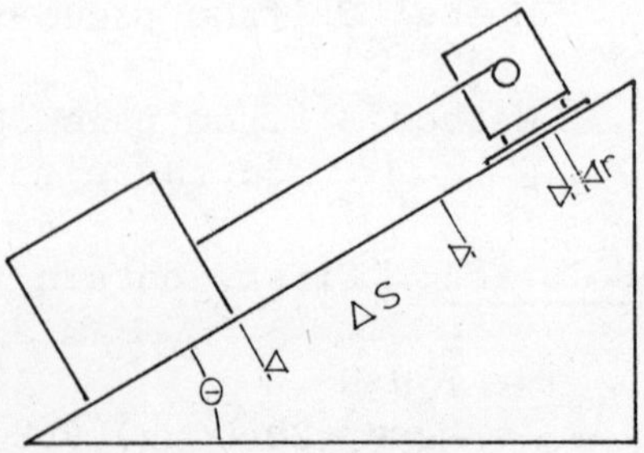


Figure 7.6

Determine the following:

| | |
|---|---|
| 1. Work done by gravity as the crate moves 10.0 m up the ramp | $W_g = F_g\Delta s \cos 120°$ <br> $= (100\ N)(10.0\ m)(-0.5) = -500\ J$ |
| 2. Force of friction as the crate moves up the ramp | $F_N = W \cos 30° = (100\ N)(0.866) = 86.6\ N$ <br> $f = \mu F_N = (0.200)(86.6\ N) = 17.3\ N$ |

| | |
|---|---|
| 3. Work done by friction as the crate moves up the ramp | $W_f = f\Delta s \cos 180°$ <br> $= (17.3\ N)(10.0\ m)(-1) = -173\ J$ |
| 4. Unbalanced force acting on the crate as it moves up the ramp | $F_u = Ma$ <br> since v = constant, a = 0 and $F_u = 0$ |
| 5. Net work done on the object as it moves up the ramp | $W_{Net} = F_u\Delta s = 0$ |
| 6. Work done on the crate by the motor | $W_{Net} = W_{motor} + W_g + W_f$ <br> $0 = W_{motor} - 500\ J - 173\ J$ <br> $W_{motor} = 673\ J$ |
| 7. Time it takes the motor to pull the crate 10 m up the ramp | $\Delta t = \frac{\Delta s}{v} = \frac{10.0\ m}{0.100\ m/s} = 100\ s$ |
| 8. Power supplied by the motor as it pulls the crate up the ramp | $P = \frac{W}{\Delta t} = \frac{673\ J}{100\ s} = 6.73\ W$ |
| 9. The force that the motor must supply in order to pull the crate up the ramp | $F_u = F_{motor} + F_{g\parallel} + f = 0$ <br> $F_{motor} = -F_{g\parallel} - f = -(-50.0\ N) - (-17.3\ N)$ <br> $= 67.3\ N$ |
| 10. Power supplied by the motor as it pulls the crate up the ramp (using a method other than that used in step 8) | $P = F_{\parallel} v = F_{motor}v$ <br> $= (67.3\ N)(0.100\ m) = 6.73\ W$ |
| 11. Torque the motor must supply in order to pull the crate up the ramp | $\tau_{motor} = rF_{motor}$ <br> $= (0.100\ m)(67.3\ N) = 6.73\ N\cdot m$ |
| 12. Angular speed of the rotating motor shaft | $\omega = \frac{v}{r} = \frac{0.100\ m/s}{0.100\ m} = 1.00\ rad/s$ |

| | |
|---|---|
| 13. Power supplied by the motor as it pulls the crate up the ramp (using a method other than that used in steps 8 and 10) | $P = \tau_{motor}\,\omega$<br>$= (6.73\ \text{Nm})(1.00\ \text{rad/sec}) = 6.73\ \text{W}$ |

Example 7.2. A crewmember of a racing shell can exert an average force of 20.0 N on his oar, working at the rate of 20.0 strokes/min. If his hands move back a distance of 0.700 m with each stroke, at what rate (in hp) is he working?

Given: F = 200 N, n = 20.0 strokes/min, $\Delta s$ = 0.700 m.

Determine: The rate (in hp) at which the crewmember is working.

Strategy: Knowing the force and distance of each stroke, we can determine the work done by the crewmember per stroke. We can multiply the work done per stroke by the number of strokes per minute to obtain the rate at which he is working. This rate can be converted from joules per minute to hp.

Solution:

$$W/\text{stroke} = F(\Delta s/\text{stroke}) = (200\ \text{N})(0.700\ \text{m/stroke}) = 140\ \text{J/stroke}$$

$$W/\text{min} = (140\ \text{J/stroke})(20.0\ \text{strokes/min}) = 2.80 \times 10^3\ \text{J/min}$$

$$P = (2.80 \times 10^3\ \frac{\text{J}}{\text{min}})(\frac{\text{min}}{60.0\ \text{s}})(\frac{\text{W}}{\text{J/s}})(\frac{\text{hp}}{746\ \text{w}}) = 6.26 \times 10^{-2}\ \text{hp}$$

Related Text Problems: 7-9 through 7-15, 7-24 through 7-29.

## 3. Work-Energy Theorem (Section 7.5)

Review: The work-energy theorem states that the work done by the resultant (i.e., the net, or unbalanced) force acting on a body is equal to the change in the kinetic energy of the body. Stated in equation form,

$$W_{Net} = \Delta KE \quad \text{where} \quad KE = mv^2/2$$

Practice:

Figure 7.7

$v_o = 0$
$F = 100$ N
$W = 100$ N
$\Delta s = 10.0$ m
$\theta = 30°$
$u = 0.100$

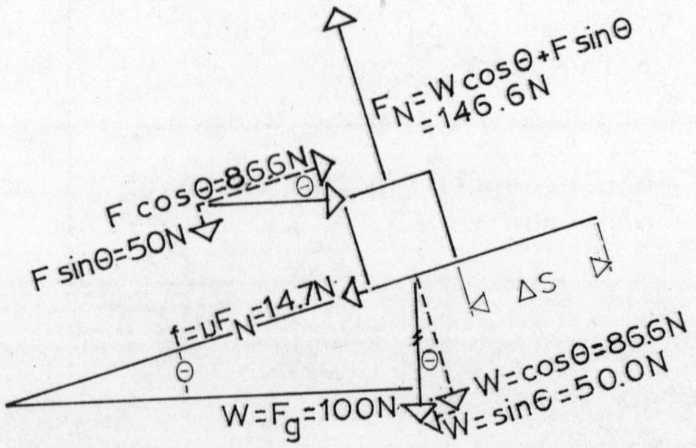

Determine the following:

| | |
|---|---|
| 1. The net, or unbalanced, force parallel to the incline that moves the object along the incline | $F_{u\parallel} = F_{\parallel} + f_{\parallel} + W_{\parallel}$<br>$= (F\cos\theta) - f - (W\sin\theta)$<br>$= (100\ N)\cos 30° - 14.7\ N - (100\ N)\sin 30°$<br>$= 86.6\ N - 14.7\ N - 50.0\ N = +21.9\ N$ |
| 2. The acceleration of the object up the incline | $a = F_u/M = 21.9\ N/10.2\ kg = +2.15\ m/s^2$ |
| 3. The square of the speed of the object after it has been pushed 10.0 m up the ramp | $v^2 - v_o^2 = 2a\Delta s$<br>$v^2 = 2(2.15\ m/s^2)(10.0\ m)$<br>$v^2 = 43.0\ m^2/s^2$ |
| 4. The change in the kinetic energy of the object as it is pushed up the incline | $\Delta KE = K_f - K_o = Mv^2/2$<br>$= (10.2\ kg)(43.0\ m^2/s^2)/2 = 219\ J$ |
| 5. The net work done by the unbalanced force in the process of pushing the object up the incline | $W_{Net} = F_{u\parallel}\Delta s$<br>$= (21.9\ N)(10.0\ m)$<br>$= 219\ J$ |

Note: We have determined (a) the change in the kinetic energy of the object as it is pushed 10.0 m up the incline (step 4), and (b) the net work done by the unbalanced force in the process of pushing the oject up the incline (step 5), and they are equal. Consequently, now that we are confident that $W_{Net} = \Delta KE$, in the future we will calculate whichever of these two quantities is the easier and know that they are equal.

Example 7.3. Figure 7.8 shows a hill that children use for sledding. One child pushes another for 5.00 m (A→B). The child and sled then travel down the 100-m incline (B→C) and onto a 600-m level stretch that terminates with a brick wall. If the strongest child can exert a 400-N force, is it safe for this child to push another child and sled that have a combined weight of 450 N (i.e., will the child on the sled hit the wall)? The coefficient of friction between the sled and snow is 0.100.

Given:

$u = 0.100$

$W_{c+s} = 450\ N$

$F_{max} = 400\ N$

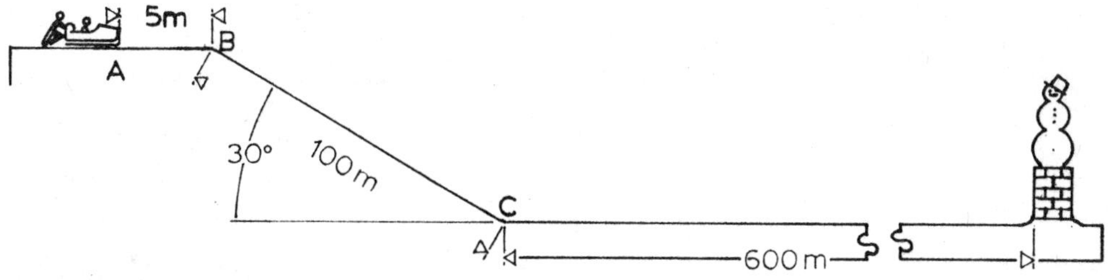


Figure 7.8

Determine: The distance the child and sled travel along the 600-m level stretch when pushed off by the strongest child ($F_{max}$ = 400 N).

Strategy: Determine the net work done on the child and sled while they are being pushed from A to B by $F_{max}$. Determine the net work done on the child and sled as they travel down the incline from B to C. Using the net work done on the sled from A to B and B to C, we can determine the kinetic energy of the child and sled at C. The child and sled will travel along the level stretch until the kinetic energy is zero, i.e., until the kinetic energy is changed by an amount equal to its value at C. This change in kinetic energy will equal the net work done on the child and sled by friction as the sled slides across the final horizontal section. Knowing the net work and that the unbalanced force is friction, we can determine the distance friction must act to do an amount of work equal to the change in kinetic energy. If this distance is less than 600 m, it is safe for the strongest child to do the pushing.

Solution: The net work done on the child and sled as they are pushed from A to B by the strongest child is:

$$W_{Net\ A\to B} = F_{u\parallel}\Delta s = (F_{max} - f)\Delta s = (F_{max} - \mu W_{c+s})\Delta s = 1.78 \times 10^3\ J$$

The net work done on the child and sled as they travel down the incline from B to C is:

$$W_{Net\ B\to C} = F_{u\parallel}\Delta s = [(W_{c+s}\ \sin 30°) - (\mu W_{c+s}\ \cos 30°)]\Delta s = 1.86 \times 10^4\ J$$

The kinetic energy of the child and sled at C are:

$$KE_C = \Delta KE_{A\to B} + \Delta KE_{B\to C} = W_{Net\ A\to B} + W_{Net\ B\to C} = 2.04 \times 10^4\ J$$

As the child and sled glide across the level stretch, we want to know how far they must go until the kinetic energy is decreased by 20.4 x $10^3$ J. Stated algebraically, we want

$$\Delta KE_{level} = W_{Net\ level} = -2.04 \times 10^4\ J$$

$$W_{Net\ level} = F_{u\parallel}\Delta s = -f\Delta s = -\mu W_{c+s}\Delta s \quad \text{or} \quad \Delta s = -\frac{W_{Net\ level}}{W_{c+s}} = 450\ m$$

Since the level stretch has 600 m before the brick wall and since the sled travels only 450 m along the level stretch when pushed by the strongest child, this hill is safe for the children to play on.

Related Text Problems: 7-19, 7-20, 7-21b.

## 4. Gravitational Potential Energy (Section 7.6)

Review: The change in gravitational potential energy is the negative of the work done by gravity. This may be stated in equation form as

$$\Delta PE = -W_g \quad \text{or} \quad \Delta PE = Mg\Delta h$$

where $\Delta h$ is the change in vertical position. When doing gravitational potential energy problems, we can choose zero gravitational potential energy at any convenient place. Most students choose the lowest vertical position for this zero point. If you locate the level of zero gravitational potential energy in this manner, then when an object moves downward, the value for h decreases, $\Delta h$ is negative, and $\Delta PE$ is negative, which means that the gravitational potential energy has decreased. In like manner, when an object moves upward, the value for h increases, $\Delta h$ is positive, and $\Delta PE$ is positive, which means that the gravitational potential energy has increased.

Practice: Consider the following two situations.

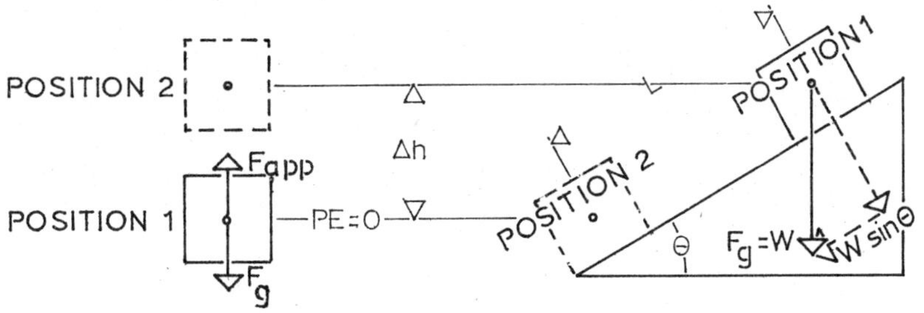


M = mass of the object
$F_{app}$ = the applied force required to move the object from position 1 to position 2 at a constant speed

M = mass of the object
L = length of the incline
θ = angle of the incline
μ = 0
$F_{app}$ = the applied force required to move the object from position 1 to position 2 at a constant speed

Determine the following for each case:

1. The work done by the force of gravity $F_g$ as the object moves from position 1 to position 2

| | |
|---|---|
| $W_g = F_{g\parallel}\Delta s = (-W)\Delta h = -Mg\Delta h$ | $W_g = F_{g\parallel}\Delta s = (W \sin\theta)L = MgL \sin\theta$ |

2. The change in the gravitational potential energy of the object as it moves from position 1 to position 2

| | |
|---|---|
| $\Delta PE = -W_g = -(-Mg\Delta h) = Mg\Delta h$ | $\Delta PE = -W_g = -MgL \sin\theta$ |

3. The gravitational potential energy of the object at the upper level, assuming that zero gravitational potential energy is at the level shown.

| | |
|---|---|
| $\Delta PE = PE_f - PE_i$ , $PE_i = 0$<br>$PE_f = \Delta PE = Mg\Delta h$ | $\Delta PE = PE_f - PE_i$ , $PE_f = 0$<br>$PE_i = -\Delta PE = -(-MgL \sin\theta) = MgL \sin\theta$ |

Example 7.4. A 400-kg crate slides down a 4.00-m frictionless ramp inclined at an angle of 20° as shown in Fig. 7.8. Determine the change in gravitational potential energy of the crate as it slides down the ramp.

Given and Diagram:

Figure 7.8

$M = 400$ kg
$L = 4.00$ m
$\theta = 20°$
$\theta = 20°$

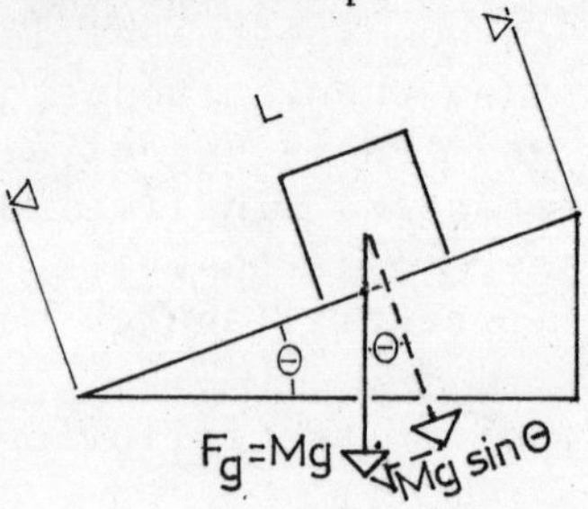


Determine: The change in the gravitational potential energy of the crate as it slides down the ramp.

Strategy: We can easily obtain the force of gravity on the crate and hence the work that gravity does on a crate as it slides down the ramp. The change in the gravitational potential energy is just the negative of the work done by gravity.

Solution:

$$W_g = F_{g\parallel}\Delta s = (Mg\ \sin\theta)(L) = (400\ \text{kg})(9.80\ \text{m/s}^2)(\sin 20°)(4.00\ \text{m}) = 5.36 \times 10^3\ \text{J}$$

$$\Delta PE = -W_g = -5.36 \times 10^3\ \text{J}$$

The gravitational potential energy of the crate is decreased by $5.36 \times 10^3$ J as it slides down the ramp.

Related Text Problems: 7-22, 7-23, 7-26 through 7-28, 7-30, 7-36, 7-38.

## 5. Work Done on an Applied Force (Section 7.7)

Review: The work done on an object by an applied force may accomplish any one or a combination of the following:

1. Change the object's kinetic energy $\Delta KE$.
2. Change the object's potential energy $\Delta PE$.
3. Do work against friction.

In Fig. 7.9, the applied force moves the object up an incline of length L.

Figure 7.9

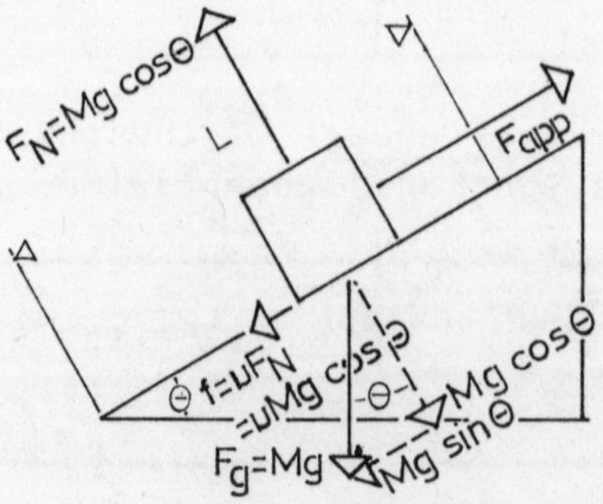

The forces doing work on the object are:

1. the applied force $F_{app}$ (supplied by you or some other object)
2. the force of gravity $F_g = Mg$
3. the force of friction $f = \mu Mg \cos\theta$

The net work done on the object by these forces is

$$W_{Net} = W_{by\ F_{app}} + W_{by\ F_g} + W_{by\ f}$$

Now recall the following:

1. The net work done on an object is equal to the change in kinetic energy of the object, hence $W_{Net} = \Delta KE$
2. The work done on an object by gravity is equal to the negative of the change in potential energy of the object, hence $W_{by\ F_g} = -\Delta PE$
3. The work done against friction is the negative of the work done by friction, hence $W_{against\ f} = -W_{by\ f}$

When the above are put into our equation for net work, we obtain:

$$\Delta KE = W_{by\ F_{app}} - \Delta PE - W_{against\ f} \quad \text{or} \quad W_{by\ F_{app}} = \Delta KE + \Delta PE + W_{against\ f}$$

This means that the work done on the object by the applied force may accomplish any one or a combination of the following:

1. Change the object's kinetic energy $\Delta KE$.
2. Change the object's potential energy $\Delta PE$.
3. Do work against friction $W_{against\ f}$.

<u>Practice:</u> Consider the following situations.

**Situation**

The object is pulled across a horizontal surface by $F_{app}$

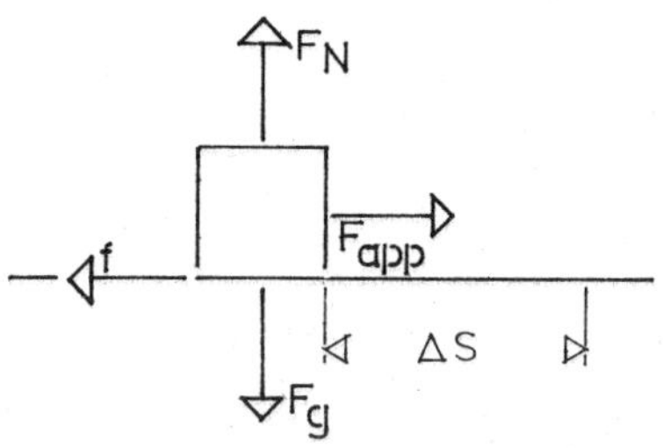


The object is pulled up an incline a distance by $F_{app}$

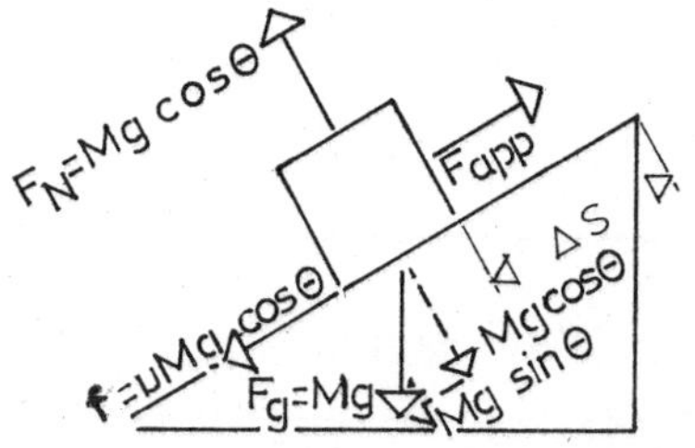


---

**Data**

$M$ = 10.2 kg, mass of object
$F_g$ = $Mg$ = 100 N, force of gravity
$\mu$ = 0.100, coefficient of kinetic friction
$F_{app}$ = 100 N, applied force

Determine the following for each situation:

1. Normal force on the object

| | |
|---|---|
| $F_N = Mg = 100$ N | $F_N = Mg \cos 30° = 86.6$ N |

2. Force of friction on the object

| | |
|---|---|
| $f = \mu F_N = 10.0$ N | $f = \mu F_N = 86.6$ N |

3. The net or unbalanced force acting on the object

| | |
|---|---|
| $F_u = F_{app} - f = 90.0$ N | $F_u = F_{app} - (Mg \sin\theta) - f = 41.3$ N |

4. The net work done on the object by $F_u$

| | |
|---|---|
| $W_{Net} = F_u \Delta s = 900$ J | $W_{Net} = F_u \Delta s = 413$ J |

5. Work done on the object by the force of gravity $F_g$

| | |
|---|---|
| $W_{by\ F_g} = F_{g\parallel} \Delta s = 0$ | $W_{by\ F_g} = F_{g\parallel} \Delta s$ <br> $= -Mg \sin\theta\ \Delta s = -500$ J |

6. Work done on the object by the force of friction

| | |
|---|---|
| $W_{by\ f} = -f \Delta s = -100$ J | $W_{by\ f} = -f \Delta s = -86.6$ J |

7. Work done against friction

| | |
|---|---|
| $W_{against\ f} = -W_{by\ f} = 100$ J | $W_{against\ f} = -W_{by\ f} = 86.6$ J |

8. Change in kinetic energy of the object as it moves the distance $\Delta s$

| | |
|---|---|
| $\Delta KE = W_{Net} = 900$ J | $\Delta KE = W_{Net} = 413$ J |

9. Change in gravitational potential energy of the object as it moves the distance $\Delta s$

| | |
|---|---|
| $\Delta PE = -W_{by\ F_g} = 0$ | $\Delta PE = -W_{by\ F_g} = 500$ J |

10. Work done by the applied force

| | |
|---|---|
| $W_{app} = F_{app}\,\Delta s = 1000$ J | $W_{app} = F_{app}\,\Delta s = 1000$ J |
| Note:<br><br>$W_{app} = \Delta KE + W_{against\ f}$<br><br>The applied force does 1000 J of work: 900 J goes into changing the kinetic energy and 100 J is used in doing work against friction. | $W_{app} = \Delta KE + \Delta PE + W_{against\ f}$<br><br>The applied force does 1000 J of work: 413.4 J goes into changing the kinetic energy of the object, 500 J goes into changing the potential energy of the object and 86.6 J is used in doing work against friction. |

Example 7.5. A dump truck hauling a heavy crate loses its brakes just before a long grade that has a dangerous curve at the bottom. The truck can just barely negotiate this curve at 20.0 m/s. The codriver climbs out of the cab, into the bed of the truck, attaches a strong chain to the crate, climbs back into the cab, and tells the driver to dump the load. The driver dumps the crate just as they start down the 1500-m-long, 10° grade and notes that they are traveling at 15.0 m/s. The coefficient of kinetic friction between the crate and the road is 0.263. Are they able to negotiate the curve? Ignore rolling friction of the truck.

Given: $M_T = 5.00 \times 10^3$ kg $\quad \theta = 10°$ $\quad v_{max} = 20.0$ m/s $\quad L = 1500$ m
$M_C = 1.00 \times 10^4$ kg $\quad \mu = 0.263$ $\quad v_{top} = 15.0$ m/s

Determine: Whether or not the truck can negotiate the curve at the bottom of the grade. That is, determine the speed of the truck and crate at the bottom of the grade and compare it to 20.0 m/s.

Strategy: As the truck and crate travel down the grade, gravity is speeding them up and friction is slowing them down. We can determine the force of friction and the work done by friction. We can determine the force of gravity and the work done by gravity and, hence, the change in gravitational potential energy. We can then use the work-energy principle to determine the $\Delta KE$ of the crate and truck as they travel down the grade. Knowing $\Delta KE$, we can determine $\Delta v$, which may be added to $v_{top}$ to obtain $v_{bottom}$.

Solution:

$$W_{by\ f} = f_{\parallel}\Delta s = -\mu(M_C g \cos\theta)\Delta s = -38.85 \times 10^6 \text{ J}$$

$$W_{by\ g} = W_g = F_{g\parallel}\Delta s = (M_T + M_C)g \sin\theta\ \Delta s = 39.07 \times 10^6 \text{ J}$$

$$W_{app} = \Delta PE + \Delta KE + W_{against\ f}$$

Recall that $W_{app} = 0$, since the brakes are not functioning

$$W_{against\ f} = -W_{by\ f} = +38.85 \times 10^6 \text{ J}$$

$$\Delta PE = -W_{by\ g} = -39.07 \times 10^6 \text{ J}$$

Hence the change in kinetic energy of the crate and truck as they travel down the 10° grade is

$$\Delta KE = W_{app} - \Delta PE - W_{against\ f}$$

$$\Delta KE = 0 - (-39.07 \times 10^6 \text{ J}) - (38.85 \times 10^6 \text{ J}) = 2.20 \times 10^5 \text{ J}$$

The kinetic speed of the crate and truck at the bottom of the grade is

$$K_f = K_i + \Delta KE = \frac{1}{2}(M_C + M_T)v_{top}^2 + \Delta KE = 2.22 \times 10^5 \text{ J}$$

Finally, the speed of the crate and truck at the bottom of the grade may be calculated.

$$K_f = \frac{1}{2}(M_C + M_T)v_f^2 \quad \text{or} \quad v_f = [2K_f/(M_C + M_T)]^{1/2} = 17.3 \text{ m/s}$$

The truck can safely negotiate the curve at the bottom of the grade.

Related Text Problems: 7-30, 7-32 through 7-36.

## 6. Conservation of Mechanical Energy (Section 7.8)

Review: From the previous principal concept, we know that the work done on an object by an applied force can change the kinetic energy of the object, change the potential energy of the object and do work against kinetic friction. We can state this algebraically as

$$W_{by\ F_{app}} = \Delta KE + \Delta PE + W_{against\ f}$$

If neither you nor anyone else exerts a force on the object, $F_{app}$ is zero and

$$W_{by\ F_{app}} = 0$$

If no friction acts on the object, the work done against friction is zero.

$$W_{against\ f} = 0$$

When the preceding algebraic statements are combined we have

$$\Delta KE + \Delta PE = 0 \quad \text{or} \quad \Delta KE = -\Delta PE$$

This last algebraic statement tells us that mechanical energy is conserved. That is, the total mechanical energy is constant. If the energy is constant, a decrease in potential energy must be accompanied by an equivalent increase in kinetic energy; and an increase in potential energy must be accompanied by an equivalent decrease in kinetic energy.

Practice: A box rests on a shelf 2.00 m off the floor. The shelf is bumped and the box falls to the floor.

Fig. 7.10 $m = 1.00$ kg
$h_A = 2.00$ m
$h_B = 1.50$ m

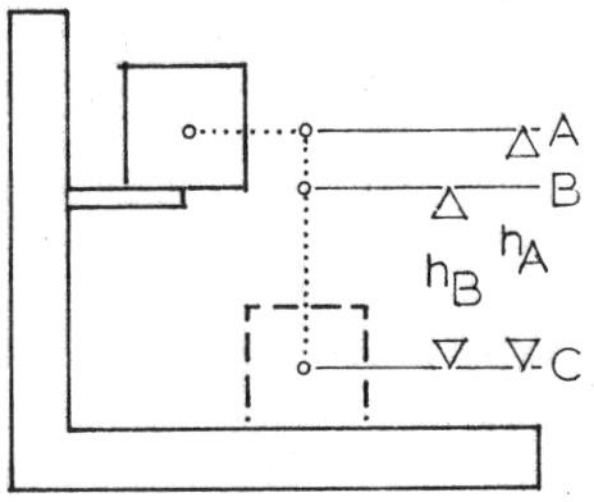


Determine the following:

| | |
|---|---|
| 1. Kinetic, potential and total energy of the box at A | $KE_A = 0$ (box is at rest)<br>$PE_A = mgh_A = 19.6$ J<br>$E_A = KE_A + PE_A = 19.6$ J |
| 2. Applied force and force of friction acting on the box as it falls | $F_{app} = 0$ and $f = 0$ |
| 3. Work done by $F_{app}$ and f as the box falls from A to C | $W_{F_{applied}} = 0$ and $W_f = 0$ |
| 4. Total, potential and kinetic energy of the box at B | $E_B = E_A = 19.6$ J<br>$PE_B = mgh_B = 14.7$ J<br>$KE_B = E_B - PE_B = 4.90$ J |
| 5. The kinetic energy of the box at B using your knowledge of kinematics | $v^2 - v_o^2 = 2a\Delta s$<br>$v_o = v_A = 0,\ v = v_B$<br>$a = -g = -9.80\ m/s^2,\ \Delta s = -0.500$ m<br>$v_B^2 = 2(-9.80\ m/s^2)(-0.500\ m) = 9.80\ m^2/s^2$<br>$K_B = mv_B^2/2 = 4.90$ J |

| | |
|---|---|
| 6. The change in potential, kinetic and total energy of the box as it falls from A to B | $\Delta KE = KE_B - KE_A = 4.90$ J<br>$\Delta PE = PE_B - PE_A = -4.90$ J<br>$\Delta E = \Delta KE + \Delta PE = 0$ J |
| 7. Total, potential and kinetic energy of the box at C | $E_C = E_B = E_A = 19.6$ J<br>$PE_C = 0$ J<br>$KE_C = E_C - PE_C = 19.6$ J |
| 8. The change in potential, kinetic and total energy of the box as it falls from B to C | $\Delta KE = KE_C - KE_B = 14.7$ J<br>$\Delta PE = PE_C - PE_B = -14.7$ J<br>$\Delta E = \Delta KE = \Delta PE = 0$ J |
| 9. Speed of the box at C | $KE_C = mv_C^2/2$<br>$v_C = (2KE_C/m)^{1/2} = 6.26$ m/s |

Example 7.6.

Figure 7.11

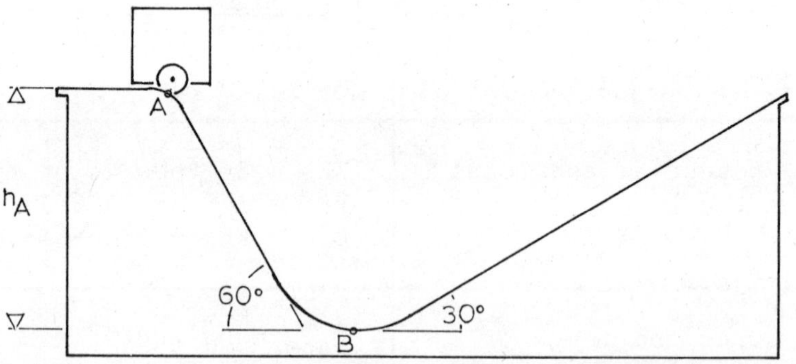


A 10.0-kg box initially at rest at position A on the track shown in fig. 7.11 is given a nudge to start it sliding along the track. The track is frictionless and position A is 3.00 m higher than position B. (a) What is the speed of the box at B? (b) What distance will the box travel up the 30° incline?

Given: $m = 10.0$ kg, $h_A = 3.00$ m, $\mu = 0$
the incline angles are 60° and 30°

Determine: Speed of the box at B and the distance the box travels up the 30° incline.

Strategy: First, we need to establish zero gravitational potential energy. Since we can choose zero gravitational potential energy any place we wish, we might as well locate it where it is the most convenient. It is convenient to choose zero gravitational potential energy at the lowest position of the box. The convenience of this choice is that all potential energies are either zero or positive. Most students find it cumbersome to work with negative energies. Since no applied force is acting and the track is frictionless, mechanical energy is conserved. Using the given information, we can determine the potential, kinetic and total energy of the box at A. Since energy is conserved, the total energy at B is the same as the total energy at A. Knowing the total energy at B, we can determine the kinetic energy and the speed of the box at B. Since energy is conserved, the box will slide up the 30° incline until all its kinetic energy is converted into potential energy. Knowing this, we can determine the distance the box travels up the 30° ramp.

<u>Solution</u>: Choose zero potential energy to be the level of position B. The potential, kinetic and total energy at position A are then as follows:

$$PE_A = mgh_A = 294 \text{ J}$$

$$KE_A = mv_A{}^2/2 = 0 \quad (v_A = 0)$$

$$E_A = PE_A + KE_A = 294 \text{ J}$$

$$E_B = PE_B + KE_B \quad \text{but} \quad PE_B = 0 \quad \text{(by choice)}$$
$$E_B = E_A = 294 \text{ J} \qquad \text{(conservation of energy)}$$

Combining these we have

$$KE_B = E_B = E_A = 294 \text{ J}$$

We also know that

$$KE_B = mv_B{}^2/2 \quad \text{or}$$

$$v_B = (2KE_B)/m)^{1/2} = 7.67 \text{ m/s}$$

The box will travel up the incline until ita final energy ($E_F$) is entirely potential. Knowing this, we can write

$$E_F = E_A, \quad E_F = PE_F + KE_F, \quad KE_F = 0, \quad PE_F = mgh_F = mg\ell_F\sin\theta$$

Combining these, we obtain

$$mg\ell_F\sin\theta = E_A \quad \text{or} \quad \ell_F = E_A/mg\ \sin\theta = 6.00 \text{ m}$$

<u>Related Text Problems</u>: 7-37, 7-39, 7-40.

---

PRACTICE TEST

Take and grade this practice test. Doing so will allow you to determine any weak spots in your understanding of the concepts taught in this chapter. The following section prescribes what you should study further to strengthen your understanding.

A $1.00 \times 10^3$ N crate is sitting at position A, as shown in the figure. You push the crate from A to B with a constant 400-N force. At B you release the crate, and it slides down the incline and across the horizontal section until it stops. Let's agree to choose zero gravitational potential energy at the level of the horizontal section.

$W = 1.00 \times 10^3$ N

$\mu_k = 0.100$

$g = 9.80$ m/s$^2$

$F_{push} = 400$ N

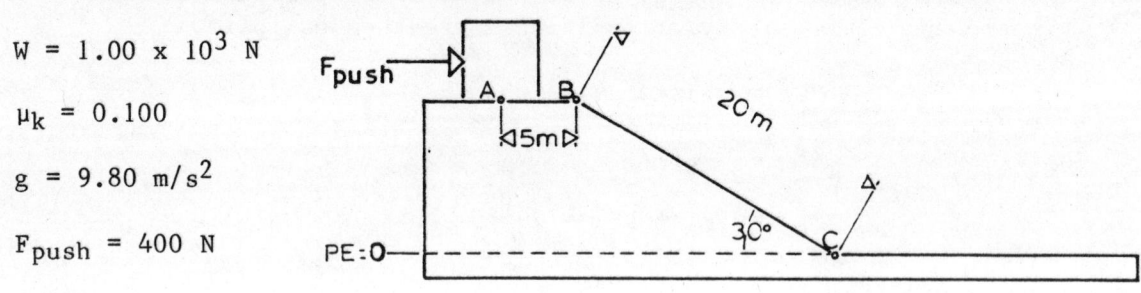

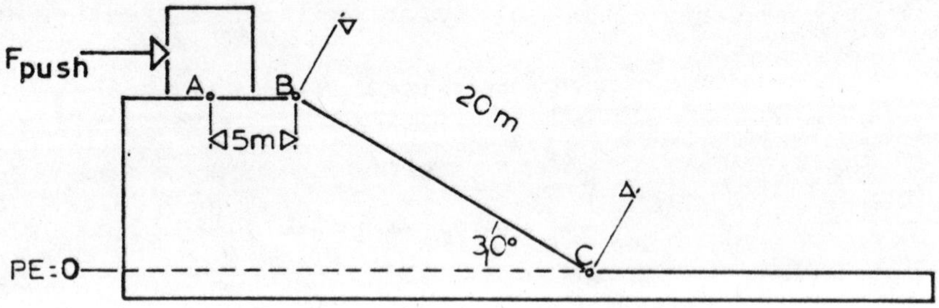


Determine the Following:

_________1. The gravitational potential energy at position A.
_________2. The work done on the crate by you as you push it from A to B.
_________3. The work done on the crate by gravity as you push it from A to B.
_________4. The change in gravitational potential energy as you push the crate from A to B.
_________5. The work done on the crate by friction as you push it from A to B.
_________6. The net work done on the crate as you push it from A to B.
_________7. The change in kinetic energy of the crate as you push it from A to B.
_________8. The speed of the crate as it passes B.
_________9. The total energy of the crate at B.
_________10. The work done on the crate by gravity as it slides down the incline.
_________11. The change in gravitational potential energy of the crate as it slides down the incline.
_________12. The work done on the crate by friction as it slides down the incline.
_________13. The net work done on the crate as it slides down the incline.
_________14. The change in kinetic energy of the crate as it slides down the incline.
_________15. The kinetic energy of the crate at c.
_________16. The decelerating force acting on the crate as it slides across the final horizontal section.
_________17. The deceleration of the crate as it slides across the final horizontal section.
_________18. The distance the crate slides along the final horizontal section before coming to rest.

(See Appendix I for answers.)

## PRINCIPAL CONCEPTS AND EQUATIONS PRESCRIPTION

Your score on the practice is an excellent measure of your understanding of this chapter. You should now use the following chart to write your own prescription for curing any of your physics ills. Look down the leftmost column to the number of the question(s) you answered incorrectly, read across that row to see which section(s) of the study guide you should return to for further study, and then do the suggested text problems to gain additional experience in working with the particular concept.

| Practice Test Question | Concepts and Equations | Prescription | |
|---|---|---|---|
| | | Principal Concept | Text Problems |
| 1 | Gravitational potential energy | 4 | 7-22,26 |
| 2 | Work | 1 | 7-1,3 |
| 3 | Work | 1 | 7-2,4 |
| 4 | Gravitational potential energy | 4 | 7-36,38 |
| 5 | Work | 1 | 7-2,7 |
| 6 | Work | 1 | 7-4,5 |
| 7 | Work-energy theorem | 3 | 7-19,20 |
| 8 | Kinetic energy | 3 | 7-18,19 |
| 9 | Work-energy theorem | 3 | 7-20,21 |
| 10 | Work | 1 | 7-1,21 |
| 11 | Gravitational potential energy | 4 | 7-22,27 |
| 12 | Work | 1 | 7-2,7 |
| 13 | Work | 1 | 7-4,5 |
| 14 | Work-energy theorem | 3 | 7-19,20 |
| 15 | Work-energy theorem | 3 | 7-20,21 |
| 16 | Friction | 7 of Ch. 4 | 4-37,39 |
| 17 | Newton's second law | 2 of Ch. 4 | 4-3,4 |
| 18 | Work-energy theorem | 3 | 7-19,21 |

# 8 Impulse and Linear Momentum

RECALL FROM PREVIOUS CHAPTERS

| Previously learned concepts and equations frequently used in this chapter | Text Section | Study Guide Page |
|---|---|---|
| Analyzing translational motion | 2.4 | 22 |
| Resolving vectors into components | 3.2 | 32 |
| Adding vectors | 3.2 | 32 |
| Treatment of two-dimensional motion | 3.5,6 | 37 |
| Newton's second law of motion: $\vec{F}_u = m\vec{a}$ | 4.3 | 47 |
| Kinetic energy: $KE = mv^2/2$ | 7.5 | 108 |
| Gravitational potential energy: $\Delta PE = -W_g = mg\Delta h$ | 7.6 | 110 |

NEW IDEAS IN THIS CHAPTER

| Concepts and equations introduced | Text Section | Study Guide Page |
|---|---|---|
| Linear momentum: $\vec{P} = m\vec{v}$ | 8.2 | 123 |
| Impulse: Impulse = $\vec{F}_{av}\Delta t$ | 8.2 | 123 |
| Impulse-momentum theorem: Impulse = $\vec{F}_{av}\Delta t = \Delta\vec{P}$ | 8.2 | 123 |
| Conservation of linear momentum: If $\vec{F}_{ext} = 0$, then $\Delta\vec{P} = 0$ and $\vec{P}_i = \vec{P}_f$ | 8.3 | 125 |
| Coefficient of restitution: $\varepsilon = \dfrac{\text{relative velocity of separation}}{\text{relative velocity of approach}}$ | 8.6 | 129 |
| In elastic collisions, momentum is conserved, kinetic energy is conserved, and $\varepsilon = 1$. | 8.4,5 | 129 |
| In perfectly inelastic collisions, momentum is conserved, kinetic energy is not conserved, and $\varepsilon = 0$. The objects stick together after collision. | 8.6 | 129 |
| In inelastic collisions (not perfectly inelastic), momentum is conserved, kinetic energy is not conserved, and $\varepsilon < 1$. | 8.6 | 129 |

## PRINCIPAL CONCEPTS AND EQUATIONS

### 1. The Impulse & Momentum Theorem (Section 8.2)

Review:

Impulse: $\vec{F}_{av}\Delta t = \vec{F}\Delta t$ (if the force is constant)

Linear momentum: $\vec{P} = m\vec{v}$

Change in linear momentum: $\Delta\vec{P} = m\vec{v}_f - m\vec{v}_i$

The impulse-momentum theorem states that impulse is equal to the change in linear momentum. In equation form, this may be written as:

$$\text{Impulse} = \vec{F}_{av}\Delta t = \Delta\vec{P} = m\vec{v}_f - m\vec{v}_i$$

Practice: A 50.0 g golf ball bounces off a cement floor as shown in Fig. 8.1. The ball hits the floor with a speed of 50.0 m/s and bounces off with no loss in speed. The ball and the floor are in contact for 0.010 s.

Figure 8.1

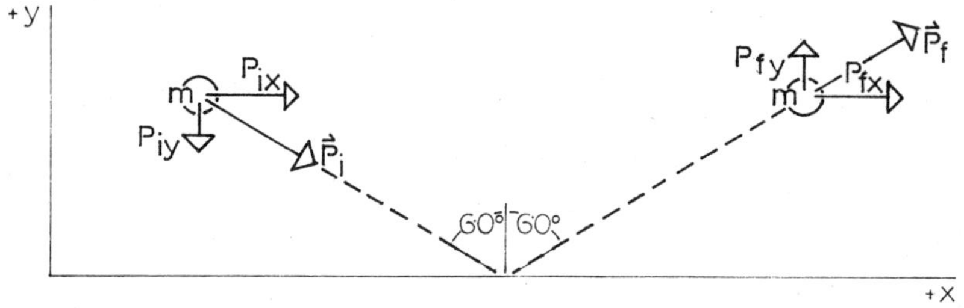


Determine the following:

| | |
|---|---|
| 1. Magnitude of the initial momentum of the ball | $P_i = mv_i = 2.50$ kg·m/s |
| 2. Magnitude of the final momentum of the ball | $P_f = mv_f = 2.50$ kg·m/s |
| 3. The x component of the initial momentum of the ball | $P_{ix} = P_i \sin 60° = 2.17$ kg·m/s |
| 4. The x component of the final momentum of the ball | $P_{fx} = P_f \sin 60° = 2.17$ kg·m/s |
| 5. Change in the x component of the momentum of the ball | $\Delta P_x = P_{fx} - P_{ix} = 0$ |
| 6. The y component of the initial momentum of the ball | $P_{iy} = -P_i \cos 60° = -1.25$ kg·m/s |

| | |
|---|---|
| 7. The y component of the final momentum of the ball | $P_{fy} = +P_f \cos 60° = +1.25\ kg \cdot m/s$ |
| 8. Change in the y component of the momentum of the ball | $\Delta P_y = P_{fy} - P_{iy}$<br>$= +1.25\ kg \cdot m/s - (-1.25\ kg \cdot m/s)$<br>$= +2.50\ kg \cdot m/s$<br>The plus sign tells us that the change in momentum is in the +y direction. |
| 9. Vector diagram showing $\vec{P}_i$, $\vec{P}_f$ and $\Delta\vec{P}$ | 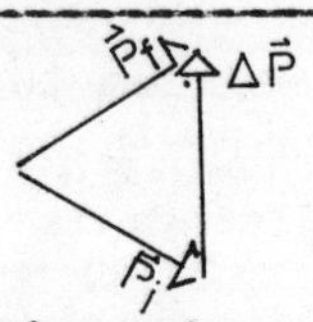<br>Notice that $\vec{P}_f = \vec{P}_i + \Delta\vec{P}$ |
| 10. Impulse imparted to the ball during the collision | $Impulse_b = \Delta\vec{P}_b = +2.50\ kg \cdot m/s$<br>The impulse imparted to the ball is in the +y direction. |
| 11. Impulse imparted to the floor during the collision | $Impulse_f = -Impulse_b = -2.50\ kg \cdot m/s$<br>The impulse imparted to the floor is in the −y direction. |
| 12. Average force exerted on the ball during impact | $\vec{F}_b = \Delta\vec{P}_b/\Delta t = +250\ N$<br>The plus sign tells us that $\vec{F}_b$ is in the positive y direction. |
| 13. Average force exerted on the floor during impact | $\vec{F}_f = -\vec{F}_b = -250\ N$<br>The minus sign tells us that $\vec{F}_f$ is in the negative y direction |

<u>Example 8.1.</u> Two students are engaged in a snowball fight. One student throws a 0.250-kg snowball with a velocity of 10.0 m/s at the other student. The snowball sticks to the student and the impact takes place over a 0.010 s time interval. Determine the collision impulse and the average impulse force experienced by the student.

<u>Given:</u> $m = 0.250\ kg$ , $v_i = 10.0\ m/s$ , $v_f = 0\ m/s$ , $\Delta t = 0.0100\ s$

<u>Determine:</u> The collision impulse and the average impulse force experienced by the student.

<u>Strategy:</u> Call the initial direction of the snowball positive. Knowing the mass and initial speed of the snowball, determine its initial momentum. Since the final momentum of the snowball is zero, its change in momentum and hence its impulse are equal in magnitude but opposite in direction to its initial

momentum. The impulse acting on the student is equal in magnitude to that acting on the snowball, but acts in the opposite direction. The average impulse force acting on the student can be determined from the impulse acting on the student and the time interval during which the impulse occurs. Since this problem occurs in one dimension, it will be convenient to drop the vector notation and use "+" and "-" to indicate the direction.

Solution:

| | |
|---|---|
| Initial momentum of the snowball | $P_i = mv_i = +2.50$ kg·m/s |
| Final momentum of the snowball | $P_f = mv_f = 0$ kg·m/s |
| Change in momentum of the snowball | $\Delta P_{sb} = P_f - P_i = -2.50$ kg·m/s |
| Impulse delivered to the snowball | $\text{Impulse}_{sb} = \Delta P_{sb} = -2.50$ kg·m/s |
| Impulse delivered to the student | $\text{Impulse}_{st} = -\text{Impulse}_{sb} = +2.50$ kg·m/s |
| Impulse force acting on the student | $F_{st} = \Delta P_{st}/\Delta t = +\ 250$ N |

The positive sign tells us that the force acting on the student is in the same direction as the initial momentum of the snowball.

Related Text Problems: 8-1 through 8-11, 8-13, 8-35, 8-41.

## 2. Conservation of Linear Momentum (Section 8.3)

Review: From the previous section, we have $\vec{F}\Delta t = \Delta\vec{P} = \vec{P}_f - \vec{P}_i$, where $\vec{F}$ represents an external force that acts for a time $\Delta t$ to change the momentum by an amount $\Delta\vec{P}$. Notice that when $\vec{F} = 0$, $\Delta\vec{P} = 0$ and $\vec{P}_i = \vec{P}_f$, that is momentum is conserved. The law of conservation of momentum states that the total linear momentum of a system remains constant if the resultant of all outside forces is negligible. That is, the resultant momentum before the collision equals the resultant momentum after the collision. Since most of our work will be in one dimension, it will be convenient to drop the vector notation and use "+" and "-" to indicate the direction.

Practice: A 0.020-kg bullet penetrates a 2.00-kg block of wood initially at rest on a horizontal frictionless surface. The bullet is traveling 300 m/s before it hits the block and 200 m/s after it emerges from the block. Let's agree to call the direction of the bullet positive. Since this problem occurs in one dimension, it will be convenient to drop the vector notation and use "+" and "-" to indicate direction.

Determine the following:

| | |
|---|---|
| 1. Initial momentum of the bullet | $P_{bi} = m_b v_{bi} = 6.00$ kg·m/s |
| 2. Final momentum of the bullet | $P_{bf} = m_b v_{bf} = 4.00$ kg·m/s |
| 3. Resultant external force acting on the bullet and block as the bullet travels through the block | $F_{ext} = 0$. As the bullet penetrates the block, the bullet and block exert an internal force on each other, but no external forces exist. |

| | | |
|---|---|---|
| 4. | Whether or not momentum is conserved | Yes, since $F_{ext} = 0$, then $\Delta P = 0$<br>$P_f = P_i$ |
| 5. | Momentum of the block after it is hit by the bullet<br><br>Use the notation $P_{bi}$ and $P_{bf}$ for the initial and final momentum of the bullet, and use $P_{wi}$ and $P_{wf}$ for the initial and final momentum of the wood block. | Given conservation of momentum, we write<br>$P_i = P_f$<br>$P_{bi} = P_{bf} + P_{wf}$ (recall that $P_{wi} = 0$)<br>$P_{wf} = P_{bi} - P_{bf} = 2.00$ kg·m/s |
| 6. | Speed of the block after it is hit by the bullet | $P_{wf} = m_w v_{wf}$<br>$v_{wf} = P_{wf}/m_w = 1.00$ m/s |

Example 8.2. A child runs and jumps onto a stationary grocery cart. If the cart has a mass 2/3 that of the child, find the ratio of the final speed of the cart and child to that of the initial speed of the child.

Given: $m_g = (2/3)m_c$, where $m_c$ and $m_g$ represent the mass of the child and grocery cart respectively.

Determine: The ratio of the final speed of the grocery cart and child ($v_{g+c}$) to that of the initial speed of the child ($v_c$). That is, determine $v_{g+c}/v_c$.

Strategy: The time it takes the child to land on the grocery cart is so short that we can ignore external forces and consequently say that momentum is conserved. A statement of conservation of momentum contains both the initial speed of the child $v_c$ and the final speed of the grocery cart and child $v_{g+c}$. Hence the ratio of $v_{g+c}/v_c$ can be determined from a statement of conservation of momentum.

Solution: The initial momentum is just that of the child, $P_i = m_c v_c$ .

The final momentum is that of the grocery cart and child:

$$P_f = (m_g + m_c)v_{g+c} = [(2/3)m_c + m_c]v_{g+c} = (5/3)m_c v_{g+c}$$

Equating the initial and final momentum and solving for the ratio of the speeds, we obtain

$$v_{g+c}/v_c = 3/5$$

Example 8.3. A 60.0-kg astronaut has been spacewalking outside her spacecraft and realizes that she is stranded (stationary) 10.0 m from the spacecraft, as shown in Fig. 8.2. In order to get back, she throws a 1.00-kg hammer directly away from the ship with a speed of 6.00 m/s. How long will it take her to reach the ship?

Given and Diagram:

Figure 8.2 $\quad d_a = 10.0$ m
$v_h = 6.00$ m/s
$m_a = 60.0$ kg
$m_h = 1.00$ kg

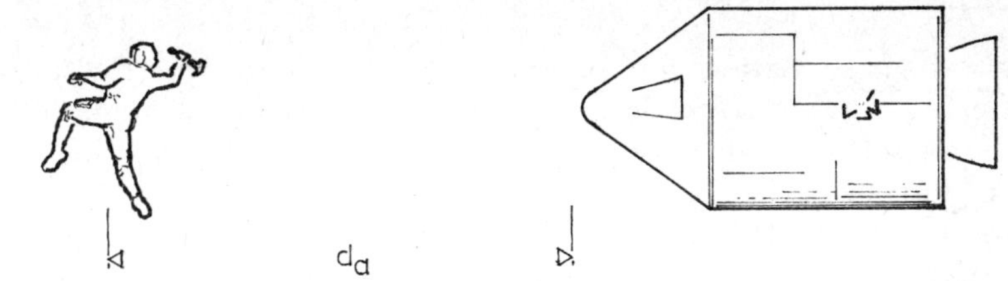


Determine: The time it takes the astronaut to return to the spacecraft.

Strategy: Take the positive direction to be radially outward from the ship and the negative direction to be radially inward. Since the initial radial momentum for the hammer and astronaut is zero and external forces are negligible, momentum in the radial direction is conserved. If the astronaut gives the hammer a radially outward momentum (positive), she will receive the same inward momentum (negative). From such a momentum statement, we can obtain the radially inward velocity (negative) of the astronaut. Knowing velocity of the astronaut and her displacement (negative), we can determine the time of travel.

Solution: $\vec{P}_a + \vec{P}_h = 0$ (conservative of momentum)

$$\vec{P}_a = -\vec{P}_h$$

$$m_a\vec{v}_a = -m_h\vec{v}_h$$

$$\vec{v}_a = -m_h\vec{v}_h/m_a = -0.100 \text{ m/s}$$

$$t_a = \vec{d}_a/\vec{v}_a = (-10.0 \text{ m})/(-0.100 \text{ m(s)}) = 100 \text{ s}$$

Example 8.4. A cannon with a certain elevation has a range R when fired on level ground (Fig. 8.3a). If a shell explodes at the top of the trajectory in such a manner that 1/3 of the mass falls straight down and the other 2/3 continues onward (Fig. 8.3b), how far (in terms of R) will the 2/3 fragment land from the cannon?

Given and Diagram:

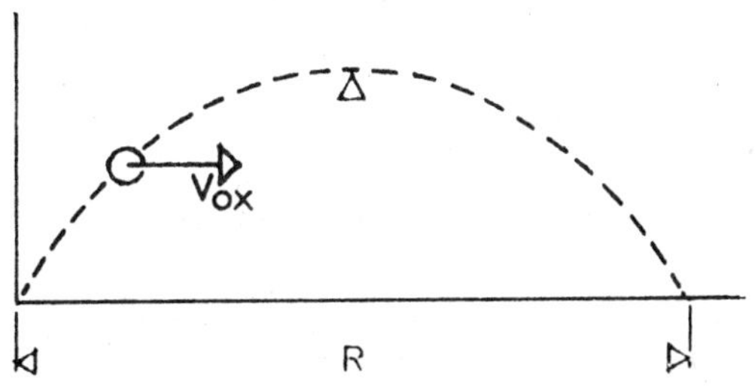


$m_s$ = mass of the shell
$m_1 = (1/3)\ m_s$
$m_2 = (2/3)\ m_s$

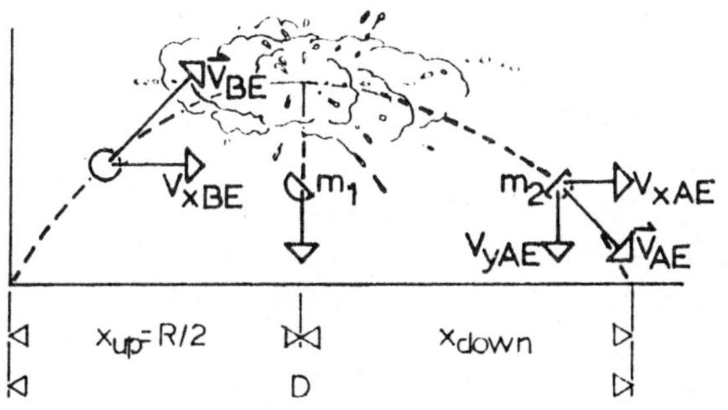


Before explosion (BE) $\qquad$ After Explosion (AE)

(a) $\qquad$ (b)

Figure 8.3

Determine: How far (in terms of R) the mass $m_2$ lands from the cannon.

Strategy: Since no external forces act in the x direction (even during the explosion), momentum in the x direction is conserved. From a statement of conservation of momentum in the x direction, we can determine the x component of the velocity of $m_2$ after the explosion in terms of the x component of the velocity of the entire shell before the explosion. Knowing the range of an unexploded shell, we can determine the time of flight of an unexploded shell in terms of the x component of its velocity. We can determine the time of flight for the exploded fragments from the time of flight of an unexploded shell. Knowing the x component of the fragment's velocity and the time of flight of the fragment, we can determine how far the fragment travels.

Solution: Since momentum is conserved in the x direction, we can equate the x component of the momentum before the explosion to the x component of the momentum after the explosion.

$$m_s v_{xBE} = m_2 v_{xAE_2} + m_1 v_{xAE_1}$$

Since $m_2 = (2/3)m_s$ and $v_{xAE_1} = 0$, this becomes

$$m_s v_{xBE} = (2/3)m_s v_{xAE_2} \quad \text{or} \quad v_{xAE_2} = (3/2)v_{xBE}$$

The x component of the velocity before the explosion ($v_{xBE}$) is equal to the x component of the velocity of an unexploding shell, $v_{ox}$. Hence the previous equation can be written as

$$v_{xAE_2} = (3/2)v_{ox}$$

The x component of the velocity, the range, and the time of flight of an unexploding shell are related by

$$R = v_{ox}T_F \quad \text{or} \quad T_F = R/v_{ox}$$

In this case, the time it takes the exploding shell to get to the top of its trajectory is one half the time of flight of the unexploded shell ($T_{up} = T_F/2$). The x component of the exploding shell's position when it explodes ($x_{up}$) is half the range of the unexploded shell ($x_{up} = R/2$).

As the shell fragments fall to the ground, they experience the same vertical acceleration and travel the same vertical distance as the entire shell on its way to the top. Consequently, we know that

$$T_{down} = T_{up} = T_F/2 = R/(2v_{ox})$$

Combining the above, we obtain

$$x_{down} = v_{xAE_2}T_{down} = (3v_{ox}/2)(R/2v_{ox}) = 3(R/4)$$

Finally

$$D = x_{up} + x_{down} = R/2 + 3(R/4) = 5(R/4)$$

Related Text Problems: 8-14, 8-15, 8-20, 8-21, 8-25, 8-31, 8-34, 8-36, 8-37.

## 3. Collisions In One Dimension (Sections. 8.4, 8.5)

Review: Collisions may be elastic or inelastic. In an elastic collision, both momentum and kinetic energy are conserved. Hence we may write

(a) $m_1v_{1i} + m_2v_{2i} = m_1v_{1f} + m_2v_{2f}$ and

(b) $$\frac{1}{2}m_1v_{1i}^2 + \frac{1}{2}m_2v_{2i}^2 = \frac{1}{2}m_1v_{1f}^2 + \frac{1}{2}m_2v_{2f}^2$$

When these two equations are solved simultaneously for the two unknowns ($v_{1f}$ and $v_{2f}$), we obtain

(c) $$v_{1f} = \frac{m_1 - m_2}{m_1 + m_2}v_{1i} + \frac{2m_2}{m_1 + m_2}v_{2i}$$ and

(d) $$v_{2f} = \frac{2m_1}{m_1 + m_2}v_{1i} + \frac{m_2 - m_1}{m_1 + m_2}v_{2i}$$

In the process of solving equations (a) and (b) for $v_{1f}$ and $v_{2f}$, we also discovered that, for an elastic collision, the relative velocity of approach for the two objects is equal to the relative velocity of separation. That is

$$v_{2i} - v_{1i} = v_{1f} - v_{2f}$$

This allows us to define a coefficient of restitution as:

$$\varepsilon = \frac{\text{relative velocity of separation}}{\text{relative velocity of approach}} = \frac{v_{1f} - v_{2f}}{v_{2i} - v_{1i}}$$

Note that $\varepsilon = 1$ for an elastic collision.

In an inelastic collision, momentum is conserved but kinetic energy is not. There is some loss of kinetic energy in an inelastic collision; hence the relative velocity of separation is less than the relative velocity of approach. That is, $\varepsilon < 1$ for inelastic collisions.

If the collision is perfectly inelastic, the objects stick together, the relative velocity of separation is zero, and $\varepsilon = 0$.

Practice: In Fig. 8.4, object 1 is fired from the gun with a speed v. The collision betwen objects 1 and 2 is perfectly inelastic, and object 1 embeds itself in object 2 in a very short interval. The surface is frictionless.

Figure 8.4

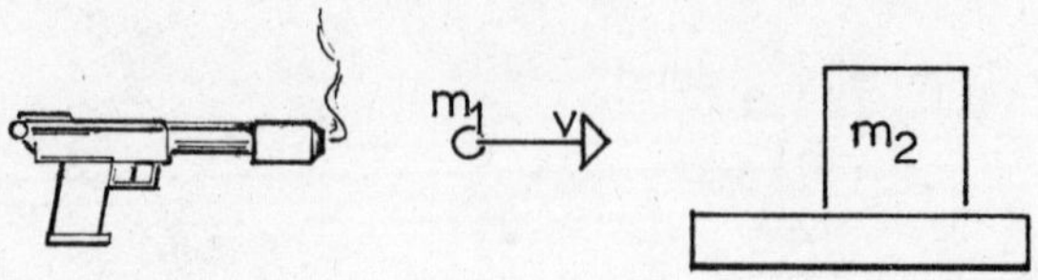


$m_1 = m_2$

$v_{1i} = v$

$\mu = 0$

| | |
|---|---|
| 1. Determine the velocity $v_{12f}$ of objects 1 and 2 after 1 embeds itself in 2. | Since momentum is conserved, we may write<br>$m_1v_{1i} + m_2v_{2i} = (m_1 + m_2)v_{12f}$<br>Knowing that $m_1 = m_2$, $v_{1i} = v$ and $v_{2i} = 0$, we obtain<br>$m_1v = 2m_1v_{12f}$ or $v_{12f} = v/2$ |

In Fig. 8.5, objects 1 and 2 collide in an elastic manner.

Figure 8.5

$m_1 = m_2$ $v_{2i} = 0$

$v_{1i} = v$ $\mu = 0$

| | |
|---|---|
| 2. Determine the velocity of objects 1 and 2 after the collision. | Since the collision is elastic, we may use equations (c) and (d) to determine $v_{1f}$ and $v_{2f}$. When the given information is inserted into equations (c) and (d), we obtain<br>$v_{1f} = v_{2i} = 0$ and $v_{2f} = v_{1f} = v$.<br>Note that the two objects have traded velocities. |

In Fig. 8.6, objects 1 and 2 collide in an elastic manner.

Figure 8.6

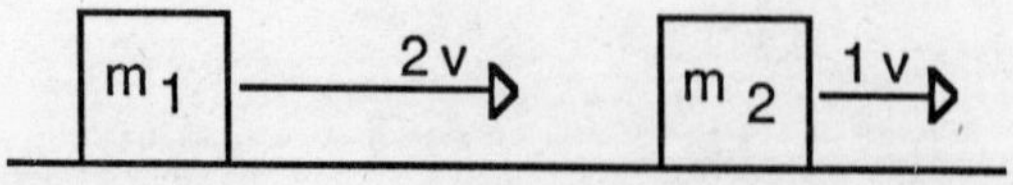


$m_1 = m_2$ $v_{2i} = 0$

$v_{1i} = 2v$ $\mu = 0$

| | |
|---|---|
| 3. Determine the velocity of objects 1 and 2 after the collision. | Insert the given information into equations (c) and (d) to obtain $v_{1f} = v_{2i} = v$ and $v_{2f} = v_{1i} = 2v$. Note that the two objects have traded velocities. |

Note: Any time two objects of equal mass undergo an elastic collision, they simply just trade velocities. (See steps 2 and 3 of this practice section.)

In Fig. 8.7, objects 1 and 2 collide in an elastic manner.

Figure 8.7

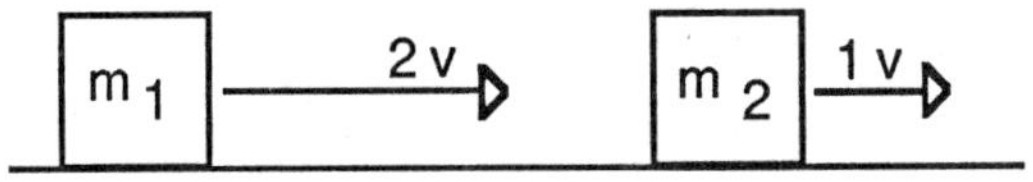


$m_1 = 2m \qquad v_{1i} = 2v$

$m_2 = m \qquad v_{2i} = v$

$\mu = 0$

| | |
|---|---|
| 4. Determine the velocity of objects 1 and 2 after the collision. | Insert the given information into equations (c) and (d) to obtain<br>$v_{1f} = \frac{m}{3m}(2v) + \frac{2m}{3m}v = \frac{4}{3}v$<br>$v_{2f} = \frac{4m}{3m}(2v) - \frac{m}{3m}v = \frac{7}{3}v$<br>Both objects continue moving to the right after the collision; however, object 1 is moving slower and object 2 is moving faster than before the collision. |

In Fig. 8.8, objects 1 and 2 collide in an elastic manner.

Figure 8.8

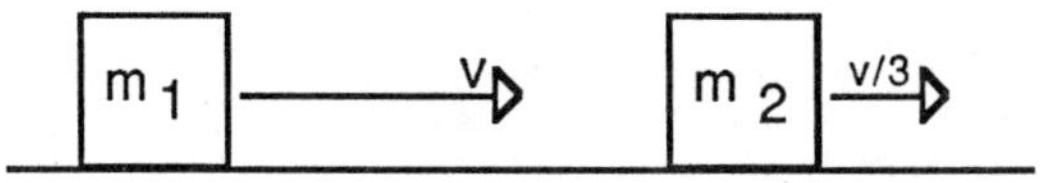


$m_1 = m \qquad v_{1i} = v$

$m_2 = 4m \qquad v_{2i} = v/3$

$\mu = 0$

| | |
|---|---|
| 5. Determine the velocity of objects 1 and 2 after the collision. | Insert the given information into equations (c) and (d) to obtain $v_{1f} = \frac{-3m}{5m}(v) + \frac{8m}{5m}(\frac{1}{3}v) = -\frac{1}{15}v$ $v_{2f} = \frac{2m}{5m}(v) + \frac{3m}{5m}(\frac{1}{3}v) = +\frac{3}{5}v$ After the collision, $m_1$ moves slowly to the left and $m_2$ continues moving to the right but with an increased speed. |

In Fig. 8.9, objects 1 and 2 collide in an elastic manner.

Figure 8.9

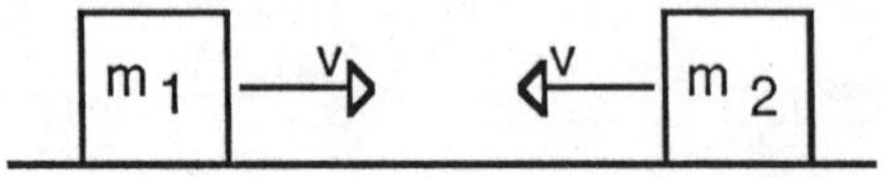


$m_1 = m \qquad v_{1i} = +v$

$m_2 = 2m \qquad v_{2i} = -v$

$\mu = 0$

| | |
|---|---|
| 6. Determine the velocity of objects 1 and 2 after the collision. | Insert the given information into equations (c) and (d) to obtain: $v_{1f} = \frac{-m}{3m}(v) + \frac{4m}{3m}(-v) = -\frac{5}{3}v$ $v_{2f} = \frac{2m}{3m}(v) + \frac{m}{3m}(-v) = +\frac{1}{3}v$ After the collision, object 1 moves to the left and object 2 moves to the right. |

In Fig. 8.10, objects 1 and 2 collide in an inelastic manner.

Figure 8.10

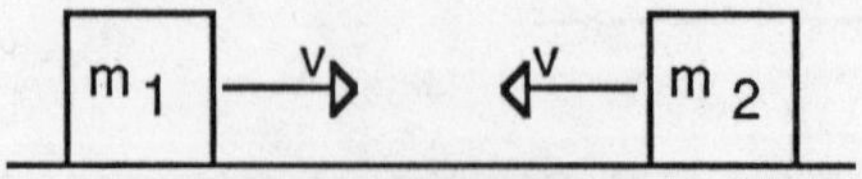


$m_1 = 2m \qquad m_2 = m$

$v_{1i} = +v \qquad v_{2i} = -v$

$\mu = 0 \qquad \varepsilon = 1/2$

| | |
|---|---|
| 7. Determine the velocity of objects 1 and 2 after the collision. | Since momentum is conserved, we can write<br>$m_1v_{1i} + m_2v_{2i} = m_1v_{1f} + m_2v_{2f}$<br><br>From the definition of the coefficient of restitution, we can write<br>$\varepsilon = (v_{1f} - v_{2f})/(v_{2i} - v_{1i})$<br><br>Inserting values into the momentum statement, we obtain<br>($\alpha$) $v = 2v_{1f} + v_{2f}$<br><br>Inserting values into the definition for the coefficient of restitution, we obtain<br>($\beta$) $-v = v_{1f} - v_{2f}$<br><br>Add ($\alpha$) and ($\beta$) to obtain $v_{1f} = 0$<br><br>Subtract ($\beta$) from ($\alpha$) to obtain<br>$v_{2f} = v$<br><br>After the collision, $m_1$ is stationary and $m_2$ moves to the right with a speed v. |

Example 8.5. A 0.010-kg projectile is fired horizontally into and becomes embedded in a suspended block of wood of mass 0.990 kg, as shown in Fig. 8.11. The block and the embedded projectile swing upward until the center of mass of the combination is raised by 0.500 m. (a) What is the velocity of the block and embedded projectile immediately after collision? (b) What is the initial velocity of the projectile? (c) What fraction of the initial kinetic energy is lost during the collision?

Diagram and Given:

Figure 8.11

$m_p = 0.010$ kg
$m_b = 0.990$ kg
$h = 0.500$ m

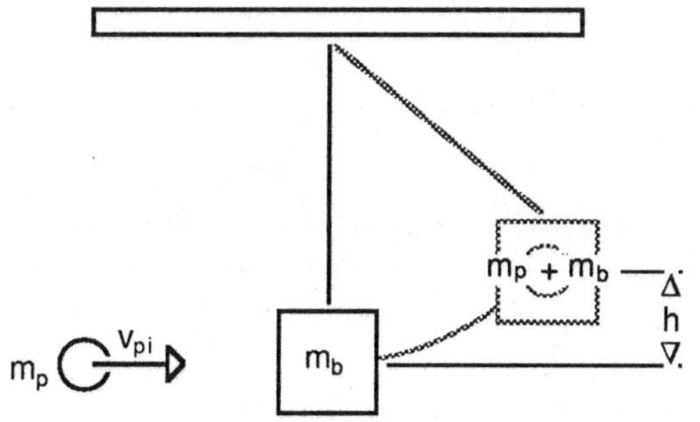

Determine:

(a) The velocity of the block and embedded projectile ($v_{pbf}$) immediately after collision
(b) The velocity of the projectile ($v_{pi}$) before collision.
(c) The fraction of the kinetic energy lost during the collision.

Strategy:

(a) Knowing that all of the kinetic energy of the block and embedded projectile at position 1 gets changed into gravitational potential energy at position 2, we can determine $v_{pbf}$.
(b) Now that $v_{pbf}$ is known, we can use the fact that momentum is conserved during the collision to determine $v_{pi}$.
(c) Once all of the masses and velocities are known, we can determine the initial and final levels of kinetic energy, the change in kinetic energy, and hence, the fraction of kinetic energy lost during the collision.

Solution:

(a) $\frac{1}{2}(m_p + m_b)\, v_{pbf}^2 = (m_p + m_b)gh$ or $v_{pbf} = \sqrt{2gh} = 3.16$ m/s

(b) $m_p v_{pi} = (m_p + m_b)\, v_{pbf}$ or $v_{pi} = (m_p + m_b)v_{pbf}/m_p = 316$ m/s

(c) The initial and final kinetic energies are, respectively

$KE_i = m_p v_{pi}^2/2 = 499$ J and $KE_f = (m_p + m_b)\, v_{pbf}^2/2 = 50.0$ J

The change in the kinetic energy during the collision is obtained by

$$\Delta KE = KE_f - KE_i = -449 \text{ J}$$

The fraction of the energy lost is $f_{lost} = |\Delta KE/KE_i| = 0.900$

Example 8.6. A block of mass $m_1 = 500$ g slides with a velocity of 10.0 cm/s across a frictionless horizontal surface and makes an elastic head-on collision with a block of mass $m_2$, which is initially at rest as shown in Fig. 8.12. After the collision, the 500-g block travels 5.00 cm/s in a direction opposite to its initial motion. Determine the mass of the second block and its velocity after the collision.

Diagram and Given:

$m_1 = 500$ g
$v_{1i} = +10.0$ cm/s
$v_{1f} = -5.00$ cm/s
$v_{2i} = 0$ cm/s

Figure 8.12

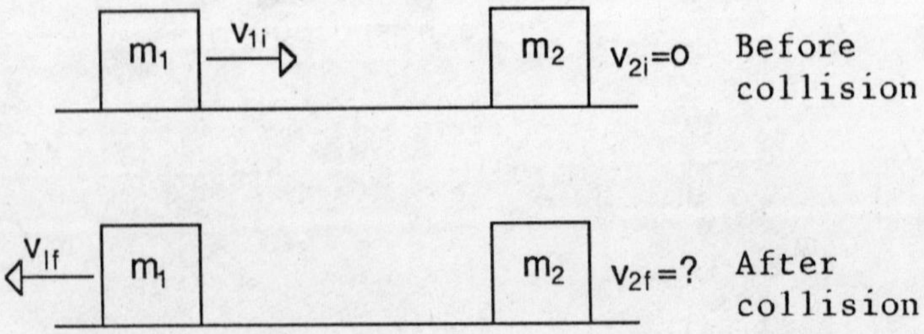

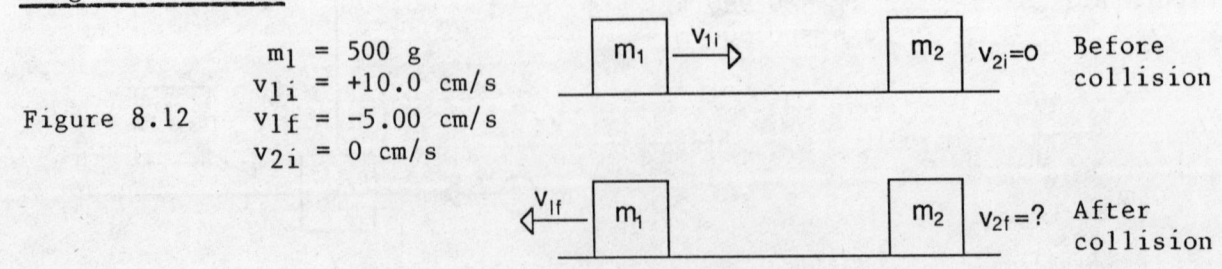

Determine:

a) The mass $m_2$ of the second block.
b) The velocity $v_{2f}$ of the second block after the collision.

Strategy: Since the collision is elastic, we know that momentum and kinetic energy are conserved. A statement of conservation of momentum will involve two unknowns, $m_2$ and $v_{2f}$. A statement of conservation of kinetic energy will involve these same two unknowns. Thus we can solve these two equations.

Solution: A statement of conservation of momentum for this situation is

$$m_1v_{1i} + m_2v_{2i} = m_1v_{1f} + m_2v_{2f}$$

Recalling that $v_{ii} = 0$ for this example, we can write

This may be rewritten as

$$(\alpha) \qquad m_1(v_{1i} - v_{1f}) = m_2v_{2f}$$

A statement of conservation of kinetic energy is

$$\frac{1}{2}(m_1v_{1i}^2) + \frac{1}{2}(m_2v_{2i}^2) = \frac{1}{2}(m_1v_{1f}^2) + \frac{1}{2}(m_2v_{2f}^2)$$

Since $v_{2i} = 0$, for this case we can write

$$m_1(v_{1i}^2 - v_{1f}^2) = m_2v_{2f}^2$$

or

$$(\beta) \qquad m_1(v_{1i} + v_{1f})(v_{1i} - v_{1f}) = m_2v_{2f}^2$$

Notice that we now have two equations, ($\alpha$) and ($\beta$), and two unknowns, $m_2$ and $v_{2f}$. Eliminate $m_2$ by dividing ($\beta$) by ($\alpha$):

$$\frac{(\beta)}{(\alpha)} = \frac{m_1(v_{1i} + v_{1f})(v_{1i} - v_{1f})}{m_1(v_{1i} - v_{1f})} = \frac{m_2\, v_{2f}^2}{m_2\, v_{2f}}$$

Solving for $v_{2f}$, we obtain

$$v_{2f} = v_{1i} + v_{1f} = +5.00 \text{ cm/s}$$

The value for $v_{2f}$ can be inserted into either equation ($\alpha$) or ($\beta$) to obtain $m_2$. Since equation ($\alpha$) looks easier, use it to obtain $m_2$.

$$m_1(v_{1i} - v_{1f})/v_{2f} = 2.00 \times 10^3 \text{ kg}$$

You can check this by inserting $v_{2f}$ into equation ($\beta$) to obtain the same value for $m_2$.

Example 8.7. A golfball rolls off a table 1.00 m high onto a concrete floor, as shown in Fig. 8.13. On the first bounce it rebounds to a height of 0.900 m. Determine the coefficient of restitution for the golfball and concrete collision and the height of the ball on the second bounce.

Diagram and Given:

Figure 8.13

$h_o = 1.00$ m

$h_1 = 0.900$ m

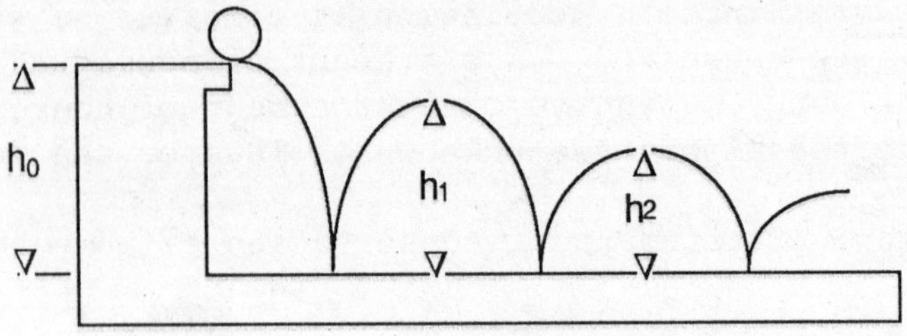


Determine: The coefficient of restitution ($\varepsilon$) for collisions between the golfball and the concrete floor. The height the ball bounces ($h_2$) after its second collision with the floor.

Strategy: Knowing how far the ball falls and the acceleration due to gravity, we can determine the speed of the ball when it hits the floor and hence the relative velocity of approach for the ball and floor. Knowing how high the ball travels after the collision and the acceleration due to gravity, we can determine the speed of the ball after it hits the floor and hence the relative velocity of separation for the ball and the floor. Knowing the relative velocity of approach and separation, we can determine the coefficient of restitution.

The relative velocity of separation after the first collision is equal to the negative of the relative velocity of approach for the second collision. Knowing the relative velocity of approach for the second collision and the coefficient of restitution for the collision, we can determine the relative velocity of separation and hence the upward speed of the ball after the collision. Knowing the upward speed of the ball after the collision, we can determine how high the ball will bounce.

Solution: The velocity of the ball before the first collision can be determined by using

$$v^2 - v_o^2 = 2a\Delta s$$

where $v_o = 0$, $a = -g = -9.80 \text{ m/s}^2$, $\Delta s = -h_o = -1.00$ m, and v is the velocity of the ball just before collision. Solving for v, we obtain

$$v = -\sqrt{2a\Delta s} = -4.43 \text{ m/s}$$

The negative root indicates that the velocity is downward. Since the floor is stationary, the relative velocity of approach before the first collision is

$$v_{rel\ approach} = 4.43 \text{ m/s}$$

The velocity of the ball after the first collision with the floor can be determined by using

$$v^2 - v_o^2 = 2a\Delta s$$

where $v = 0$ (velocity of the ball at the top of the bounce), $a = -g = -9.80\ m/s^2$, $\Delta s = +h_1 = +0.900$ m, and $v_o$ is the initial upward speed of the ball after the first collision. Solving for $v_o$, we obtain

$$v_o = +\sqrt{-2a\Delta s} = +4.20\ m/s$$

The positive root indicates that the velocity is upward. Since the floor is stationary, the relative velocity of separation of the ball and floor after the first collision is

$$v_{rel\ separation} = 4.20\ m/s$$

The coefficient of restitution is

$$\varepsilon = v_{rel\ separation}/v_{rel\ approach} = 0.948$$

For the second collision, the relative velocity of approach is 4.20 m/s, since the ball returns to the floor with a speed equal to its initial upward speed. The relative velocity of separation can then be determined by

$$v_{rel\ separation} = \varepsilon\ (v_{rel\ approach}) = 3.98\ m/s$$

Since the floor is stationary, the initial upward velocity is equal to the relative velocity of separation. Hence

$$v_o = v_{rel\ separation} = 3.98\ m/s$$

The height $h_2$ reached by the ball after its second bounce (after two collisions with the floor) can be determined by using

$$v^2 - v_o^2 = 2a\Delta s$$

where $v = 0$ (velocity of the ball at the top of the bounce), $\Delta s = +h_2$, $a = -g = -9.80\ m/s^2$ and $v_o = 3.98$ m/s. Solving for $\Delta s$, we obtain

$$\Delta s = h_2 = -v_o^2/2a = 0.808\ m$$

Related Text Problems: 8-14 through 8-18, 8-22 through 8-28.

---

## PRACTICE TEST

Take and grade this practice test. Doing so will allow you to determine any weak spots in your understanding of the concepts taught in this chapter. The following section prescribes what you should study further to strengthen your understanding.

A 100-g ball moving at a speed of 40.0 m/s strikes a wall at an angle of 60° with respect to the normal. There is no change in the component of the velocity parallel to the wall. The coefficient of restitution for the collision is 0.600. The ball and wall are in contact for 0.100 s. Determine the following:

______ 1. Magnitude of the ball's initial momentum

______ 2. Component of the ball's initial velocity perpendicular to the wall

__________ 3. Component of the ball's final velocity perpendicular to the wall

__________ 4. Angle with which the ball leaves the wall

__________ 5. Magnitude of the ball's final momentum

__________ 6. Magnitude of the impulse imparted to the ball

__________ 7. Magnitude of the average force exerted on the ball

$m_1$ is released from rest and collides elastically with $m_2$. The objects have the same mass, the horizontal surface is frictionless, the cord attached to $m_1$ is 1.00 m long, and θ is 30.0.

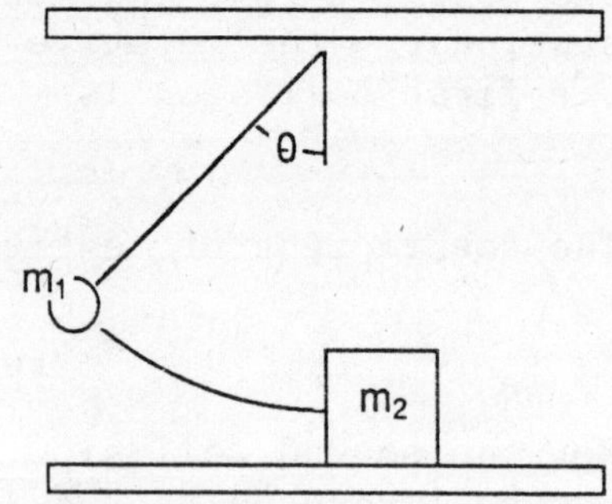


Determine the following:

__________ 8. Speed of $m_1$ the instant before the collision

__________ 9. Speed of $m_1$ after the collision

__________ 10. Speed of $m_2$ after the collision

$m_1$ and $m_2$ are traveling as shown at the right. $m_1$ slides down the ramp, overtaking and colliding elastically with $m_2$. The surface is frictionless.

$m_1 = 400$ g  $h = 1.00$ m  $\mu = 0$

$m_2 = 200$ g  $v_{1i} = v_{2i} = 1.00$ m/s

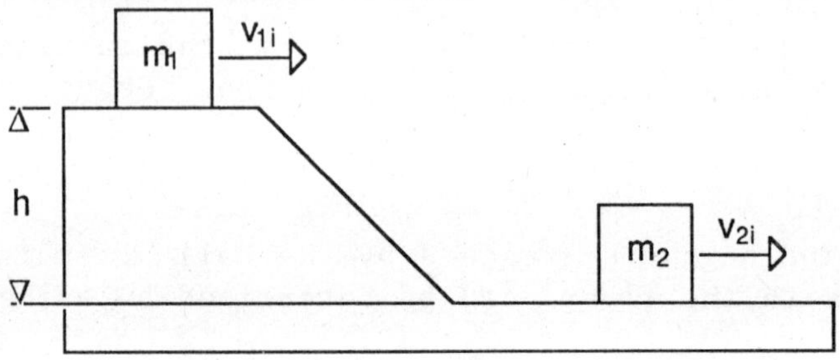


Determine the following:

__________ 11. Speed of $m_1$ the instant before the collision

__________ 12. Speed of $m_1$ after the collision

__________ 13. Speed of $m_2$ after the collision

$m_1$ and $m_2$ travel toward each other before colliding inelasltically, as shown in the figure at the right. The surface is frictionless.

$m_1 = 400$ g  $v_{1i} = +1$ m/s  $\varepsilon = 0.800$

$m_2 = 200$ g  $v_{2i} = -3$ m/s  $\mu = 0$

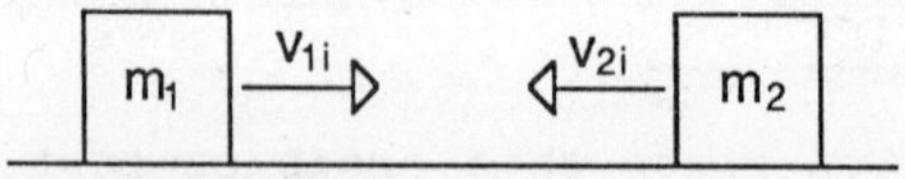


Determine the following:

__________ 14. Velocity of $m_1$ after the collision

__________ 15. Velocity of $m_2$ after the collision

$m_1$ is fired into $m_2$ which is initially at rest on a rough surface. The collision is totally inelastic. The composite object ($m_1$ + $m_2$) travels 5.00 m across the surface before coming to rest.

$m_1 = 20.0$ g  $v_{1i} = 100$ m/s

$m_2 = 200$ g  $v_{ii} = 0.00$ m/s

$\Delta s = 5.00$ m

Determine the following:

__________ 16. Speed of the composite object the instant after collision
__________ 17. Acceleration of the composite object as it slides
__________ 18. Frictional force decelerating the composite object
__________ 19. Coefficient of friction for the composite object on the rough surface

(See Appendix I for answers.)

## PRINCIPAL CONCEPTS AND EQUATIONS PRESCRIPTION

Your score on the practice is an excellent measure of your understanding of this chapter. You should now use the following chart to write your own prescription for curing any of your physics ills. Look down the leftmost column to the number of the question(s) you answered incorrectly, read across that row to see which section(s) of the study guide you should return to for further study, and then do the suggested text problems to gain additional experience in working with the particular concept.

| Practice Test Question | Concepts and Equations | Prescription | |
|---|---|---|---|
| | | Principal Concept | Text Problems |
| 1 | Momentum | 1 | 8-2,3 |
| 2 | Components of a vector | 6 of Ch. 3 | 3-3,8 |
| 3 | Coefficient of restitution | 3 | 8-22,23 |
| 4 | Coefficient of restitution | 3 | 8-22,23 |
| 5 | Inelastic collisions | 3 | 8-29,30 |
| 6 | Impulse momentum theorem | 1 | 8-1,5 |
| 7 | Impulse | 1 | 8-6,7 |
| 8 | Conservation of mechanical energy | 6 of Ch. 7 | 7-37,39 |
| 9 | Elastic collision | 3 | 8-16,17 |
| 10 | Elastic collision | 3 | 8-18,19 |
| 11 | Conservation of mechanical energy | 6 of Ch. 7 | 7-39,40 |
| 12 | Elastic collision | 3 | 8-16,17 |
| 13 | Elastic collision | 3 | 8-16,17 |
| 14 | Inelastic collision | 3 | 8-26,27 |
| 15 | Inelastic collision | 3 | 8-26,27 |
| 16 | Conservation of momentum | 2 | 8-15,25 |
| 17 | $v^2 - v_o^2 = 2a\Delta s$ | 5 of Ch. 2 | 2-20,24 |
| 18 | Newton's second law | 2 of Ch. 4 | 4-3,4 |
| 19 | Coefficient of friction | 7 of Ch. 4 | 8-24 |

# 9 Rotational Dynamics

## RECALL FROM PREVIOUS CHAPTERS

| Previously learned concepts and equations frequently used in this chapter | Text Section | Study Guide Page |
|---|---|---|
| Analyzing tanslational motion | 2.4 | 22 |
| Newton's second law: $\vec{F} = m\vec{a}$ | 4.3 | 47 |
| Relationship between linear and rotational quantities $\Delta s = r\Delta\theta$, $v = r\omega$, $a = r\alpha$ | 5.2,5.3,4 | 68 |
| Analyzing rotational motion | 5.4 | 71 |
| Torque: $\tau = rF \sin\theta$ | 6.2 | 87 |
| Work: $W = F_{\parallel}\Delta s$ | 7.3 | 100 |
| Translational kinetic energy: $KE = mv^2/2$ | 7.5 | 108 |
| Gravitational potential energy: $PE = mgh$ | 7.6 | 110 |

## NEW IDEAS IN THIS CHAPTER

| Concepts and equations introduced | Text Section | Study Guide Page |
|---|---|---|
| Moment of inertia: $I = \sum_i m_i r_i^2$ | 9.2 | 140 |
| Torque and acceleration: $\tau = I\alpha$ | 9.2 | 143 |
| Parallel-axis theorem: $I_p = I_o + md^2$ | 9.2 | 145 |
| Work and torque: $W = \tau\Delta\theta$ | 9.3 | 146 |
| Rotational kinetic energy: $KE_R = I\omega^2/2$ | 9.3 | 146 |
| Work and rotational KE: $W_{net} = \Delta KE_R$ | 9.3 | 146 |
| Angular momentum: $L = I\omega$ | 9.5 | 152 |
| Angular impulse: $\Delta L = \tau\Delta t$ | 9.5 | 152 |
| Conservation of angular momentum: $L_f = L_i$ | 9.5 | 154 |

## PRINCIPAL CONCEPTS AND EQUATIONS

### 1. Moment of Inertia (Section 9.2)

Review: The moment of inertia for an extended body about an axis P can be determined by the expression

$$I = \sum_{i=1}^{n} m_i r_i^2$$

This quantity may be thought of as the body's resistance to angular acceleration. It is important to note that I depends not only on the mass but also on its distribution about the axis of rotation. For most extended objects, it is difficult to obtain an expression for I without using calculus. Thus the text authors have provided a list of expressions (Table 9-1) for I for a variety of extended bodies. Table 9-1 gives an expression for I about an axis through the center of mass of the body.

Practice: Refer to Fig. 9-1 to determine the desired quantities.

Figure 9.1 $m_1$, $m_2$, and $m_3$ are tiny mass points.
$m_1 = m_2 = m_3 = m$

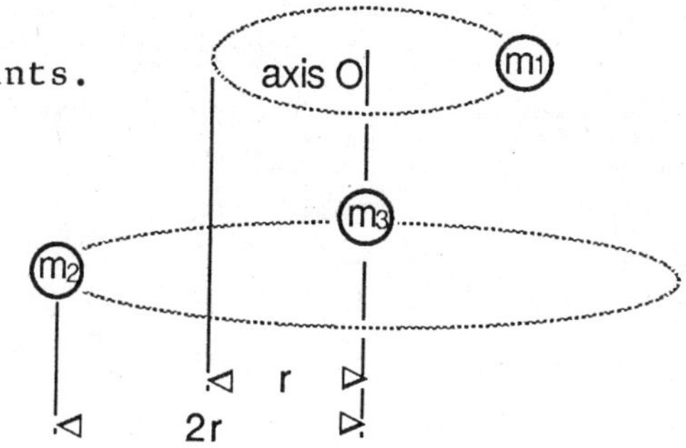


Determine the following:

| | |
|---|---|
| 1. Moment of inertia of $m_1$ about 0 | $I_1 = m_1r_1^2 = mr^2$ |
| 2. Moment of inertia of $m_2$ about 0 | $I_2 = m_2r_2^2 = m(2r)^2 = 4mr^2$ |
| 3. Moment of inertia of $m_3$ about 0 | $I_3 = m_3r_3^2 = m_3(0) = 0$ |
| 4. Total moment of inertia about 0 | $I_T = \sum_{i=1}^{3} m_ir_i^2$ <br> $= m_1r_1^2 + m_2r_2^2 + m_3r_3^2$ <br> $= I_1 + I_2 + I_3 = 5mr^2$ |

A child is riding on a merry-go-round (m-g-r) at a position that is one-half the distance to the rim.

Figure 9.2
$r$ = radius of the m-g-r
$r_c = r/2$ = position of child
$m_m$ = mass of the m-g-r
$m_c$ = mass of the child

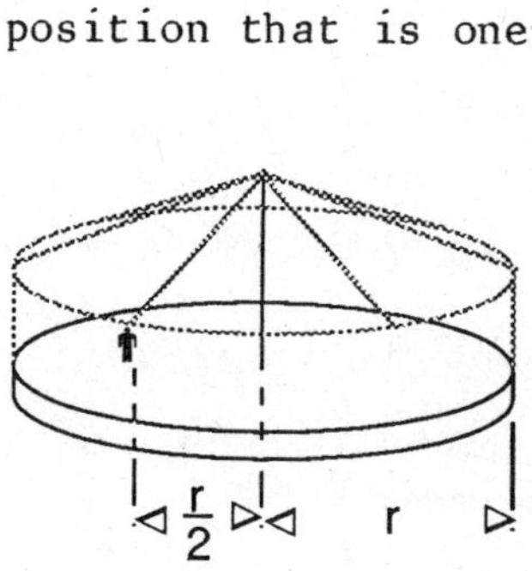

Determine the following:

| | |
|---|---|
| 5. Moment of inertia of the m-g-r about O | $I_m = m_m r^2/2$ |
| 6. Moment of inertia of the child about O | $I_c = m_c r_c^2 = m_c(r/2)^2 = m_c r^2/4$ |
| 7. Total moment of inertia about O | $I_T = I_m + I_c = (2m_m + m_c)r^2/4$ |

Example 9.1. A cylinder of length $L = 1.00$ m, radius $R = 1.00 \times 10^{-2}$ m, and mass $m_{cy} = 5.00 \times 10^{-1}$ kg is mounted on an axle of radius $r = 5.00 \times 10^{-3}$ m and negligible mass, as shown in Fig. 9.3. Two identical weights, each of mass $m_w = 1.00$ kg, are mounted on the cylinder $2.50 \times 10^{-1}$ m from the axle. Where on the cylinder should the two weights be relocated if we wish to double the system's inertia about the axle.

Diagram and Given:

Figure 9.3

$$
\begin{aligned}
L &= 1.00 \text{ m} \\
m_{cy} &= 5.00 \times 10^{-1} \text{ kg} \\
m_w &= 1.00 \text{ kg} \\
r_{axle} &= 5.00 \times 10^{-3} \text{ m} \\
r_{cy} &= 1.00 \times 10^{-2} \text{ m} \\
I_{cy} &= m_{cy}L^2/12 \text{ (Table 9-1)} \\
d &= 2.50 \times 10^{-1} \text{ m}
\end{aligned}
$$

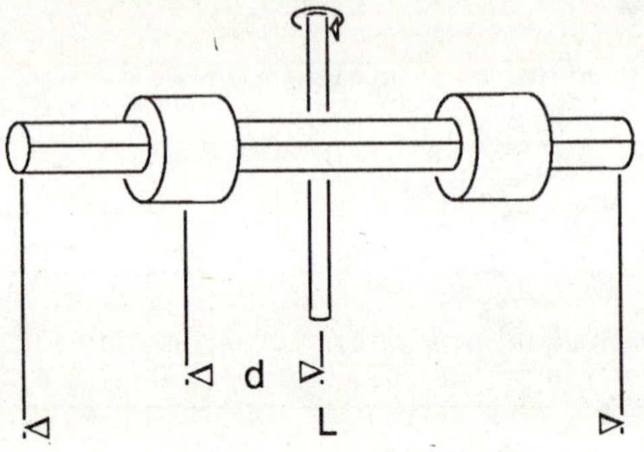


Determine: Where on the cylinder the two weights should be relocated in order to double the system's moment of inertia.

Strategy: With the given information, we can determine the moment of inertia of the cylinder about the axis, the moment of inertia of each weight (at its present position) about the axis, and hence the moment of inertia of the present system about the axis. We can then double this to determine the new moment of inertia of the system. Knowing the new moment of inertia of the system and the moment of inertia of the cylinder about the axis, we can determine the moment of inertia of the weights in their new location. Knowing the moment of inertia of the weights in their new location and the mass of the weights, we can determine the new location of the weights.

Solution: The moment of inertia of the cylinder about the axis is

$$I_{cy} = m_{cy}L^2/12 = (5.00 \times 10^{-1} \text{ kg})(1.00 \text{ m}^2)/12 = 4.17 \times 10^{-2} \text{ kg}\cdot\text{m}^2$$

The moment of inertia of each weight in its present position about the axis is

$$I_w = m_w d^2 = (1.00 \text{ kg})\ (2.50 \times 10^{-1} \text{ m})^2 = 6.25 \times 10^{-2} \text{ kg}\cdot\text{m}^2$$

The total moment of inertia of the system about the axis with the weights in their present positions is $I_{Total} = I_{cy} + 2I_w = 16.7 \times 10^{-2}$ kg·m$^2$

The new moment of inertia of the system is to be twice this value for $I_{total}$.

$$I_{total\ new} = 2I_{total} = 33.4 \times 10^{-2}\ kg\cdot m^2$$

The new total moment of inertia of the system about the axis has a contribution from the cylinder and from the weights in their new position.

$$I_{total\ new} = I_{cy} + 2I_{w\ new}$$

$$I_{w\ new} = (I_{total\ new} - I_{cy})/2 = 14.6 \times 10^{-2}\ kg\cdot m^2$$

Finally, we can determine the new location of the weights.

$$I_{w\ new} = m_w d_{new}^2 \qquad d_{new} = 3.82 \times 10^{-1}\ m$$

If the weights are moved from $d = 2.50 \times 10^{-1}$ m to $d = 3.82 \times 10^{-1}$ m the moment of inertia of the system will double.

Related Text Problems: 9-2, 9-3, 9-4, 9-11a, 9-35a.

## 2. The Relationship Between Net External Torque Acting on a Body and Resulting Angular Acceleration (Section 9.2)

Review: If an object has a rotational moment of inertia I about an axis and experiences a net torque $\tau_{net}$ about the axis, it will experience an angular acceleration $\alpha$ about that axis. The physical quantities are related by the expression $\tau_{net} = I\alpha$.

Note: The expression $\tau_{net} = I\alpha$ is the rotational analog of the expression $F_{net} = ma$.

Practice: Shown in Fig. 9.4 is a solid disk mounted on an axle. A constant tangential force is applied by means of a string that has been wrapped around the circumference of the disk. The disk is initially at rest, and it experiences a constant frictional torque while it is rotating.

Figure 9.4

$m = 1.00$ kg = mass of the disk
$r = 2.00 \times 10^{-1}$ m = radius of the disk
$F = 2.00$ N = constant force
$\tau_f = 1.00 \times 10^{-1}$ N·m = frictional torque
$\omega_o = 0$ = initial angular speed
The axle is so small that we can ignore the hole made in the disk for the axle.

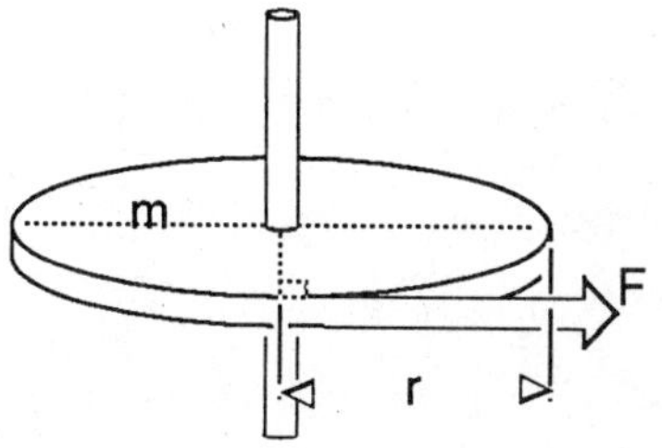


Determine the following:

| | |
|---|---|
| 1. Moment of inertia of the disk about the axle | $I = mr^2/2 = 2.00 \times 10^{-2}\ kg\cdot m^2$ |

| | |
|---|---|
| 2. Accelerating torque about the axle as a result of the constant force F | $\tau = rF \sin 90° = 4.00 \times 10^{-1}$ N·m |
| 3. Net torque about the axle | $\tau_{net} = \tau - \tau_f = 3.00 \times 10^{-1}$ N·m |
| 4. Angular acceleration of the disk about the axle | $\alpha = \tau_{net}/I = 15.0$ rad/s$^2$ |

Example 9.2. Three 1.00-kg spheres are held in a horizontal equilateral triangle configuration by a Y-shaped frame, as shown in Fig. 9.5. Each branch of the frame is $2.00 \times 10^{-2}$ m long. This configuration is mounted on a vertical axis through the center of the frame, and forces of 1.00 N, 2.00 N, and 3.00 N are applied to the spheres at right angles to the branches of the Y. Determine the resulting angular acceleration of the configuration about the axis.

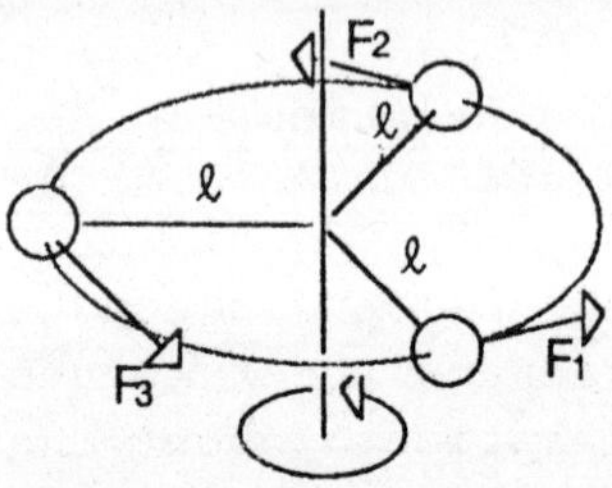


Given and Diagram:

Figure 9.5 $m = 1.00$ kg
$\ell = 2.00 \times 10\text{-}2$ m
$F_1 = 1.00$ N, $F_2 = 2.00$ N, $F_3 = 3.00$ N

Determine: The resulting angular acceleration of the system about the axis.

Strategy: Knowing the mass of each sphere and its distance from the axis of rotation, we can determine the moment of inertia of the the triangular configuration about the axis. Knowing the magnitude, direction, and moment arm for each force, we can determine the net torque acting on the system. Now that $\tau_{net}$ and I about the axis have been determined, we can obtain the angular acceleration $\alpha$ about that axis.

Solution: The moment of inertia of the system about the axis is

$$I = 3m\ell^2 = 12.0 \times 10^{-4} \text{ kg·m}^2$$

The net torque about the axis is

$$\tau_{net} = +\tau_1 - \tau_2 + \tau_3 = \ell(F_1 - F_2 + F_3) = 4.00 \times 10^{-2} \text{ N·m}$$

The resulting angular acceleration about the axis is

$$\alpha = \tau_{net}/I = 33.0 \text{ rad/s}^2$$

Related Text Problems: 9-1, 9-6, through 9-13, 9-21 through 9-24, 9-30, 9-35.

## 3. Parallel-Axis Theorem (Section 9.2)

Review: The moment of inertia of an object about any axis is equal to the moment of inertia about a parallel axis through the center of mass plus the product of the mass of the object and the square of the distance between the two axes.

Figure 9.6 $I_P = I_O + md^2$

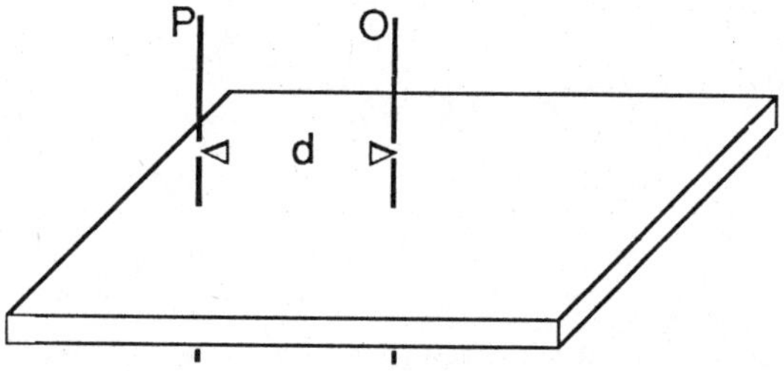


Practice: Consider the long cylindrical rod of mass m, radius r, and length $\ell$ shown in Fig. 9.7.

Figure 9.7

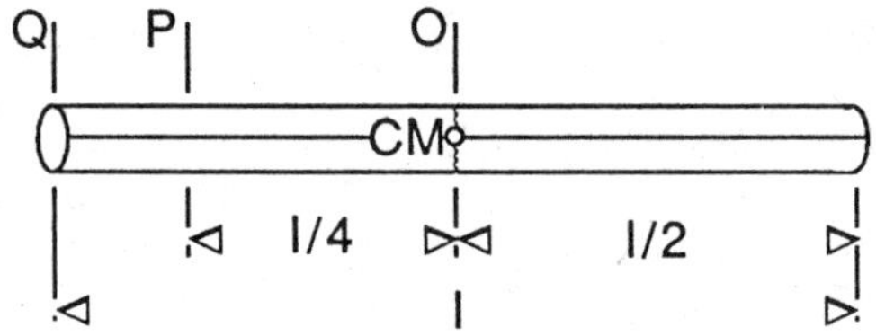


Determine an expression for the following:

| | |
|---|---|
| 1. Moment of inertia of the rod about the axis through its center of mass | From Table 9-1 of the text: $I_O = m\ell^2/12$ |
| 2. Moment of inertia of the rod about the axis P | Using the parallel-axis theorem: $I_P = I_O + md^2 = m\ell^2/12 + m(\ell/4)^2 = 7m\ell^2/48$ |
| 3. Moment of inertia of the rod about the axis Q | Using the parallel-axis theorem: $I_Q = I_O + md^2 = m\ell^2/12 + m(\ell/2)^2 = m\ell^2/3$ |

Example 9.3. Determine the moment of inertia of a flat circular disk about the axes parallel to the axis through the CM. Axis P is one-half the distance to the rim, and axis Q is a distance 2r from the CM. The mass of the disk is 2.00 kg and its radius is $1.00 \times 10^{-1}$ m.

Given and Diagram:

Figure 9.8 $m = 2.00$ kg
$r = 1.00 \times 10^{-1}$ m

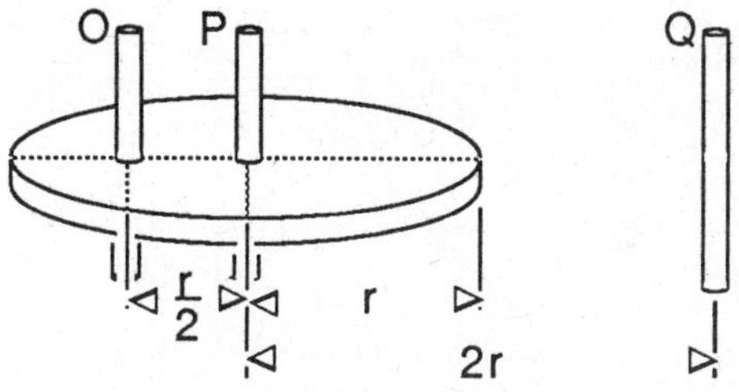

Determine: The moment of inertia of the flat circular disk about axes P and Q.

Strategy: Knowing m and r, we can determine the moment of inertia of the disk about the axis O through its center of mass. Since the distances to the other two axes are known, we can use the parallel-axis theorem to determine the moment of inertia about them.

Solution: The moment of inertia about axis O through the center of mass of the disk is

$$I_O = mr^2/2 = (2.00\ \text{kg})(1 \times 10^{-1}\ \text{m})^2/2 = 1 \times 10^{-2}\ \text{kg}\cdot\text{m}^2$$

Using the parallel-axis theorem, obtain the moment of inertia about the axis P.

$$I_P = I_O + md^2 = I_O + m(r/2)^2 = I_O + mr^2/4 = 1.5 \times 10^{-2}\ \text{kg}\cdot\text{m}^2$$

Using the parallel-axis theorem, obtain the moment of inertia about the axis Q.

$$I_Q = I_O + md^2 = I_O + m(2r)^2 = I_O + 4mr^2 = 9.00 \times 10^{-2}\ \text{kg}\cdot\text{m}^2$$

Related Text Problems: 9-5, 9-23, 9-25.

## 4. Work and Rotational Kinetic Energy (Section 9.3).

Review: If a torque $\tau$ is applied to a body, causing it to rotate through an angle $\Delta\theta$ about a fixed axis, the work done by the applied torque is $W = \tau\Delta\theta$.

Note: $W = \tau\Delta\theta$ is the rotational analog of $W = F\Delta s$.

The rotational kinetic energy of a body about a fixed axis is $KE_R = I\omega^2/2$.

Note: $KE_R = I\omega^2/2$ is the rotational analog of $KE_T = mv^2/2$.

The net work done on a body by the net torque is equal to the change in the rotational kinetic energy of the body. This may be stated algebraically as

$$W_{net} = \Delta KE_R$$

Note: $W_{net} = \Delta KE_R$ is the rotational analog of $W_{net} = \Delta KE_T$.

When all the above are combined into one algebraic statement, we have

$$W_{net} = \tau_{net}\Delta\theta = \Delta KE_R = KE_{Rf} - KE_{Ri} = I\omega_f^2/2 - I\omega_i^2/2$$

Practice: A solid cylindrical disk of mass m and radius r is mounted on a small axle with a frictionless bearing, as shown in Fig. 9.9. A lightweight cord is attached to and wrapped around the rim of the disk several times. An object of mass m is attached to the free end of the cord, released from rest, and allowed to fall a distance h.

Figure 9.9

$M$ = mass of disk
$m$ = mass of hanging object
$r$ = radius of disk
$h$ = distance object falls
$v_o = 0$ = initial linear speed of m
$\omega_o = 0$ = initial angular speed of M

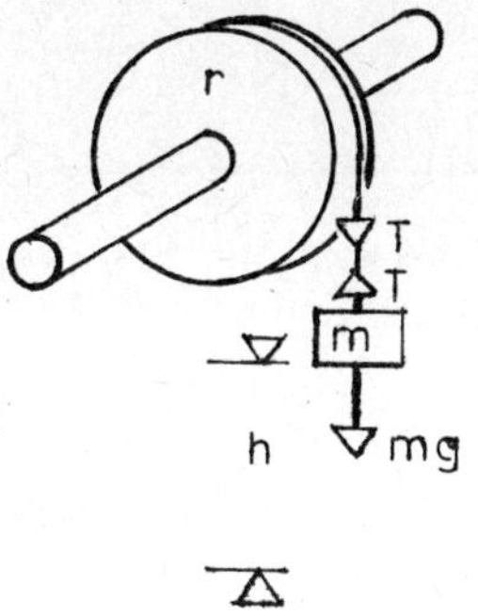


Determine an expression for the following:

| | |
|---|---|
| 1. Work done on the system (disk plus hanging mass) by gravity | $W_g = F_g \Delta s \cos\theta = mgh \cos 0° = mgh$ |
| 2. Linear acceleration of the hanging mass | A statement of Newton's second law for the hanging mass is $ma = mg - T$. From this, we may write tension as $T = m(g - a)$. Then, using torque, obtain $\tau = rT = I\alpha$ or $rm(g - a) = (Mr^2/2)(a/r)$, which can be solved for acceleration to obtain $a = 2mg/(M + 2m)$ |
| 3. Final linear speed of the hanging mass | $v_f^2 - v_o^2 = 2a\Delta s$<br>$v_f = (2ah)^{1/2} = [4mgh/(M + 2m)]^{1/2}$ |
| 4. Final angular speed of the disk | $\omega_f = v_f/r = [4mgh/(M + 2m)]^{1/2}/r$ |
| 5. Change in translational kinetic energy of the hanging mass | $\Delta KE_T = KE_{Tf} = mv_f^2/2 = 2m^2gh/(M + 2m)$ |
| 6. Change in rotational kinetic energy of the disk | $\Delta KE_R = KE_{Rf} = I\omega_f^2/2 = mMgh/(M + 2m)$ |
| 7. Total change in kinetic energy (translational plus rotational) | $\Delta KE = \Delta KE_T + \Delta KE_R = mgh$<br>Note that the total change in kinetic energy is equal to the net work done on the system (step 1). |
| 8. Net force acting on the hanging mass | $F_{net} = ma = 2m^2g/(M + 2m)$ |

| | |
|---|---|
| 9. Net work done on hanging mass | $W_{net\ m} = F_{net\ m}\Delta s \cos\theta$<br>$= 2m^2gh/(M + 2m)$<br>Note that $W_{net\ m} = \Delta KE_T$, that is, the net work done on the hanging mass is equal to its change in translational kinetic energy. |
| 10. Tension in the cord | Recall from step 2 that $T = m(g - a)$ and $a = 2mg/(M + 2m)$.<br>Combining these, we obtain<br>$T = mMg/(M + 2m)$ |
| 11. Net torque acting on the disk | $\tau_{net} = rT = mMgr/(M + 2m)$ |
| 12. Angular displacement of the disk | $\theta = h/r$ |
| 13. Net work done on the disk | $W_{net\ M} = \tau_{net}\theta = mMgh/(M + 2m)$<br>Note that $W_{net\ M} = \Delta KE_R$, that is, the net work done on the disk is equal to its change in rotational kinetic energy. |
| 14. Net work done on the system (hanging mass and disk) | $W_{net} = W_{net\ m} + W_{net\ M} = mgh$ |

Note: Gravity does an amount of work (mgh) on the system. This work is distributed between the hanging mass and the rotating disk ($mgh = W_{net\ m} + W_{net\ M}$). The net work done on the hanging mass goes into changing its translational kinetic energy ($W_{net\ m} = \Delta KE_T$), and the net work done on the rotating disk goes into changing its rotational kinetic energy ($W_{net\ M} = \Delta KE_R$).

Example 9.4. A child pushes a disk-shaped merry-go-round (m-g-r) with a 400-N tangential force for 10.0 s. The m-g-r has a mass of 250 kg and a radius of 1.50 m. If the it is initially at rest, determine its rotational kinetic energy at the end of the 10.0 s.

Given and Diagram:

Figure 9.10

$F = 100$ N = tangential force
$t = 5.00$ s = time force acts
$m = 250$ kg = mass of m-g-r
$R = 1.50$ m = radius of m-g-r
$\omega_o = 0$ = initial angular speed of m-g-r

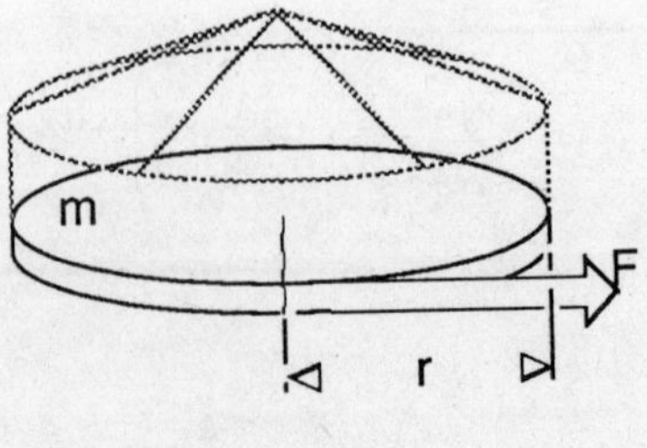

Determine: The rotational kinetic energy of the m-g-r after the child has been pushing for 10.0 s with a 100-N tangential force.

Strategy: Knowing m, r, and F, we can determine the torque $\tau$ acting on the m-g-r and its moment of inertia I. From I and $\tau$, we can determine the angular acceleration $\alpha$. Knowing the angular acceleration $\alpha$ and time t for that acceleration, we can determine the angular displacement $\Delta\theta$. From $\tau$ and $\Delta\theta$, we can determine the net work on the m-g-r and hence its rotational kinetic energy.

Solution: The net torque on the m-g-r about the cylindrical axis is

$$\tau = rF \sin 90° = (1.50 \text{ m})(1.00 \times 10^2 \text{ N}) = 1.50 \times 10^2 \text{ N}\cdot\text{m}$$

The moment of inertia of the m-g-r about the cylindrical axis is

$$I = mr^2/2 = (2.50 \times 10^2 \text{ kg})(1.50 \text{ m})^2/2 = 281 \text{ kg}\cdot\text{m}^2$$

The angular acceleration of the m-g-r is

$$\alpha = \tau/I = 1.50 \times 10^2 \text{ N}\cdot\text{m}/281 \text{ kg}\cdot\text{m}^2 = 2.14 \text{ rad/s}$$

The angular displacement of the m-g-r during the push time is

$$\Delta\theta = \alpha t^2/2 = (2.14 \text{ rad/s}^2)(5.00 \text{ s})^2/2 = 26.8 \text{ rad}$$

The net work done on the m-g-r is

$$W_{net} = \tau\Delta\theta = (1.50 \times 10^2 \text{ N}\cdot\text{m})(26.8 \text{ rad}) = 4.02 \times 10^3 \text{ J}$$

The rotational kinetic energy at the end of the 10.0 s push is

$$KE_{R_{final}} = \Delta KE_R = W_{net} = 4.02 \times 10^3 \text{ J}$$

Example 9.5. A section of a large, thin-walled steel pipe (m = 1.00 kg and $r = 5.00 \times 10^{-1}$ m) is rolling across a horizontal floor at a speed of 2.00 m/s. Determine the work that must be done in order to stop the rolling pipe.

Given and Diagram:

Figure 9.11

$m = 1.00$ kg
$r = 5.00 \times 10^{-1}$ m
$v = 2.00$ m/s

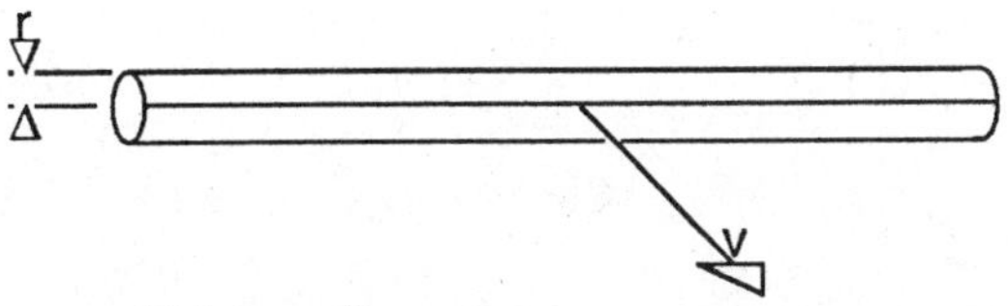


Determine: The amount of work to stop the rolling pipe.

Strategy: From the given information, we can determine the translational kinetic energy ($KE_T$), the angular speed ($\omega$), the moment of inertia (I), and the rotational kinetic energy ($KE_R$) of the pipe. From $KE_T$ and $KE_R$, we can determine the total change in kinetic energy and hence the work required to stop the pipe.

Solution: The translational kinetic energy of the pipe is

$$KE_T = mv^2/2 = (1.00 \times 10^2 \text{ kg})(2.00 \text{ m/s})^2/2 = 2.00 \times 10^2 \text{ J}$$

The angular speed of the pipe is

$$\omega = v/r = (2.00 \text{ m/s})/5.00 \times 10^{-1} \text{ m} = 4.00 \text{ rad/s}$$

The rotational moment of inertia of the pipe is

$$I = mr^2 = (1.00 \times 10^2 \text{ kg})(5.00 \times 10^{-1} \text{ m})^2 = 25.0 \text{ kg}\cdot\text{m}^2$$

The rotational kinetic energy is

$$KE_R = I\omega^2/2 = (2.50 \text{ kg}\cdot\text{m}^2)(4.00 \text{ rad/s})^2/2 = 2.00 \times 10^{-2} \text{ J}$$

The total change in kinetic energy is

$$\Delta KE = \Delta KE_R + \Delta KE_T = KE_R + KE_T = 4.00 \times 10^2 \text{ J}$$

The net work required to stop the rolling section of pipe is

$$W_{net} = \Delta KE = 4.00 \times 10^2 \text{ J}$$

Related Text Problems: 9-13 through 9-16.
9-31.

## 5. Translational and Rotational Motion Combined (Section 9.4)

Review: The motion of an object undergoing both translational and rotational motion can be considered from two points of view. The first considers the translational motion of the center of mass and the rotational motion about the center of mass. The second considers the instantaneous rotational motion about the instantaneously stationary point of contact between the rolling object and the surface on which it is rolling.

View 1
Translation of O and rotation about O

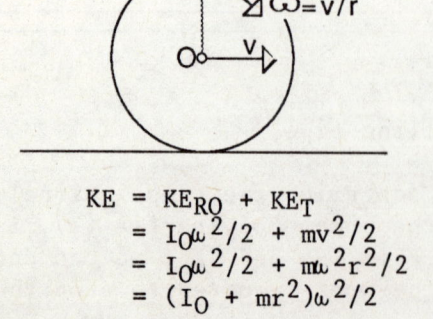

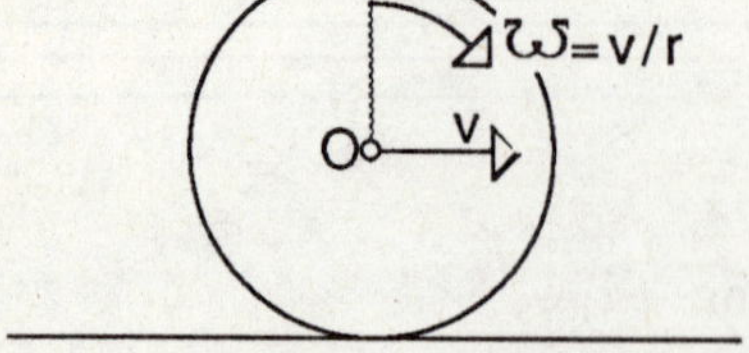


$$\begin{aligned} KE &= KE_{RO} + KE_T \\ &= I_O\omega^2/2 + mv^2/2 \\ &= I_O\omega^2/2 + m\omega^2r^2/2 \\ &= (I_O + mr^2)\omega^2/2 \end{aligned}$$

Figure 9.12

View 2
Rotation about P

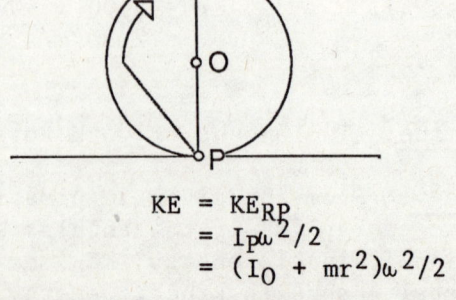

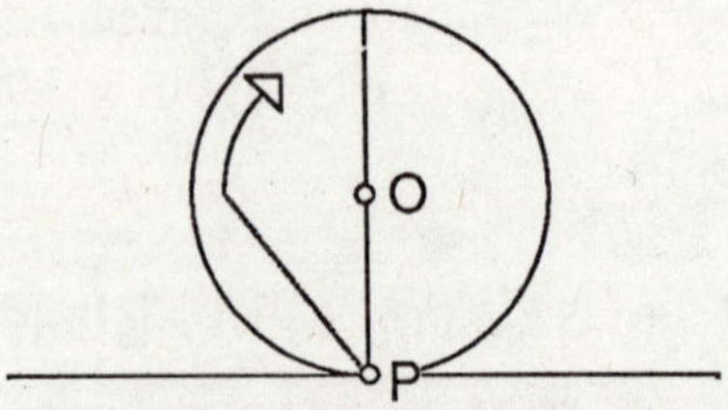


$$\begin{aligned} KE &= KE_{RP} \\ &= I_P\omega^2/2 \\ &= (I_O + mr^2)\omega^2/2 \end{aligned}$$

Figure 9.13

Notice that both viewpoints lead to the same answer.

Practice: A sphere of mass m and radius r rolls without slipping down an incline of length L and inclination θ, as shown in Fig. 9.14.

Figure 9.14

m = mass of sphere
r = radius of sphere
L = length of incline
θ = angle of inclination

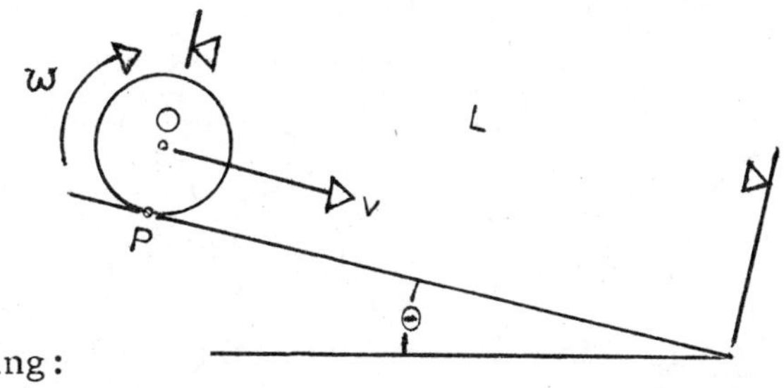


Determine an expression for the following:

| | |
|---|---|
| 1. Rotational moment of inertia of the sphere about O | $I_O = 2mr^2/5$ |
| 2. Rotational moment of inertia of the sphere about P | $I_P = I_O + mr^2 = 7mr^2/5$ |
| 3. Relationship between the linear speed of the sphere's CM and its angular speed about its CM | $v = \omega r$ |
| 4. Kinetic energy of the sphere at any time according to view 1 | $KE = KE_{RO} + KE_T$<br>$KE = I_O\omega^2/2 + mv^2/2 = 7mv^2/10$ |
| 5. Kinetic energy of the sphere at any time according to view 2 | $KE = KE_{RP} = I_P\omega^2/2 = 7mv^2/10$ |
| 6. Linear speed of the sphere's CM at the bottom of the incline | The change in KE of the sphere is equal to the work done on it by gravity. This work is the negative of the change in the sphere's potential energy.<br>$\Delta KE = KE_f - KE_i$, $KE_i = 0$<br>$\Delta KE = W_g = -\Delta PE = mgL \sin\theta$<br>$\Delta KE = 7mv_f^2/10$ by step 4 or 5<br>Combining the above, obtain<br>$v_f = [10gL \sin\theta/7]^{1/2}$ |
| 7. Time it takes the sphere to reach the bottom of the incline | $v_{av} = (v_i + v_f)/2 = L/t$<br>$t = 2L/v_f = [14L/5g \sin\theta]^{1/2}$ |

Example 9.6. The solid cylinder in Fig. 9.15 is supported by two cords wrapped tightly around it and attached to the ceiling. If the cylinder is released from rest, determine the linear acceleration of the CM.

Given and Diagram:

Figure 9.15 $v_o = 0$
$g = 9.80\ m/s^2$

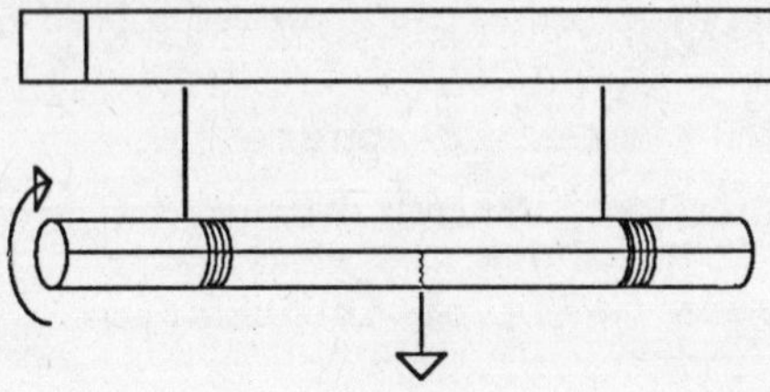

Determine: The linear acceleration of the cylinder's CM as it falls.

Strategy: Using view 1, we can state that the kinetic energy at any time is the translational kinetic energy of the CM plus the rotational kinetic energy about the CM. Using the work-energy theorem, we can also state that the change in kinetic energy of the cylinder is equal to the work done on it by gravity. These two statements can be combined to obtain the speed of the CM at any time. Finally we can use our knowledge of kinematics to determine the acceleration of the CM.

Solution: Using view 1, we write the kinetic energy at any time as

$$KE = mv^2/2 + I_O\omega^2/2 = mv^2/2 + (mr^2/2)(v/r)^2/2 = 3\ mv^2/4$$

The work done by gravity at any time is $W_g = F_gL = mgL$, where L is the distance that the gravitational force $F_g$ has acted at any time. According to the work-energy theorm, the net work done on the cylinder is equal to the change in its kinetic energy.

$$W_g = \Delta KE;\ \Delta KE = KE_f - KE_i;\ KE_f = 3mv_f^2/4;\ KE_i = 0;\ W_g = mgL$$

Substituting, we obtain

$$mgL = 3mv_f^2/4 \quad \text{or} \quad v_f^2 = 4gL/3$$

Finally, from our knowledge of kinematics, we write

$$v_f^2 - v_i^2 = 2a\Delta s;\ v_i = 0;\ v_f^2 = 4gL/3;\ \Delta s = L$$

which when combined gives

$$4gL/3 = 2aL \quad \text{or} \quad a = 2g/3 = 6.53\ m/s^2$$

Related Text Problems: 9-14 through 9-16, 9-19, 9-26, 9-27, 9-29 through 9-31.

## 6. Angular Impulse and Angular Momentum (Section 9.5)

Review: The angular momentum of a rotating system is given by $L = I\omega$. If a torque $\tau$ acts for a time $\Delta t$ on a rotating system, it imparts an angular impuse $\tau\Delta t$. The impulse-momentum theorem allows us to write

$$\tau\Delta t = \Delta L = \Delta(I\omega) = (I\omega)_f - (I\omega)_i$$

Practice: A child pushes with a force F tangential to the rim of a merry-go-round (m-g-r) of mass m and radius r for a time interval $\Delta t$ (Figure 9.16). The m-g-r has an initial angular speed of $\omega_i$.

Figure 9.16

$m$ = mass of m-g-r
$r$ = radius of m-g-r
$F$ = tangential force
$\Delta t$ = time of push
$\omega_i$ = initial angular speed of m-g-r

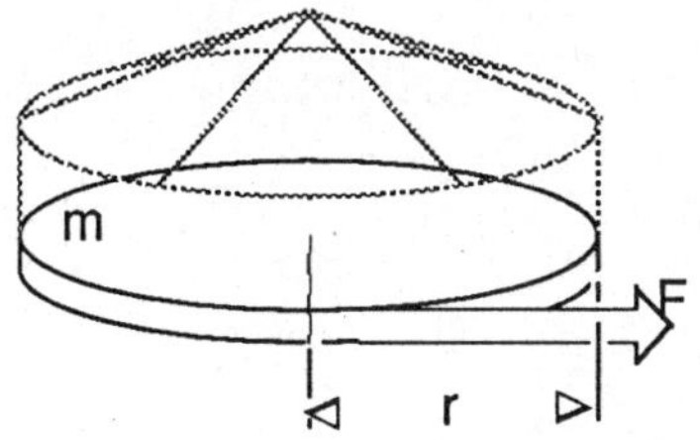


Determine an expression for the following:

| | |
|---|---|
| 1. Torque exerted on m-g-r by the child | $\tau = rF$ |
| 2. Angular impulse applied to the m-g-r by child | angular impulse $= \tau\Delta t = rF\Delta t$ |
| 3. Rotational moment of inertia of the m-g-r | $I = mr^2/2$ |
| 4. Initial angular momentum of the m-g-r | $L_i = I\omega_i = mr^2\omega_i/2$ |
| 5. Change in angular momentum of the m-g-r as a result of the child's push | $\Delta L$ = angular impulse $= rF\Delta t$ |
| 6. Final angular momentum of the m-g-r | $L_f = L_i + \Delta L$<br>$= 1/2mr^2\omega_i + rF\Delta t$ |
| 7. Final angular speed of the m-g-r | $L_f = I\omega_f = mr^2\omega_f/2$<br>$L_f = 1/2mr^2\omega_i + rF\Delta t$<br>Equating these two statements, we obtain<br>$\omega_f = \omega_i + 2F\Delta t/mr$ |

We can check our result by using a different approach to obtain $\omega_f$.

| | |
|---|---|
| 8. Angular acceleration of the m-g-r while the child is pushing | $\alpha = \tau/I = 2F/mr$ |
| 9. Final angular speed of the m-g-r | $\omega_f = \omega_i + \alpha\Delta t = \omega_i + 2F\Delta t/mr$ |

Note: In steps 7 and 9 of the preceding practice section, we obtained the same expression for the final angular speed using two different methods.

Example 9.7. A 1.00-kg disk with a 0.100-m radius is mounted as shown in Fig. 9.17. A lightweight cord is wrapped around the rim of the disk and pulled with a constant force of 10.0 N for 10.0 s. If the disk is initially at rest, determine its final angular speed.

Given and Diagram:

$m = 1.00$ kg
$r = 0.100$ m
Figure 9.17 $F = 10.0$ N
$\Delta t = 10.0$ s

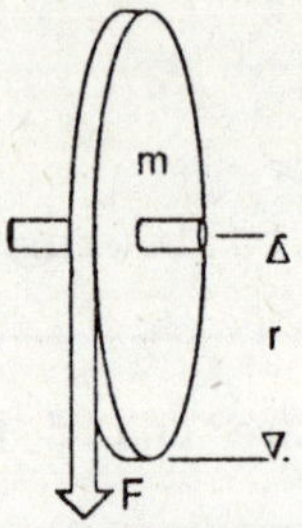


Determine: The angular speed of the rotating system after the 10.0-N force is applied for 10.0 s.

Strategy: Knowing r and F, we can determine the torque acting on the disk. Knowing the torque and the time it acts, we can determine the impulse applied to the disk and hence the change in angular momentum. Knowing M and r, we can determine the rotational moment of inertial of the system. We can determine the final angular speed from the rotational moment of inertia and the change in angular momentum.

Solution:

$$\tau = rF = (0.100 \text{ m})(10.0 \text{ N}) = 1.00 \text{ N}\cdot\text{m}$$

$$\text{Impulse} = \tau\Delta t = (1.00 \text{ N}\cdot\text{m})(10.0 \text{ s}) = 10.0 \text{ N}\cdot\text{m}\cdot\text{s} = \Delta L$$

$$I = mr^2/2 = (1.00 \text{ kg})(0.100 \text{ m})^2/2 = 5.00 \times 10^{-3} \text{ kg}\cdot\text{m}^2$$

$$\Delta L = \Delta(I\omega) = I\Delta\omega = I(\omega_f - \omega_i) = I\omega_f$$

$$\text{or } \omega_t = \Delta L/I = 10.0 \text{ N}\cdot\text{m}\cdot\text{s}/5.00 \times 10^{-3} \text{ kg}\cdot\text{m}^2 = 2.00 \times 10^3 \text{ rad/s}^2$$

Related Text Problem: 9-43.

## 7. Conservation of Angular Momentum (See 9.5)

Review: The rotational analog of the impulse-momentum theorem is given by $\tau\Delta t = \Delta L$. Notice when $\tau = 0$ then $\Delta L = 0$ and $L_i = L_f$, that is, angular momentum is conserved. The law of conservation of angular momentum states that the angular momentum of the system remains constant if the net external torque is negligible.

Practice: A merry-go-round, which can be considered to be a large flat disk of mass M and radius r, rotates about an axis through its center of mass with an angular speed $\omega_i$. A child of mass m stands watching and then jumps radially onto the edge of the m-g-r, as shown in Fig. 9.18.

Figure 9.18

$M = 150$ kg
$m = 25.0$ kg
$r = 1.50$ m
$\omega_i = 0.200$ rad/s

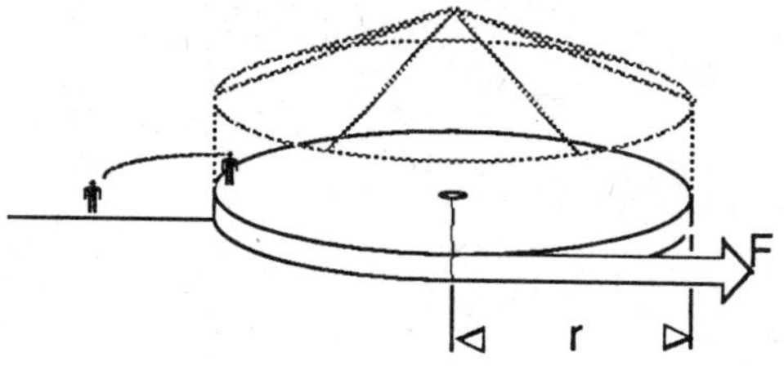


Determine the following:

| | |
|---|---|
| 1. Moment of inertia of the m-g-r about 0 | $I_M = Mr^2/2 = 169$ kg·m$^2$ |
| 2. Initial angular momentum of the m-g-r about 0 | $L_{Mi} = I_M\omega_i = 33.8$ kg·m$^2$/s |
| 3. Initial angular momentum of the child about 0 | $L_{ci} = 0$ since $\omega_{ci} = 0$ |
| 4. Initial angular momentum of system (m-g-r + child) about 0 | $L_i = L_{Mi} + L_{ci} = 33.8$ kg·m$^2$ |
| 5. Net torque the child exerts on the m-g-r as she jumps onto it | $\tau_{net} = 0$ |
| 6. Change in angular momentum of the system (m-g-r + child) | Since $\tau_{net} = 0$, then $\Delta L = 0$ and $L_f = L_i = 33.8$ kg·m$^2$/s |
| 7. Moment of inertia of the child after she jumps onto the m-g-r | $I_c = mr^2 = 56.3$ kg.m$^2$ |
| 8. Total moment of inertia of the final rotating system (m-g-r + child) | $I_f = I_M + I_c = 225$ kg·m$^2$ |
| 9. Final angular speed of the rotating system (m-g-r + child) | $\omega_t = L_t/I_t = 0.150$ rad/s |

Example 9.8. The 0.100-kg puck in Fig. 9.19 is sliding with a speed of 2.00 m/s in a circular orbit of radius 1.00 m on the horizontal frictionless surface. The centripetal force is provided by a student holding onto the cord. If the student pulls steadily downward on the string until the radius of the orbit is 0.500 m, determine the final speed of the puck and the work done by the student.

Given and Diagram:

Figure 9.19

$m = 1.00$ kg
$r_1 = 1.00$ m
$r_2 = 0.500$ m
$v_1 = 2.00$ m/s

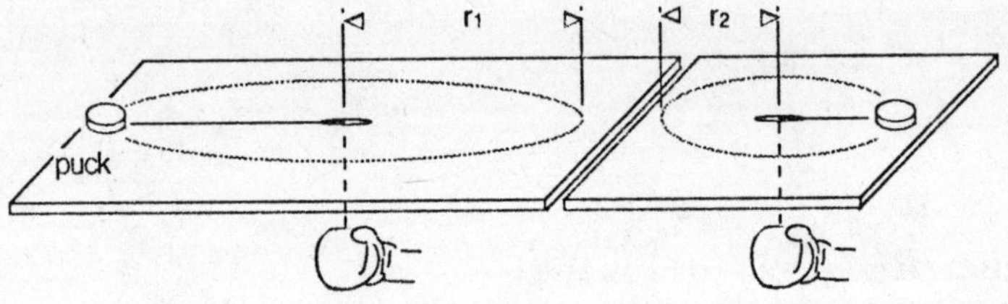


Determine: (a) the linear speed $v_2$ of the puck after the student pulls on the cord until $r_2 = r_1/2$ and (b) the work done by the student.

Strategy: The force acts in such a manner that no external torque is applied to the system; consequently, angular momentum is conserved. Knowing that angular momentum is conserved, the definition of angular momentum, and the relationship between linear and angular speed, we can determine $v_2$. Knowing the initial and final speed of the puck, we can determine its initial and final kinetic energy, the change in its kinetic energy, and hence the work done on it by the student.

Solution: Conserving angular momentum, we can write $L_2 = L_1$.

Substituting $L = I\omega$, $I = mr^2$, and $\omega = v/r$ we obtain

$$v_2 = r_1 v_1 / r_2 = 2v_1 = 4.00 \text{ m/s}$$

The work done by the student is equal to the change in kinetic energy of the puck.

$$W = \Delta KE = KE_2 - KE_1 = mv_2^2/2 - mv_1^2/2 = 3\,mv_1^2/2 = 0.600 \text{ J}$$

Related Text Problems: 9-37 through 9-42.

## 8. Both Energy and Torque Methods Can be Used to Solve Rotational Motion Problems (Sections 9.2, 9.3)

Review: In the review section for principal concept 2 of this chapter, we saw that $\tau_{net} = I\alpha$. In the review section for principal concept 4, we saw that $W_{net} = \Delta KE$. The concept introduced here is that both torque and energy methods can be used to solve rotational motion problems.

Practice: Two objects and a pulley, all of equal mass, are arranged as shown in Fig. 9.20. The horizontal surface is frictionless.

Figure 9.20

$m_1 = m_2 = m_p = m$
r = radius of pulley

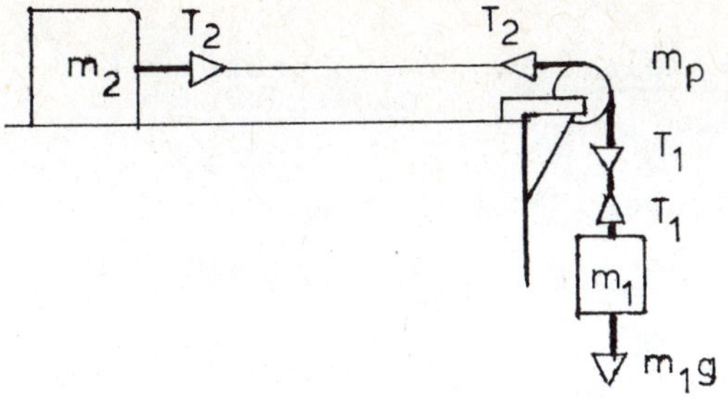


Determine an expression for the following:

| | |
|---|---|
| 1. Kinetic energy of the system at any time | $KE = m_1v^2/2 + m_2v^2/2 + I_p\omega^2/2$<br>$I_p = m_pr^2/2$, $\omega = v/r$<br>$KE = 5mv^2/4$ |
| 2. Work done by gravity when $m_1$ has been lowered by some distance h | $W_g = m_1gh = mgh$ |
| 3. Translational speed of $m_1$ and $m_2$ after $m_1$ has been lowerd by some distance h | $W_g = \Delta KE$; $W_g = mgh$; $\Delta KE = 5mv^2/4$<br>Substituting, we obtain<br>$v = 2[gh/5]^{1/2}$ |
| 4. Translational acceleration of $m_1$ and $m_2$ at any time | $v_f^2 - v_i^2 = 2a\Delta s$<br>$v_i = 0$; $v_f = 2[gh/5]^{1/2}$; $\Delta s = h$<br>Substituting, we obtain $a = 2g/5$ |

Note: The preceding expression foe acceleration was obtained using energy methods.

| | |
|---|---|
| 5. Tension $T_1$ | By Newton's second law<br>$m_1a = m_1g - T_1$ or $T_1 = m_1(g - a)$ |
| 6. Tension $T_2$ | By Newton's second law, $m_2a = T_2$ |
| 7. Net torque on $m_p$ | $\tau_{net} = r(T_1 - T_2) = rm(g - 2a)$ |
| 8. Translational acceleration of $m_1$ and $m_2$ at any time | $\tau_{net} = I\alpha$<br>$\tau_{net} = Rm(g - 2a)$; $I = mR^2/2$<br>$\alpha = a/R$<br>Substituting and solving for a, we obtain $a = 2g/5$ |

Note: The preceding expression for acceleration was obtained using torque methods. It is exactly the same as the Step 4 value obtained using energy methods.

Example 9.9. Two objects and a pulley are arranged as shown in Fig. 9-21.

Figure 9.21 $m_1 = 5.00$ kg $v_i = 0$
$m_2 = 2.00$ kg $\mu = 0.100$
$m_p = 1.00$ kg $g = 9.80$ m/s$^2$

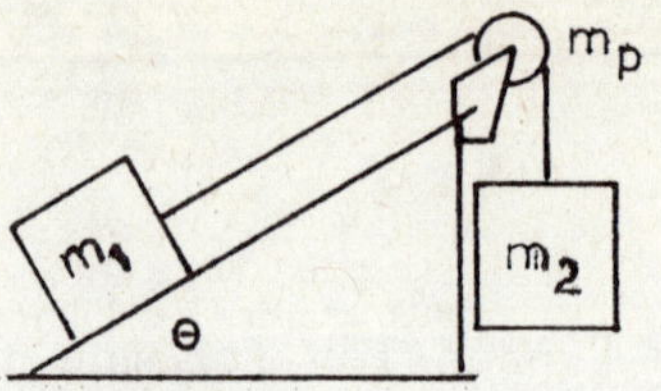


Calculate the translational acceleration of $m_1$ and $m_2$.

Determine: The translational acceleration of $m_1$ and $m_2$.

Strategy: Method I (energy method). First, obtain an expression for the kinetic energy of the system at any time. Second, obtain an expression for the net work done on the system when it moves some distance h. Finally, use the work-energy theorem to obtain an expression for v and your knowledge of kinematics to obtain an expression for the acceleration.

Method II (torque method). First, use Newton's second law to obtain expressions for the tensions $T_1$ and $T_2$. Second, obtain an expression for the net torque acting on the system. Finally, use $\tau = I\alpha$ to obtain an expression for v and your knowledge of kinematics to obtain an expression for the acceleration.

Solution: Method I. Obtain an expression for the kinetic energy of the system.

$$\Delta KE = KE_f - KE_i = KE_f = m_1v^2/2 + m_2v^2/2 + I_p\omega^2/2 = [m_1 + m_2 + m_p/2]\ v^2/2$$

Obtain an expression for the net work done on the system when it moves some distance h.

$$W_{net} = W_g + W_f = m_1gh - (m_2gh \sin\theta) - (\mu m_2gh \cos\theta) = [m_1 - m_2 \sin\theta - \mu m_2 \cos\theta]gh$$

Now equate $W_{net}$ and $\Delta KE$ and solve for $v^2$. Then use $v_f^2 - v_i^2 = 2a\Delta s$ to obtain

$$a = [m_1 - m_2 \sin\theta - \mu m_2 \cos\theta]g/[m_1 + m_2 + m_p/2] = 0.511g = 5.01 \text{ m/s}^2$$

Method II. Use Newton's second law to obtain an expression for the tensions $T_1$ and $T_2$.

$$m_1a = m_1g - T_1 \quad \text{or} \quad T_1 = m_1(g - a)$$

$$m_2a = T_2 - m_2g \sin\theta - \mu m_2g \cos\theta \quad \text{or} \quad T_2 = m_2(a + g \sin\theta + \mu g \cos\theta)$$

The net torque acting on the pulley is

$$\tau_{net} = r(T_1 - T_2) = r[m_1(g - a) - m_2(a + g \sin\theta + \mu g \cos\theta)]$$

We can obtain another expression for $\tau_{net}$:

$$\tau_{net} = I\alpha = (m_pr^2/2)a/r = m_par/2$$

Equating these expressions for $\tau_{net}$ and solving for the acceleration we obtain

$$a = [m_1 - m_2 \sin\theta - \mu m_2 \cos\theta]g/[m_1 + m_2 + m_p/2] = 5.01 \text{ m/s}^2$$

Related Text Problems: 9-13 through 9-16, 9-21, 9-22, 9-27, 9-30.

---

PRACTICE TEST

Take and grade this practice test. Doing so will allow you to determine any weak spots in your understanding of the concepts taught in this chapter. The following section prescribes what you should study further to strengthen your understanding.

A disk of radius r and mass m (Fig. 9.22) is rotating at an initial angular speed of $\omega_i$. A sphere of putty of mass m is dropped from a height h onto the disk at a location of r/2. The putty sticks to the disk at the point of contact.

Figure 9.22

$\omega_i$ = initial angular speed
m = mass of putty = mass of disk
r = radius of disk
h = height from which putty is dropped

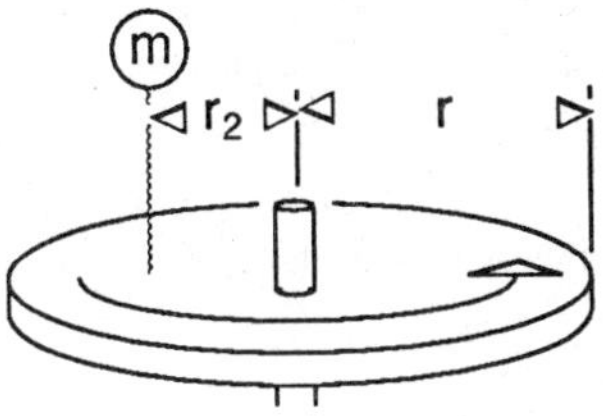


Develop expressions for the following physical quantities in terms of the known information only.

______ 1. The initial moment of inertia of the rotating system
______ 2. The initial angular momentum of the rotating system
______ 3. The initial kinetic energy of the rotating system
______ 4. The moment of inertia of the putty after it is dropped onto the disk
______ 5. The final moment of inertia of the rotating system
______ 6. The final angular momentum of the rotating system
______ 7. The final angular speed of the rotating system
______ 8. The final kinetic energy of the rotating system
______ 9. The change in the kinetic energy of the rotating system as the putty falls onto the disk
______ 10. What happened to this decrease in kinetic energy

A rigid body of irregular shape is mounted on an axle of radius r and negligible mass, as shown in Fig. 9.23. A string is wrapped around the axle, and a block of mass m is attached to it. The system is released from rest, and the block descends a distance h in a time t.

Figure 9.23

m = mass of block
r = radius of axle
h = distance block descends in time t
t = time block takes to descend the distance h
g = acceleration due to gravity

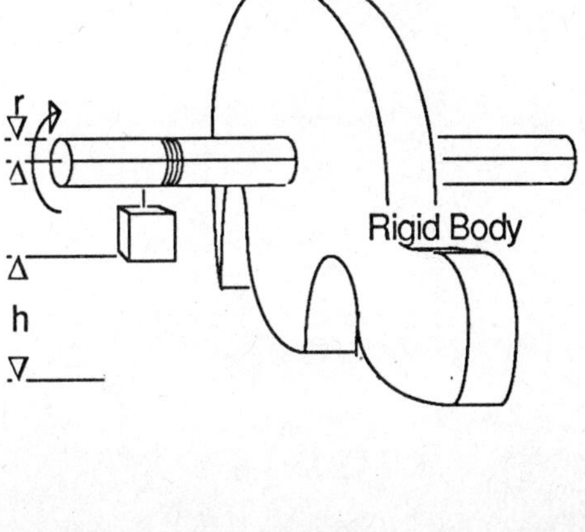

Determine expressions for the following physical quantities in terms of the known information only.

______ 11. The final linear speed of the block (that is, its speed after it has fallen the distance h)
______ 12. The final angular speed of the rigid body
______ 13. The final translational kinetic energy of the block
______ 14. The decrease in gravitational potential energy of the block as it descends the distance h
______ 15. The final rotational kinetic energy of the rigid body
______ 16. The moment of inertia of the rigid body obtained by using energy principles
______ 17. The linear acceleration of the block
______ 18. The angular acceleration of the rigid body
______ 19. The tension in the string that supports the block
______ 20. The torque tending to rotate the rigid body
______ 21. The moment of inertia of the rigid body obtained by using torque principles

(See Appendix I for answers)

## PRINCIPAL CONCEPTS AND EQUATIONS PRESCRIPTION

Your score on the practice is an excellent measure of your understanding of this chapter. You should now use the following chart to write your own prescription for curing any of your physics ills. Look down the leftmost column to the number of the question(s) you answered incorrectly, read across that row to see which section(s) of the study guide you should return to for further study, and then do the suggested text problems to gain additional experience in working with the particular concept.

| Practice Test Question | Concepts and Equations | Prescription: Principal Concept | Prescription: Text Problems |
|---|---|---|---|
| 1 | Moment of inertia | 1 | 9-2a,3 |
| 2 | Angular momentum | 6 | 9-32,33 |
| 3 | Rotational kinetic energy | 4 | 9-13,14 |
| 4 | Moment of inertia | 1 | 9-3,11a |
| 5 | Moment of inertia | 1 | 9-4,35a |
| 6 | Conservation of angular momemtum | 6 | 9-37,38 |
| 7 | Angular momentum | 6 | 9-33,34 |
| 8 | Rotational kinetic energy | 4 | 9-15,16 |
| 9 | Rotational kinetic energy | 4 | 9-19,21 |
| 10 | Inelastic collision | 3 of Ch. 8 | 8-35 |
| 11 | Equations used to analyze translational motion | 5 of Ch. 2 | 2-19,20 |
| 12 | Relationship between linear and angular quantities | 2 of Ch. 5 | 5-5,11 |
| 13 | Translational kinetic energy | 3 of Ch. 7 | 7-18,19 |
| 14 | Gravitational potential energy | 4 of Ch. 7 | 7-22,27 |
| 15 | Rotational kinetic energy | | |
| 16 | Work-energy theorem | 3 of Ch. 7 | 7-19,20 |
| 17 | Equations used to analyze translational motion | 5 of Ch. 2 | 2-19,20 |
| 18 | Relationship between linear and angular quantities | 2 of Ch. 5 | 5-16c,17a |
| 19 | Newton's second law | 2 of Ch. 4 | 4-3,6 |
| 20 | Torque | 2 of Ch. 6 | 6-12,16 |
| 21 | Relationship between torque and angular acceleration | 2 | 9-6,7 |

# 10 Simple Harmonic Motion

RECALL FROM PREVIOUS CHAPTERS

| Previously learned concepts and equations frequently used in this chapter | Text Section | Study Guide Page |
|---|---|---|
| The relationship between linear and angular speed: $v = r\omega$ | 5.3 | 68 |
| Centripetal acceleration: $a_c = v^2/r$ | 5.5 | 73 |
| Kinetic energy: $KE = mv^2/2$ | 7.5 | 108 |

NEW IDEAS IN THIS CHAPTER

| Concepts and equations introduced | Text Section | Study Guide Page |
|---|---|---|
| The condition for SHM: $\vec{a} = -k\vec{x}/m$ | 10.3 | 162 |
| Relationships between angular frequency $\omega$, linear frequency f, and period of oscillation $\tau$: $\omega = (k/m)^{1/2}$; $f = \omega/2\pi = (k/m)^{1/2}/2\pi$; $\tau = 1/f = 2\pi/\omega = 2\pi(m/k)^{1/2}$ | 10.4 | 163 |
| Equations to describe SHM: $x = A\cos(\omega t + \theta_o)$; $v = -\omega A\sin(\omega t + \theta_o)$; $a = -\omega^2 A\cos(\omega t + \theta_o)$ | 10.4 | 166 |
| Mechanical energy in SHM: Kinetic energy: $KE = mv^2/2$; Potential energy: $PE = kx^2/2$; Total energy: $E = KE + PE = kA^2/2$ | 10.5 | 173 |

PRINCIPAL CONCEPTS AND EQUATIONS

1. Condition for Simple Harmonic Motion (Sections 10.3, 10.4)

Review: Figure 10.1a shows a mass m resting on a horizontal frictionless surface and attached to an unstretched spring that has a spring constant k. In Fig. 10.1b, the mass is displaced by an amount $+\vec{x}$. Since the spring is stretched an amount $\vec{x}$, it exerts a restoring force $\vec{F} = -k\vec{x}$. The negative sign is an explicit reminder that the restoring force and the displacement are in opposite directions. According to Newton's second law, the unbalanced force acting on the mass is equal to the product of the mass and its acceleration (i.e. $\vec{F}_u = m\vec{a}$). In this case, the unbalanced force acting on the mass is the restoring force of the spring ($\vec{F}_u = -k\vec{x}$). Combining these expressions for the unbalanced force and solving for the acceleration, we obtain $\vec{a} = -k\vec{x}/m$. Anytime an object's motion is such that the acceleration is proportional to and in the opposite direction of the displacement, that object is executing simple harmonic motion.

Figure 10.1

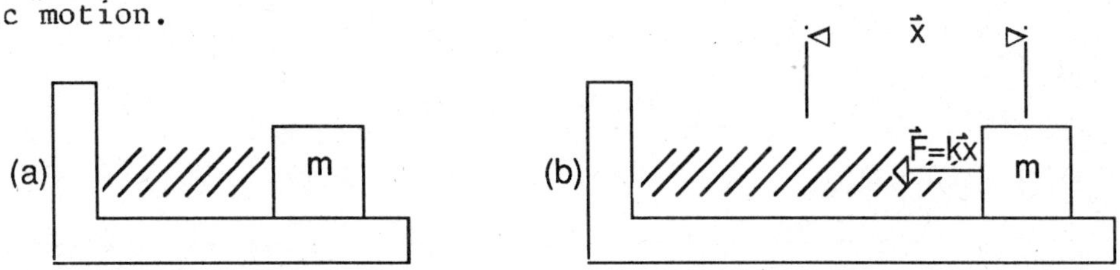


If it is determined that an object of mass m attached to a spring of constant k is executing SHM, its angular frequency $\omega$, linear frequency f, and period of oscillation $\tau$ are given by the following expressions.

angular frequency: $\omega = (k/m)^{1/2}$

linear frequency: $f = \omega/2\pi = (k/m)^{1/2}/2\pi$

period of oscillation: $\tau = 1/f = 2\pi/\omega = 2\pi(m/k)^{1/2}$

Practice: Figure 10.2a shows a 2.00-kg mass attached to a 0.200-m-long spring and resting on a horizontal frictionless surface. Figure 10.2b shows that when you pull on the mass-spring system with a 40.0-N force, you can stretch the spring by 0.100 m.

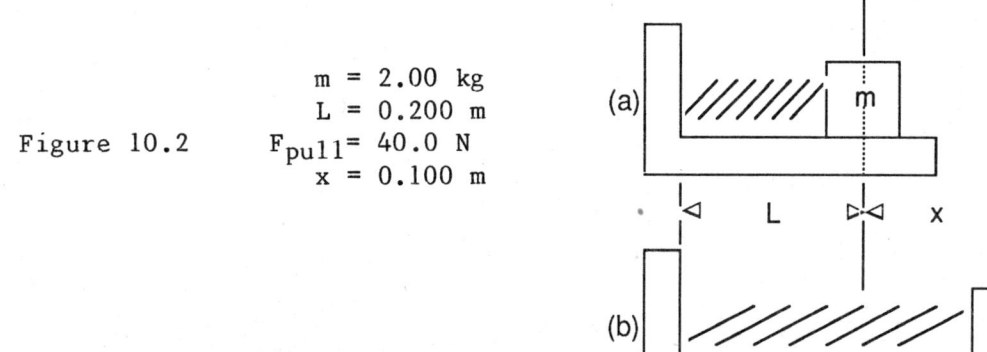

Figure 10.2

$m = 2.00$ kg
$L = 0.200$ m
$F_{pull} = 40.0$ N
$x = 0.100$ m

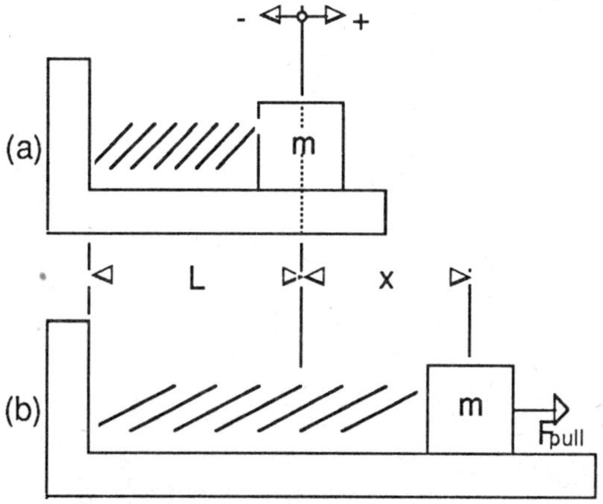


Determine the following:

| | |
|---|---|
| 1. The restoring force $F_r$ exerted on the mass by the spring while you are exerting the 40.0-N force on the mass | Since the mass is in equilibrium (as long as you hold onto it), we can write $F_r = -F_{pull} = -40.0$ N. The minus sign tells us that the restoring force is to the left. |
| 2. The spring constant | $k = -F_r/x$<br>$= -(-40.0\ \text{N})/(0.100\ \text{m}) = 400$ N/m |

| | |
|---|---|
| 3. Acceleration of the mass the instant it is released | $a = -kx/m = -20.0\ m/s^2$ |
| 4. Angular frequency of the oscillating system | $\omega = (k/m)^{1/2} = 14.1$ rad/s |
| 5. Linear frequency of the oscillating system | $f = \omega/2\pi = 2.24$ Hz |
| 6. Period of oscillation | $\tau = 1/f = 0.446$ s |
| 7. Time for 10 complete oscillations | $t = 10\ \tau = 4.46$ s |
| 8. Force you would have to supply to compress the spring to a length of 0.050 m | For this case, $x = -0.150$ m<br><br>$F_{pull} = -F_r = -(-kx) = -60.0$ N<br><br>The minus sign confirms that you would have to push to the left in order to compress the spring. |
| 9. Acceleration of the mass when released from the situation described in step 8 | $a = -kx/m = +30.0\ m/s^2$ or<br>$a = F_r/m = +30.0\ m/s^2$<br><br>The positive sign tells us that the mass will accelerate to the right when released. |
| 10. Period of oscillation of the system when released from the situation described in step 8. | $\tau = 0.446$ s<br>Same as in step 6. Since m and k are not changed, the period is not changed [$\tau = 2\pi(m/k)^{1/2}$]. |

Example 10.1. A 0.400-kg mass suspended from a spring stretches the spring 0.100 m. A 0.600-kg mass is attached to the 0.400-kg mass, and the spring is set into oscillation on a horizontal frictionless surface by stretching it 0.150 m and then releasing it. Determine the maximum horizontal acceleration of the combined mass on the end of the spring and the period of oscillation.

Given: $m_1 = 0.400$ kg $m_2 = 0.600$ kg $x_1 = 0.100$ m $x_2 = 0.150$ m

Determine: The maximum acceleration and period of oscillation of the 1.00-kg mass $(m_1 + m_2)$ on the end of the spring after the spring is stretched horizontally 0.150 m and then released.

Strategy: Knowing $m_1$, we can determine the force needed to stretch the spring a distance $x_1$. When we know this force, we also know the restoring force exerted by the spring when it is stretched a distance $x_1$. Knowing the restoring force and the stretch, we can determine the spring constant k. Knowing that k does not change when the spring is placed in a horizontal position, the total mass attached to the spring, and the maximum horizontal stretch, we can determine the maximum acceleration of the combined mass and its period of oscillation.

Solution: Figure 10.3 shows the different situations involved in the problem.

Figure 10.3

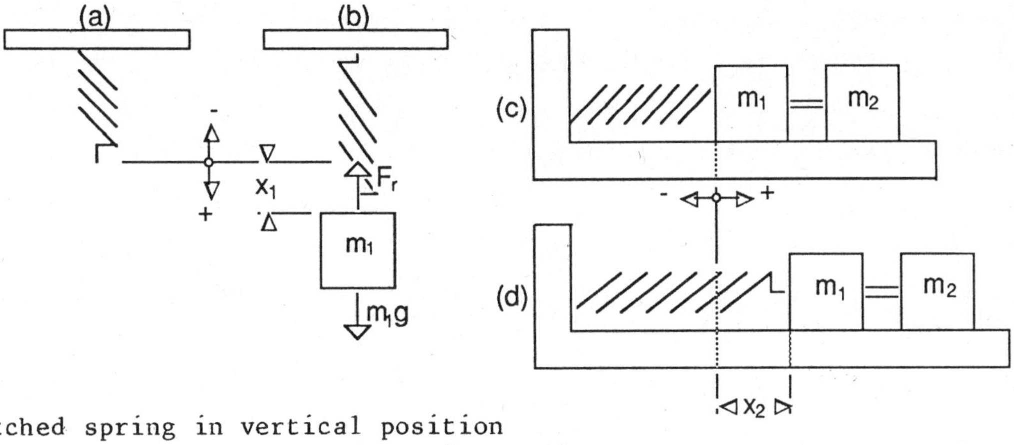


(a) Unstretched spring in vertical position
(b) Suspended mass $m_1$ stretches the spring a distance $x_1$
(c) Unstretched spring in a horizontal position with mass $m_1 + m_2$ attached
(d) Maximum horizontal displacement from equilibrium with mass $m_1 + m_2$ attached

From figure 10.3b, we see that after $m_1$ is suspended from the spring, a new equilibrium is established. As a result, we can write

$$F_r = -m_1g \quad \text{and} \quad F_r = -kx_1$$

Combining these expressions for the restoring force and solving for k, we obtain

$$k = m_1g/x_1 = (0.400 \text{ kg})(9.80 \text{ m/s}^2)/(0.100 \text{ m}) = 39.2 \text{ N/m}$$

From Fig. 10.3d, we see that the maximum displacement from equilibrium is $x_2$. Hence the maximum acceleration of the combined mass is

$$a = -kx_2/(m_1 + m_2) = -(39.2 \text{ N/m})(0.150 \text{ m})/(1.00 \text{ kg}) = -5.88 \text{ m/s}^2$$

The period of oscillation is

$$\tau = 2\pi[(m_1 + m_2)/k]^{1/2} = 2\pi(1.00 \text{ kg}/39.2 \text{ N/m})^{1/2} = 1.00 \text{ s}$$

Related Text Problems: 10-6a, 10-7b, c, 10-9, 10-10, 10-15, 10-16.

## 2. Describing Simple Harmonic Motion (Section 10.4)

Review: A complete description of an object's motion specifies its displacement, velocity, and acceleration at all times. Figure 10.4 summarizes the use of a reference circle to obtain expressions for the displacement, velocity, and acceleration of a mass undergoing SHM. Since this motion takes place in one dimension, we use plus and minus to denote the direction of these vector quantities.

If the angular speed of a reference particle on the reference circle is related to the spring-mass system by $\omega^2 = k/m$, the following is true about the x component of the motion of the the reference particle.

The displacement of the mass in SHM is the same as the x component of the displacement of the particle traveling on the reference circle.

$$x = A\cos\theta = A\cos\omega t$$

The velocity of the mass is the same as the x component of the velocity of the reference particle.

$$v = V_x = -V\sin\theta = -\omega A\sin\omega t$$

The acceleration of the mass is the same as the x component of the centripetal acceleration of the reference particle.

$$a = a_{cx} = -a_c\cos\theta = -\omega^2 A\cos\omega t$$

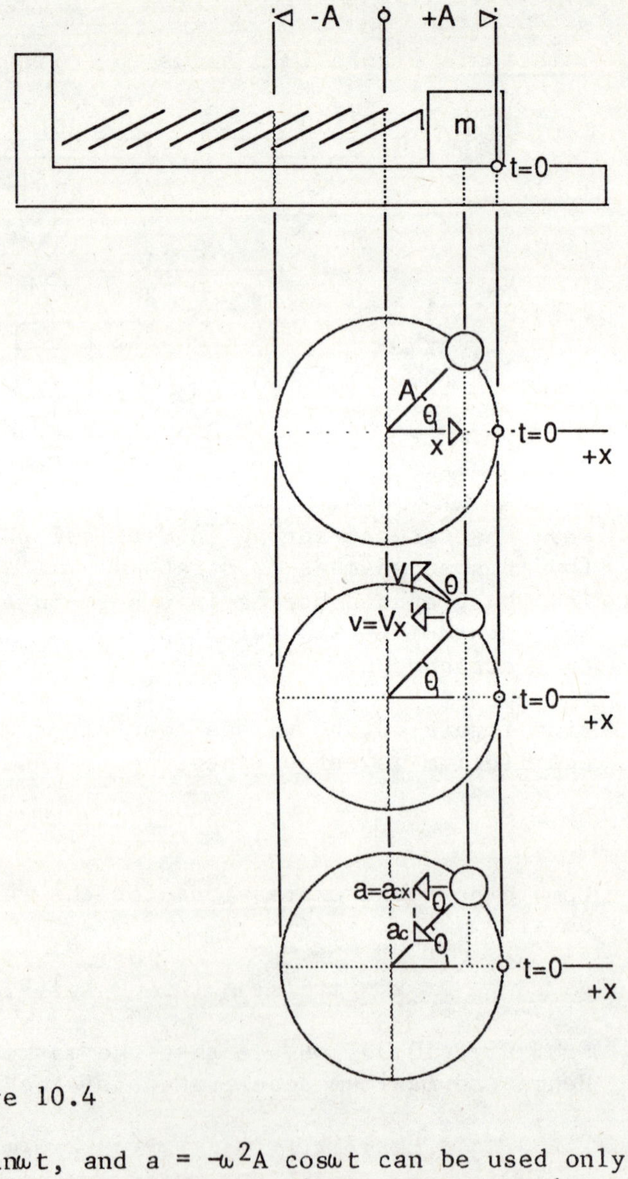


Figure 10.4

The expressions $x = A\cos\omega t$, $v = -\omega A\sin\omega t$, and $a = -\omega^2 A\cos\omega t$ can be used only for the case shown in Fig. 10.4, (that is, only when the displacement has its positive maximum value at $t = 0$). These expressions for x, v, and a can be made more general (that is, modified to allow $t = 0$ at any position on the reference circle) by incorporating an initial phase angle $\theta_o$:

$x = A\cos(\omega t + \theta_o)$ $\qquad v = -\omega A\sin(\omega t + \theta_o)$ $\qquad a = -\omega^2 A\cos(\omega t + \theta_o)$

where $\theta_o$ locates the particle on the reference circle relative to the positive x axis at $t = 0$.

Practice: Figure 10.5a shows a 2.00-kg mass resting on a horizontal frictionless surface and attached to a spring of constant $k = 400$ N/m and length $L = 0.200$ m. Figure 10.5b shows the object displaced +0.100 m and released at $t = 0$.

Figure 10.5

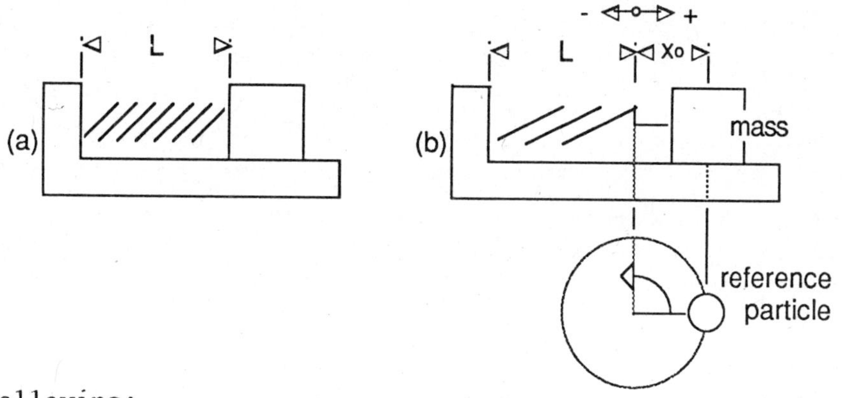


Determine the following:

| | |
|---|---|
| 1. Radius of the reference circle | $A = x_{max} = 0.100$ m |
| 2. Angular speed of the reference particle | $\omega^2 = k/m = 200\ rad^2/s^2$<br>$\omega = 14.1$ rad/s |
| 3. Linear speed of the reference particle | $V = \omega A = 1.41$ m/s |
| 4. Oscillation frequency of the mass and rotation frequency of the reference particle | $\omega = 2\pi f$<br>$f = \omega/2\pi = 2.25$ Hz |
| 5. Period of oscillation of the mass and period of rotation of the reference particle | $\tau = 1/f = 0.444$ s |
| 6. Magnitude of the reference particle's displacement, velocity, and acceleration at any time | $r = A = 0.100$ m<br>$V = \omega A = 1.41$ m/s<br>$a_c = \omega^2 A = 20.0$ m/s |

| | |
|---|---|
| 7. x component of the reference particle's displacement, velocity, and acceleration at t = 0 | $x = +0.100$ m<br>$V_x = 0$<br>$a_{cx} = -\omega^2 A = -20.0$ m/s<br>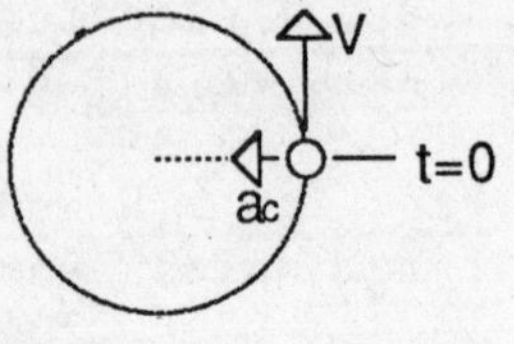<br> |
| 8. Displacement, velocity and acceleration of the mass at t = 0 | $x = A\cos\omega t = +A = +0.100$ m<br>$v = -\omega A\sin\omega t = 0$<br>$a = -\omega^2 A\cos\omega t = -\omega^2 A = -20.0\ \text{m/s}^2$ |
| 9. x component of the reference particle's displacement, velocity, and acceleration at t = T/4 | $x = 0$<br>$V_x = -\omega A = -14.1$ m/s<br>$a_{cx} = 0$<br>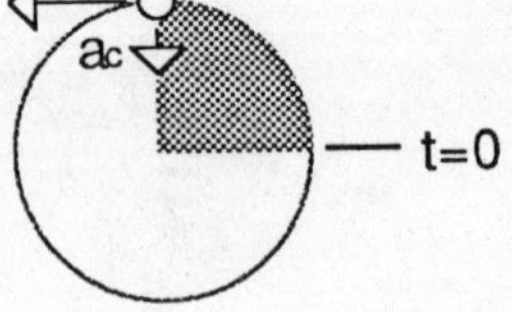<br> |
| 10. Displacement, velocity, and acceleration of the mass at t = T/4 | $x = A\cos\omega t = A\cos\frac{2\pi}{T}\frac{T}{4} = 0$<br>$v = -\omega A\sin\omega t = -\omega A\sin\frac{2\pi}{T}\frac{T}{4} = -1.41$ m/s<br>$a = -\omega^2 A\cos\omega t = -\omega^2 A\cos\frac{2\pi}{T}\frac{T}{4} = 0$ |
| 11. x component of the reference particle's displacement, velocity, and acceleration at t = T/2 | $x = -A = -0.100$ m<br>$V_x = 0$<br>$a_{cx} = \omega^2 A = 20.0\ \text{m/s}^2$<br>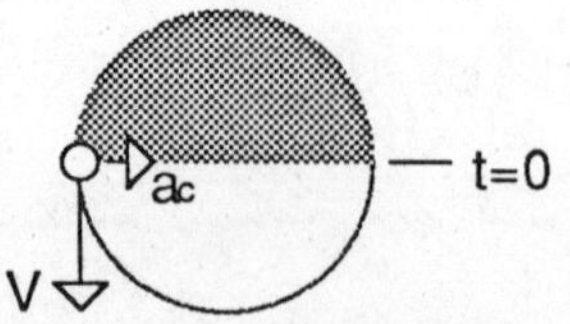<br> |
| 12. Displacement, velocity, and acceleration of the mass at t = T/2 | $x = A\cos\omega t = A\cos\frac{2\pi}{T}\frac{T}{2} = -0.100$ m<br>$v = -\omega A\sin\omega t = -\omega A\sin\frac{2\pi}{T}\frac{T}{2} = 0$<br>$a = -\omega^2 A\cos\omega t = -\omega^2 A\cos\frac{2\pi}{T}\frac{T}{2} = +20.0\ \text{m/s}^2$ |
| 13. x component of the reference particle's displacement, velocity, and acceleration at t = 3T/4 | $x = 0$<br>$V_x = +\omega A = +1.41$ m/s<br>$a_{cx} = 0$<br>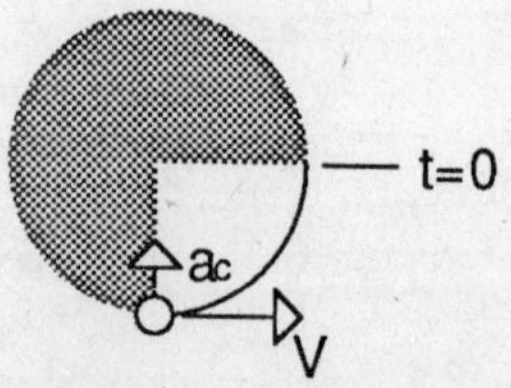<br> |

| | |
|---|---|
| 14. Displacement, velocity, and acceleration of the mass at $t = 3T/4$ | If $t = 3T/4$, then $\omega t = \frac{2\pi}{T}\frac{3T}{4} = \frac{3\pi}{2}$<br>$x = A\cos\omega t = A\cos 3\pi/2 = 0$<br>$v = -\omega A\sin\omega t = -\omega A\sin 3\pi/2 = 1.41$ m/s<br>$a = -\omega^2 A\cos\omega t = -\omega^2 A\cos 3\pi/2 = 0$ |
| 15. Displacement, velocity, and acceleration of the mass at $t = 0.05$ s | $x = A\cos\omega t$<br>$= (0.100\text{ m})\cos[(14.1\text{ rad/s})(0.050\text{ s})]$<br>$= 0.072$ m<br><br>$v = -\omega A\sin\omega t$<br>$= -(1.41\text{ m/s})\sin[(14.1\text{ rad/s})0.0500\text{ s})]$<br>$= -0.914$ m/s<br><br>$a = -\omega^2 A\cos\omega t$<br>$= -(20.0\text{ m/s}^2)\cos[(14.1\text{ rad/s})(0.050\text{ s})]$<br>$= -15.2\text{ m/s}^2$ |
| 16. Displacement, velocity, and acceleration of the mass when the angle on the reference circle is 120° | $\omega t = \theta = 120° = 2\pi$ rad/3<br>$x = A\cos\omega t = A\cos(2\pi\text{ rad/3})$<br>$= -5.00 \times 10^{-2}$ m<br>$v = -\omega A\sin\omega t = -\omega A\sin(2\pi\text{ rad/3})$<br>$= -1.22$ m/s<br>$a = -\omega^2 A\cos\omega t = -\omega^2 A\cos(2\pi\text{ rad/3})$<br>$= +10.0\text{ m/s}^2$ |

Note: The preceding practice section shows that the motion of an oscillating mass is the same as the x component of the motion of a particle on the reference circle. For example, compare your answer to step 7 with your answer to step 8, step 9 with step 10, 11 with 12, and 13 with 14.

Figure 10.6a shows a 2.00-kg object resting on a horizontal frictionless surface and attached to a spring of constant $k = 400$ N/m and length $L = 0.200$ m. The object is displaced +0.100 m (Fig. 10.6b) and released. After its release, the object undergoes SHM, and the particle on the reference circle has an angular speed of $\omega \sqrt{k/m} = 14.1$ rad/s.

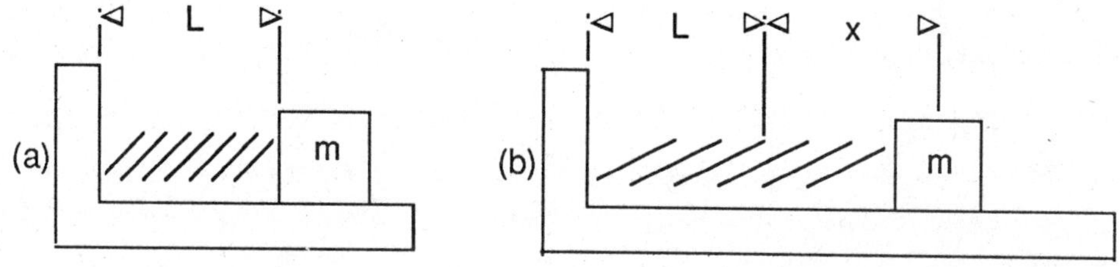

Figure 10.6

Case I. The time record is started ($t = 0$) the first time the oscillating object goes through the equilibrium position. Determine the following:

| | |
|---|---|
| 17. The initial phase angle | $\theta_o = 90° = \pi$ rad/2 |
| 18. The x component of the displacement, velocity, and acceleration of a particle on the reference circle | $x = 0$<br>$V_x = -\omega A = (k/A)^{1/2}A = -1.41$ m/s<br>$a_{cx} = 0$ |
| 19. The displacement, velocity, and acceleration of the oscillating object at t = 0 | $x = A\cos(\omega t + \theta_o) = A\cos 90° = 0$<br>$v = -\omega A\sin(\omega t + \theta_o) = -\omega A\sin 90° = -1.41$ m/s<br>$a = -\omega^2 A\cos(\omega t + \theta_o) = -\omega^2 A\cos 90° = 0$ |
| 20. The x component of the displacement, velocity, and acceleration of the reference particle at t = 3T/4 | $x = +A = +0.100$ m<br>$V_x = 0$<br>$a_{cx} = -\omega^2 A = -20.0$ m/s$^2$ |
| 21. The displacement, velocity and acceleration of the oscillating object at t = 3T/4 | If $t = 3T/4$, then $\omega t = \frac{2\pi}{T}\frac{3T}{4} = 3\pi/2$<br>If $\theta_o = \pi/2$, then $\omega t + \theta_o = 2\pi$<br><br>$x = A\cos(\omega t + \theta_o) = A\cos 2\pi = +0.100$ m<br>$v = -\omega A\sin(\omega t + \theta_o) = -\omega A\sin 2\pi = 0$<br>$a = -\omega^2 A\cos(\omega t + \theta_o) = -\omega^2 A\cos 2\pi = -20.0$ m/s$^2$ |

Case II. The time record is started (t = 0) when the oscillating object has its maximum negative displacement. Determine the following:

| | |
|---|---|
| 22. The initial phase angle | $\theta_o = 180° = \pi$ rad |
| 23. The x component of the displacement, velocity, and acceleration of the reference particle at t = 0 | $x = -A = -0.100$ m<br>$V_x = 0$<br>$a_{cx} = \omega^2 A = +20.0$ m/s$^2$ |
| 24. The displacement, velocity, and acceleration of the oscillating object at t = 0 | $x = A\cos(\omega t + \theta_o) = A\cos\pi = 0.100$ m<br>$v = -\omega A\sin(\omega t + \theta_o) = -\omega A\sin\pi = 0$<br>$a = -\omega^2 A\cos(\omega t + \theta_o) = -\omega^2 A\cos\pi = +20.0$ m/s$^2$ |

| | |
|---|---|
| 25. The x component of the displacement, velocity, and acceleration of the reference particle at t = T/4 | $x = 0$<br>$V_x = +\omega A = +1.41$ m/s<br>$a_{cx} = 0$ |
| 26. The displacement, velocity and acceleration of the oscillating object at t = T/4 | If $t = T/4$, then $\omega t = \frac{2\pi}{T}\frac{T}{4} = \frac{\pi}{2}$<br>If $\theta_o = \pi$, then $\omega t + \theta_o = 3\pi/2$<br>$x = A\cos(\omega t + \theta_o) = A\cos(3\pi/2) = 0$<br>$v = -\omega A\sin(\omega t + \theta_o) = -\omega A\sin(3\pi/2)$<br>$= 1.41$ m/s<br>$a = -\omega^2 A\cos(\omega t + \theta_o) = -\omega^2 A\cos(3\pi/2)$<br>$= 0$ |

Example 10.2. The SHM of a 0.200-kg mass attached to a spring and oscillating horizontally on a frictionless surface is described by $x = (0.200\text{ m})\cos(2\pi t + \pi/2)$. At time t = 0.125 s, what is the mass's (a) displacement; (b) velocity; (c) acceleration?

Given: $x = (0.200\text{ m})\cos(2\pi t + \pi/2)$ $\quad\omega = 2\pi$ rad/s $\quad A = 0.200$ m
$m = 0.200$ k $\quad\theta_o = (\pi/2)$rad $\quad t = 0.125$ s

Note: Values of A, $\omega$, and $\theta_o$ are determined by comparing the expression for x with $x = A\cos(\omega t + \theta_o)$.

Determine: The (a) displacement; (b) velocity; (c) acceleration of the mass at t = 0.125 s.

Strategy: First, by using $\theta = (\omega t + \theta_o)$, we can determine the location of the particle on the reference circle and get some idea of what to expect for values of x, v, and a. Second, we can insert the value for $(2\pi t + \pi/2)$ and other given quantities into the appropriate expressions to determine x, v, and a. Finally, we can verify our work by checking the signs on x, v, and a to see if they are correct.

Solution: Let's first determine the location of the particle on the reference circle in order to get some idea of what to expect for values of x, v, and a.

$\theta = (\omega t + \theta_o)$
$\theta = [(2\pi \text{ rad/s})(0.125\text{ s}) + (\pi/2)\text{ rad}]$
$\theta = (3\pi/4)$ rad

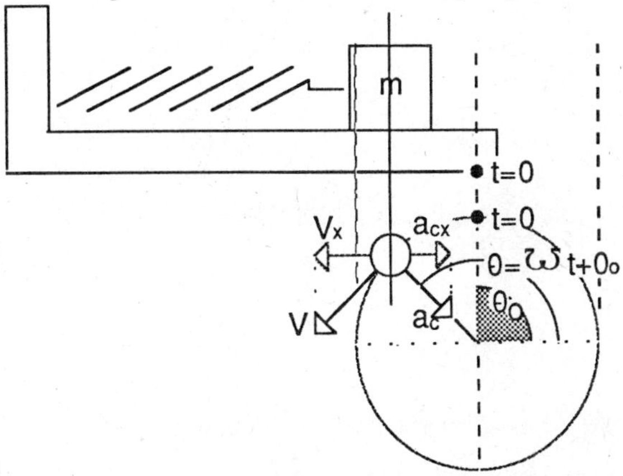


Figure 10.7

From Fig. 10.7, we see that at $t = 0.125$ s, x and v are negative and a is positive.

Second, let's insert the value for $(2\pi t + \pi/2)$ and other given quantities into expressions for x, v, and a.

(a) $x = (0.200\text{ m})\cos(2\pi t + \pi/2) = (0.200\text{ m})\cos(3\pi/4) = -0.141\text{ m}$
(b) $v = -\omega A\sin(2\pi t + \pi/2) = -(2\pi)(0.200\text{ m})\sin(3\pi/4) = -0.889\text{ m/s}$
(c) $a = -\omega^2 A\cos(2\pi t + \pi/2) = -(2\pi)^2(0.200\text{ m})\cos(3\pi/4) = +5.58\text{ m/s}^2$

Notice that the signs on x, v, and a agree with what we established using the reference circle. Finally as, a further check on our work, let's calculate the acceleration using another expression.

$$a = -\omega^2 x = -(2\pi)^2(-0.141\text{ m}) = +5.58\text{ m/s}^2$$

Example 10.3. A mass resting on a horizontal frictionless surface is connected to a fixed spring. The mass is displaced +0.200 m from its equilibrium position and released. The first time the oscillating mass goes through the equilibrium position with a positive speed, a record of time is started (i.e., $t = 0$). If at $t = 0.500$ s the displacement is $x = +0.100$ m, what is the period of the oscillating mass?

Given: $A = 0.200$ m
$x = +0.100$ m at $t = 0.500$ s
$t = 0$ when $x = 0$ m and $v = +\text{max}$

Determine: The period of the oscillating mass

Note: The value for t is derived from the fact that the time record starts the first time the oscillating mass goes through the equilibrium position with a positive speed.

Strategy: Use a reference circle to establish the initial phase angle $\theta_o$. Use the expression for the position of the oscillating mass at any time to determine the value for $\omega t$. Once the value for $\omega t$ is known, the value for $\omega$ and, subsequently, the value for $\tau$ can be determined.

Solution: Establish a value for $\theta_o$ with the aid of a reference circle.

Figure 10.8 $\theta_o = (3\pi/2)$ rad

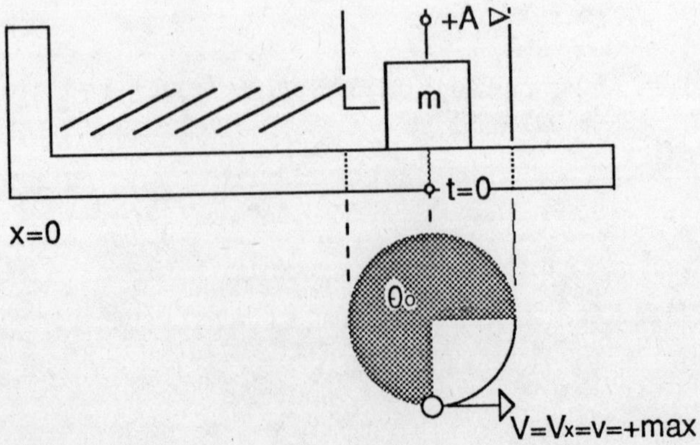

Use the expression $x = A\cos(\omega t + \theta_o)$ to determine a value for $\cos(\omega t + 3\pi/2)$.

$$0.100 \text{ m} = (0.200 \text{ m})\cos(\omega t + 3\pi/2) \quad \text{or} \quad \cos(\omega t + 3\pi/2) = +0.500$$

Obtain a value for $(\omega t + 3\pi/2)$ by taking the arccos of this last expression.

$$(\omega t + 3\pi/2) = \pi/3 \quad \text{or} \quad 5\pi/3$$

We choose the value $5\pi/3$ because the value $\pi/3$ results in a negative value for $\omega$ and hence $\tau$. Knowing a value for $(\omega t + \pi/2)$, we can determine a value for $\omega t$ and hence $\tau$.

$$\omega t + (3\pi/2) \text{ rad} = (5\pi/3) \text{ rad} \quad \text{or} \quad \omega t = (\pi/6) \text{ rad}$$

using $t = 0.500$ s, we obtain $\omega = (\pi/3)$rad/s

using $\omega = 2\pi/\tau$, we obtain $\tau = 6.00$ s

Related Text Problems: 10-1 through 10-6, 10-8, 10-12, 10-13, 10-14, 10-17, 10-18.

## 3. Mechanical Energy in SHM (Section 10.5)

Review: Figure 10.9 shows a mass-spring system at rest on a horizontal frictionless surface at $t = 0$. When the mass is released, it undergoes SHM about the equilibrium position.

Figure 10.9

$m$ = mass of oscillating object
$k$ = spring constant
$A$ = amplitude of oscillation
$\theta_o$ = initial phase angle

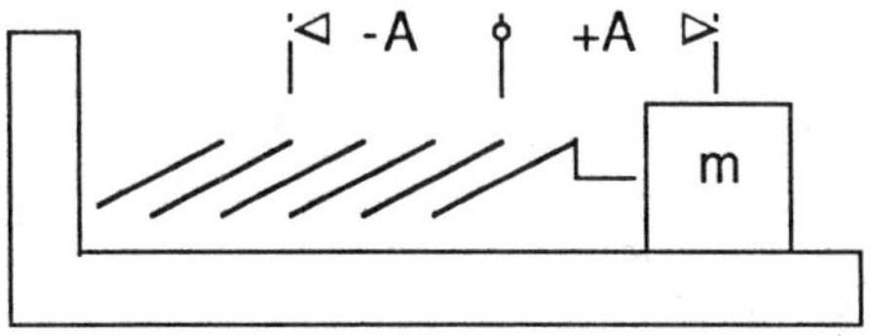


The position and speed of the mass at any time are $x = A\cos(\omega t + \theta_o)$ and $v = -\omega A\sin(\omega t + \theta_o)$, and its kinetic energy at any time is

$$KE = \frac{1}{2}mv^2 = \frac{1}{2}m\omega^2A^2\sin^2(\omega t + \theta_o) = \frac{1}{2}kA^2\sin^2(\omega t + \theta_o)$$

The potential energy of the spring at any time is

$$PE = \frac{1}{2}kx^2 = \frac{1}{2}kA^2\cos^2(\omega t + \theta_o)$$

Notice that the potential energy varies in time from 0 to $kA^2/2$. The total mechanical energy of the mass-spring system is the sum of the KE of the mass and the PE of the spring.

$$E = KE + PE = \frac{1}{2}kA^2\sin^2(\omega t + \theta_o) + \frac{1}{2}kA^2\cos^2(\omega t + \theta_o) = \frac{1}{2}kA^2$$

Notice that the total energy of the system is a constant in time.

Practice: A 2.00-kg object attached to the end of a spring with a 400 N/m spring constant is undergoing SHM on a horizontal frictionless surface. To start the system oscillating, the spring was compressed by 0.100 m and then released. The time record was started ($t = 0$) the first time the object went through the equilibrium position.

Determine the following:

| | |
|---|---|
| 1. Maximum kinetic energy | $KE_{max} = kA^2/2 = 2.00$ J |
| 2. Maximum potential energy | $PE_{max} = kA^2/2 = 2.00$ J |
| 3. Total mechanical energy at any time | $E = kA^2/2 = 2.00$ J |
| 4. Potential energy when $x = A/2$ | $PE = kx^2/2 = kA^2/8 = 0.500$ J |
| 5. Kinetic energy when $x = A/2$ | $KE = E - PE = 1.50$ J |
| 6. Initial phase angle | $\theta_o = 270° = (3\pi/2)$ rad |
| 7. Angular speed of a particle on the associated reference circle | $\omega = (k/m)^{1/2} = 14.1$ rad/s |
| 8. Position of the oscillating object at $t = 0.100$ s | $\omega t + \theta_o = 1.41$ rad $+ 4.71$ rad $= 6.12$ rad<br>$x = A\cos(\omega t + \theta_o) = +0.0987$ m |
| 9. Speed of the object at $t = 0.100$ s | $\omega t + \theta_o = 612$ rad |
| 10. Potential energy of the system at $t = 0.100$ s | $PE = kx^2/2 = 1.95$ J |
| 11. Kinetic energy of the system at $t = 0.100$ s | $KE = mv^2/2 = 0.05$ J |
| 12. Total energy of the system at $t = 0.100$ s | $E = KE + PE = 2.00$ J, or<br>$E = kA/2 = 2.00$ J |

| | |
|---|---|
| 13. Position of the object when its KE is equal to the PE of the spring | $KE + PE = E$, $E = kA^2/2$, $PE = KE$<br>$2PE = kA^2/2$, $PE = kx^2/2$<br>$2(kx^2/2) = kA^2/2$, or $x = 0.0707$ m |

Example 10.4. A 0.500-kg mass is attached to the end of a spring that has a spring constant of 50.0 N/m. This system is set into SHM on a horizontal frictionless surface by displacing it 0.100 m from equilibrium and then releasing it. The record of time ($t = 0$) is started when the mass has a maximum negative displacement. At $t = 0.100$ s, what is the system's (a) potential energy, (b) kinetic energy, (c) total energy?

Given: $m = 0.500$ kg, $k = 50.0$ N/m, $A = 0.100$ m, $t = 0.100$ s,
$t = 0$ when $x = -A$ and $v_c$

Determine: The potential, kinetic, and total energy of the oscillating system 0.100 s after the clock is started.

Strategy: By using a reference circle, we can determine the initial phase and then expressions for x and v at any time. We can then use these expressions to determine PE, KE, and E. Finally, we can check our work by using another expression to calculate E.

Solution: Let's first get a clear picture of what is happening by looking at Fig. 10.10.

Mass spring system at the equilibrium position

Mass displaced 0.100 m from the equilibrium position

Mass at the maximum negative displacement ($t = 0$)

Mass a short time after it is at the maximum negative displacement

Reference particle on the reference circle

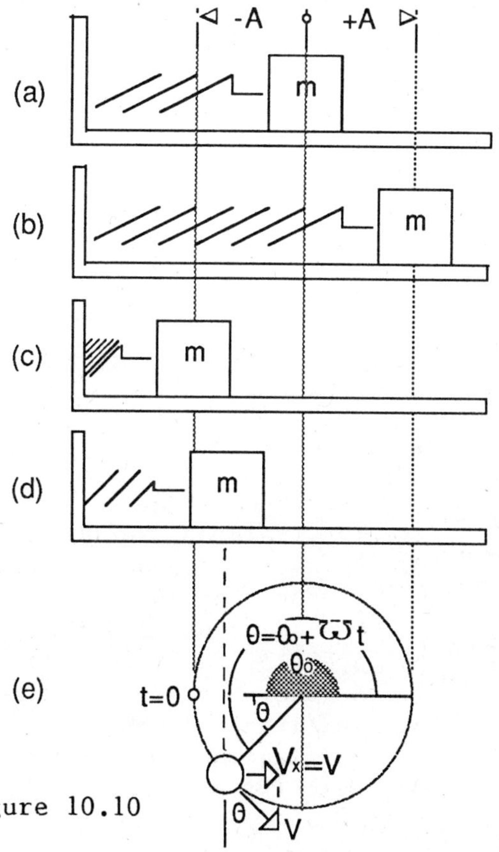


Figure 10.10

From Fig. 10.10, we can establish that $\theta_o = \pi$ rad. From the given information, we know that $\omega\sqrt{k/m}$ = 10 rad/s. Then at t = 0.100 s, we have $\omega t + \theta_o$ = 4.14 rad. The values for x and v at this time are

$$x = A\cos(\omega t + \theta_o) = (0.100 \text{ m})\cos(4.14 \text{ rad}) = -0.0542 \text{ m}$$
$$v = -\omega A\sin(\omega t + \theta_o) = -(10.0/\text{s})(0.100 \text{ m})\sin 4.14 \text{ rad}) = +0.841 \text{ m/s}$$

Knowing x and v t = 0.100 s, we can determine PE, KE, and E at that time.

$$PE = kx^2/2 = (50.0 \text{ N/m})(-0.0542 \text{ m})^2/2 = 0.0734 \text{ J}$$
$$KE = mv^2/2 = (0.500 \text{ kg})(+0.841 \text{ m/s})^2/2 = 0.177 \text{ J}$$
$$E = PE + KE = 0.0734 \text{ J} + 0.177 \text{ J} = 0.250 \text{ J}$$

As a check on our work, we can determine E using another expression.

$$E = kA^2/2 = (50.0 \text{ N/m})(0.100 \text{ m})^2/2 = 0.250 \text{ J}$$

Related Text Problems: 10-6e, 10-7, 10-18 through 10-21.

---

## PRACTICE TEST

Take and grade this practice test. Doing so will allow you to determine any weak spots in your understanding of the concepts taught in this chapter. The following section prescribes what you should study further to strengthen your understanding.

A 2.00-kg mass is attached to the end of a spring. The mass-spring system is placed on a horizontal frictionless surface, as shown in Fig. 10.11, A force of 20.0 N is required to stretch the spring 0.100 m. You start the system oscillating by compressing the spring 0.200 m and then releasing it. You start your record of time (t = 0) the first time the oscillating mass goes through equilibrium.

Figure 10.11

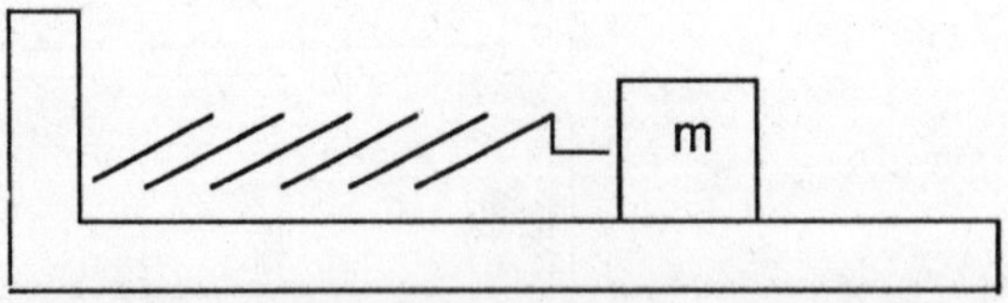


Determine the following:

_______ 1. Spring constant
_______ 2. Amplitude of oscillation
_______ 3. Angular frequency of a particle on the reference circle
_______ 4. Linear frequency of the oscillating mass
_______ 5. Period of oscillation
_______ 6. Speed of the reference particle
_______ 7. Acceleration of the reference particle
_______ 8. Initial phase angle
_______ 9. Displacement of the mass after one-half period
_______ 10. Velocity of the mass after one-half period

_______ 11. Acceleration of the mass after one-half period
_______ 12. Force exerted by the spring on the mass after one-half period
_______ 13. Kinetic energy of the mass after one-half period
_______ 14. Potential energy of the spring after one-half period
_______ 15. Total mechanical energy of the mass-spring system
_______ 16. Displacement of the mass at t = 0.300 s
_______ 17. Velocity of the mass at t = 0.300 s
_______ 18. Acceleration of the mass at t = 0.300 s
_______ 19. Potential energy of the system at t = 0.300 s
_______ 20. Kinetic energy of the system at t = 0.300 s

(See Appendix I for answers)

## PRINCIPAL CONCEPTS AND EQUATIONS PRESCRIPTION

Your score on the practice is an excellent measure of your understanding of this chapter. You should now use the following chart to write your own prescription for curing any of your physics ills. Look down the leftmost column to the number of the question(s) you answered incorrectly, read across that row to see which section(s) of the study guide you should return to for further study, and then do the suggested text problems to gain additional experience in working with the particular concept.

| Practice Test Question | Concepts and Equations | Prescription | |
|---|---|---|---|
| | | Principal Concept | Text Problems |
| 1 | Spring constant | 1 | 10-2,4 |
| 2 | Amplitude of oscillation | 2 | 10-4,8 |
| 3 | Angular frequency | 1 | 10-2,6 |
| 4 | Linear frequency | 1 | 10-12,15 |
| 5 | Period of oscillation | 1 | 10-9,10 |
| 6 | Speed of Reference particle | 2 | 10-4,8 |
| 7 | Acceleration of reference particle | 2 | 10-13,14 |
| 8 | Phase angle | 2 | 10-4,18 |
| 9 | Describing SHM (displacement) | 2 | 10-1,5 |
| 10 | Describing SHM (velocity) | 2 | 10-6,13 |
| 11 | Describing SHM (acceleration) | 2 | 10-8,14 |
| 12 | Condition for SHM (force) | 1 | 10-4,13 |
| 13 | Kinetic energy in SHM | 3 | 10-18,21 |
| 14 | Potential energy in SHM | 3 | 10-19,20 |
| 15 | Total Mechanical energy in SHM | 3 | 10-18,20 |
| 16 | Describing SHM (displacement) | 2 | 10-3,4 |
| 17 | Describing SHM (velocity) | 2 | 10-6,12 |
| 18 | Describing SHM (acceleration) | 2 | 10-8,13 |
| 19 | Potential energy in SHM | 3 | 10-20-21 |
| 20 | Kinetic energy in SHM | 3 | 10-7,18 |

# 11 Mechanical Waves and Sound

## RECALL FROM PREVIOUS CHAPTERS

Your success with this chapter will be affected by your understanding of previous chapters. However, you should be able to proceed with this chapter without reviewing any previous concepts and equations.

## NEW IDEAS IN THIS CHAPTER

| Concepts and equations introduced | Text Section | Study Guide Page |
|---|---|---|
| Mechanical wave motion | 11.1 | 178 |
| Traveling waves: $y = y_o \sin[\frac{2\pi}{\lambda}(x - vt)]$ | 11.2 | 180 |
| Speed of propagation of a transverse traveling wave in a string: $v = [T/\rho_L]^{1/2}$ | 11.3 | 182 |
| Standing transverse waves in a string fixed at both ends: $\lambda_n = 2L/n$; $f_n = nv/2L$; $n = 1,2,3,...$ | 11.5 | 182 |
| Standing longitudinal waves in pipes<br>Closed: $\lambda_n = \frac{4L}{2n-1}$; $f_n = \frac{(2n-1)v}{4L}$; $n = 1,2,3,..$<br>Open: $\lambda_n = 2L/n$; $f_n = nv/2L$; $n = 1,2,3,..$ | 11.6 | 185 |
| Sound intensity level: $SIL = 10 \log(I/I_o)$ | 11.8 | 187 |
| Doppler effect: $f_o = (v - v_o)f_s / (v - v_s)$ | 11.9 | 188 |

## PRINCIPAL CONCEPTS AND EQUATIONS

### 1. Mechanical Wave Motion (Section 11.1)

Review: A mechanical oscillator moving in simple harmonic motion in a medium creates a sinusoidal disturbance. This disturbance propagates through the medium (away from the source) with a constant velocity and is called a traveling wave. Fig. 11.1 shows a section of a transverse wave traveling to the right. It is important to appreciate that the particles of the medium oscillate about their equilibrium position while the disturbance is traveling through the medium.

Figure 11.1

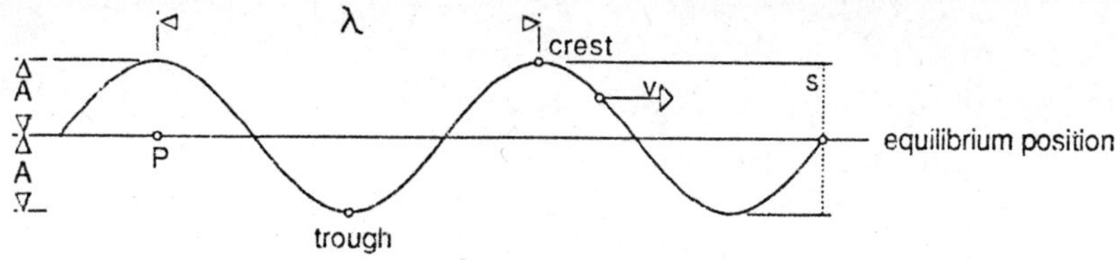


λ = wavelength = minimum distance over which wave repeats itself
A = amplitude = maximum displacement of medium from equilibrium
f = frequency = number of waves passing some arbitrary point P per second
τ = period = time for one wavelength to pass some point P; time for a particle of the medium to undergo a complete cycle of its motion
v = speed of propagation = speed at which wave moves through medium
s = path of a medium particle oscillating about the equilibrium position

Two important relationships for these physical quantities are

$$\tau = 1/f \quad \text{and} \quad v = f\lambda .$$

Practice: Figure 11.2 shows a series of three rapid succession photos of a traveling wave in front of a grid. The grid is 0.150 m x 0.150 m

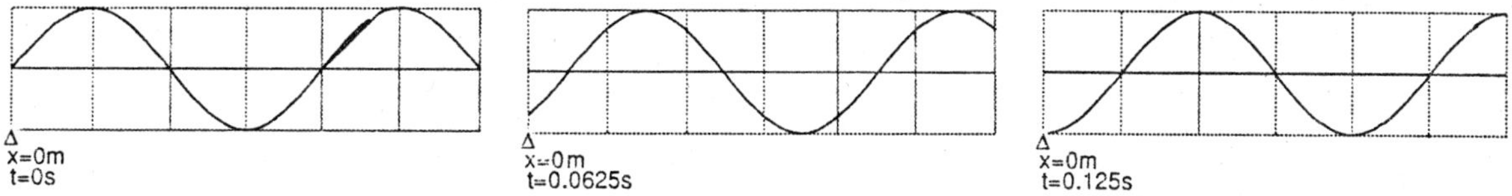


Figure 11.2

Determine the following for the traveling wave:

| | |
|---|---|
| 1. Amplitude | A = 0.150 m |
| 2. Wavelength | λ = 0.600 m |
| 3. Period | Since the wave advances λ/4 in 0.125 s,<br>τ = 4(0.125 s) = 0.500 s |
| 4. Frequency | f = 1/τ = 2.00 Hz |
| 5. Speed of propagation | Since the wave advances Δx = λ = 0.600 m in Δt = τ = 0.500 s,<br>v = Δx/Δt = (0.600 m)/(0.500 s) = 1.20 m/s<br>v = fλ = (2.00 Hz)(0.600 m) = 1.20 m/s |

Example 11.1. A cello string vibrates with a frequency of 500 Hz. If the speed of sound in air is 340 m/s, calculate (a) the number of times the string

vibrates wihle the sound travels 100 m and (b) the wavelength of the sound wave.

Given: $f_c$ = 500 Hz = frequency of vibration of the cello string
v = 340 m/s = speed of sound in air
d = 100 m = distance the sound travels

Determine: (a) The number of times the string vibrates while the sound travels 100 m. (b) The wavelength of the sound wave.

Strategy: Knowing d and v, we can determine the time it takes the sound to travel the distance. Knowing the string's frequency of vibration, we can determine the period $\tau$. From t and $\tau$, we can determine the number of times the string vibrates while the sound travels the distance d. The frequency of the sound in air is the same as the vibrational frequency of the string. From f and v for air, we can determine the wavelength of the sound wave.

Solution: The time it takes the sound to travel the 100 m is

$$t = d/v = (100 \text{ m})/(340 \text{ m/s}) = 0.294 \text{ s}$$

The period of oscillation for the vibrating string is

$$\tau = 1/f = 1/500 \text{ Hz} = 2.00 \times 10^{-2} \text{ s}$$

The period $\tau$ is the time for one vibration of the string. The time t is the total time the string has to vibrate. If during the time t the string vibrates N times, we can write $t = N\tau$. Then the number of vibrations of the string while the sound travels the 100 m is

$$N = t/\tau = (0.294 \text{ s})/(2.00 \times 10^{-2} \text{ s}) = 147$$

The vibrational frequency of the sound in air is the same as the vibrational frequency of the sound source, namely the cello string. The wavelength is

$$\lambda = v/f = (340 \text{ m/s})/ (500 \text{ Hz}) = 0.680 \text{ m}$$

Related Text Problems: 11-1 to 11-4, 11-12.

## 2. Traveling Waves (Section 11.2)

Review: The following expression describes a wave traveling to the right in a medium.

$$y = y_o \sin \left[\frac{2\pi}{\lambda} (x - vt)\right]$$

where $y_o$ = amplitude of the wave
v = speed of propagation of the wave
$\lambda$ = wavelength of the wave
x = distance from the source
t = time we consider medium at distance x from the source
y = transverse displacement of the traveling wave at position x and time t

In order to avoid cluttering this expression, we usually do not show units of the physical quantities involved. It is understood, however that if $y_o$, $\lambda$, and x are in meters, v in meters/second and t in seconds, then y is in meters. If you pick a value for x, the transverse motion of the medium at that location is sinusoidal in time. If you pick a value for t, the transverse displacement of the medium is sinusoidal in x.

Practice: Given the following expression for a traveling transverse wave

$$y = (0.200\ \text{m}) \sin[50\pi\ (x - 4.00t)]$$

Determine the following:

| | |
|---|---|
| 1. Amplitude of the wave | $y_o = 0.200$ m |
| 2. Wavelength | $2\pi/\lambda = 50\pi$ or $\lambda = 0.0400$ m |
| 3. Speed of propagation | $vt = 4.00\ t$ or $v = 4.00$ m/s |
| 4. Frequency | $f = v/\lambda = (4.00\ \text{m/s})/(0.0400\ \text{m}) = 100$ Hz |
| 5. Period | $\tau = 1/f = 1.00 \times 10^{-2}$ s |
| 6. The appearance of this traveling wave between x = -0.040 m and x = 1.20 m at t = 0 s | $y = (0.200\ \text{m})\sin[50\pi(x - 4.00t)]$<br>At t = 0, this becomes<br>$y = (0.200\ \text{m})\sin(50\pi x)$.<br>When this is plotted for values of x between x = -0.0400 m and x = 0.120 m, we obtain the following figure<br>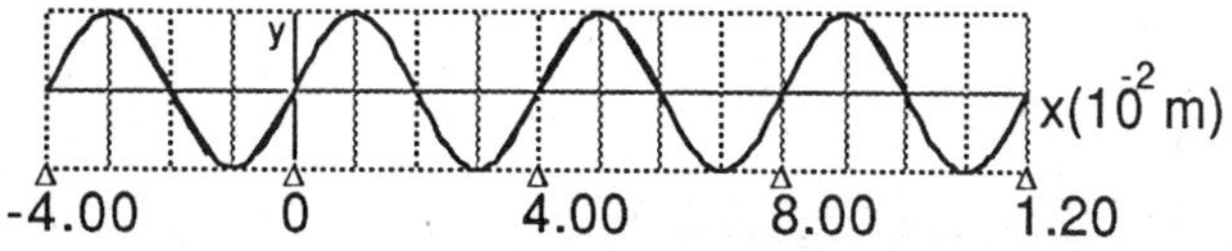<br> |
| 7. The time t when the medium at x = 0 has a transverse displacement y = 0.200 m for the first time after the record of time is started | Examining the traveling wave at t = 0 (step 6), we see that the medium at x = 0 has a displacement of y = 0.200 m in $3\tau/4$ or 0.00750 s. We can also obtain this result analytically by starting with the expression for the traveling transverse wave<br>$y = (0.200\ \text{m})\sin[50\pi(x - 4.00t)]$<br>Inserting y = 0.200 m and x = 0 m, we obtain<br>$0.200\ \text{m} = (0.200\ \text{m})\sin(-2\pi\pi t)$, |

| | |
|---|---|
| | which reduces to<br>$-1 = \sin 200\pi t$<br>This condition is satisfied for the first time when |
| 8. The time when the medium at $x = 1.00$ m has a transverse displacement of $y = -0.100$ m for the first time after the record of time is started | $y = (0.200\text{ m})\sin[50\pi(x - 4.00t)]$<br>Insert $x = 1.00$ m and $y = -0.100$ m to get<br>$-0.100\text{ m} = (0.200\text{ m})\sin[50\pi(1.00 - 4.00t)]$<br>or $-0.500 = \sin[50\pi(1.000 - 4.00t)]$<br>This is true for the first time when<br>$50\pi(1.00 - 4.00t) = 7\pi/6$ or $t = 0.244$ s |
| 9. The transverse displacement of the medium at $x = 0.750$ m and $t = 0.150$ s | $y = (0.200\text{ m})\sin[50\pi(x - 4.00t)]$<br>Insert x and t to obtain<br>$y = (0.200\text{ m})\sin[50\pi(0.150)] = -0.200$ m |

Example 11.2. A traveling transverse wave has an amplitude of 0.150 m, a frequency of 200 Hz, and a speed of propagation of 100 m/s. For this wave, write the equation for the displacement of any point in the medium at any time.

Given: $y_o = 0.150$ m, $f = 200$ Hz, $v = 100$ m/s

Determine: An expression for the transverse displacement of any point in the medium at any time.

Strategy: Insert the given quantities into the general expression that describes a transverse wave traveling through a medium.

Solution: Combine v and f to obtain λ: $\lambda = v/f = 0.500$ m

Insert $y_o$, λ, and v into $y = y_o \sin[\frac{2\pi}{\lambda}(x - vt)]$ to obtain

$$y = (0.150\text{ m})\sin[4\pi(x - 100t)]$$

Related Text Problems: 11-6 through 11-9, 11-15, 11-19.

## 3. Transverse Waves in a String (Sections 11.3 and 11.5)

Review: The speed of propagation of a transverse disturbance through a string is determined by the tension (T) and the linear mass density ($\rho_L$) of the string: $v = (T/\rho_L)^{1/2}$.

If the ends of the string are fixed, the transverse disturbance is reflected back into the string. For a string with a given tension T, linear mass density $\rho_L$, and length L, we can find transverse disturbance frequencies that cause standing waves. Standing transverse waves can be set up in a string (fixed at both ends), as shown in Fig. 11.3

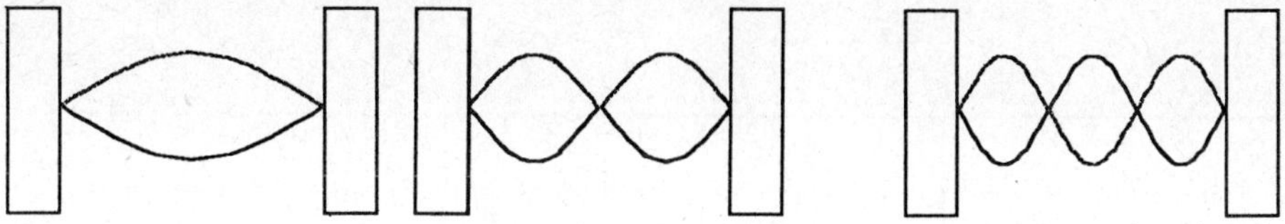

general pattern

| | | | |
|---|---|---|---|
| $n = 1$ | $n = 2$ | $n = 3$ | $n = 1,2,3, \ldots$ |
| $L = \lambda_1/2$ | $L = 2(\lambda_2/2) = \lambda_2$ | $L = 3(\lambda_3/2)$ | $L = n\lambda_n/2$ |
| $\lambda_1 = 2L$ | $\lambda_2 = L$ | $\lambda_3 = 2L/3$ | $\lambda_n = 2L/n$ |
| $f_L = v/2L$ | $f_2 = v/L$ | $f_3 = 3v/2L$ | $f_n = nv/2L$ |
| Fundamental | 1st Overtone | 2nd Overtone | n - 1 Overtone |
| 1st Harmonic | 2nd Harmonic | 3rd Harmonic | n Harmonic |

Figure 11.3

Practice: A lightweight string is attached to a vibrating tuning fork at one end, a block of mass M is suspended from the other end, as shown in Fig. 11.4.

Figure 4.
$L = 1.00$ m = length of cord
$m = 0.00100$ kg = mass of cord
$M = 0.500$ kg = mass of block

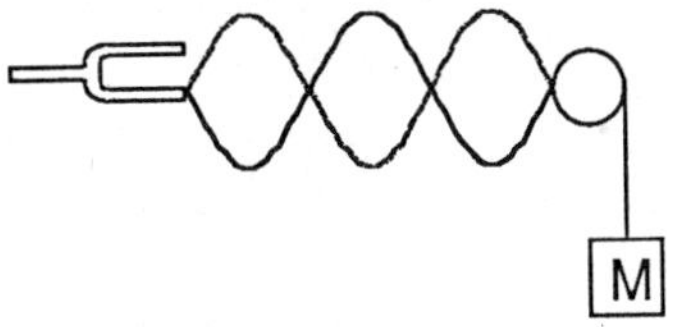


Determine the following:

| | |
|---|---|
| 1. Tension in the string | $T = Mg = 4.90$ N |
| 2. Linear mass density of the string | $\rho_L = m/L = 2.00 \times 10^{-3}$ kg/m |
| 3. Speed of the transverse wave as it travels down the string | $v = (T/\rho_L)^{1/2} = 49.5$ m/s |
| 4. Wavelength of the transverse wave in the string | $\lambda = 2L/3 = 0.667$ m |
| 5. Frequency of the transverse wave in the string | $f = v/\lambda = 74.2$ Hz This is also the frequency of the tuning fork. |
| 6. The mass M' that will cause the string to vibrate at its fundamental frequency | $\rho_L$, f, v, and L remain the same.<br>M and $\lambda$ change and are related by<br>$(\alpha)\ v = f\lambda = (T/\rho_L)^{1/2} = (Mg/\rho_L)^{1/2}$<br>For the new situation<br>$(\beta)\ v = f\lambda' = (T'/\rho_L)^{1/2} = (M'g/\rho_L)^{1/2}$<br>Divide $(\beta)$ by $(\alpha)$ to obtain $M' = M(\lambda'/\lambda)$<br>$\lambda = 0.667$ m (step 4) and $\lambda' = 2L = 2.00$ m<br>$M' = 4.50$ kg |

| | |
|---|---|
| 7. The length L' of the string that would vibrate at its fundamental frequency with the 0.500-kg mass attached | $\rho_L$, f, v, and T remain the same. L and $\lambda$ change and are related by $v = f\lambda = (T/\rho_L)^{1/2}$ and $\lambda = 2L/n$; thus $f(2L/n) = (T/\rho_L)^{1/2}$ For the present case, this expression gives $2fL/3 = (T/\rho_L)^{1/2}$ and for the fundamental case, it gives $2fL'/1 = (T/\rho_L)^{1/2}$ Divide the second equation by the first to obtain $L' = L/3 = 0.333$ m |

Example 11.3. The strings of a guitar are 0.600 m long. A particular string has a linear mass density of $6.00 \times 10^{-2}$ kg/m and is stretched to a tension of 50.0 N. (a) What is the fundamental frequency? (b) How far from the end should the string be pressed in order to cause it to vibrate in its third harmonic?

Given: $L = 0.600$ m, $\rho_L = 6.00 \times 10^{-2}$ kg/m, $T = 50.0$ N

Determine: (a) The fundamental frequency of the string. (b) Where to press in order to cause it to vibrate in the third harmonic (i.e., with a frequency three times that of the fundamental).

Strategy: Knowing T and $\rho_L$, we can determine v. Knowing L, we can determine $\lambda$ for the fundamental. The frequency for the fundamental can be determined from v and $\lambda$. Once the fundamental frequency is known, the frequency of the third harmonic can be determined. Knowing the frequency of the third harmonic and that v is still the same, we can determine the wavelength and hence where to press the string.

Solution: (a) The speed of propagation of the transverse waves down the string is

$$v = (T/\rho_L)^{1/2} = (50.0 \text{ N}/6.00 \times 10^{-2} \text{ kg/m})^{1/2} = 91.3 \text{ m/s}$$

The wavelength for the fundamental is $\lambda = 2L = 1.20$ m. The fundamental frequency is

$$f_1 = v/\lambda = (91.3 \text{ m/s})/(1.20 \text{ m}) = 76.1 \text{ Hz}$$

(b) The frequency of the third harmonic is $f_3 = 3f_1 = 228$ Hz.

The associated wavelength is $\lambda_3 = v/f_3 = (91.3 \text{ m/s})/(228 \text{ Hz}) = 0.400$ m.

If the string is 0.600 m long, it will vibrate in the third harmonic (have a wavelength of 0.400 m) if touched 0.400 m from one end, as shown in Fig. 11.5.

Figure 11.5

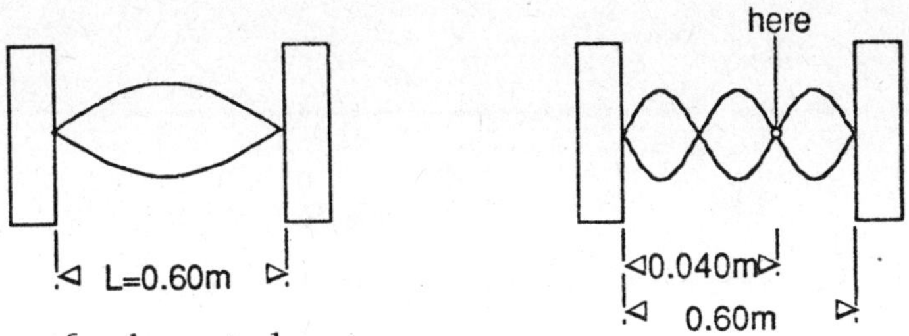


fundamental

Related Text Problems: 11-10, 11-11, 11-13, 11-14, 11-18, 11-20 through 11-25.

## 4. Standing Longitudinal Waves in a Pipe (Section 11.6)

Review: Standing longitudinal waves can be set up in both closed and open pipes, as shown in Fig. 11.6 and 11.7.

| | | | general pattern |
|---|---|---|---|
| $n = 1$ | $n = 2$ | $n = 3$ | $n = 1,2,3,...$ |
| $L = \lambda_1/4$ | $L = 3\lambda_2/4$ | $L = 5\lambda_3/4$ | $L = (2n - 1)\lambda_n/4$ |
| $\lambda_1 = 4L$ | $\lambda_2 = 4L/3$ | $\lambda_3 = 4L/5$ | $\lambda_n = 4L/(2n - 1)$ |
| $f_1 = v/4L$ | $f_2 = 3(v/4L) = 3f_1$ | $f_3 = 5(v/4L) = 5f_1$ | $f_n = (2n - 1)v/4L$ |
| Fundamental | 1st Overtone | 2nd Overtone | n - 1 Overtone |
| 1st Harmonic | 3rd Harmonic | 5th Harmonic | (2n - 1) Harmonic |

Figure 11.6. Closed Pipes

| | | | general pattern |
|---|---|---|---|
| $n = 1$ | $n = 2$ | $n = 3$ | $n = 1,2,3,...$ |
| $L = \lambda_1/2$ | $L = 2\lambda_2/2$ | $L = 3\lambda_3/2$ | $L = n\lambda_n/2$ |
| $\lambda_1 = 2L$ | $\lambda_2 = L$ | $\lambda_3 = 2L/3$ | $\lambda_n = 2L/n$ |
| $f_1 = v/2L$ | $f_2 = v/L = 2f_1$ | $f_3 = 3(v/2L) = 5\ f_1$ | $f_n = nv/2L$ |
| Fundamental | 1st Overtone | 2nd Overtone | n - 1 Overtone |
| 1st Harmonic | 2nd Harmonic | 3rd Harmonic | n Harmonic |

Figure 11.7. Open Pipes

Practice: Figure 11.8 shows a closed pipe of length $L_c = 1.00$ m and an open pipe of adjustable length $L_o$. The speed of sound in air is 340 m/s.

Figure 11.8

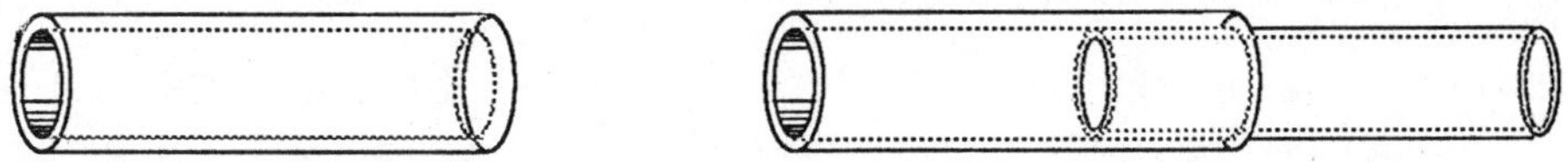

$L_c = 1.00$ m $\qquad$ $L_o$ = adjustable

Determine the following:

| | |
|---|---|
| 1. Fundamental frequency of the closed pipe | $f_n = (2n - 1)v/4L,\ n = 1$<br>$f_1 = v/4L_c = 86.3$ Hz |
| 2. Wavelength associated with the fundamental of the closed pipe | $\lambda_n = 4\ L_c/2n - 1,\ n = 1$<br>$\lambda_1 = 4\ L_c = 4.00$ m |
| 3. Frequency of the fifth harmonic of the closed pipe | Harmonics are multiples of the fundamental frequency. For the 5th harmonic, $2n - 1 = 5$, $n = 3$, and $f_3 = 5f_1 = 432$ Hz |
| 4. Wavelength associated with the 3rd overtone of the closed pipe | $\lambda_n = 4\ L_c/(2n - 1),\ n = 4$<br>$\lambda_4 = 4L_c/7 = 0.571$ m |
| 5. Length of the open pipe in order for a closed-pipe fundamental to cause an open-pipe second harmonic | $f_{2o} = f_{1c} = 86.3$ Hz<br>$f_n = nv/2L$ for an open pipe<br>$n_o = 2$ for the 2nd harmonic<br>$L_o = n_o/2f_{2o} = 4.00$ m |
| 6. Length of the open pipe in order for a closed-pipe third harmonic to cause an open-pipe second overtone | For the second overtone in the open pipe, $n_o = 3$. For the third harmonic in the closed pipe, $n_c = 2$.<br>$f_{2c} = (2n_c - 1)v/4\ L_c = 259$ Hz<br>$f_{3o} = f_{2c} = 259$ Hz<br>$f_{3o} = n_o v/2L_o$<br>$L_o = n_o v/2f_{3o} = 2.00$ m |

Example 11.4. If you have a closed pipe that is twice as long as an open pipe, can the two ever create sound of the same frequency?

Given: $L_c = 2L_o$, the closed pipe is twice as long as the open pipe; $v_c = v_o$ because the speed of sound in air is the same for both the closed and open pipes.

Determine: Whether or not these two pipes can create sound of the same frequency.

Strategy: Equate expressions for the frequency of sound from the closed and open pipe; insert the fact that $L_c = 2L_o$; then see if it is possible to satisfy the resulting relationship.

Solution: The expressions for the frequency of sound from a closed and an open pipe, respectively, are: $f_{nc} = (2n_c - 1)v/4L_c$ and $f_{no} = n_o v/2L_o$

Equating these expressions, we obtain $(2n_c - 1)v/4L_c = n_o v/2L_o$.

Cancelling v and inserting $L_c = 2L_o$, we find that the condition that must be satisfied in order to obtain the same frequency sound from both pipes is

$$2n_c - 1 = 4n_o$$

Since $n_c$ and $n_o$ can be any positive integer value, the left side of this expression is always an odd integer and the right side is an even integer. Since the equality cannot be satisfied, it is evident that we cannot obtain sound of the same frequency from the two pipes when $L_c = 2L_o$.

Related Text Problems: 11-26 through 11-30.

## 5. Sound Intensity Level (Section 11.8)

Review: Intensity of sound is the energy per unit time that the sound wave carries across an imaginary unit area placed perpendicular to the sound. Intensity is measured in $(J/s)/m^2$ or Watts per square meter ($W/m^2$). The lowest intensity that the ear can detect is $I_o = 1.00 \times 10^{-12}$ $W/m^2$, and sound becomes painful at an intensity of 1.00 $W/m^2$. Sound intensity level (SIL) for sound of intensity I is defined as follows:

$$SIL = 10 \log(I/I_o)$$

SIL has no units, but we give the name decibel (dB) to sound levels reported in this manner. The reference intensity $I_o$ is the threshold of hearing, or $10^{-12}$ $W/m^2$.

Practice: A rock concert is held outside. The slightly elevated stage holds a number of speakers that have a total power output of $400\pi$ W. Assume the sound travels outward in a spherical shell and any sound that hits the crowd sitting on the ground is totally absorbed.

Determine the following:

| | |
|---|---|
| 1. The sound intensity heard by observer 1 sitting 100 m away | $I_1 = (E/t)/A_1 = P/4\pi r_1^2 = 10^{-2}$ $W/m^2$ where $A_1 = 4\pi r_1^2$ is the area of a sphere of radius $r_1$. The sound is distributed uniformly over a spherical shell and the bottom half is absorbed. |
| 2. The SIL heard by observer 1 | $SIL_1 = 10 \log(I_1/I_o)$<br>$= 10 \log(10^{-2}/10^{-12}) = 100$ dB |
| 3. The distance of observer 2 from the speakers in order for her to hear sound 1/16 the intensity of that heard by observer 1 | $I = (E/t)/A = P/A$ or $P = IA$<br>$P$ = constant; hence $P_1 = P_2$ or<br>$I_1 A_1 = I_2 A_2$<br>Insert $A_1 = 4\pi r_1^2$, $A_2 = 4\pi r_2^2$ and $I_2 = I_1/16$ to obtain<br>$r_2 = r_1[I_1/(I_1/16)]^{1/2} = 400$ m |

| | |
|---|---|
| 4. The SIL heard by observer 2 | $SIL_2 = 10 \log (I/I_o)$<br>$= 10 \log[(I_1/16)/I_o]$<br>$= 10 \log(6.25 \times 10^{-4}/10^{-12})$<br>$= 10 \log(6.25 \times 10^8)$ dB<br>$= 10 (\log 6.25 + 8 \log 10)$<br>$= 88.0$ dB |
| 5. The distance of observer 3 from the speakers in order for him to hear sound 0.80 the SIL of that heard by observer 1 | $SIL_3 = 0.80\ SIL_1 = 80.0$ dB<br>$SIL_3 = 10 \log(I_3/I_o)$ or<br>$I_3 = I_o \log^{-1}(SIL_3/10) = 10^{-4}$ W/m$^2$<br>$I_3 = P/A_3 = P/4\pi r_3^2$ or<br>$r_3 = (P/4\pi I_3)^{1/2} = 10^3$ m |

Example 11.5: How far from a 100π-W speaker should an usher seat a person with a 20.0-dB hearing impairment if that person wants to hear sound at the 60.0-dB level?

Given: P = 100π W = power of speaker
20.0 dB = hearing impairment
SIL = 60.0 dB = desired sound intensity level

Determine: The distance from a 100π-W speaker that a person with a 20.0 dB hearing impairment must be located in order to get a SIL of 60.0 dB.

Strategy: From the desired SIL and the hearing impairment, we can determine the SIL at the person's seat. Knowing the SIL, we can determine the sound intensity at the seat. Finally, we can obtain the distance of the seat from the sound intensity and speaker power.

Solution: SIL at the seat: SIL = 60.0 dB + 20.0 dB = 80.0 dB

Sound intensity at the seat

$$SIL = 10 \log(I/I_o) \quad \text{or} \quad I = I_o \log^{-1}(SIL/10) = 10^{-4} \text{ W/m}^2$$

Distance from seat to speaker

$$I = P/A = P/4\pi r^2 \quad \text{or} \quad r = [P/4\pi I]^{1/2} = 500 \text{ m}$$

Related Text Problems: 11-32, 11-33, 11-34.

## 6. The Doppler Effect (Section 11.9)

Review: If the frequency of sound emitted by a source is $f_s$, the frequency of sound heard by the observer is

$$f_o = (v - v_o)f_s / (v - v_s)$$

where $v$ = speed of sound in air

$v_o$ = speed of the observer (positive if in the direction from the source to the observer and negative if in the opposite direction)

$v_s$ = speed of the source (positive if in the direction from the source to the observer and negative if in the opposite direction)

<u>Practice</u>: Shown below are six possible situations for a sound source (train whistle) and an observer. The frequency of the source is 100 Hz. The source speed is 30.0 m/s, and that of the observer is 20.0 m/s when moving. The speed of sound in air is 340 m/s. Determine $f_o$ for each of the following situations:

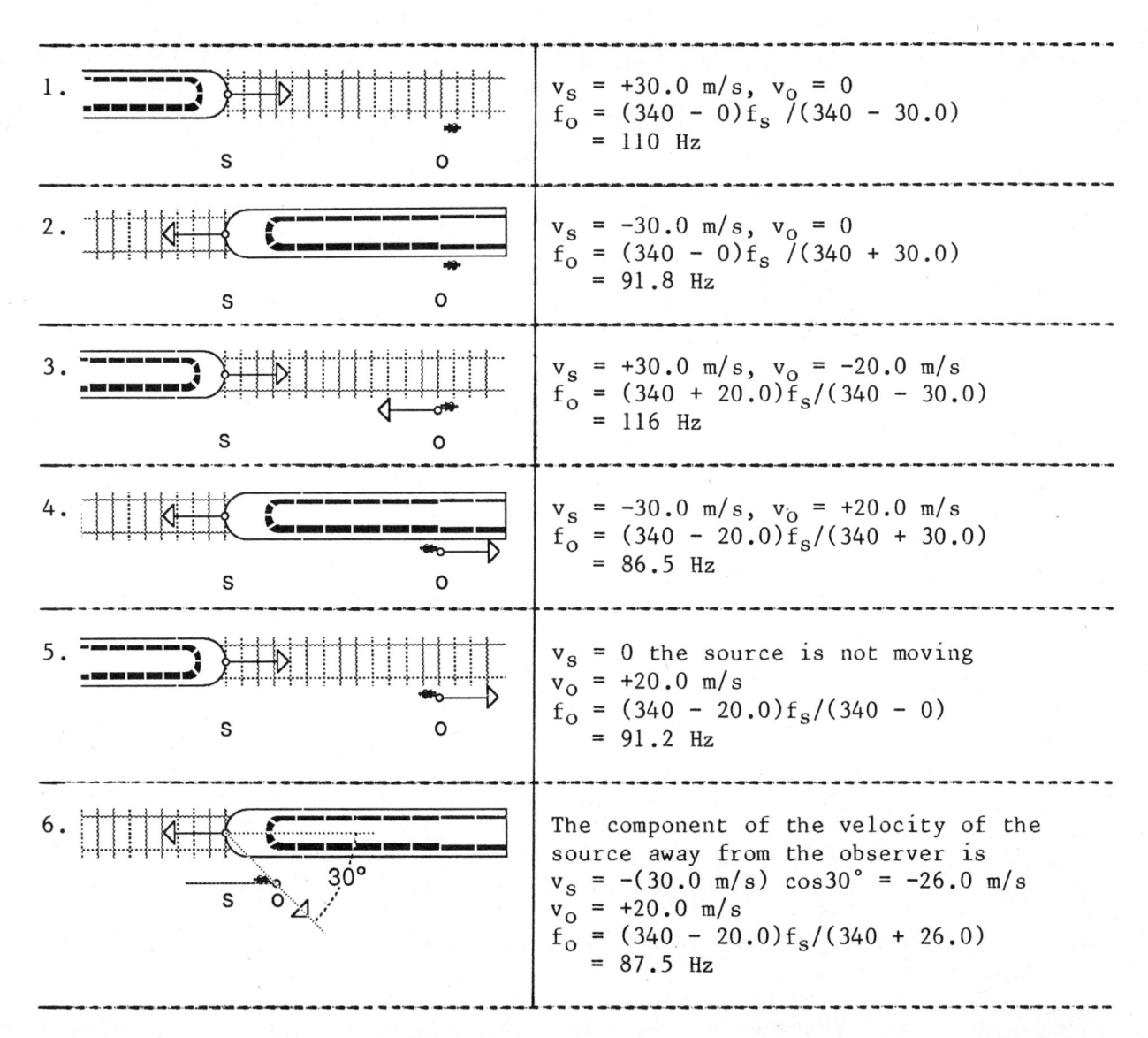

1. 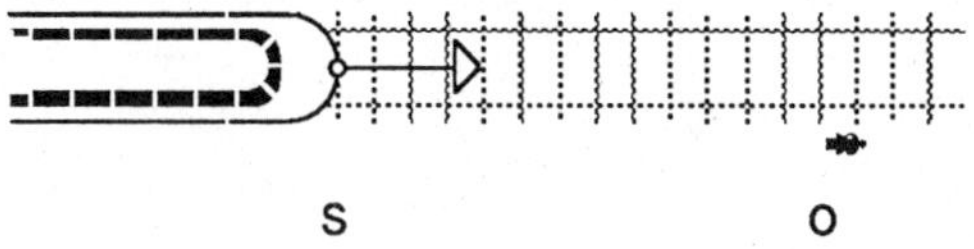


$v_s$ = +30.0 m/s, $v_o$ = 0
$f_o = (340 - 0)f_s/(340 - 30.0)$
= 110 Hz

2. 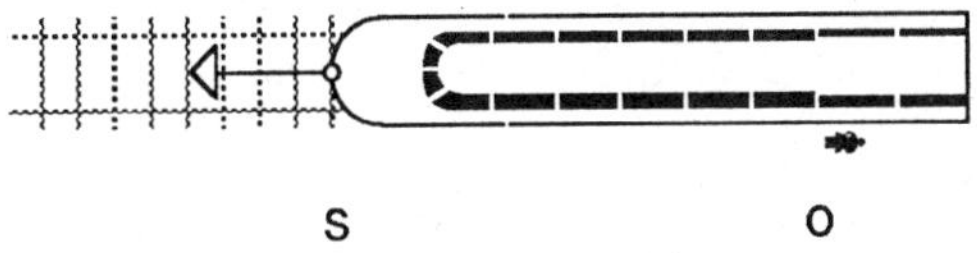


$v_s$ = −30.0 m/s, $v_o$ = 0
$f_o = (340 - 0)f_s/(340 + 30.0)$
= 91.8 Hz

3. 

$v_s$ = +30.0 m/s, $v_o$ = −20.0 m/s
$f_o = (340 + 20.0)f_s/(340 - 30.0)$
= 116 Hz

4. 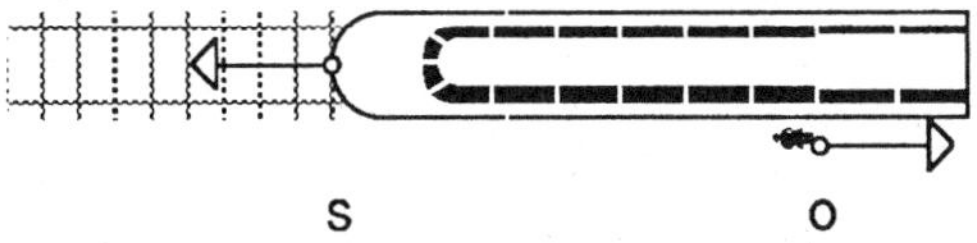


$v_s$ = −30.0 m/s, $v_o$ = +20.0 m/s
$f_o = (340 - 20.0)f_s/(340 + 30.0)$
= 86.5 Hz

5. 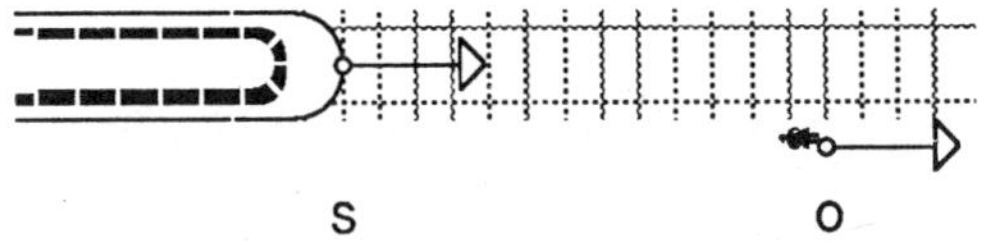


$v_s$ = 0 the source is not moving
$v_o$ = +20.0 m/s
$f_o = (340 - 20.0)f_s/(340 - 0)$
= 91.2 Hz

6. 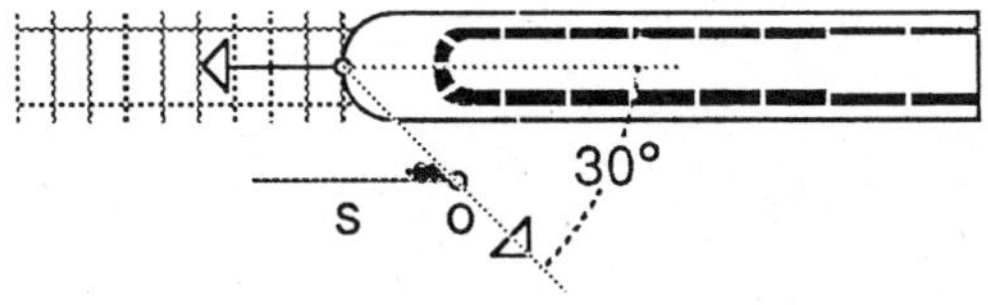


The component of the velocity of the source away from the observer is
$v_s = -(30.0\ \text{m/s})\cos 30° = -26.0$ m/s
$v_o$ = +20.0 m/s
$f_o = (340 - 20.0)f_s/(340 + 26.0)$
= 87.5 Hz

<u>Example 11.6.</u> A woman standing beside a railroad track measures a drop of 15 Hz in the pitch of the whistle of a train as it passes her. if the actual frequency of the whistle is 120 Hz, how fast was the train moving?

Given: $\Delta f = 15$ Hz, $f_s = 120$ Hz, $v = 340$ m/s

Determine: The speed of the train, $v_s$

Strategy: Write expressions for the frequency of sound heard by the woman as the train approaches and recedes. Set the difference between these two expressions equal to $\Delta f$ and solve for $v_s$.

Solution: Frequency heard by the observer as the train approaches ($f_{oa}$)

$$f_{oa} = (340 - 0)f_s/(340 - v_s)$$

Frequency heard by the observer as the train recedes ($f_{or}$)

$$f_{or} = (340 - 0)f_s/(340 + v_s)$$

We know that

$$\Delta f = f_{oa} - f_{or} = [340\ f_s/(340 - v_s)] - [340\ f_s/(340 + v_s)]$$

Since $F_s = 120$ Hz and $\Delta f = 15$ Hz, this reduces to

$$v_s^2 + 5440v_s - 340^2 = 0 \quad \text{or} \quad v_s = 21.2 \text{ m/s}$$

Related Text Problems: 11-35, 11-36, 11-37.

---

## PRACTICE TEST

Take and grade this practice test. Doing so will allow you to determine any weak spots in your understanding of the concepts taught in this chapter. The following section prescribes what you should study further to strengthen your understanding.

Water waves approach a buoy in the ocean. There is a distance of 10.0 m between adjacent crests, and a crest reaches the buoy every 5.00 s. The buoy bobs up and down in such a manner that its vertical position varies by 2.00 m. Determine the following:

__________ 1. Amplitude of the waves

__________ 2. Frequency of the waves

__________ 3. Wavelength of the waves

__________ 4. Speed of propagation of the waves

A traveling transverse wave is described by $y = (0.100 \text{ m})\sin[2\pi(x - 2.00t)]$. Determine the following:

__________ 5. Speed of propagation of the wave

__________ 6. Frequency of the wave

__________ 7. Transverse displacement of the medium at $x = 20.0$ m at $t = 10.0$ s

__________ 8. First time the medium at $x = 1.00$ m has a transverse displacement of $y = -0.100$ m

A stainless steel cable is stretched between two posts. The cable is 20.0 m long and has a mass of 4.00 kg. The wire is struck with a hammer at one end, and the pulse returns in 0.200 s. Determine the following:

__________ 9. Linear mass density of the cable
__________ 10. Speed of propagation of the disturbance
__________ 11. Tension in the wire

A string on an instrument is 1.00 m long, has a mass of $2.00 \times 10^{-2}$ kg, and is vibrating in the second harmonic with a frequency of 300 Hz. Determine the following:

__________ 12. Wavelength in the string
__________ 13. Speed of propagation of the transverse waves in the string
__________ 14. Tension in the string
__________ 15. Tension in the string that would cause it to vibrate in the fundamental mode with a freequency of 300 Hz

An open organ pipe is 1.00 m long. The frequency for its third harmonic is 510 Hz. Determine the following:

__________ 16. Wavelength of the standing longitudinal wave set up in the pipe when the air in it is vibrating in its third harmonic
__________ 17. Fundamental frequency for this pipe
__________ 18. Wavelength for the fifth harmonic
__________ 19. Length of a closed pipe that has a fundamental frequency equal to the frequency of the third harmonic of the open pipe

(See Appendix I for answers)

---

## PRINCIPAL CONCEPTS AND EQUATIONS PRESCRIPTION

Your score on the practice is an excellent measure of your understanding of this chapter. You should now use the following chart to write your own prescription for curing any of your physics ills. Look down the leftmost column to the number of the question(s) you answered incorrectly, read across that row to see which section(s) of the study guide you should return to for further study, and then do the suggested text problems to gain additional experience in working with the particular concept.

| Practice Test Question | Concepts and Equations | Prescription | |
|---|---|---|---|
| | | Principal Concept | Text Problems |
| 1 | Mechanical waves: amplitude | 1 | 11-6,9 |
| 2 | Mechanical waves: frequency | 1 | 11-5,6 |
| 3 | Mechanical waves: wavelength | 1 | 11-5,6 |
| 4 | Mechanical waves: speed of propagation | 1 | 11-1,2 |
| 5 | Traveling transverse wave: speed | 2 | 11-6 |
| 6 | Traveling transverse wave: frequency | 2 | 11-6 |
| 7 | Traveling transverse wave: displacement | 2 | 11-7,8 |
| 8 | Traveling transverse wave: displacement | 2 | 11-7,8 |
| 9 | Transverse wave in a string: density | 3 | 11-11,23 |
| 10 | Transverse wave in a string: speed | 3 | 11-4,22 |
| 11 | Transverse wave in a string: tension | 3 | 11-22,23 |
| 12 | Standing waves in a string: wavelength | 3 | 11-24,25 |
| 13 | Standing waves in a string: speed | 3 | 11-20,21 |
| 14 | Transverse wave in a string: tension | 3 | 11-18,23 |
| 15 | Transverse wave in a string: tension | 3 | |
| 16 | Standing waves in a pipe: wavelength | 4 | 11-26,27 |
| 17 | Standing waves in a pipe: frequency | 4 | 11-26,27 |
| 18 | Standing waves in a pipe: wavelength | 4 | 11-26,27 |
| 19 | Standing waves in a pipe: frequency | 4 | 11-29,30 |

# 12 Some Properties of Materials

## RECALL FROM PREVIOUS CHAPTERS

Your success with this chapter will be affected by your understanding of previous chapters. However, you should be able to proceed with this chapter without reviewing any previous concepts and equations.

## NEW IDEAS IN THIS CHAPTER

| Concepts and equations introduced | Text Section | Study Guide Page |
|---|---|---|
| Stress: Stress = F/A | 12.2 | 193 |
| Strain: Tensile or compressive strain = $\Delta L/L$ | 12.2 | 196 |
| Shear strain = $\Delta S/L$ | | |
| Elastic moduli: | 12.3 | 198 |
| Young's modulus $Y = (F/A)/(\Delta L/L)$ | | |
| Shear modulus $S = (F/A)/(\Delta S/L)$ | | |
| Bulk modulus $B = P/(\Delta V/V)$ | | |

## PRINCIPAL CONCEPTS AND EQUATIONS

### 1. Stress (Section 12.2)

Review: Stress is the force per unit area of cross section. In equation form, we write:

$$\text{Stress} = F/A$$

Figure 12.1 illustrates three types of stress applied to a solid rod with a rectangular cross section.

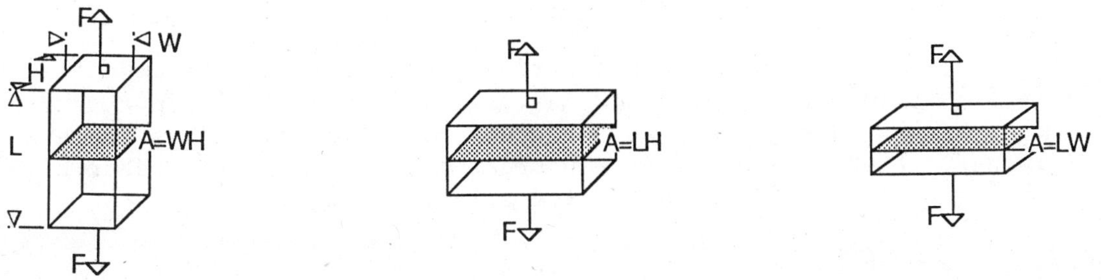


Stress = F/A = F/WH (i)    F/LH (ii)    F/LW (iii)

(a) Tensile stress

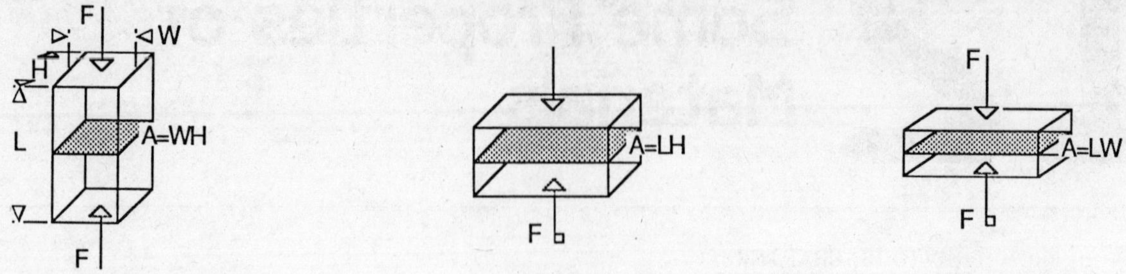


Stress = F/A = F/WH (i) | F/LH (ii) | F/LW (iii)

(b) Compressive stress

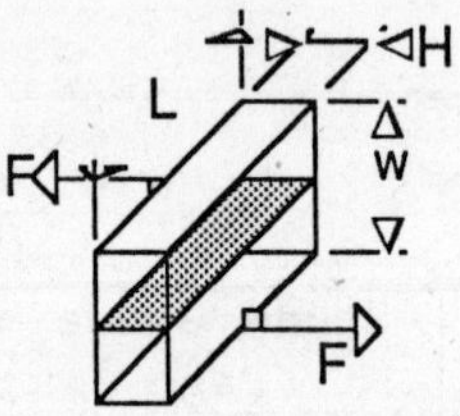

Stress = F/A = F/HL (i) | F/WL (ii) | F/HL (iii)

(c) Shear stress

Figure 12.1

Practice: The dimensions of the rod in Fig. 12.1 are L = 0.200 m, W = 0.100 m, and H = 0.0500 m. The magnitude of the force used is 400 N.

Determine the following:

| | |
|---|---|
| 1. Tensile stress for case a-i | $F/A = F/WH = 8.00 \times 10^4\ N/m^2$ |
| 2. Compressive stress for case b-ii | $F/A = F/LH = 4.00 \times 10^4\ N/m^2$ |
| 3. Shear stress for case c-ii | $F/A = F/LW = 2.00 \times 10^4\ N/m^2$ |

Example 12.2. The ultimate tensile strength of steel used to make elevator cable is $5.50 \times 10^8\ N/m^2$. Find the maximum upward acceleration that can be given to a $5.00 \times 10^3$ kg elevator supported by a 1.00 cm radius cable while maintaining a safety factor of 6. In order to have a safety factor of six, the cable must be able to withstand a force six times the load.

Given: $m = 5.00 \times 10^3$ kg = mass of elevator
$r = 1.00 \times 10^{-2}$ m = radius of cable
$(F/A)_{max} = 5.50 \times 10^8\ N/m^2$ = ultimate tensile strength
$f = 6.00$ = safety factor

Determine: The maximum upward acceleration that can be given to the elevator while maintaining a safety factor of 6.

Strategy: Knowing the ultimate tensile strength and the safety factor, we can determine the maximum tensile stress allowed for the cable. Knowing this stress and the cross-sectional area, we can determine the maximum tension allowed. Finally, we can use this tension and the mass of the elevator to determine the maximum allowed acceleration.

Solution: Maximum tensile stress allowed in the cable is

$$\text{max stress allowed} = \frac{\text{ultimate stress}}{\text{safety factor}} = 9.17 \times 10^7\ \text{N/m}^2$$

$$\text{max tension allowed} = (\text{max stress allowed})(\pi r^2) = 2.88 \times 10^4\ \text{N}$$

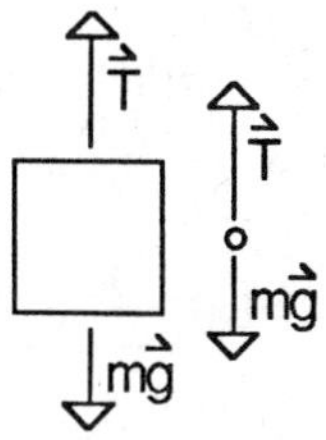


Figure 12.2

Using the free-body force diagram shown in Fig. 12.2, we can write an expression for $F_{net}$ and then determine a.

The net force on the elevator is

$$F_{net} = T - mg = ma \text{ or } a = (T - mg)/m = 3.80\ \text{m/s}^2$$

Example 12.2. A contractor wishes to support a sagging floor with a steel pipe. The ultimate compressive strength of steel is $4.10 \times 10^8$ N/m$^2$. The pipe has an outside radius of 12.0 cm and inside radius of 10.0 cm. Determine the maximum load this pipe can support before buckling.

Given: $S_{max} = 4.10 \times 10^8$ N/m$^2$ = maximum compressive stress
$r_o = 12.0 \times 10^{-2}$ m = outside radius of pipe
$r_i = 10.0 \times 10^{-2}$ m = inside radius of pipe

Determine: The maximum load the pipe can support before buckling.

Strategy: Using $r_o$ and $r_i$, we can determine the cross-sectional area of the pipe. Knowing this area and the maximum compressive stress of the pipe, we can determine the maximum load the pipe can support before buckling.

Solution: The cross-sectional area of the pipe is

$$A = \pi(r_o^2 - r_i^2) = 1.38 \times 10^{-2}\ \text{m}^2$$

The maximum load the pipe can support is $F = S_{max}A = 5.66 \times 10^6$ N

Example 12.3: A punching press that exerts a $2.50 \times 10^4$ N force is employed to punch circular holes 1.00 cm in diameter in a sheet of metal. If the metal can withstand a shear stress of $5.00 \times 10^8$ N/m$^2$, find the maximum thickness of a sheet of this metal that can be used.

Given and Diagram:

$F = 2.50 \times 10^4$ N
$d = 1.00 \times 10^{-2}$ m
shear stress $= 5.00 \times 10^8$ N/m$^2$

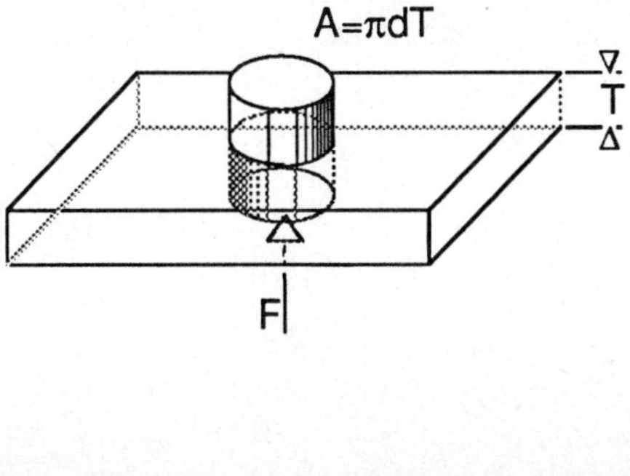


Figure 12.3

Determine: The maximum thickness T of a sheet of metal through which the press can punch a 1.00-cm-diameter hole.

Strategy: Using the shearing stress the metal can endure and the force exerted by the press, we can determine the maximum area the press can shear. Knowing this area and the diameter of the punch, we can determine the thickness T.

Solution: The area the press can shear is

$$A = F/\text{shear stress} = 5.00 \times 10^{-5}\ \text{m}^2$$

The thickness of the metal may be determined by

$$A = \pi dT \quad \text{or} \quad T = A/\pi d = 1.59 \times 10^{-3}\ \text{m}$$

Related Text Problems: 12-1, 12-3, 12-4a, 12-6a, 12-7a.

## 2. Strain (Section 12.2)

Review: The strain or an object is a measure of its deformation under stress. Figure 12.4 illustrates three types of strain.

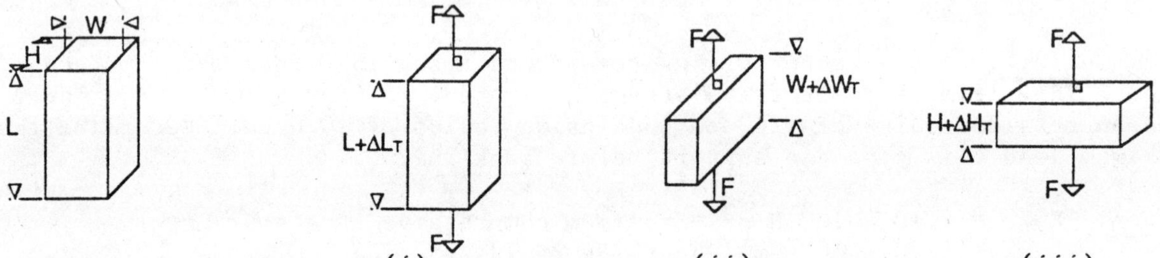

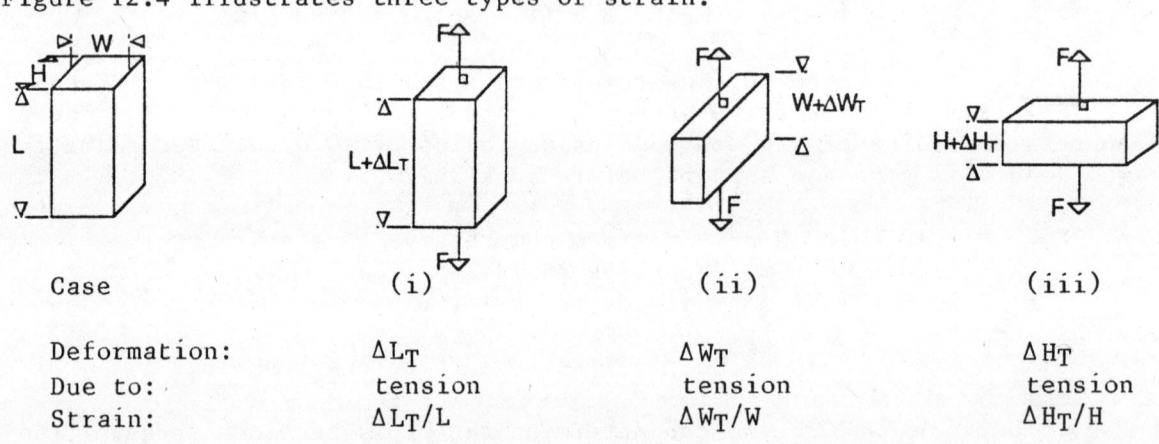

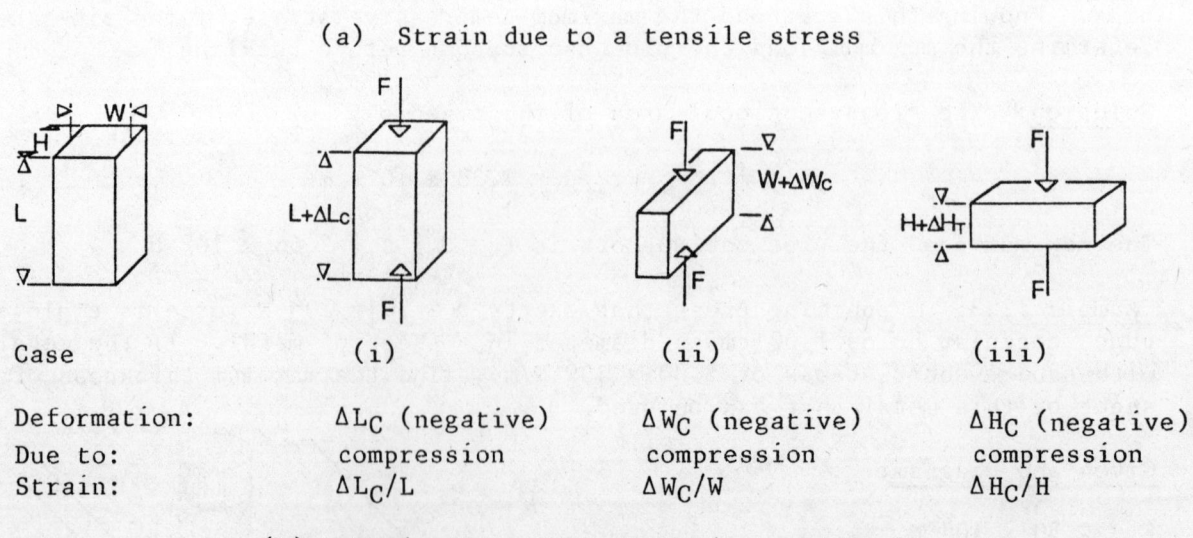

| Case | (i) | (ii) | (iii) |
|---|---|---|---|
| Deformation: | $\Delta L_T$ | $\Delta W_T$ | $\Delta H_T$ |
| Due to: | tension | tension | tension |
| Strain: | $\Delta L_T/L$ | $\Delta W_T/W$ | $\Delta H_T/H$ |

(a) Strain due to a tensile stress

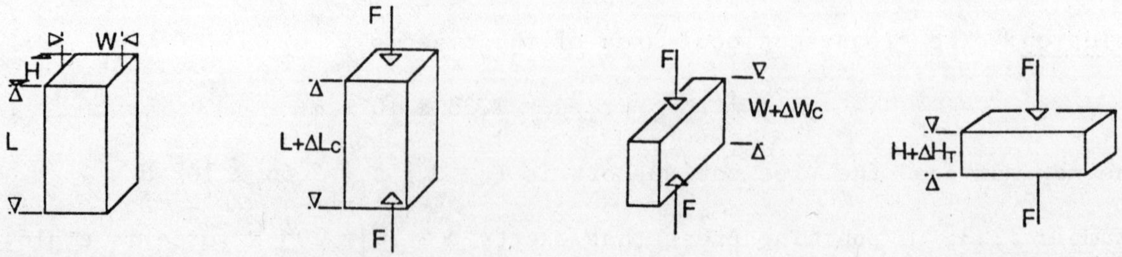


| Case | (i) | (ii) | (iii) |
|---|---|---|---|
| Deformation: | $\Delta L_C$ (negative) | $\Delta W_C$ (negative) | $\Delta H_C$ (negative) |
| Due to: | compression | compression | compression |
| Strain: | $\Delta L_C/L$ | $\Delta W_C/W$ | $\Delta H_C/H$ |

(b) Strain due to a compressive stress

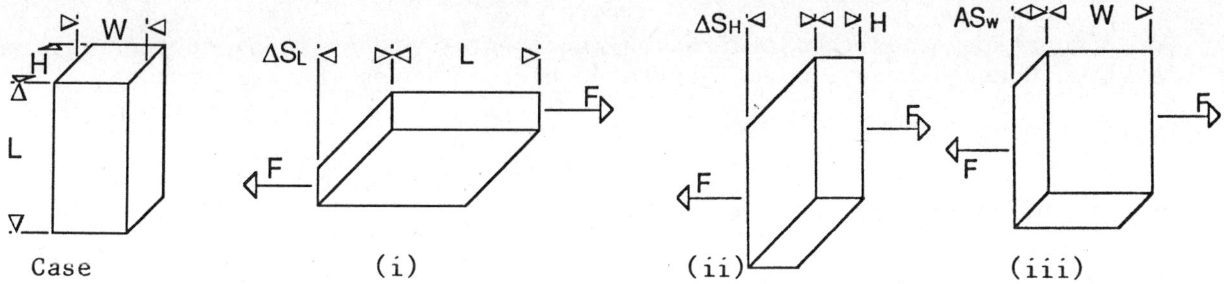

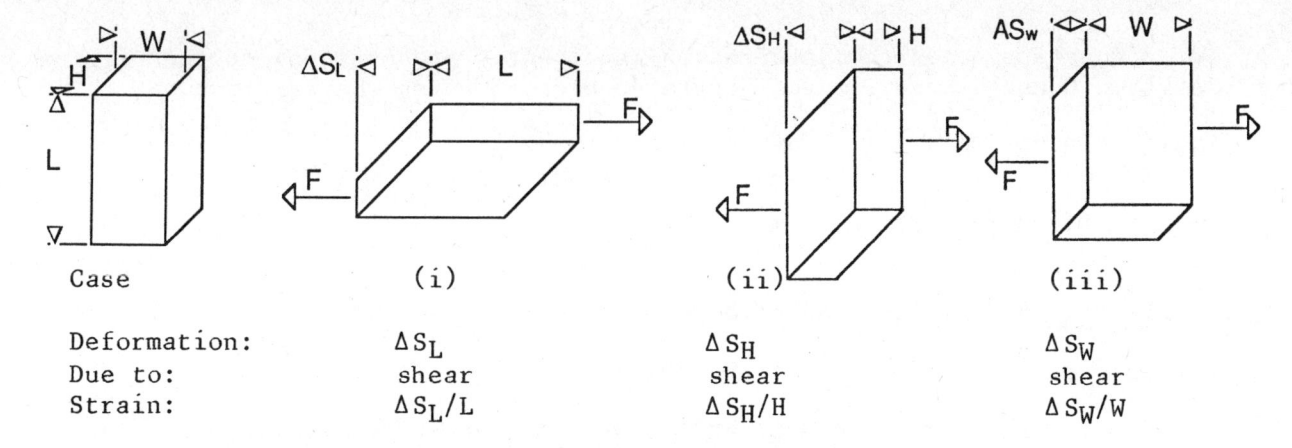

| Case | (i) | (ii) | (iii) |
|---|---|---|---|
| Deformation: | $\Delta S_L$ | $\Delta S_H$ | $\Delta S_W$ |
| Due to: | shear | shear | shear |
| Strain: | $\Delta S_L/L$ | $\Delta S_H/H$ | $\Delta S_W/W$ |

(c) Strain due to a shearing stress

Figure 12.4

Practice: The data for the object shown in Fig. 12.4 are as follows:

| | | |
|---|---|---|
| $L = 2.00 \times 10^{-1}$ m | $\Delta L_T = 8.00 \times 10^{-8}$ m | $\Delta L_C = 6.00 \times 10^{-8}$ m |
| $W = 1.00 \times 10^{-1}$ m | $\Delta W_T = 4.00 \times 10^{-8}$ m | $\Delta W_C = 3.00 \times 10^{-8}$ m |
| $H = 2.00 \times 10^{-2}$ m | $\Delta H_T = 1.00 \times 10^{-8}$ m | $\Delta H_C = 7.50 \times 10^{-9}$ m |
| $\Delta S_L = 2.00 \times 10^{-7}$ m | $\Delta S_H = 2.50 \times 10^{-8}$ m | $\Delta S_W = 2.50 \times 10^{-8}$ m |

Determine the following:

| | |
|---|---|
| 1. Tensile strain for case a-i | Strain $= \Delta L_T/L = 4.00 \times 10^{-7}$ |
| 2. Compressive strain for case b-ii | Strain $= \Delta W_C/W = 3.00 \times 10^{-7}$ |
| 3. Shear strain for case c-iii | Strain $= \Delta S_W/W = 2.50 \times 10^{-7}$ |

Example 12.4. A steel wire 1.00 m long supports a load and is stretched $1.00 \times 10^{-2}$ m. What is the strain in this wire, and how much would a 2.00-m steel wire stretch under the same load?

Given: $L = 1.00$ m, $\Delta L_T = 1.00 \times 10^{-2}$ m, $L' = 2.00$ m

Determine: The strain for the 1.00-m wire and the stretch $\Delta L'_T$ for the 2.00-m wire under the same load.

Strategy: Knowing the length L of the wire and the change in length $\Delta L_T$, we can determine the strain. Knowing that the deformation is directly proportional to the original length (if the load is constant), we can determine $\Delta L'_T$.

Solution: The strain in the 1.00-m wire is

$$\text{strain} = \Delta L_T/L = 1.00 \times 10^{-2}$$

For a constant load, the deformation is directly proportional to the original length. Consequently, if the length doubles, so does the stretch.

Since $L' = 2L$, then $\Delta L'_T = 2\Delta L_T = 2.00 \times 10^{-2}$ m

Related Text Problems: 12-2, 12-4b, 12-7b.

## 3. Elastic Moduli (Section 12.3)

Review: The elastic moduli is the ratio of stress to strain

$$\text{Young's modulus } Y = \frac{\text{tensile or compressive stress}}{\text{tensile or compressive strain}} = \frac{F/A}{\Delta L/L}$$

$$\text{shear modulus } S = \frac{\text{shear stress}}{\text{shear strain}} = \frac{F/A}{\Delta L/L}$$

$$\text{bulk modulus } B = \frac{\text{volume stress}}{\text{volume strain}} = \frac{F/A}{\Delta V/V} = \frac{P}{\Delta V/V}$$

A quantity related to bulk modulus is compressibility K.

$$K = 1/B$$

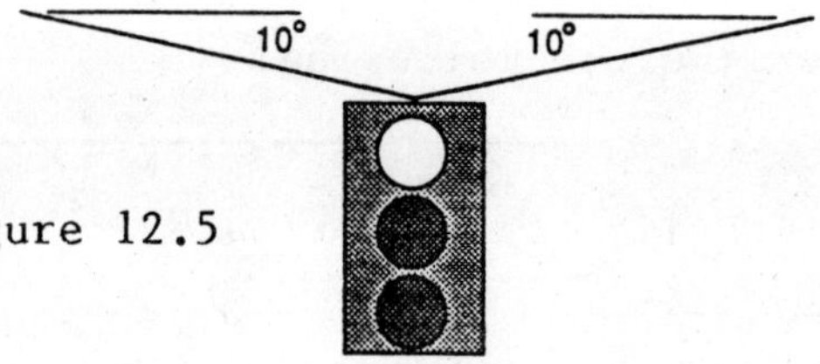


Figure 12.5

Practice: A 50.0-kg traffic light is suspended from a $5.00 \times 10^{-3}$ m radius cable, as shown in Fig. 12.5. The cable hangs 10° below the horizontal due to the weight of the light.

Determine the following

| | |
|---|---|
| 1. Weight of the light | $w = mg = 490$ kg |
| 2. Tension in the cable | Construct a free-body diagram, choose a coordinate system, and resolve the forces into components. 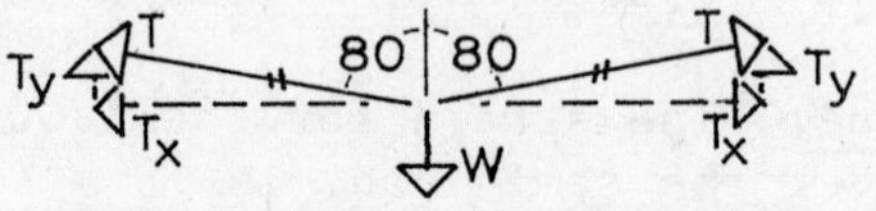  $2T_y = 2T \sin 10° = w$, or $T = (w/2) \sin 10° = 1410$ N |
| 3. Tensile stress | $A = \pi r^2 = 7.85 \times 10^{-5}$ m$^2$<br>Stress $= F/A = T/A = 1.80$ N/m$^2$ |

| | |
|---|---|
| 4. Young's modulus for steel | $Y = 21.0 \times 10^{10}$ N/m$^2$ (Table 12-1 of the text) |
| 5. Tensile strain | Y = stress/strain<br>strain = stress/Y = $8.57 \times 10^{-5}$ |
| 6. Fractional change in length of the cable | $\Delta L/L$ = strain = $8.57 \times 10^{-5}$ |

Example 12.5: Figure 12.6 shows a rectangular slab of jello subjected to a shearing force at its upper surface. The force of static friction acting on the bottom surface is sufficient to keep the jello stationary. Calculate the shear stress, shear strain, and shear modulus for jello.

Figure 12.6

$L = 1.00 \times 10^{-1}$ m  $\Delta S_L = 1.00 \times 10^{-2}$ m
$W = 5.00 \times 10^{-2}$ m  $F = 5.00 \times 10^{-1}$ N
$H = 3.00 \times 10^{-2}$ m

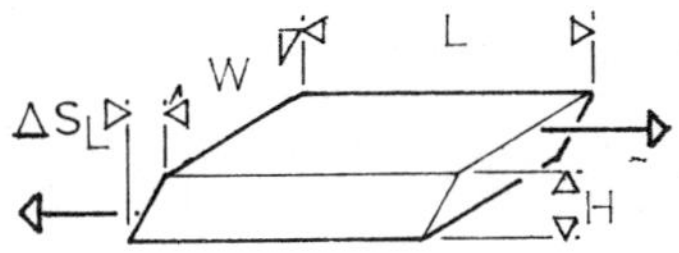


Determine: Shear stress, shear strain, and shear modulus.

Given: L, W, H, F, and $\Delta S_L$

Strategy: The shearing force F and the area LW can be used to obtain the shear stress. The deformation $\Delta S_L$ due to the shearing force and the length of the slab can be used to obtain the shear strain. Finally, the shear modulus can be obtained from the stress and strain.

Solution:

Shear stress = F/A = F/LW = $1.00 \times 10^{-4}$ N/m$^2$
Shear strain = $\Delta S_L/L$ = $1.00 \times 10^{-1}$
Shear modulus = stress/strain = $1.00 \times 10^{-3}$ N/m$^2$

Example 12.6. A brass sphere of radius 50.0-cm is subjected to a uniform pressure of $1.00 \times 10^{10}$ N/m$^2$. What is the radius of the sphere while under pressure?

Given: $r = 5.00 \times 10^{-1}$ m = radius of the sphere
$P = 1.00 \times 10^{10}$ N/m$^2$ = pressure on the sphere
$B = 6.80 \times 10^{10}$ N/m$^2$ = bulk modulus for brass

Determine: The radius of the sphere while under pressure.

Strategy: We can determine the volume V of the sphere from the radius. We can determine the change in volume and the new volume V' from P, B, and V. Knowing the new volume V', we can determine the new radius r'.

Solution: The volume of the sphere is

$$V = 4\pi r^3/3 = 5.24 \times 10^{-1}\ m^3$$

The change in volume of the sphere under pressure is

$$\Delta V = PV/B = 7.71 \times 10^{-2}\ m^3$$

The volume of the sphere under pressure is

$$V' = V - \Delta V = 4.47 \times 10^{-1}\ m^3$$

The radius of the sphere under pressure is

$$r' = (3V'/4\pi)^{1/3} = 4.75 \times 10^{-1}\ m$$

Related Text Problems: 12-4c, 12-5, 12-6b, 12-7c, 12-8 through 12-12.

---

PRACTICE TEST

Take and grade this practice test. Doing so will allow you to determine any weak spots in your understanding of the concepts taught in this chapter. The following section prescribes what you should study further to strengthen your understanding.

When a 10.0 kg mass is suspended from a wire of length 2.00 m and radius $2.00 \times 10^{-3}$ m the wire stretches $1.50 \times 10^{-3}$ m.
Determine the following:

__________ 1. Stress in the wire
__________ 2. Strain in the wire
__________ 3. Young's modulus for the wire

Two metal bars 0.500 cm thick are held together by these rivets, as shown in Fig. 12.7. The rivets have a radius of $2.00 \times 10^{-8}$ m, and the maximum shear stress they can withstand is $5.00 \times 10^8\ N/m^2$.
Determine the following:

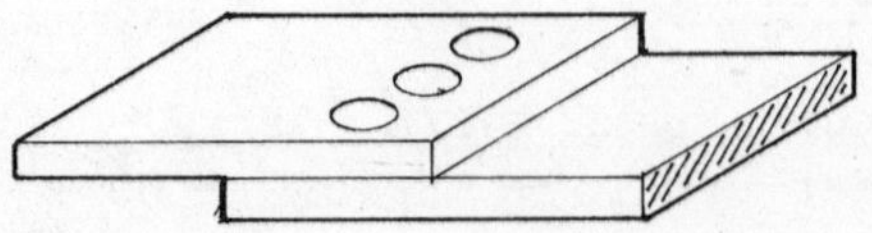

Figure 12.7

__________ 4. The area that must be sheared in order to separate the bars by applying a force parallel to them.

__________ 5. The force applied parallel to the bars that will shear the three rivets.

A rectangular slab of jello length 1.00 x $10^{-1}$ m, width 5.00 x $10^{-2}$ m, and height 3.00 x $10^{-2}$ m is subjected to a shearing force of 5.00 x $10^{-1}$ N, as shown in Fig. 12.8. The angle $\phi$ is 10.0°
Determine the following:

__________ 6. Shear stress on the jello
__________ 7. Shear strain on the jello
__________ 8. Shear modulus for the jello

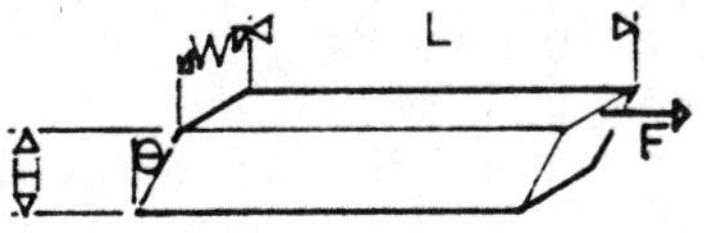


Fig. 12.8

Lead blocks of length 2.00 x $10^{-1}$ m, width 1.00 x $10^{-1}$ m, and height 2.00 x $10^{-2}$ m are stacked ten high. The mass of each block is 4.40 kg, and Young's Modulus for lead is 1.6 x $10^{10}$ N/m$^2$.
Determine the following:

__________ 9. Stress on the bottom block
__________ 10. Strain on the bottom block
__________ 11. Compression of the bottom block

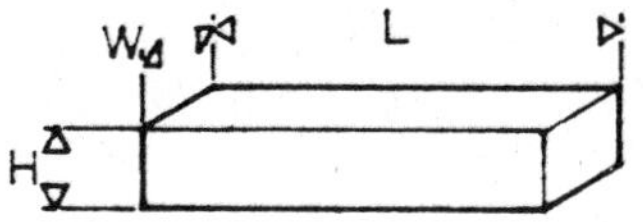


A solid brass sphere with a radius of 5.00 x $10^{-2}$ m is placed in a vacuum chamber, and the pressure is reduced to 1.00 x $10^{-4}$ percent of atmospheric pressure. The bulk modulus for brass is 6.80 x $10^{10}$ N/m$^2$.
Determine the following:

__________ 12. Change in volume of the brass sphere
__________ 13. Change in radius of the brass sphere

(See Appendix I for answers)

## PRINCIPAL CONCEPTS AND EQUATIONS PRESCRIPTION

Your score on the practice is an excellent measure of your understanding of this chapter. You should now use the following chart to write your own prescription for curing any of your physics ills. Look down the leftmost column to the number of the question(s) you answered incorrectly, read across that row to see which section(s) of the study guide you should return to for further study, and then do the suggested text problems to gain additional experience in working with the particular concept.

| Practice Test Question | Concepts and Equations | Prescription | |
|---|---|---|---|
| | | Principal Concept | Text Problems |
| 1 | Tensile stress | 1 | 12-1,3 |
| 2 | Tensile strain | 2 | 12,2-4 |
| 3 | Young's modulus | 3 | 12,4-7 |
| 4 | Shear stress | 1 | 12,14,15 |
| 5 | Shear stress | 1 | 12,15-16 |
| 6 | Shear stress | 1 | 12-14,15 |
| 7 | Shear strain | 2 | 12-14,15 |
| 8 | Shear modulus | 3 | 12-15 |
| 9 | Compressive stress | 1 | 12-6,8 |
| 10 | Young's modulus | 3 | 12-4,5 |
| 11 | Compressive strain | 2 | 12-8,9 |
| 12 | Bulk modulus | 3 | 12-18,19 |
| 13 | Bulk modulus | 3 | 12-17,18 |

# 13 Mechanics of Fluids

RECALL FROM PREVIOUS CHAPTERS

| Previously learned concepts and equations frequently used in this chapter | Text Section | Study Guide Page |
|---|---|---|
| Adding vectors analytically | 3.2 | 32 |
| Free-body force diagrams | 4.2 | 44 |
| First condition of equilibrium: $\sum_i \vec{F}_i = 0$ | | |
| Work: $W = F_{\parallel}\Delta s$ | 7.3 | 100 |
| Kinetic energy: $KE = mv^2/2$ | 7.5 | 108 |
| Work-energy theorem: $W_{net} = \Delta KE$ | 7.5 | 108 |
| Potential energy: $\Delta PE = mg\Delta h$ | 7.6 | 110 |

NEW IDEAS IN THIS CHAPTER

| Concepts and equations introduced | Text Section | Study Guide Page |
|---|---|---|
| Mass density: $\rho = m/V$ | 13.2 | 203 |
| Pressure: $P = F/A$ | 13.3 | 205 |
| Pressure due to a fluid: $P = \rho gh$ | 13.3 | 205 |
| Absolute pressure: $P_{abs} = P_A + \rho gh$; $P_{abs} = P_A + P_G$ | 13.3 | 205 |
| Archimedes' principle: $F_B = w_{fd} = m_{fd}g = \rho_f V_{fd} g$ | 13.5 | 208 |
| Continuity equation: $A_1 V_1 = A_2 V_2$ | 13.6 | 211 |
| Bernoulli's equation: $P + \rho gh + \rho v^2/2 = \text{constant}$ | 13.7 | 213 |

PRINCIPLE CONCEPTS AND EQUATIONS

1. Mass Density (Section 13.2)

Review: The density $\rho$ of a homogeneous substance is the mass of the material per unit volume.

$$\rho = m/V$$

Practice: Consider the three objects and the given information in Figure 13.1

Figure 13.1

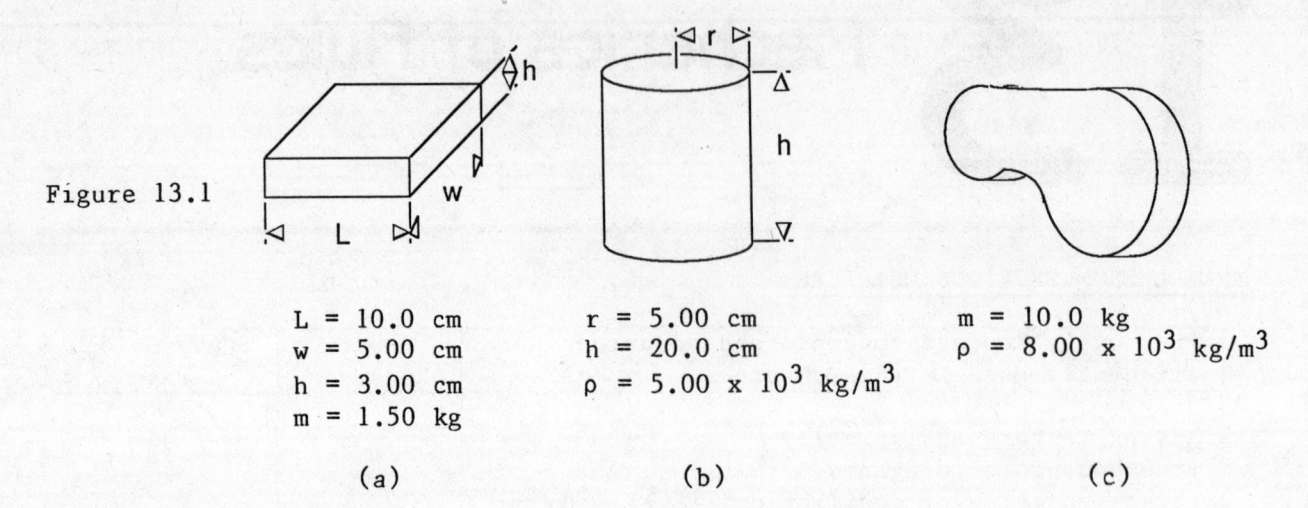

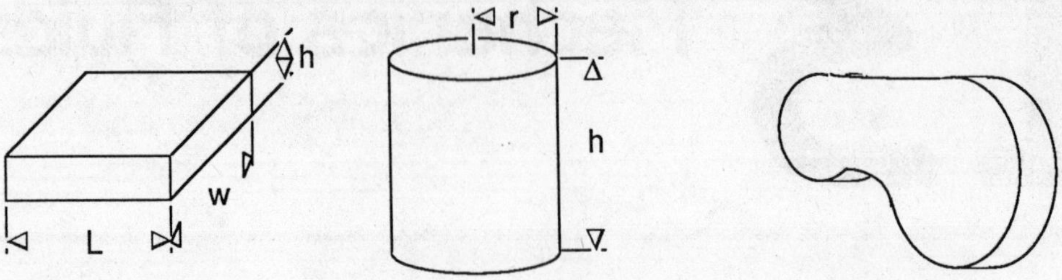


| (a) | (b) | (c) |
|---|---|---|
| $L = 10.0$ cm<br>$w = 5.00$ cm<br>$h = 3.00$ cm<br>$m = 1.50$ kg | $r = 5.00$ cm<br>$h = 20.0$ cm<br>$\rho = 5.00 \times 10^3$ kg/m$^3$ | $m = 10.0$ kg<br>$\rho = 8.00 \times 10^3$ kg/m$^3$ |

Determine the following:

| | |
|---|---|
| 1. Mass density of the object in Fig. 13.1a | $V = Lwh = 1.50 \times 10^{-4}$ m$^3$<br>$\rho = m/V = 1.00 \times 10^4$ kg/m$^3$ |
| 2. Mass of the object in Fig. 13.1b | $V = \pi r^2 h = 1.57 \times 10^{-3}$ m$^3$<br>$m = \rho V = 7.85$ kg |
| 3. Volume of the object in Figure 13.1c | $V = m/\rho = 1.25 \times 10^{-3}$ m$^3$ |

Example 13.1. Which weighs more, 2.00 m$^3$ of brass or 1.50 m$^3$ of mercury?

Given: $V_b = 2.00$ m$^3$ = volume of brass
$V_{Hg} = 1.50$ m$^3$ = volume of mercury
We also know that density of brass and mercury (Table 13.1 of text)
$\rho_b = 8.67 \times 10^3$ kg/m$^3$
$\rho_{Hg} = 13.6 \times 10^3$ kg/m$^3$

Determine: Which of these two weighs more?

Strategy: Knowing the volume and density of each sample, we can determine the mass and hence the weight.

Solution: $w_b = m_b g = \rho_b V_b g = 1.70 \times 10^5$ N
$w_{Hg} = m_{Hg} g = \rho_{Hg} V_{Hg} g = 2.00 \times 10^5$ N

Related Problems: 13-1, 13-3, 13-4.

## 2. Fluid Pressure (Sections 13.3, 13.4)

Review: Pressure is the magnitude of the force per unit area.

$$P = F/A$$

The pressure due to a height h of a fluid or at a depth h in a fluid of mass density $\rho$ may be written as

$$P = \rho gh$$

The absolute pressure at a depth h in a liquid is equal to the sum of the pressure due to the liquid ($\rho gh$) and atmospheric pressure ($P_A$).

$$P_{abs} = P_A + \rho gh$$

Most pressure-measuring devices (called gauges) are calibrated to read zero when the pressure is equal to atmospheric pressure. That is, gauges indicate the pressure difference between inside and outside the container. Consequently, the absolute pressure inside the container is given by

$$P_{abs} = P_A + P_G$$

Pascal's principle states that when a change in pressure is applied to an enclosed fluid, the change is transmitted undiminished to every point in the fluid and to the walls of the container.

Practice: Figure 13.2a shows a cylindrical container equipped with a movable air-tight piston lid and a U-tube mercury manometer. Figure 13.2b shows that when the cylinder is half-filled with an unknown liquid, the difference between the level of mercury in the two sides of the U-tube is 1.47 cm. Figure 13.2c shows the same situation as Figure 13.2b, except that the piston lid ($m_1$ = 1.00 kg) has been installed and a 9.00-kg mass ($m_2$) has been placed on it.

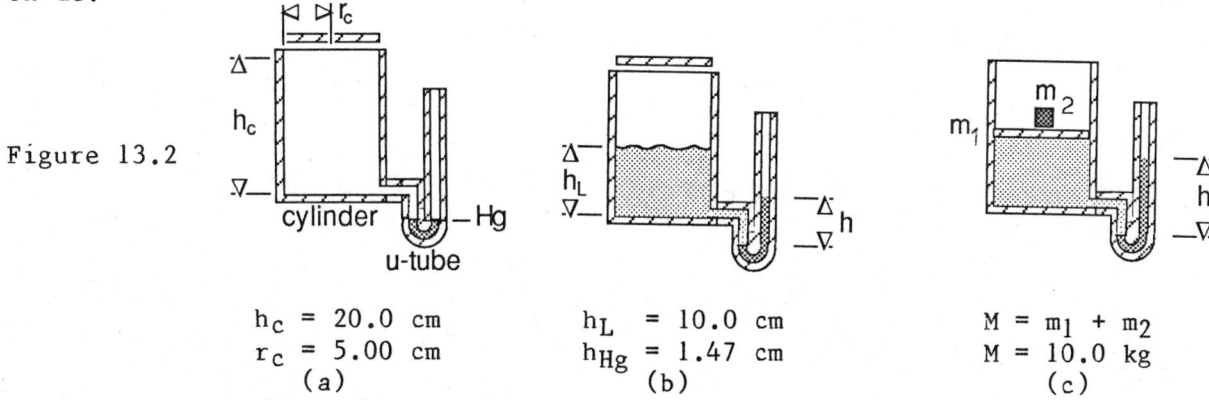

Figure 13.2

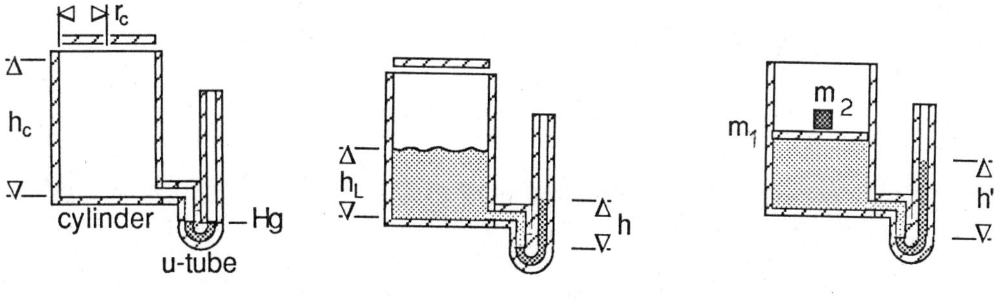


| (a) | (b) | (c) |
|---|---|---|
| $h_c$ = 20.0 cm | $h_L$ = 10.0 cm | $M = m_1 + m_2$ |
| $r_c$ = 5.00 cm | $h_{Hg}$ = 1.47 cm | M = 10.0 kg |

Determine the following:

| | |
|---|---|
| 1. Gauge pressure of the cylinder in Fig. 13.2a. | $P_G = 0$ |

| | |
|---|---|
| 2. Absolute pressure of the cylinder in Fig. 13.2a | $P_{abs} = P_A = 1.01 \times 10^5$ Pa |
| 3. Force on the inside bottom of the cylinder in Fig. 13.2a | $F = P_A A = P_A \pi r_c^2 = 793$ N |
| 4. Gauge pressure of the cylinder in Fig. 13.2b | $P_G = \rho_{Hg} g h_{Hg} = 1.96 \times 10^3$ Pa |
| 5. Pressure at the bottom of the cylinder due to the liquid in Fig. 13.2b | $P_{L\ bottom} = P_G = 1.96 \times 10^3$ Pa |
| 6. Pressure due to the liquid at a depth of 5.00 cm in Fig. 13.2b | Let h represent the depth<br>$P_L = \rho_L g h$<br>since $h = h_L/2$<br>$P_h = P_{L\ bottom}/2 = 9.8 \times 10^2$ Pa |
| 7. Density of the liquid in the cylinder | $\rho_L = P_{L\ bottom}/g h_L$<br>$= 2.00 \times 10^3\ kg/m^3$ |
| 8. Force on the bottom of the cyliner due to the liquid in Fig. 13.2b | $F_L = P_{L\ bottom} A = 15.4$ N or<br>$F_L = w_L = m_L g = \rho_L V_L g$<br>$= \rho_L \pi r^2 h_L g = 15.4$ N |
| 9. Absolute pressure at the bottom of cylinder in Fig. 13.2b | $P_{abs} = P_{L\ bottom} + P_A$<br>$= 1.96 \times 10^3$ Pa $+ 1.0 \times 10^5$ Pa<br>$= 1.02 \times 10^5$ Pa |
| 10. Pressure on the liquid due to the mass M in Fig. 13.2c | $F = Mg = 98.0$ N<br>$P = F/A = 1.25 \times 10^4$ Pa |
| 11. Pressure on the bottom of the cylinder due to the mass M in Fig. 13.2c | According to Pascal's principle, the pressure everywhere inside the container is increased by $1.25 \times 10^4$ Pa.<br>$P_{M\ bottom} = 1.25 \times 10^4$ Pa |
| 12. Gauge pressure of the cylinder in Fig. 13.2c | $P_G = P_L + P_M = 1.45 \times 10^4$ Pa |

| | |
|---|---|
| 13. Difference in the height of the mercury in the sides of the U-tube in Fig. 13.2c | $P_G = \rho_{Hg} g h_{Hg}$<br>$h_{Hg} = P_G/\rho_{Hg} g = 1.09 \times 10^{-1}$ m |

Example 13.2. The hydraulic system in Figure 13.3 is used to test the ultimate compressive stress of various samples. Stress is created on the sample by setting weights on piston 1.

Figure 13.3.

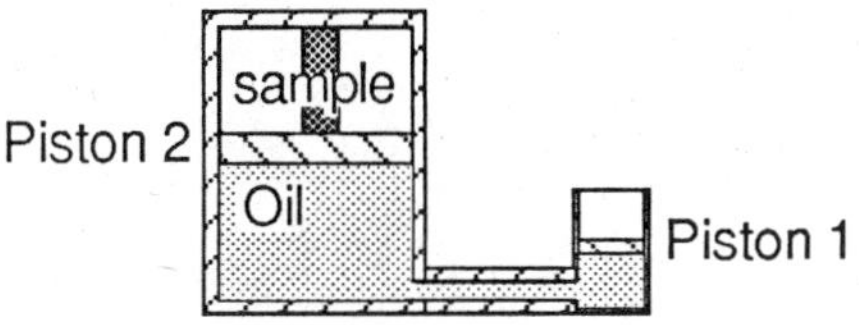


The samples are cylinders with a radius of 1.00 cm and a length of 10.0 cm. Determine the mass that must be placed on piston 1 in order to compress an aluminum sample by $1.00 \times 10^{-4}$ m.

Given:
$A_1 = 1.00 \times 10^{-3}\ m^2$ = area of piston 1
$A_2 = 1.00 \times 10^{-1}\ m^2$ = area of piston 2
$Y_{Al} = 7.00 \times 10^{10}\ N/m^2$ = Young's Modulus for compression of Aluminum
$L = 1.00 \times 10^{-1}$ m = length of sample
$\Delta L = 1.00 \times 10^{-5}$ m = compression of the sample
$r = 1.00 \times 10^{-2}$ m = radius of sample

Determine: Mass to be placed on piston 1 to compress the sample by $1.00 \times 10^{-5}$ m.

Strategy: Knowing Y, $\Delta L$, L, and r, we can determine the force ($F_s$) needed to compress the aluminum sample. Knowing $F_s$ and the area of piston 2 ($A_2$), we can determine the necessary fluid pressure. From the fluid pressure and the area of piston 1 ($A_1$), we can determine the force and hence the mass needed at piston 1.

Solution: The expression for Young's modulus is $Y = (F/A)/(\Delta L/L)$ This may be solved for the force ($F_s$) needed to compress the sample

$$F_s = YA(\Delta L/L) = Y(\pi r^2)\Delta L/L = 2.20 \times 10^3 \text{ N}$$

Since this force is to be supplied by piston 2, we obtain

$$F_2 = F_s = 2.20 \times 10^3 \text{ N}, \quad \text{and} \quad P_2 = F_2/A_2 = 2.20 \times 10^4 \text{ N/m}^2$$

According to Pascal's law, $P_1 = P_2$. By the definition of pressure, $P_1 = F_1/A$. Combining these, we obtain

$$F_1 = P_2 A_1 = (2.20 \times 10^4 \text{ N/m}^2)(1.00 \times 10^{-3} \text{ m}^2) = 22.0 \text{ N}$$

The mass required to supply this force is $m_1 = F_1/g = 2.24$ kg

Related Text Problems: 13-6 through 13-11.

## 3. Archimedes' Principle (Section 13.5)

Review: Archimedes' principle states that a body immersed in a fluid is buoyed up by a force equal to the weight of the fluid displaced by the body.

$$F_B = w_{fd} = m_{fd}g = \rho_f V_{fd} g$$

Where $w_{fd}$, $m_{fd}$, and $V_{fd}$, are the weight, mass, and volume of the fluid displaced; $\rho_f$ is the density of the fluid; and g is the acceleration due to gravity.

Practice: Figure 13.4a shows a block of wood, its length, cross-sectional area, and mass. Figure 13.4b shows the block held under water by a weight. Figure 13.4c shows the block half submerged in an unknown liquid (called liquid 1). Figure 13.4d shows the result of pouring another liquid (liquid 2) on top of liquid 1. Note that liquids 1 and 2 do not mix.

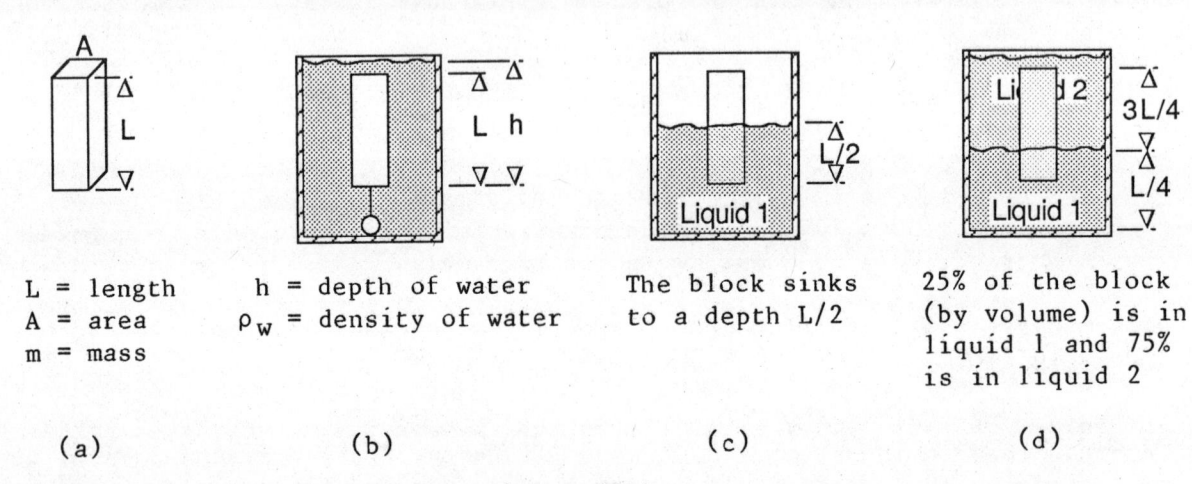

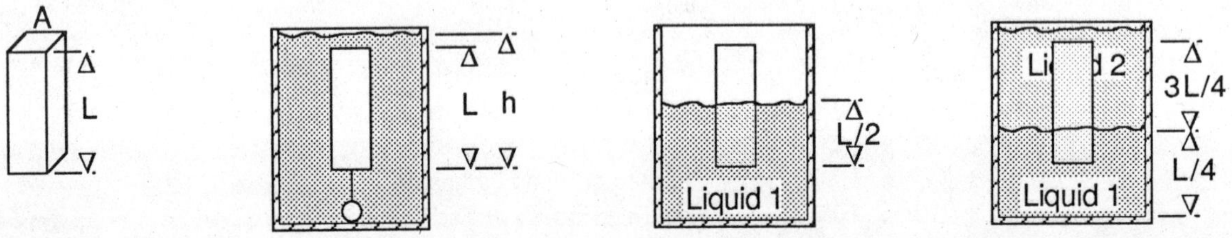


(a) L = length, A = area, m = mass

(b) h = depth of water, $\rho_w$ = density of water

(c) The block sinks to a depth L/2

(d) 25% of the block (by volume) is in liquid 1 and 75% is in liquid 2

Figure 13.4

Determine an expression for the following in terms of given quantities:

| | |
|---|---|
| 1. Volume of the block | $V_b = AL$ |
| 2. Density of the block | $\rho_b = m/V_b = m/AL$ |
| 3. Pressure at the bottom of the block due to the water in Fig. 13.4b | $P_{bottom} = \rho_w gh$ |
| 4. Pressure at the top of the block due to the water in Fig. 13.4b | $P_{top} = \rho_w g(h - L)$ |

| | |
|---|---|
| 5. Pressure difference between the bottom and top of the block due to the water in Fig. 13.4b | $\Delta P = P_{bottom} - P_{top}$ <br> $= \rho_w gh - \rho_w g(h - L) = \rho_w gL$ |
| 6. Upward (buoyant) force on the block due to the water in Fig. 13.4b | $F = \Delta PA = \rho_w gLA$ |
| 7. Weight of water displaced by the block in Fig. 13.4b | $w_{wd} = m_{wd}g = \rho_w V_{wd}g = \rho_w LAg$ |

Note: The expression for the buoyant force in step 6 is the same as the expression for the weight of water displaced in step 7. You have just discovered the same thing that Archimedes discovered.

| | |
|---|---|
| 8. Mass density of liquid 1 | Since the block is in equilibrium, the buoyant force due to liquid 1 must equal the weight of the block <br> $F_{B1} = w_b$ <br> $w_b = m_b g$ <br> $F_{B1} = w_{1d} = m_{1d}g = \rho_1 V_{1d}g = \rho_1 ALg/2$ <br> $m_b g = \rho_1 ALg/2$ <br> $\rho_1 = 2m_b/AL$ |
| 9. Mass density of liquid 2 | $w_b = F_{B1} + F_{B2}$ <br> $w_b = m_b g$ <br> $F_{B1} = \rho_1 ALg/4 = m_b g/2$ ($\rho_1$ from step 8) <br> $F_{B2} = 3\rho_2 ALg/4$ <br> $m_b g = m_b g/2 + 3\rho_2 ALg/4$, or <br> $\rho_2 = 2m_b/3AL$ |

<u>Example 13.3</u>. An object of volume $3.00 \times 10^{-4} m^3$ hangs by a cord from a spring balance, as shown in Fig. 13.5. When the object hangs in air, (figure 13.5a) the balance reads 1.00 kg. When the object hangs in a liquid of unknown density, the balance reads 0.550 kg. Determine the density of the liquid.

Figure 13.5

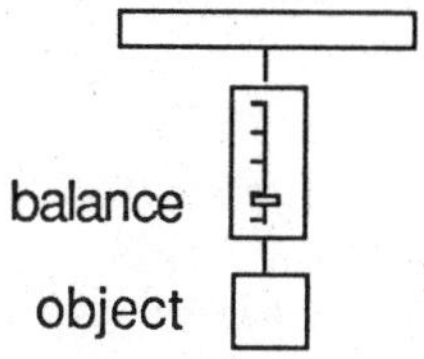


Balance reading
m = 1.00 kg
(a)

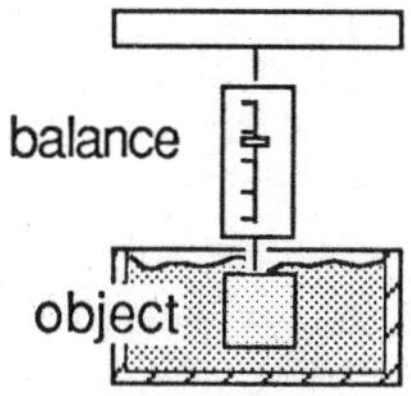


Balance reading
m' = 0.550 kg
(b)

Given: $m = 1.00$ kg = balance reading with the object suspended in air
$m' = 0.550$ kg = balance reading with the object suspended in liquid
$V = 3.00 \times 10^{-4}$ m$^3$ = volume of oject

Determine: Density $\rho_\ell$ of the liquid

Strategy: Since the object is in equilibrium, we can write a summation-of-forces statement that includes the tension in the cord, the buoyant force, and the weight of the object. We can determine the weight of the object, the tension in the cord, and hence the buoyant force from the given information. Once the buoyant force is known, we can use Archimedes' principle to determine the density of the liquid.

Solution: A free-body diagram for the object is shown in Fig. 13.6. Since the object is in equilibrium, we can write the following summation-of-forces statement:

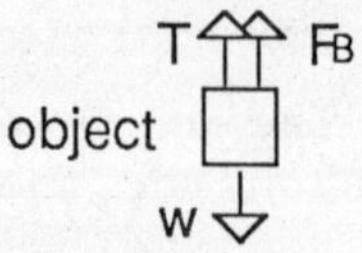


Figure 13.6

$$T + F_B = w$$

where $T = m'g$ = cord tension when the object is in the liquid
$w = mg$ = weight of the object
$F_B = w_{\ell d} = \rho_\ell V_{\ell d} g$ = buoyant force

When expressions for T, w, and $F_B$ are inserted into the summation-of-forces statement and it is solved for $\rho_\ell$, we obtain

$$\rho_\ell = (m - m')/V_{\ell d}$$

Noting that $V_{\ell d} = V$ and inserting values, we obtain

$$\rho_\ell = 1.50 \times 10^3 \text{ kg/m}^3$$

Example 13.4. A flat-bottomed barge has a length of 50.0 m, width of 10.0 m, and a depth of 2.00 m. The barge weighs $8.00 \times 10^5$ N, and we want it to float 0.500 m out of the water when loaded. What volume of coal ($\rho_c = 1.80 \times 10^3$ kg/m$^3$) can be loaded on the barge?

Given: $L = 50.0$ m = length of barge
$W = 10.0$ m = width of barge
$h = 2.00$ m = depth of barge
$d = 1.50$ m = depth of barge in water
$w_b = 8.00 \times 10^5$ N = weight of barge
$\rho_c = 1.80 \times 10^3$ kg/m$^3$ = density of coal
$\rho_w = 1.00 \times 10^3$ kg/m$^3$ = density of water

Determine: The volume of coal that can be loaded on to the barge with the barge floating 0.500 m out of the water.

Strategy: Knowing L, w, and d, we can determine the volume of water displaced ($V_{wd}$) by the barge. Knowing $V_{wd}$, $\rho_w$, and g, we can determine the weight of water displaced and hence the buoyant force $F_B$. Knowing that $F_B$ must support the weight of the barge $w_b$ and the weight of the coal $w_c$, we can determine

$w_c$. Knowing $w_c$ and $\rho_c$ we can determine the volume of coal $V_c$ that can be loaded onto the barge.

Solution: The volume of water displaced is $V_{wd} = Lwd$. The weight of water displaced and hence the buoyant force are given by

$$F_B = w_{wd} = m_{wd}g = \rho_w V_{wd} g = \rho_w Lwdg$$

This buoyant force must support the barge and coal, and so

$$F_B = w_b + w_c \qquad \text{or} \qquad w_c = F_B - w_b$$

Finally, the volume of the coal can be determined by

$$w_c = m_c g = \rho_c V_c g$$

$$V_c = \frac{w_c}{\rho_c g} = \frac{F_B - w_b}{\rho_c g} = \frac{\rho_w Lwdg - w_B}{\rho_c g} = 371 \text{ m}^3$$

Related Problems: 13-12 through 13-15.

4. Continuity Equation (Section 13.6)

Review: Figure 13.7 shows a rigid pipe full of an incompressible fluid of density $\rho$.

Figure 13.7

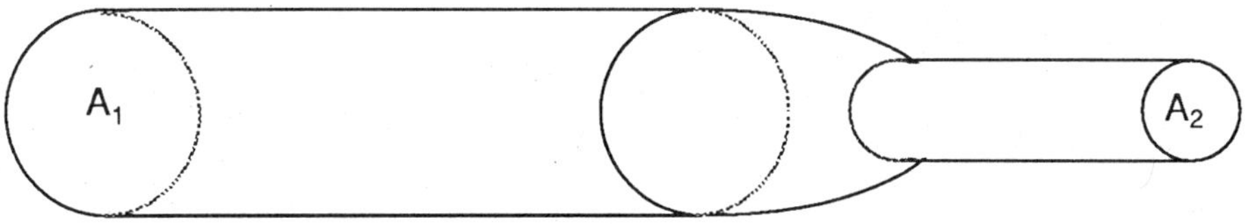


The volume flow rate is given by $Q = Av$. Since the pipe is rigid and full of an incompressible fluid, the rate at which fluid flows past $A_1$ must equal the rate at which it flows past $A_2$. That is

$$Q_1 = Q_2 \qquad \text{or} \qquad A_1 v_1 = A_2 v_2$$

This expression is the continuity equation.

Practice: Water enters a building in a pipe of inside diameter $5.00 \times 10^{-2}$ m and is piped to various locations in pipes of inside diameter $2.00 \times 10^{-2}$ m (Fig. 13.8). At one of the faucets, it is discovered that a $2.00 \times 10^{-2}$ m$^3$ container can be filled in 50.0 s.

$d_f = 2.00 \times 10^{-2}$ m = inside diameter of the pipe at the faucet

$d_b = 5.00 \times 10^{-2}$ m = inside diameter of the pipe entering the building

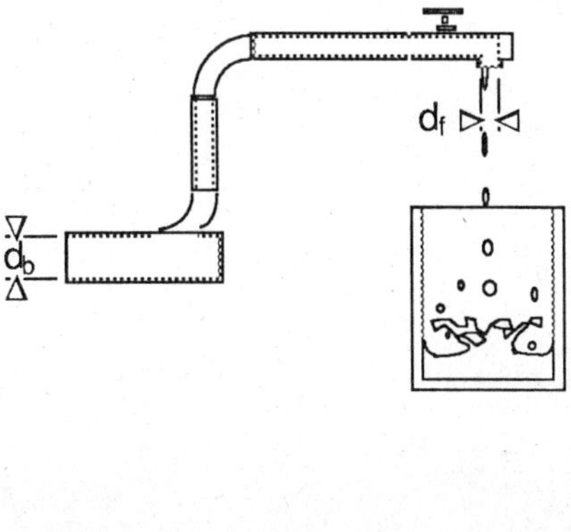


Figure 13.8

Determine the following:

| | |
|---|---|
| 1. Volume flow rate out of the faucet | $Q_f$ = volume flow/time<br>$Q_f = 2.00 \times 10^{-2}\ m^3/50.0\ s$<br>$= 4.00 \times 10^{-4}\ m^3/s$ |
| 2. Speed at which water leaves the faucet | $A_f = \pi d_f^2/4 = 3.14 \times 10^{-4}\ m^2$<br>$v_f = Q_f/A_f = 1.27\ m/s$ |
| 3. Volume flow rate at which water enters the building | $Q_b = Q_f = 4.00 \times 10^{-4}\ m^3/s$ |
| 4. Speed at which water enters the building | $A_b = \pi d_b^2/4 = 19.6 \times 10^{-4}\ m^2$<br>$v_b = Q_b/A_b = 0.204\ m/s$ |

Example 13.5. A pipe 0.250 m in diameter, has a constriction 0.100 m in diameter. The pipe is filled to capacity with water, and the water flows through the constricted section of pipe at the rate of 500 kg/s. Determine (a) the volume flow rate for the pipe and (b) the speed of the water in each section of the pipe.

Given: $d_1 = 0.250$ m = diameter of large pipe
$d_2 = 0.100$ m = diameter of small pipe
$m/t = 500$ kg/s = mass flow rate of water through constriction

Determine: (a) The volume flow rate; (b) the speed at which water flows through each section of pipe.

Strategy: Knowing the mass flow rate (m/t) and the density of water, we can determine the volume flow rate Q. Knowing Q and the pipe diameters, we can determine the speed at which water flows through each section.

Solution: Since the pipe is filled to capacity, the volume flow rate is the same for both sections and can be determined as follows:

$$Q = V/t = (m/\rho)/t = (m/t)/\rho = (5.00 \times 10^2\ kg/s)/(10^3\ kg/m^3) = 5.00 \times 10^{-1}\ m^3/s$$

The speed of flow through each section of pipe can be determined as follows:

$$v_1 = Q/A_1 = Q/(\pi d_1^2/4) = 10.2\ m/s$$

$$v_2 = Q/A_2 = Q/(\pi d_2^2/4) = 63.8\ m/s$$

Related Text Problems: 13-16, 13-17a, 13-18, 13-19a, 13-22b.

## 5. Bernoulli's Equation (Sections. 13.7, 13.8)

Review: Bernoulli's equation states that if a rigid pipe is full of a flowing incompressible fluid, then

$$P + \rho gh + \rho v^2/2 = \text{constant}$$

Practice: Consider the section of pipe and the given information in Fig. 13.9. The pipe is full of water.

Figure 13.9

$r_1 = 1.00 \times 10^{-1}$ m
$r_2 = 3.00 \times 10^{-2}$ m
$v_2 = 2.00$ m/s
$P_2 = 1.00 \times 10^4$ Pa
$\rho_w = 1.00 \times 10^3$ kg/m$^3$

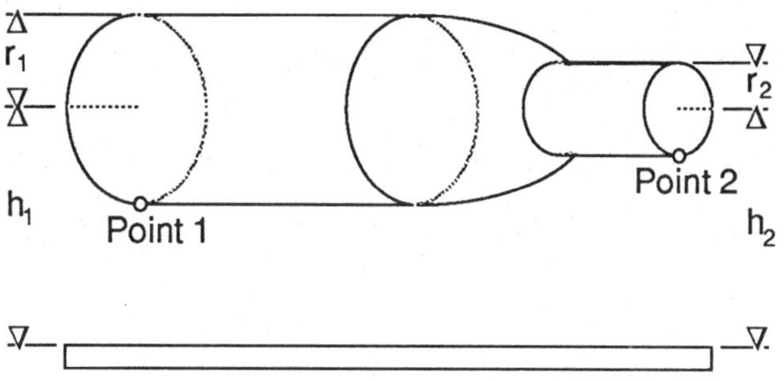


Note that $h_1 = h_2$ for this situation.

Determine the following:

| | |
|---|---|
| 1. Volume flow rate at point 2 | $Q_2 = A_2v_2 = \pi r_2^2v_2 = 5.65 \times 10^{-3}$ m$^3$/s |
| 2. Volume flow rate at point 1 | $Q_1 = Q_2 = 5.65 \times 10^{-3}$ m$^3$/s |
| 3. Water speed at point 1 | $v_1 = Q_1/A_1 = Q_1/\pi r_1^2 = 0.222$ m/s<br>or<br>$A_1v_1 = A_2v_2$ which gives<br>$v_1 = (A_2/A_1)v_2 = (r_2/r_1)^2v_2 = v_2/9$<br>$= 0.222$ m/s |
| 4. Water pressure at point 1 | $P_1 + \rho gh_1 + \rho v_1^2/2 = P_2 + \rho gh_2 + \rho v_2^2/2$<br>For this case, $h_1 = h_2$; hence<br>$P_1 = P_2 + \rho(v_2^2 - v_1^2)/2 = 1.20 \times 10^4$ Pa |

Consider the section of pipe and the given information in Fig. 13.10. The pipe is full of water with a mass flow rate of 4.00 kg/s.

Figure 13.10

$A_1 = A_2 = 3.00 \times 10^{-4}$ m$^2$
$h_1 = 0$
$h_2 = 4.00$ m
$m/t = 4.00$ kg/s

Notice that $v_1 = v_2$ for this situation.

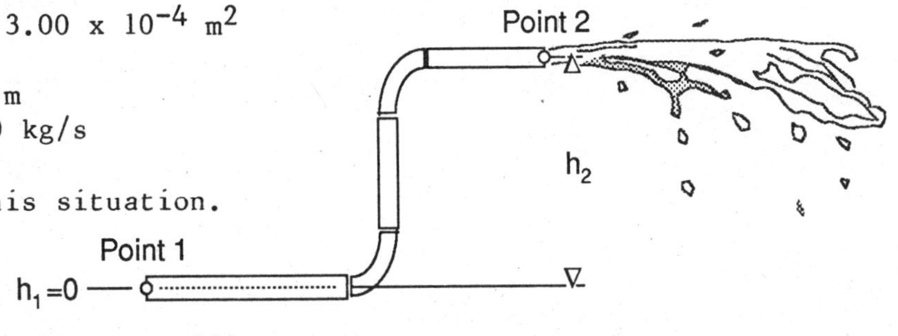

Determine the following:

| | |
|---|---|
| 5. Volume flow rate | $Q = V/t = (m/\rho)/t = (m/t)/\rho$<br>$= 4.00 \times 10^{-3}\ m^3/s$ |
| 6. Water speed | $v = Q/A = 13.3\ m/s$ |
| 7. Gauge pressure at point 1 | $P_1 + \rho g h_1 + \rho v_1^2/2 = P_2 + \rho g h_2 + \rho v_2^2/2$<br>For this case, $v_1 = v_2$; hence<br>$P_1 = P_2 + \rho g\ (h_2 - h_1)$<br>$h_1 = 0$<br>$P_2 = P_A$ = atmospheric pressure<br>$P_1 = P_{1G} + P_A$<br>$P_{1G} = \rho g h_2 = 3.92 \times 10^4$ Pa |

Figure 13.11 shows a large cylindrical water tank open at the top and with a small hole in the bottom.

Figure 13.11 $\quad r_1 = 2.00$ m, $r_2 = 2.00 \times 10^{-2}$ m, $h_1 = 5.00$ m

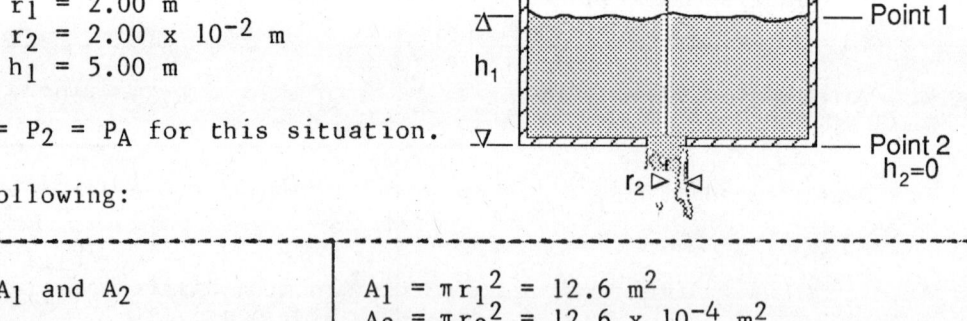

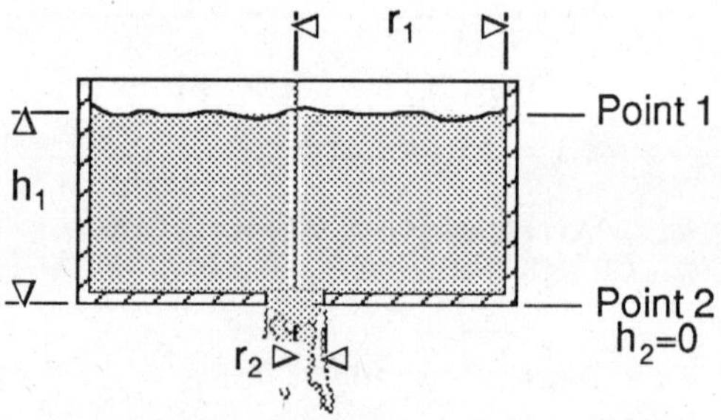


Notice that $P_1 = P_2 = P_A$ for this situation.

Determine the following:

| | |
|---|---|
| 8. The areas $A_1$ and $A_2$ | $A_1 = \pi r_1^2 = 12.6\ m^2$<br>$A_2 = \pi r_2^2 = 12.6 \times 10^{-4}\ m^2$ |
| 9. Any valid assumptions about $v_1$ | Since $A_1 = 10^4 A_2$, a valid assumption is that $v_1$ is negligibly small relative to $v_2$. For this reason, let's agree to set $v_1 = 0$. |
| 10. Speed of water leaving tank | $P_1 + \rho g h_1 + \rho v_1^2/2 = P_2 + \rho g h_2 + \rho v_2^2/2$<br>For this case, $P_1 = P_2$ and $v_1 = 0$; hence<br>$v_2 = (2gh_1)^{1/2} = 9.90\ m/s$ |
| 11. Volume of water leaving tank in 100 s | $Q_2 = A_2 v_2 = 1.25 \times 10^{-2}\ m^3/s$<br>$Q_2 = V_2/t$<br>$V_2 = Q_2 t = 1.25\ m^3$ |

Example 13.6. The design of a particular airplane calls for a lift of 800 N per square meter of wing area. If the flow velocity is 100 m/s past the lower wing surface, what flow velocity past the upper wing surface will give the required lift? The density of air is 1.25 kg/m$^3$.

Given: Lift = $8.00 \times 10^2$ N/m$^2$ = pressure difference between upper and lower surfaces

$v_\ell$ = $1.00 \times 10^2$ m/s = flow velocity past lower surface

$\rho_a$ = 1.25 kg/m$^3$ = density of air

Determine: Flow velocity past upper surface ($v_u$) to create the required lift.

Strategy: Since the wing thickness is small relative to the other distances involved, it is a good approximation to set $h_u = h_\ell$. The lift information gives us the required pressure difference between the lower and upper wing surfaces. We can use the pressure difference and $v_\ell$ to determine $v_u$.

Solution: The lift information gives us the pressure difference between the two surfaces.

$$\text{lift} = P_\ell - P_u = 8 \times 10^2 \text{ N/m}^2$$

Inserting this information into Bernoulli's equation

$$P_\ell + \rho_a g h_\ell + \rho_a v_\ell^2/2 = P_u + \rho_a g h_u + \rho_a v_u^2/2$$

we obtain

$$v_u = [2(P_\ell - P_u)/\rho_a + v_\ell^2]^{1/2} = 106 \text{ m/s}$$

Related Problems: 13-17b, 13-19b, 13-21 through 13-35.

---

PRACTICE TEST

Take and grade the practice test. Then use the chart in the next section to determine any weak areas and to write a prescription which will allow you to strengthen your understanding of physics in these areas.

Figure 13.12 shows a cylindrical water tank with an attached mercury U-tube manometer.

Figure 13.12

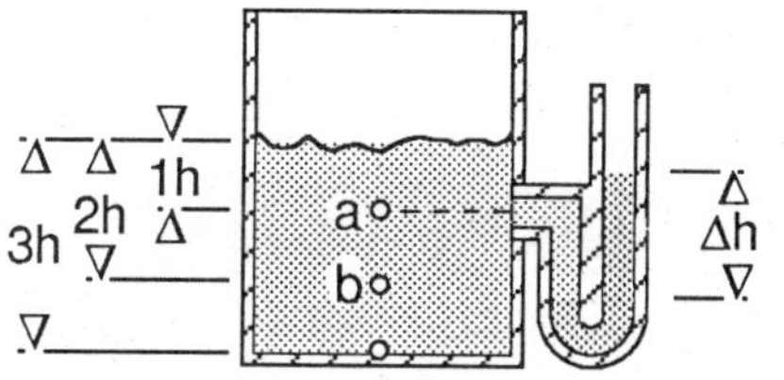


$\Delta h$ = 0.368 m of mercury

$\rho_{Hg}$ = $13.6 \times 10^3$ kg/m$^3$

$\rho_{water}$ = $1.00 \times 10^3$ kg/m$^3$

Determine the following:

______________ 1. Pressure due to water at level a

______________ 2. Pressure due to water at level b

______________ 3. Absolute pressure at bottom of tank

______________ 4. Distance h below surface where manometer is attached

Figure 13.13 shows a simple hydraulic system

$A_1 = 8.00 \times 10^{-2}\ m^2$ = area of piston 1
$A_2 = 40.0 \times 10^{-2}\ m^2$ = area of piston 2
$m_1 = 5.00$ kg = the mass on piston 1

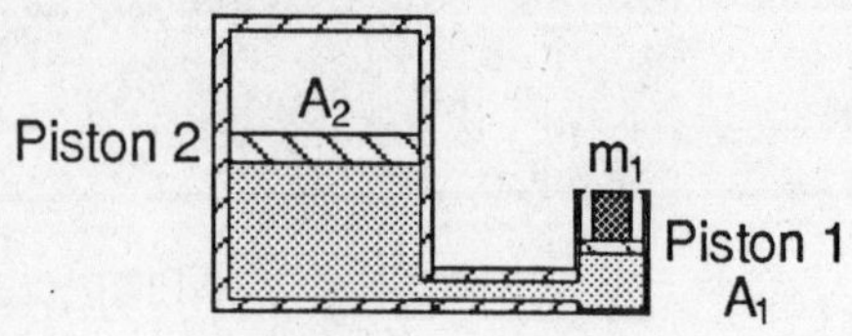


Figure 13.13

Determine the following:

_____________ 5. Pressure on fluid due to $m_1$
_____________ 6. Pressure on piston 2
_____________ 7. Upward force at piston 2 due to $m_1$
_____________ 8. Mass that could be lifted at piston 2 by $m_1$

The cylindrical object shown in Figure 13.14a floats in liquid 1 and hangs suspended by a cord in liquid 2. The tension in the cord is represented by T.

Figure 13.14

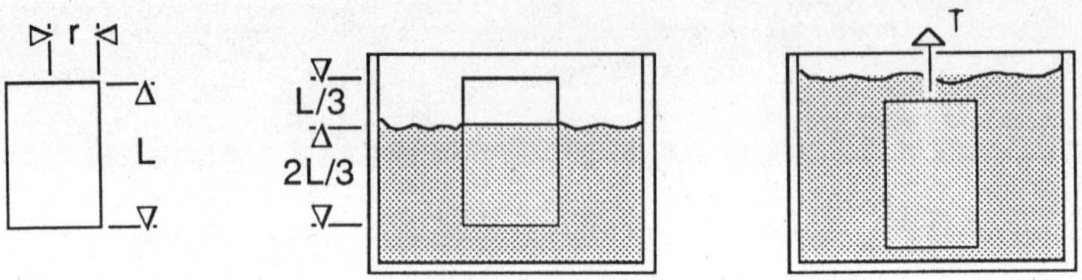


$L = 1.00 \times 10^{-1}$ m
$r = 2.00 \times 10^{-2}$ m
$m = 8.80 \times 10^{-2}$ kg
(a)

object floats in liquid 1
(b)

object suspended in liquid 2
$T = 2.45 \times 10^{-1}$ N
(c)

Determine the following:

_____________ 9. Density of object
_____________10. Weight of object
_____________11. Buoyant force on object in liquid 1
_____________12. Weight of liquid 1 displaced by object
_____________13. Density of liquid 1
_____________14. Buoyant force on object in liquid 2
_____________15. Weight of liquid 2 displaced by object
_____________16. Density of liquid 2

Figure 13.15 shows a rigid pipe full of a flowing incompressible fluid and an attached mercury manometer.

$d_1 = 1.00 \times 10^{-1}$ m = diameter of large pipe
$d_2 = 2.50 \times 10^{-2}$ m = diameter of small pipe
$\rho_f = 1.20 \times 10^{+3}\ kg/m^3$ = mass density of fluid
$m/t = 9.42$ kg/s = mass flow rate of fluid

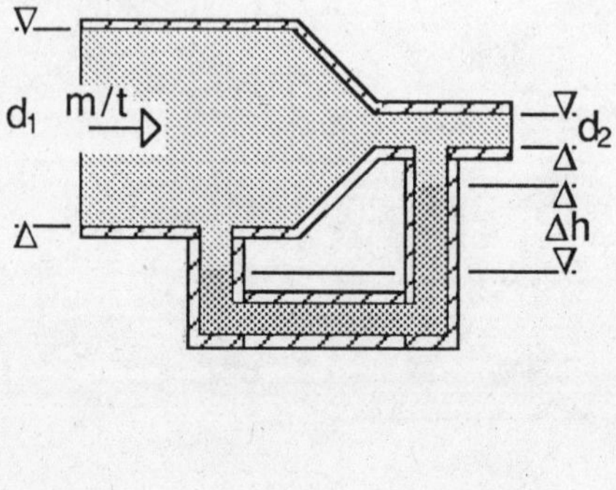


Figure 13.15

Determine the following:

_____ 17. Volume flow rate of fluid
_____ 18. Flow speed through large pipe
_____ 19. Flow speed through small pipe
_____ 20. Difference in pressure between the two pipes
_____ 21. Difference in mercury level in the sides of the U-tube

(See Appendix I for answers.)

## PRINCIPAL CONCEPTS AND EQUATIONS PRESCRIPTION

Your score on the practice test is an excellent measure of your understanding of this chapter. You should now use the following chart to write your own prescription for curing any of your physics ills. Look down the leftmost column to the number of the question(s) you answered incorrectly, read across that row to see which section(s) of the study guide you should return to for further study, and then do the suggested text problems to gain additional experience in working with the particular concept.

| Question Number | Concepts and Equations | Prescription | |
|---|---|---|---|
| | | Principal Concept | Text Problems |
| 1 | Pressure due to a fluid | 2 | 13-10,11 |
| 2 | Pressure due to a fluid | 2 | 13-9,10 |
| 3 | Absolute pressure | 2 | 13-7,8 |
| 4 | Pressure due to a fluid | 2 | 13-6,9 |
| 5 | Pressure | 2 | 13-5,7 |
| 6 | Pascal's principle | 2 | 13-6 |
| 7 | Pressure | 2 | 13-5,7 |
| 8 | Weight and mass | 2 of Ch. 4 | 4-27,28 |
| 9 | Density | 1 | 13-1,3 |
| 10 | Weight and mass | 2 of Ch. 4 | 4-27,28 |
| 11 | First condition of equilibrium | 1 of Ch. 6 | 6-1,2 |
| 12 | Archimedes' principle | 3 | 13-2,3 |
| 13 | Archimedes' principle | 3 | 13-12,13 |
| 14 | First condition of equilibrium | 1 of Ch. 6 | 6-4,5 |
| 15 | Archimedes' principle | 3 | 13-14,15 |
| 16 | Density | 1 | 13-1,3 |
| 17 | Volume flow rate | 1 | 13-16,17 |
| 18 | Continuity equation | 4 | 13-17,18 |
| 19 | Continuity equation | 4 | 13-19,22 |
| 20 | Bernoulli's equation | 5 | 13-17,19 |
| 21 | Pressure due to a fluid | 2 | 13-27,29 |

# 14 Temperature, Gases, and Kinetic Theory

## RECALL FROM PREVIOUS CHAPTERS

| Previously learned concepts and equations frequently used in this chapter | Text Section | Study Guide Page |
|---|---|---|
| Average speed: $v_{avg} = \Delta s/\Delta t$ | 2.3 | 19 |
| Newton's third Law: $\vec{F}_{12} = -\vec{F}_{21}$ | 4.5 | 57 |
| Kinetic energy: $KE = mv^2/2$ | 7.5 | 108 |
| Momentum: $p = mv$ | 8.2 | 123 |
| Stress: Stress = F/A | 12.1 | 193 |
| Strain: Strain = $\Delta L/L$ | 12.2 | 196 |
| Young's modulus: $Y = \text{stress/strain} = (F/A)/(\Delta L/L)$ | 12.3 | 198 |
| Pressure: $P = F/A$ | 13.3 | 205 |

## NEW IDEAS IN THIS CHAPTER

| Concepts and equations introduced | Text Section | Study Guide Page |
|---|---|---|
| Relationship between temperature scales: $t_F = (9/5)t_C + 32$; $t_C = (5/9)(t_F - 32)$; $T = t_C + 273$ | 14.3 | 219 |
| Thermal expansion: Linear $\Delta L = \alpha L\Delta T$; Area $\Delta A = 2\alpha A\Delta T$; Volume $\Delta V = 3\alpha V\Delta T$ | 14.4 | 222 |
| Gas laws: Boyle's law $P_1V_1 = P_2V_2$ if $T_1 = T_2$; Charles' law $V_1/T_1 = V_2/T_2$ if $P_1 = P_2$; Combined $P_1V_1/T_1 = P_2V_2/T_2$; Ideal gas law $PV = nRT$ | 14.5,6 | 224 |
| Kinetic theory of gases | 14.8 | 227 |
| Root-mean-square velocity: $v_{rms} = (3kT/m)^{1/2}$ | 14.8 | 227 |

PRINCIPAL CONCEPTS AND EQUATIONS

## 1. Temperature Scales (Section 14.3)

Review: The temperature scales in common use, the degree abbreviation, the values for melting ice and boiling water, and the functional relationships are as follows:

| Temperature Scale | Degree Abbr. | Value for Melting Ice | Value for Boiling Water | Functional Relationship |
|---|---|---|---|---|
| Fahrenheit | °F | 32°F | 212°F | $t_F = (9/5)t_C + 32$ |
| Celsius | °C | 0°C | 100°C | $t_C = (5/9)(t_F - 32)$ |
| Kelvin | K | 273 K | 373 K | $T = t_C + 273$ |

Practice: Figure 14.1 shows five thermometers, each having different numerical values for the temperatures of melting ice and boiling water.

Figure 14.1

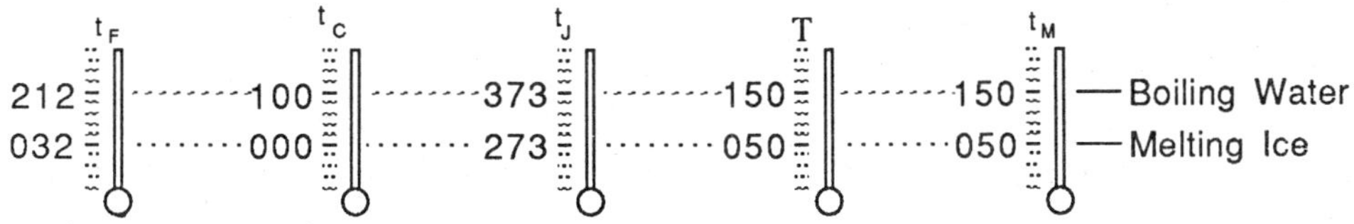


| Scale | Fahrenheit | Celsius | Kelvin | Jo | Mo |
|---|---|---|---|---|---|
| Degree Abbr. | °F | °C | K | °J | °M |

Note: Before starting the practice section, it is essential that we agree on notation.

$t_C = 10°C$ stands for ten degrees Celcius, as specific value on the Celcius scale.

$\Delta t_C = 10C°$ stands for a change in temperature of ten degrees on the Celcius scale (i.e. ten Celcius degrees).

When considering the Celcius and Fahrenheit scales, express the fact that 100 Celcius degrees is equal to 180 Fahrenheit degrees as

$$100\ C° = 180\ F°$$

Express the fact that 0° on the Celcius scale is equal to 32° on the Fahrenheit scale as

$$0°C = 32°F$$

Determine the following:

| | |
|---|---|
| 1. The relationship between $t_J$ and $t_C$ | 100 C° = 100 J°, or 1 C° = 1 J°<br>Every time the Celsius scale advances 1 C°, Jo's scale advances 1 J°. Since Jo starts out 50° higher, we write<br>$t_J = t_C + 50$<br>Notice that values of $t_C$ equal to 0°C and 100°C, give values of $t_J$ equal to 50°J and 150°J, respectively. |
| 2. The relationship between $t_M$ and $t_C$ | 100 C° = 300 M°, or 1 C° = 3 M°<br>Every time the Celsius scale advances 1 C°, Mo's scale advances 3 M°. Since Mo starts out 150° lower, we write<br>$t_M = 3t_C - 150$<br>Check: $t_C$ = 0°C → $t_M$ = -150°M<br>$t_C$ = 100°C → $t_M$ = 150°M |
| 3. The relationship between $t_J$ and $t_C$ | We already know $t_M$ as a function of $t_C$ (step 2) and $t_C$ as a function of $t_J$ (step 1). We can combine these to obtain $t_M$ as a function of $t_J$.<br>$t_M = 3t_C - 150$<br>$t_M = 3(t_J - 50) - 150 = 3t_J - 300$<br>(Perform the check.) |
| 4. The relationship between $t_F$ and $t_C$ | 100 C° = 180 F° or 1 C° = (9/5) F°<br>Every time the Celsius scale advances 1 C°, the Fahrenheit scale advances (9/5) F°. Since the Fahrenheit scale starts out 32° ahead, we write<br>$t_F = (9/5)t_C + 32$<br>(Perform the check.) |
| 5. The relationship between $t_C$ and T | 100 C° = 100 K° or 1 C° = 1 K°<br>Every time the Celsius scale advances 1 C°, the Kelvin scale advances 1 K°. Since the Kelvin scale starts out 273° ahead, we write $T = t_C + 273$<br>(Perform the check.) |

|  |  |
| --- | --- |
| 6. The relationship between $t_F$ and $T$ | $t_F = (9/5)t_C + 32$ (step 4)<br>$t_C = T - 273$ (step 5)<br>Combining these, we obtain<br>$t_F = (9/5)(T-273) + 32$<br>$t_F = (9/5)T - 459.4$<br>(Perform the check.) |
| 7. The scale reading when $t_F = t_C$ | $t_F = (9/5)t_C + 32$, set $t_F = t_C$<br>$t_C = (9/5)t_C + 32$ or $t_C = -40°C$<br>Check: Insert $t_C = -40°$ into the top expression to obtain $t_F = -40°$ |
| 8. The scale reading when $t_J = t_M$ | First notice in Fig. 14.1 that $t_J = t_M$ at 150°. However, let's prove this algebraically as follows:<br>$t_M = 3t_J - 300$, set $t_M = t_J$<br>$t_J = 3t_J - 300$ or $t_J = 150°J$ |

Example 14.1. Using their thermometers, inhabitants of planet X report that ice melts at -100°X and water boils at 150°X. We wish to send them a container of liquid oxygen (which boils at -183°C) with instructions about its properties. What should we tell them the boiling point is in °X?

Given: Melting point of ice = -100°X
Boiling point of water = 150°X
Boiling point of oxygen = -183°C

Determine: The boiling point of oxygen in °X.

Strategy: Determine the number of X° equal to 1 C°, compare the melting points of ice in °X and °C, and then develop a functional relationship between $t_X$ and $t_C$. Insert the boiling point of oxygen in °C to determine its value in °X.

Solution: 100 C° = 250 X° or 1 C° = 2.5 X°
Every time the Celsius scale advances by 1 C°, the X-scale will advance 2.5 X°. Since the X scale starts out 100° behind, we write

$$t_X = 2.5t_C - 100$$

Check: $t_C = 0°C \rightarrow t_X = -100°X$, and $t_C = 100°C \rightarrow t_X = 150°X$

Since the check agrees with the calibration information (values for melting ice and boiling water), we are confident our functional relationship is correct. Finally, we insert $t_C = -183°C$ to obtain $t_X = -558°X$.

Related Text Problems: 14-1 to 14-5.

## 2. Thermal Expansion (Section 14.4)

Review: Most objects expand when heated. Figure 14.2 shows an object with dimensions L, W, and H at temperature T and the change in these dimensions when the temperature is increased an amount $\Delta T$.

Figure 14.2

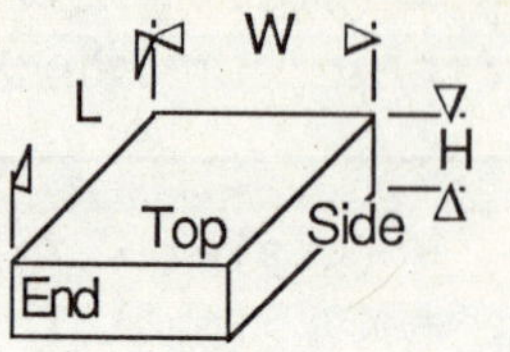


Temperature T

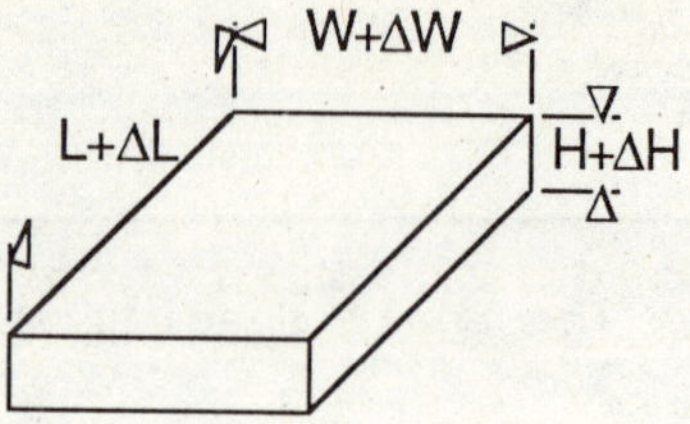


Temperature T + $\Delta T$

The linear dimensions are changed as follows:

L is changed by an amount $\Delta L = \alpha L \Delta T$
W is changed by an amount $\Delta W = \alpha W \Delta T$
H is changed by an amount $\Delta H = \alpha H \Delta T$

The quantity $\alpha$ (linear coefficient of thermal expansion) tells us the change in length (expansion or contraction) per unit length per degree change in temperature.

An area is changed by an amount

$$\Delta A = \gamma A \Delta T, \text{ where } \gamma = 2\alpha$$

The quantity $\gamma$ is the area coefficient of thermal expansion.

The volume is changed by an amount

$$\Delta V = \beta V \Delta T, \text{ where } \beta = 3\alpha$$

The quantity $\beta$ is the volume coefficient of thermal expansion. When you work text problem 14-17, you will learn why $\gamma = 2\alpha$ and $\beta = 3\alpha$.

Practice: The following is known about the object shown in Fig. 14.2:

$L = 2.00 \times 10^{-1}$ m, $W = 5.00 \times 10^{-2}$ m, $H = 2.00 \times 10^{-2}$ m, $\alpha = 2.50 \times 10^{-5}\ K^{-1}$ and $\Delta T = 100°C$

Determine the following:

| | |
|---|---|
| 1. The change in the linear dimensions of the object | $\Delta L = \alpha L \Delta T = 5.00 \times 10^{-4}$ m<br>$\Delta W = \alpha W \Delta T = 1.25 \times 10^{-4}$ m<br>$\Delta H = \alpha H \Delta T = 5.00 \times 10^{-5}$ m |

| | |
|---|---|
| 2. The change in the area of the top of the object | $A = LW$, $\gamma = 2\alpha$<br>$\Delta A = \gamma A \Delta T = 2\alpha LW \Delta T = 5.00 \times 10^{-5}\ m^2$ |
| 3. The change in volume of the object | $V = LWH$, $\beta = 3\alpha$<br>$\Delta V = \beta V \Delta T = 3\alpha LWH \Delta T = 1.50 \times 10^{-6}\ m^3$ |

A walkway connecting two buildings is supported by two steel I-beams that are slipped into place with a perfect fit on a cold day ($t_C = -10°C$). Consider a day when the steel temperature reaches 90°C.

Figure 14.3

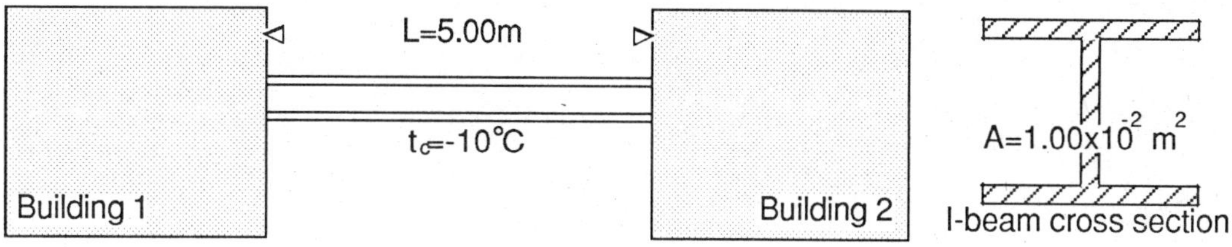


Determine the Following:

| | |
|---|---|
| 4. The stress needed on each I-beam to prevent expansion | $Y$ = stress/strain<br>stress = $(Y)$(strain) $= Y\Delta L/L = Y\alpha\Delta T$<br>stress $= 2.52 \times 10^8\ N/m^2$ |
| 5. The force needed on each I-beam to prevent expansion | stress $= F/A = 2.52 \times 10^8\ N/m^2$<br>$A = 1.00 \times 10^{-2}\ m^2$<br>$F$ = (stress)$(A) = 2.52 \times 10^6$ N |
| 6. The total force the walls of the building must be able to withstand to prevent expansion of the beams | $F_T = 2F = 5.04 \times 10^6$ N. Since normal wall construction cannot stand up to forces of this magnitude, an expansion joint is highly recommended. |

Example 14.2. The outside diameter of a wagon wheel is 1.00 m. An iron tire for this wheel has an inside diameter of 0.995 m at 25°C. To what temperature must the tire be heated in order for it to just slip onto the wheel?

Given: $d_w = 1.00$ m, $d_t = 0.995$ m at $t_C = 25°C$, $\alpha_{iron} = 12 \times 10^{-6}/K$

Determine: The temperature to which the tire must be raised so that its thermal expansion will allow it to slip onto the wheel.

Strategy: The solution to this problem depends on the knowledge that a cavity in a body expands or contracts with a change in temperature exactly as a solid object of the same size and composition would. We can treat this as either an area problem or a linear problem. In the area solution, we determine by how much we must change the area of the tire so that it is equal to the area of the wheel. Knowing $\Delta A$, we can determine $\Delta T$. In the linear solution, we determine how much we must change the diameter of the tire so it is equal to the diameter of the wheel. Knowing $\Delta d_t$, we can determine $\Delta T$.

Solution:

Area method
$$A_w = \pi r_w^2 \qquad A_t = \pi r_t^2$$
$$\Delta A_t = A_w - A_t = \pi(r_w^2 - r_t^2) = \gamma A_t \Delta T = 2\alpha\pi r_t^2 \Delta T$$
$$\Delta T = [(r_w/r_t)^2 - 1]/2\alpha = 420°C$$
$$T_f = T_i + \Delta T = 445°C$$

Linear Method
$$d_w = 1.00 \text{ m} \qquad d_t = 0.995 \text{ m}$$
$$\Delta d_t = d_w - d_t = \alpha d_t \Delta T$$
$$\Delta T = [(d_w/d_t) - 1]/\alpha = 420°C$$
$$T_f = T_i + \Delta T = 445°C$$

Related Problems: 14-6 through 14-17.

## 3. Gas Laws (Sections. 14.5, 14.6)

Review:

Boyle's law: $PV$ = constant or $P_1V_1 = P_2V_2$ at constant T
Charles' law: $V/T$ = constant or $V_1/T_1 = V_2/T_2$ at constant P
Charles' and Boyle's laws combined: $PV/T$ = constant or $P_1V_1/T_1 = P_2V_2/T_2$
Ideal gas law: $PV = nRT$, where R (= 8.314 J/mol·K) is a constant for all gases and n is the number of moles of gas. One mole of a gas contains Avogadro's number $N_A$ (= $6.02 \times 10^{23}$) of molecules and at STP occupies a volume of $2.24 \times 10^{-2}$ m$^3$.

Practice: The following is known about a quantity of an ideal gas.

$V_1 = 1.00 \times 10^{-2}$ m$^3$, $T_1 = 27.0°C$, $P_1 = 2.00 \times 10^5$ Pa, $\rho = 12.8 \times 10^{-1}$ kg/m$^3$

Determine the following:

| | |
|---|---|
| 1. Volume of the gas if the temperature changes to $T_2 = 227°C$ and the pressure remains constant | $V_1/T_1 = V_2/T_2$ at constant P<br>$V_2 = V_1T_2/T_1$<br>$V_2 = (1.00 \times 10^{-2}\text{ m}^3)(500\text{ K}/300\text{ K})$<br>$V_2 = 1.67 \times 10^{-2}$ m$^3$ |
| 2. Volume of the gas if the pressure changes to $P_2 = 5.00 \times 10^5$ Pa and the temperature remains constant | $P_1V_1 = P_2V_2$ at constant T<br>$V_2 = P_1V_1/P_2$<br>$V_2 = 4.00 \times 10^{-3}$ m$^3$ |
| 3. Pressure of the gas if the temperature changes to $T_2 = 227°C$ and the volume remains constant | $P_1V_1/T_1 = P_2V_2/T_2$<br>If $V_1 = V_2$, this reduces to<br>$P_2 = P_1(T_2/T_1) = 3.33 \times 10^5$ Pa |

| | |
|---|---|
| 4. Temperature of the gas in °C if the pressure changes to $P_2 = 3.00 \times 10^5$ Pa and the volume changes to $1.50 \times 10^{-2}$ m$^3$ | $P_1V_1/T_1 = P_2V_2/T_2$<br>$T_2 = P_2V_2T_1/P_1V_1 = 675$ K<br>$t_C = T - 273 = 402$°C |
| 5. Number of moles of the gas | $PV = nRT$<br>$n = PV/RT = 0.802$ mole |
| 6. Number of molecules of the gas | Every mole of the gas contains $N_A$ molecules.<br>$N = nN_A = 4.83 \times 10^{23}$ molecules |
| 7. Mass of the gas | $\rho = M_g/V$<br>$M_g = \rho V = 12.8 \times 10^{-3}$ kg |
| 8. Mass of one molecule | We just determined that N molecules have a mass M, and so the mass per molecule is<br>$m = M_g/N = 2.65 \times 10^{-26}$ kg/molecule |
| 9. Molar mass of the gas | The molar mass is the mass of one mole or $N_A$ molecules<br>$M = mN_A = 1.60 \times 10^{-2}$ kg/mole |

Example 14.3 A scuba diver's $1.00 \times 10^{-2}$ m$^3$ tank is filled with air at a gauge pressure of $150 \times 10^5$ Pa. If the diver uses $2.50 \times 10^{-2}$ m$^3$ of air per minute at the same pressure as the water pressure at her depth below the surface, how long can she remain under water at a depth of 20.0 m in a fresh-water lake? The temperature of the water falls from 25°C at the top to 20°C as she dives.

Given:
$V_1 = 10.0 \times 10^{-3}$ m$^3$ = volume of the tank
$P_{1G} = 1.50 \times 10^7$ Pa = initial gauge pressure of the tank
$r = 2.50 \times 10^{-2}$ m$^3$/60 s $= 4.17 \times 10^{-4}$ m$^3$/s = rate of use of air
$h = 20.0$ m = depth of scuba diver
$P_{2G}$ = pressure at which the diver uses oxygen = same as the water pressure at her depth
$T_1 = 25$°C = water temperature at the surface
$T_2 = 20$°C = water temperature at 20 m
$\rho_w = 1.00 \times 10^3$ kg/m$^3$ = density of water

Determine: The length of time the diver can remain under water (at a depth of 20.0 m) with this tank of air.

Strategy: From the given information, we can determine the absolute pressure and temperature at the surface and at 20.0 m. Since the volume at the surface is given, we can determine the volume available to the diver at 20 m. Knowing the volume available and the rate of consumption, we can determine the time the tank of air will last.

Solution: First, let's determine the absolute pressure at the surface and at a depth of 20.0 m.

Surface $P_1 = P_{1G} + P_A = 150 \times 10^5$ Pa $+ 1.01 \times 10^5$ Pa $= 151 \times 10^5$ Pa
Depth of 20 m $P_2 = \rho gh + P_A = 1.96 \times 10^5$ Pa $+ 1.01 \times 10^5$ Pa $= 2.97 \times 10^5$ Pa

Next, let's determine the absolute temperature at the surface and at 20.0 m.

Surface $T_1 = t_{C1} + 273 = 298$ K
Depth of 20 m $T_2 = t_{C2} + 273 = 293$ K

Using the ideal gas law, we can determine the amount of air available at 20.0 m.

$$P_1V_1/T_1 = P_2V_2/T_2 \quad \text{or} \quad V_2 = P_1V_1T_2/P_2T_1 = 0.500 \text{ m}^3$$

Because the tank is still full of air when it is all "used up" (no more air is available to the diver when the pressure inside the tank is the same as that outside the tank), the volume of air available is

$$V_{available} = V_2 - V_1 = 0.490 \text{ m}^3$$

The time this volume of air will last when being consumed at the rate r is

$$\text{Amount} = \text{rate} \times \text{time} \quad \text{or} \quad t = V_{available}/r = 1.18 \times 10^3 \text{ s}$$

Example 14.4. The volume of an oxygen tank is $20.0 \times 10^{-3}$ m$^3$. As oxygen is withdrawn from the tank, the reading on a pressure gauge drops from $10.0 \times 10^5$ Pa to $2.00 \times 10^5$ Pa and the temperature of the gas in the tank drops from 30°C to 15°C. (a) How many kilograms of oxygen were originally in the tank? (b) How many kilograms were withdrawn? (c) What volume would the withdrawn oxygen occupy at STP?

Given: $V_1 = 20.0 \times 10^{-3}$ m$^3$ $P_{1G} = 10.0 \times 10^5$ Pa $T_1 = 30$°C
$P_{2G} = 2.00 \times 10^5$ Pa $T_2 = 15$°C

Determine: (a) The mass of oxygen originally in the tank.
(b) The mass of oxygen withdrawn from the tank.
(c) The volume of the withdrawn oxygen at STP.

Strategy: Using the ideal gas law, we can determine the initial and final number of moles of oxygen in the tank. Knowing the number of moles and the molecular mass for oxygen, we can determine the initial and final mass of oxygen in the tank. Knowing the initial and final mass of oxygen in the tank, we can determine the mass withdrawn. Knowing the mass withdrawn, we can determine the number of moles withdrawn. Finally, we can use the ideal gas law to determine the volume of the withdrawn oxygen at STP.

Solution: We can find the initial and final number of moles in the tank by using the ideal gas law.

$P_1 = P_{1G} + P_A = 11.0 \times 10^5$ Pa; $T_1 = 30°C = 303$ K
$P_2 = P_{2G} + P_A = 3.01 \times 10^5$ Pa ; $T_2 = 15°C = 288$ K
$n_1 = P_1V_1/RT_1 = 8.74$ moles and $n_2 = P_2V_2/RT_2 = 2.52$ moles

The molecular mass M (that is, the mass of one mole) of oxygen is $32.0 \times 10^{-3}$ kg (recall that oxygen is diatomic $O_2$). The initial and final masses of oxygen gas in the tank are

$$M_1 = n_1M = 2.80 \times 10^{-1} \text{ kg} \quad \text{and} \quad M_2 = n_2M = 8.06 \times 10^{-2} \text{ kg}$$

The mass of gas withdrawn is $M_w = M_1 - M_2 = 1.94 \times 10^{-1}$ kg

The number of moles withdrawn is $n_w = n_1 - n_2 = 6.22$ moles

The volume this amount of oxygen occupies at STP is

$V_w = n_wRT/P$ ; $T = 0°C = 273$ K, $P = 1.01 \times 10^5$ Pa, hence $V_w = 1.40 \times 10^{-1}$ m$^3$

Related Problems: 14-18, 14-31 through 14-35.

## 4. Kinetic Theory of Gases (Sections 14.8)

Review: If n moles of a gas of molecular mass M are placed in a container of volume V at a temperature T, then the following is true:

$M_g = nM = Nm$ = total mass of gas
$m = M/N_A$ = mass of a single molecule of gas
$N = nN_A$ = number of molecules of gas
$\langle v^2 \rangle = 3kT/m$ = average of the square of the speeds of the molecules
$v_{rms} = [\langle v^2 \rangle]^{1/2}$ = root mean square of the speeds of the molecules
$\langle KE \rangle/\text{molecule} = m\langle v^2 \rangle/2 = 3kT/2$ = average kinetic energy per molecule
$P = NkT/V$ = pressure in the container
$k = R/N_A$ = relationship between constants

Practice: Six moles of helium gas are placed in a $2.00 \times 10^{-3}$ m$^3$ container at a temperature of 27°C.

Determine the following:

| | |
|---|---|
| 1. Molecular mass of helium | $M = 4.00 \times 10^{-3}$ kg/mole |
| 2. Mass of one helium molecule | $m = M/N_A = 6.64 \times 10^{-27}$ kg |
| 3. Number of molecules of helium in the container | $N = nN_A = 3.61 \times 10^{24}$ molecules |

| | |
|---|---|
| 4. Average of the square of the speeds of the molecules | $\langle v^2 \rangle = 3kT/m = 1.87 \times 10^6\ m^2/s^2$ |
| 5. Root-mean-square speed of the molecules | $v_{rms} = [\langle v^2 \rangle]^{1/2} = 1.37 \times 10^3$ m/s |
| 6. Average kinetic energy per molecule | $\langle KE \rangle = m\langle v^2 \rangle/2 = 6.21 \times 10^{-21}$ J<br>$\langle KE \rangle = 3kT/2 = 6.21 \times 10^{-21}$ J |
| 7. Total kinetic energy of the gas | $KE = N\langle KE \rangle = 2.24 \times 10^4$ J |
| 8. Pressure in the container | $P = NkT/V = 7.48 \times 10^6\ N/m^2$<br>$P = nRT/V = 7.48 \times 10^6\ N/m^2$ |

Example 14.5. At what temperature will the $v_{rms}$ of nitrogen molecules be equal to that of helium molecules at 27°C?

Given: Helium gas at 27°C and nitrogen gas

Determine: The temperature at which nitorgen gas molecules will have the same $v_{rms}$ as helium gas molecules at 27°C.

Strategy: We can write an expression for $v_{rms}$ as a function of T for both gases. Since we want $v_{rms\ N_2}$ to equal $v_{rms\ He}$, we can equate these expressions and solve for $T_{N_2}$.

Solution:

$$v_{rms\ N_2} = v_{rms\ He}$$

$$3kT_{N_2}/m_{N_2} = 3kT_{He}/m_{He} \quad \text{or} \quad T_{N_2} = T_{He}(m_{N_2}/m_{He})$$

Since $m_{N_2} = M_{N_2}/N_A$ and $m_{He} = M_{He}/N_A$, we can write

$$T_{N_2} = T_{He_2}(M_{N_2}/M_{He}) = (300\ K)(28/4) = 2100\ K$$

Related Text Problems: 14-36 through 14-40.

---

PRACTICE TEST Take and grade the practice test. Doing so will allow you to determine any weak spots in your understanding of the concepts taught in this chapter. The following section prescribes what you should study further to stregthen your understanding.

Sue decides to create her own temperature scale. She places an uncalibrated mercury thermometer into an ice-water bath and calls the mercury level 20°S. She then places the uncalibrated thermometer into boiling water and calls the mercury level 170°S. Finally, she divides the distance between these two levels into 150 equal lengths. Determine the following:

__________ 1. The reading on a Celsius thermometer when Sue's thermometer reads 95°S.

__________ 2. The reading on Sue's thermometer when a Fahrenheit thermometer reads 80°F.

__________ 3. The Kelvin temperature when Sue's thermometer reads 125°S.

A steel cable 4.00 m long is stretched tightly across a driveway on a day when the temperature is 30°C. Determine the following on a day when the temperature is -10°C:

__________ 4. The additional strain on the wire

__________ 5. The additional stress on the wire

At room temperature (25°C), a brass sphere 3.0000 cm in diameter is $5.0000 \times 10^{-4}$ cm larger than the inside diameter of a steel ring.

__________ 6. Determine the single temperature of both sphere and ring at which the sphere just slips through the ring.

A 150-$cm^3$ glass test tube is filled to the brim with acetone at 0°C.

__________ 7. Determine the amount of acetone that will overflow if the test tube and contents are heated to 100°C.

An ideal gas is pumped into a $1.00 \times 10^{-2}$ $m^3$ container at 27°C until the gauge pressure reads $8.99 \times 10^5$ Pa. Determine the following:

__________ 8. Absolute pressure of the gas

__________ 9. Number of moles of the gas in the container

__________ 10. Pressure of the gas if the temperature is doubled

__________ 11. Pressure of the gas if the temperature is doubled and the volume of the container is reduced by a factor of three.

A $1.00 \times 10^{-3}$ $m^3$ container is filled with $1.60 \times 10^{-1}$ kg of $O_2$ gas having a $v_{rms}$ of 500 m/s. Determine the following:

__________ 12. Molecular mass of $O_2$

__________ 13. Mass of one molecule of $O_2$

__________ 14. Number of moles of $O_2$

__________ 15. Number of molecules of $O_2$

__________ 16. Temperature of the $O_2$

__________ 17. Pressure of the $O_2$

__________ 18. Kinetic energy of the $O_2$

(See Appendix I for answers.)

## PRINCIPAL CONCEPTS AND EQUATIONS PRESCRIPTION

Your score on the practice is an excellent measure of your understanding of this chapter. You should now use the following chart to write your own prescription for curing any of your physics ills. Look down the leftmost column to the number of the question(s) you answered incorrectly, read across that row to see which section(s) of the study guide you should return to for further study, and then do the suggested text problems to gain additional experience in working with the particular concept.

| Practice Test Question | Concepts and Equations | Prescription | |
|---|---|---|---|
| | | Principal Concept | Text Problems |
| 1 | Temperature scales | 1 | 14-4,5 |
| 2 | Temperature sclaes | 1 | 15-1,2 |
| 3 | Temperature scales | 1 | 14-3,4 |
| 4 | Linear thermal expansion: $\Delta L = \alpha L \Delta T$ | 2 | 14-6,8 |
| 5 | Young's modulus: Y = stress/strain | 3 of Ch. 12 | 12-4,7 |
| 6 | Area thermal expansion: $\Delta A = 2\alpha A \Delta T$ | 2 | 14-10,17 |
| 7 | Volume thermal expansion: $\Delta V = 3\alpha V \Delta T$ | 2 | 14-11,13 |
| 8 | Absolute pressure: $P = P_G + P_A$ | 2 of Ch. 13 | 14-22 |
| 9 | Ideal gas law: $PV = nRT$ | 3 | 14-24,25 |
| 10 | Ideal gas law: $PV = nRT$ | 3 | 14-26,28 |
| 11 | Ideal gas law: $PV = nRT$ | 3 | 14-23,27 |
| 12 | Molecular mass M | 3 | 14-29,30 |
| 13 | Mass of one molecule: $m = M/N_A$ | 3 | |
| 14 | Number of moles: $M_g = nM$ | 3 | 14-26,34 |
| 15 | Number of molecules: $N = nN_A$ | 3 | 14-31 |
| 16 | $v_{rms} = [3kT/m]^{1/2}$ | 4 | 14-37,38 |
| 17 | Ideal gas law: $PV = nRT$ | 3 | 14-25 |
| 18 | Kinetic theory: $KE = N\langle KE\rangle = M\langle v^2\rangle/2$ | 4 | 14-39,40 |

# 15 Heat and Heat Transfer

## RECALL FROM PREVIOUS CHAPTERS

| Previously learned concepts and equations frequently used in this chapter | Text Section | Study Guide Page |
|---|---|---|
| Power: $P = \Delta W/\Delta t$ | 7.4 | 106 |
| Kinetic energy: $KE = mv^2/2$ | 7.5 | 108 |
| Potential energy: $\Delta PE = mg\Delta h$ | 7.6 | 110 |
| Temperature scales: $T = t_C + 273$ | 14.3 | 219 |

## NEW IDEAS IN THIS CHAPTER

| Concepts and equations introduced | Text Section | Study Guide Page |
|---|---|---|
| Mechanical equivalent of heat | 15.1 | 231 |
| Specific heat capacity: $c = \Delta Q/m\Delta T$ | 15.2 | 234 |
| Latent heat: $L = \Delta Q/m$ | 15.3 | 235 |
| Calorimetry: $\Delta Q_{Net} = 0$ | 15.3 | 238 |
| Conduction of heat: $\Delta Q/\Delta t = kA\Delta T/d$ | 15.4 | 240 |
| Convection of heat: $\Delta Q/\Delta t = hA\Delta T$ | 15.5 | 240 |
| Radiation of heat: $\Delta Q/\Delta t = \varepsilon\sigma AT^4$ | 15.6 | 240 |
| Relative humidity: $RH = (\rho/\rho_s)100\%$ | 15.7 | 248 |

## PRINCIPAL CONCEPTS AND EQUATIONS

### 1. Mechanical Equivalent of Heat (Section 15.1)

Review: Heat is a form of energy. The temperature of a system increases when heat is added and decreses when heat is removed. Traditional (but not SI) units of heat are

calorie (cal) = amount of heat necessary to raise the temperature of one gram of water by one Celsius degree

Calorie (Cal) = 1000 cal

British Thermal Unit (Btu) = amount of heat necessary to raise the temperature of one pound of water by one Fahrenheit degree

In order to use the quantity heat in our calculations, we must first express it in standard SI energy units (i.e., in Joules). Conversion factors have been determined experimentally and are

$$1.00 \text{ cal} = 4.184 \text{ J}$$
$$1.00 \text{ Cal} = 1000 \text{ cal} = 4184 \text{ J}$$
$$1.00 \text{ Btu} = 252 \text{ cal} = 1054 \text{ J}$$

Practice: Consider a 100-W light bulb in a desk lamp. Assume that 90% of the energy is converted into heat.

Determine the following:

| | |
|---|---|
| 1. Energy per unit time consumed by the light bulb | $E/t = P = 100 \text{ W} = 100 \text{ J/s}$ |
| 2. Energy per unit time converted into heat | $(E/t)_{heat} = 0.900(E/t) = 90.0 \text{ J/s}$ |
| 3. Amount of heat energy (in J) produced in 2 h | $E_{heat} = (E/t)_{heat}t$<br>$= (90.0 \text{ J/s})(2 \text{ h})(3600 \text{ s/h})$<br>$= 6.48 \times 10^5 \text{ J}$ |
| 4. Amount of heat energy (in cal) produced in 2 h | $E_{heat} = (6.48 \times 10^5 \text{ J})(1 \text{ cal}/4.184 \text{ J})$<br>$= 1.55 \times 10^5 \text{ cal}$ |

A man proposes to work off a large serving of blueberry pie and ice cream with a 25-kg barbell. Each lift of the barbell is 2.10 m, and his body is 12% efficient under these circumstances. The pie and ice cream combination contains 1500 Cal.

Determine the following:

| | |
|---|---|
| 5. Total number of calories consumed | $N = (1500 \text{ Cal})(1000 \text{ cal/Cal})$<br>$= 1.50 \times 10^6 \text{ cal}$ |
| 6. Amount of external work (in J) the man must do (at 12% efficiency) to work off the consumed calories | $W = (0.120)(1.50 \times 10^6 \text{ cal})(4.184 \text{ J/cal})$<br>$= 7.53 \times 10^5 \text{ J}$ |
| 7. Amount of external work required to lift the barbell once | $W = \Delta PE = mg\Delta h$<br>$= (25.0 \text{ kg})(9.80 \text{ m/s}^2)(2.10 \text{ m})$<br>$= 5.15 \times 10^2 \text{ J}$ |

| | |
|---|---|
| 8. Number of times he must lift the barbell to work off the desert | $N = W_{total}/W_{each\ time}$<br><br>$= 7.53 \times 10^5\ J/5.15 \times 10^2\ J$<br>$= 1.46 \times 10^3$ |

Example 15.1 A 1000-N crate is pulled 10.0 m across a rough surface, as shown in Fig. 15.1.

$w$ = 1000 N = weight of crate
$F$ = 400 N = force pulling crate
$\theta$ = 30° = angle shown
$s$ = 10.0 m = distance crate is pulled
$\mu$ = 0.200 = coefficient of kinetic friction

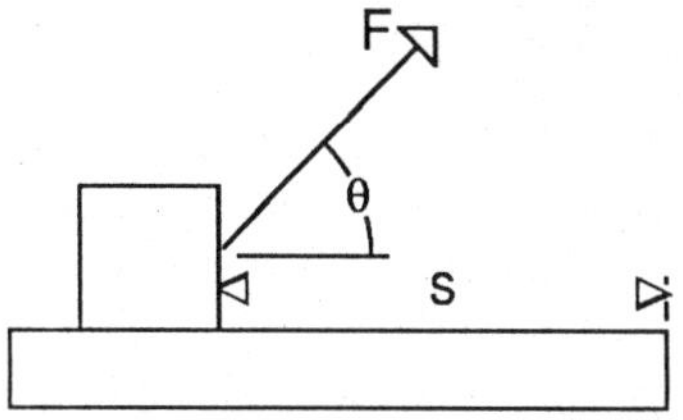


Figure 15.1

If all the work done on the crate by friction is converted into heat, how many calories of heat are generated?

Given: $W$ = weight of crate
$F$ = force on crate
$\theta$ = angle between F and S
$s$ = distance moved
$\mu$ = coefficient of kinetic friction

Determine: The number of calories of heat generated if all the work done on the crate by friction is converted into heat.

Strategy: Using the given information, we can determine first the normal force N, then the force of friction, and finally the amount of work done by friction. Work in Joules may be converted to heat in calories.

Solution: Figure 15.2 shows the crate and a free-body force diagram for it.

Figure 15.2

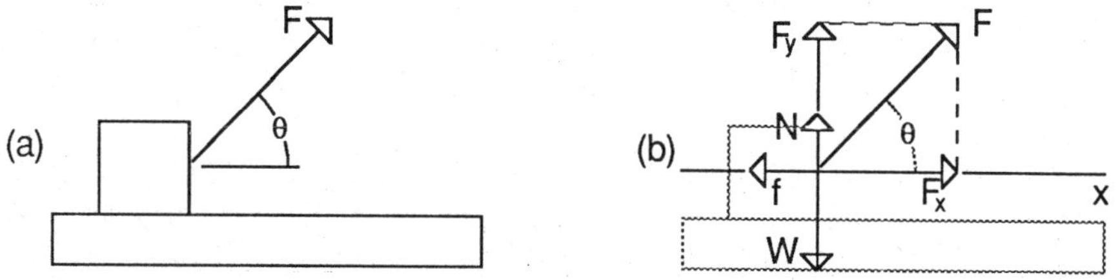


The normal force acting on the crate is obtained by realizing that the crate is in vertical equilibrium, and hence the sum of the forces in the y direction is zero.

$$\sum F_y = F_y + N_y + W_y = 0, \quad \text{where} \quad F_y = F\sin\theta, \quad N_y = +N, \quad W_y = -W$$

$$F\sin\theta + N - W = 0 \quad \text{or} \quad N = W - F\sin\theta = 800\ \text{N}$$

The frictional force acting on the crate is $f = \mu N = 160$ N
The work done on the crate by friction is $W_f = f_{\parallel} s = (-160\ \text{N})(10.0\ \text{m}) = -1600\ \text{J}$

The heat produced by this work is $Q = (-1600\ J)(cal/4.184\ J) = -382\ cal$
The minus sign in the work calculation tells us that friction is causing a loss of energy. The minus sign in the heat calculation tells us that the heat is lost.

Related Text Problems: 15-1 through 15-6.

## 2. Specific Heat Capacity (Section 15.2)

Review: The specific heat capacity, or simply the specific heat, of a substance is defined as the amount of heat required to raise a unit mass a unit temperature interval. This may be stated in equation form as

$$c = \Delta Q/m\Delta T$$

where c is the specific heat of a body of mass m and $\Delta Q$ is the heat required to increase the temperature by an amount $\Delta T$.

Practice: A 40.0-kg steel engine expends energy at the rate of 200 W while iding. Twenty percent of this energy goes into heating the engine by internal friction. The design is such that the engine radiates 90% of all the heat produced. The specific heat of steel is 449 J/kg·K.

Determine the following:

| | |
|---|---|
| 1. Rate (in J/s) at which energy is being consumed by friction | $(E/t)_f = 0.20(E/t)$<br>$= 0.20(200\ W) = 40.0\ J/s$ |
| 2. Rate (in cal/s) at which heat is produced due to friction | $(Q/t)_f = (E/t)_f(1\ cal/4.184\ J)$<br>$= (40.0\ J/s)(1\ cal/4.184\ J)$<br>$= 9.56\ cal/s$ |
| 3. Total amount of heat (in cal) produced due to friction in 10.0 min. | $Q_{total\ f} = (Q/t)_f t$<br>$= (9.56\ cal/s)(10.0\ min)$<br>$= 5.74 \times 10^3\ cal$ |
| 4. Amount of heat (in cal) available to raise the temperature of the engine during the 10.0 min | Since the engine radiates 90% of the heat produced, 10% is avaailable to heat the engine.<br>$Q_{heat} = 0.10Q_{total\ f} = 5.74 \times 10^2\ cal$ |
| 5. Change in temperature of the engine during the 10.0 min. | $m = 40.0\ kg,\ c = 449\ J/kg\cdot K$<br>$Q_{heat} = (5.74 \times 10^2\ cal)(4.184\ J/cal)$<br>$= 2.40 \times 10^3\ J$<br>$\Delta T = Q_{heat}/mc = 0.134\ K = 0.134°C$ |

| | |
|---|---|
| 6. Amount of water that could undergo a temperature change of 100°C during the 10.0 min if all the radiated heat was available for heating water | $Q_{rad} = 0.90Q_{total\ f} = 2.16 \times 10^4$ J<br>$\Delta T = 100$ K, $c = 4.18 \times 10^3$ J/kg·K<br>$m = Q_{rad}/c\Delta T = 0.517$ kg |

Example 15.2. A lead sphere is dropped 10.0 m into a large tank of water. When the sphere enters the water, its kinetic energy is quickly dissipated through friction. If 80% of the work done by friction goes into heating the sphere, what is the change in temperature of the sphere?

Given: $\Delta h = 10.0$ = distance sphere falls before entering water
0.80 = fraction of work done by friction that goes into heating sphere

Determine: The change in temperature of the sphere as its kinetic energy is dissipated through friction.

Strategy: Knowing the distance the sphere falls, we can write an expression for its decrease in gravitational potential energy. This decrease in gravitational potential energy is equal to the increase in kinetic energy, which in turn is equal to the amount of work done by friction on the sphere. Knowing that 80% of this work done by friction goes into heating the sphere, we can determine the heat supplied to the sphere and hence its temperature change.

Solution: Since the sphere falls a distance h before hitting the water, an expression for its change in gravitational potential energy is $\Delta PE = -mg\Delta h$. Since this decrease in gravitational potential energy is equal to the increase in kinetic energy, which is dissipated by doing work against friction, we can write

$$W_f = \Delta KE = -\Delta PE = mg\Delta h$$

Since 80% of this energy goes into heating the sphere, we can write

$$Q = 0.8W_f = 0.8mg\Delta h$$

We can then determine the temperature change as follows:

$$Q = mc\Delta T = 0.8mg\Delta h \quad \text{or} \quad \Delta T = 0.8g\Delta h/c = 0.613 \text{ K} = 0.613^\circ\text{C}$$

Related Text Problems: 15-3, 15-4, 15-5, 15-7, 15-8.

## 3. Latent Heat of Phase Change (Section 15.3)

Review: In general, the temperature of a substance increases when heat is added. When the heat added causes a phase change, however, the temperature does not increase. During a phase change, the heat energy added goes into the work of breaking molecular bonds rather than increasing the temperature. The heat energy involved in a phase change is called latent heat and can be expressed as

$$\Delta Q = mL$$

The latent heat can be latent heat of fusion $L_f$: solid to liquid (melt) or liquid to solid (freeze) or latent heat of vaporization $L_v$: liquid to gas (vaporize) or gas to liquid (condense).

For water, the latent heats are

$L_f = \Delta Q/m = 80$ cal/g $= 3.35 \times 10^5$ J/kg = amount of heat required to convert 1.00 g of ice at 0°C into 1.00 g of water at 0°C.

$L_v = \Delta Q/m = 540$ cal/g $= 2.26 \times 10^6$ J/kg = amount of heat required to convert 1.00 g of water at 100°C into 1.00 g of steam at 100°C.

Practice: Figure 15.3 shows a plot of temperature vs. heat energy supplied to 1.00 kg of some substance. The substance is a solid at temperatures lower than -10.0°C and a gas at temperatures above 20.0°C.

Figure 15.3

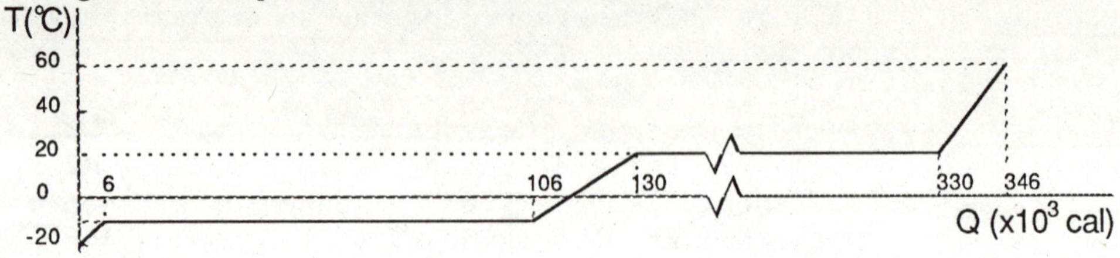


Determine the following:

| | |
|---|---|
| 1. Amount of heat energy needed to raise the temperature of the substance from -20°C to -10°C | $\Delta Q = Q_f - Q_i = 6.00 \times 10^3$ cal |
| 2. Specific heat of the substance as a solid | $\Delta Q = 6.00 \times 10^3$ cal $= 2.51 \times 10^4$ J<br>$\Delta T = 10.0$°C $= 10.0$ K, $m = 1.00$ kg<br>$c_{solid} = \Delta Q/m\Delta T = 2.51 \times 10^3$ J/kg·K |
| 3. Melting point for the substance | $T_{melt} = -10.0$°C |
| 4. Amount of heat required to melt the substance | $\Delta Q = Q_f - Q_i = 100 \times 10^3$ cal |
| 5. Latent heat of fusion for the substance | $\Delta Q = 1.00 \times 10^5$ cal $= 4.18 \times 10^5$ J<br>$m = 1.00$ kg<br>$L_f = \Delta Q/m = 4.18 \times 10^5$ J/kg |
| 6. Specific heat of the substance as a liquid | $\Delta Q = 24.0 \times 10^3$ cal $= 1.00 \times 10^5$ J<br>$\Delta T = 30.0$°C $= 30.0$ K, $m = 1.00$ kg<br>$c_{liquid} = \Delta Q/m\Delta T = 3.33 \times 10^3$ J/kg·K |

| | |
|---|---|
| 7. Latent heat of vaporization for the substance | $\Delta Q = 200 \times 10^3$ cal $= 8.37 \times 10^5$ J<br>$m = 1.00$ kg<br>$L_v = \Delta Q/m = 8.37 \times 10^5$ J/kg |
| 8. Specific heat of the substance as a gas | $\Delta Q = 16.0 \times 10^3$ cal $= 6.69 \times 10^4$ J<br>$\Delta T = 40°C = 40$ K, $m = 1.00$ kg<br>$c_{gas} = \Delta Q/m\Delta T = 1.67 \times 10^3$ J/kg·K |

Example 15.3
A passive-solar house is to have storage facilities for $1.00 \times 10^8$ J of heat energy, to be stored in a salt with the following physical properties.

$T_{melt} = T_{freeze} = 35.0°C$, $L_f = 3.00 \times 10^5$ J/kg, $\rho = 1.50 \times 10^3$ kg/m$^3$
$c_s = 2.00 \times 10^3$ J/kg·K = specific heat as a solid
$c_\ell = 3.00 \times 10^3$ J/kg·K = specific heat as a liquid

The minimum and maximum temperatures of the salt are, respectively, 25.0°C and 55.0°C. Determine the minimum storage space required.

Given: $T_{melt}$, $L_f$, $\rho$, $c_s$, $c_\ell$, $Q_{total}$, $T_i$, and $T_f$

Determine: The minimum volume of storage space required for the salt.

Strategy: Under the most severe conditions, the salt will start the day as a solid at 25.0°C. During the day, it will absorb heat to (a) warm as a solid to its melting point, (b) melt (i.e., undergo a phase change from a solid to a liquid), and (c) warm as a liquid to 55.0 °C. We can write an expression for the amount of heat required for each step and then equate the sum of these expressions to the total amount of heat to be stored. The only unknown in this equation is the mass of the salt. Once the mass of salt is known, we can use its density to determine the volume of salt and hence the minimum storage requirements.

Solution: Expressions for the heat required to warm the salt as a solid, melt it, and warm it as a liquid are

$$\Delta Q_s = mc_s\Delta T_s \qquad \Delta Q_{melt} = mL_f \qquad \Delta Q_\ell = Mc_\ell\Delta T$$

The sum of these three expressions gives the total heat to be stored:

$$Q_{total} = mc_s\Delta T_s + mL_f + mc_\ell\Delta T_\ell$$

$$m = Q_{total}/(c_s\Delta T_s + L_f + c_\ell\Delta T_\ell) = 2.63 \times 10^2 \text{ kg}$$

The volume of salt and hence the maximum storage requirement is

$$V = m/\rho = (2.63 \times 10^2 \text{ kg})/(1.50 \times 10^3 \text{ kg/m}^3) = 0.175 \text{ m}^3$$

Related Text Problems: 15-9 through 15-14, 15-16.

## 4. Calorimetry by the Method of Mixture (Section 15.3)

Review: Figure 15.4 shows a hot copper cylinder and a calorimeter containing water. After the copper is placed in the calorimeter, the water and copper reach an equilibrium temperature $T_e$.

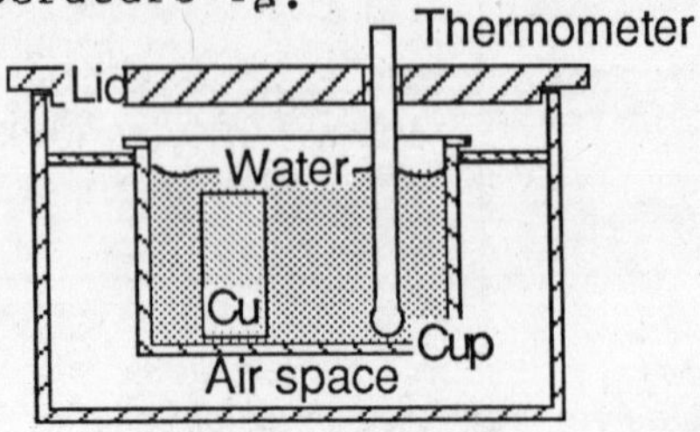


Figure 15.4.

The copper, calorimeter cup, and water are the thermal system of interest, and everything else is the environment. The calorimeter design is such that essentially no heat energy is exchanged between the system and the environment. Let's call the heat loss of the copper $\Delta Q_{Cu}$, the heat gain of the calorimeter cup $\Delta Q_{cup}$ and the heat gain of the water $\Delta Q_w$. Summarizing, we have

$$\text{Heat loss} = \Delta Q_{Cu}$$
$$\text{Heat gain} = \Delta Q_{cup} + \Delta Q_w$$

Since the system is isolated from the environment, the net flow of thermal energy for the system is zero. Thermal energy is transferred (the heat lost by the copper is gained by the cup and water); however, the net flow of thermal energy is zero. This allows us to write

$$\text{Net flow of thermal energy} = 0$$
$$\text{Heat loss + heat gain} = 0$$
$$\Delta Q_{Cu} + \Delta Q_{cup} + \Delta Q_w = 0$$

Practice: Consider the situation illustrated in Fig. 15.4 and the following data:

| | | |
|---|---|---|
| $m_{Cu}$ = 100 g | $c_{Cu}$ = 384 J/kg·K | $T_{iw} = T_{i\ cup}$ = 20.0°C |
| $m_{cup}$ = 50 g | $c_{cup}$ = 902 J/kg·K | $T_{iCu}$ = 100°C |
| $m_w$ = 250 g | $c_w$ = 4180 J/kg·K | |

Let $T_e$ represent the final equilibrium temperature.

Determine the following:

| | |
|---|---|
| 1. An expression for the heat lost by the copper | $\Delta T_{Cu} = T_{final} - T_{initial} = T_e - T_{i\ Cu}$<br>$\Delta Q_{Cu} = m_{Cu}c_{Cu}(T_e - T_{i\ Cu})$ |
| 2. An expression for the heat gain by the calorimeter cup | $\Delta T_{cup} = T_e - T_{i\ cup}$<br>$\Delta Q_{cup} = m_{cup}c_{cup}(T_e - T_{i\ cup})$ |

| | |
|---|---|
| 3. An expression for the heat gain by the water | $\Delta T_w = T_e - T_{iw}$<br>$\Delta Q_w = m_w c_w (T_e - T_{iw})$ |
| 4. An expression for the net heat flow for the system (copper, cup, and water) | $\Delta Q_{net} = \Delta Q_{Cu} + \Delta Q_{cup} + \Delta Q_w = 0$ |
| 5. Equilibrium temperature | $\Delta Q_{Cu} + \Delta Q_{cup} + \Delta Q_w = 0$<br>$m_{Cu} c_{Cu} (T_e - T_{i\ cu})$<br>$+ m_{cup}\ c_{cup} (T_e - T_{i\ cup})$<br>$+ m_w c_w (T_e - T_i) = 0$<br>Insert values to obtain $T_e = 22.7°C$ |

Example 15.4. How many 20.0-g ice cubes at -10.0°C must be placed in a 2.00-kg iron skillet to lower its temperature from 150°C to 50°C? Assume no heat loss to the environment.

Given:
$m_{ice\ cube} = 2.00 \times 10^{-2}$ kg
$T_{i\ ice} = -10.0°C$
$m_s = 2.00$ kg
$T_{is} = 150°C$
$c_{ice} = 2094$ J/kg·C°
$c_w = 4180$ J/kg·C°
$c_s = 449$ J/kg·C°
$L_f = 3.35 \times 10^5$ J/kg
$T_e = 50.0°C$

Determine: The number of ice cubes needed to lower the temperature of the skillet from 150°C to 50.0°C.

Strategy: From the given information, we can determine the amount of heat to be lost by the skillet. Knowing that all of this heat must be gained by the ice, we can determine the mass of ice that can be warmed from -10.0°C, melted, and the resulting water warmed to 50.0°C by the available amount of heat. Knowing the mass of ice and the mass of each ice cube,. we can determine the number of cubes needed.

Solution: The heat lost by the skillet is

$$\Delta Q_s = m_s c_s \Delta T_s = m_s c_s (T_{fs} - T_{is}) = -8.98 \times 10^4 \text{ J}$$

We are assuming no heat loss to the environment, hence

$$\Delta Q_{ice} + \Delta Q_s = 0 \quad \text{or} \quad \Delta Q_{ice} = -\Delta Q_s = 8.98 \times 10^4 \text{ J}$$

Since this amount of heat is available to warm the ice from -10.0°C to 0°C, to melt the ice, and to warm the resulting water from 0°C to 50.0°C, we can write

$$m_{ice} c_{ice} [0°C - (-10.0°C)] + m_{ice} L_f + m_w c_w (50.0°C - 0°C) = 8.98 \times 10^4 \text{ J}$$

Note that $m_w = m_{ice}$ because the water is the result of melting ice. Inserting values and solving for $m_{ice}$, we obtain $m_{ice} = 1.59 \times 10^{-1}$ kg.

The number of ice cubes needed to supply this amount of ice is

$$N = m_{ice}/m_{ice\ cube} = 7.95$$

Related Text Problems: 15-17 through 15-22.

## 5. Heat Transfer (Sections. 15.4, 15.5, 15.6)

Review: Heat may be transferred from one body to another by conduction, convection, or radiation.

In conduction, which is heat transfer through a substance, energy is passed from one layer of moleucles to another. The equation that describes the rate at which this process occurs is

$$\Delta Q/\Delta t = kA(T_2 - T_1)/d$$

Figure 15.5

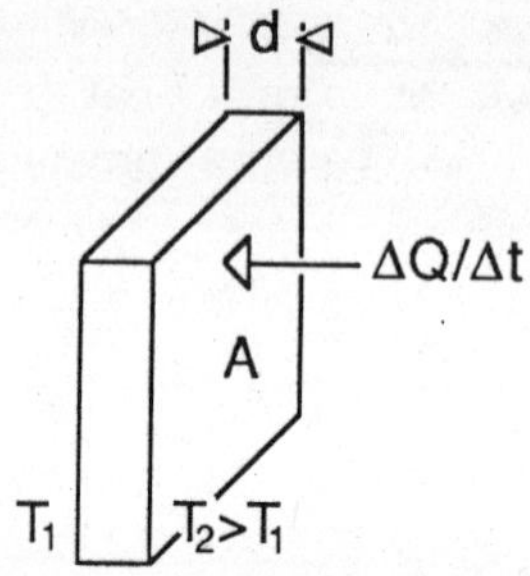


where $T_1$ and $T_2$ are temperatures, A is the cross-sectional area perpendicular to the heat transfer, d is the material thickness, k is the thermal conductivity, and $\Delta Q/\Delta t$ is the rate of conduction of heat. If the heat is conducted through more than one type of material, this equation is modified to

$$\Delta Q/\Delta t = \frac{A(T_2 - T_1)}{d_1/k_1 + d_2/k_2 + d_3/k_3}$$

Figure 15.6

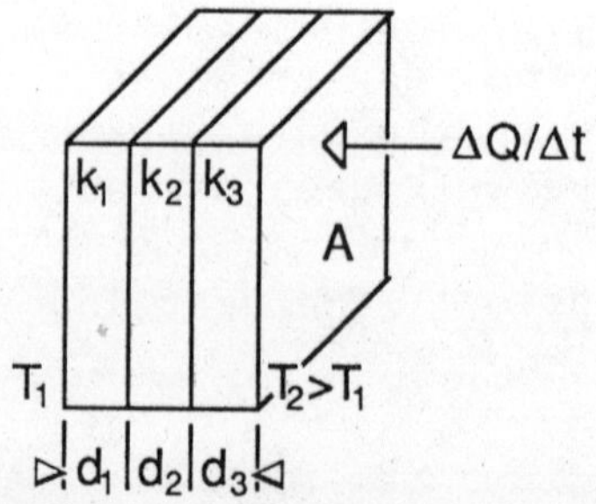


Convection is heat transfer from one point to another by the transfer of a warm fluid. Convection may be forced or natural. Figure 15.7 shows a situation where object 2 is heated by the forced convection of air past hot object 1. The convection takes place because object 1 heats the air, the air is forced to object 2, and the air heats object 2.

Figure 15.7

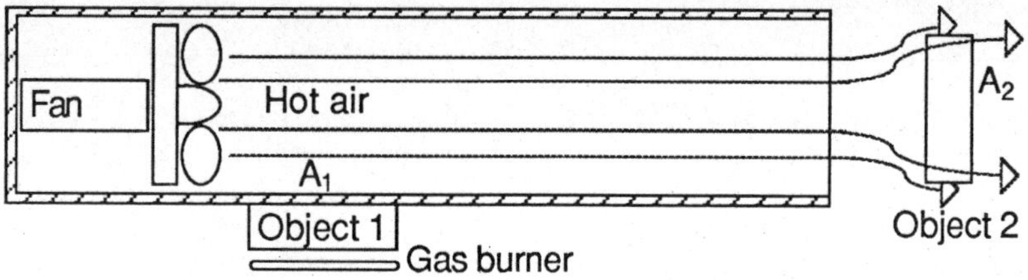


The rate at which heat is transferred from object 1 to the air is

$$(\Delta Q/\Delta t)_{1\rightarrow air} = h_1A_1(T_1 - T_{air\ cold})$$

The rate at which heat is transferred from air to object 2 is

$$(\Delta Q/\Delta t)_{air\rightarrow 2} = h_2A_2(T_{air\ hot} - T_2),\ \text{where}$$

$T_1$ = temperature of object 1
$T_{air\ cold}$ = temperature of the cold air heated by object 1
$T_{air\ hot}$ = temperature of the hot air heating object 2
$T_2$ = temperature of object 2
$A_1$ and $A_2$ = surface area of objects 1 and 2, respectively
$h_1$ and $h_2$ = convection coefficient for surfaces 1 and 2, respectively

Radiation is heat transfer in the form of electromagnetic waves. All bodies are both emitting and absorbing radiant energy all the time. The rate at which a body emits radiant energy to its environment is

$$(\Delta Q/\Delta t)_{emit} = \varepsilon\sigma A T^4_{obj}$$

The rate at which a body absorbs radiant energy from the environment is

$$(\Delta Q/\Delta t)_{abs} = \varepsilon\sigma A T^4{}_{envir}$$

The net rate at which a body acquires energy is

$$(\Delta Q/\Delta t)_{net} = (\Delta Q/\Delta t)_{abs} - (\Delta Q/\Delta t)_{emit} = \varepsilon\sigma A(T^4_{envir} - T^4_{obj})$$

The net gain of energy will be positive if $T_{envir} > T_{obj}$ and negative if $T_{envir} < T_{obj}$.

Practice (conduction): Single pane (SP) glass is 1/16 in. thick, and thermal pane (TP) glass is two single panes with an air gap of 1/4 in. A 1.00-$m^2$ window cost \$15 for SP and \$60 for TP. The building the window is to be placed in buys heat at the rate of \$0.10/kW·h. We wish to investigate the cost effectiveness of TP on a day when the temperature inside the building is 25°C and the temperature outside is -15°C. To assume that the outer surface of the window is -15°C and the inner surface is 25°C is entirely erroneous, as may be verified by touching the inner surface of a window pane on a cold day. Suppose experimentation determines that on such a day the temperature difference between the inside and outside of the window is 2.00°C. Thermal conductivity for glass is $k_g$ = 1.00 W/(m·°c).

Determine the following:

| | |
|---|---|
| 1. Rate at which heat is lost through SP | $d_g = (1/16 \text{ in})(1 \text{ m}/39.4 \text{ in})$ <br> $= 1.59 \times 10^{-3}$ m <br> $(\Delta Q/\Delta t)_{SP} = k_g A(T_1 - T_2)/d = 1.26 \times 10^3$ W |
| 2. Amount of heat lost through SP in one day | $\Delta Q_{SP} = (\Delta Q/\Delta t)_{SP} t = (1.26 \times 10^3 \text{ W})(1 \text{ day})$ <br> $= 1.09 \times 10^8$ J |
| 3. Rate at which heat is lost through TP | $(\Delta Q/\Delta t)_{TP} = A(T_1 - T_2)/[2d_g/k_g) + (d_a/k_a)]$ <br> $= 624$ W |
| 4. Amount of heat lost through TP in one day | $\Delta Q_{TP} = (\Delta Q/\Delta t)_{TP} t = (624 \text{ W})(1 \text{ day})$ <br> $= 5.39 \times 10^7$ J |
| 5. Difference in the amount of heat lost in one day | $\Delta Q_{diff} = \Delta Q_{SP} - \Delta Q_{TP} = 5.51 \times 10^7$ J |
| 6. Cost of heat energy in dollar per joule | $\text{Cost} = \left(\frac{\$0.10}{\text{kW}\cdot\text{h}}\right)\left(\frac{1 \text{ h}}{3600 \text{ s}}\right)\left(\frac{1 \text{ kW}}{1000}\right)\left(\frac{1 \text{ W}}{1 \text{ J/S}}\right)$ <br> $= \$2.78 \times 10^{-8}/\text{J}$ |
| 7. Cost to supply difference in heat loss for one day | $\text{Cost} = (5.51 \times 10^7 \text{ J})(\$2.78 \times 10^{-8}/\text{J})$ <br> $= \$1.53$ |
| 8. Cost difference between purchase of TP and SP | Cost difference = \$60.0 - \$15.0 = \$45.0 |
| 9. Number of days like the one being considered before TP would pay for itself | Cost/day = \$1.53/day <br> Cost difference = \$45.00 <br> N = \$45.00/(\$1.53/day) = 29.4 days |

Example 15.5 (conduction). The outer walls of a house have a layer of wood 2.00 cm thick and a layer of styrofoam insulation 3.00 cm thick. The thermal conductivity of the wood and styrofoam, respectively, are $1.40 \times 10^{-1}$ W/m·C° and $1.00 \times 10^{-2}$ W/m·C°. The temperature of the interior wall is 20°C, and that of the exterior wall -10°C. (a) How much heat flows through 1.00 $m^2$ of the wall in 1.00 h? (b) What is the temperature at the styrofoam-wood interface?

Given: $d_w$ = 2.00 x $10^{-2}$ m = thickness of wood
$d_s$ = 3.00 x $10^{-2}$ m = thickness of styrofoam
$k_w$ = 1.40 x $10^{-1}$ W/m·C° = thermal conductivity of wood
$k_s$ = 1.00 x $10^{-2}$ W/m·C° = thermal conductivity of styrofoam
$T_{in}$ = 20°C = temperature of inside surface of styrofoam
$T_{out}$ = -10°C = temperature of outside surface of wood
t = 1.00 h

Determine: (a) The amount of heat energy lost through 1.00 $m^2$ of the wall in 1.00 h. (b) The temperature of the styrofoam-wood interface.

Strategy: (a) From the given information, we can determine the rate at which heat is conducted per square meter of the composite wall. Knowing the rate at which the heat is conducted and that the conduction time is 1.00 h, we can determine the amount of heat conducted through 1.00 square meter in 1.00 h. (b) Let's represent the temperature at the wood-styrofoam interface by T and then write an expression for the rate of transfer of heat per square meter for the wood and styrofoam separately. The rate of transfer of heat per square meter through either the styrofoam or the wood can be no different from that through the composite wall, and this latter value has been determined. Thus we can use either expression to obtain T.

Solution: (a) The rate at which heat is conducted per square meter of the composite wall is

$$(\Delta Q/\Delta t)/A = (T_1 - T_2)/[d_w/k_w) + (d_s/k_s] = 9.55 \text{ W/m}^2$$

(b) Represent the temperature at the interface by T, and write an expression for the rate of transfer of heat per square meter for each material.

$$\left(\frac{\Delta Q/\Delta t}{A}\right)_w = k_w[T - (-10°C)]/d_w \qquad \left(\frac{\Delta Q/\Delta t}{A}\right)_s = k_s(25°C - T)/d_s$$

The rate of transfer of heat per square meter through either the styrofoam or the wood is the same as that through the composite wall, which was determined in (a). We can use either of the preceding expressions to obtain T. Using the expression for wood, we obtain

$$T = [\left(\frac{\Delta Q/\Delta t}{A}\right) d_w/k_w] - 10°C = 3.64°C$$

You should use the expression for styrofoam to check our results.

Related Text Problems (conduction): 15-23, 15-25, through 15-31a.

Practice (convection): A steam pipe of 10.0-cm outside diameter has an uninsulated section 1.00 m long. The outer surface of the pipe has a constant temperatuer 95°C, and the temperature of the surrounding air is 25°C. For this situation, h = 6.78 W/$m^2$·C°

Determine the following:

| | |
|---|---|
| 1. Area of the pipe that can lose heat to air in the room | $A = 2\pi r\ell = 3.14 \times 10^{-1}\ m^2$ |
| 2. Rate at which heat is lost to the room by natural convection | $\Delta Q/\Delta t = hA\Delta T = 1.49 \times 10^2\ J/s$ |
| 3. Amount of heat lost to room by natural convection per day | $\Delta Q = (\Delta Q/\Delta t)t = 1.29 \times 10^7\ J$ |

Example 15.6 (convection). Convection coefficients in $W/m^2 \cdot C°$ for the natural convection of air are

| | |
|---|---|
| Horizontal plate, facing upward | $h = 2.49(\Delta T)^{1/4}$ |
| Horizontal plate, facing downward | $h = 1.31(\Delta T)^{1/4}$ |
| Vertical plate | $h = 1.77(\Delta T)^{1/4}$ |

A metal cube 1.00 m on a side is maintained at 95.0°C in a 25.0°C room. Determine the amount of heat transferred to the room from the cube by natural convection in 1.00 h.

Given: $T_c$ = 95.0°C = temperature of the cube
$T_r$ = 25.0°C = temperature of the room
$\ell$ = 1.00 m =length of one side of cube
$t$ = 1.00 h = time for convection
Formulas for various convection coefficients

Determine: The amount of heat transferred from the cube to the room by natural convection in 1.00 h.

Strategy: From the given information, we can determine $\Delta T$ and then the various convection coefficients. Once the convection coefficients are known, we can determine the rate at which heat is lost from the cube by natural convection. Knowing the rate of convection, we can determine the amount of heat lost in 1.00 h.

Solution: Determine $\Delta T$ and A.

$$\Delta T = T_c - T_r = 70.0°C \qquad A_{top} = A_{bottom} = A_{side} = 1.00\ m^2$$

Determine the convection coefficients.

$$\text{Top } h = 2.49(70)^{1/4}\ W/m^2 \cdot C° = 7.20\ W/m^2 \cdot C°$$
$$\text{Bottom } h = 1.31(70)^{1/4}\ W/m^2 \cdot C° = 3.79\ W/m^2 \cdot C°$$
$$\text{Side } h = 1.77(70)^{1/4}\ W/m^2 \cdot C° = 5.12\ W/m^2 \cdot C°$$

Determine the rate at which heat is lost by natural convection from the top, bottom, and sides of the cube.

$$(\Delta Q/\Delta t)_{top} = h_{top} A_{top} \Delta T = 504 \text{ W}$$
$$(\Delta Q/\Delta t)_{bottom} = h_{bottom} A_{bottom} \Delta T = 265 \text{ W}$$
$$(\Delta Q/\Delta t)_{sides} = 4 h_{side} A_{side} \Delta T = 1430 \text{ W}$$

Determine the rate at which heat is lost by natural convection from the entire cube.

$$(\Delta Q/\Delta t)_{total} = (\Delta Q/\Delta t)_{top} + (\Delta Q/\Delta t)_{bottom} + (\Delta Q/\Delta t)_{sides} = 2200 \text{ W}$$

Determine the amount of heat lost by the cube by natural convection in 1.00 h.

$$\Delta Q = (\Delta Q/\Delta t)_{total}\, t = (2.20 \times 10^3 \text{ J/s})(1.00 \text{ h})(3600 \text{ s/h}) = 7.92 \times 10^6 \text{ J}$$

Related Text Problems (convection): 15-31, 15-32, 15-33.

Practice (radiation): A student in a bathing suit lies in the sun, and the following information is known.

$T_{stu}$ = 38°C = 311 K = surface temperature of skin
$T_{sun}$ = $5.80 \times 10^3$ K = surface temperature of sun
$r_{sun}$ = $6.96 \times 10^8$ m = radius of sun
$d$ = $1.50 \times 10^{11}$ m = distance from sun to earth
$\sigma$ = $5.67 \times 10^{-8}$ W/m$^2\cdot$K$^4$ = Stefan-Boltzmann constant
$\varepsilon_{sun}$ = 1.00 = emissivity of sun
$\varepsilon_{stu}$ = 0.980 = emissivity of student
$A_{stu}$ = 1.50 m$^2$ = exposed area of student
$L_{per}$ = $2.43 \times 10^6$ J/kg = heat of vaporization for perspiration
$(\Delta Q/\Delta t)_{metab}$ = $1.00 \times 10^3$ W = rate at which student's metabolism produces heat

Determine the following:

| | |
|---|---|
| 1. Rate at which the sun is emitting energy | $(\Delta Q/\Delta t)_{sun} = \sigma\varepsilon_{sun}A_{sun}T_{sun}^4 = 3.90 \times 10^{26}$ W |
| 2. Area of the sun | $A_{sun} = 4\pi r^2_{sun} = 6.08 \times 10^{18}$ m$^2$ |
| 3. Rate at which energy is emitted per square meter of the sun's surface | $\dfrac{(\Delta Q/\Delta t)_{sun}}{A_{sun}} = 6.41 \times 10^7$ W/m$^2$ |

| | | |
|---|---|---|
| 4. | Rate at which energy from the sun arrives at a large sphere centered at the sun and with a radius equal to the distance from the sun to the earth | $(\frac{\Delta Q}{\Delta t})_{sphere} = (\frac{\Delta Q}{\Delta t})_{sun} = 3.90 \times 10^{26}$ W<br>Energy arrives at this sphere at the same rate that it is emitted by the sun. |
| 5. | Rate per square meter at which energy arrives at the earth | The $3.90 \times 10^{26}$ W is spread uniformly over the sphere of radius $d = 1.5 \times 10^{11}$m. Consequently, the energy per square meter arriving at all points on this sphere is<br>$(\Delta Q/\Delta t)_{sphere} = 3.90 \times 10^{26}$ W<br>$A_{sphere} = 4\pi d^2 = 2.83 \times 10^{23}$ m$^2$<br>$(\frac{\Delta Q/\Delta t}{A})_{sphere} = (\frac{\Delta Q/\Delta t_{sphere}}{A_{sphere}}) = 1380$ W/m$^2$<br>Since the earth is on this large sphere of radius d, the rate per square meter at which energy arrives at the earth is also<br>$(\frac{\Delta Q/\Delta t}{A})_{earth} = 1.38 \times 10^3$ W/m$^2$ |
| 6. | Rate at which student absorbs radiant energy from the sun | $(\Delta Q/\Delta t)_{abs} = (\frac{\Delta Q/\Delta t}{A})_{earth} A_{stu} = 2070$ W |
| 7. | Rate at which student radiates energy | $(\Delta Q/\Delta t)_{emit} = \sigma \varepsilon_{stu} A_{stu} T^4_{stu} = 780$ W |
| 8. | Net rate at which heat is supplied to student via radiation | $(\Delta Q/\Delta t)_{rad} = (\Delta Q/\Delta t)_{abs} - (\Delta Q/\Delta t)_{emit}$<br>$= 1.29 \times 10^3$ W |
| 9. | Total rate at which heat is supplied to student via radiation and metabolism | $(\Delta Q/\Delta t)_{total} = (\Delta Q/\Delta t)_{rad} + (\Delta Q/\Delta t)_{metab}$<br>$= 2.29 \times 10^3$ W |

| | |
|---|---|
| 10. The rate at which perspiration must evaporate in order for body heat to remain constant | $Q = mL_{per}$<br>$(\Delta Q/\Delta t) = (\Delta m/\Delta t)L_{per}$<br>$\Delta m/\Delta t = (\Delta Q/\Delta t)/L_{per} = 9.42 \times 10^{-4}$ kg/s |

Example 15.7 (radiation). A glass of soda warms from 6.00°C to 10.0°C in 5.00 min when the air temperature is 35.0°C. How long will it take to warm from 10.0°C to 15.0°C?

Given: $T_1 = 6.00°C = 279$ K $\quad T_3 = 15.0°C = 288$ K $\quad \Delta t = 5.00$ min
$T_2 = 10.0°C = 283$ K $\quad T_a = 35.0°C = 308$ K

Determine: The length of time required for the soda to warm from $T_2$ to $T_3$.

Strategy: Write an expression for the net rate at which the soda absorbs energy by radiation while warming from $T_1$ to $T_2$ and again while warming from $T_2$ to $T_3$. These two expressions can be manipulated in such a manner as to eliminate all unknown quantities except the time to warm the soda from $T_2$ to $T_3$.

Solution: The rate at which the soda absorbs heat from the room while warming from $T_1$ to $T_2$ is

$$(\Delta Q/\Delta t)_{abs} = \sigma \varepsilon A T_a^4$$

The rate at which it loses heat to the room while warming from $T_1$ to $T_2$ is

$$(\Delta Q/\Delta t)_{emit} = \sigma \varepsilon A T_e^4$$

where $T_e = 8.00°C = 280$ K is halfway between $T_1$ and $T_2$.

The net rate at which the soda gains energy by radiation while warming from $T_1$ to $T_2$ is

$$(\Delta Q/\Delta t)_{net} = (\Delta Q/\Delta t)_{abs} - (\Delta Q/\Delta t)_{emit} = \sigma \varepsilon A(T_a^4 - T_e^4)$$

Since all of this heat goes into heating the soda, we can write

$$\Delta Q = mc\Delta T \quad \text{or} \quad \Delta Q/\Delta t = mc\Delta T/\Delta t \text{ , hence}$$

(α) $(\Delta Q/\Delta t) = \sigma \varepsilon A(T_a^4 - T_e^4) = mc\Delta T/\Delta t$, where $T_a = 308$ K, $T_e = 280$ K, $\Delta T = T_2 - T_1 = 4.00$ K, and $\Delta t = 5.00$ min

In like manner, as the soda warms from $T_2$ to $T_3$ we can write

(β) $\Delta Q/\Delta t = \sigma \varepsilon A(T_a^4 - T_e'^4) = mc\Delta T'/\Delta t'$, where $T_a = 308$ K, $T_e' = 285.5$ K, $\Delta T' = T_3 - T_2 = 5.00$ K, and $\Delta t'$ is to be determined

Divide ($\alpha$) by ($\beta$) and solve for $\Delta t'$.

$$\Delta t' = \frac{(T_a^4 - T_e^4)}{(T_a^4 - T_e'^4)} (\frac{\Delta T'}{\Delta T}) \Delta t = \frac{(308^4 - 280^4)}{(308^4 - 285.5^4)} (\frac{5}{4}) (5.00 \text{ min}) = 7.57 \text{ min}$$

Related Text Problems (radiation): 15-34 through 15-37.

## 6. Relative Humidity (Sec. 15.7)

Review: Relative humidity is the ratio of the density of water vapor present in air at a given temperature to the density of water vapor when the air is saturated at that same temperature. Since relative humidity is usually reported as a percentage, we write

$$RH = (\rho/\rho_s) \times 100\%$$

Practice: The volume of a house is 450 $m^3$. The house is open on a day when the temperature and relative humidity, respectively, are 28°C and 65%. The house and is closed the air conditioner turned on, and after some time the temperature and relative humidity, respectively, are 22°C and 45%.

Determine the following:

| | |
|---|---|
| 1. The density of water vapor in saturated air at 28°C and 22°C | $\rho_s(28°C) = 26.9\ g/m^3$ (from Table 15.4 in the text). To obtain $\rho_s$ (22°C), we can interpolate 22°C between the values at 20°C and 24°C. $\rho_s(22°C) = \rho_s(20°C) + (0.5)\,[\rho_s(24°C) - \rho_s(20°C)]$ $\rho_s\ (22°C) = 19.3\ g/m^3$ |
| 2. The initial density of water vapor, when T = 28°C and RH = 65% | $RH = (\rho_i/\rho_s) \times 100\%$ $\rho_i = RH\ \rho_s/100\%$ $= (65\%)(26.9\ g/m^3)/100\% = 17.5\ g/m^3$ |
| 3. The final density of water vapor, when T = 22°C and RH = 45% | $RH = (\rho_f/\rho_s) \times 100\%$ $\rho_f = RH\ \rho_s/100\%$ $= (45\%)(19.3\ g/m^3)/100\% = 8.69\ g/m^3$ |
| 4. The mass of water that was removed from each cubic meter of air | $\Delta\rho = \lvert\rho_f - \rho_i\rvert = 8.81\ g/m^3$ $\rho = m/V$ or $\Delta m/V = \Delta\rho = 8.81\ g/m^3$ |

| | |
|---|---|
| 5. The total mass $M_T$ and volume $V_T$ of water removed | $M_T = (\Delta m/V)V_{house}$<br>$= (8.81\ g/m^3)(4.50 \times 10^2\ m^3) = 3.96\ kg$<br>$V_T = M_T/\rho_{water} = (3.96\ kg)/(10^3\ kg/m^3)$<br>$= 3.96\ kg \times 10^{-3}\ m^3$ |

Example 15.8. A glass of water is in equilibrium with air in a room at 28°C. Ice is added to the glass, and water vapor begins to condense on the glass when the temperature of the water reaches 20.0°C. Determine the relative humidity of air in the room.

Given: $T_r$ = 28.0°C = temperature of air in room and initial temperature of water
$T_c$ = 20.0°C = temperature of water when condensation starts

Determine: The relative humidity of air in the room.

Strategy: The water vapor is condensing on the glass at 20.0°C, and so the air is saturated at 20.0°C. We can use Table 15-4 from the text to determine the mass of water per cubic meter of saturated air at 20.0°C. We can use the gas laws to determine the volume occupied by the air containing this mass of water at 28.0°C. Knowing the mass of the water and the volume of the air at 28.0°C, we can determine the density of the water vapor in saturated air, Knowing the density of the water vapor at 28.0°C and using Table 15-4 to obtain the water vapor density of saturated air at 28.0°C, we can determine RH for air in the room.

Solution: The density of water vapor in saturated air, and hence the mass of water per cubic meter of air at 20.0°C, are $\rho_s(20.0°C) = 17.1\ g/m^3$. The volume occupied by this mass of water vapor plus air at 28.0°C can be obtained from Charles' law.

$$\frac{V_1}{T_1} = \frac{V_2}{T_2} \quad \text{or} \quad V_1 = V_2\left(\frac{T_1}{T_2}\right) = 1.00\ m^3\left(\frac{301\ K}{293\ K}\right) = 1.03\ m^3$$

The density of the water vapor at 28.0°C is

$$\rho(28.0°C) = m/V_1 = (17.1\ g)/(1.03\ m^3) = 16.6\ g/m^3$$

The water vapor density of saturated air at 28.0°C is obtained from Table 15-4.

$$\rho_s(28.0°C) = 26.9\ g/m^3$$

The relative humidity of the air in the room is

$$RH = (\rho/\rho_s)\ 100\% = [(16.6\ g/m^3)/(26.9\ g/m^3)](100\%) = 61.7$$

Related Text Problems: 15-39, 15-40, 15-41.

PRACTICE TEST

Take and grade this practice test. Doing so will allow you to determine any weak spots in your understanding of concepts taught in this chapter. The following section prescribes what you should study further to strengthen your understanding.

In the experiment shown in Fig. 15.8 a 1.00-kg mass is allowed to fall 5.00 m at a constant speed while turning the paddles in water. Insulation prevents any heat from escaping from the system. The mass of water in the container is 0.500 kg.

Figure 15.8 $m = 1.00$ kg
$m_w = 0.500$ kg
$h = 5.00$ m

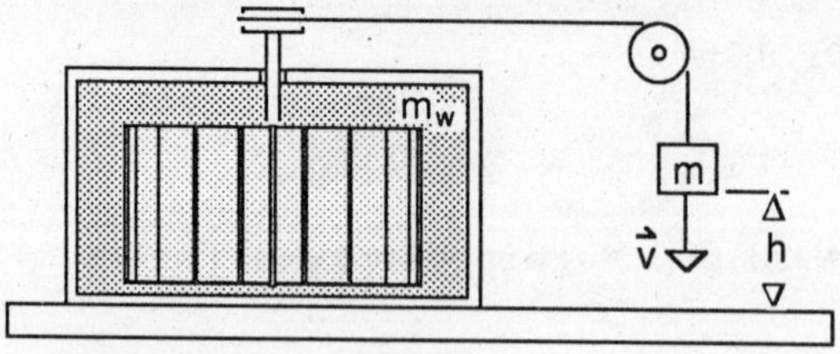


Determine the following:

_____________ 1. Energy (in J) delivered to water
_____________ 2. Energy (in cal) delivered to water
_____________ 3. Temperature change (in °C) of water

A 75.0-g aluminum calorimeter cup contains 25.0 g of water at 25.0°C. Three 30.0-g ice cubes at -10.0°C are placed in the cup. Determine the following:

_____________ 4. Amount of heat (in cal) needed to warm ice from -10.0°C to 0°C
_____________ 5. Amount of heat (in cal) needed to melt ice
_____________ 6. Amount of heat (in cal) available from water and cup to warm and melt ice
_____________ 7. Final equilibrium temperature
_____________ 8. Amount of ice remaining in cup after thermal equilibrium is established.

Rods of aluminum, steel, and copper are welded together to form a Y-shaped object (Figure 15.9). The free end of the copper is maintained at 100°C, and the free ends of the aluminum and steel are maintained at 10.0°C. The rods are all 1.00 m long, have a cross-sectional area of 5.00 $cm^2$, and are isulated so that essentially no heat is lost from the surface. Determine the Following:

_____________ 9. Temperature at the junction
_____________ 10. Rate (in cal/s) at which heat flows along copper rod
_____________ 11. Temperature at midpoint of aluminum rod

Figure 15.9

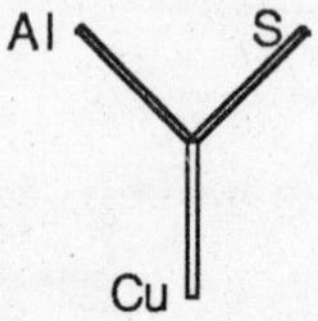


A water heater has an outside surface area of 3.00 $m^2$, a convection coefficient of 2.25 $W/m^2 \cdot °C$ and an emissivity of 0.500. The surface temperature of the heater is 30°C and it is in a room where the temperature is 25°C. The cost of heating water with this tank is \$0.05/kW·h. Determine the following:

_____________ 12. Net rate of heat loss (in cal/s) due to radiation

____________ 13. Rate of heat loss (in cal/s) due to convection
____________ 14. Cost of heat loss due to convection and radiation per day

A physics laboratory is 18.0 m x 12.0 m x 6.0 m. On a day when students are determining the latent heat of vaporization of water, the temperature and relative humidity in the room are T = 27.0°C and RH = 60.0%. Determine the following:

____________ 15. Water vapor density of saturated air in room
____________ 16. Density of water vapor in room
____________ 17. Mass of water in air in room

(See Appendix I for Answers.)

---

PRINCIPAL CONCEPTS AND EQUATIONS PRESCRIPTION

Your score on the practice is an excellent measure of your understanding of this chapter. You should now use the following chart to write your own prescription for curing any of your physics ills. Look down the leftmost column to the number of the question(s) you answered incorrectly, read across that row to see which section(s) of the study guide you should return to for further study, and then do the suggested text problems to gain additional experience in working with the particular concept.

| Question Test Number | Concepts and Equations | Prescription | |
|---|---|---|---|
| | | Principal Concept | Text Problems |
| 1 | Gravitational potential energy: $\Delta PE = -W_g = mg\Delta h$ | 4 of Ch. 7 | 15-2,3 |
| 2 | Mechanical equivalent of heat | 1 | 15-1,4 |
| 3 | Specific heat capacity: $c = \Delta Q/m\Delta T$ | 2 | 15-3,4 |
| 4 | Specific heat capacity: $c = \Delta Q/m\Delta T$ | 2 | 15-5,7 |
| 5 | Latent heat: $L = \Delta Q/m$ | 3 | 15-9,10 |
| 6 | Specific heat capacity: $c = \Delta Q/m\Delta T$ | 2 | 15-4,8 |
| 7 | Calorimetry: $\Delta Q_{net} = 0$ | 4 | 15-17,18 |
| 8 | Latent heat: $L = \Delta Q/m$ | 3 | 15-20,21 |
| 9 | Conduction: $\Delta Q/\Delta t = kA\Delta T/d$ | 5 | 15-23,25 |
| 10 | Conduction: $\Delta Q/\Delta t = kA\Delta T/d$ | 5 | 15-26,27 |
| 11 | Conduction: $\Delta Q/\Delta t = kA\Delta T/d$ | 5 | 15-28,29 |
| 12 | Radiation: $\Delta Q/\Delta t = \varepsilon\sigma AT^4$ | 5 | 15-34,35 |
| 13 | Convection: $\Delta Q/\Delta t = hA\Delta T$ | 5 | 15-31,32 |
| 14 | Radiation and convection | 5 | 15-23,32 |
| 15 | Relative humidity: Table 15-4 | 6 | 15-39,4 |
| 16 | Relative humidity: $RH = (\rho/\rho_s)x100\%$ | 6 | 15-40,41 |
| 17 | Mass density: $\rho = m/V$ | 1 of Ch. 13 | 15-39,41 |

---

# 16 Thermodynamics

## RECALL FROM PREVIOUS CHAPTERS

| Previously learned concepts and equations frequently used in this chapter | Text Section | Study Guide Page |
|---|---|---|
| Temperature scales: $T = t_C + 273$ | 14.3 | 219 |
| Ideal gas law: $PV = nRT$ | 14.6 | 224 |
| Specific heat capacity: $c = \Delta Q/m\Delta T$ | 15.2 | 234 |
| Latent heat: $L = \Delta Q/m$ | 15.3 | 235 |

## NEW IDEAS IN THIS CHAPTER

| Concepts and equations introduced | Text Section | Study Guide Page |
|---|---|---|
| First law of thermodynamics: $\Delta U = \Delta Q - W$ | 16.2 | 252 |
| Molar heat capacity: $\Delta Q = nC_V\Delta T$ and $\Delta Q = nC_P\Delta T$ | 16.3 | 255 |
| Isovolumetric process: $\Delta V = 0$, $W = 0$, $\Delta Q = \Delta U = nC_V\Delta T$ | 16.4 | 255 |
| Isobaric process: $\Delta P = 0$, $\Delta Q = nC_P\Delta T$, $\Delta U = nC_V\Delta T$, $W = P\Delta V = nR\Delta T$ | 16.4 | 255 |
| Isothermal process: $\Delta T = 0$, $\Delta U = 0$, $\Delta Q = W = nRT\ln(V_2/V_1)$ | 16.4 | 255 |
| Adiabatic process: $\Delta Q = 0$, $\gamma = C_P/C_V$, $PV^\gamma = \text{constant}$, $W = -\Delta U = -nC_V\Delta T$ | 16.4 | 255 |
| Heat engines: $\varepsilon = W/Q_H = 1 - (Q_L/Q_H)$ | 16.5 | 261 |
| Refrigerators: $COP = Q_L/W = Q_L/(Q_H - Q_L)$ | 16.5 | 261 |
| Carnot cycle: $\varepsilon = W/Q_H = 1 - (T_L/T_H)$ | 16.7 | 263 |
| Carnot refrigerator: $COP = T_L/(T_H - T_L)$ | 16.7 | 263 |
| Entropy: $\Delta S = \Delta Q/T$ | 16.8 | 267 |
| Second law of thermodynamics: $\Delta S \geq 0$ | 16.8 | 267 |

## PRINCIPAL CONCEPTS AND EQUATIONS

### 1. The First Law of Thermodynamics (Section 16.2)

Review: A mathematical statement of the first law of thermodynamics is

$$\Delta U = \Delta Q - W$$

where $\Delta U$ = internal energy = $\begin{cases} + \text{ for increase in internal energy} \\ - \text{ for decrease in internal energy} \end{cases}$

$\Delta Q$ = heat = $\begin{cases} + \text{ for heat added} \\ - \text{ for heat released} \end{cases}$

$W$ = work = $\begin{cases} + \text{ for work done by the system} \\ - \text{ for work done on the system} \end{cases}$

The first law of thermodynamics says that, if you put heat energy ($\Delta Q$) into a system and get work ($W$) out, the internal energy changes by an amount $\Delta U = \Delta Q - W$. This is a statement of conservation of energy.

If a system goes from state 1 ($P_1$, $V_1$, $T_1$) to state 2 ($P_2$, $V_2$, $T_2$), $\Delta U$ is a constant independent of how the system goes from state 1 to state 2. If a system starts out in state 1, makes several successive changes, and then returns to its original state, then $\Delta U = 0$.

Practice: Figure 16.1 shows two (path a and path b) of the numerous possible ways of taking an ideal gas from state 1 ($P_1$, $V_1$, $T_1$) to state 2 ($P_2$, $V_2$, $T_2$) and one (path c) of the numerous possible ways of taking a gas from state 2 to state 1.

Figure 16.1

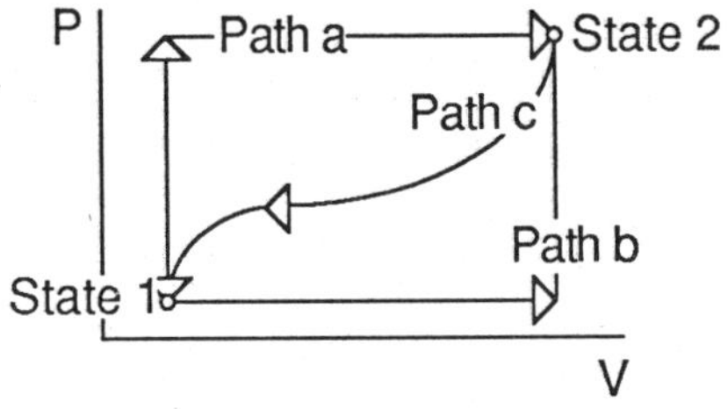


When the gas goes from state 1 to state 2 along path a, it absorbs 100 J of heat and does 50.0 J of work.

When the gas goes from state 1 to state 2 along path b, 80.0 J of heat flows into the system.

In order to return the gas to state 1 from state 2 along path c, 40.0 J of work must be done on the gas.

Determine the following:

| | |
|---|---|
| 1. Change in internal energy of the gas as it goes from state 1 to state 2 along path a | $\Delta U = \Delta Q - W$<br>$\Delta Q = +100$ J and $W = +50.0$ J (given)<br>$\Delta U = +100 \text{ J} - (+50.0 \text{ J}) = +50.0$ J<br>The internal energy increases by 50.0 J |
| 2. Change in internal energy of the gas as it goes from state 1 to state 2 along path b | $\Delta U$ = constant between any two states regardless of path; hence<br>$\Delta U_b = \Delta U_a = +50.0$ J |

| | |
|---|---|
| 3. Change in internal energy of the gas for a complete cycle (state 1 to state 2 back to state 1) | For a complete cycle, the final state of the gas is the same as the initial state; hence<br>$\Delta U_{1\to2\to1} = 0$ |
| 4. Change in internal energy of the gas as it goes from state 2 to state 1 along path c | No matter how the gas goes from state 1 to state 2,<br>$\Delta U_{1\to2} = +50$ J<br>For a complete cycle,<br>$\Delta U_{1\to2\to1} = \Delta U_{1\to2} + \Delta U_{2\to1} = 0$<br>consequently,<br>$\Delta U_{2\to1} = -\Delta U_{1\to2} = -50$ J<br>No matter how the gas goes from state 2 to state 1,<br>$\Delta U_{2\to1} = -50$ J<br>hence $\Delta U_c = -50$ J |
| 5. Work done by the gas in going from state 1 to state 2 along path b | $\Delta U_b = +50.0$ J (step 2)<br>$\Delta Q_b = +80.0$ J (given)<br>$W_b = \Delta Q_b - \Delta U_b = +30.0$ J |
| 6. Change in heat of the gas in going from state 2 to state 1 along path c | $\Delta U_c = -50.0$ J (step 4)<br>$W_c = -40.0$ J (given)<br>$\Delta Q_c = \Delta U_c + W_c = -90.0$ J |
| 7. Net work for a complete cycle along paths a and c | $W_a = +50.0$ J and $W_c = -40.0$ J (given)<br>$W_{net} = W_a + W_c = +10.0$ J |
| 8. Net change in heat for a complete cycle along paths a and c | $\Delta Q_a = +100.0$ J (given)<br>$\Delta Q_c = -90.0$ J (step 6)<br>$\Delta Q_{net} = \Delta Q_a + \Delta Q_c = +10.0$ J<br>Note $\Delta U_{net} = \Delta Q_{net} - W_{net}$<br>$\Delta U_{net} = 0$ (step 3)<br>Hence $W_{net} = \Delta Q_{net}$<br>which agrees with step 7 |

Example 16.1. During a thermodynamic process, 300 cal of heat is added to a system and its internal energy decreases by 500 J. How much work is done on or by this system?

Given: $\Delta Q = +300$ cal; $\Delta U = -500$ J

Determine: The amount of work done on or by this system.

Strategy: We can use the given information and the first law of thermodynamics to determine the work W. The sign of W will allow us to determine whether the system is doing work or work is being done on it.

Solution: $\Delta U = -500\ J$

$\Delta Q = (+300\ cal)(4.184\ J/cal) = +1260\ J$

$W = \Delta Q - \Delta U = +1260\ J - (-500\ J) = +1760\ J$

The amount of work done by the system is 1760 J.

Related Text Problems: 16-1 through 16-5.

## 2. Reversible Thermodynamics Processes (Sections 16.3, 16.4)

Review: A reversible process is one that takes place in such a way that the system is always in thermal equilibrium. The important reversible thermodynamic processes and their consequences can be treated graphically and analytically. The graphical treatment is made possible by the fact that reversible processes may be represented by curves on P-V and P-T graphs. The graphical treatment is especially useful because it helps us visualize what is taking place.

Isovolumetric process (constant-volume)

$\Delta V = 0$

$W = P\Delta V = 0$

$\Delta Q = nC_V\Delta T$

$\Delta U = \Delta Q$

Isovolumetric is frequently shortened to isometric.

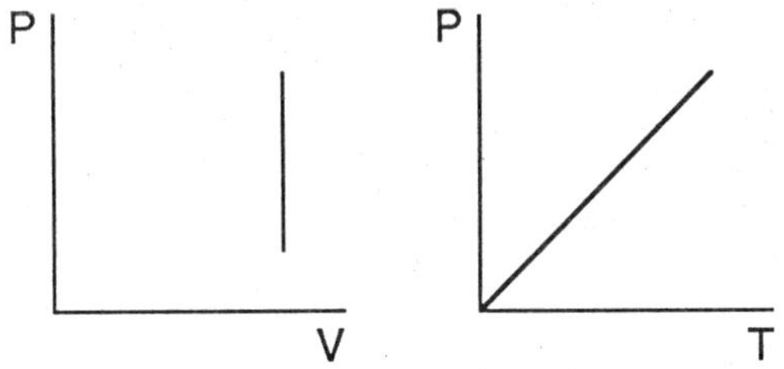


Isobaric process (constant-pressure)

$\Delta P = 0$

$W = P\Delta V = nR\Delta T$

$\Delta Q = nC_P\Delta T$

$\Delta U = nC_V\Delta T$

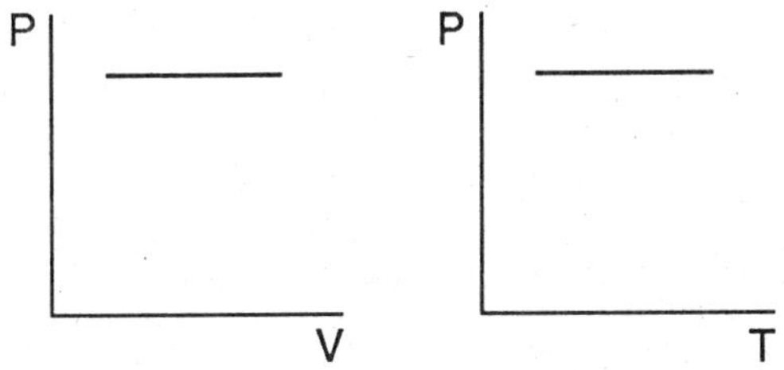


Isothermal process (constant-temperature)

$\Delta T = 0$

$\Delta U = nC_V\Delta T = 0$

$\Delta Q = W = nRT\ln(V_2/V_1)$

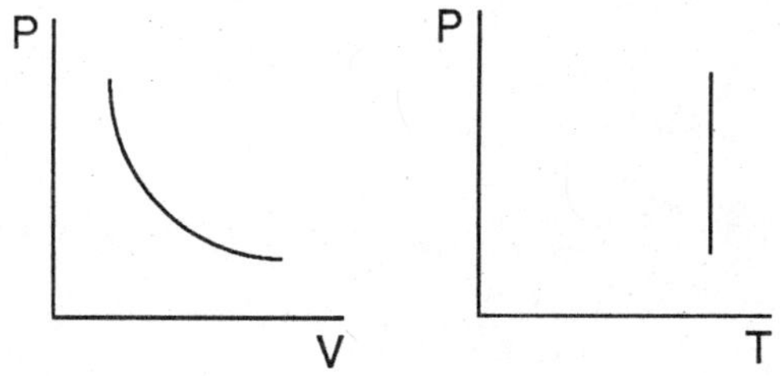

Adiabatic process (no heat change)

$\Delta Q = 0$

$\Delta U = nC_V \Delta T$

$W = -\Delta U = -nC_V \Delta T$

$PV^\gamma$ = constant; $\gamma = C_P/C_V$

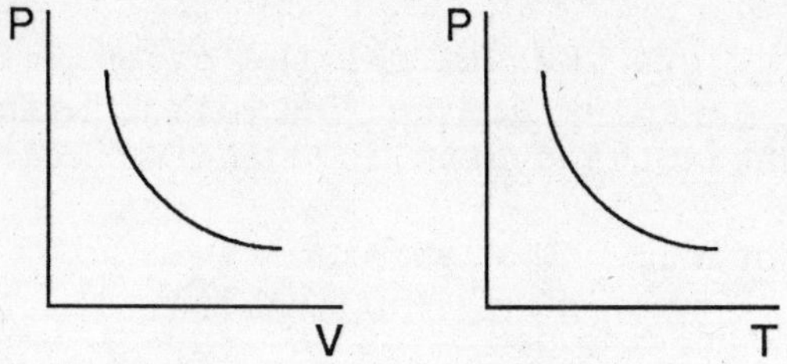


Note: $\Delta U = nC_V \Delta T$, always

$C_P - C_V = R$

Practice: A cylinder contains 2.00 mole of helium gas ($C_P$ = 20.77 J/mol·K; $C_V$ = 12.46 J/mol·K). The P-V and P-T plots in Fig. 16.2 show the thermodynamic processes and information for this gas as it goes through a complete cycle (1→2→3→4→1).

Figure 16.2.

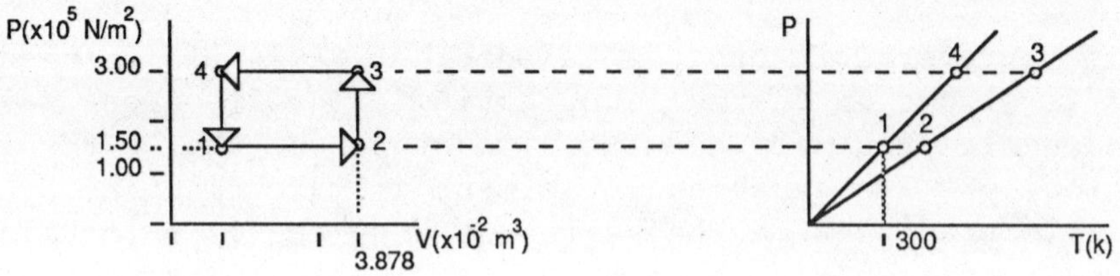


(a) P vs. V (b) P vs. T

Determine the following:

| | |
|---|---|
| 1. Thermodynamic process the gas undergoes as it goes from state 1 to state 2 | Isobaric: $\Delta P = 0$ |
| 2. Pressure, temperature, and volume of the gas in state 1 | $P_1 = 1.50 \times 10^5$ N/m$^2$ (P-V plot)<br>$T_1$ = 300 K (P-T plot)<br>$P_1V_1 = nRT_1$<br>$V_1 = nRT_1/P_1 = 3.324 \times 10^{-2}$ m$^3$ |
| 3. Pressure, temperature, and volume of gas in state 2 | $P_2 = P_1 = 1.50 \times 10^5$ N/m$^2$ (P-V plot)<br>$V_2 = 3.878 \times 10^{-2}$ m$^3$ (P-V plot)<br>$P_2V_2 = nRT_2$<br>$T_2 = P_2V_2/nR$ = 350 K |
| 4. Work done by the gas as it goes from state 1 to state 2 | $\Delta T = T_2 - T_1$ = 50.0 K<br>$P = 1.50 \times 10^5$ N/m$^2$<br>$\Delta V = V_2 - V_1 = 5.54 \times 10^{-3}$ m$^3$<br>$W_{12} = nR\Delta T = 8.31 \times 10^2$ J<br>$W_{12} = P\Delta V = 8.31 \times 10^2$ J<br>Note that $W_{12}$ is the area under the curve on the P-V plot. |

| | |
|---|---|
| 5. Heat supplied to the gas as it goes from state 1 to state 2 | $\Delta T = 50.0$ K<br>$\Delta Q_{12} = nC_p\Delta T_{12} = 2.08 \times 10^3$ J |
| 6. Change in internal energy of the gas as it goes from state 1 to state 2 | $\Delta T_{12} = 50.0$ K<br>$\Delta U_{12} = nC_V\Delta T_{12} = 1.25 \times 10^3$ J<br>$\Delta U_{12} = \Delta Q_{12} - W_{12} = 1.25 \times 10^3$ J |
| 7. Thermodynamic process the gas as it undergoes as it goes from state 2 to state 3 | Isovolumetric: $\Delta V = 0$ |
| 8. Pressure, temperature, and volume of the gas in state 3 | $P_3 = 3.00 \times 10^5$ N/m$^2$ (P-V plot)<br>$V_3 = V_2 = 3.878 \times 10^{-2}$ m$^3$ (P-V plot)<br>$P_3V_3 = nRT_3$; $T_3 = P_3V_3/nR = 700$ K<br>We may also obtain $T_3$ from<br>$P_3/T_3 = P_2/T_2$; $T_3 = T_2P_3/P_2 = 700$ K |
| 9. Work done by or on the gas as it goes from state 2 to state 3 | $\Delta V_{23} = 0$<br>$W_{23} = P\Delta V_{23} = 0$ |
| 10. Heat supplied to or taken from the gas as it goes from state 2 to state 3 | $\Delta T_{23} = 350$ K<br>$\Delta Q_{23} = nC_V\Delta T_{23} = 8.72 \times 10^3$ J<br>This heat is supplied to the gas. |
| 11. Change in internal energy of the gas as is goes from state 2 to state 3 | $W_{23} = 0$; $\Delta Q_{23} = 8.72 \times 10^3$ J<br>$\Delta U_{23} = \Delta Q_{23} - W_{23} = 8.72 \times 10^3$ J |
| 12. Pressure, temperature, and volume of the gas in state 4 | $P_4 = P_3 = 3.00 \times 10^5$ N/m$^2$ (P-V plot)<br>$V_4 = V_1 = 3.324 \times 10^{-2}$ m$^3$ (step 2)<br>$P_4V_4 = nRT_4$; $T_4 = P_4V_4/nR = 600$ K<br>We may also obtain $T_4$ from<br>$V_3/T_3 = V_4/T_4$; $T_4 = T_3V_4/V_3 = 600$ K |
| 13. Work done on the gas as it goes from state 3 to state 4 to state 4 | $\Delta V_{34} = V_4 - V_3 = -5.54 \times 10^{-3}$ m$^3$<br>$\Delta T_{34} = T_4 - T_3 = -100$ K<br>$W_{34} = P_4\Delta V_{34} = -1.66 \times 10^3$ J<br>$W_{34} = nR\Delta T = -1.66 \times 10^3$ J |

| | | |
|---|---|---|
| 14. | Heat supplied to or taken from the gas as it goes from state 3 to state 4 | $\Delta T_{34} = -100$ K<br>$\Delta Q_{34} = nC_p\Delta T_{34} = -4.15 \times 10^3$ J<br>This heat is taken from the gas. |
| 15. | Change in internal energy of the gas as it goes from state 3 to state 4 | $\Delta T_{34} = -100$ K<br>$\Delta U_{34} = nC_V\Delta T_{34} = -2.49 \times 10^3$ J<br>$\Delta U_{34} = \Delta Q_{34} - W_{34} = -2.49 \times 10^3$ J |
| 16. | Work done on or by the gas as it goes from state 4 to state 1 | $\Delta V_{41} = 0$<br>$W_{41} = P\Delta V_{41} = 0$ |
| 17. | Heat supplied to or taken from the gas as it goes from state 4 to state 1 | $\Delta T_{41} = -300$ K<br>$\Delta Q_{41} = nC_V\Delta T_{41} = -7.48 \times 10^3$ J |
| 18. | Change in internal energy of the gas as it goes from state 4 to state 1 | $\Delta T_{41} = -300$ K<br>$\Delta U_{41} = nC_V\Delta T = -7.48 \times 10^3$ J<br>$\Delta U_{41} = \Delta Q_{41} - W_{41} = -7.48 \times 10^3$ J |
| 19. | Net change in energy for the complete cycle | $\Delta U_{net} = 0$<br>$\Delta U_{net} = \Delta U_{12} + \Delta U_{23} + \Delta U_{34} + \Delta U_{41} = 0$ |
| 20. | Net heat change for the complete cycle | $\Delta Q_{net} = \Delta Q_{12} + \Delta Q_{23} + \Delta Q_{34} + \Delta Q_{41}$<br>$\Delta Q_{net} = -8.30 \times 10^2$ J |
| 21. | Net work done for the complete cycle | $W_{net} = W_{12} + W_{23} + W_{34} + W_{41}$<br>$W_{net} = -8.30 \times 10^2$ J |

A cylinder contains 1.00 mol of neon gas ($C_p = 20.77$ J/mol·K; $C_V = 12.46$ J/mol·K). The P-V and P-T plots in Fig. 16.3 show the thermodynamic processes and information for this gas as it goes through a complete cycle ($1 \rightarrow 2 \rightarrow 3 \rightarrow 4 \rightarrow 1$).

Figure 16.3.

(a) P vs. V

(b) P vs. T

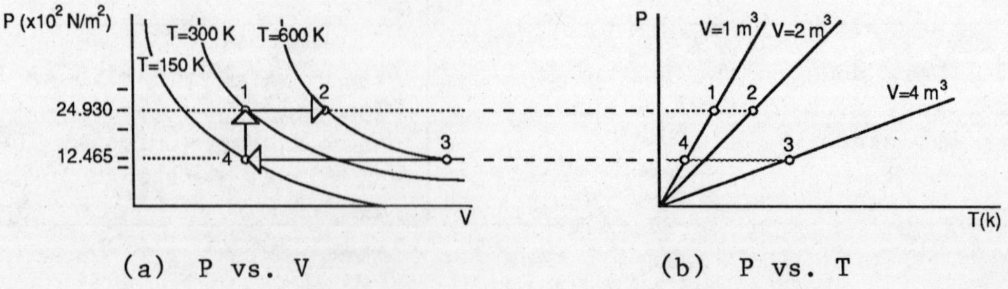

Determine the following:

| | |
|---|---|
| 1. $P_1$, $V_1$, $T_1$ | $P_1 = 24.93 \times 10^2\ N/m^2$ (P-V or P-T plot)<br>$V_1 = 1.00\ m^3$ (P-T plot)<br>$T_1 = 300\ K$ (P-V plot) |
| 2. $P_2$, $V_2$, $T_2$ | $P_2 = P_1 = 24.93 \times 10^2\ N/m^2$ (P-V plot)<br>$V_2 = 2.00\ m^3$ (P-T plot)<br>$T_2 = 600\ K$ (P-V plot) |
| 3. Thermodynamic process the gas undergoes as it goes from state 1 to state 2 | Isobaric: $\Delta P_{12} = 0$ |
| 4. $W_{12}$ | $\Delta V_{12} = 1.00\ m^3$; $\Delta T_{12} = 300\ K$<br>$W_{12} = nR\Delta T_{12} = 2.49 \times 10^3\ J$<br>$W_{12} = P_1\Delta V_{12} = 2.49 \times 10^3\ J$ |
| 5. $\Delta Q_{12}$ | $\Delta T_{12} = 300\ K$<br>$\Delta Q_{12} = nC_p\Delta T_{12} = 6.23 \times 10^3\ J$ |
| 6. $\Delta U_{12}$ | $\Delta T_{12} = 300\ K$<br>$\Delta U_{12} = nC_v\Delta T_{12} = 3.74 \times 10^3\ J$<br>$\Delta U_{12} = \Delta Q_{12} - W_{12} = 3.74 \times 10^3\ J$ |
| 7. Thermodynamic process the gas undergoes as it goes from state 2 to state 3 | Isothermal: $\Delta T_{23} = 0$ |
| 8. $\Delta U_{23}$ | $\Delta U_{23} = nC_v\Delta T_{23} = 0$ |
| 9. $W_{23}$ | $W_{23} = nRT_2 \ln(V_3/V_2) = 3.46 \times 10^3\ J$ |
| 10. $\Delta Q_{23}$ | $\Delta U_{23} = \Delta Q_{23} - W_{23}$; $\Delta U_{23} = 0$<br>$\Delta Q_{23} = W_{23} = 3.46 \times 10^3\ J$ |
| 11. Thermodynamic process the gas undergoes as it goes from state 3 to state 4 | Isobaric: $\Delta P_{34} = 0$ |

| | |
|---|---|
| 12. $W_{34}$ | $\Delta V_{34} = -3.00\ m^3$; $\Delta T_{34} = -450$ K<br>$W_{34} = P_3 \Delta V_{34} = -3.74 \times 10^3$ J<br>$W_{34} = nR\Delta T_{34} = -3.74 \times 10^3$ J |
| 13. $\Delta Q_{34}$ | $\Delta T_{34} = -450$ K<br>$\Delta Q_{34} = nC_p \Delta T_{34} = -9.35 \times 10^3$ J |
| 14. $\Delta U_{34}$ | $\Delta T_{34} = -450$ K<br>$\Delta U_{34} = nC_V \Delta T_{34} = -5.61 \times 10^3$ J<br>$\Delta U_{34} = \Delta Q_{34} - W_{34} = -5.61 \times 10^3$ J |
| 15. Thermodynamic process the gas undergoes as it goes from state 4 to state 1 | Isovolumetric: $\Delta V_{41} = 0$ |
| 16. $W_{41}$ | $W_{41} = 0$ since $\Delta V_{41} = 0$ |
| 17. $\Delta Q_{41}$ | $\Delta T_{41} = 150$ K<br>$\Delta Q_{41} = nC_V \Delta T_{41} = 1.87 \times 10^3$ J |
| 18. $\Delta U_{41}$ | $\Delta U_{41} = \Delta Q_{41} - W_{41} = 1.87 \times 10^3$ J |
| 19. $\Delta U_{net}$ | $\Delta U_{net} = \Delta U_{12} + \Delta U_{23} + \Delta U_{34} + \Delta U_{41} = 0$ |
| 20. $\Delta Q_{net}$ | $\Delta Q_{net} = \Delta Q_{12} + \Delta Q_{23} + \Delta Q_{34} + \Delta Q_{41}$<br>$= 2.21 \times 10^3$ J |
| 21. $W_{net}$ | $W_{net} = W_{12} + W_{23} + W_{34} + W_{41}$<br>$= 2.21 \times 10^3$ J<br>also $\Delta U_{net} = \Delta Q_{net} - W_{net} = 0$<br>$W_{net} = \Delta Q_{net}$<br>This agrees with the results of step 20 |

Example 16.2. Six moles of neon gas at of 301 K and $1.50 \times 10^7\ N/m^2$ undergoes an adiabatic expansion to half this pressure. What are the final pressure, volume, and temperature of the gas?

Given: $n = 6$ mol; $T_1 = 301$ K; $P_1 = 1.50 \times 10^7\ N/m^2$; $P_2 = P_1/2$; $\Delta Q = 0$

Determine: The final pressure, volume, and temperature after the gas expands adiabatically to one half the initial pressure.

Strategy: Knowing that the gas expands to half the initial pressure, we can determine the final pressure. We can use the given information to determine the initial volume ($V_1$) of the gas. Knowing that the expansion is adiabatic, we can determine the final volume and then the final temperature of the gas.

Solution: $P_2 = P_1/2 = 7.50 \times 10^6$ N/m$^2$

The initial volume may be determined by $P_1V_1 = nRT_1$ or $V_1 = nRT_1/P_1 = 1.00 \times 10^{-3}$ m$^3$. For an adiabatic process, we know that $PV^\gamma$ = constant, and so we can determine the final volume:

$$P_1V_1^\gamma = P_2V_2^\gamma; \quad (V_2/V_1)^\gamma = P_1/P_2 = 2.00; \quad V_2/V_1 = (2.00)^{1/\gamma}$$

$$V_2 = V_1(2.00)^{1/\gamma} = (1.00 \times 10^{-3} \text{ m}^3)(2.00)^{1/1.67} = 1.51 \times 10^{-3} \text{ m}^3$$

Knowing that the expansion is adiabatic and treating the neon as an ideal gas, we can determine the final temperature:

$$P_1V_1^\gamma = P_2V_2^\gamma; \quad P_1V_1 = nRT_1; \quad P_2V_2 = nRT_2$$

Eliminating the pressure, we obtain

$$T_1V_1^{\gamma-1} = T_2V_2^{\gamma-1}; \quad T_2 = T_1(V_1/V_2)^{\gamma-1} = 228 \text{ K}$$

Related Text Problems: 16-6 through 16-21.

## 3. Heat Engines and Refrigerators (Section 16.5)

Review: A heat engine and a refrigerator are shown schematically in Fig. 16.4.

Figure 16.4

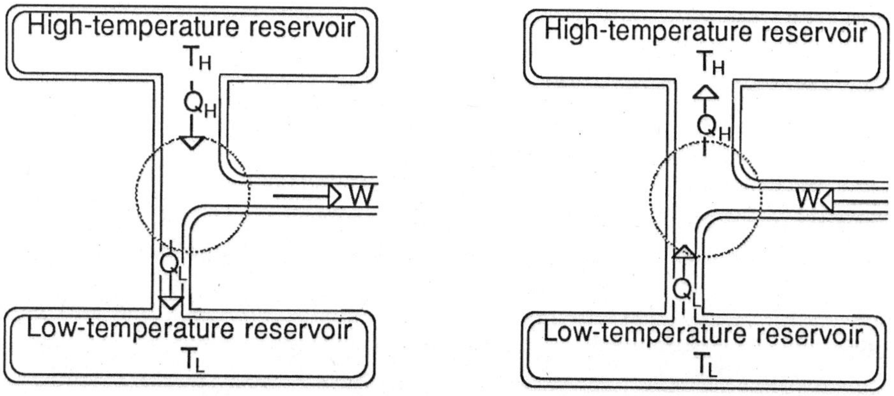


$$\varepsilon = W/Q_H$$
$$W = Q_H - Q_L$$
$$\varepsilon = (Q_H - Q_L)/Q_H$$
$$\varepsilon = 1 - (Q_L/Q_H)$$

(a) Heat Engine

$$COP = Q_L/W$$
$$W = Q_H - Q_L$$
$$COP = Q_L/(Q_H - Q_L)$$

(b) Refrigerator

Practice: A heat engine absorbs 2000 J of heat from a high-temperature reservoir and exhausts 1600 J to a low-temperature reservoir. When the same engine is run in reverse as a refrigerator between the same two reservoirs, 600 J of work is required to extract 1400 J of heat from the low-temperature reservoir.

Determine the following:

| | |
|---|---|
| 1. Amount of work done by the heat engine | $W = Q_H - Q_L = 2000\text{ J} - 1600\text{ J} = 400\text{ J}$ |
| 2. Efficiency of the heat engine | $\varepsilon = W/Q_H = (400\text{ J})/(2000\text{ J}) = 0.200$ |
| 3. Coefficient of performance for the refrigerator | $COP = Q_L/W = (1400\text{ J})/(600\text{ J}) = 2.33$ |
| 4. Amount of heat exhausted to the high-temperature reservoir by the refrigerator | $W = Q_H - Q_L$<br>$Q_H = W + Q_L = 600\text{ J} + 1400\text{ J} = 2000\text{ J}$ |

Note: Heat-engine problems involve the four quantities $Q_H$, $Q_L$, $W$, and $\varepsilon$. You will usually be given two of them and asked to solve for the other two. Refrigerator problems involve the four quantities $Q_L$, $W$, $Q_H$, and COP. You will usually be given two of them and asked to solve for the other two.

Example 16.3. A heat engine working with a thermodynamic efficiency of 0.400 exhausts $1.50 \times 10^8$ J of heat to a low-temperature reservoir. Calculate the work done by the engine.

Given: $\varepsilon = 0.400$, $Q_L = 1.50 \times 10^8$ J

Determine: Work done by the engine

Strategy: Knowing the efficiency and the heat exhausted, we can determine the heat accepted from the high temperature reservoir and then the work done by the engine.

Solution: The heat accepted from the high-temperature reservoir is

$$\varepsilon = 1 - (Q_L/Q_H);\ Q_H = Q_L/(1 - \varepsilon) = 2.50 \times 10^8\text{ J}$$

The work done by the engine is

$$W = Q_H - Q_L = 2.50 \times 10^8\text{ J} - 1.50 \times 10^8\text{ J} = 1.00 \times 10^8\text{ J}$$

Example 16.4. An ice-cube tray holding 400 mℓ of water at 20.0°C is placed in a freezer. If the unit is powered by a 0.333 hp motor and the coefficient of performance is 2.00, how long does it take the water to freeze? (Ignore the thermal effects of the tray).

Given: $V_w$ = 400 mℓ, P = 0.333 hp, $T_{wi}$ = 20°C, COP = 2.00

Determine: Time required to freeze the water

Strategy: Knowing the volume and initial temperature of the water, we can determine the amount of heat that must be removed from the low-temperature reservoir in order to freeze the water. We can determine the rate at which work is done by the refrigerator from the known horsepower rating of the motor. We can then use this rate and the coefficient of performance to determine the rate at which heat is removed from the low-temperature reservoir. Knowing the amount of heat that must be removed and the rate at which it is being removed, we can determine the time needed to freeze the the water.

Solution: First, let's determine the amount of heat that must be removed to freeze the water.

$$m_w = V_w\rho_w = (400\ \text{m}\ell)(10^{-3}\ \ell/\text{m}\ell)(10^{-3}\ \text{m}^3/\ell)(10^3\ \text{kg/m}^3) = 4.00 \times 10^{-1}\ \text{kg}$$
$$Q_L = Q\ (\text{cool water to } 0°\text{C}) + Q\ (\text{freeze water})$$
$$Q_L = m_w c_w \Delta T_w + m_w L_f = m_w(c_w \Delta T_w + L_f)$$
$$Q_L = (4.00 \times 10^{-1}\ \text{kg})[(4180\ \text{J/kg}\cdot\text{K})(20\ \text{K}) + 3350\ \text{J/kg}] = 3.48 \times 10^4\ \text{J}$$

Second, let's determine the rate at which heat is removed from the low-temperature reservoir.

$$P = W/t = (0.333\ \text{hp})(746\ \text{W/hp})[(\text{J/s})/\text{W}] = 249\ \text{J/s}$$

COP = $Q_L/W$; $Q_L$ = (COP)W, which may be divided by t to obtain

$$Q_L/t = \text{COP}(W/t) = 2.00(249\ \text{J/s}) = 498\ \text{J/s}$$

Finally, let's determine the time required to freeze (that is, to remove the $3.48 \times 10^4$ J of heat from) the water.

$$t = Q_L/(Q_L/t) = (3.48 \times 10^4\ \text{J})/(4.98 \times 10^2\ \text{J/s}) = 6.99\ \text{s}$$

This number is unrealistically small because we have assumed that the motor is 100% efficient and that all of its work output goes into freezing the ice.

Related Text Problems: Heat engines-16-21 through 16-24; refrigerators-16-25 through 16-30.

## 4. The Carnot Cycle (Section 16.7)

Review: The Carnot cycle is shown in Figure 16.5.

Figure 16.5

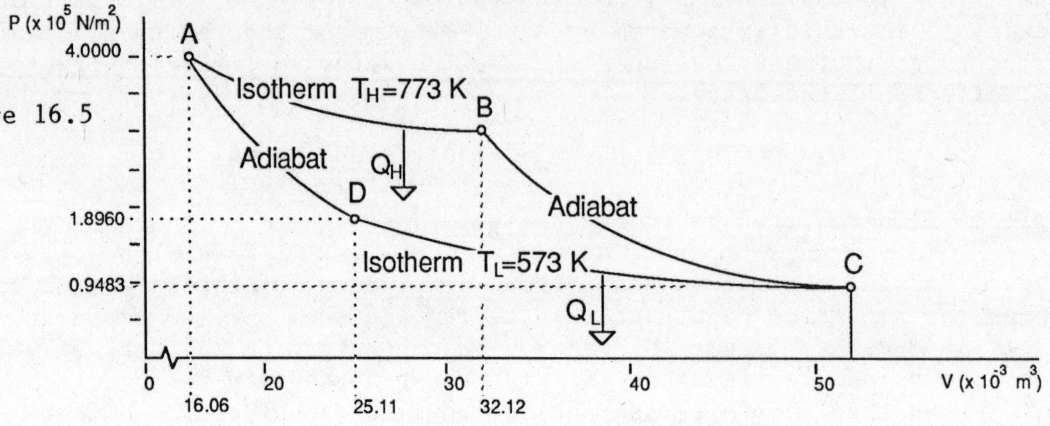


The efficiency of a heat engine operating in a Carnot cycle is

$$\varepsilon = 1 - (T_L/T_H)$$

The coefficient of performance of a Carnot refrigerator is

$$COP = T_L/(T_H - T_L)$$

Practice: The working substance for the heat engine undergoing the Carnot cycle in Fig. 16.5 is 1 mol of He gas ($C_P$ = 20.77 J/mol·K; $C_V$ = 12.46 J/mol·K).

Determine the following:

| | |
|---|---|
| 1. Work done by the gas as it expands isothermally from A to B | n = 1 mol    $V_A = 16.06 \times 10^{-3}\ m^3$<br>R = 8.31 J/mol·K    $V_B = 32.12 \times 10^{-3}\ m^3$<br>T = 773 K<br>$W_{AB} = nRT\ln(V_B/V_A) = 4.453 \times 10^3$ J |
| 2. The change in internal energy of the gas as it goes from A to B | $\Delta T_{AB} = 0$<br>$\Delta U_{AB} = nC_V\Delta T_{AB} = 0$ |
| 3. Heat absorbed from the high-temperature reservoir as the gas goes from A to B | $Q_H = \Delta Q_{AB}$<br>$\Delta U_{AB} = \Delta Q_{AB} - W_{AB} = 0$<br>$Q_H = \Delta Q_{AB} = W_{AB} = 4.453 \times 10^3$ J |
| 4. $\Delta Q_{BC}$ | $\Delta Q_{BC} = 0$ since it is an adiabatic process |

| | |
|---|---|
| 5. $\Delta U_{BC}$ | $\Delta T_{BC} = -200$ K<br>$\Delta U_{BC} = nC_V\Delta T_{BC} = -2.492 \times 10^3$ J |
| 6. $W_{BC}$ | $\Delta U_{BC} = \Delta Q_{BC} - W_{BC}$; $\Delta Q_{BC} = 0$<br>$W_{BC} = -\Delta U_{BC} = 2.492 \times 10^3$ J |
| 7. $W_{CD}$ | $V_C = 50.22 \times 10^{-3}$ $m^3$<br>$V_D = 25.11 \times 10^{-3}$ $m^3$<br>$T_C = T_D = 573$ K<br>$W_{CD} = nRT\ln(V_D/V_C) = -3.301 \times 10^3$ J |
| 8. $\Delta U_{CD}$ | $\Delta U_{CD} = 0$ since $\Delta T_{CD} = 0$ |
| 9. $Q_L$ | $\Delta U_{CD} = \Delta Q_{CD} - W_{CD} = 0$<br>$Q_L = \Delta Q_{CD} = W_{CD} = -3.301 \times 10^3$ J |
| 10. $\Delta Q_{DA}$ | $\Delta Q_{DA} = 0$ since it is an adiabatic process |
| 11. $\Delta U_{DA}$ | $\Delta T_{DA} = +200$ K<br>$\Delta U_{DA} = nC_V\Delta T_{DA} = +2.492 \times 10^3$ J |
| 12. $W_{DA}$ | $\Delta U_{DA} = \Delta Q_{DA} - W_{DA}$; $\Delta Q_{DA} = 0$<br>$W_{DA} = -\Delta U_{DA} = -2.494 \times 10^3$ J |
| 13. $W_{net}$ | $W_{net} = W_{AB} + W_{BC} + W_{CD} + W_{DA}$<br>$= 1.152 \times 10^3$ J<br>$W_{net} = Q_H - Q_L = 1.152 \times 10^3$ J |
| 14. Efficiency ($\varepsilon$) for this Carnot cycle | $Q_H = 4.453 \times 10^3$ J (step 3)<br>$W_{net} = 1.152 \times 10^3$ J (step 13)<br>$T_H = 773$ K; $T_L = 573$ K<br>$\varepsilon = W_{net}/Q_H = 0.259$<br>$\varepsilon = 1 - (T_L/T_H) = 0.259$ |
| 15. COP of the associated Carnot refrigerator | $COP = T_L/(T_H - T_L) = 2.87$ |

Example 16.5. A Carnot engine takes in heat from a 327°C reservoir and has an efficiency of 0.250. If the exhaust temperature remains the same and the efficiency increases to 0.400, what is the new temperature of the hot reservoir?

Given: $T_{H1}$ = 327°C = temperature of the hot reservoir
$\varepsilon_1$ = 0.250 = efficiency when $T_{H1}$ = 327°C
$\varepsilon_2$ = 0.400 = efficiency when reservoir is at $T_{H2}$

Determine: The temperature $T_{H2}$ of the hot reservoir that will give this Carnot engine an efficiency of 0.400.

Strategy: Knowing $T_{H1}$ and $\varepsilon_1$, we can determine $T_{L1}$. Knowing $\varepsilon_2$ and that $T_{L2} = T_{L1}$, we can determine $T_{H2}$.

Solution: $\varepsilon_1 = 1 - (T_{L1}/T_{H1})$; $T_{L1} = (1 - \varepsilon_1)T_{H1} = 450$ K
$T_{L2} = T_{L1} = 450$ K
$\varepsilon_2 = 1 - (T_{L2}/T_{H2})$; $T_{H2} = T_{L2}/(1 - \varepsilon_2) = 750$ K

Example 16.6. A Carnot refrigerator operates between reservoirs at 23°C and -10°C. If it works long enough to convert a 200-g aluminum ice cube tray containing 400 g of water at 20°C to ice at -10°C, determine (a) the net amount of heat absorbed from the water and tray, (b) the coefficient of performance, (c) the work supplied to the refrigerator, and (d) the amount of heat exhausted to the room.

Given: $T_H$ = 23°C = temperature of hot reservoir
$T_L$ = -10°C = temperature of cold reservoir
$M_{Al}$ = 200 g = mass of aluminum tray
$M_w$ = 400 g = mass of water
$T_i$ = 20°C = initial temperature of water and tray

Determine: $Q_L$ = heat absorbed from cold reservoir (freezer)
COP = coefficient of performance
W = work supplied to refrigerator
$Q_H$ = amount of heat exhausted to hot reservoir (room)

Strategy: (a) Knowing the masses, temperatures, specific heats, and heat of fusion, we can determine the amount of heat absorbed from the water and tray. (b) Knowing the temperature of the hot and cold reservoirs, we can determine the COP. (c) Knowing COP and $Q_L$, we can determine W. (d) Knowing W and $Q_L$, we can determine $Q_H$.

Solution: (a) The amount of heat absorbed from the cold reservoir ($Q_L$) is

$$Q_L = \Delta Q \text{ (cool and freeze water)} + \Delta Q \text{ (cool tray)}$$
$$Q_L = M_w c_w(20\text{ K}) + M_w L_f + M_{ice}c_{ice}(10\text{ K}) + M_{Al}c_{Al}(30\text{ K}) = 1.81 \times 10^5 \text{ J}$$

(b) The coefficient of performance is

$$COP = T_L/(T_H - T_L) = (263\text{ K})/(33\text{ K}) = 7.97$$

(c) The amount of work supplied to the refrigerator is

$$COP = Q_L/W; \quad W = Q_L/COP = 2.27 \times 10^4 \text{ J}$$

(d) The heat exhausted to the hot reservoir is

$$W = Q_H - Q_L; \quad Q_H = W + Q_L = 20.4 \times 10^4 \text{ J}$$

Related Text Problems: 16-33 through 16-41.

## 5. Entropy (Section 16.8)

Review: If a system absorbs a quantity of heat $\Delta Q$ at an absolute temperature T, its change in entropy is defined as

$$\Delta S = \Delta Q/T$$

The second law of thermodynamics states that the only processes an isolated system can undergo are those during which the net entropy either increases or remains constant. In symbols,

$$\Delta S \geq 0 \text{ (for an isolated system)}$$

Entropy is a measure of the internal energy of a system that is unavailable to do work. As natural processes in isolated systems proceed toward higher degrees of randomness (disorder), the entropy increases.

Practice: One kilogram of water at 6°C is mixed with 1 kg of water at 0°C.

Determine the following:

| | |
|---|---|
| 1. Final (equilibrium) temperature of the mixture | $\Delta Q = Q_{gain} + Q_{loss} = 0$<br>$Q_{gain} = m_w c_w(T_E - 0°C)$<br>$Q_{loss} = m_w c_w(T_E - 6°C)$<br>$m_w c_w(T_E - 0°C) + m_w c_w(T_E - 6°C) = 0$<br>$2T_E - 6°C = 0; \; T_E = 3°C$ |
| 2. Entropy change of the water as it cools from 6°C to 5°C | Use an average value for T<br>$T = 5.5°C = 278.5$ K<br>$\Delta T = -1.0$ K<br>$\Delta Q = m_w c_w \Delta T = -4180$ J<br>$\Delta S = \Delta Q/T = -15.0$ J/K |
| 3. Entropy change of the water as it cools from 5°C to 4°C | Use $T = 4.5°C = 277.5$ K<br>$\Delta T = -1.0$ K<br>$\Delta Q = m_w c_w \Delta T = -4180$ J<br>$\Delta S = \Delta Q/T = -15.1$ J/K |

| | |
|---|---|
| 4. Entropy change of the water as it cools from 4°C to 3°C | Use $T = 3.5°C = 276.5\ K$<br>$\Delta T = -1.0\ K$<br>$\Delta Q = m_w c_w \Delta T = -4180\ J$<br>$\Delta S = \Delta Q/T = -15.1\ J/K$ |
| 5. Entropy change of the water as it cools from 6°C to 3°C | $\Delta S = \Delta S_{6-5} + \Delta S_{5-4} + \Delta S_{4-3} = 45.2\ J/k$<br>Use $T = 4.5°C = 277.5\ K$<br>$\Delta T = -3.0\ K$<br>$\Delta Q = m_w c_w c \Delta T = -12{,}540\ J$<br>$\Delta S = \Delta Q/T = -45.2\ J/K$ |

Note: We have obtained essentially the same answer for the total change in entropy using both small steps (1°C) and one large step (3°C) to the equilibrium temperature. For convenience, we will use just one step in future work.

| | |
|---|---|
| 6. Entropy change of the water as it warms from 0°C to 3°C | Use $T = 1.5°C = 274.5\ K$<br>$\Delta T = 3.0\ K$<br>$Q = m_w c_w \Delta T = 12{,}540\ J$<br>$\Delta S = \Delta Q/T = 45.7\ J/K$ |
| 7. Total entropy change as the system comes to equilibrium | $\Delta S_T = \Delta S_{cool} + \Delta S_{warm}$<br>$\Delta S_T = -45.2\ J/K + 45.7\ J/K = +0.500\ J/K$ |

Example 16.7. A system of gas undergoes an isothermal process at 127°C. During this process, the gas does $3.00 \times 10^4$ J of work and the internal energy is lowered by $2.00 \times 10^4$ J. Determine the entropy change of the gas.

Given: $T = 127°C$, $W = 3.00 \times 10^4\ J$, $\Delta U = -2.00 \times 10^4\ J$

Determine: The entropy change ($\Delta S$) of the gas

Strategy: Knowing $\Delta U$ and W, we can determine $\Delta Q$. Once $\Delta Q$ and T are known, we can determine the entropy change.

Solution: Using the first law of thermodynamics, we can obtain $\Delta Q$.

$$\Delta U = \Delta Q - W; \quad \Delta Q = \Delta U + W = +1.00 \times 10^4\ J$$

The entropy change is

$$\Delta S = \Delta Q/T = (1.00 \times 10^4\ J)/(400\ K) = 25.0\ J/K$$

Related Text Problems: 16-42 through 16-45.

PRACTICE TEST

Take and grade the practice test. Then use the chart in the next section to determine any weak areas and to write a prescription which will allow you to strengthen your understanding of physics in these areas.

Figure 16.6 shows the P-V and P-T diagrams for a system of He gas as it undergoes the complete cycle 1→2→3→4→5→1. As the gas goes from state 2 to state 3, it absorbs $11.09 \times 10^3$ J of heat. As it goes from state 3 to state 4, it does $5.168 \times 10^3$ J of work.

Figure 16.6

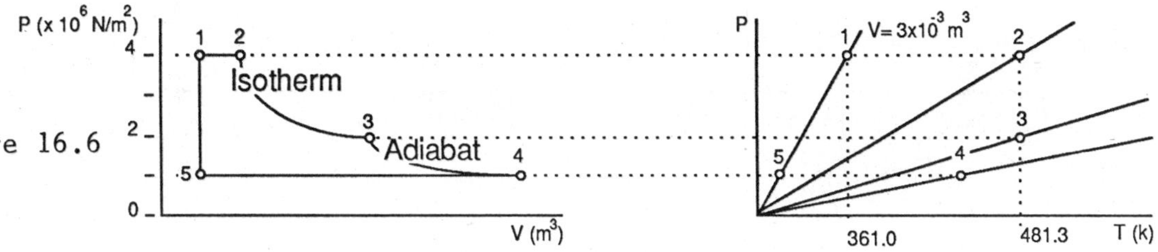


Determine the following:

______ 1. Number of moles of gas in the system
______ 2. Volume of the gas in state 2
______ 3. Work done by the gas as it goes from state 1 to state 2
______ 4. Heat absorbed by the gas as it goes from state 1 to state 2
______ 5. Change in internal energy of the gas as it goes from state 1 to state 2
______ 6. Volume of the gas in state 3
______ 7. Work done by the gas as it goes from state 2 to state 3
______ 8. Change in entropy of the gas as it goes from state 2 to state 3
______ 9. Volume of the gas in state 4
______ 10. Temperature of the gas in state 4
______ 11. Change in entropy of the gas as it goes from state 3 to state 4
______ 12. Change in internal energy of the gas as it goes from state 3 to state 4
______ 13. Work done on the gas as it goes from state 4 to state 5
______ 14. Change in internal energy of the gas as it goes from state 4 to state 5
______ 15. Heat absorbed by the gas as it goes from state 4 to state 5
______ 16. Change in internal energy of the system for a complete cycle
______ 17. Change in heat of the gas for a complete cycle
______ 18. Net work done on or by the gas for a complete cycle

(See Appendix I for answers.)

PRINCIPAL CONCEPTS AND EQUATIONS PRESCRIPTION

Your score on the practice is an excellent measure of your understanding of this chapter. You should now use the following chart to write your own prescription for curing any of your physics ills. Look down the leftmost column to the number of the question(s) you answered incorrectly, read across that row to see which section(s) of the study guide you should return to for further study, and then do the suggested text problems to gain additional experience in working with the particular concept.

| Question Test Number | Concepts and Equations | Prescription | |
|---|---|---|---|
| | | Principal Concept | Text Problems |
| 1 | $P_1V_1 = nRT_1$ | 3 of Ch. 14 | 16-11,12 |
| 2 | $P_2V_2 = nRT_2$ | 3 of Ch. 14 | 16-11,13 |
| 3 | $W_{12} = P_1\Delta V_{12}$; $W_{12} = nR\Delta T_{12}$ | 2 | 16-7,8 |
| 4 | $\Delta Q_{12} = nC_p\Delta T_{12}$ | 2 | 16-6 |
| 5 | $\Delta U_{12} = nC_v\Delta T_{12}$; $\Delta U_{12} = \Delta Q_{12} - W_{12}$ | 1,2 | 16-6,14 |
| 6 | $\Delta Q_{23} = nRT_2\ln(V_3/V_2)$ | 2 | 16-12 |
| 7 | $\Delta U_{23} = \Delta Q_{23} - W_{23}$ | 1 | 16-2,3 |
| 8 | $\Delta S_{23} = \Delta Q_{23}/T_2$ | 5 | 16-42,43 |
| 9 | $P_3V_3{}^\gamma = P_4V_4{}^\gamma$ | 2 | 16-16,17 |
| 10 | $T_3V_3{}^{\gamma-1} = T_4V_4{}^{\gamma-1}$ | 2 | 16-13 |
| 11 | $\Delta S_{34} = \Delta Q_{34}/T$ | 5 | 16-44,45 |
| 12 | $\Delta U_{34} = nC_v\Delta T_{34}$ | 1 | 16-6,18 |
| 13 | $W_{45} = P_4\Delta V_{45}$; $W_{45} = nR\Delta T_{45}$ | 2 | 16-8,11 |
| 14 | $\Delta U_{45} = nC_v\Delta T_{45}$ | 2 | 16-4,18 |
| 15 | $\Delta U_{45} = \Delta Q_{45} - W_{45}$; $\Delta Q_{45} = nC_p\Delta T_{45}$ | 1 | 16-1,6 |
| 16 | $\Delta U_{11} = 0$ for a complete cycle | 2 | 16-1,2 |
| 17 | $\Delta Q_{11} = \sum \Delta Q$ | 2 | 16-1,3 |
| 18 | $W_{net} = \sum W$; $\Delta U_{11} = \Delta Q_{11} - W_{net}$ | 2 | 16-1,8 |

# 17 Electrostatic Forces

## RECALL FROM PREVIOUS CHAPTERS

| Previously learned concepts and equations frequently used in this chapter | Text Section | Study Guide Page |
|---|---|---|
| Analyzing translational motion | 2.4 | 22 |
| Resolving vectors into components | 3.2 | 32 |
| Adding vectors | 3.2 | 32 |
| Newton's second law of motion: $\vec{F}_u = m\vec{a}$ | 4.3 | 47 |
| The first condition of equilibrium: $\sum_i \vec{F}_i = 0$ | 6.1 | 83 |

## NEW IDEAS IN THIS CHAPTER

| Concepts and equations introducted | Text Section | Study Guide Page |
|---|---|---|
| Coulomb's law: $F = k_e q_1 q_2/r^2$ | 17.4 | 271 |
| Electric field: $E = F/q$ and $E = k_e Q/r^2$ | 17.5 | 277 |

## PRINCIPAL CONCEPTS AND EQUATIONS

1. Coulomb's Law (Section 17.4)

Review: Figure 17.1 shows several charged objects in various configurations, along with information about the magnitude and direction of the forces involved.

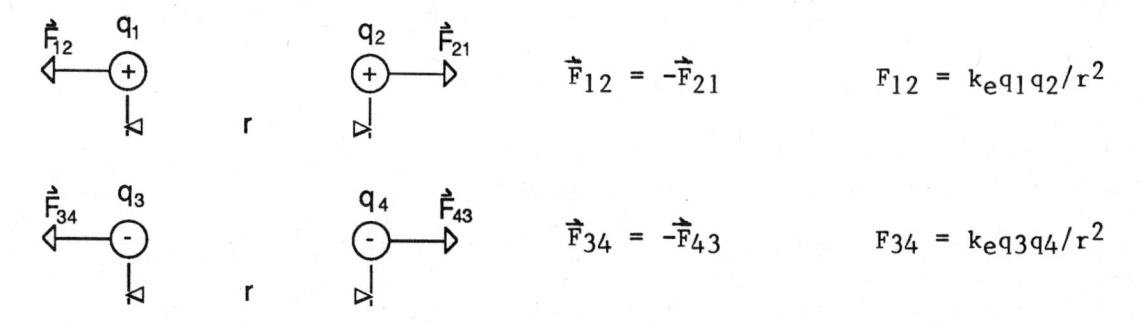

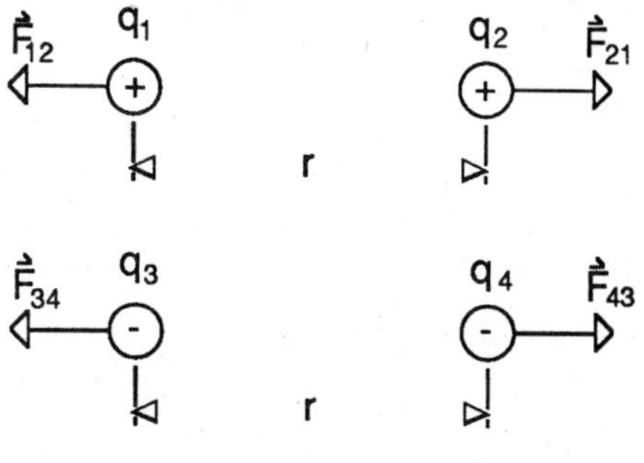


$\vec{F}_{12} = -\vec{F}_{21}$ $\qquad F_{12} = k_e q_1 q_2/r^2$

$\vec{F}_{34} = -\vec{F}_{43}$ $\qquad F_{34} = k_e q_3 q_4/r^2$

Figure 17.1

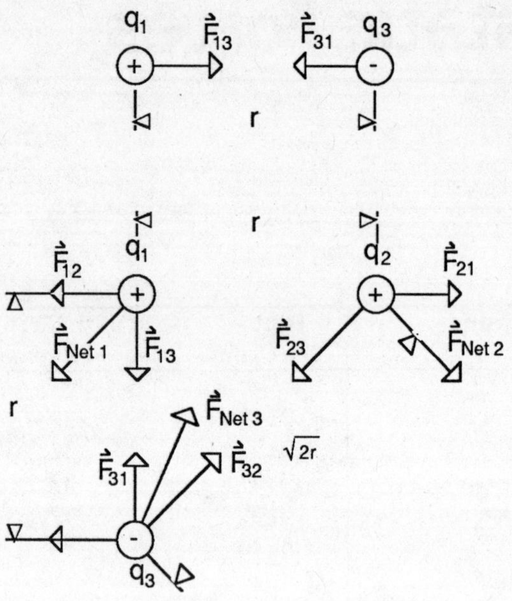


$$\vec{F}_{13} = -\vec{F}_{31} \qquad F_{13} = k_e q_1 q_3/r^2$$

$$\vec{F}_{12} = -\vec{F}_{21} \qquad F_{12} = k_e q_1 q_2/r^2$$

$$\vec{F}_{13} = -\vec{F}_{31} \qquad F_{13} = k_e q_1 q_3/r^2$$

$$\vec{F}_{23} = -\vec{F}_{32} \qquad F_{23} = k_e q_2 q_3/2r^2$$

$$\vec{F}_{net1} = \vec{F}_{12} + \vec{F}_{13}$$

$$\vec{F}_{net2} = \vec{F}_{23} + \vec{F}_{21}$$

$$\vec{F}_{net3} = \vec{F}_{31} + \vec{F}_{32}$$

Figure 17.1 (continued)

We can summarize this information by making the following statements:

1. Like charges repel one another.
2. Unlike charges attract one another.
3. The force of repulsion or attraction is proportional to the product of the charges.
4. The force of repulsion or attraction is inversely proportional to the square of the distance between the charges.

<u>Practice:</u> Figure 17.2 shows two positive charges separated by a distance $r = 1.00$ m.

Figure 17.2

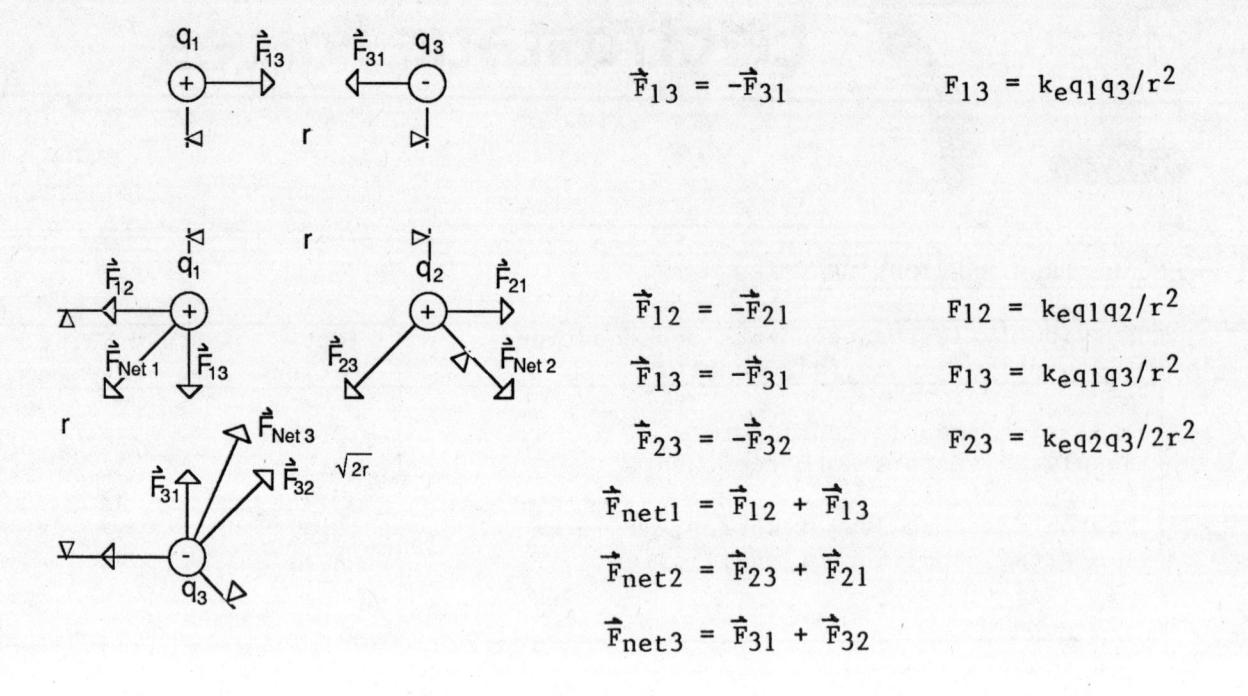

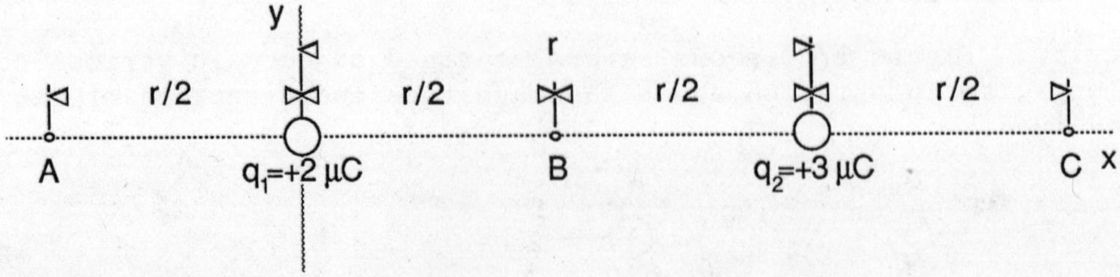


Determine the following:

| | |
|---|---|
| 1. Magnitude and direction of the force on $q_2$ due to $q_1$ | $F_{21} = k_e q_1 q_2/r^2 = 5.40 \times 10^{-2}$ N<br>Since $q_2$ is repelled by $q_1$, this force is in the +x direction. |

| | |
|---|---|
| 2. Magnitude and direction of the force on $q_1$ due to $q_2$ | $\vec{F}_{12} = -\vec{F}_{21} = 5.40 \times 10^{-2}$ N in the $-x$ direction |

A third charge, $q_3 = -1.00\ \mu C$, is placed at position A.

| | |
|---|---|
| 3. Magnitude and direction of the force on $q_1$ due to $q_2$ | $\vec{F}_{12} = 5.40 \times 10^{-2}$ N in the $-x$ direction. This is the same answer as in step 2. |

Note: The force one charge exerts on another is not affected by the introduction of other charges.

| | |
|---|---|
| 4. The magnitude and direction of the force on $q_1$ due to $q_3$ | $\vec{F}_{13} = k_e q_1 q_3/r^2 = 7.20 \times 10^{-2}$ N in the $-x$ direction |
| 5. The net force on $q_1$ |   $\vec{F}_{net1} = \vec{F}_{12} + \vec{F}_{13} = 1.26 \times 10^{-1}$ N in the $-x$ direction |
| 6. The net force on $q_3$ if it is repositioned at B | 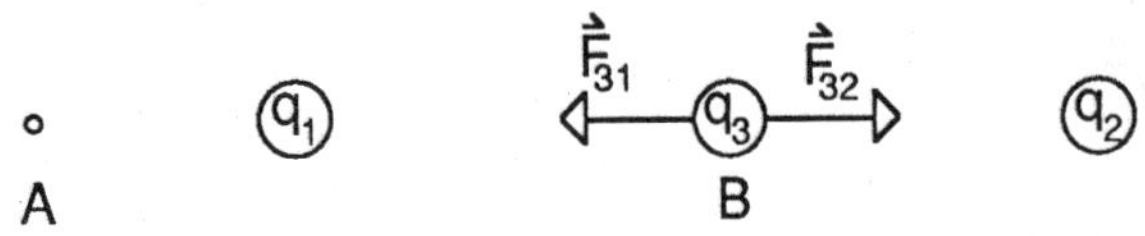  $\vec{F}_{31} = k_e q_3 q_1/r^2 = 7.2 \times 10^{-2}$ N, $-x$ direction<br>$\vec{F}_{32} = k_e q_3 q_2/r^2 = 10.8 \times 10^{-2}$ N, $+x$ direction<br>$\vec{F}_{Net3} = 3.6 \times 10^{-2}$ N, $+x$ direction |
| 7. The location of $q_3$ in order for $F_{net3}$ to equal zero | If $q_3$ is placed to the left of $q_1$, both $q_1$ and $q_2$ force it in the $-x$ direction. If $q_3$ is placed to the right of $q_2$, both $q_1$ and $q_2$ force it in the $+x$ direction. Only when $q_3$ is between $q_1$ and $q_2$ can the forces cancel. We want $F_{net3} = 0$, or $F_{31} = F_{32}$. Let's say this happens when $q_3$ is a distance s from $q_1$, hence 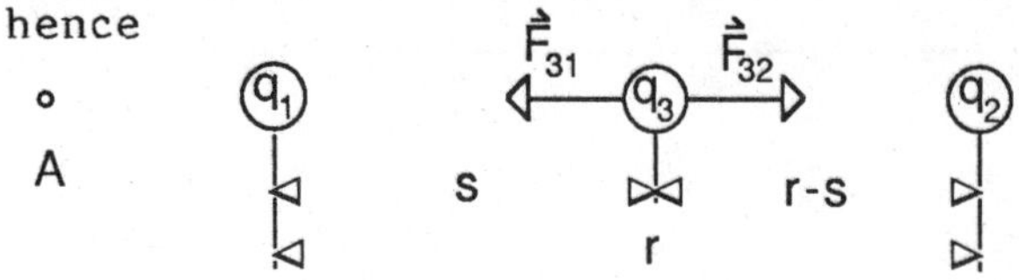  |

$$k_e q_1 q_3/s^2 = k_e q_2 q_3/(r - s)^2$$

or $$s = r/[1 \pm (q_2/q_1))^{1/2}]$$

$$s = 0.449 \text{ m and } -4.45 \text{ m}$$

Of these two mathematical results, only s = 0.449 m is physically correct. A positive value for s puts $q_3$ to the right of $q_1$ and between $q_1$ and $q_2$. We have already agreed that the forces cancel only in this region.

Note: The fact that $q_3$ canceled in the previous step tells us that the answer is independent of $q_3$. A charge of any size or sign will give the same result.

Note: If charges $q_1$ and $q_2$ have the same sign, the net force on a third charge $q_3$ (regardless of size or sign) is be zero only when $q_3$ is placed between $q_1$ and $q_2$. Also $q_3$ will be located closer to the weaker of the $q_1 - q_2$ pair.

Note: If $q_1$ and $q_2$ have different signs, the net force on a third charge $q_3$ (regardless of size or sign) is zero only when $q_3$ is placed outside the region between $q_1$ and $q_2$. Also, $q_3$ will be located closer to the weaker of the $q_1 - q_2$ pair.

$q_1$, $q_2$, and $q_3$ aranged as shown in Figure 17.3.

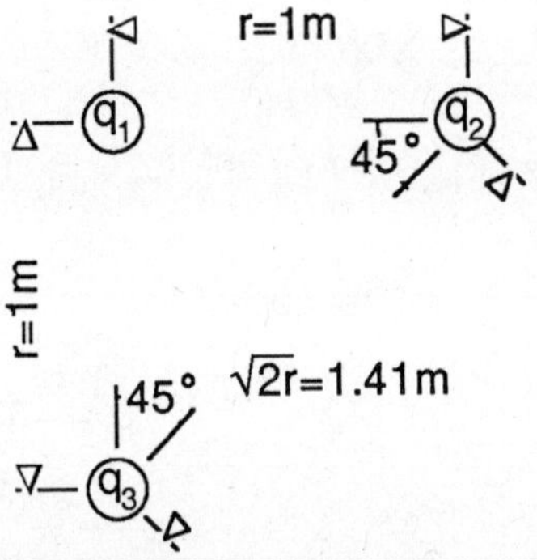


Figure 17.3

$q_1 = +2.00\ \mu C$ $\qquad r = 1.00$ m

$q_2 = +3.00\ \mu C$

$q_3 = -1.00\ \mu C$

8. The magnitude and direction of the net force on $q_1$

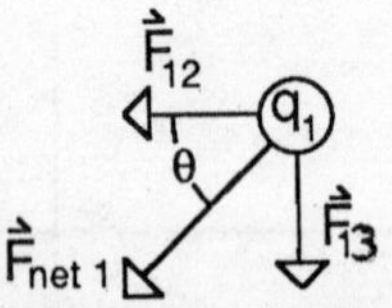


$$F_{12} = k_e q_1 q_2/r^2 = 5.40 \times 10^{-2} \text{ N}$$
$$F_{13} = k_e q_1 q_3/r^2 = 1.80 \times 10^{-2} \text{ N}$$

$$F_{net1} = [F_{12}{}^2 + F_{13}]^{1/2} = 5.69 \times 10^{-2} \text{ N}$$
$$\theta = \tan^{-1} (F_{13}/F_{12}) = 18.4°$$

| | |
|---|---|
| 9. The magnitude and direction of the net force on $q_2$ | 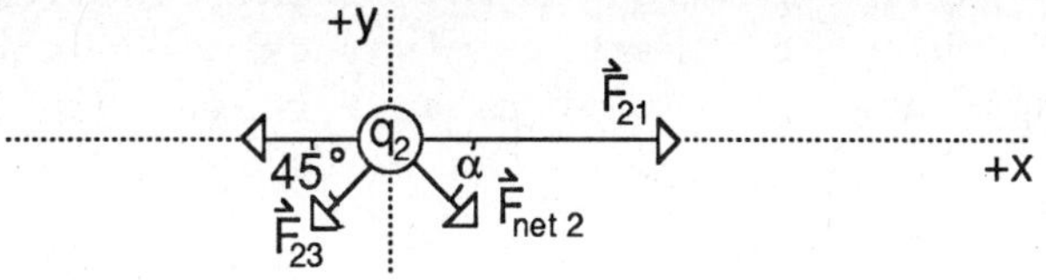<br>$F_{21x} = k_e q_2 q_1/r^2 = 5.40 \times 10^{-2}$ N<br>$F_{23} = k_e q_2 q_3/2r^2 = 1.36 \times 10^{-2}$ N<br>$F_{23x} = -F_{23} \cos 45° = -0.962 \times 10^{-2}$ N<br>$F_{23y} = -F_{23} \sin 45° = -0.962 \times 10^{-2}$ N<br><br>$F_{net2x} = +4.44 \times 10^{-2}$ N<br>$F_{net2y} = -0.962 \times 10^{-2}$ N<br><br>$F_{net2} = [F^2_{net2y} + F^2_{net2x}]^{1/2} = 0.0454$ N<br><br>$\alpha = \tan^{-1}(F_{net2y}/F_{net2x}) = 12.2°$ |
| 10. The magnitude and direction of the net force on $q_3$ | 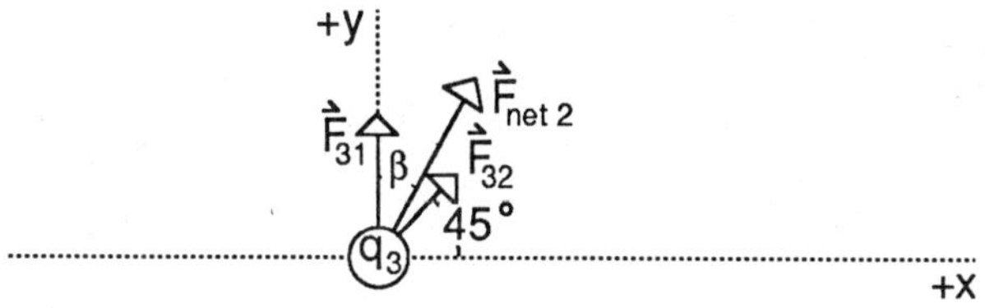<br>$F_{31y} = k_e q_3 q_1/r^2 = 1.80 \times 10^{-2}$ N<br>$F_{32} = k_e q_3 q_2/2r^2 = 1.36 \times 10^{-2}$ N<br>$F_{32x} = F_{32} \cos 45° = 0.962 \times 10^{-2}$ N<br>$F_{32y} = F_{32} \sin 45° = 0.962 \times 10^{-2}$ N<br><br>$F_{net3x} = 0.962 \times 10^{-2}$ N<br>$F_{net3y} = 2.76 \times 10^{-2}$ N<br><br>$F_{net3} = [F_{net3x}^2 + F_{net3y}^2]^{1/2} = 2.92$ N<br><br>$\beta = \tan^{-1}(F_{net3x}/F_{net3y}) = 19.2°$ |

Example 17.1.

Two small spheres, each with mass $1.50 \times 10^{-2}$ g, are given identical charges, attached to fine silk threads 1.00 m long, and hung from a common point. The separation of the spheres is 0.250 m. Determine the tension in each thread and the charge on each sphere.

Given: $m = 1.50 \times 10^{-2}$ g, $r = 0.250$ m, $L = 1.00$ m, and each sphere has the same charge q

Determine: The tension in each string and the charge on each sphere.

Strategy: First, let's draw a free-body diagram for one of the spheres, showing the tension in the string, the force of gravity, and the repulsive Coulomb force. Next, let's pick a convenient coordinate system and find the components of each of these forces. Finally, using the fact that the sphere is in equilibrium, we can write summation-of-forces equations to determine T and q.

Solution 17.4. Figure 17.4a depicts the original situation, and Fig. 17.4b is a free-body diagram for one sphere, showing all the forces acting on it, a convenient coordinate system, and the components of the forces along the coordinate system.

Figure 17.4

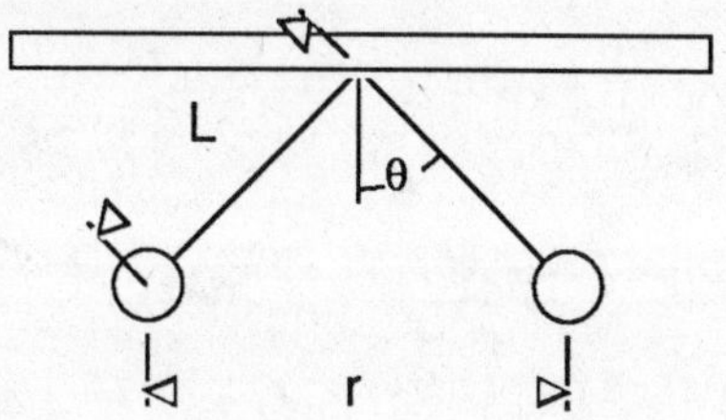


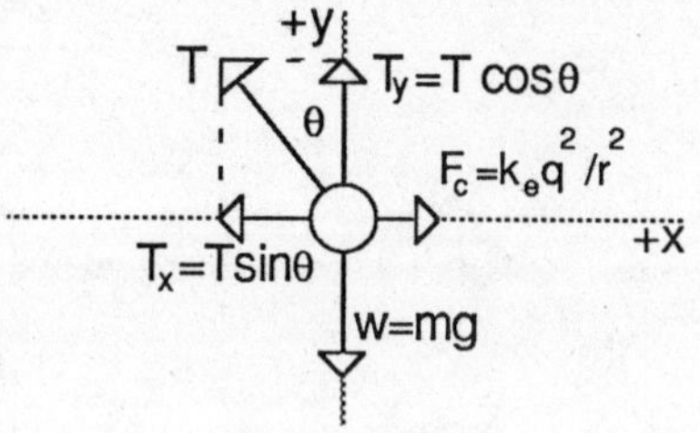


We can obtain the angle θ by

$$\theta = \sin^{-1}\left(\frac{r/2}{L}\right) = \sin^{-1}(0.125/1.00) = 7.18°$$

Summing the y components of the forces, we have

$$\sum F_y = T\cos\theta - mg = 0; \quad T = mg/\cos\theta = 1.48 \times 10^{-4}\ \text{N}$$

Summing the x components of the forces, we have

$$\sum F_x = (k_e q^2/r^2) - T\sin\theta = 0; \quad q = \pm\, r(T\sin\theta/k_e)^{1/2} = \pm 1.13 \times 10^{-8}\ \text{C}$$

The ± tells us that the change can be either positive or negative.

Example 17.2. Suppose we could place $10^{15}$ electrons in each of two 2.00-kg containers. If we then fastened one container securely to the earth and connected the two containers with a strong, lightweight cord 2.00 m long, determine the tension in the cord and the initial acceleration of the free container when the cord is cut.

Given: $m_c$ = 2.00 kg = mass of each container
$m_e$ = 9.11 x $10^{-31}$ kg = mass of electron
e = 1.60 x $10^{-19}$ C = charge on electron
N = $10^{15}$ = number of electrons
L = 2.00 m = length of cord connecting the two containers

Determine: T = tension in the cord
a = acceleration of free container when cord is cut

Strategy: First, using the given information about mass, charge, and number of electrons, we need to determine the charge and mass on each container. Next, let's draw a free-body diagram for the container not attached firmly to the earth. This diagram should show the Coulomb force, the force of gravity, and the tension in the cord. Knowing that the container is in equilibrium, we can determine the tension from a summation-of-forces statement. Knowing all of the forces acting on the container and the mass of the container and electrons, we can determine the net force on the container and hence its acceleration when the cord is cut.

Solution: The charge on each container is

$$Q = Ne = (10^{15} \text{ electrons})(1.60 \times 10^{-19} \text{ C/electron}) = 1.60 \times 10^{-4} \text{ C}$$

The mass of electrons in each container is

$$m_e = (10^{15} \text{ electrons})(9.11 \times 10^{-31} \text{ kg/electron}) = 9.11 \times 10^{-16} \text{ kg}$$

The mass of the container and its electrons is

$$M = m_c + m_e = 2.00 \text{ kg} + 9.11 \times 10^{-16} \text{ kg} \approx 2.00 \text{ kg}$$

Figure 17.5a depicts the original situation, Fig. 17.5b is a free-body diagram for the container, and Fig. 17.4c is a free-body diagram for the container after the cord is cut.

Figure 17.5

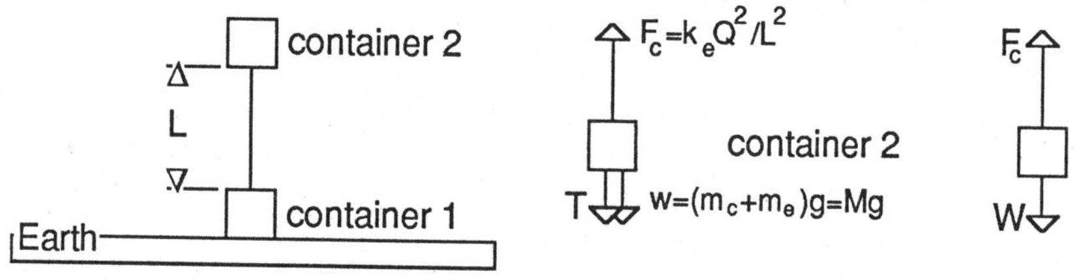


(a) (b) (c)

Since the container is in equilibrium, we can write

$$\sum F = +k_e Q^2/\ell^2 - T - Mg = 0$$

$$T = k_e Q^2/\ell^2 - Mg = 38.0 \text{ N}$$

When the cord is cut, the net force on the free container and electrons is

$$F_{net} = k_e Q^2/\ell^2 - Mg = 38.0 \text{ N}$$

and the subsequent acceleration is

$$a = F_{net}/M = 19.0 \text{ m/s}^2$$

Related Text Problems: 17-1 through 17-12.

## 2. The Electric Field (Section 17.5)

Review: Figure 17.6 shows an object with a charge +Q surrounded by three test charges $+q_{o1}$, $+q_{o2}$, and $+q_{o3}$. The test charges are all a distance r from the center of the charged object.

Figure 17.6

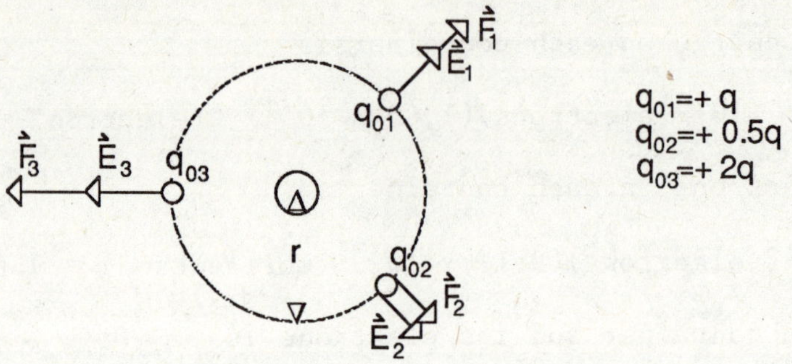


The magnitudes of the electrostatic forces and electric fields felt by each test charge are

$F_1 = k_eQq_{o1}/r^2 = k_eQq/r^2$ $\qquad E_1 = F_1/q_{o1} = k_eQ/r^2$

$F_2 = k_eQq_{o2}/r^2 = k_eQ(q/2)/r^2$ $\qquad E_2 = F_2/q_{o2} = k_eQ/r^2$

$F_3 = k_eQq_{o3}/r^2 = k_eQ(2q)/r^2$ $\qquad E_3 = F_3/q_{o3} = k_eQ/r^2$

Notice that the magnitude of the electrostatic force at position r is directly proportional to the magnitude of the test charge. For example, when the magnitude of the test charge doubles, the magnitude of the electrostatic force doubles. However, the magnitude of the electric field (i.e., the electric force per unit charge) is independent of the magnitude of the test charge. The magnitude of the electric field at any point in space depends only on the magnitude of the charge that created the field and on the distance of the point from the charge. The magnitude of the electric field is defined as the force per unit charge:

$$E = F/q_o = (k_eQq_o/r^2)/q_o = k_eQ/r^2$$

The direction of the electric field at any point in space is the same as the direction of the force felt by a positive test charge at that point.

Practice: A point charge $q_1 = +50.0\ \mu C$ is secured to a point in space.

Determine the following:

| | |
|---|---|
| 1. The electric field 1.00 m from $q_1$ | $\vec{E} = k_eq_1/r^2 = 4.50 \times 10^5$ N/C directed radially outward |

2. How far from $q_1$ we would find a 5-N/C electric field

$E = k_e q_1/r^2$ or
$r = \pm (k_e q_1/E)^{1/2} = \pm 300$ m, use $r = +300$ m
Since all measurements are radially outward, a negative value has no meaning.

3. Where we should place a point charge $q_2 = +100\ \mu C$ in order to obtain a zero electric field 1.00 m from $q_1$

Our knowledge of electrostatics allows us to conclude that the fields cancel only at some point between the two charges. This allows us to draw

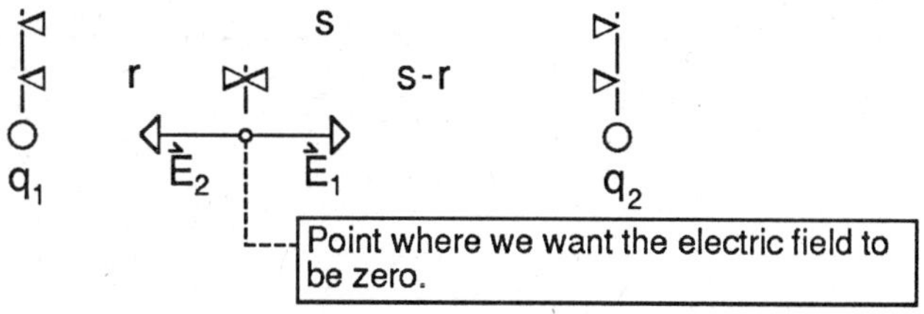


$$E_1 = E_2$$

$$k_e q_1/r^2 = k_e q_2/(s-r)^2$$
$$s = r[1 \pm (q_2/q_1)^{1/2}]$$
$$= 2.41 \text{ m and } -0.41 \text{ m}$$

Place $q_2$ 2.41 m from $q_1$. Since all measurementes are radially outward from $q_1$, a negative value has no meaning.

4. Where we should place a point charge $q_3 = -100\ \mu C$ in order to obtain a zero electric field 1.00 m from $q_1$

Our knowledge of electrostatics allows us to conclude that the fields cancel for the following arrangement:

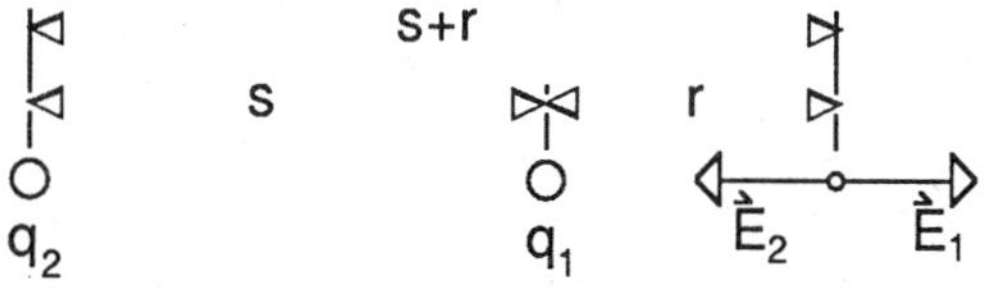


$$E_1 = E_3$$
$$k_e q_1/r^2 = k_e q_3/(S+r)^2$$
$$s = -r[1 \mp (q_3/q_1)^{1/2}]$$
$$s = 0.414 \text{ m and } -2.41 \text{ m}$$

Of these two mathematical results, only the value $s = 0.414$ m is physically correct

A particle of mass $1.00 \times 10^{-2}$ kg and charge 100 μC is released from rest in a uniform electric field. It achieves a speed of 10.0 m/s after being accelerated by the electric field for 1.00 m. Figure 17.7 shows before and after pictures.

Figure 17.7

(a) Before (b) After

Determine the following:

| | |
|---|---|
| 5. Acceleration of the particle | $v = 10.0$ m/s, $v_o = 0$ m/s, $s = 1.00$ m<br>$v^2 - v_o^2 = 2as$<br>$a = (v^2 - v_o^2)/2s = 50.0\ m/s^2$ |
| 6. Force on the particle | $F = ma = 0.500$ N |
| 7. Magnitude of the electric field intensity | $E = F/q = 5.00 \times 10^3$ N/C |
| 8. Change in kinetic energy of the particle | $\Delta KE = KE - KE_o = mv^2/2 - 0 = 0.500$ J |
| 9. Work done on the particle by the electric field | $W = Fs \cos\theta = Fs \cos 0° = 0.500$ N |

Note: All the work done on the particle by the electric field goes into increasing the particle's kinetic energy.

---

Example 17.3. Two objects of charge $q_1 = +20.0\ \mu C$ and $q_3 = -30.0\ \mu C$ have negligible mass and are connected by a lightweight cord 1.00 m long. The object $q_2$ is then securely attached to a point in space. Determine the magnitude of the electric field that will create a 2.00-N tension in the cord.

Given: $q_1 = +20.0\ \mu C$, $q_2 = -30.0\ \mu C$, $L = 1.00$ m, $T = 2.00$ N

Determine: The magnitude of the electric field that will create a 2.00 N tension in the cord.

Strategy: There is tension in the cord because $q_1$ is forced away from $q_2$ by the electric field. Since $q_1$ is in equilibrium, we can sum the forces and determine the magnitude of the electric force needed to create the desired tension. Knowing the magnitude of the electric force and the charge $q_1$, we can determine the magnitude of the electric field needed to create the desired tension.

Solution: Figure 17.8a shows $q_2$ attached to a point in space, $q_1$ attached to $q_2$ by a cord of length L, and the electric field. Figure 17.8b is a free-body diagram for $q_1$.

Figure 17.8

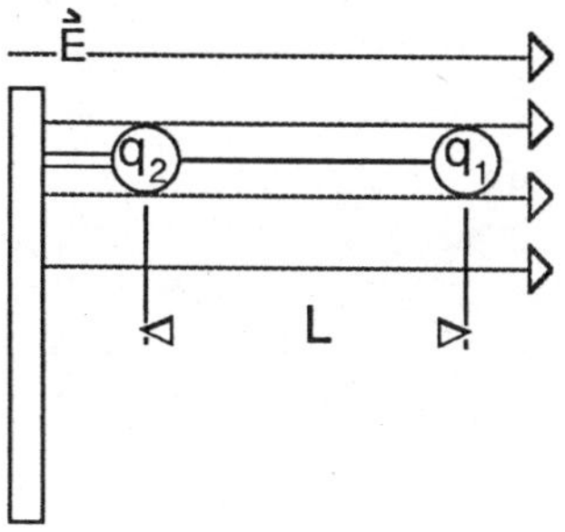


(a)

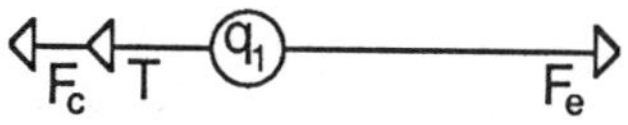


$T = 2.00$ N = tension
$F_c = k_e q_1 q_2/\ell^2 = 5.40$ N = Coulomb force
$F_E$ = electric field force

(b)

Since $q_1$ is in equilibrium, we can write

$$\sum F = F_E - T - F_c = 0; \quad F_E = T + F_c = 7.40 \text{ N}$$

The magnitude of the electric field is given by

$$E = F_E/q_1 = 3.70 \times 10^5 \text{ N/C}$$

Example 17.4. An electron traveling horizontally at $4.00 \times 10^6$ m/s enters a region where a vertically upward, 500-N/C electric field exists. If the region in which the electric field exists is 0.100 m long, determine the vertical displacement of the electron as it crosses this region.

Given: An electron with $q = 1.60 \times 10^{-19}$ C and $m = 9.11 \times 10^{-31}$ kg
$v = 4.00 \times 10^6$ m/s = speed of electron as it enters the E field
$E = 500$ N/C = magnitude of E field
$L = 0.100$ m = length of region of E field

Determine: The vertical displacement of the electron as it crosses the electric field.

Strategy: Knowing the magnitude of the electric field and the charge on the electron, we can determine the vertical force on the electron. Knowing this vertical force and the mass of the electron, we can determine its vertical acceleration. Knowing the horizontal speed of the electron and the length of the region in which the electric field exists, we can determine the time the electron spends in the field. Knowing the vertical acceleration and the time the electron undergoes this acceleration, we can determine the vertical displacement.

Solution: Figure 17.9 shows the details of the problem.

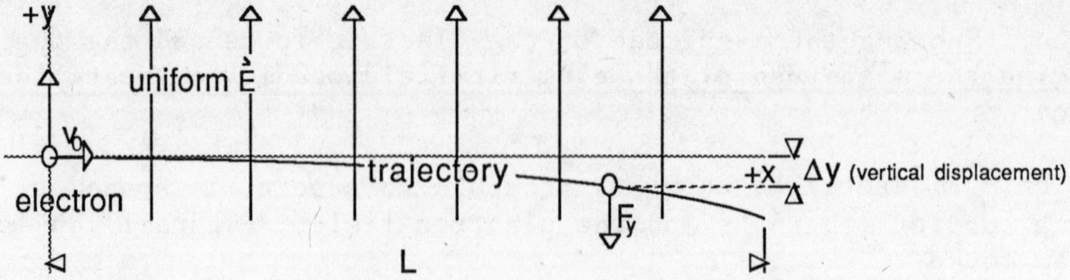


Figure 17.9

Let's work the problem algebraically and then insert values.

The vertical force on the electron is $F_y = -qE$

The vertical acceleration of the electron is $a_y = F_y/m = -qE/m$

The time it takes the electron to cross the electric field can be obtained as follows:

$x = x_o + v_{ox}t + a_xt^2/2 = v_ot \quad (x_o = 0,\ v_{ox} = v_o,\ a_x = 0)$
$x = L$ when $t = t_E$ (time in the electric field)
$L = v_ot_E;\ t_E = L/v_o$

The y displacement can be obtained as follows:

$y = y_o + v_{oy}t + a_yt^2/2 = -gEt^2/2m \quad (y_o = 0,\ v_{oy} = 0,\ a_y = -qE/m)$
$y = y_E$ (displacement due to the electric field) when $t = t_E = L/v_o$
$y_e = -(qE/2m)(L^2/v_o^2) = -2.74 \times 10^{-2}$ m

Related Text Problems: 17-13 through 17-25.

---

PRACTICE TEST

Take and grade the practice test. Doing so will allow you to determine any weak spots in your understanding of the concepts taught in this chapter. The following section prescribes what you should study further to strengthen your understanding.

Figure 17.10 shows three charges and four positions in space.

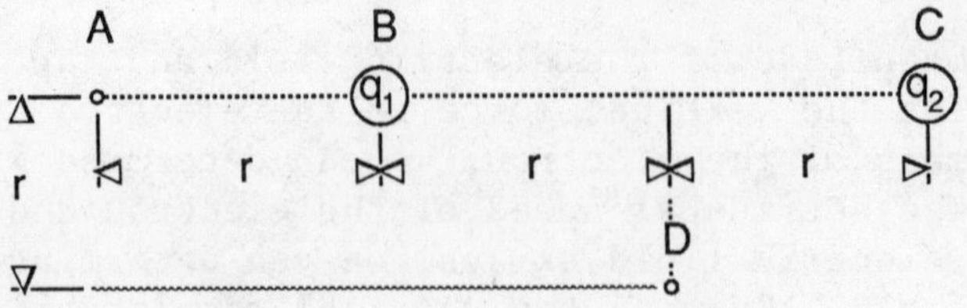


Figure 17.10

$q_1 = +200\ \mu C$
$q_2 = -300\ \mu C$
$q_3 = +100\ \mu C$
$r = 1.00$ m

Determine the following:

________ 1. Force on $q_2$ due to $q_1$ ($\vec{F}_{21}$)
________ 2. Force on $q_1$ due to $q_2$ ($\vec{F}_{12}$)

Now place $q_3$ at A

_____________ 3. Force on $q_3$ due to $q_1$ ($\vec{F}_{31}$)
_____________ 4. Force on $q_3$ due to $q_2$ ($\vec{F}_{32}$)
_____________ 5. Net force on $q_3$ ($\vec{F}_{net3}$)
_____________ 6. Electric field at A ($\vec{E}_A$)

Now place $q_3$ at D

_____________ 7. Force on $q_3$ due to $q_1$ ($\vec{F}_{31}$)
_____________ 8. Force on $q_3$ due to $q_2$ ($\vec{F}_{32}$)
_____________ 9. Net force on $q_3$ ($\vec{F}_{net3}$)
_____________ 10. Electric field at D ($\vec{E}_D$)
_____________ 11. Location where $q_3$ experiences no net force
_____________ 12. Point on a line passing through A, B, and C where $\vec{E}$ is zero

Figure 17.11 shows an object of mass m and charge q suspended by a weightless string of length L in a uniform electric field $\vec{E}$.

Figure 17.11

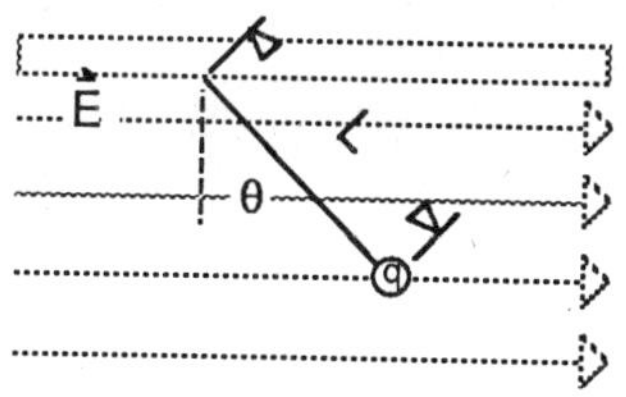


$\theta = 10.0°$
$m = 1.00 \times 10^{-2}$ kg
$q = 100$ μC
$L = 1.00$ m

Determine the following:

_____________ 13. Tension in the string
_____________ 14. Magnitude of $\vec{E}$

Figure 17.12 shows an electron traveling at 100 m/s and entering a region in which there is a 500-V/m electric field.

Figure 17.12

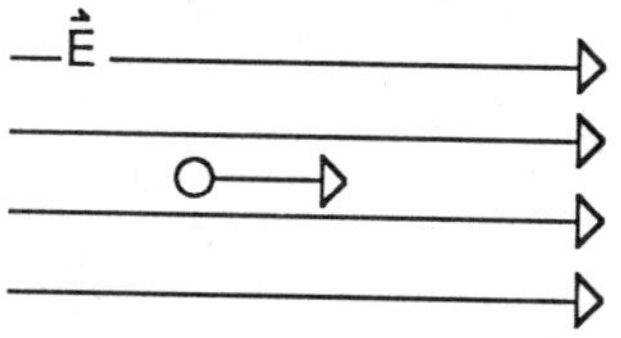


Determine the following:

_____________ 15. Force on electron
_____________ 16. Acceleration of electron
_____________ 17. Distance electron travels in field before coming to rest
_____________ 18. Time electron travels in field before coming to rest

(See Appendix I for Answers.)

## PRINCIPAL CONCEPTS AND EQUATIONS PRESCRIPTION

Your score on the practice is an excellent measure of your understanding of this chapter. You should now use the following chart to write your own prescription for curing any of your physics ills. Look down the leftmost column to the number of the question(s) you answered incorrectly, read across that row to see which section(s) of the study guide you should return to for further study, and then do the suggested text problems to gain additional experience in working with the particular concept.

| Practice Test Question | Concepts and Equations | Prescription: Principal Concept | Prescription: Text Problems |
|---|---|---|---|
| 1 | Coulomb's law: $F = k_e q_1 q_2/r^2$ | 1 | 17-1,3 |
| 2 | Coulomb's law | 1 | 17-3,4 |
| 3 | Coulomb's law | 1 | 17-4,5 |
| 4 | Coulomb's law | 1 | 17-5,6 |
| 5 | Adding vectors | 4 of Ch. 3 | 17-7,8 |
| 6 | Electric field: $E = F/q_o = kQ/r^2$ | 2 | 17-13,25 |
| 7 | Coulomb's law | 1 | 17-3,5 |
| 8 | Coulomb's law | 1 | 17-1,6 |
| 9 | Adding vectors | 4 of Ch. 3 | 17-11,12 |
| 10 | Electric field | 2 | 17-14,18 |
| 11 | Coulomb's law | 1 | 17-9 |
| 12 | Electric field | 2 | 17-9,22 |
| 13 | First condtion of equilibrium $\sum_i F_i = 0$ | 1 of Ch. 6 | 17-10 |
| 14 | Electric field | 2 | 17-10,18 |
| 15 | Electric field | 2 | 17-16,20 |
| 16 | Newton's second law: $F = ma$ | 2 of Ch. 4 | 17-15,16 |
| 17 | Translational motion: $v^2 - v_o^2 = 2as$ | 5 of Ch. 2 | 17-15,17 |
| 18 | Translational motion: $v = v_o + at$ | 5 of Ch. 2 | 17-16,17 |

# 18 Electrostatic Energy and Capacitance

RECALL FROM PREVIOUS CHAPTERS

| Previously learned concepts and equations frequently used in this chapter | Text Section | Study Guide Page |
|---|---|---|
| Work: $W = F_{\parallel}\Delta s = F\Delta s \cos\theta$ | 7.3 | 100 |
| Kinetic Energy: $KE = mv^2/2$ | 7.5 | 108 |
| Gravitational potential energy: $\Delta PE = -W_g$ | 7.6 | 110 |
| Coulomb's law: $F = k_e q_1 q_2/r^2$ | 17.4 | 260 |
| Electric field: $E = F/q$ | 17.5 | 267 |

NEW IDEAS IN THIS CHAPTER

| Concepts and equations introduced | Text Section | Study Guide Page |
|---|---|---|
| Change in electric potential energy: $\Delta PE_{ab} = k_e Q q_o[(1/r_b) - (1/r_a)]$ | 18.2 | 285 |
| Electric potential difference: $V_{ab} = \Delta PE_{ab}/q_o = k_e Q[(1/r_b) - (1/r_a)]$ | 18.2 | 289 |
| Absolute potential: $V = k_e Q/r$ | 18.3 | 292 |
| Potential difference and $\vec{E}$ field: $E = -V/\Delta s$ | 18.4 | 294 |
| Capacitance: $C = q/V = \varepsilon_o A/d$ | 18.6 | 298 |
| Energy stored in a capacitor: $W = CV^2/2 = qV/2 = \varepsilon_o AdE^2/2$ | 18.7 | 298 |
| Energy per unit volume: $\mu = W/Ad = \varepsilon_o E^2/2$ | 18.7 | 298 |
| Dielectric coefficient: $K = C/C_o$ | 18.8 | 298 |

PRINCIPAL CONCEPTS AND EQUATIONS

1. Change in Electric Potential Energy (Section 18.2)

Review: Since the Coulomb force is conservative, the change in electric potential energy when a test charge $q_o$ is moved in the electric field of a source charge Q is the negative of the work done by the Coulomb force. Figure 18.1 shows a source charge and its associated electric field, a test charge being moved from some initial position $r_a$ to some final position $r_b$, and the Coulomb force that Q exerts on $q_o$.

Figure 18.1

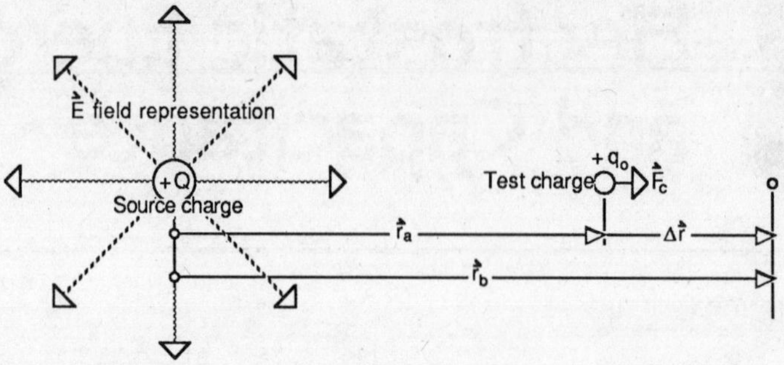


$$\Delta PE_{ab} = -W \text{ (by the Coulomb force } F_c)$$

The change in electric potential energy is

$$\Delta PE_{ab} = k_e Q q_o[(1/r_b) - (1/r_a)]$$

This is a mutual change in electric potential energy. It is not just a change in the electric potential energy of $q_o$ or Q. It is the change in electric potential energy of the system.

Practice: Figure 18.2 shows a source charge Q = +100 μC, its associated electric field, and two test charges $q_{o1}$ = +1.00 μC and $q_{o2}$ = −1.00 μC. We are interested in moving $q_{o1}$ and $q_{o2}$ around in this electric field and determining the subsequent changes in electric potential energy.

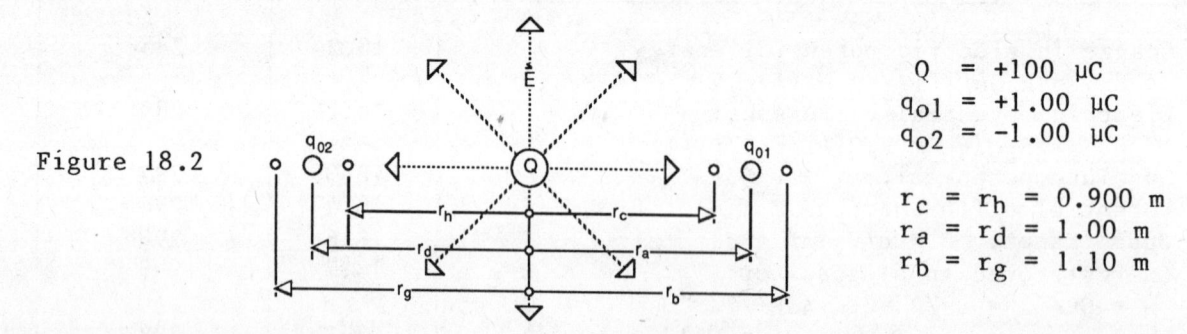

Figure 18.2

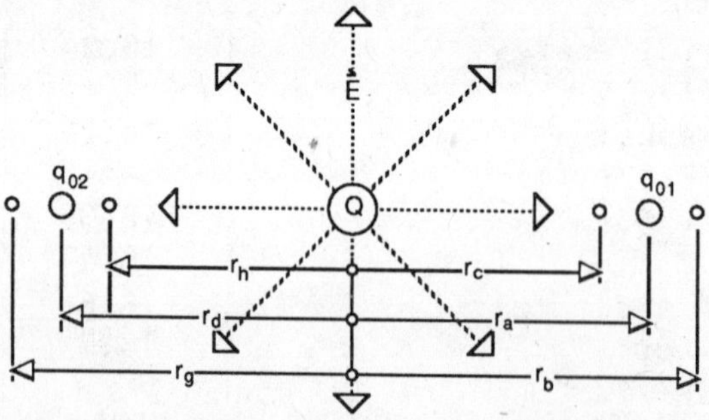


Q = +100 μC
$q_{o1}$ = +1.00 μC
$q_{o2}$ = −1.00 μC

$r_c = r_h$ = 0.900 m
$r_a = r_d$ = 1.00 m
$r_b = r_g$ = 1.10 m

Determine the following:

1. The sign on the work done by the Coulomb force as $q_{o1}$ moves from
   (a) $r_a$ to $r_b$
   (b) $r_a$ to $r_c$

(a)

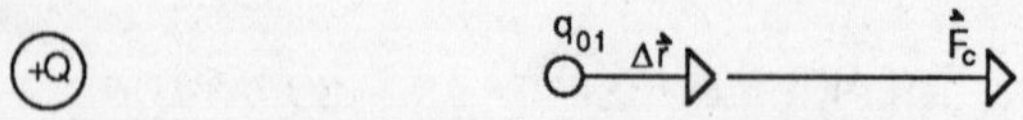


As $q_{o1}$ goes from $r_a$ to $r_b$, the angle between $F_c$ and $\Delta r$ is zero, hence

$$W_{ab} = F_c \Delta r \cos\theta = F_c \Delta r \cos 0° = +$$

(b)

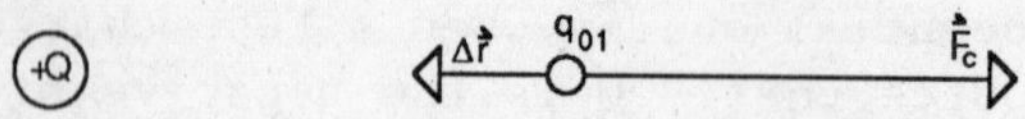


$$W_{ac} = F_c \Delta r \cos\theta = F_c \Delta r \cos 180° = -$$

| | |
|---|---|
| 2. The sign on the change in the electric potential energy as $q_{o1}$ moves from<br>(a) $r_a$ to $r_b$<br>(b) $r_a$ to $r_c$ | The change in electric potential energy is the negative of the work done by the Coulomb force.<br><br>(a) $\Delta PE_{ab} = -W_{ab}$ = (negative)<br>(b) $\Delta PE_{ac} = -W_{ac}$ = (positive)<br><br>We can also investigate the expression<br><br>$\Delta PE_{ab} = k_eQq_o[(1/r_b) - (1/r_a)]$<br><br>(a) Q and $q_{o1}$ are positive, but since $r_a < r_b$, $\Delta PE_{ab}$ is negative.<br><br>(b) Q and $q_{o1}$ are positive, but since $r_c < r_a$, $\Delta PE_{ac}$ is positive. |
| 3. The change in electric potential energy as $q_{o1}$ moves from<br>(a) $r_a$ to $r_b$<br>(b) $r_a$ to $r_c$ | $\Delta PE_{ab} = k_eQq_{o1}[(1/r_b) - (1/r_a)]$<br><br>(a) $\Delta PE_{ab} = -8.18 \times 10^{-2}$ J<br>(b) $\Delta PE_{ac} = +10.0 \times 10^{-2}$ J<br><br>Note: The signs for these values are consistent with our findings in step 2. |
| 4. The work done by the Coulomb force on $q_{o1}$ as it is moved from<br>(a) $r_a$ to $r_b$<br>(b) $r_a$ to $r_c$ | (a) $W_{ab} = -\Delta PE_{ab} = +8.18 \times 10^{-2}$ J<br>(b) $W_{ac} = -\Delta PE_{ac} = -10.0 \times 10^{-2}$ J<br><br>Note: The signs for these values are consistent with our findings in step 1. |

Note: If the source and test charge have the same sign, positive work is done by the Coulomb force and the mutual electric potential energy decreases as $q_o$ moves away from Q; negative work is done by the Coulomb force and the mutual electric potential energy increases as $q_o$ moves towards Q.

| | |
|---|---|
| 5. The sign on the work done by the Coulomb force and the sign on the change in electric potential energy as $q_{o2}$ goes from<br>(a) $r_d$ to $r_h$<br>(b) $r_d$ to $r_g$ | (a) $\theta = 0$; $W_{dh} = F_c\Delta r \cos 0° = +$<br>$\Delta PE_{dh} = -W_{dh}$ (negative)<br><br>(b) $\theta = 180°$; $W_{dg} = F_c\Delta r \cos 180° = -$<br>$\Delta PE_{dg} = -W_{dg}$ (positive) |

| | |
|---|---|
| 6. The change in mutual electric potential energy and the work done on $q_{o2}$ by the Coulomb force as it moves from (a) $r_d$ to $r_h$ (b) $r_d$ to $r_g$ | $\Delta PE_{dh} = k_e Q q_{o2} [(1/r_h) - (1/r_d)]$ <br> (a) $\Delta PE_{dh} = -10.0 \times 10^{-2}$ J <br> $W_{dh} = -\Delta PE_{dh} = +10.0 \times 10^{-2}$ J <br> (a) $\Delta PE_{dg} = +8.18 \times 10^{-2}$ J <br> $W_{dg} = -\Delta PE_{dg} = -8.18 \times 10^{-2}$ J |

Note: If the source and the test charge have opposite signs, negative work is done by the Coulomb force and the mutual electric potential energy increases as $q_o$ moves away from Q; positive work is done by the Coulomb force and the mutual electric potential energy decreases as $q_o$ moves toward Q.

Note: If you need more practice, redo the preceding calculations using a negative source charge.

Example 18.1. Figure 18.3 shows two charges $q_1$ and $q_2$ on the y axis an equal distance from the origin. A third charge $q_3$ is an infinite distance away on the x axis.

Figure 18.3

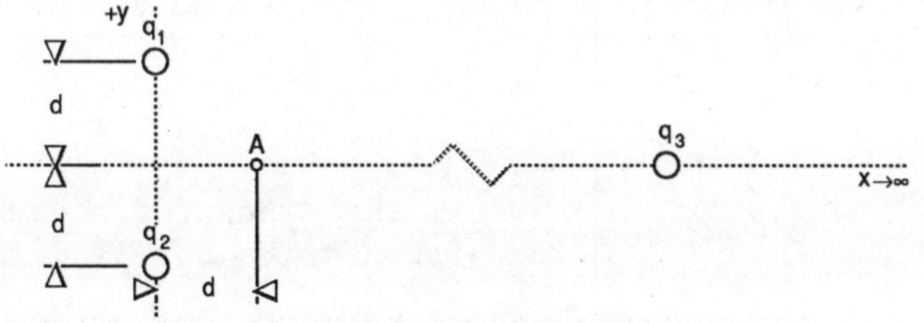


$q_1 = +2.00$ μC
$q_2 = -4.00$ μC
$q_3 = -6.00$ μC
$d = 0.500$ m

Determine the change in electric potential energy as $q_3$ is moved along the x axis to point A.

Given: $q_1 = +2.00$ μC, $q_2 = -4.00$ μC, $q_3 = -6.00$ μC, $d = 0.500$ m

Determine: The change in electric potential energy as $q_3$ is moved along the x axis from infinitely far away to point A.

Strategy: As $q_3$ approaches $q_2$, negative work is done by the Coulomb force and the electrical potential energy increases. We can determine this increase in electric potential energy from the given information. As $q_3$ approaches $q_1$, positive work is done by the Coulomb force and the electric potential energy decreases. We can determine this decrease in electric potential energy from the given information. Finally, we can obtain the total change in electric potential energy by adding the increase and decrease in electric potential energy.

Solution: First, let's obtain the increase in electric potential energy as $q_3$ approaches $q_2$.

$$r_i = \infty \text{ and } r_f = \sqrt{2}d = 0.707 \text{ m}$$

$$\Delta PE_{if} = k_e q_3 q_2 \, [(1/r_f) - (1/r_i)] = 3.06 \times 10^{-1} \text{ J}$$

Second, let's obtain the decrease in electric potential energy as $q_3$ approaches $q_1$.

$$r_i = \infty \text{ and } r_f = \sqrt{2}d = 0.707 \text{ m}$$

$$\Delta PE_{if} = k_e q_1 q_3 \, [(1/r_f) - (1/r_i)] = -1.53 \times 10^{-1} \text{ J}$$

Finally, let's obtain the total change in electric potential energy by adding the increase and decrease in electric potential energy.

$$\Delta PE_{total} = (+3.06 \times 10^{-1} \text{ J}) + (-1.53 \times 10^{-1} \text{ J}) = +1.53 \times 10^{-1} \text{ J}$$

Overall, the electric potential energy is increased as $q_3$ approaches $q_1$ and $q_2$.

Related Text Problems: 18-7, 18-8, 18-9, 18-19.

## 2. Electric Potential Difference (Section 18.2)

Review: When a test charge $q_o$ moves from an initial position $r_a$ to a final position $r_b$ in the electric field of a source charge Q, the mutual change in electric potential energy is given by

$$\Delta PE_{ab} = k_e Q q_o \, [(1/r_b) - (1/r_a)]$$

From this expression, it is obvious that ΔPE depends not only on Q but also on $q_o$. We would like to have a quantity that describes the space around the source charge and depends only on the source charge and the distance from it. We can obtain such a quantity by dividing out the charge $q_o$. The new quantity is the electric potential difference and is given by

$$V_{ab} = \Delta PE_{ab}/q_o = k_e Q \, [(1/r_b) - (1/r_a)]$$

The electric potential difference between two points in space is the change in electric potential per unit charge.

Practice: Consider the source charge Q, its associated electric field and the test charges $q_{o1}$ and $q_{o2}$, as shown in Fig. 18.4.

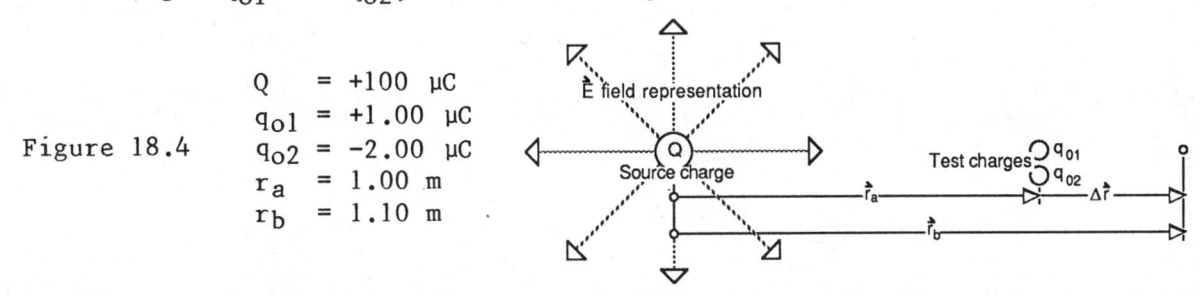

Figure 18.4

$Q = +100 \ \mu C$
$q_{o1} = +1.00 \ \mu C$
$q_{o2} = -2.00 \ \mu C$
$r_a = 1.00$ m
$r_b = 1.10$ m

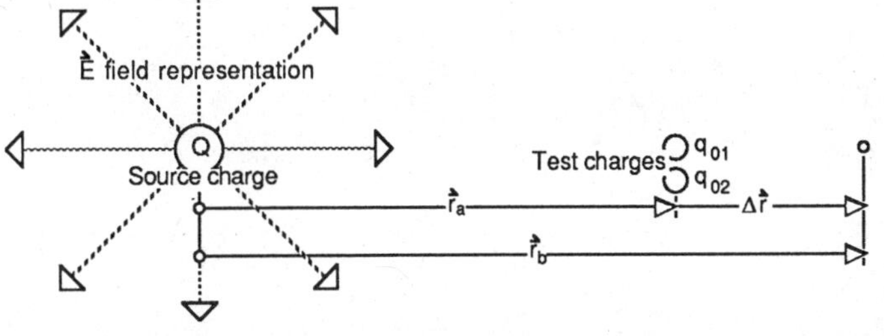

Determine the following:

| | |
|---|---|
| 1. The change in electric potential energy as $q_{o1}$ and $q_{o2}$ are moved from $r_a$ to $r_b$ | $\Delta PE = k_e Q q_o[(1/r_b) - (1/r_a)]$<br>$q_{o1}$: $\Delta PE_{ab} = -8.18 \times 10^{-2}$ J<br>$q_{o2}$: $\Delta Pe_{ab} = +16.4 \times 10^{-2}$ J |

Note: The difference in mutual electric potential energy when a test charge is moved from one point to another in the electric field of a source charge depends on the magnitude and sign of the test charge. (See step 1.)

| | |
|---|---|
| 2. The electric potential difference between $r_a$ and $r_b$ for $q_{o1}$ and $q_{o2}$ | $q_{o1}$: $V_{ab} = \Delta PE_{ab}/q_{o1} = -8.18 \times 10^4$ V<br>$V_{ab} = kQ[(1/r_b) - (1/r_a)]$<br>$= -8.18 \times 10^4$ V<br>$q_{o2}$: $V_{ab} = \Delta PE_{ab}/q_{o2} = -8.18 \times 10^4$ V<br>$V_{ab} = kQ[(1/r_b) - (1/r_a)]$<br>$= -8.18 \times 10^4$ V |

Note: The electric potential difference between any two points in space around a source charge Q depends only on Q and the distance from it. The electric potential difference is independent of the magnitude and sign of the test charge being moved. (See step 2.)

Note: If you need more practice, repeat these calculations for a negative source charge and do a case where $r_b < r_a$.

Example 18.2. A charge $q_1 = +300$ μC is placed in space. What is the electric potential difference between two points at distances $r_a = 1.20$ m and $r_b = 1.50$ m away? How much work would it require to push a charge $q_2 = +200$ μC from $r_b$ to $r_a$?

Given: $q_1 = +300$ μC, $q_2 = +200$ μC, $r_a = 1.20$ m, $r_b = 1.50$ m

Determine: The electric potential difference between the two points a distance $r_a$ and $r_b$ away. The amount of work required to push $q_2$ from $r_b$ to $r_a$.

Strategy: Knowing $q_1$, $r_a$, and $r_b$, we can determine the electric potential difference between $r_a$ and $r_b$. knowing the electric potential difference between $r_a$ and $r_b$ and the charge $q_2$, we can determine the change in electric potential energy and hence the work required to push $q_2$ from $r_b$ to $r_a$.

Solution: The electric potential difference between $r_a$ and $r_b$ is

$$V_{ab} = k_e q_1[(1/r_b) - (1/r_a)] = -4.50 \times 10^5 \text{ V}$$

The electric potential difference between $r_b$ and $r_a$ is

$$V_{ba} = k_e q_1[(1/r_a) - (1/r_b)] = +4.50 \times 10^5 \text{ V}$$

This allows us to conclude that the electric potential difference between $r_a$ and $r_b$ is $4.50 \times 10^5$ V and that point $r_a$ is at the higher electric potential.

The change in mutual electric potential energy when $q_2$ is pushed from $r_b$ to $r_a$ is

$$\Delta PE_{ba} = V_{ba}q_2 = (+4.50 \times 10^5 \text{ V})(200\ \mu\text{C}) = 90.0 \text{ J}$$

If the mutual electric potential energy increases by 90.0 J when $q_2$ is moved from $r_b$ to $r_a$, 90.0 J of work is required to push $q_2$ from $r_b$ to $r_a$.

Example 18.3. Two protons are initially at rest 5.00 cm apart. If they are released, what will their speed be when they are 25.0 cm apart?

Given: $m_p = 1.67 \times 10^{-27}$ kg, $q = 1.60 \times 10^{-19}$ C, $v_o = 0$, $r_o = 5.00$ cm, $r_f = 25.0$ cm.

Determine: Speed of the protons when they are 25.0 cm apart.

Strategy: The protons exert a force on each other, but no outside dissipative forces act on the system. As a result, energy is conserved ($E_i = E_f$) and, as the protons move, the change in electric potential energy plus the change in kinetic energy must sum to zero. We can determine the decrease in electric potential energy from the given information. Knowing that the increase in kinetic energy is equal to the decrease in electric potential energy, we can determine the final speed of the protons.

Solution: The initial and final situation are shown in Fig. 18.5.

Figure 18.5

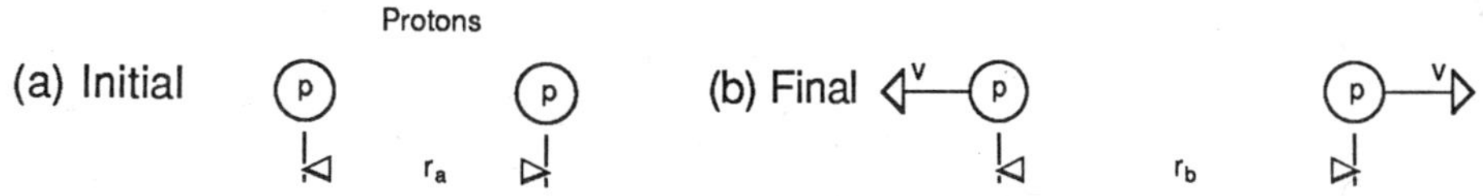


The change in electric potential energy is

$$\Delta PE_{ab} = k_e q^2[(1/r_b) - (1/r_a)] = -3.69 \times 10^{-27} \text{ J}$$

This decrease in electric potential energy goes into increasing the kinetic energy of the protons.

$$\Delta KE_{ab} = -\Delta PE_{ab} = 3.69 \times 10^{-27} \text{ J}$$

Half of this change belongs to each proton, hence

$$m_p v^2/2 = KE_{ab}/2 \qquad v = [KE_{ab}/m_p]^{1/2} = 1.05 \text{ m/s}$$

Related Text Problems: 18-1 through 18-6, 18-10, 18-15 through 18-18, 18-20, 18-21, 18-23.

## 3. Absolute Potential (Section 18.3)

Review: The electric potential difference between two points in the electric field of a source charge Q is

$$V_{ab} = k_eQ[(1/r_b) - (1/r_a)]$$

This can be written as

$$V_{ab} = k_eQ/r_b - k_eQ/r_a$$

Since $V_{ab}$ is the electric potential difference, we can write

$$V_{ab} = V_b - V_a$$

This leads us to define the electric potentials at $r_b$ and $r_a$, respectively, as

$$V_b = k_eQ/r_b \qquad V_a = k_eQ/r_a$$

If we let $r_a$ be infinitely far from the source charge Q, then $V_a = 0$ and $V_b$ is the electric potential at a distance $r_b$ from Q with respect to the zero level. Hence if we agree to make zero of the absolute potential the absolute potential due to a point charge an infinite distance away, the absolute potential at a distance r from the source charge Q is

$$V = k_eQ/r$$

We call this expression the absolute electric potential at a point, but it is really the electric potential difference between that point and another point infinitely far away.

The absolute potential at a point due to several charges is equal to the algebraic sum of the potentials due to each of the charges:

$$V = V_1 + V_2 + V_3 + \ldots V_n = \sum_{i=1}^{n} V_i = k_e \sum_{i=1}^{n} (q_i/r_i)$$

Practice: Consider the two charges $Q_1$ and $Q_2$ and the points in space A, B, C and D shown in Fig. 18.6.

Figure 18.6

$Q_1 = +100\ \mu C$
$Q_2 = -200\ \mu C$
$Q_3 = +300\ \mu C$
$r_B = r_C = 1.00$ m
$r_D = 0.500$ m

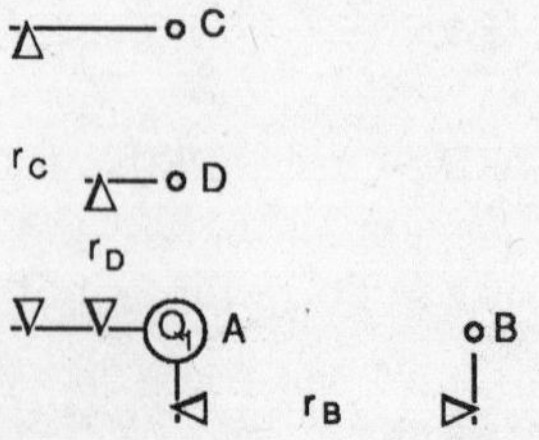

Determine the following:

| | |
|---|---|
| 1. The electric potential difference between a point infinitely far from $Q_1$ and point B | $V_{\infty B} = k_eQ_1[(1/r_B) - (1/r_\infty)] = 9.00 \times 10^5$ V |
| 2. The absolute electric potential at point B | $V_B = k_eQ_1/r_B = 9.00 \times 10^5$ V |

Note: The absolute electric potential at B is the same as the electric potential difference between the point B and a point an infinite distance away.

| | |
|---|---|
| 3. The absolute electric potential at point C with $Q_1$ at A and $Q_2$ at B | The contribution due to $Q_1$ is $(V_C)_1 = k_eQ_1/r_C = 9.00 \times 10^5$ V<br>The contribution due to $Q_2$ is<br>$(V_C)_2 = k_eQ_2/[r_C^2 + r_B^2]^{1/2} = -12.7 \times 10^5$ V<br>The total is<br>$V_C = (V_C)_1 + (V_C)_2 = -3.70 \times 10^5$ V |
| 4. The electric potential difference between points D and C when $Q_1$ is at A and $Q_2$ at B | $V_C = -3.70 \times 10^5$ V (from step 3)<br>$V_D = k_e[Q_1/r_D + Q_2/(r_D^2 + r_B^2)^{1/2}]$<br>$= 1.90 \times 10^5$ V<br>$V_{CD} = V_D - V_C = 5.60 \times 10^5$ V |
| 5. The mutual change in electric potential energy if $Q_3$ is moved from C to D while $Q_1$ is at A and $Q_2$ at B | $\Delta PE_{CD} = Q_3V_{CD} = 168$ J |
| 6. The work done by the Coulomb force as $Q_3$ is moved from C to D while $Q_1$ is at A and $Q_2$ at B | $W_{CD} = -\Delta PE_{CD} = -168$ J |

Example 18.4. Calculate (a) the absolute potential at a point 1.00 m from a 1.00-C charge, (b) the change in electric potential energy as a proton is brought from a very large distance to that point, and (c) the work done by the Coulomb force as the proton is brought from a very large distance to that point.

Given: $Q = 1.00$ C = source charge
$q = 1.6 \times 10^{-19}$ C = proton charge
$r = 1.00$ m = distance from source charge

Determine: (a) The absolute potential at a point 1.00 m from a 1.00-C charge, (b) the change in electric potential energy as a proton is brought from infinity to the point, (c) the work done by the Coulomb force as the proton is brought from infinity to the point.

Strategy; Knowing the magnitude and sign of the source charge and the distance from the charge, we can determine the absolute potential. Knowing the absolute potential and the magnitude and sign of the charge being moved toward the source charge, we can determine the change in electric potential energy and then the work done by the Coulomb force.

Solution: (a) The absolute potential at a point 1.00 m from a 1.00-C charge is

$$V = k_e Q/r = (9.00 \times 10^9 \text{ N}\cdot\text{m}^2/\text{C}^2)(1.00 \text{ C})/(1.00 \text{ m}) = 9.00 \times 10^9 \text{ V}$$

(b) The change in electric potential energy as a proton is moved from infinity to a point 1.00 m from the 1.00-C charge is

$$\Delta PE = Vq_p = (9.00 \times 10^9 \text{ V})(1.6 \times 10^{-19} \text{ C}) = 1.44 \times 10^{-9} \text{ J}$$

(c) The work done by the Coulomb force is

$$W = -\Delta PE = -1.44 \times 10^{-9} \text{ J}$$

Related Text Problems: 18-11, 18-12, 18-13, 18-28.

## 4. Potential Difference and Electric Field (Section 18.4)

Review: Consider the positive test charge $q_o$ in the uniform electric field $\vec{E}$ shown in Fig. 18.7.

Figure 18.7

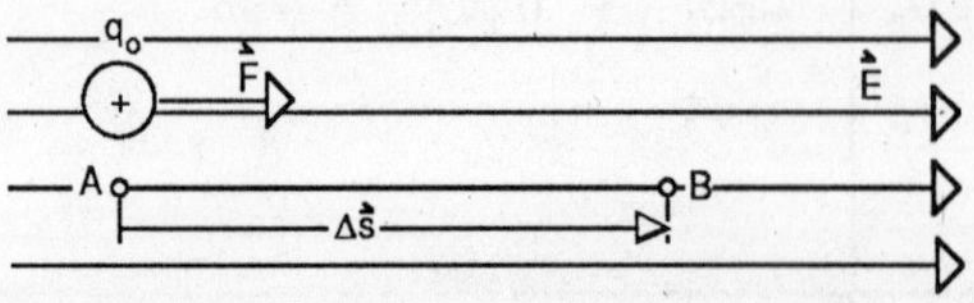


The electric field exerts a force on $q_o$ during the displacement $\Delta\vec{s}$. The electric field does work on $q_o$, changing the electric potential energy.

$F = q_o E$ = force $\vec{E}$ exerts on $q_o$
$W = F\Delta s \cos 0° = q_o E \Delta s$ = work $\vec{E}$ does on $q_o$
$\Delta PE = -W = -q_o E \Delta s$ = change in electric potential energy of $q_o$
$V = \Delta PE/q_o = -E\Delta s$ = electric potential difference between points A and B

Practice: Figure 18.8 shows a uniform electric field created between two metal plates connected to a 50.0-V power supply and a test charge $q_o = +1.00$ electronic charge.

Figure 18.8

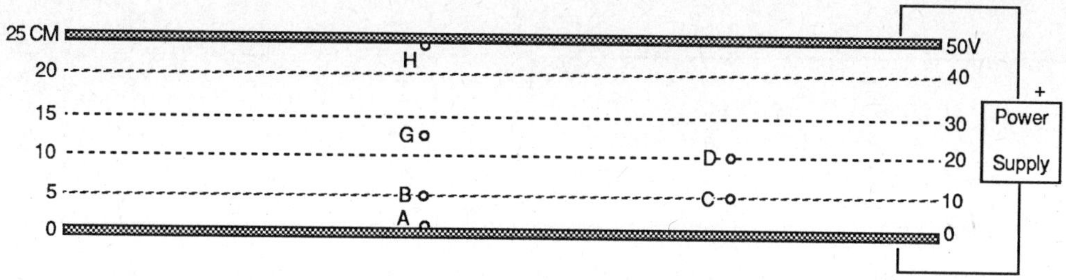


Determine the following:

| | |
|---|---|
| 1. The electric potential difference between the two plates | $V = 50.0$ V |
| 2. The electric field intensity in the region between the plates | $E = -V/\Delta s$<br>$= (-50.0\ \text{V})/(25.0 \times 10^{-2}\ \text{m}) = -200\ \text{V/m}$<br>The minus sign tells us that the direction of $\vec{E}$ is opposite that of the positive change (i.e., the increase) in V. V increases toward the top plate; hence $\vec{E}$ is directed downward. |
| 3. The force needed to move $q_o$ uniformly to any position in the $\vec{E}$ field | $F = q_oE$<br>$= (1.60 \times 10^{-19}\ \text{C})(2.00 \times 10^{2}\ \text{V/m})$<br>$= 3.20 \times 10^{-17}$ N |
| 4. Work the $\vec{E}$ field does on $q_o$ as it is moved from A to B | $W = F\Delta s \cos\theta$<br>$= q_oE\Delta s \cos 180°$<br>$= (1e)(2.00 \times 10^{2}\ \text{V/m})(5.00 \times 10^{-2}\ \text{m})(-1)$<br>$= -10.0$ eV |

Note: Recall that $W = q_oV$. To find the work done in moving a 1-e charge through a potential difference of 1 V, we have two options

$$W = q_oV = (1.60 \times 10^{-19}\ \text{C})(1.00\ \text{V}) = 1.60 \times 10^{-19}\ \text{J}$$

$$W = q_oV = (1\text{-e})(1.00\ \text{V}) = 1.00\ \text{eV}$$

The second option involves less effort (but the same amount of work) on our part; hence we use it as a matter of convenience. We can easily convert from electron volts to joules by

$$1\ \text{eV} = 1.6 \times 10^{-19}\ \text{J}$$

The unit electron volts is discussed in Section 18.3 of your text.

| | |
|---|---|
| 5. Change in electric potential energy as $q_o$ is moved from A to B | $\Delta PE = -W_{field} = +10.0$ eV |

Note: We have the freedom to choose zero potential energy in such a manner as to make the problem easy for us to solve. With this in mind, let's agree to call the negative plate zero electric potential energy.

| | |
|---|---|
| 6. The electric potential energy of $q_o$ at B | $PE_B = +10.0$ eV, since we have agreed that A is at zero electric potential energy |
| 7. Work done by the $\vec{E}$ field as $q_o$ is moved from B to C | $W_{BC} = F\Delta s \cos 90° = 0$ |
| 8. Change in electric potential energy of $q_o$ as it moved from B to C | $\Delta PE_{BC} = 0$ |

Note: The electric potential energy of $q_o$ is the same, regardless of where it is placed on the dotted line containing B and C. This is an equipotential line. It is a 10.0-V equipotential line, and since $q_o = 1\text{-e}$, its electric potential energy is 10.0-eV any place on this line.

| | |
|---|---|
| 9. Work done on $q_o$ by the $\vec{E}$ field as it moves from C to D | $W_{CD} = F\Delta s \cos 180°$<br>$= -q_oE\Delta s$<br>$= -10.0$ eV |
| 10. Change in electric potential energy of $q_o$ as it moves from C to D | $\Delta PE_{CD} = -W_{CD} = +10.0$ eV |
| 11. Total work done on $q_o$ by the $\vec{E}$ field in moving from A to D | $W_{AD} = W_{AB} + W_{BC} = -20.0$ eV |
| 12. Electric potential energy of $q_o$ at D | $PE_D = 20.0$ eV<br>D is on a 20.0-V equipotential line |

| | |
|---|---|
| 13. Magnitude of the electric field intensity at G | 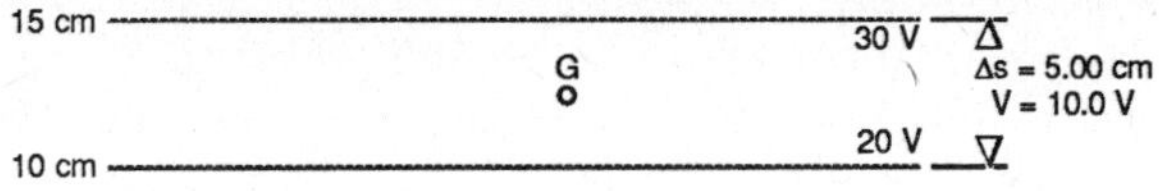<br>$V = 10.0$ V, $\Delta s = 5.00 \times 10^{-2}$ m<br>$\lvert E \rvert = \lvert V/\Delta s \rvert = 200$ V/m<br>Note: We obtained this same value in step 2. Obviously the E field is uniform. |
| 14. Electric potential energy of $q_o$ at H | $PE_H = q_oV = 50.0$ eV |
| 15. The kinetic energy at B if $q_o$ is released at rest from H | $\Delta PE_{HB} + \Delta KE_{HB} = 0$<br>$\Delta KE_{HB} = -\Delta PE_{HB} = -q_oV$<br>$= -(1e)(-40.0\text{ V}) = 40.0$ eV |

Example 18.5. Two metal plates are separated by 5.00 cm and connected across a power supply. An electron is released from rest at the negative plate and strikes the positive plate at a speed of $1.325 \times 10^7$ m/s. Calculate (a) the potential difference between the plates and (b) the magnitude and direction of the electric field intensity.

Given: $d = 5.00 \times 10^{-2}$ m = plate separation
$v_o = 0$ = initial speed of electron
$v_f = 1.325 \times 10^7$ m/s = final speed of electron
$q = 1e = 1.60 \times 10^{-19}$ C = charge on electron
$m = 9.11 \times 10^{-31}$ kg = mass of electron

Determine: (a) The potential difference V between the plates and (b) the magnitude and direction of the electric field between the plates.

Strategy: Knowing the initial and final speeds of the electron, we can determine its change in kinetic energy. Knowing the change in kinetic energy of the electron, we can determine its change in electric potential energy and hence the electric potential difference it has experienced. The electric potential difference experienced by the electron is just the potential difference between the plates. Knowing the potential difference between the plates and the plate separation, we can establish the magnitude and direction of the electric field.

Solution: The change in kinetic energy of the electron as it travels from the negative to the positive plate is

$$\Delta KE = KE_f - KE_i = mv_f^2/2 - mv_i^2/2 = 8.00 \times 10^{-17}\text{ J}$$

The change in the electron's electric potential energy is

$$\Delta PE = -\Delta KE = -8.00 \times 10^{-17}\ J$$

The electric potential difference experienced by the electron is the potential difference between the plates and is calculated by

$$V = \Delta PE/q = (-8.00 \times 10^{-17}\ J)/(-1.60 \times 10^{-19}\ C) = 500\ V$$

The magnitude of the electric field intensity is

$$E = V/d = 500\ V/5.00 \times 10^{-2}\ m = 1.00 \times 10^{4}\ V/m$$

The direction of the electric field intensity is opposite the direction of increasing V. Since V increases from the negative to the positive plate, E goes from the positive to the negative plate.

Related Text Problems: 2-24, 2-25, 2-27.

## 5. Capacitors (Sections 18.6, 18.7, 18.8)

Review: A capacitor is a device that stores a charge. The capacitance C of a capacitor tells us the amount of charge q the capacitor can store for every volt of potential difference V across the plates:

$$C = q/V$$

The charge-storing ability, or capacitance, of a capacitor depends on the geometry of its construction. That is, C is dependent on the area of the plates and the distance between them:

$$C = \varepsilon_o A/d \quad \text{where } \varepsilon_o = 1/4\pi k_e$$

In the process of charging a capacitor, charges are moved through a potential difference (i.e., work is done on them). The work that must be done in order to charge the capacitor is stored as potential energy. We can describe this work, or stored energy, as

$$W = CV^2/2 = qV/2$$

Since we also say that the energy is stored in the electric field, it is convenient to express W in terms of E:

$$W = \varepsilon_o AdE^2/2$$

We can also obtain the energy per unit volume μ of the capacitor as

$$\mu = W/Ad = \varepsilon_o E^2/2$$

If a dielectric material is placed between the plates of a capacitor, the capacitance increases from $C_o$ to C. The dielectric coefficient K of a material is a number that tells you the factor by which the capacitance is multiplied when the dielectric material is inserted; hence

$$K = C/C_o$$

If a capacitor has a capacitance $C_o$ without a dielectric, its capacitance with a dielectric is

$$C = KC_o = K\varepsilon_o A/d = \varepsilon A/d, \text{ where } \varepsilon = K\varepsilon_o$$

Practice: The plates of a parallel-plate capacitor each have an area A, and their initial separation is $d_o$. We have a battery with a terminal potential difference $V_o$ available for charging the capacitor. We also have sheets of dielectric with any desired thickness and a dielectric constant K.

Determine expressions for the following:

| | |
|---|---|
| 1. The initial capacitance $C_o$ | $C_o = \varepsilon_o A/d_o$ |

Now let's connect the capacitor to the battery.

| | |
|---|---|
| 2. The charge $q_o$ on each plate | $q_o = C_o V_o$ |
| 3. The magnitude of the electric field intensity $E_o$ | $E_o = V_o/d_o$ |
| 4. The energy $W_o$ stored in the capacitor $W_o$ | $W_o = C_o V_o^2/2 = q_o V_o/2$ |
| 5. The energy per unit volume $\mu_o$ | $\mu_o = W_o/Ad_o = \varepsilon_o E_o^2/2$ |

Now disconnect the battery and pull the plates apart until $d = 2d_o$.

| | |
|---|---|
| 6. The new capacitance C' | $C' = \varepsilon_o A/d' = \varepsilon_o A/2d_o = C_o/2$ |
| 7. The charge q' on each plate | Since the plates have been disconnected, no charge can flow either on or off them. Therefore $q' = q_o$ |

| | |
|---|---|
| 8. The new potential difference $V'$ across the plates | $V' = q'/C' = q_o/(C_o/2) = 2V_o$ |
| 9. The new electric field $E'$ | $E' = V'/d' = 2V_o/2d_o = E_o$ |
| 10. The new energy $W'$ stored in the capacitor | $W' = C'V'^2/2 = (C_o/2)(2\ V_o)^2/2 = 2W_o$<br>$W' = q'V'/2 = q_o(2V_o)/2 = 2W_o$ |
| 11. The source of this increase in energy | As you pull the plates apart, you add your work to the system, thus giving it energy. |

Let's now go back to our initial capacitor, charge it, disconnect the battery, and insert a sheet of the dielectric that has a thickness $d_o$.

| | |
|---|---|
| 12. The new capacitance $C_d$ | $C_d = KC_o$ |
| 13. The charge $q_d$ on each plate | Since the plates have been disconnected, no charge can flow on or off them<br>$q_d = q_o$ |
| 14. The new potential difference $V_d$ across the plates | $V_d = q_d/C_d = q_o/KC_o = V_o/K$ |
| 15. The new electric field $E_d$ | $E_d = V_d/d_d = (V_o/K)/d_o = E_o/K$ |
| 16. The new energy $W_d$ stored in the capacitor | $W_d = C_dV_d{}^2/2 = (KC_o)(V_o/K)^2/2 = W_o/K$<br>$W_d = q_dV_d/2 = q_o(V_o/K)/2 = W_o/K$ |
| 17. The cause of this decrease in energy | As the dielectric material approaches the capacitor, the dipoles are lined up by the electric field. This gives rise to a net induced charge on the sides of the dielectric material causing it to be pulled into the region between the plates. The energy to do this comes out of the energy stored in the capacitor. |

Example 18.6. The capacitance of a capacitor is 50.0 μF when filled with a dielectric of K = 5.00. If this capacitor is connected to a 25.0-V battery, charged, and the dielectric removed, calculate
(a) the charge on the plates while the dielectric is between the plates
(b) the energy stored with the dielectric between the plates
(c) the charge on the plates after the dielectric is removed
(d) the energy stored after the dielectric is removed.

Given: C = 50.0 μF = the capacitance of the capacitor when the dielectric is in place
K = 5.00 = the dielectric coefficient
V = 25.0 V = terminal potential difference of the battery

Determine: (a) Q = charge on the plates with the dielectric in place, (b) W = energy stored with the dielectric in place, (c) $Q_o$ = charge on the plates with the dielectric removed, and (d) $W_o$ = energy stored with the dielectric removed.

Strategy: Knowing C and V, we can determine Q and W. Knowing C and K and recognizing that $V_o = V$, we can determine $Q_o$ and $W_o$.

Solution: We can determine Q and W by setting up the following equations.

$$Q = CV = (50.0 \times 10^{-6}\ F)(25.0\ V) = 1.25 \times 10^{-3}\ C$$

$$W = CV^2/2 = (50 \times 10^{-6}\ F)(25.0\ V)^2/2 = 1.56 \times 10^{-2}\ J$$

Since the battery remains connected, $V_o = V$

Since the dielectric is removed, $C_o = C/K$

We can determine the new charge $Q_o$ and energy stored $W_o$ by setting up the following equations.

$$Q_o = C_oV_o = (C/K)V = Q/K = (1.25 \times 10^{-3}\ C)/5 = 2.5 \times 10^{-4}\ C$$

$$W_o = C_oV_o^2/2 = (C/K)V^2/2 = W/K = 1.56 \times 10^{-2}\ J/5 = 3.12 \times 10^{-3}\ J$$

Related Text Problems: 18-29 through 18-43

PRACTICE TEST

Take and grade this practice test. Doing so will allow you to determine any weak spots in your understanding of the concepts taught in this chapter. The following section prescribes what you should study further to strengthen your understanding.

Consider the source charge Q and the test charge $q_o$ shown in Fig. 18.9. The test charge is initially a large distance from the source charge.

Figure 18.9

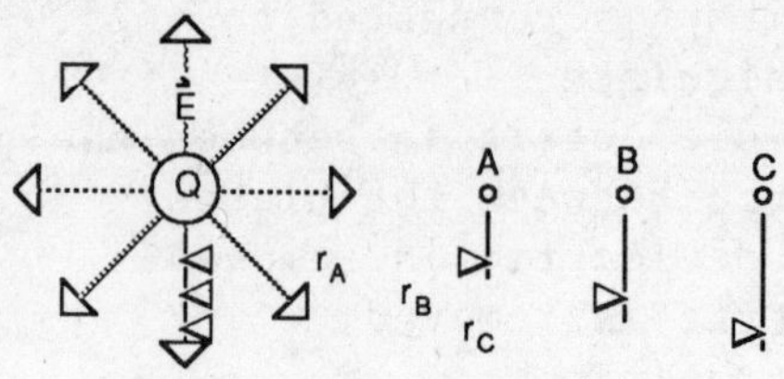


$Q = +100\ \mu C$ $\quad r_A = 1.00$ m

$q_o = +1.00\ \mu C$ $\quad r_B = 1.10$ m

$r_C \rightarrow \infty$

Determine the following:

__________ 1. Change in electric potential energy when $q_o$ is moved from $r_C$ to $r_B$

__________ 2. Change in electric potential energy when $q_o$ is moved from $r_B$ to $r_A$

__________ 3. Electric potential energy of $q_o$ at A

__________ 4. Electric potential difference between A and B

__________ 5. Absolute potential at A

Consider the electrode arrangement and resulting equipotential lines shown in Fig. 18.10.

Figure 18.10.

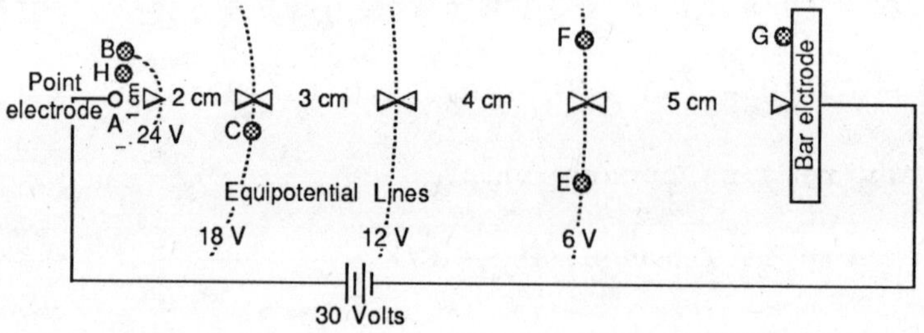

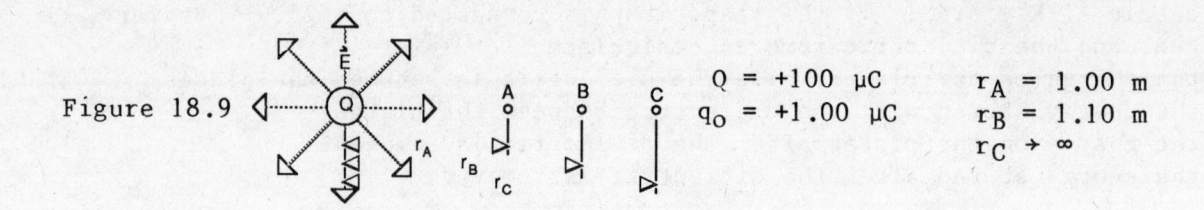

Let's agree to assign zero electric potential energy to the bar electrode. Determine the following:

__________ 6. Electric potential energy in eV of a +2 electronic (+2e) charge particle when it is at F

__________ 7. Change in electric potential energy in eV of a +2e charge particle in going from F to E

__________ 8. Work done on the +2e charge particle in eV by the electric field as it moves from F to E

__________ 9. Work done on the +2e charge particle in eV by the electric field as it moves from E to C

__________ 10. Electric potential energy of the +2e charge particle in eV when it is at C

__________ 11. Electric field strength in V/m at H

__________ 12. Force (in N) required to hold the +2e charge particle stationary at H

Now let's release the +2e charge particle at rest from A.

__________ 13. Kinetic energy of the +2e charge particle when it gets back to the 12.0-V equipotential line.

__________ 14. Speed in m/s of the +2e charge particle when it crashes into the bar electrode. M of the particle is $6.40 \times 10^{-19}$ kg

Figure 18.11 summarizes a sequence of events involving a capacitor, battery, and slab of dielectric material.

Figure 18.11 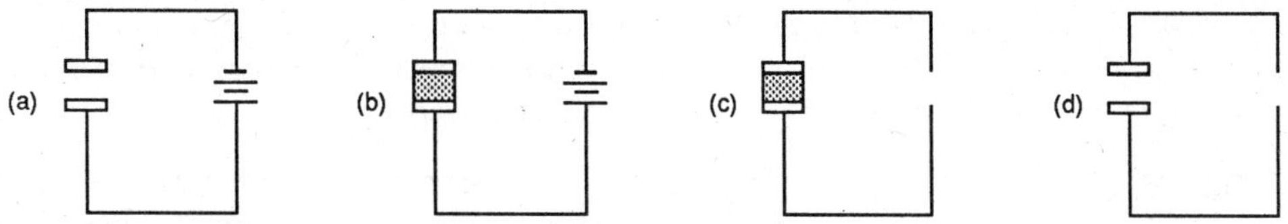


(a) The capacitor is connected to the battery and charged.
(b) While the battery is connected, a dielectric slab is inserted.
(c) The battery is disconnected.
(d) The dielectric is withdrawn.

The following information is known:

$A = 1.00 \times 10^{-2}$ $m^2$ = area of the plates
$d = 2.00 \times 10^{-2}$ m = separation of the plates
$V_o = 24.0$ V = size of the battery
$K = 5.00$ = dielectric constant

Determine the following:

__________ 15. Capacitance of the capacitor in case (a)
__________ 16. Charge on each plate of the capacitor in case (a)
__________ 17. Capacitance of the capacitor in case (b)
__________ 18. Charge on each plate of the capacitor in case (b)
__________ 19. Charge on each plate in case (d)
__________ 20. Potential difference across the plates in case (d)

(See Appendix I for answers.)

PRINCIPAL CONCEPTS AND EQUATIONS PRESCRIPTION

Your score on the practice is an excellent measure of your understanding of this chapter. You should now use the following chart to write your own prescription for curing any of your physics ills. Look down the leftmost column to the number of the question(s) you answered incorrectly, read across that row to see which section(s) of the study guide you should return to for further study, and then do the suggested text problems to gain additional experience in working with the particular concept.

| Practice Test Question | Concepts and Equations | Prescription: Principal Concept | Prescription: Text Problems |
|---|---|---|---|
| 1 | Change in electric PE: $\Delta PE_{CB} = k_e Q q_o[(1/r_B) - (1/r_C)]$ | 1 | 18-7,8 |
| 2 | Change in electric PE: $\Delta PE_{BA} = k_e Q q_o[(1/r_A) - (1/r_B)]$ | 1 | 18-9,10 |
| 3 | Electric potential energy: $PE_A = \Delta PE_{CB} + \Delta PE_{BA}$ | 1 | 18-3,7 |
| 4 | Electric pot. diff: $V_{BA} = \Delta PE_{BA}/q_o$ | 2 | 18-3,4 |
| 5 | Absolute potential: $V_A = V_{\infty A} = \Delta PE_{\infty A}/q_o$ | 3 | 18-11,12 |
| 6 | Electric Potential Energy: $PE_F = \Delta PE_{GF} = q_o V_{GF}$ | 2 | 18-3,16 |
| 7 | Change in electric PE: $\Delta PE_{FE} = q_o V_{FE}$ | 2 | 18-3,15 |
| 8 | Work: $W_{FE} = \Delta PE_{FE} = q_o V_{FE}$ | 1 | 18-2,5 |
| 9 | Work: $W_{EC} = \Delta PE_{EC} = q_o V_{EC}$ | 2 | 18-1,11 |
| 10 | Electric potential energy: $PE_C = \Delta PE_{GC} = V_{GC}/q_o$ | 2 | 18-11,22 |
| 11 | Electric field strength: $E = -V/\Delta s$ | 4 | 18-24,25 |
| 12 | Electric field strength: $E = F/q_o$ | 2 of Ch. 17 | 18-27 |
| 13 | Conservation of energy: $\Delta KE = -\Delta PE$ | 3 of Ch. 7 | 18-15,16 |
| 14 | Kinetic energy: $KE = mv^2/2$ | 3 of Ch. 7 | 18-17,18 |
| 15 | Capacitance: $C = \varepsilon_o A/d$ | 5 | 18-29,33 |
| 16 | Capacitance: $C = q/V$ | 5 | 18-31,32 |
| 17 | Dielectric coefficient: $K = C/C_o$ | 5 | 18-30,31 |
| 18 | Capacitance: $C = q/V$ | 5 | 18-32,41 |
| 19 | Dielectric coefficient: $K = C/C_o$ | 5 | 18-31,32 |
| 20 | Capacitance: $C = q/V$ | 5 | 18-38,39 |

# 19 Electric Current, Resistance, EMF

RECALL FROM PREVIOUS CHAPTERS

| Previously learned concepts and equations frequently used in this chapter | Text Section | Study Guide Page |
|---|---|---|
| Power: $P = \Delta W/\Delta t$ | 7.4 | 106 |
| Electric potential difference: $V = W/q$ | 18.2 | 289 |

NEW IDEAS IN THIS CHAPTER

| Concepts and equations introducted | Text Section | Study Guide Page |
|---|---|---|
| Current: $I = \Delta q/\Delta t = \eta qAv_d$ | 19.2 | 305 |
| Ohm's law: $R = V/I$ | 19.3 | 308 |
| Emf, terminal potential difference, and internal resistance: $\varepsilon = V + Ir$ | 19.4,5 | 310 |
| Resistivity: $R = \rho L/A$ | 19.6 | 313 |
| Temperature coefficient of resistivity: | 19.6 | 313 |
| $\rho = \rho_o(1 + \alpha\Delta T)$ and $R = R_o(1 + \alpha\Delta T)$ | 19.6 | 313 |
| Power: $P = IV = I^2R = V^2/R$ | 19.7 | 315 |

PRINCIPAL CONCEPTS AND EQUATIONS

## 1. Electric Current (Section 19.2)

Review: We define electric current as the time rate of flow of positive charge past a section of conductor.

$$I = \Delta q/\Delta t$$

We can also express current as

$$I = \eta qAv_d$$

where:

$\eta$ = number of charge carriers per unit volume
$q$ = magnitude of the charge on each carrier
$A$ = cross-sectional area of the conductor
$v_d$ = drift velocity of the charge carriers

If there are n different kinds of charge carriers, each with its own charge and drift velocity we write

$$I = A \sum_{i=1}^{n} \eta_i q_i v_{di}$$

<u>Practice:</u> Figure 19.1 shows a device with positive and negative charge carriers.

Figure 19.1

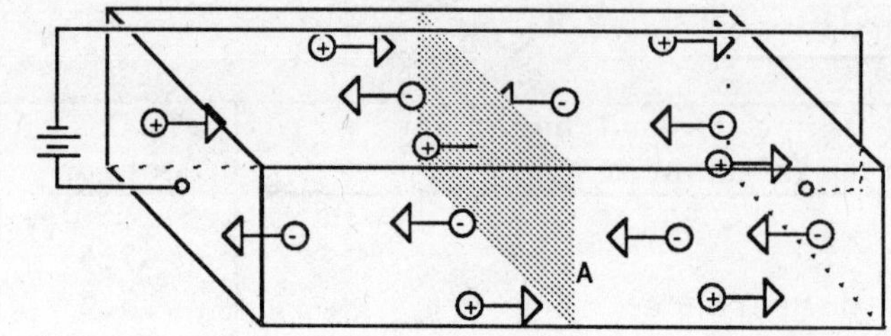


$q = 1.60 \times 10^{-19}$ C = magnitude of the charge on both positive and negative charge carriers

$A = 4.00 \times 10^{-4}\ m^2$ = cross-sectional area of the device

$\eta_+ = 6.00 \times 10^{24}/m3$ = positive charge carriers per unit volume

$\eta_- = 4.00 \times 10^{26}/m^3$ = negative charge carriers per unit volume

$v_{d+} = 3.00 \times 10^{-5}$ m/s = drift velocity of the positive charge carriers

$v_{d-} = 1.00 \times 10^{-4}$ m/s = drift velocity of the negative charge carriers

Determine the following:

| | |
|---|---|
| 1. Number of positive and negative charge carriers crossing A per unit time | $N_+/t = \eta_+ A v_{d+} = 7.20 \times 10^{16}/s$<br>$N_-/t = \eta_- A v_{d-} = 1.60 \times 10^{19}/s$ |
| 2. Amount of positive and negative charge crossing A per unit time | $q_+/t = (N_+/t)q = 1.15 \times 10^{-2}$ C/s<br>$q_-/t = (N_-/t)q = -2.56$ C/s |
| 3. Net flow of charge across A per unit time | $Q_{Net}/t = (q_+/t) + (q_-/t) = -2.55$ C/s |
| 4. Current (magnitude and direction) across A | $I = Q_{Net}/t = 2.55$ C/S left to right. From step 3 and Fig. 19.1 we realize that a net charge of -2.55 C crosses A each second from right to left. Negative movement to the left is the same as positive movement to the right. Conventional current is in the direction of motion of positive charge carriers. |

| | |
|---|---|
| 5. Time for a net charge of 10.0 C to drift across A | $I = \Delta q/\Delta t$<br>$\Delta t = \Delta q/I = 3.92$ s |

Example 19.1. An An electron is traveling at a speed of $2.00 \times 10^6$ m/s in a small, evacuated, circular glass tube. The radius of the tube is $1.00 \times 10^{-3}$ m and the radius of the electron's orbit is $1.00 \times 10^{-2}$ m. Determine the current due to the electron's motion.

Given and Diagram:

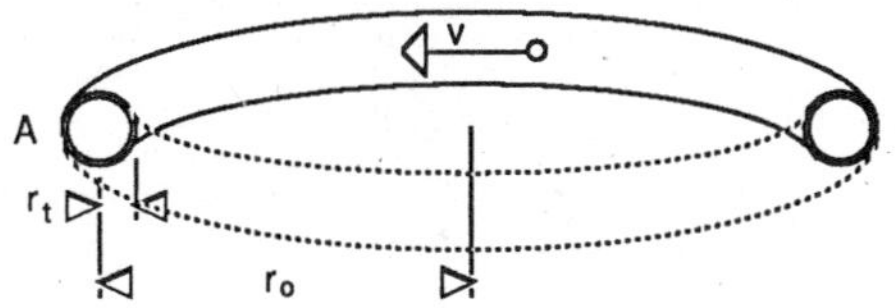


$r_o = 1.00 \times 10^{-2}$ m = radius of electron orbit
$r_t = 1.00 \times 10^{-3}$ m = radius of circular tube
$v = 2.00 \times 10^6$ m/s = speed of the electron
$q = 1.60 \times 10^{-19}$ C = charge on the electron

Figure 19.2

Determine: The current due to the electrons motion

Strategy: Knowing the electron's radius of orbit and speed, we can determine the time for one orbit. Knowing that an electron passes point A once every orbit and the time of the orbit, we can determine the current due to the electron's motion.

Solution: The length of the electron's orbit is

$$C = 2\pi r_o = 6.28 \times 10^{-2} \text{ m}$$

The time for the electron to make one orbit is

$$T = C/v = 3.14 \times 10^{-8} \text{ s}$$

Charge $\Delta q = 1.60 \times 10^{-19}$ C passes A every $\Delta t = 3.14 \times 10^{-8}$ s, hence

$$I = \Delta q/\Delta t = 5.1 \times 10^{-12} \text{ A}$$

Example 19.2. A wire carries a current of 1.00 A. How many electrons must pass through a cross-sectional area of the wire in 2.00 s to produce this current?

Given: $I = 1.00$ A = current in the wire
$\Delta t = 2.00$ S = time of current flow
$q_e = 1.60 \times 10^{-19}$ C = charge on an electron

Determine: The number of electrons which must pass through a cross-sectional area of the wire every 2.00 s in order to have a current of 1.00 A.

Strategy: Knowing the current and the time, we can determine the total charge that must pass A in order to sustain the current. Knowing the total charge and the charge on each charge carrier, we can determine the number of charge carriers needed.

Solution: The total charge that must pass A is

$$\Delta q = I\Delta t = 2.00 \text{ C}$$

The number of electrons needed to supply this charge is

$$N = \Delta q/q_e = 1.25 \times 10^{19}$$

Related Text Problems: 19-1 through 19-9.

## 2. Ohm's Law (Section 19.3)

Review: Figure 19.3 shows a 10.0-Ω resistor connected to first one cell (a), then two cells (b), and finally three cells (c). The ammeter measures the current in the circuit. The voltmeter measures the potential difference across the resistor, which is the same as the terminal potential difference of the battery.

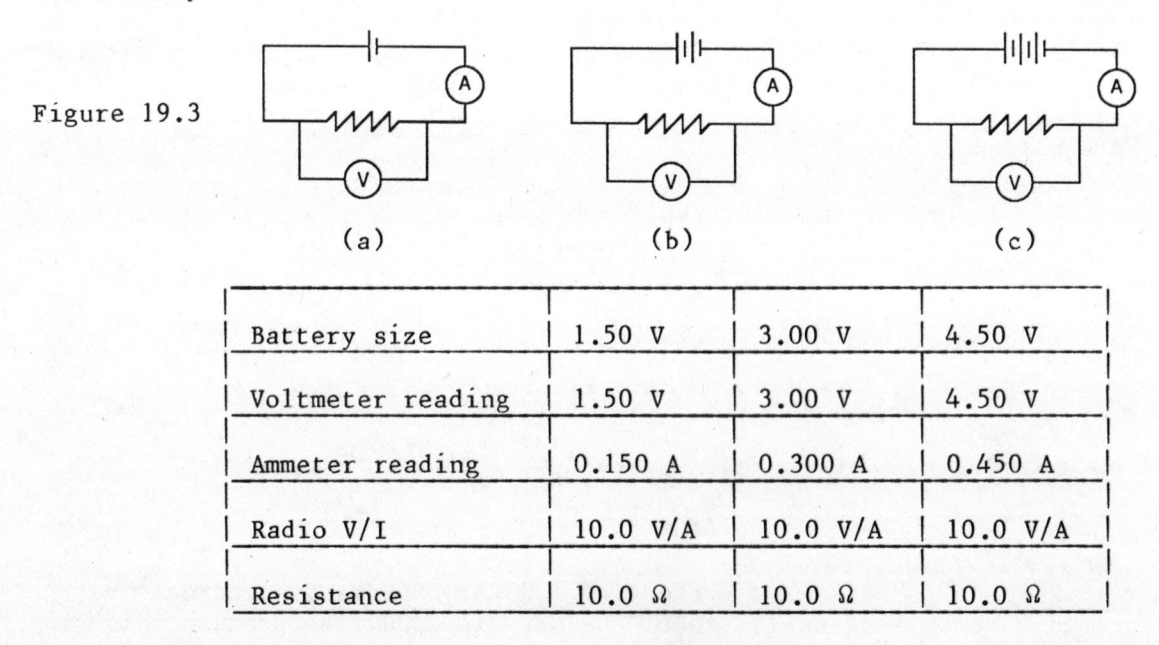

Figure 19.3

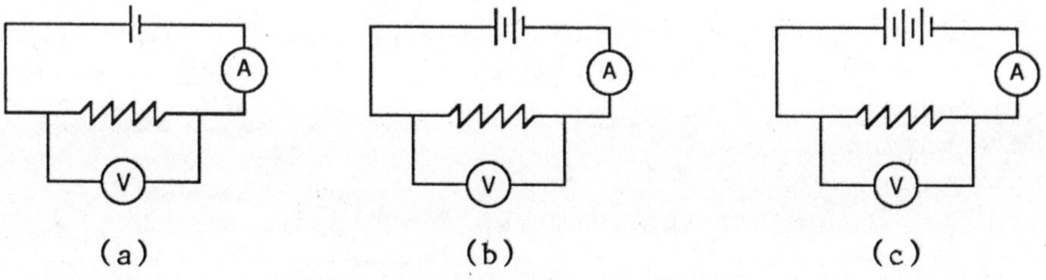


| | | | |
|---|---|---|---|
| Battery size | 1.50 V | 3.00 V | 4.50 V |
| Voltmeter reading | 1.50 V | 3.00 V | 4.50 V |
| Ammeter reading | 0.150 A | 0.300 A | 0.450 A |
| Radio V/I | 10.0 V/A | 10.0 V/A | 10.0 V/A |
| Resistance | 10.0 Ω | 10.0 Ω | 10.0 Ω |

Notice that as V increases, I also increases. That is I is proportional to V.

$$I \propto V$$

Notice also that the ratio of V/I is a constant, the resistance of the resistor. Stated algebraically

$$R = V/I$$

This is the most commonly used form of Ohm's law.

Practice: Figure 19.4 shows several simple circuits and the given information.

Figure 19.4

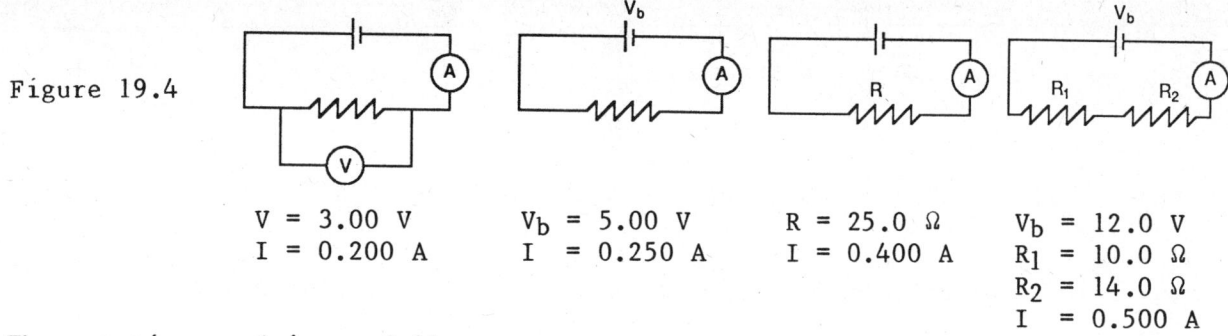

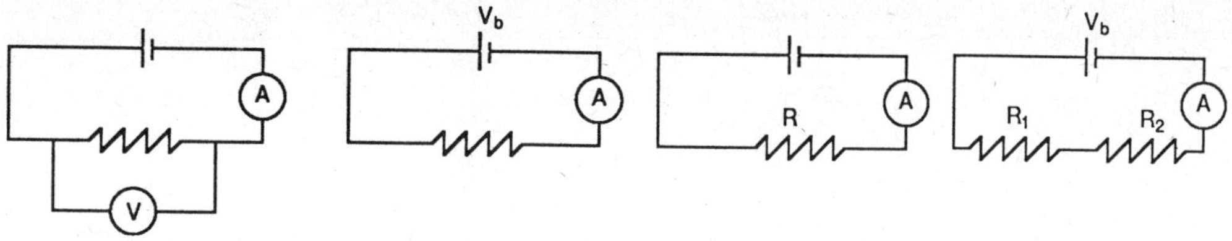


| (a) | (b) | (c) | (d) |
|---|---|---|---|
| $V = 3.00$ V<br>$I = 0.200$ A | $V_b = 5.00$ V<br>$I = 0.250$ A | $R = 25.0\ \Omega$<br>$I = 0.400$ A | $V_b = 12.0$ V<br>$R_1 = 10.0\ \Omega$<br>$R_2 = 14.0\ \Omega$<br>$I = 0.500$ A |

The notation used is as follows:

$V$ = voltmeter reading = potential difference across the resistor
$I$ = ammeter reading = current in the circuit
$R$ = size of the resistor
$V_b$ = battery size = terminal potential difference of the battery

Determine the following:

| | |
|---|---|
| 1. Terminal potential difference of the battery for case (a) | $V_b = V = 3.00$ V |
| 2. Resistance of the resistor for case (a) | $R = V/I = 15.0\ \Omega$ |
| 3. Potential difference across the resistor for case (b) | $V = V_b = 5.00$ V |
| 4. Resistance of the resistor for case (b) | $R = V/I = 20.0\ \Omega$ |
| 5. Potential difference across the resistor for case (c) | $V = IR = 10.0$ V |
| 6. Terminal potential difference of the battery for case (c) | $V_b = V = 10.0$ V |
| 7. Potential difference across $R_1$ for case (d) | $V_1 = IR_1 = 5.00$ V |
| 8. Potential difference across $R_2$ for case (d) | $V_2 = IR_2 = 7.00$ V |

| | |
|---|---|
| 9. The total potential difference across $R_1$ and $R_2$ for case (d) | $V_b = V_T = V_1 + V_2 = 12.0$ V |

Example 19.3. A small flashlight uses a bulb that has a resistance of 10.0 Ω and two 1.50-V dry cells. The dry cells can last 3.00 h under continuous use. Determine (a) the current through the bulb, (b) the charge that drifts through the bulb filament each second, (c) the number of electrons that drift through the filament each second, (d) the work done by the dry cells each second, and (e) the total amount of work done by the dry cells.

Given: $R = 10.0\ \Omega$ = resistance of the bulb filament
$V = (1.50\ \text{V/cell})(2\ \text{cells}) = 3.00$ V = potential difference across the bulb filament
$T = 3.00$ h = lifetime of the cells when used continuously in the flashlight
$q_e = 1.60 \times 10^{-19}$ C = charge on an electron

Determine:

(a) $I$ = the current through the bulb
(b) $\Delta q/\Delta t$ = the charge that drifts through the bulb each second
(c) $N/\Delta t$ = the number of electrons that drift through the bulb each second
(d) $W/\Delta t$ = work done by the dry cells each second
(e) $W_T$ = total amount of work done by the dry cells

Strategy:

(a) Knowing $V$ and $R$, we can determine $I$
(b) Knowing $I$, we also know $\Delta q/\Delta t$
(c) Knowing $\Delta q/\Delta t$ and $q_e$, we can determine $N/\Delta t$
(d) Knowing $\Delta q/\Delta t$ and $V$, we can determine $W/\Delta t$
(e) Knowing $W/\Delta t$ and $T$, we can determine $W_T$

Solution:

(a) $I = V/R = 0.300$ A
(b) $\Delta q/\Delta t = I = 0.300$ C/s
(c) $\Delta q/\Delta t = Nq_e/\Delta t$ or $N/\Delta t = (\Delta q/\Delta t)/q_e = 1.88 \times 10^{18}$/s
(d) $W/\Delta t = V(\Delta q/\Delta t) = 0.900$ J/s
(e) $W_T = (W/\Delta t)T = 9.72 \times 10^3$ J

Related Text Problems: 19-10 through 19-15.

## 3. Emf, Terminal Potential Difference, and Internal Resistance of a Battery (Sections 19.4, 19.5)

Review: Figure 19.5 shows a cell of emf ε and internal resistance r delivering a current I to an external resistor R.

Figure 19.5

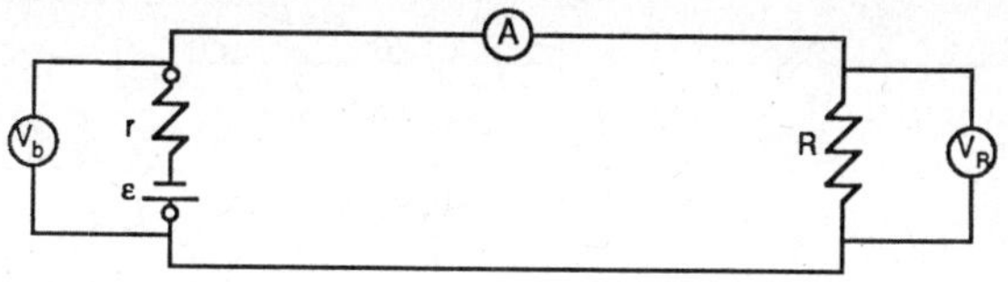


Notation: $\varepsilon$ = emf of the cell
R = external resistance
r = internal resistance
I = circuit current measured by A
$V_R$ = potential difference across R
$V_b$ = potential difference across the cell terminals

The cell does work on the charge carriers as they move through it from terminal to terminal. The energy given to the charge carriers by the cell is used up as they work their way through the external resistance R and the internal resistance r. If $\varepsilon$ represents the emf of the cell or the energy given to each charge carrier by the cell, $V_R$ represents the potential difference across the external resistance or the energy loss by each charge carrier as it passes through R, and $V_r$ represent the energy loss by each charge carrier due to the internal resistance of the cell, then we can write

$$\varepsilon = V_R + V_r$$

This says that the energy supplied to the charge carriers by the cell is consumed as they go through circuit resistance (external and internal). Using Ohm's law, we can write $V_R$ = IR and $V_r$ = Ir.

The energy loss by the wires and ammeter is negligible, hence

$$V_b = V_R$$

Combining the preceding equations we obtain an expression for the potential difference across the terminals of the cell when it is delivering a current I.

$$V_b = \varepsilon - Ir$$

In order to charge the cell, we must force a current (i.e., charge carriers) through the cell in the reverse direction. To do this, we must give the charge carriers enough energy to overcome the work done on them by the cell ($\varepsilon$) and the energy loss due to the internal resistance ($V_r$ = Ir). Hence, the amount of energy we must supply each charge carrier, or the potential difference we must put across the terminals of the cell to be charged, is

$$V_b = \varepsilon + Ir$$

Practice: The same cell is connected across two different resistors, with voltmeter and ammeter readings as shown in Fig. 19.6.

Figure 19.6

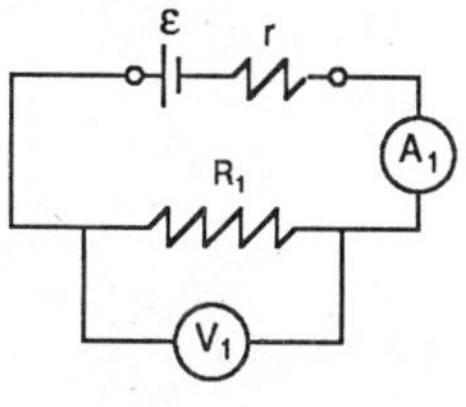


$V_1$ = 15.0 V
$I_1$ = 0.500 A

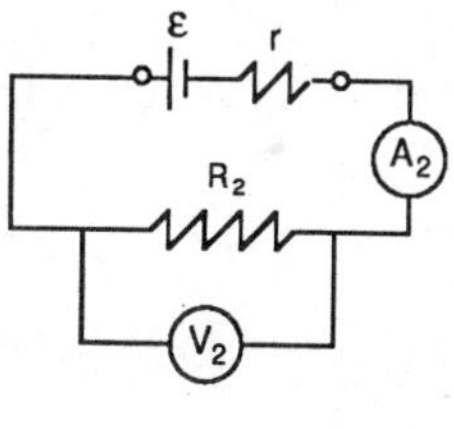


$V_2$ = 15.5 V
$I_2$ = 0.250 A

Determine the following:

| | |
|---|---|
| 1. Resistance of $R_1$ and $R_2$ | $R_1 = V_1/I_1 = 30.0\ \Omega$<br>$R_2 = V_2/I_2 = 62.0\ \Omega$ |
| 2. Internal resistance of the cell | $\varepsilon = V_1 + I_1r$ and $\varepsilon = V_2 + I_2r$<br>Since it is the same cell, we can equate these expressions for $\varepsilon$.<br>$V_1 + I_1r = V_2 + I_2r$ or<br>$r = (V_2 - V_1)/(I_1 - I_2) = 2.00\ \Omega$ |
| 3. Emf of the cell | $\varepsilon = V_1 + I_1r = 16.0$ V or<br>$\varepsilon = V_2 + I_2r = 16.0$ V |
| 4. Potential difference needed to charge the cell at 1.00 A | $V_b = \varepsilon + Ir = 18.0$ V |

Example 19.4. When switch S is open, the voltmeter V, connected across the terminals of the dry cell in Fig. 19.7, reads 1.48 V. When the switch is closed, the voltmeter reading drops to 1.33 V, and the ammeter A reads 1.25 A. Find the emf and internal resistance of the cell. Assume the meters have no effect on the circuit.

Figure 19.7

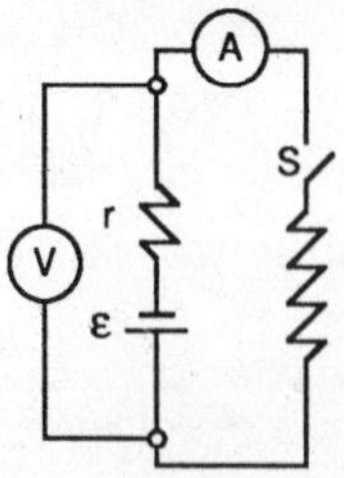


Given:

| | Switch open | Switch closed |
|---|---|---|
| V | 1.48 V | 1.33 V |
| I | 0.00 A | 1.25 A |

Determine: The emf $\varepsilon$ and internal resistance r of the cell.

Strategy: We can obtain the emf from the switch-open data and the internal resistance from the switch-closed data.

Solution: When the switch is open, I = 0.00 A and $\varepsilon = V + Ir = V = 1.48$ V.

When the switch is closed, I = 1.25 A and $\varepsilon = V + Ir$ or $r = (\varepsilon - V)/I = 0.125\ \Omega$

Related Text Problems: 19-16 through 19-22.

## 4. Resistivity (Section 19.6)

Review: The resistance of a wire at some temperature can be calculated with the expression

$$R = \rho L/A$$

From this expression, we see that the resistance of a wire is

1. directly proportional to its length L
2. inversely proportional to its cross-sectional area A, and
3. a function of the material (characterized by $\rho$)

The resistivity constant $\rho$ is different for each material. The resistivity and hence the resistance depends on the temperature, as shown here:

$$\rho = \rho_o\ (1 + \alpha\Delta T) \text{ and } R = R_o(1 + \alpha\Delta T)$$

Notice that $\Delta R = R - R_o = R_o\alpha\Delta T$ or $\alpha = \Delta R/R_o\Delta T$

The constant $\alpha$ (different for each material) is called the temperature coefficient of resistivity. It tells us the change in resistance for every ohm of resistance and degree rise in temperature.

Practice: A nichome wire with the following characteristics is used as a heating element.

| | | |
|---|---|---|
| $\rho = 1.00 \times 10^{-6}\ \Omega\cdot m$ | $L = 1.00$ m | $T = 20°C$ |
| $\alpha = 4.50 \times 10^{-4}/C°$ | $A = 1.00 \times 10^{-8}\ m^2$ | |

When the element is connected to a 120-V outlet, the current stabilizes at 0.900 A.

Determine the following:

| | |
|---|---|
| 1. Resistance of the heating element at T = 20°C | $R_{20} = \rho L/A = 100\ \Omega$ |
| 2. Resistance of the element at T = 0°C | $R_{20} = R_o[1 + \alpha(20°C)]$<br>$R_o = R_{20}/[1 + \alpha(20°C)] = 99.11\ \Omega$ |
| 3. Resistance of the element at T = 50°C | $R_{50} = R_o[1 + \alpha(50°C)]$<br>Using $R_o$ from step 2, obtain<br>$R_{50} = 101.3\ \Omega$<br>Or using the expression for $R_o$ from step 2, obtain<br>$R_{50} = R_{20}\ [1 + \alpha(50°C)]/[1 + \alpha(20°C)]$<br>$R_{50} = 101.3\ \Omega$ |

| | | |
|---|---|---|
| 4. | Resistance of the element at $T = -20^\circ C$ | $R_{-20} = R_o[1 + \alpha(-20^\circ C)] = 98.22\ \Omega$ or<br>$R_{-20} = R_{20}[1 + \alpha(-20^\circ C)]/[1 + \alpha(20^\circ C)]$<br>$R_{-20} = 98.22\ \Omega$ |
| 5. | Final resistance of the wire (i.e., its resistance while hot) | $R_f = V/I = 133.3\ \Omega$ |
| 6. | Final temperature of the wire | $R_f = R_o(1 + \alpha T)$ or<br>$T = [(R_f/R_o) - 1]/\alpha = 767^\circ C$ |

Example 19.5. A copper wire of length $L = 2.00$ m and cross-sectional area $A = 5.00 \times 10^{-8}\ m^2$ has a potential difference across it sufficient to produce a 5.00 A current at 30°C. If the temperature of the wire is increased to 300°C, by how much must the potential difference across the wire change in order to maintain the 5.00 A current?

Given:
$L = 2.00$ m = length of the wire
$A = 5.00 \times 10^{-8}\ m^2$ = cross-sectional area of the wire
$I = 5.00$ A = current in the wire
$T_i = 30^\circ C$ = initial temperature of the wire
$T_f = 30^\circ C$ = final temperature of the wire
$\rho = 1.67 \times 10^{-8}\ \Omega \cdot m$ = resistivity of copper
$\alpha = 6.80 \times 10^{-3}/C^\circ$ = temperature coefficient of resistivity for copper

Determine: The amount by which the potential difference across the wire must change ($\Delta V$) in order to maintain a 5.00 A current.

Strategy: Knowing the length, cross-sectional area, and type of material, we can calculate the resistance at the initial temperature. Knowing the resistance at the initial temperature, we can determine the resistance at the final temperature. We can determine the initial and final potential difference across the resistor from the current and the initial and final resistance. Finally, knowing the initial and final potential difference, we can determine the amount by which the potential difference must change in order to maintain a 5.00-A current.

Solution: The initial resistance of the wire is

$$R_{30} = \rho L/A = 0.668\ \Omega$$

The final resistance of the wire can be determined as follows:

$$R_{30} = R_o\ [1 + \alpha(30^\circ C)] \text{ or } R_o = R_{30}/[1 + \alpha(30^\circ)]$$

$$R_{300} = R_o\ [1 + \alpha(300^\circ C)] = R_{30}\ [1 + \alpha(300^\circ C)]/[1 + \alpha(30^\circ C)] = 1.69\ \Omega$$

The initial and final potential difference are

$$V_{30} = IR_{30} = 3.34\ V \quad \text{and} \quad V_{300} = IR_{300} = 8.45\ V$$

Finally, the amount by which the potential difference across the wire must be changed in order to maintain a 5.00-A current is

$$\Delta V = V_{300} - V_{30} = 5.11\ V$$

Related Text Problems: 19-23 through 19-29.

## 5. Energy and Power In Electric Circuits (Section 19.7)

Review: Figure 19.8 shows a simple circuit consisting of a cell, resistor, switch, voltmeter, and ammeter.

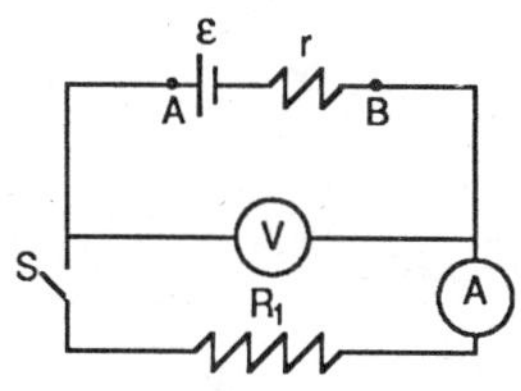

Figure 19.8

Notations
$\varepsilon$ = emf of the cell
$r$ = internal resistance of the cell
$I$ = current measured by A
$V$ = potential difference measured by V
$R$ = external resistance
$V_R$ = potential difference across R
$V_{AB}$ = terminal potential difference of the cell
$S$ = switch

Using the given notation, we can write:

$P_\varepsilon = I\varepsilon$ = power delivered by the emf
$P_r = I^2r$ = power consumed by the internal resistance of the cell
$P_R = I^2R = IV_R = V_R^2/R$ = power consumed by the external resistor
$P_\varepsilon = P_r + P_R$ = the rate at which energy is supplied by the cell is equal to the rate at which energy is consumed by r and R

Practice: The data below were collected with the circuit shown in Fig. 19.8.

| | Switch open | Switch closed |
|---|---|---|
| V | 3.00 V | 2.50 V |
| I | 0.00 A | 1.00 A |

Determine the following:

| | |
|---|---|
| 1. Emf of the cell | $\varepsilon = V_{AB} + Ir$<br>When the switch is open, $I = 0$ and<br>$\varepsilon = V_{AB} = V = 3.00\ V$ |

| | | |
|---|---|---|
| 2. | Internal resistance of the cell | $\varepsilon = V_{AB} + Ir$<br>$r = (\varepsilon - V_{AB})/I = 0.500\ \Omega$ |
| 3. | External resistance | $R = V_R/I = V/I = 2.50\ \Omega$ |
| 4. | Energy supplied by the cell in 100 s | $\Delta U_c = \varepsilon\Delta q = \varepsilon I\Delta t = 300$ J |
| 5. | Power supplied by the cell | $P_c = \Delta U_c/\Delta t = 3.00$ W or<br>$P_c = \varepsilon I = 3.00$ W |
| 6. | Power consumed by R in 100 s | $\Delta U_R = V_R\Delta q = V_R I\Delta t = 250$ J or<br>$\Delta U_R = I^2R\Delta t = 250$ J |
| 7. | Power consumed by R | $P_R = \Delta U_R/\Delta t = 2.50$ W or<br>$P_R = V_R I = 2.50$ W or<br>$P_R = I^2R = 2.50$ W |
| 8. | Energy consumed by r in 100 s | $\Delta U_r = (\varepsilon - V)\Delta q = (\varepsilon - V)I\Delta T = 50.0$ J<br>or $= (Ir)I\Delta t = I^2\Delta t = 50.0$ J |
| 9. | Power consumed by r | $P_r = \Delta U_r/\Delta t = 0.500$ W or<br>$P_r = (\varepsilon - V)I = 0.500$ W or<br>$P_r = I^2r = 0.500$ W |
| 10. | Total power consumed by R and r | $P_R = 2.50$ W (Step 7)<br>$P_r = 0.500$ W (Step 9)<br>$P_T = P_R + P_r = 3.00$ W<br>Note that $P_T$ consumed is equal to the power supplied by the cell (Step 5). |
| 11. | Efficiency of the cell | $\varepsilon = P(\text{available for use})/P_T$<br>$\varepsilon = P_R/P_T = 0.833$ |

Example 19.6. Household circuits operate at 120 V and are commonly fused with 15-A fuses. A bathroom circuit consists of a single-bulb light fixture and a wall outlet. The wire for the entire circuit is 15.0 m of 18-gauge ($1.02 \times 10^{-3}$ m diameter) copper wire. (a) Determine the maximum wattage light bulb we can use and still run a 1650 W hair dryer without melting a fuse. (b) If we buy energy at the rate of 10¢ per kilowatt-hour (kW·h), how much does it cost us to run the hair dryer and a 60.0-W light for 0.250 h?

Given: $V = 120$ Volts
$I = 15.0$ A
$L = 15.0$ m
$A = \pi r^2 = 8.17 \times 10^{-7}\ m^2$
$P_{dryer} = 1650$ W
$\rho = 1.67 \times 10^{-8}\ \Omega\cdot m$
$C = 10¢/kW\cdot h$

Determine: (a) the maximum wattage lightbulb we can use and not burn a fuse. (b) The cost to run the hair dryer and a 60.0-W bulb for 0.250 h.

Strategy: (a) From the given information, we can determine the resistance of the circuit wire and hence the power consumed by the wire when the current is 15.0 A. Knowing I and V, we can determine the maximum power available. Knowing the maximum power available, the power consumed by the circuit wire, and the power consumed by the hair dryer, we can determine the maximum wattage lightbulb we can use and not burn a fuse. (b) Knowing expressions for the total power supplied and the power loss to the wiring and values for the power consumed by the dryer and the bulb, we can determine the current in the circuit when a 60-W bulb is used. Knowing the current and the line voltage we can determine the total power supplied to the circuit (wiring, dryer, and bulb). Knowing the total power supplied, the time it is supplied, and the cost of energy, we can determine the cost to run the dryer and 60-W bulb for 0.250 h.

Solution:

(a)
$$R_{wire} = \rho L/A = 3.07 \times 10^{-1}\ \Omega$$
$$P_{wire} = I^2 R_{wire} = 69.1\ W$$
$$P_{dryer} = 1650\ W$$
$$P_{total} = IV = 1.8 \times 10^3\ W$$
$$P_{light} = P_{total} - P_{dryer} - P_{wire} = 80.9\ W$$

(b) $P_{total} = IV$, $P_{wire} = I^2 R_{wire}$, $R_{dryer} = 1650$ W, $P_{light} = 60$ W

$$P_{total} = P_{wire} + P_{dryer} + P_{light}$$

$$IV = I^2 R_{wire} + P_{dryer} + P_{light}$$

$$I(120\ V) = I^2(3.07 \times 10^{-1}) + 1650\ W + 60\ W$$

When this quadratic is solved for I we obtain

$$I = 14.8\ A$$
$$P_{total} = IV = (14.8\ A)(120\ V) = 1.78 \times 10^3\ W$$
$$E_{total} = P_{total}\Delta t = (1.78 \times 10^3\ W)(0.250\ h) = 445\ W\cdot h$$
$$\text{Cost} = CE_{total} = (10¢/kW\cdot h)(445\ W\cdot h)(kW/1000\ W) = 4.45¢$$

Related Text Problems: 19-30 through 19-39.

PRACTICE TEST

Take and grade this practice test. Doing so will allow you to determine any weak spots in your understanding of the concepts taught in this chapter. The following section prescribes what you should study further to strengthen your understanding.

Figure 19.9 shows a circuit and data collected with that circuit.

Figure 19.9

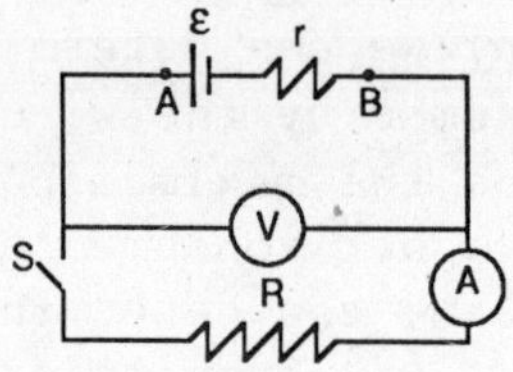


| | S open | S closed |
|---|---|---|
| V | 3.00 V | 2.00 V |
| I | 0.00 A | 1.00 A |

Determine the following:

______ 1. Emf of the cell
______ 2. Terminal potential difference of the cell with S closed
______ 3. Potential difference across R with S closed
______ 4. Resistance of R
______ 5. Internal resistance of the cell
______ 6. Charge drifting through R in 10.0 s
______ 7. Power delivered by the cell
______ 8. Power consumed by R
______ 9. Power consumed by r
______ 10. Efficiency of the cell
______ 11. Work done by the cell in 10.0 s

A heating coil designed to be used with a standard 120-V wall outlet has the following characteristics.

$$\rho = 1.00 \times 10^{-6}\ \Omega\cdot\text{m at } 25^\circ\text{C}$$
$$\alpha = 5.00 \times 10^{-4}/\text{C}^\circ$$
$$A = 2.00 \times 10^{-4}\ \text{m}^2$$

If the current to the coil is monitored, one finds that when the coil is at room temperature (25.0°C), the current to the coil is 1.00 A and that after the coil has been on for a long time, the final current is 0.900 A.

Determine the following:

______ 12. Resistance of the coil at 25.0°C
______ 13. Length of wire used in the coil
______ 14. Resistance of the coil at 50.0°C
______ 15. Resistivity of the coil material at -10.0°C
______ 16. Resistance of the coil when it has been on for a long time
______ 17. Final temperature of the coil
______ 18. Power concumption of the coil after it has been on a long time

(See Appendix I for answers.)

PRINCIPAL CONCEPTS AND EQUATIONS PRESCRIPTION

Your score on the practice is an excellent measure of your understanding of this chapter. You should now use the following chart to write your own prescription for curing any of your physics ills. Look down the leftmost column to the number of the question(s) you answered incorrectly, read across that row to see which section(s) of the study guide you should return to for further study, and then do the suggested text problems to gain additional experience in working with the particular concept.

| Practice Test Question | Concepts and Equations | Prescription | |
|---|---|---|---|
| | | Principal Concept | Text Problems |
| 1 | Emf: $\varepsilon = V + Ir$ | 3 | 19-19,22 |
| 2 | Terminal potential difference | 3 | 19-16,19 |
| 3 | Potential difference | 3 | 19-20,22 |
| 4 | Ohm's law: $R = V/I$ | 2 | 19-10,12 |
| 5 | Internal resistance: $\varepsilon = V + Ir$ | 3 | 19-17,18 |
| 6 | Current: $I = \Delta q/\Delta t$ | 1 | 19-1,2 |
| 7 | Power: $P = I\varepsilon$ | 5 | 19-32,39 |
| 8 | Power: $P = I^2R = IV = V^2/R$ | 5 | 19-30,31 |
| 9 | Power: $P = I^2r$ | 5 | 19-32,33 |
| 10 | Efficiency: $\varepsilon = P_{out}/P_{total}$ | 5 | |
| 11 | Work: $W = qV$ or $W = Pt$ | 2 of Ch. 18 | 19-35,36 |
| 12 | Ohm's law: $R = V/I$ | 2 | 19-11,13 |
| 13 | Resistivity: $R = \rho L/A$ | 4 | 19-23,24 |
| 14 | Temperature coefficient $\rho = \rho_o(1 + \alpha\Delta T)$ | 4 | 19-26,27 |
| 15 | Temperature coefficient | 4 | 19-28,29 |
| 16 | Ohm's law: $R = V/I$ | 2 | 19-14,15 |
| 17 | Temperature coefficient | 4 | 19-26,27 |
| 18 | Power: $P = I^2R = IV = V^2/R$ | 5 | 19-34,35 |

# 20 Direct-Current Circuits

RECALL FROM PREVIOUS CHAPTERS

| Previously learned concepts and equations frequently used in this chapter | Text Section | Study Guide Page |
|---|---|---|
| Electric potential difference: $V = W/q$ | 18.2 | 289 |
| Capacitance: $C = q/V$ | 18.6 | 298 |
| Current: $I = \Delta q/\Delta t$ | 19.2 | 305 |
| Ohm's Law: $R = V/I$ | 19.3 | 308 |
| Power: $P = IV = I^2R = V^2/R$ | 19.7 | 315 |

NEW IDEAS IN THIS CHAPTER

| Concepts and equations introduced | Text Section | Study Guide Page |
|---|---|---|
| Resistors in series: $I_1 = I_2 = I_3 = I_T$; $V_1 + V_2 + V_3 = V_T$; $R_1 + R_2 + R_3 = R_S$ | 20.3 | 321 |
| Resistors in parallel: $I_1 + I_2 + I_3 = I_T$; $V_1 = V_2 = V_3 = V_T$; $1/R_1 + 1/R_2 + 1/R_3 = 1/R_P$ | 20.3 | 321 |
| Kirchhoff's Rules: $\sum_i I_i = 0$; $\sum_i V_i = 0$ | 20.4 | 316 |
| Galvanometer to ammeter: $R_S = I_mR_m/(I - I_m)$ | 20.5 | 320 |
| Galvanometer to voltmeter: $R_{series} = (V - I_mR_m)/I_m$ | 20.5 | 320 |
| Capacitors in parallel: $V_1 = V_2 = V_3 = V_T$; $q_1 + q_2 + q_3 = q$; $C_1 + C_2 + C_3 = C_P$ | 20.6 | 323 |
| Capacitors in series: $V_1 + V_2 + V_3 = V_T$; $q_1 = q_2 = q_3 = q$; $1/C_1 + 1/C_2 + 1/C_3 = 1/C_S$ | 20.6 | 323 |
| Resistors and capacitors in dc circuits | 20.7 | 325 |
| Charging $q = q_{max}(1-e^{-tR_c/C})$; $I = I_{max}e^{-t/R_cC}$; $V_{R_c} = Ve^{-t/R_cC}$; $V_c = V(1-e^{-t/R_cC})$ | | |
| Discharging: $q = q_oe^{-t/R_dC}$; $I = -I_{max}e^{-t/R_dC}$; $V_{R_d} = -Ve^{-t/R_dC}$; $V_C = Ve^{-t/R_dC}$ | | |

PRINCIPAL CONCEPTS AND EQUATIONS

## 1. Resistors In Series and Parallel (Section 20.3)

Review: Figure 20.1 illustrates a series circuit.

Figure 20.1

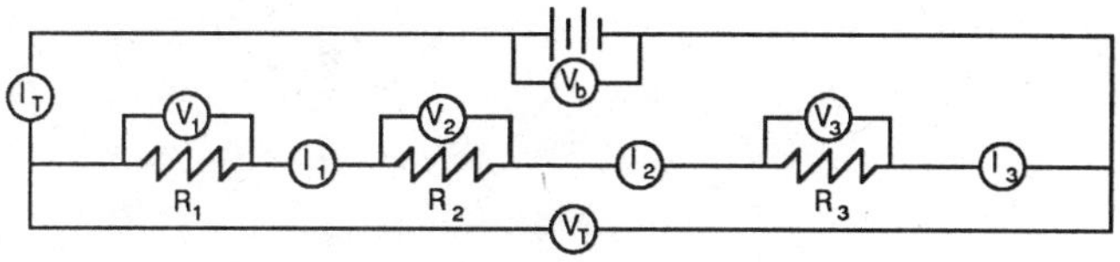


For this series circuit, the following is true:

| | |
|---|---|
| $I_1 = I_2 = I_3 = I_T$ | The current through each resistor is the same, and it is equal to the total current in the circuit. |
| $V_1 + V_2 + V_3 = V_T = V_b$ | The sum of the losses in electric potential difference across the resistors is equal to the total loss in electric potential difference and to the gain in electric potential difference from the battery. |
| $R_1 + R_2 + R_3 = R_S$ | Resistors add for a series circuit. |
| $V_1 = I_1R_1,\ V_2 = I_2R_2$<br>$V_3 = I_3R_3,\ V_T = I_TR_S$ | Ohm's law can be applied to each resistor individually, to any combination of resistors, and to the entire circuit. |

Practice: The following data are given for the circuit in Fig. 20.1.

$$I_T = 2.00 \text{ A},\ V_1 = 2.00 \text{ V},\ V_2 = 3.00 \text{ V, and } V_3 = 4.00 \text{ V}$$

Determine the following:

| | |
|---|---|
| 1. Current through each resistor | $I_1 = I_2 = I_3 = I_T = 2.00$ A |
| 2. Resistance of each resistor | $R_1 = V_1/I_1 = 1.00\ \Omega$<br>$R_2 = V_2/I_2 = 1.50\ \Omega$<br>$R_3 = V_3/I_3 = 2.00\ \Omega$ |
| 3. Total potential difference across the three resistors | $V_T = V_1 + V_2 + V_3 = 9.00$ V |
| 4. Potential difference for the battery | $V_b = V_T = 9.00$ V |

| | |
|---|---|
| 5. Total resistance for the circuit | $R_S = R_1 + R_2 + R_3 = 4.50\ \Omega$ or <br> $R_S = V_T/I_T = 4.50\ \Omega$ |
| 6. Power loss by each resistor | $P_1 = I_1V_1 = I_1^2R_1 = V_1^2/R_1 = 4.00$ W <br> $P_2 = I_2V_2 = I_2^2R_2 = V_2^2/R_2 = 6.00$ W <br> $P_3 = I_3V_3 = I_3^2R_3 = V_3^2/R_3 = 8.00$ W |
| 7. Power loss by the entire circuit | $P_T = P_1 + P_2 + P_3 = 18.0$ W or <br> $P_T = I_TV_T = I_T^2R_S = V_T^2/R_S = 18.0$ W |
| 8. Power delivered by the battery | $P_b = P_T = 18.0$ W or <br> $P_b = I_TV_b = 18.0$ W |

Review: Figure 20.2 illustrates a parallel circuit.

Figure 20.2

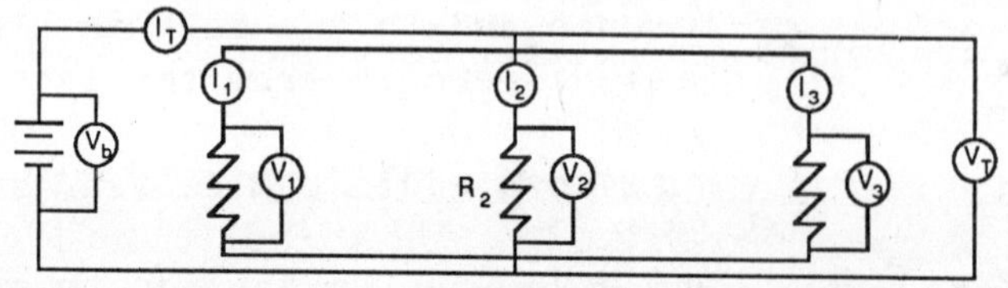


For this parallel circuit, the following is true:

$I_1 + I_2 + I_3 = I_T$ — The sum of the currents through each branch is equal to the total current in the circuit.

$V_1 = V_2 = V_3 = V_T = V_b$ — The potential difference is the same across each resistor and is equal to the total potential difference across the combination of resistors and the potential difference of the battery.

$$\frac{1}{R_1} + \frac{1}{R_2} + \frac{1}{R_3} = \frac{1}{R_p}$$ Resistors add in a reciprocal manner for parallel circuits.

$V_1 = I_1R_1$, $V_2 = I_2R_2$
$V_3 = I_3R_3$, $V_T = I_TR_T$ — Ohm's law can be applied to each resistor individually, to any combination of resistors, and to the entire circuit.

The following data are given for the circuit in Fig. 20.2.

$$V_T = 2.00\ \text{V},\ I_T = 2.00\ \text{A},\ R_1 = 2.00\ \Omega,\ R_2 = 3.00\ \Omega$$

Determine the following:

| | |
|---|---|
| 1. Total resistance of the circuit | $R_P = V_T/I_T = 1.00\ \Omega$ |
| 2. Resistance of $R_3$ | $\frac{1}{R_P} = \frac{1}{R_1} + \frac{1}{R_2} + \frac{1}{R_3}$<br>$\frac{1}{R_3} = \frac{1}{R_P} - \frac{1}{R_1} - \frac{1}{R_2} = \frac{1}{6}\ \Omega$ or $R_3 = 6.00\ \Omega$ |
| 3. Potential difference across each resistor | $V_1 = V_2 = V_3 = V_T = 2.00$ V |
| 4. Current through each resistor | $I_1 = V_1/R_1 = 1.00$ A<br>$I_2 = V_2/R_2 = 0.667$ A<br>$I_3 = V_3/R_3 = 0.333$ A<br>Note: $I_1 + I_2 + I_3 = I_T$ |
| 5. Power loss by each resistor | $P_1 = I_1V_1 = I_1^2R_1 = V_1^2/R_1 = 2.00$ W<br>$P_2 = I_2V_2 = I_2^2R_2 = V_2^2/R_2 = 1.33$ W<br>$P_3 = I_3V_3 = I_3^2R_3 = V_3^2/R_3 = 0.666$ W |
| 6. Power loss by the entire circuit | $P_T = P_1 + P_2 + P_3 = 4.00$ W or<br>$P_T = I_TV_T = I_T^2R_P = V_T^2/R_P = 4.00$ W |

Review: Figure 20.3 shows a series-parallel combination circuit redrawn into successively simpler circuits.

(a) Series-parallel combination
$I_2 = I_3;\ I_2 + I_4 = I_1;\ I_1 + I_5 = I_T$
$V_2 + V_3 = V_4;\ V_1 + V_4 = V_5 = V_T = V_b$

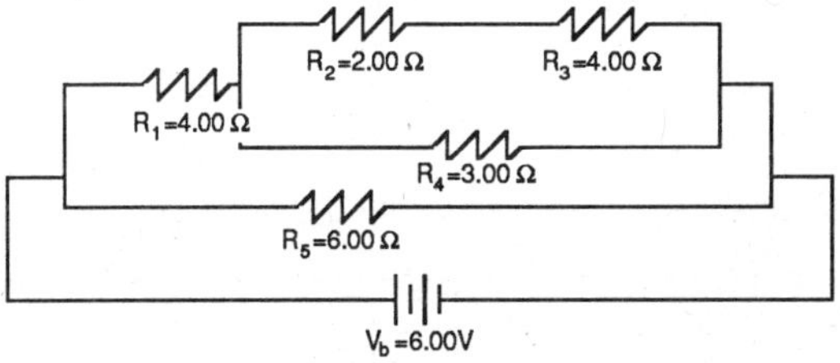


(b) Since $R_2$ and $R_3$ are in series, we can replace them with the eqivalent resistor $R_S$

$I_S + I_4 = I_1;\ I_1 + I_5 = I_T$
$V_S = V_4;\ V_1 + V_4 = V_5 = V_T = V_b$
$R_S = R_2 + R_3$

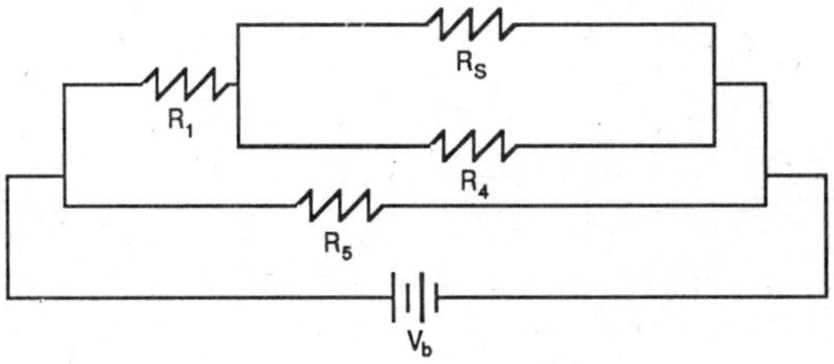


Figure 20.3

(c) Since $R_S$ and $R_4$ are in parallel, we can replace them with the equivalent resistor $R_P$.

$I_P = I_1$; $I_1 + I_5 = I_T$
$V_1 + V_P = V_5 = V_T = V_b$
$1/R_P = 1/R_S + 1/R_4$

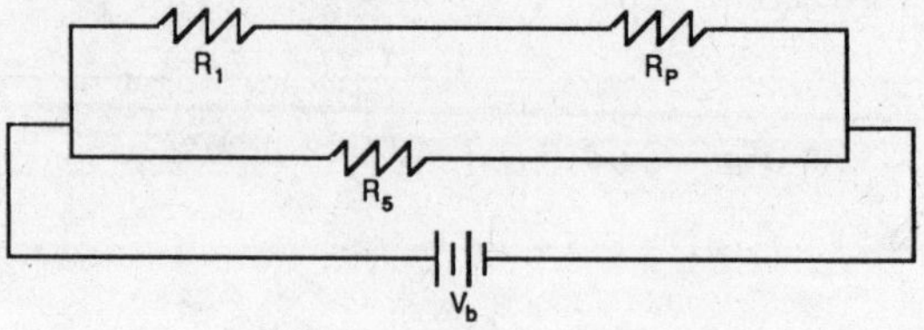


(d) Since $R_1$ and $R_P$ are in series, we can replace them with the equivalent resistor $R_S'$.

$I_S' + I_5 = I_T$
$V_S' = V_5 = V_T = V_b$
$R_S' = R_1 + R_P$

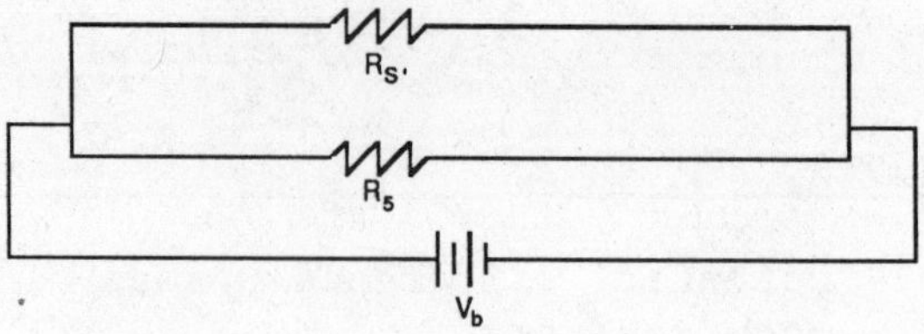


(e) Since $R_S'$ and $R_5$ are in parallel, we can replace them with an equivalent resistor. Since this will be the total resistance, let's call it $R_T$

$V_T = V_b$
$1/R_T = 1/R_S' + 1/R_5$

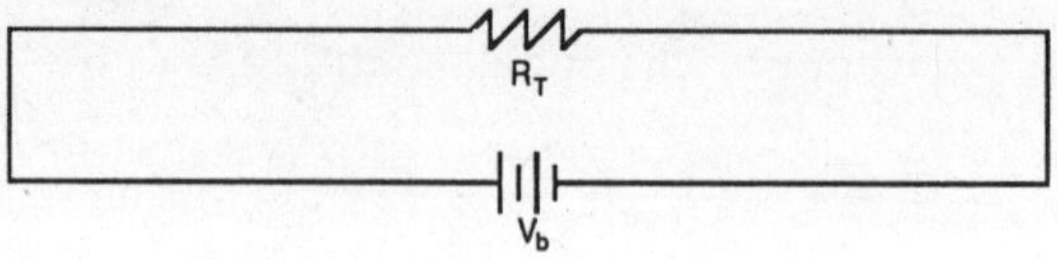


Figure 20.3 (continued from page 323)

Determine the following:

| | |
|---|---|
| 1. Total resistance for the circuit | $R_S = R_2 + R_3 = 6.00\ \Omega$ (Figs. 20.3 a and b)<br>$\frac{1}{R_P} = \frac{1}{R_S} + \frac{1}{R_4}$ or $R_P = 2.00\ \Omega$ (Figs. 20.3 b and c)<br><br>$R_S' = R_1 + R_P = 6.00\ \Omega$ (Figs. 20.3 c and d)<br>$\frac{1}{R_P} = \frac{1}{R_S'} + \frac{1}{R_5}$ or $R_T = 3.00\ \Omega$ (Figs. 20.3 d and e) |
| 2. Total current for the circuit | $V_T = V_b = 6.00$ V (Fig. 20.3 e)<br>$I_T = V_T/R_T = 2.00$ A |
| 3. Potential difference and current for $R_5$ and $R_S'$ | $V_S' = V_5 = V_T = V_b = 6.00$ V (Figs. 20.3 e and d)<br>$I_S' = V_S'/R_S' = 1.00$ A<br>$I_5 = V_5/R_5 = 1.00$ A<br>As a check, note that<br>$I_S' + I_5 = I_T = 2.00$ A |

| | |
|---|---|
| 4. Potential difference and current for $R_1$ and $R_P$ | $I_1 = I_P = I_S' = 1.00$ A (Figs. 20.3 d and c)<br>$V_1 = I_1R_1 = 4.00$ V<br>$V_P = I_PR_P = 2.00$ V<br>As a check, note that<br>$V_1 + V_P = V_5 = V_T = 6.00$ V |
| 5. Potential difference and current for $R_S$ and $R_4$ | $V_S = V_4 = V_P = 2.00$ V (Figs. 20.3 c and b)<br>$I_S = V_S/R_S = 0.333$A<br>$I_4 = V_4/R_4 = 0.667$ A<br>As a check,note that<br>$I_S + I_4 = I_1 = 1.00$ A |
| 6. Potential difference and current for $R_2$ and $R_3$ | $I_2 = I_3 = I_S = 0.333$ A (Figs. 20.3 b and a)<br>$V_2 = I_2R_2 = 0.666$ V<br>$V_3 = I_3R_3 = 1.33$ V<br>As a check, note that<br>$V_2 + V_3 = V_4 = 2.00$ V |
| 7. Power loss by each resistor | $P_1 = I_1V_1 = I_1^2R_1 = V_1^2/R_1 = 4.00$ W<br>$P_2 = I_2V_2 = I_2^2R_2 = V_2^2/R_2 = 0.222$ W<br>$P_3 = I_3V_3 = I_3^2R_3 = V_3^2/R_3 = 0.443$ W<br>$P_4 = I_4V_4 = I_4^2R_4 = V_4^2/R_4 = 1.33$ W<br>$P_5 = I_5V_5 = I_5^2R_5 = V_5^2/R_5 = 6.00$ W |
| 8. Total power loss by the resistors | $P_T = P_1 + P_2 + P_3 + P_4 + P_5 = 12.0$ W or<br>$P_T = I_TV_T = I_T^2R_T = V_T^2/R_T = 12.0$ W |

The results of analyzing this circuit are summarized in the following table.

| Poten. diff volts | Current amps | Resistance ohms | Power watts | |
|---|---|---|---|---|
| $V_1 = 4.00$ | $I_1 = 1.00$ | $R_1 = 4.00$ | $P_1 = 4.00$ | Quantities you know or wish to know |
| $V_2 = 0.666$ | $I_2 = 0.333$ | $R_2 = 2.00$ | $P_2 = 0.222$ | |
| $V_3 = 1.33$ | $I_3 = 0.333$ | $R_3 = 4.00$ | $P_3 = 0.443$ | |
| $V_4 = 2.00$ | $I_4 = 0.667$ | $R_4 = 3.00$ | $P_4 = 1.33$ | |
| $V_5 = 6.00$ | $I_5 = 1.00$ | $R_5 = 6.00$ | $P_5 = 6.00$ | |
| $V_T = 6.00$ | $I_T = 2.00$ | $R_T = 3.00$ | $P_T = 12.0$ | |
| $V_S = 2.00$ | $I_S = 0.333$ | $R_S = 6.00$ | | intermediate quantities that must be determined |
| $V_P = 2.00$ | $I_P = 1.00$ | $R_P = 2.00$ | | |
| $V_S' = 6.00$ | $I_S' = 1.00$ | $R_S' = 6.00$ | | |

Note: When working problems of this type, you will find it convenient to set up a grid like the one shown above. The advantages of such a grid are as follows:

1. It helps you keep track of what you know and what you need to determine.
2. Ohm's law can be applied to any row. Since $V = IR$, the I column times the R column should equal the V column.
3. You can check to see that resistors in series have the same current.
4. You can check to see that resistors in parallel have the same potential.
5. The V column times the I column should equal the P column.
6. The I column squared times the R column should equal the P column.
7. The V column squared divided by the R column should equal the P column.

Example 20.1. Consider the circuit shown in Fig. 20.4 and then determine

(a) the potential difference across each resistor
(b) the current through each resistor
(c) the power loss for each resistor and the entire circuit

Figure 20.4.

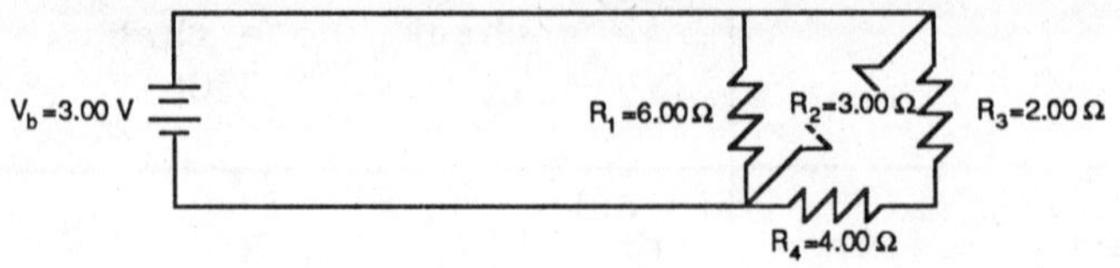


Given: $R_1 = 6.00\ \Omega$, $R_2 = 3.00\ \Omega$, $R_3 = 2.00\ \Omega$, $R_4 = 4.00\ \Omega$, and $V_b = 3.00$ V

Determine: The potential difference across each resistor, the current through each resistor, and the power loss for each resistor and the entire circuit.

Strategy: First, let's redraw the circuit so it is easier to see which resistors are in parallel and which resistors are in series. Next, we can determine $R_T$ and then $I_T$. Then, using our knowledge of series and parallel resistors, we can determine the desired quantities. We can also use our knowledge of series and parallel resistors to check our work. In order to keep track of what we know, let's construct a VIRP grid.

Solution: The circuit in Fig. 20.4 can be redrawn as shown in Fig. 20.5.

Figure 20.5.

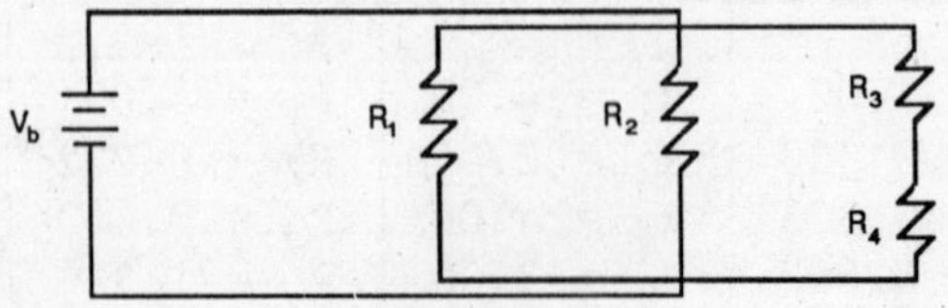


Since $R_3$ and $R_4$ are in series, we can replace them with the equivalent resistor $R_S = R_3 + R_4 = 6.00\ \Omega$. Enter this information in the grid. Next, we can redraw the circuit as shown in Fig. 20.6.

Figure 20.6.

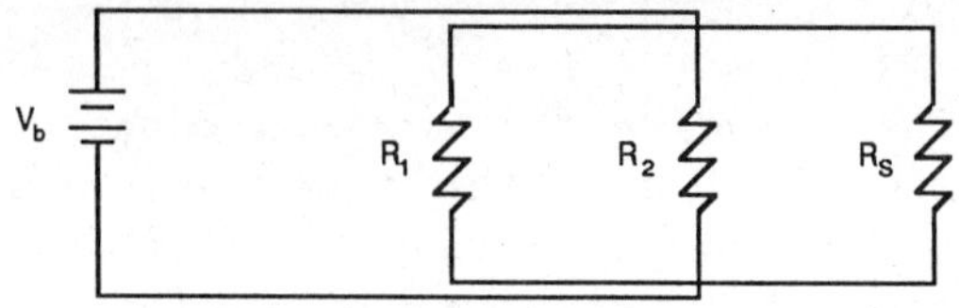


Since $R_1$, $R_2$ and $R_S$ are in parallel, we can obtain the equivalent resistance and the total current. Enter this information into the grid.

$$1/R_P = 1/R_1 + 1/R_2 + 1/R_S \quad \text{or } R_P = 1.50\ \Omega$$

$$I_T = V_T/R_P = 2.00\text{ A}$$

Using our knowledge of parallel circuits, we see that

$V_1 = V_2 = V_S = V_b = 3.00$ V
$I_1 = V_1/R_1 = 0.500$ A, $I_2 = V_2/R_2 = 1.00$ A, $I_S = V_S/R_S = 0.500$ A
As a check, note that $I_1 + I_2 + I_S = I_T$
Enter this information into the grid.

Using our knowledge of series circuits, we see that

$I_3 = I_4 = I_S = 0.500$ A
$V_3 = I_3R_3 = 1.00$ V, $V_4 = I_4R_4 = 2.00$ V
As a check, note that $V_3 + V_4 = V_S = V_b$.
Enter this information into the grid.

Finally, we can calculate the power losses.

$P_1 = I_1V_1 = I_1^2R_1 = V_1^2/R_1 = 1.50$ W
$P_2 = I_2V_2 = I_2^2R_2 = V_2^2/R_2 = 3.00$ W
$P_3 = I_3V_3 = I_3^2R_3 = V_3^2/R_3 = 0.500$ W
$P_4 = I_4V_4 = I_4^2R_4 = V_4^2/R_4 = 1.00$ W
$P_T = I_TV_T = I_T^2R_T = V_T^2/R_T = 6.00$ W
As a check, note that $P_1 + P_2 + P_3 + P_4 = P_T$
Enter this information into the grid.

At this point, the grid should appear as follows:

| V(volts) | I(amps) | R(ohms) | P(watts) |
|---|---|---|---|
| $V_1 = 3.00$ | $I_1 = 0.500$ | $R_1 = 6.00$ | $P_1 = 1.50$ |
| $V_2 = 3.00$ | $I_2 = 1.00$ | $R_2 = 3.00$ | $P_2 = 3.00$ |
| $V_3 = 1.00$ | $I_3 = 0.500$ | $R_3 = 2.00$ | $P_3 = 0.500$ |
| $V_4 = 2.00$ | $I_4 = 0.500$ | $R_4 = 4.00$ | $P_4 = 1.00$ |
| $V_b = 3.00$ | $I_T = 2.00$ | $R_T = 1.50$ | $P_T = 6.00$ |
| $V_S = 3.00$ | $I_S = 0.500$ | $R_S = 6.00$ | |

Just as a final check, note that

(a) for each row $V = IR$, $P = IV$, $P = I^2R$, and $P = V^2/R$,
(b) all resistors in series have the same current,
(c) all resistors in parallel have the same potential difference,
(d) currents for resistors in parallel add to give the total current and
(e) potential differences for resistors in series add to give the total potential difference

Related Text Problems: 20-3 through 20-15.

## 2. Kirchhoff's Rules of Electric Circuits (Section 20.4)

Review: Kirchhoff's rules can be used to solve circuit problems that cannot be solved by the direct application of Ohm's law. Such a circuit is shown in Fig. 20.7.

Figure 20.7

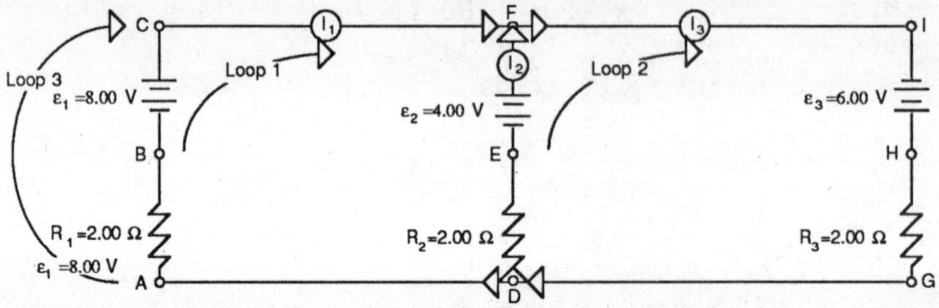


Kirchhoff's first rule states that the algebraic sum of the currents into any point is zero. Mathematically, we write

$$\sum_i I_i = I_1 + I_2 - I_3 = 0 \quad \text{for point F}$$

$$\sum_i I_i = I_3 - I_1 - I_2 = 0 \quad \text{for point D}$$

Currents directed into a point are defined as positive and those directed out of a point are defined as negative. Notice that the previous two equations are identical. Even though we have two points, we only have one point equation. In general, n points will result in n-1 different point equations.

Kirchhoff's second rule states that the algebraic sum of the voltages in traversing a closed loop is zero. Mathematically, we write

$$\sum_i V_i = \varepsilon_1 - \varepsilon_2 + I_2R_2 - I_1R_1 = 0 \quad \text{for loop \#1}$$

$$\sum_i V_i = \varepsilon_2 + \varepsilon_3 - I_3R_3 - I_2R_2 = 0 \quad \text{for loop \#2}$$

$$\sum_i V_i = \varepsilon_1 + \varepsilon_3 - I_3R_3 - I_1R_1 = 0 \quad \text{for loop \#3}$$

Notice that the previous three equations are not independent. The sum of the first two loop equations is equal to the third loop equation. Even though we have only two useful equations, the third one serves as a simple check on our work. In general, n loops will result in n-1 independent loop equations.

If the electric potential increases when traversing a circuit element, use a "+" sign. If it decreases use a "-" sign. This results in the following sign convention.

1. Traverse an emf from the negative to the positive terminal -- this results in an increase in electric potential, hence $+\varepsilon$.
2. Traverse an emf from the positive to the negative terminal -- this results in a decrease in electric potential, hence $-\varepsilon$.
3. Traverse a resistor in the same direction as the current -- the result is a decrease in electric potential, hence $-V = -IR$.
4. Traverse a resistor opposite the direction of the current -- the result is an increase in electric potential, hence $V = IR$.

If the following prescription (repeated from Section 20-4 of your text) is systematically followed, you will be able to routinely solve problems involving circuits like these.

1. Draw a circuit diagram that includes all elements of the circuit and label all known quantities.
2. Assign a current direction to each branch. Choose a different symbol for each separate current.
3. Apply Kirchhoff's rule to n-1 junctions obtaining n-1 independent equations.
4. Select enough closed loops from the circuit so that, including the equations from step 3, you have as many independent equations as currents. Be sure that every element in the circuit is included in at least one loop.
5. Apply Kirchhoff's loop rule to the closed loops selected in step 4, and obtain the rest of the needed independent equations.
6. Rewrite and organize all equations in a convenient form.
7. Solve the equations in step 6 as simultaneous equations.

<u>Practice:</u> Suppose that the following information is given for the circuit in Fig. 20.7.

$$\varepsilon_1 = 8.00 \text{ V},\ \varepsilon_2 = 4.00 \text{ V},\ \varepsilon_3 = 6.00 \text{ V},\ R_1 = R_2 = R_3 = 2.00\ \Omega$$

Determine the following:

| | |
|---|---|
| 1. The point equations for F and D | F: $I_1 + I_2 - I_3 = 0$<br>D: $I_3 - I_1 - I_2 = 0$<br><u>Note:</u> The second point equation gives us no new information |

| | |
|---|---|
| 2. Loop equations for loops 1, 2, and 3 | 1: $\varepsilon_1 - \varepsilon_2 + I_2R_2 - I_1R_1 = 0$<br>2: $\varepsilon_2 + \varepsilon_3 - I_3R_3 - I_2R_2 = 0$<br>3: $\varepsilon_1 + \varepsilon_3 - I_3R_3 - I_1R_1 = 0$<br><u>Note</u>: The third loop equation gives us no new information. We can obtain the third loop equation by adding the first two. |
| 3. Three equations which can be solved simultaneously to obtain $I_1$, $I_2$, and $I_3$ | (a) $I_1 + I_2 - I_3 = 0$<br>(b) $\varepsilon_1 - \varepsilon_2 + I_2R_2 - I_1R_1 = 0$<br>(c) $\varepsilon_2 + \varepsilon_3 - I_3R_3 - I_2R_2 = 0$<br><u>Note</u>: The point equation and any two of the three loop equations will work. |
| 4. Obtain $I_1$, $I_2$, and $I_3$ | (a) $I_1 + I_2 = I_3$<br>(b) $4.00\ V = 2I_1 - 2I_2$ or (b') $2V = I_1 - I_2$<br>(c) $10.0\ V = 2I_2 + 2I_3$ or (c') $5V = I_2 + I_3$<br>Insert $I_3$ from (a) into (c) to obtain<br>(d) $5V = 2I_2 + I_1$<br>Subtract (b') from (d) to obtain<br>$I_2 = 1.00\ A$<br>Insert $I_2$ into (c') to obtain<br>$I_3 = 4.00\ A$<br>Insert $I_2$ and $I_3$ into (a) to obtain<br>$I_1 = 3.00\ A$ |
| 5. Determine the electrical potential difference $V_{CA}$ | Let's travel from A to C with a positive charge. We go through $R_1$ in the same direction as the current, hence B is at a lower potential than A. In fact, $V_{BA} = -I_1R_1 = -6.00\ V$. We go through $\varepsilon_1$ such that we gain potential. In fact, $V_{CB} = +8.00\ V$. Putting this together, we have<br>$V_{CA} = V_{BA} + V_{CB} = -I_1R_1 + \varepsilon_1 = +2.00\ V$ |
| 6. Determine the electrical potential difference $V_{FD}$ | $V_{ED} = -I_2R_2 = -2.00\ V$<br>$V_{FE} = +\varepsilon_2 = +4.00\ V$<br>$V_{FD} = V_{ED} + V_{FE} = -I_2R_2 + \varepsilon_2 = +2.00\ V$ |
| 7. Determine the electrical potential difference $V_{IG}$ | $V_{HG} = +I_3R_3 = +8.00\ V$<br>$V_{IH} = -\varepsilon_3 = -6.00\ V$<br>$V_{IG} = V_{HG} + V_{GH} = I_3R_3 - \varepsilon_3 = +2.00\ V$ |

Note: The three branches A to C, D to F, and G to I are in parallel; subsequently, $V_{CA}$, $V_{FD}$, and $V_{IG}$ should be equal. From steps 5, 6, and 7 on the previous page, we see that this is indeed true. If any one of the currents were incorrect, we would not get $V_{CA} = V_{FD} = V_{IG}$. This simple check allows us to be certain that we have correctly determined $I_1$, $I_2$, and $I_3$.

Example 20.2. For the circuit shown in Fig. 20.8, find the current in each resistor.

Figure 20.8

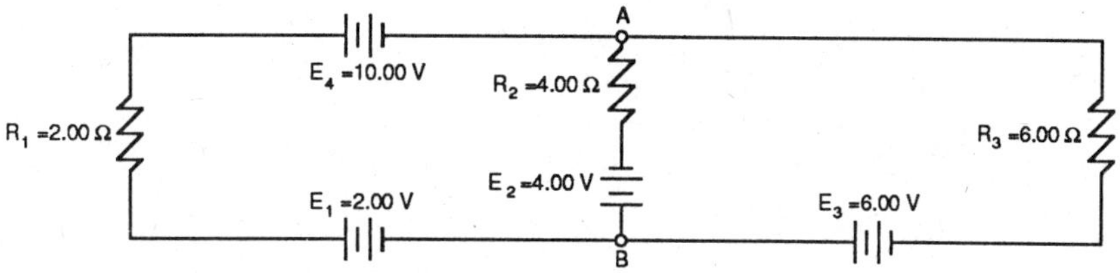


Given: $R_1 = 2.00\ \Omega$, $R_2 = 4.00\ \Omega$, $R_3 = 6.00\ \Omega$, $\varepsilon_1 = 2.00$ V, $\varepsilon_2 = 4.00$ V, $\varepsilon_3 = 6.00$ V, $\varepsilon_4 = 10.0$ V

Determine: The current in each resistor

Strategy: Assume a direction for the current in each branch and then write the point equation for A or B and two independent loop equations. These three equations contain the three unknown currents and hence can be solved simultaneously for the currents.

Solution: Assume the direction for the current in each branch as shown in Fig. 20.9.

Figure 20.9.

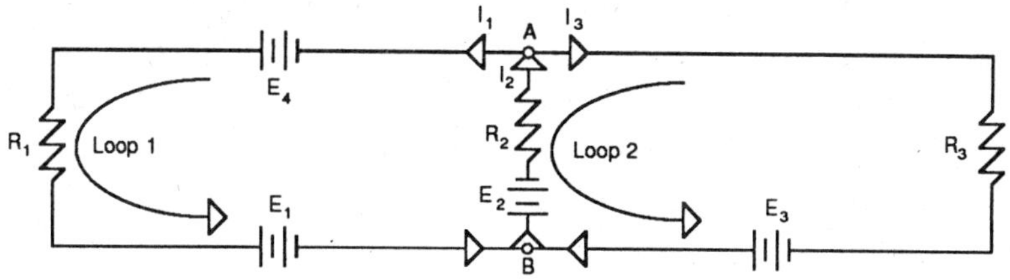


The point equation for B is

$$\text{(a)} \quad I_1 + I_3 - I_2 = 0$$

The loop equations for loops 1 and 2, respectively, are

$$\text{(b)} \quad \varepsilon_4 - I_1R_1 - \varepsilon_1 - \varepsilon_2 - I_2R_2 = 0$$

$$\text{(c)} \quad \varepsilon_2 - \varepsilon_3 + I_3R_3 + I_2R_2 = 0$$

Insert values into (b) and (c) to obtain

$$\text{(b')} \quad I_1 + 2I_2 = 2$$

$$\text{(c')} \quad 3I_3 + 2I_2 = 1$$

Solve (a) for $I_1$ and substitute it into (b') to obtain

$$\text{(b'')} \quad -I_3 + 3I_2 = 2$$

Multiply (b") by 3 to obtain

$$(b''') \quad -3I_3 + 9I_2 = 6$$

Add (b"') and (c') to eliminate $I_3$ and obtain $I_2$.

$$11I_2 = 7 \quad \text{or} \quad I_2 = 0.636 \text{ A}$$

Substitute $I_2$ into (b') and (c') to obtain, respectively, $I_1$ and $I_3$.

From (b') $\quad I_1 = 2 - 2I_2 = 0.728$ A
From (c') $\quad I_3 = (1 - 2I_1)/3 = -0.0907$ A
As a check, substitute $I_1$ and $I_3$ into (a) to see if the same value for $I_2$ is obtained.
From (a) $\quad I_1 + I_3 = 0.637$ A

Related Text Problems: 20-16 through 20-23.

## 3. Galvanometers, Voltmeters, and Ammeters (Section 20.5)

Review: A galvanometer has a fixed resistance $R_m$ and is designed for a full-scale deflection with a current $I_m$, hence it can withstand a potential difference of $V_m = I_mR_m$. A galvanometer can be used to measure currents greater than $I_m$ if a shunt resistor $R_S$ is connected in parallel with it. Such a combination is called an ammeter (Fig. 20.10).

Figure 20.10

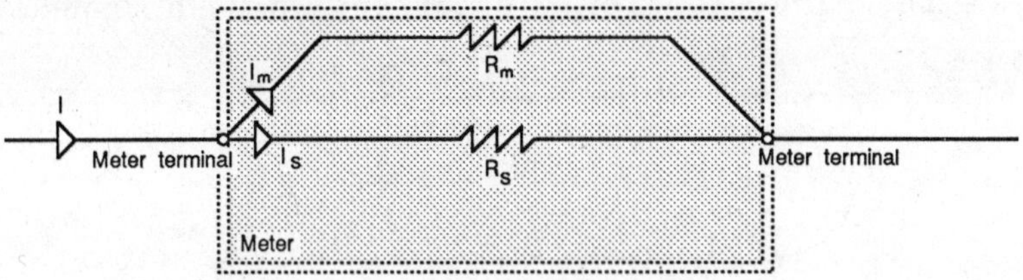


Since the resistors are in parallel, we can write

$$I_m + I_S = I$$
$$V_S = V_m \qquad \text{but } V_S = I_SR_S \text{ and } V_m = I_mR_m \text{ (by Ohm's law)}$$
$$I_SR_S = I_mR_m$$
$$(I - I_m)R_S = I_mR_m$$
$$R_S = I_mR_m/(I - I_m)$$

Generally $I_m$ and $R_m$ are known, and we need to determine the size of the shunt resistor $R_S$ needed to provide us with the capability of measuring a current I.

A galavanometer can be used to measure potential difference greater than $V_m = I_mR_m$ if a resistor is placed in series with it. Such a combination is called a voltmeter (Fig. 20.11).

Figure 20.11.

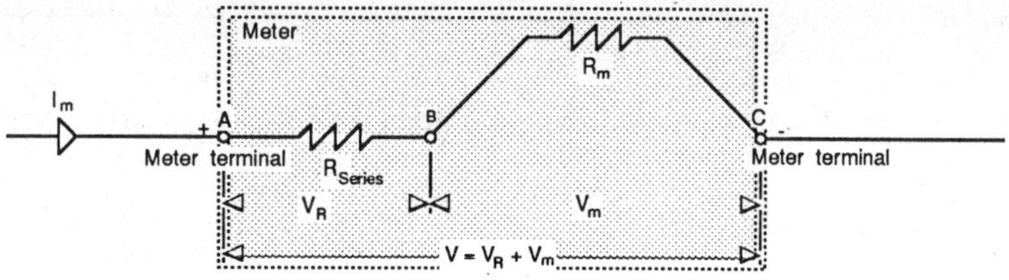


Since the resistors are in series

$$I_R = I_m$$
$$V_R + V_m = V$$
$$I_m R_{series} + I_m R_m = V$$
$$R_{series} = (V - I_m R_m)/I_m$$

Generally $I_m$ and $R_m$ are known, and we need to determine the size of the series resistor $R_{series}$ needed to provide us with the capability of measuring a potential difference V.

A galvanometer can also be used to measure resistance, as shown in Fig. 20.12.

Figure 20.12

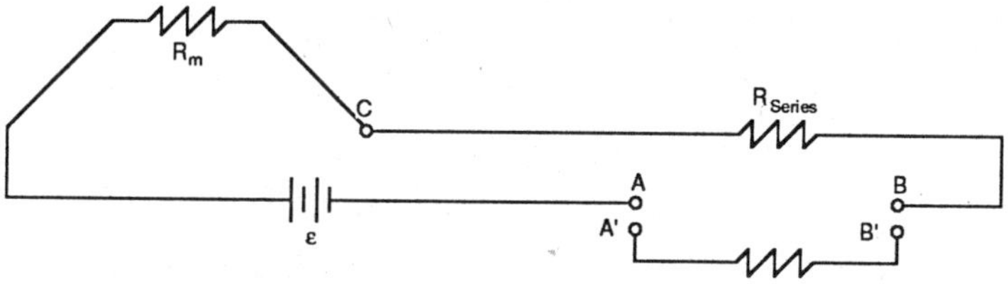


The resistor $R_{series}$ is selected so that the galvanometer reads full scale (i.e., $I = I_m$) when A and B are connected, hence zero resistance ($R_x = 0$). When the circuit is open ($R_x = \infty$), there is no current, so the galvanometer reads zero. Thus, an ohmmeter reads backwards.

$R_x = 0$ gives a full-scale deflection
$R_x = \infty$ gives no deflection

For $R_x$ in between these extremes, the meter gives a reading between 0 and full scale. Several known resistors can be used to calibrate the meter.

Practice: A galvanometer has a resistance $R_m = 30.0\ \Omega$ and is designed for a full scale deflection when $I_m = 1.00 \times 10^{-3}$ A.

Determine the following:

| | |
|---|---|
| 1. Shunt resistor needed to make a 0 to 1.00-A ammeter | $R_S = I_m R_m/(I - I_m)$<br>$R_S = 3.00 \times 10^{-2}\ \Omega$ |
| 2. Series resistor needed to make a 0 to 5.00-V voltmeter | $R_{series} = (V - I_m R_m)/I_m$<br>$R_{series} = 4.97 \times 10^{3}\ \Omega$ |

| | |
|---|---|
| 3. The current in the circuit if this galvanometer has a deflection one-half of full scale when $R_S = 6.00 \times 10^{-2}\ \Omega$ | When $R_S = 6.00 \times 10^{-2}\ \Omega$, the maximum current the meter can handle is obtained by<br>$R_S = I_mR_m/(I - I_m)$ or<br>$I = I_m[(1 + (R_m/R_S)]$<br>$I = 0.501$ A for full scale, hence<br>$I = 0.251$ A for half scale |
| 4. The potential difference being measured if this galvanometer is used with a $2.00 \times 10^3\ \Omega$ series resistor and has a one-fourth scale deflection | When $R_{series} = 2.00 \times 10^3\ \Omega$, the max potential difference the meter can handle is obtained by<br>$R_{series} = (V - I_mR_m)/I_m$ or<br>$V = I_m(R_{series} + R_m)$<br>$V = 2.03$ V for full-scale hence<br>$V = 2.03/4 = 0.508$ V for one-fourth of full scale deflection |
| 5. The size of the resistor needed to make an ohmmeter if a 9.0-V cell is used with this galvanometer | $\varepsilon = I_m(R_m + R)$ or<br>$R = (\varepsilon/I_m) - R_m = 8970\ \Omega$ |

Example 20.3. A galvanometer is converted into a 0-5.00-A ammeter with a 0.100-Ω shunt resistor and into a 0-5.00-V voltmeter with a $4.50 \times 10^3\ \Omega$ series resistor. Determine the galvanometer characteristics (i.e., $R_m$ and $I_m$).

Given: $R_S = 0.100\ \Omega$ = shunt resistor for a 0-5.00-A ammeter
$R_{series} = 900\ \Omega$ = series resistor for a 0-5.00-V voltmeter
$I = 5.00$ A = maximum current measured by the ammeter
$V = 5.00$ V = maximum potential difference measured by the voltmeter

Determine: $I_m$ and $R_m$ for the galvanometer

Strategy: Knowing $R_S$, $R_{series}$, V, and I, we can write two equations (one for the voltmeter and one for the ammeter) containing the two unknowns $I_m$ and $R_m$. These two equations can be solved simultaneously to obtain $I_m$ and $R_m$.

Solution: For the ammeter and voltmeter, respectively, we can write

(a) $R_S = I_mR_m/(I - I_m)$ and (b) $R_{series} = (V - I_mR_m)/I_m$

Inserting values into (a) and (b) we obtain

(c) $0.100 = I_mR_m/(5.00 - I_m)$ and (d) $900 = (5.00 - I_mR_m)/I_m$

Manipulating the algebra in (c) and (d) we obtain

(e) $0.500 - 0.100\ I_m = I_mR_m$ and (f) $I_mR_m = 5.00 - 900\ I_m$

We can equate these expressions for $I_m R_m$, eliminate $R_m$ and solve for $I_m$.

$$0.500 - 0.100\ I_m = 5.00 - 900\ I_m \quad \text{or} \quad I_m = 5.00 \times 10^{-3}\ \text{A}$$

We can insert $I_m$ into (e) or (f) to obtain $R_m$

(e) $0.500 - 0.100\ I_m = I_m R_m$ or $R_m = (0.500 - 0.100\ I_m)/I_m = 99.9\ \Omega$

(f) $I_m R_m = 5.00 - 900\ I_m$ or $R_m = (5.00 - 900\ I_m)/I_m = 100\ \Omega$

Related Text Problems: 20-24 through 20-28.

## 4. Capacitors In Parallel and Series (Section 20.6)

Review: Figure 20.13 shows four different combinations of three capacitors and reviews all relationships concerning potential difference, charge, and capacitance. This figure and section 20.6 of your text should be studied until you can easily write down such relations for any capacitor combination.

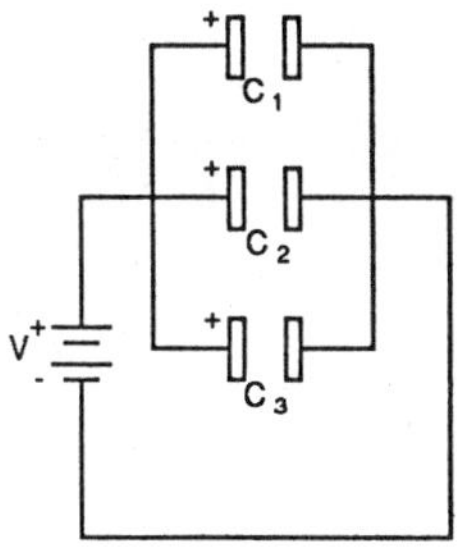


(a) Parallel combination of capacitors

$V_1 = V_2 = V_3 = V_T = V$
$q_1 + q_2 + q_3 = q$
$C_1 + C_2 + C_3 = C_P$
$C_1 = q_1/V_1;\ C_2 = q_2/V_2;\ C_3 = q_3/V_3;\ C_P = q/V$

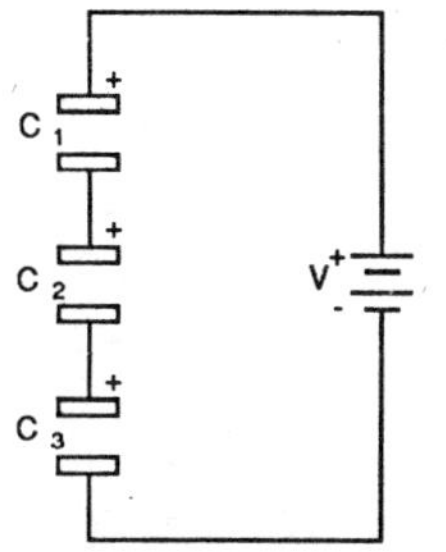


(b) Series combination of capacitors

$V_1 + V_2 + V_3 = V_T = V$
$q_1 = q_2 = q_3 = q$
$1/C_1 + 1/C_2 + 1/C_3 = 1/C_S$
$C_1 = q_1/V_1;\ C_2 = q_2/V_2;\ C_3 = q_3/V_3;\ C_S = q/V$

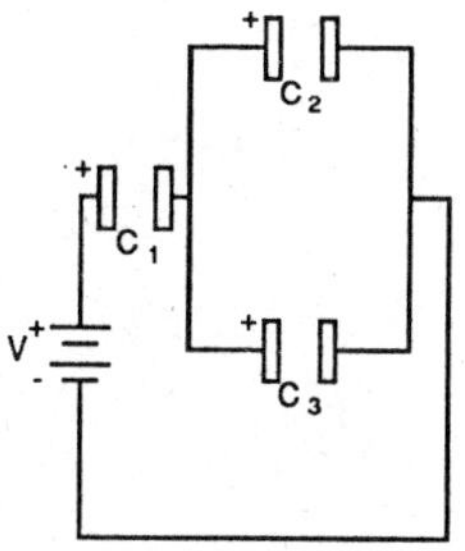


(c) Parallel and series combination of capacitors

$V_2 = V_3 = V_P\ ;\ V_1 + V_P = V$
$q_2 + q_3 = q_P\ ;\ q_1 = q_P = q$

$C_2 + C_3 = C_P\ ;\ 1/C_1 + 1/C_P = 1/C_S$

$C_1 = q_1/V_1;\ C_2 = q_2/V_2;\ C_3 = q_3/V_3$
$C_P = q_P/V_P;\ C_S = q/V$

Figure 20.13

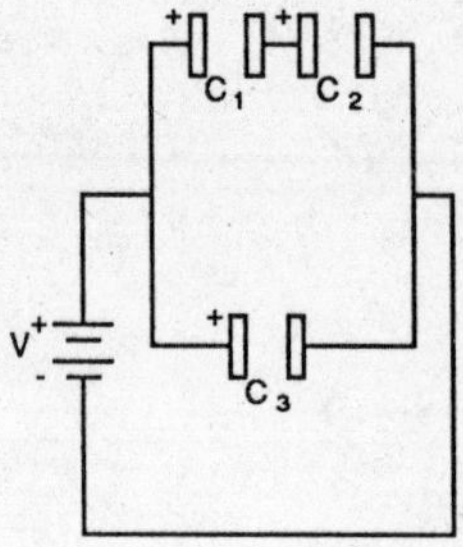


(d) Series and parallel combination of capacitors

$V_1 + V_2 = V_S = V_3 = V_P$
$q_1 = q_2 = q_3;\ q_S + q_3 = q_P = q$
$1/C_1 + 1/C_2 = 1/C_S;\ C_S + C_3 = C_P$
$C_1 = q_1/V_1,\ C_2 = q_2/V_2;\ C_3 = q_3/V_3$
$C_P = q_P/V_P;\ C_S = q/V$

Figure 20.13 (continued from page 335)

Example 20.4. For the circuit shown in Fig. 20.13d, determine (a) the equivalent capacitance and (b) the charge on each capacitor. Let $C_1 = 3.00 \times 10^{-6}$ F, $C_2 = 6.00 \times 10^{-6}$ F, $C_3 = 3.00 \times 10^{-6}$ F, and V = 10.0 V.

Given: $C_1 = 3.00 \times 10^{-6}$ F, $C_2 = 6.00 \times 10^{-6}$ F, $C_3 = 3.00 \times 10^{-6}$ F, V = 10.0 V

Determine: $C_T$ -- the equivalent capacitance
$q_1$, $q_2$, and $q_3$ -- the charge on each capacitor

Strategy: Knowing $C_1$, $C_2$, and $C_3$, we can determine $C_S$ and then $C_P$, the equivalent capacitance. Knowing $C_P$ and V, we can determine the total amount of charge and then the charge on each capacitor.

Solution: Knowing $C_1$ and $C_2$, we can obtain $C_S$

$$1/C_S = 1/C_1 + 1/C_2 \quad \text{or} \quad C_S = 2.00 \times 10^{-6}\ \text{F}$$

Knowing $C_S$ and $C_3$, we can obtain $C_P$, the equivalent capacitance

$$C_P = C_3 + C_S = 5.00 \times 10^{-6}\ \text{F}$$

Knowing $C_p$ and V, we can determine the total charge delivered to the capacitor.

$$Q = C_P V = 5.00 \times 10^{-5}\ \text{C}$$

Knowing Q and the way the capacitors are combined, we can determine the charge on each capacitor.

$$Q = Q_3 + Q_S$$
$$Q_3 = C_3 V = (3.00 \times 10^{-6}\ \text{F})(10.0\ \text{V}) = 3.00 \times 10^{-5}\ \text{C}$$
$$Q_S = Q - Q_3 = (5.00 \times 10^{-5}\ \text{C}) - (3.00 \times 10^{-5}\ \text{C}) = 2.00 \times 10^{-5}\ \text{C}$$
$$Q_1 = Q_2 = Q_S = 2.00 \times 10^{-5}\ \text{C}$$

Related Text Problems: 20-29 through 20-35.

## 5. Resistors and Capacitors In DC Circuits (Section 20.7)

Review: Figure 20.14 shows a circuit that allows us to charge a capacitor C through the resistor $R_c$ when switch $S_c$ is closed and $S_d$ is open. We can then discharge the capacitor C through the resistor $R_d$ when switch $S_c$ is opened and $S_d$ is closed.

Figure 20.14

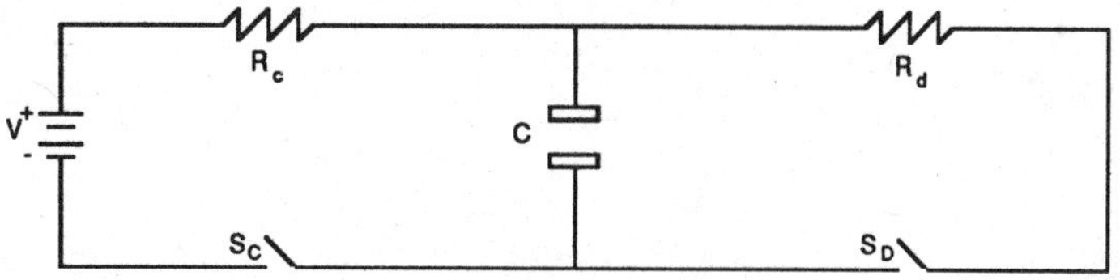


Charging the capacitor ($S_c$ closed and $S_d$ open): As the capacitor is charged, the charge on the capacitor (q), the current in the circuit (I), the potential difference across the resistor ($V_{R_c}$), and the potential difference across the capacitor ($V_c$) all vary in time as follows:

(a) $q = CV(1 - e^{-t/R_cC}) = q_{max}(1 - e^{-t/R_cC})$, where $q_{max} = CV$

Equation (a) and Fig. 20.15 show how the charge on the capacitor varies in time.

Figure 20.15

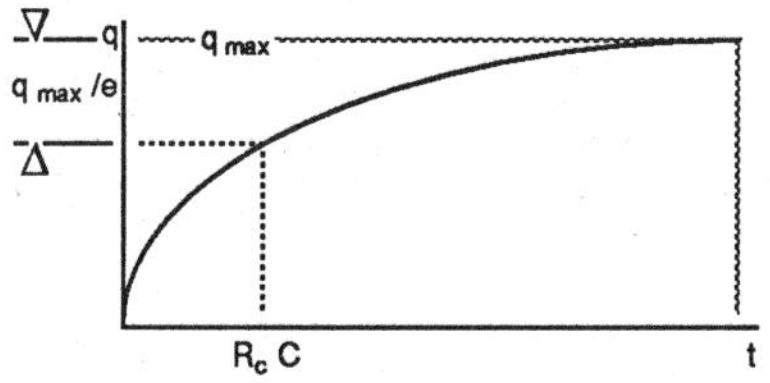


$q = 0$ at $t = 0$
$q = q_{max}(1 - e^{-1}) = 0.63q_{max}$ at $t = R_cC$
$q \to q_{max}$ as $t \to \infty$

Write Kirchhoff's loop equation for the charging circuit and solve for I to obtain:

(b) $I = -\frac{q}{R_cC} + \frac{V}{R_c} = -\frac{CV}{R_cC}(1 - e^{-t/R_cC}) + \frac{V}{R_c} = \frac{V}{R_c} e^{-t/R_cC} = I_{max} e^{-t/R_cC}$

Equation (b) and Fig. 20.16 show how the current in the circuit varies in time.

Figure 20.16

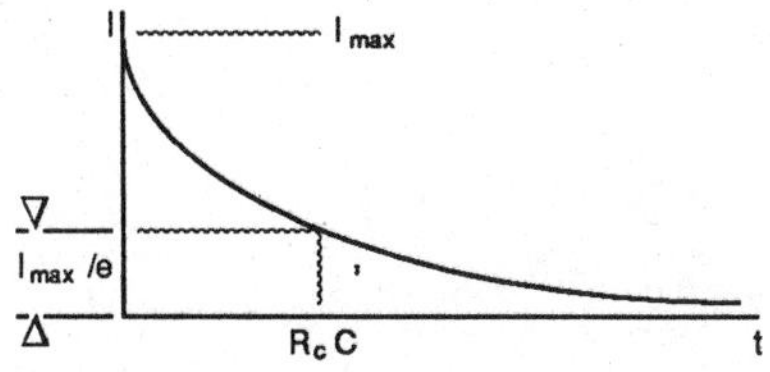


$I = I_{max} = V/R_c$ at $t = 0$
$I = I_{max}/e = 0.37I_{max}$ at $t = R_cC$
$I \to 0$ as $t \to \infty$

(c) $V_{R_c} = IR_c = \frac{V}{R_c} e^{-t/R_cC} R_c = Ve^{-t/R_cC}$

Equation (c) and Fig. 20.17 show how the potential difference across the resistor $R_c$ varies in time.

Figure 20.17

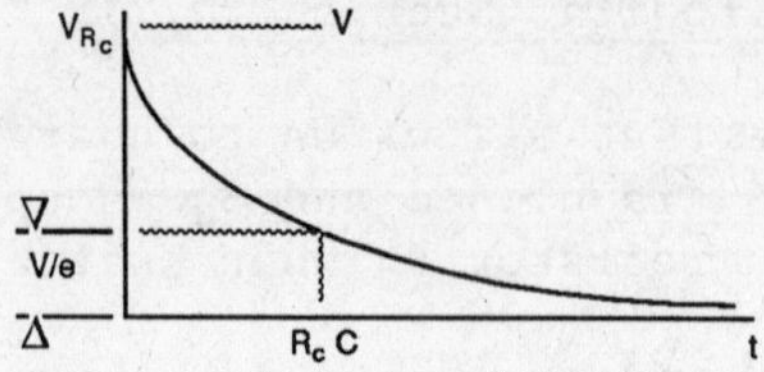


$V_{R_C} = V$ at $t = 0$

$V_{R_C} = V/e = 0.37V$ at $t = R_cC$

$V_{R_C} \rightarrow 0$ as $t \rightarrow \infty$

(d) $V = q/C = CV(1 - e^{-t/R_cC})/C = V(1 - e^{-t/R_cC})$

Equation (d) and Fig. 20.18 show how the potential difference across the capacitor varies in time.

Figure 20.18

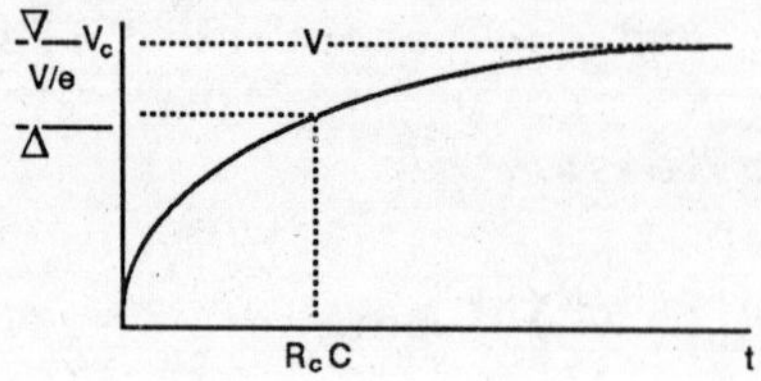


$V_C = 0$ at $t = 0$
$V_C = V(1 - e^{-1}) = 0.63V$
at $t = R_cC$
$V_C \rightarrow V$ as $t \rightarrow \infty$

Discharging the capacitor ($S_c$ open and $S_d$ closed):

Equation (e) and Fig. 20.19 show how the charge on the capacitor varies in time.

(e) $$q = q_o\, e^{-t/R_dC}$$

Figure 20.19

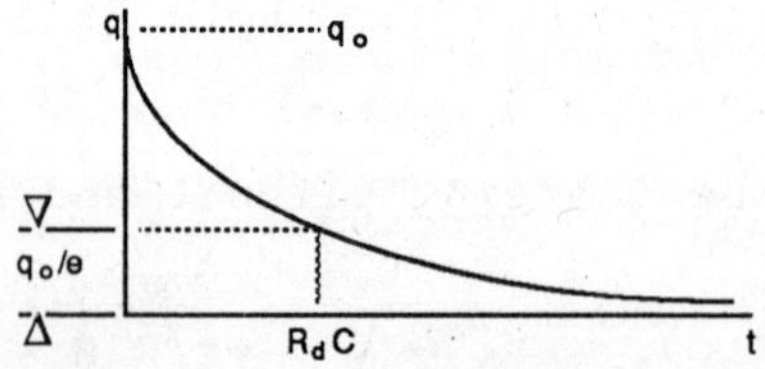


$q = q_o$ at $t = 0$
$q = q_o/e = 0.37q_o$ at $t = R_dC$
$q \rightarrow 0$ as $t \rightarrow \infty$

Equation (f) and Fig. 20.20 show how the current in the circuit varies with time.

(f) $$I = -\frac{q}{R_dC} = -\frac{q_o e^{-t/R_dC}}{R_dC} = -\frac{CV\, e^{-t/R_dC}}{R_dC} = -I_{max}\, e^{-t/R_dC}$$

The minus sign means that the current is in the direction opposite to that of the charging circuit.

Figure 20.20

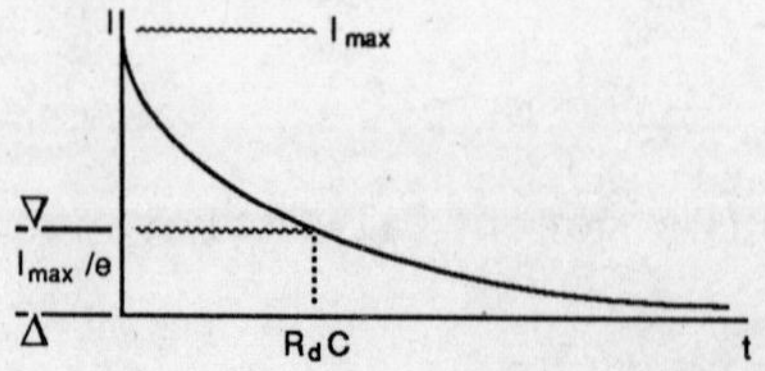


$I = I_{max} = V/R_d$ at $t = 0$
$I = I_{max}/e = 0.37I_{max}$
at $t = R_dC$
$I \rightarrow 0$ as $t \rightarrow \infty$

Equation (g) and Fig. 20.21 show how the potential difference across the resistor varies with time.

(g) $V_{R_d} = IR_d = -I_{max}(e^{-t/R_dC})R_d = -\frac{V}{R_d}(e^{-t/R_dC})R_d = -Ve^{-t/R_dC}$

Figure 20.21

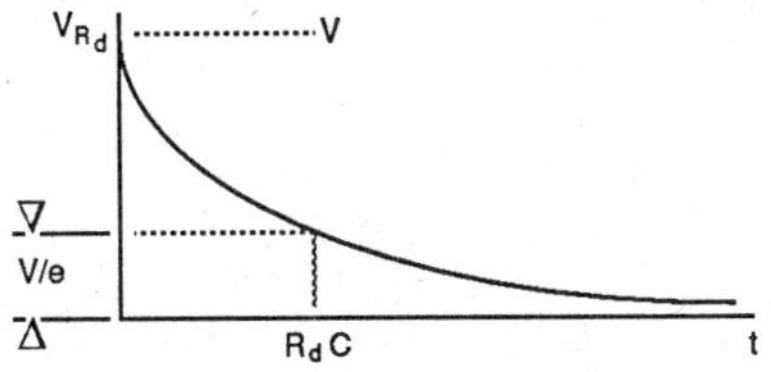


$V_{R_d} = V$ at $t = 0$

$V_{R_d} = V/e = 0.37V$ at $t = R_dC$

$V_{R_d} \to 0$ as $t \to \infty$

Equation (h) and Fig. 20.22 show how the potential difference across the capacitor varies with time.

(h) $V_C = \frac{q}{C} = \frac{q_o e^{-t/R_dC}}{C} = \frac{CV e^{-t/R_dC}}{C} = Ve^{-t/R_dC}$

Figure 20.22

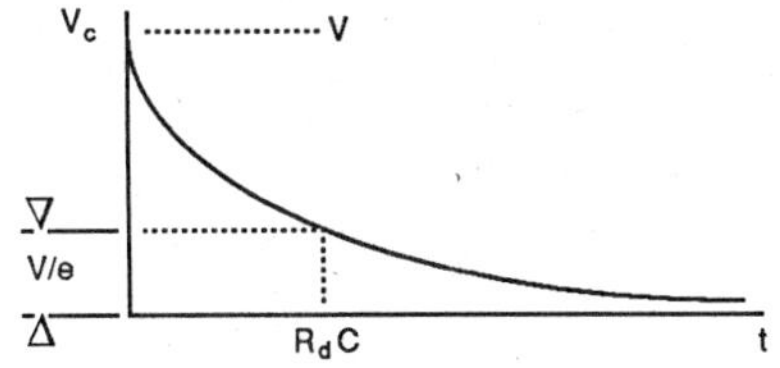


$V_C = V$ at $t = 0$
$V_C = V/e = 0.37V$ at $t = R_dC$
$V_C \to 0$ as $t \to \infty$

<u>Practice:</u> Consider Fig. 20.14 with the following values.

$R_c = 100\ \Omega$, $R_d = 50.0\ \Omega$, $C = 5.00 \times 10^{-6}$ F, $V = 10.0$ V

First, let's close $S_c$ and open $S_d$ and then determine the following:

| | |
|---|---|
| 1. $q$, $V_{R_c}$, $V_C$, and $I$ at $t = 0$<br>(i.e., at the instant the switch is closed | $q = CV(1 - e^{-t/R_cC}) = 0$<br>$I = I_{max}e^{-t/R_cC} = I_{max} = V/R_c = 0.100$ A<br>$V_{R_c} = Ve^{-t/R_cC} = V = 10.0$ V<br>$V_C = V(1 - e^{-t/R_cC})$ |
| 2. $q$, $V_{R_c}$, $V_C$, and $I$ at $t = R_cC$<br>(i.e., at a time equal to the time constant) | $q = CV(1 - e^{-1}) = 3.16 \times 10^{-5}$ C<br>$I = I_{max}/e = (V/R_c)/e = 3.68 \times 10^{-2}$ A<br>$V_{R_c} = V/e = 3.68$ V<br>$V_C = V(1 - e^{-1}) = 6.32$ V<br>Note as a check that $V_{R_c} + V_C = V$ |

| | |
|---|---|
| 3. $q$, $V_{R_c}$, $V_C$, and $I$ at a time equal to one tenth the time constant | $t = 0.100\ R_cC$<br>$q = CV(1 - e^{-0.1}) = 4.76 \times 10^{-6}$ C<br><br>$I = I_{max}\ e^{-0.1} = (V/R_C)e^{-0.1} = 9.05 \times 10^{-2}$ A<br>$V_{R_C} = V\ e^{-0.1} = 9.05$ V<br>$V_C = V(1 - e^{-0.1}) = 0.950$ V<br>Note as a check that $V_{R_c} + V_C = V$ |
| 4. Time constant for charging | $T_c = R_cC = 5.00 \times 10^{-4}$ s |
| 5. Time for the charge on the capacitor to reach one-half its maximum value | $q = q_{max}/2$<br>$q = q_{max}(1 - e^{-t/R_cC})$<br>$q_{max}/2 = q_{max}(1 - e^{-t/R_cC})$<br>$2 = e^{t/R_cC}$<br>$t = R_cC \ln 2 = 3.47 \times 10^{-4}$ s |

Now let's open $S_c$ and close $S_d$ and then determine the following:

| | |
|---|---|
| 6. Time constant for discharging | $T_d = R_dC = 2.50 \times 10^{-4}$ s |
| 7. $q$, $V_{R_d}$, $V_C$, and $I$ the instant $S_d$ is closed | $q = q_oe^{-t/R_dC} = q_o = CV = 5.00 \times 10^{-5}$ C<br>$V_{R_d} = -Ve^{-t/R_dC} = -V = -10.0$ V<br>$V_C = Ve^{-t/R_dC} = V = 10.0$ V<br>$I = I_{max}\ e^{-t/R_dC} = -I_{max} = -V/R_d$<br>$= 0.200$ A |

Example 20.5. A capacitor is placed in series with a 25.0-V battery and a $6.00 \times 10^4$-Ω resistor. It takes 10.0 s for the capacitor to charge from 0.0 V to 20.0 V. Calculate the value of the capacitor.

Given: $V = 25.0$ V, $R = 6.00 \times 10^4$ Ω, $V_C = 20.0$ V at $t = 10.0$ s

Determine: C -- the value of the capacitor

Strategy: We can write an expression for the potential difference across the capacitor as a function of time. We can then solve this expression for C and substitute known values.

Solution: The potential difference across the capacitor as a function of time can be expressed as

$$V_C = V(1 - e^{-t/RC})$$

This can be solved for C as follows

$$V_C/V = (1 - e^{-t/RC})$$

$$e^{-t/RC} = 1 - (V_C/V)$$

$$-t/RC = \ln[1 - (V_C/V)]$$

$$C = -t/R\{\ln[1 - (V_C/V)]\} = 1.04 \times 10^{-4} \text{ F}$$

Related Text Problems: 20-36 through 20-41.

---

PRACTICE TEST

Take and grade the practice test. Then use the chart in the next section to determine any weak areas and to write a prescription which will allow you to strengthen your understanding of physics in these areas.

Figure 20.23 $R_1 = 5.00\ \Omega$, $R_2 = 3.00\ \Omega$, $R_3 = 6.00\ \Omega$, $R_4 = 3.00\ \Omega$, $V_b = 2.50$ V

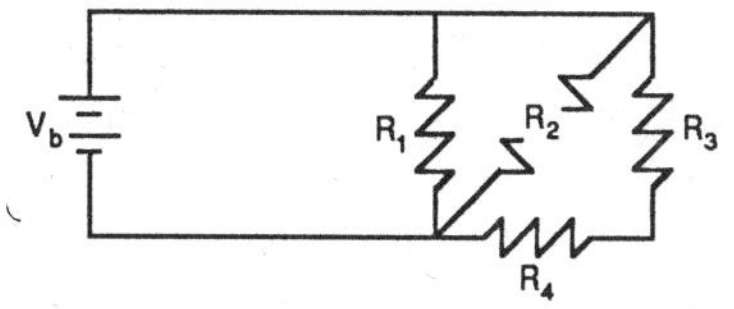


For the circuit shown in Fig. 20.23, determine the following:

__________ 1. Total resistance for the circuit
__________ 2. Current through $R_1$
__________ 3. Current through $R_4$
__________ 4. Potential difference across $R_2$

Figure 20.24 $\varepsilon_1 = \varepsilon_2 = \varepsilon_3 = 6.00$ V, $R_1 = 4.00\ \Omega$, $R_2 = 2.00\ \Omega$, $R_3 = 8.00\ \Omega$

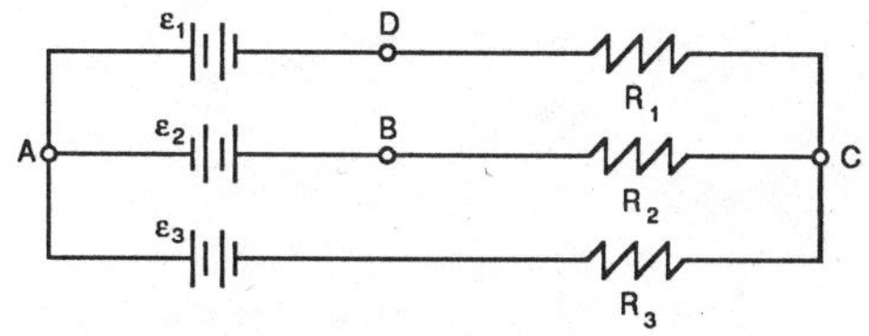


For the circuit shown in Fig. 20.24, determine the following:

__________ 5. Current through $R_1$
__________ 6. Current through $R_2$
__________ 7. Current through $R_3$

__________ 8. Potential difference $V_{AC}$

__________ 9. Potential difference $V_{BD}$

A galvanometer has a 30.0-Ω resistance and is designed for a full scale deflection at $1.00 \times 10^{-3}$ A. Determine the following:

__________ 10. Shunt resistor needed to make a 0-1-A ammeter

__________ 11. Series resistor needed to make a 0-5-V voltmeter

__________ 12. Current in a circuit if this galvanometer is used with a $6.00 \times 10^{-2}$-Ω shunt resistor as an ammeter and it has a one-half scale deflection.

Figure 20.25

$C_1 = 20.0\ \mu F$
$C_2 = 2.00\ \mu F$
$C_3 = 8.00\ \mu F$
$C_4 = 10.0\ \mu F$
$V_T = 10.0\ V$

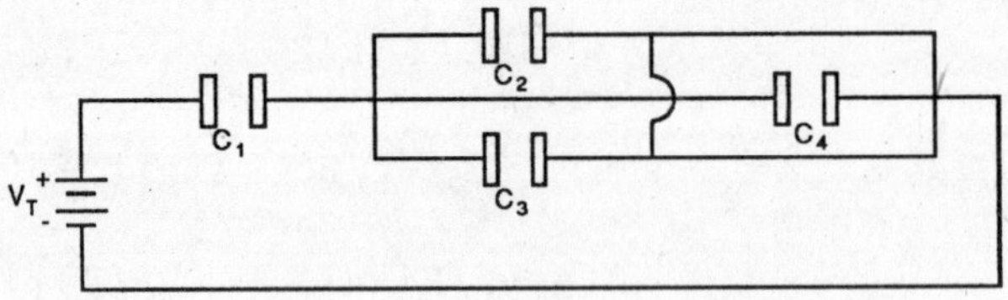


For the circuit shown in Fig. 20.25, determine the following:

__________ 13. Total capacitance

__________ 14. Potential difference across $C_2$

__________ 15. Charge on $C_3$

__________ 16. Charge on $C_4$

Figure 20.26

R = 200 Ω
C = 10.0 μF
V = 10.0 V
t = 0 when S is closed

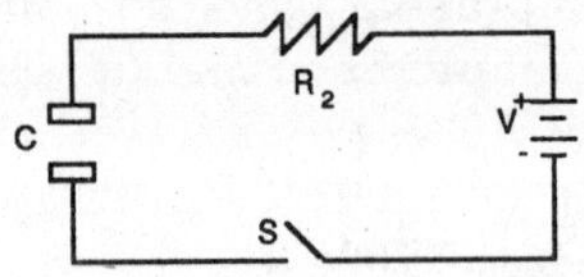


For the circuit shown in Fig. 20.26, determine the following:

__________ 17. Potential difference across R at t=0

__________ 18. Time constant

__________ 19. Current after one time constant

__________ 20. Charge on the capacitor after two time constants

(See Appendix I for answers.)

## PRINCIPAL CONCEPTS AND EQUATIONS PRESCRIPTION

Your score on the practice is an excellent measure of your understanding of this chapter. You should now use the following chart to write your own prescription for curing any of your physics ills. Look down the leftmost column to the number of the question(s) you answered incorrectly, read across that row to see which section(s) of the study guide you should return to for further study, and then do the suggested text problems to gain additional experience in working with the particular concept.

| Practice Test Question | Concepts and Equations | Prescription: Principal Concept | Prescription: Text Problems |
|---|---|---|---|
| 1 | Resistors in series and parallel | 1 | 20-3,4 |
| 2 | Resistors in series and parallel | 1 | 20-6,7 |
| 3 | Resistors in series and parallel | 1 | 20-8,9 |
| 4 | Resistors in series and parallel | 1 | 20-13,14 |
| 5 | Kirchhoff's rules | 2 | 20-16,17 |
| 6 | Kirchhoff's rules | 2 | 20-17,18 |
| 7 | Kirchhoff's rules | 2 | 29-18,19 |
| 8 | Kirchhoff's rules | 2 | 20-19,20 |
| 9 | Kirchhoff's rules | 2 | 20-20,21 |
| 10 | Galvanometer to ammeter | 3 | 20-24,26 |
| 11 | Galvanometer to voltmeter | 3 | 20-25,27 |
| 12 | Galvanometer to ammeter | 3 | 20-24,28 |
| 13 | Capacitors in series and parallel | 4 | 20-29,30 |
| 14 | Capacitors in series and parallel | 4 | 20-31,32 |
| 15 | Capacitors in series and parallel | 4 | 20-33,34 |
| 16 | Capacitors in series and parallel | 4 | 20-34,35 |
| 17 | Resistors and capacitors in dc circuits | 5 | 20-36,40 |
| 18 | Resistors and capacitors in dc circuits | 5 | 20-37,41 |
| 19 | Resistors and capacitors in dc circuits | 5 | 20-36,38 |
| 20 | Resistors and capacitors in dc circuits | 5 | 20-37,39 |

# 21 Magnetic Phenomena

## RECALL FROM PREVIOUS CHAPTERS

| Previously learned concepts and equations frequently used in this chapter | Text Section | Study Guide Page |
|---|---|---|
| Analytical treatment of vectors (components and addition) | 3.2,3,4 | 32 |
| Uniform circular motion: $F_c = mv^2/r$ | 5.5 | 73 |

## NEW IDEAS IN THIS CHAPTER

| Concepts and equations introducted | Text Section | Study Guide Page |
|---|---|---|
| Magnetic force on a moving charge: $F = qvB \sin\theta$ | 21.2 | 344 |
| Magnetic field due to current-carrying wires: | 21.3 | 350 |
| Long straight wire: $B = \mu_0 I/2\pi a$ | | |
| Flat circular coil: $B = \mu_0 NI/2r$ | | |
| Long solenoid: $B = \mu_0 NI/L$ | | |
| Toroid: $B = \mu_0 NI/2\pi r$ | | |
| Magnetic forces on and between currents: $F = I\ell B \sin\theta$ and $F = \mu_0 I_1 I_2 \ell/2\pi a$ | 21.4 | 353 |
| Motional emf: $\varepsilon = B\ell v_\perp$ | 21.6 | 358 |
| Magnetic flux: $\Phi = BA \cos\theta$ | 21.6 | 358 |
| Induced emf: $\varepsilon = -\Delta\Phi/\Delta t$ | 21.6 | 358 |

## PRINCIPAL CONCEPTS AND EQUATIONS

### 1. Magnetic Force On Moving Charges (Section 21.2)

Review: A charged particle moving through a magnetic field experiences a force described by

$$F = qvB \sin\theta$$

where:

$F$ is the magnitude of the force experienced by the charged particle
$q$ is the charge on the particle
$v$ is the speed of the particle
$B$ is the magnitude of the magnetic field strength measured in Tesla (T)
$\theta$ is the angle between the direction of $\vec{v}$ and $\vec{B}$

The direction of the force is determined by the right-hand rule. Using your right hand, point your fingers in the direction of $\vec{v}$. Now orient your hand so that you can cross your fingers from the direction of $\vec{v}$ into the direction of $\vec{B}$. Your extended thumb indicates the direction of $\vec{F}$ for a positively charged particle. If the particle is negative, proceed as if it is positive and then just reverse the direction of the force. Figure 21.1 shows several applications of the right-hand rule. In Fig. 21.1, $q = 1.60 \times 10^{-19}$ C, $v = 1.00 \times 10^3$ m/s, and $B = 1.00$ T.

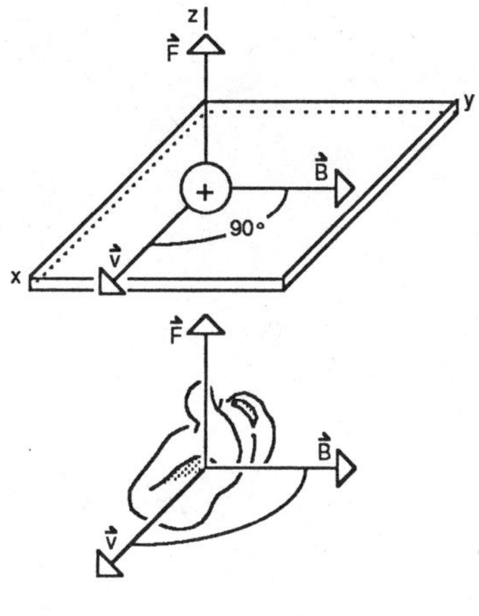


(a) Cross fingers from $\vec{v}$ to $\vec{B}$
$F = qvB \sin 90° = 1.60 \times 10^{-16}$ N

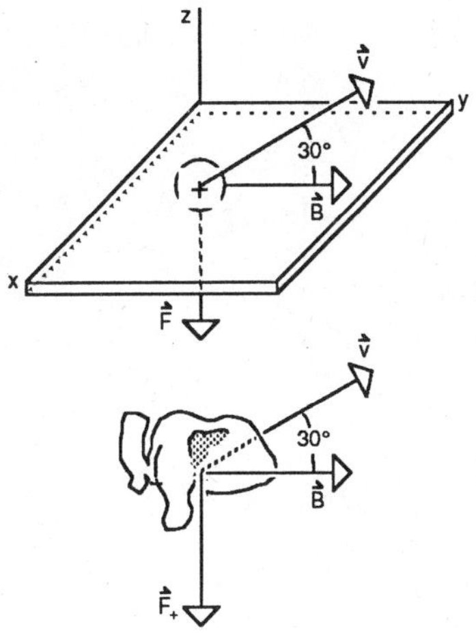


(b) Cross fingers from $\vec{v}$ to $\vec{B}$
$F = qvB \sin 30° = 8.00 \times 10^{-15}$ N

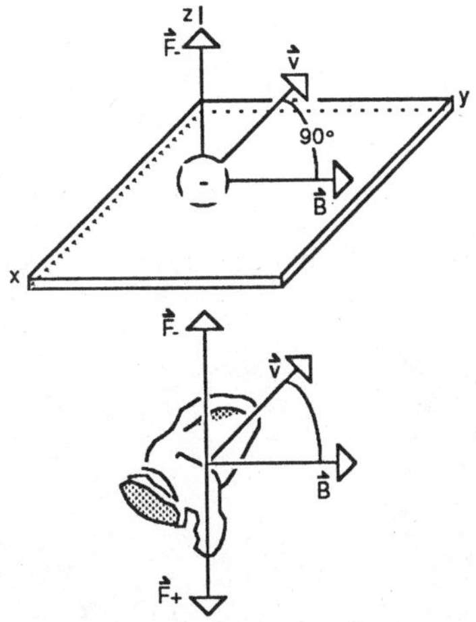


(c) Cross fingers from $\vec{v}$ to $\vec{B}$
$F = qvB \sin 90° = 1.60 \times 10^{-16}$ N

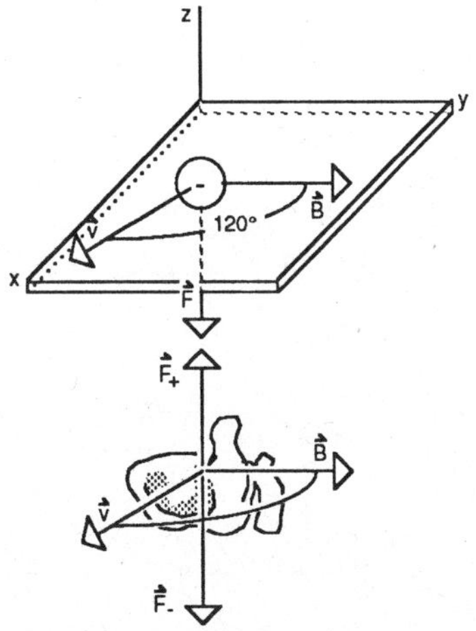


(d) Cross fingers from $\vec{v}$ to $\vec{B}$
$F = qvB \sin 120° = 1.39 \times 10^{-16}$ N

Figure 21.1

If $\vec{v}$ is parallel or anti parallel to $\vec{B}$, no magnetic force exists ($\sin 0° = \sin 180° = 0$) and the charged particle moves straight through the $\vec{B}$ field. If $\vec{v}$

is perpendicular to $\vec{B}$, the magnetic force is at right angles to both $\vec{v}$ and $\vec{B}$. You should recall from section 5.5 of the text that when this is the case, the particle travels in a circle, with the magnetic force supplying the central force. If $\vec{v}$ has components parallel and perpendicular to $\vec{B}$, the perpendicular components causes the particle to have circular motion in the field, the parallel component causes the particle to move through the field, and the result is a helical trajectory. Figure 21.2 shows a particle of charge +q entering a region where a magnetic field $\vec{B}$ is directed up out of the page. By applying the right-hand rule we see that the magnetic force supplies a central force causing the particle to move in a circle.

Figure 21.2

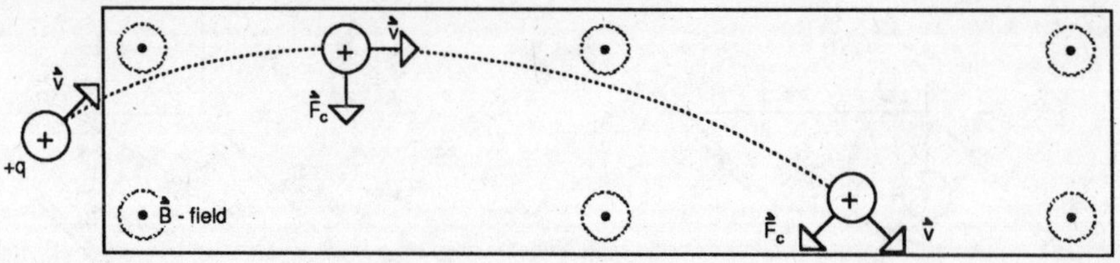


Since a particle of mass m and speed v is moving in a circle of radius r, it must be experiencing a central force of magnitude

$$F_c = mv^2/r$$

For this case, the central force is supplied by the magnetic force. Hence we can write

$$F_c = F_B = qvB \sin 90° = qvB$$

Combining these two expressions for $F_c$, we have

$$mv^2/r = qvB \quad \text{or} \quad mv = qBr$$

Note: This last expression involves five physical quantities (m, v, q, B, and r). Consequently we can determine any one of these five quantities given the other four.

Practice: Figure 21.3 shows several charged particles moving in a magnetic field.

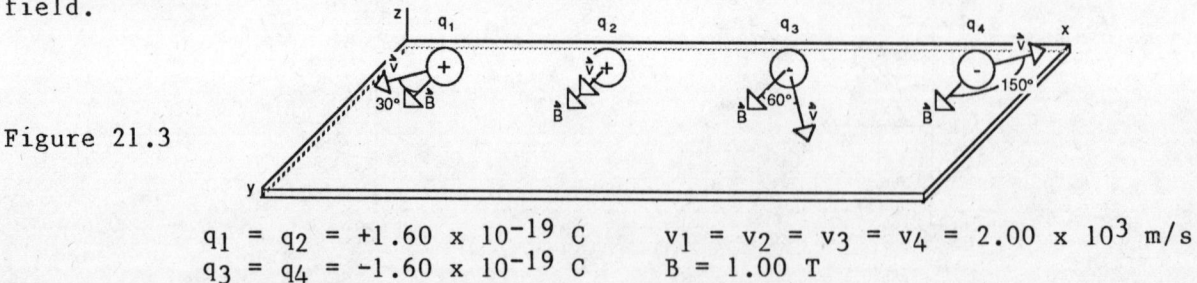

Figure 21.3

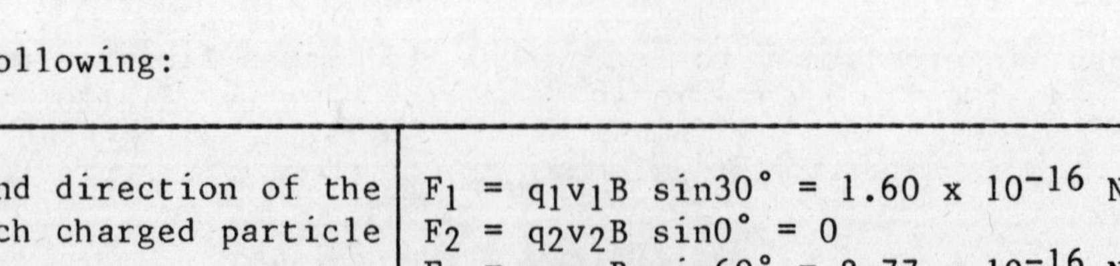


$q_1 = q_2 = +1.60 \times 10^{-19}$ C $\quad v_1 = v_2 = v_3 = v_4 = 2.00 \times 10^3$ m/s

$q_3 = q_4 = -1.60 \times 10^{-19}$ C $\quad B = 1.00$ T

Determine the following:

| | |
|---|---|
| 1. Magnitude and direction of the force on each charged particle | $F_1 = q_1v_1B \sin 30° = 1.60 \times 10^{-16}$ N (+Z)<br>$F_2 = q_2v_2B \sin 0° = 0$<br>$F_3 = q_3v_3B \sin 60° = 2.77 \times 10^{-16}$ N (+Z)<br>$F_4 = q_4v_4B \sin 120° = 2.77 \times 10^{-16}$ N (+Z) |

Figure 21.4 shows several charged particles traveling in a region where a magnetic field is directed vertically into the page.

Figure 21.4 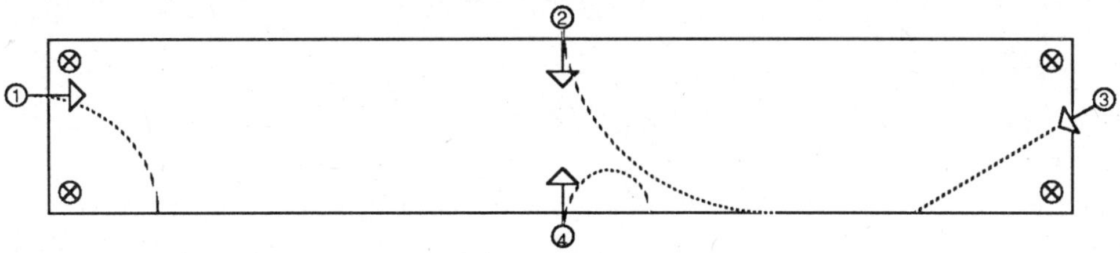

Determine the following:

| | |
|---|---|
| 2. The particle(s) with no charge | Particle 3 is not deflected, consequently it is experiencing no magnetic force. Since v, B, and $\theta$ are not zero, q must be zero |
| 3. The particle(s) with a negative charge | Particles 1 and 4 (using the right-hand rule) |
| 4. The particle(s) with a positive charge | Particle 2 (using the right-hand rule) |
| 5. If particles 1 and 2 have the same mass and magnitude of charge, which one has the greater speed? | $mv = qBr$ or $r = mv/qB$<br>Since q, B, and m are the same for each particle, a greater v results in a greater r. Subsequently, particle 2 must have the greater speed. |
| 6. If particles 2 and 4 have the same speed and mass, which one has the greater charge? | $mv = qBr$ or $r = mv/qB$<br>Since m, v, and B are the same for each particle, a greater q results in a smaller r. Subsequently, particle 4 must have the greater charge. |

Example 21.1: Figure 21.5 shows a cylindrical tube aligned with a magnetic field of 0.200 T. A proton is injected into the cylinder with a velocity of $1.00 \times 10^6$ m/s directed 30° above the direction of the field. As shown, the particle will travel in a spiral trajectory. Determine the following:

a) central force experienced by the particle,
b) radius of the spiral trajectory of the particle,
c) time for the particle to make one complete revolution, and
d) length of the tube in order for the particle to make three complete revolutions

Figure 21.5

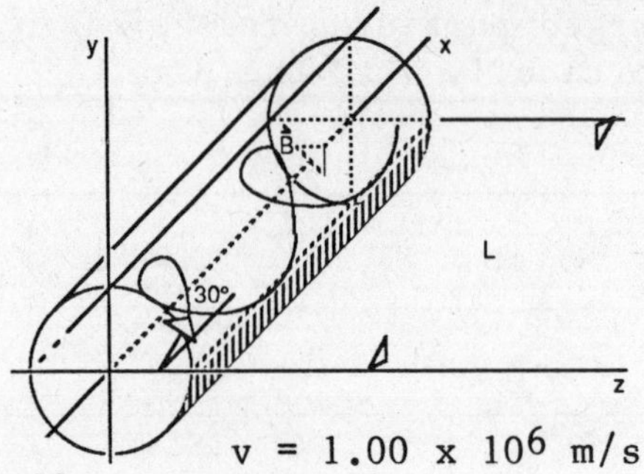


Given: B = 0.200 T    $v = 1.00 \times 10^6$ m/s    $m = 1.67 \times 10^{-27}$ kg
$q = 1.60 \times 10^{-19}$ C    $\theta = 30°$

Determine:
(a) $F_c$ -- central force experienced by the particle
(b) r -- radius of the spiral trajectory
(c) T -- time for the particle to make one complete revolution
(d) L -- length of the tube in order for the particle to make three revolutions

Strategy:
(a) Knowing q, v, θ, and B, we can determine the magnetic force and hence the central force $F_c$.
(b) Knowing $F_c$, m, v and θ, we can determine r.
(c) Knowing r, v, and θ, we can determine T.
(d) Knowing T, v, and θ, we can determine L.

Solution: First, let's determine the component of the velocity parallel and perpendicular to the magnetic field.

$v_\parallel = v \cos\theta = 8.66 \times 10^5$ -- this component moves the particle down the tube
$v_\perp = v \sin\theta = 5.00 \times 10^5$ -- this component causes the particle to experience a central force

(a) The central force is supplied by the magnetic force and can be determined by

$$F_c = F_B = qv_\perp B = 1.60 \times 10^{-14}\ \text{N}$$

(b) Now that the central force is known, the radius of the spiral trajectory can be determined by

$$F_c = mv_\perp^2/r \text{ or } r = mv_\perp^2/F_c = 2.61 \times 10^{-2}\ \text{m}$$

(c) Using the radius of curvature, the time for one revolution can be determined by

$$T = 2\pi r/v_\perp = 3.28 \times 10^{-7}\ \text{s}$$

(d) Finally, we can determine the length of tube needed for the particle to undergo three revolutions.

Speed of the particle down the tube is $v_\parallel = v \cos 30° = 8.66 \times 10^5$ m/s
Time for three revolutions is $t = 3T = 9.84 \times 10^{-7}$ s
Length of tube needed is $L = tv_\parallel = 0.852$ m

Example 21.1. Singly charged positive ions traveling at various speeds are introduced into the experimental apparatus shown in Fig. 21.6. The ions are first accelerated by the electric field established by placing a potential difference $V_1$ across the vertical parallel plates. Next, they pass through the velocity selector. Finally, they enter a region where a magnetic field of 0.200 T exists.

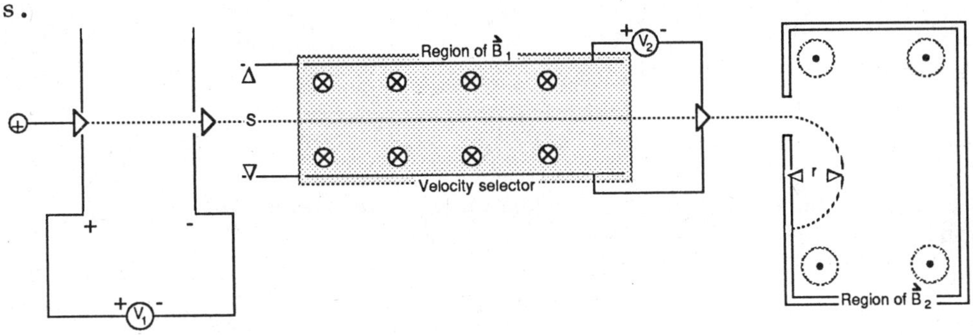


Figure 21.6

$m = 5.80 \times 10^{-26}$ kg -- mass of the ions
$q = 1.60 \times 10^{-19}$ C -- charge on the ions
$V_1 = 20.0$ V -- potential difference across the vertical parallel plates
$V_2 = 750$ V -- potential difference across the horizontal parallel plates in the velocity selector
$S = 5.00 \times 10^{-2}$ m -- separation of the plates in the velocity selector
$B_1 = 0.500$ T -- magnetic field in the velocity selector
$B_2 = 0.200$ T -- final magnetic field experienced by the ions

Calculate (a) The minimum speed of the ions as they enter the velocity selector, (b) the speed of the ions as they exit the velocity selector, and (c) the radius of curvature of the ions in the final magnetic field $\vec{B}_2$.

Given: $q$, $m$, $V_1$, $V_2$, $S$, $B_1$, and $B_2$

Determine:

(a) $v_{min}$ -- the minimum speed of ions entering the velocity selector.
(b) $v$ -- the speed of ions exiting the velocity selector
(c) $r$ -- the radius of curvature of the ions in the magnetic field $\vec{B}_2$

Strategy: Knowing the charge on the ions and the potential difference $V_1$, we can determine the change in electric potential energy of the ions as they travel through the vertical parallel plates. Knowing this change in electric potential energy, we can determine the change in kinetic energy and hence the minimum speed of ions entering the velocity selector. Knowing $V_2$ and $S$, we can determine the $\vec{E}$-field in the velocity selector. We can then use the magnitude of the $\vec{E}$-field and $\vec{B}$-field in the velocity selector to determine the exit speed of the ions. We can calculate the radius of curvature in the final $\vec{B}$-field from $B_2$, $v$, $q$, and $m$.

Solution: Since no dissipative forces are involved, total energy is conserved ($\Delta KE + \Delta PE = 0$). Since energy is conserved, the increase in kinetic energy of the ions traveling through $V_1$ is equal to the negative of the decrease in electric potential energy.

$$\Delta KE = -\Delta PE = qV_1 = 3.20 \times 10^{-18} \text{ J}$$

An ion enters the velocity selector with the minimum speed if it enters the region between the vertical parallel plates with zero KE. If this is the case, its speed can be obtained as follows:

$$\Delta KE = KE_f - KE_i = (mv^2_{min}/2 - 0) \text{ or } v_{min} = (2\Delta KE/m)^{1/2} = 1.05 \times 10^4 \text{ m/s}$$

The speed of ions exiting the velocity selector can be obtained as follows:

$$v = E/B_1 = (V_2/S)/B_1 = 3.00 \times 10^4 \text{ m/s}$$

The radius of curvature of the ions in the magnetic field $B_2$ can be obtained by combining $F_c = mv^2/r$ and $F_B = qvB_2$ to obtain

$$r = mv/qB_2 = 5.44 \times 10^{-2} \text{ m}$$

Related Text Problems: 21-1 through 21-13.

## 2. Magnetic Field Due to a Current Carrying Wire (Section 21.3)

Review: A current-carrying wire creates a magnetic field. Figure 21.7 shows several different current-carrying wires, the resulting magnetic field, and expressions for the magnitude of the $\vec{B}$-field.

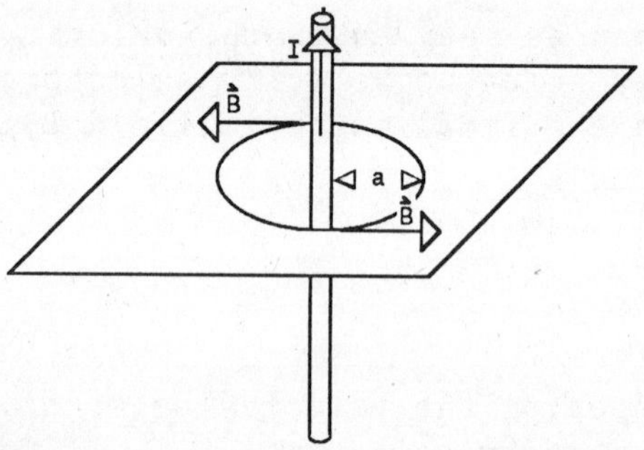


(a) $\vec{B}$-field at distance a from long straight wire
$B = \mu_o I/2\pi a$

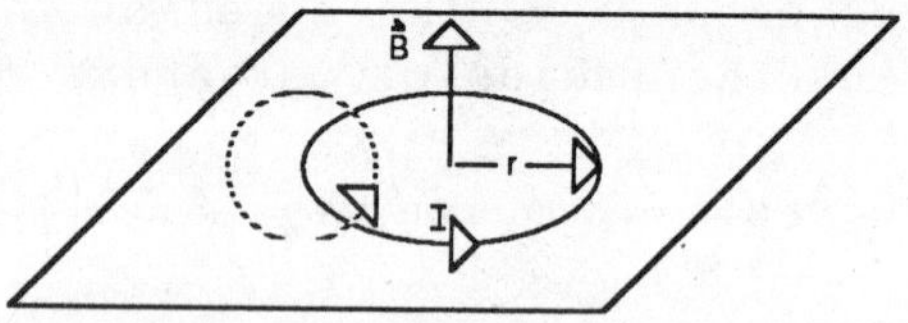


(b) $\vec{B}$-field at the center of a flat circular coil of N turns
$B = \mu_o NI/2r$

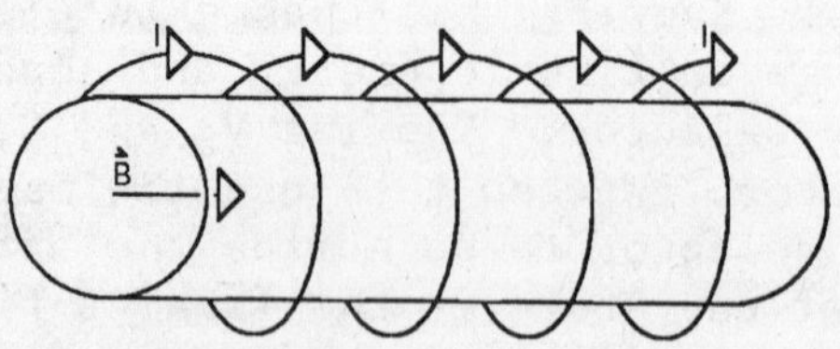


(c) $\vec{B}$-field inside a long straight solenoid
$B = \mu_o NI/L = \mu_o nI$

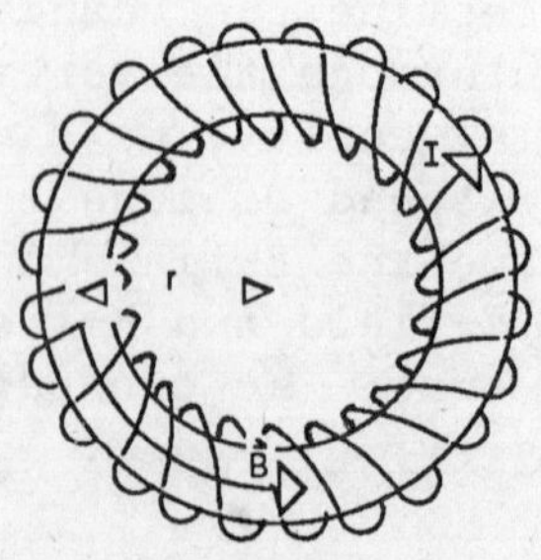


(d) $\vec{B}$-field inside a toroid
$B = \mu_o NI/2\pi r$

Figure 21.7

The direction of the magnetic field for each case shown in Fig. 21.7 is determined as follows:

(a) Grasp the current-carrying wire with your right hand in such a manner that your thumb is in the direction of the current. The fingers of your right hand curl around the wire in the direction of the $\vec{B}$-field.

(b) You can apply the preceding right-hand rule to the coil (see the small field circles) to determine that the $\vec{B}$-field at the center is upward. You can also grasp the coil in your right hand in such a manner that the fingers of your right hand curl around the coil in the direction of the current. Your thumb then indicates the direction of the $\vec{B}$-field.

(c) and (d) Apply either the right rule introduced in (a) or the right hand rule introduced in (b).

Practice: Consider the situation shown in Fig. 21.8.

Figure 21.8.

$v = 1.00 \times 10^5$ m/s
$q = 2.00 \times 10^{-10}$ C
$d = 2.00 \times 10^{-2}$ m
$I = 1.00$ A
+ = up out of page
− = down into page

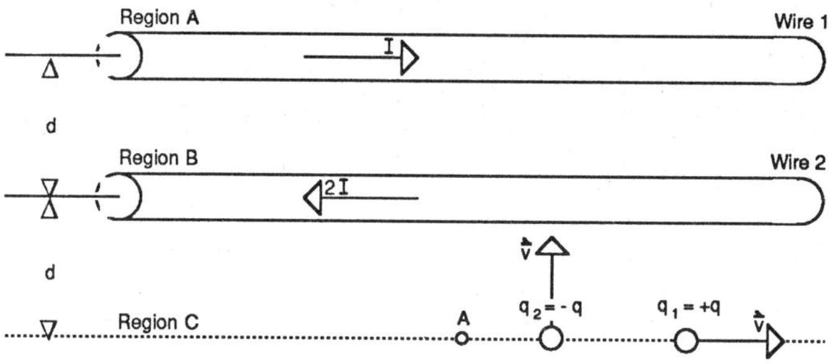


Determine the following:

| | |
|---|---|
| 1. Magnitude and direction of the magnetic field at A | The field at A due to wire 1 is $\vec{B}_1 = -\mu_0 I/2\pi(2d) = -5.00 \times 10^{-6}$ T The field at A due to wire 2 is $\vec{B}_2 = +\mu_0(2I)/2\pi d = +2.00 \times 10^{-5}$ T The net field at A is $\vec{B} = \vec{B}_1 + \vec{B}_2 = +1.50 \times 10^{-5}$ T The + sign indicates the direction |
| 2. Magnitude and direction of the magnetic field midway between the wires | $\vec{B}_1 = -\mu_0 I/2\pi(d/2) = -2.00 \times 10^{-5}$ T $\vec{B}_2 = -\mu_0(2I)/2\pi(d/2) = -4.00 \times 10^{-5}$ T The net field is $\vec{B} = \vec{B}_1 + \vec{B}_2 = -6.00 \times 10^{-5}$ T |
| 3. Magnitude and direction of the magnetic field experienced by the current in wire 1 | The magnetic field experienced by the current in wire 1 is due to the current in wire 2. $\vec{B}_2 = -\mu_0(2I)/2\pi d = -2.00 \times 10^{-5}$ T |

| | |
|---|---|
| 4. The distance from wire 1 in region A where the magnetic field is zero | In region A the $\vec{B}$-field due to wire 1 is up (+) and the $\vec{B}$-field due to wire 2 is down (-). Let x equal the distance from wire 1, then $\vec{B}_1 + \vec{B}_2 = 0$<br>$+\mu_o I/2\pi x - \mu_o(2I)/(2\pi)(d+x) = 0$ or<br>$x = d = 2.00 \times 10^{-2}$ m |
| 5. What should be done to the current in wire 1 to get zero magnetic field at A? | $\vec{B}_1 = -\mu_o I/2\pi(2d) = -\mu_o I/4\pi d$<br>$\vec{B}_2 = +\mu_o(2I)/2\pi d = +\mu_o I/\pi d$<br>Note that $B_2$ is 4 times $B_1$. Consequently, we need to increase the current in wire 1 by a factor of 4 in order to get B=0 at A. |
| 6. What should be done to the current in wire 1 to get zero magnetic field midway between the two wires? | $\vec{B}_1 = -\mu_o I/2\pi(d/2) = -\mu_o I/\pi d$<br>$\vec{B}_2 = -\mu_o(2I)/2\pi(d/2) = -2\mu_o I/\pi d$<br>We need to reverse the direction and double the magnitude of the current in wire 1 in order to get B=0 midway between the wires |
| 7. Magnitude and direction of the force on $q_1$ | $F = q_1 vB \sin 90.0°$<br>$\vec{B} = +1.50 \times 10^{-5}$ T (see step 1 of this practice)<br>$\vec{F} = 3.00 \times 10^{-10}$ N towards the bottom of the page |
| 8. Magnitude and direction of the force on $q_2$ | $F = qvB \sin 90.0°$<br>$\vec{F} = 3.00 \times 10^{-10}$ N towards the left of the page |

<u>Example 21.3</u>: Two flat circular coils are placed concentric to each other in the same plane in Fig. 21.9.

Figure 21.9

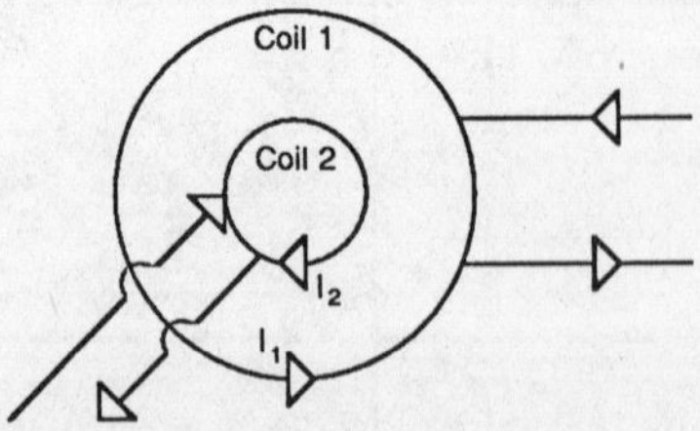


$r_1 = 2.00 \times 10^{-1}$ m, $N_1 = 5$ turns, $I_1 = 2.00$ A
$r_2 = 1.00 \times 10^{-1}$ m, $N_2 = 10$ turns, $I_2 = 4.00$ A

Determine the magnitude and direction of the magnetic field at the center of the coils.

Given: $r_1$ and $r_2$ -- radius of each flat circular coil
$N_1$ and $N_2$ -- number of turns in each coil
$I_1$ and $I_2$ -- current in each coil

Determine: Magnitude and direction of the net $\vec{B}$-field at the center of the coil.

Strategy: From the given information, we can determine the magnitude and direction of the $\vec{B}$-field at the center of the coils due to the current in each coil. We can add these results vectorially to obtain the net $\vec{B}$-field.

Solution: The magnetic field at the center due to 1 turn of wire from each coil is determined as follows:

$$\vec{B}_1 = +\mu_o I_1/2r_1 = +6.28 \times 10^{-6} \text{ T}$$

$$\vec{B}_2 = -\mu_o I_2/2r_2 = -25.1 \times 10^{-6} 0\text{T}$$

The magnetic field at the center due to the number of turns of wire (N) is

$$\vec{B}_{1T} = +N_1\vec{B}_1 = +31.4 \times 10^{-6} \text{ T and } \vec{B}_{2T} = N_2\vec{B}_2 = -251 \times 10^{-6} \text{ T}$$

The net magnetic field at the center of the coil is

$$\vec{B}_{Net} = \vec{B}_{1T} + \vec{B}_{2T} = -2.20 \times 10^{-4} \text{ T}$$

The minus sign indicates that the net field is directed downward.

Related Text Problems: 21-14 through 21-22.

## 3. Magnetic Forces On And Between Currents (Section 21.4)

Review: When a current-carrying wire is placed in a magnetic field, it experiences a force described by:

$$F = I\ell B \sin\theta$$

where:
- I -- represents the current in the wire
- B -- represents the magnetic field strength
- $\ell$ -- represents the length of the current-carrying wire that is actually in the magnetic field
- $\theta$ -- represents the angle between the direction of the current and the direction of the $\vec{B}$-field
- F -- represents the magnetic force on the current-carrying wire due to the $\vec{B}$-field

The direction of the force is obtained by using the following right-hand rule. Point the fingers of your right hand in the direction of the current and orient your hand so that your fingers can be crossed from the direction of I into the direction of $\vec{B}$. Your extended thumb then indicates the direction of the force.

Note: This right-hand rule for current-carrying wires is essentially the same as the right-hand rule for moving positively charged particles. This is not surprising when we recall that our concept of conventional current is based on the movement of positively charged particles.

Figure 21.10 shows three different cases of current-carrying wires in a magnetic field, the direction of the resulting force, how the direction of the force was obtained using the right-hand rule, and the magnitude of the magnetic force. For each case, B = 1.00 T and I = 2.00 A

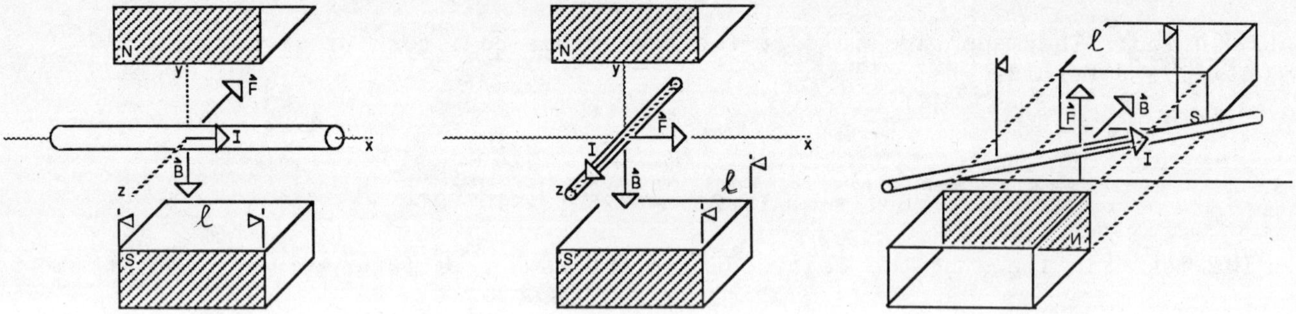


| (a) | (b) | (c) |
|---|---|---|
| $\theta = 90.0°$ | $\theta = 90.0°$ | $\theta = 30.0°$ |
| $\ell = 10.0$ cm | $\ell = 20.0$ cm | $\ell = 28.3$ cm |

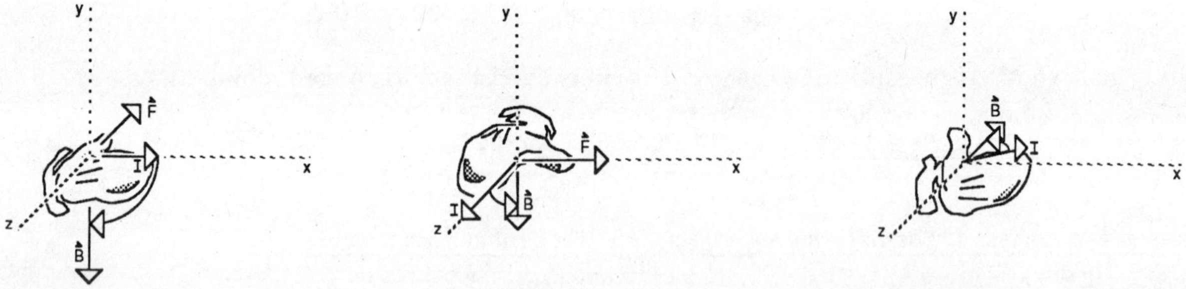


| (a) | (b) | (c) |
|---|---|---|
| Cross fingers from I to B | Cross fingers from I to B | Cross fingers from I to B |
| $F = BI\ell \sin 90.0°$ | $F = BI\ell \sin 90.0°$ | $F = BI\ell \sin 30.0°$ |
| $F = 0.200$ N | $F = 0.400$ N | $F = 0.283$ N |
| -Z direction | +X direction | +Y direction |

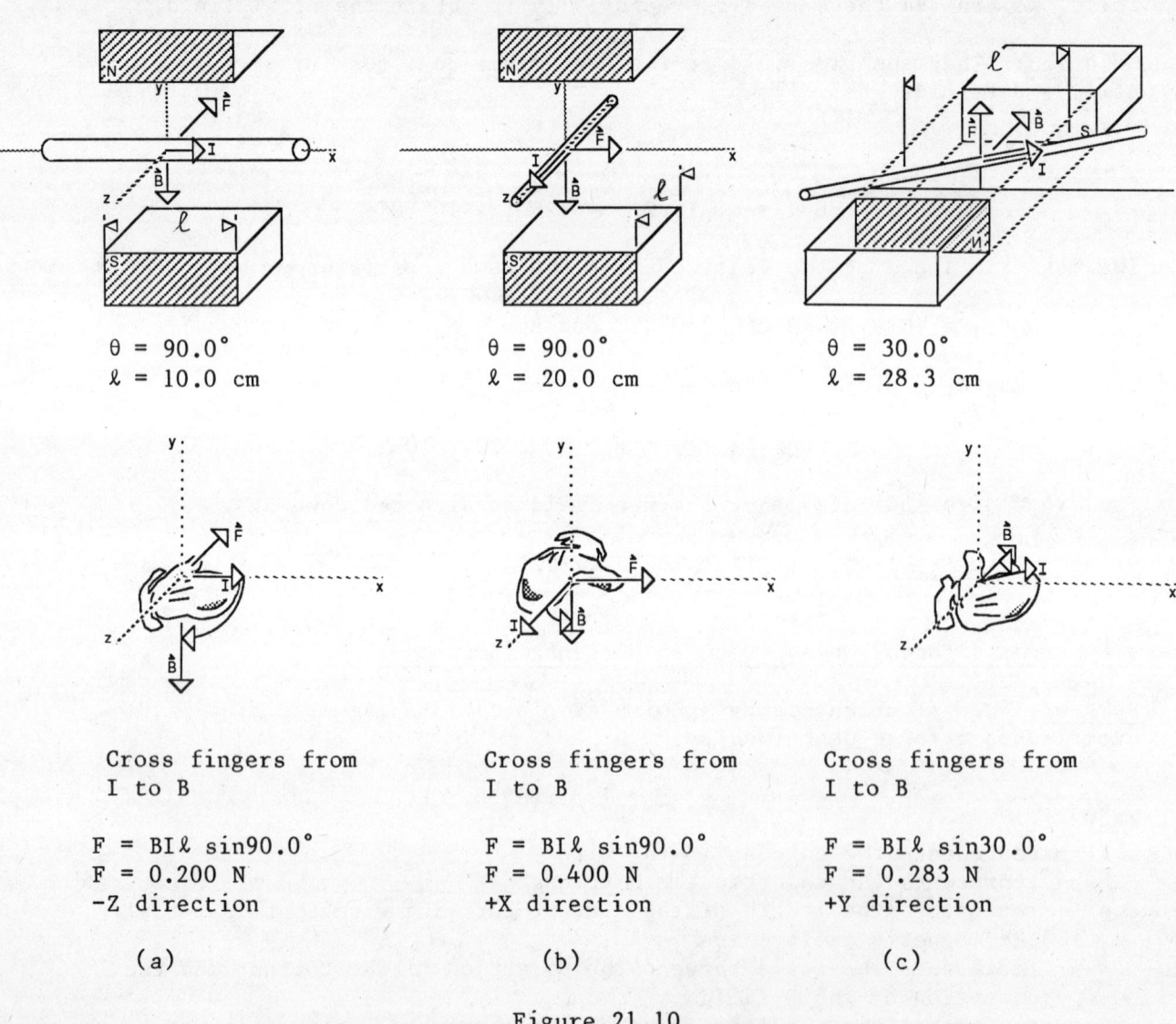

Figure 21.10

Figure 21.11 shows segments of two long parallel wires separated by distance a and carrying currents $I_1$ and $I_2$, respectively. Each wire creates a magnetic field that exerts a magnetic force on the current in the other wire.

For both cases:

$B_1 = \mu_0 I_1/2\pi a$
$B_2 = \mu_0 I_2/2\pi a$
$F_1 = B_2 I_1 \ell \sin 90.0°$
$= \mu_0 I_1 I_2 \ell/2\pi a$
$F_2 = B_1 I_2 \ell \sin 90.0°$
$= \mu_0 I_1 I_2 \ell/2\pi a$
$F/\ell = \mu_0 I_1 I_2/2\pi a$

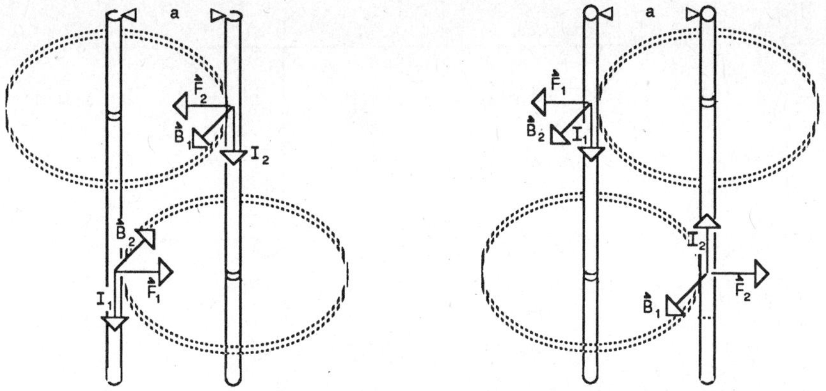


(a) Parallel currents (b) Antiparallel currents

Figure 21.11

Practice: Figure 21.12 shows three current-carrying wires mounted so that they cannot move.

Figure 21.12.
$I_1 = 1.00$ A
$I_2 = 2.00$ A
$I_3 = 3.00$ A
$d = 0.100$ m

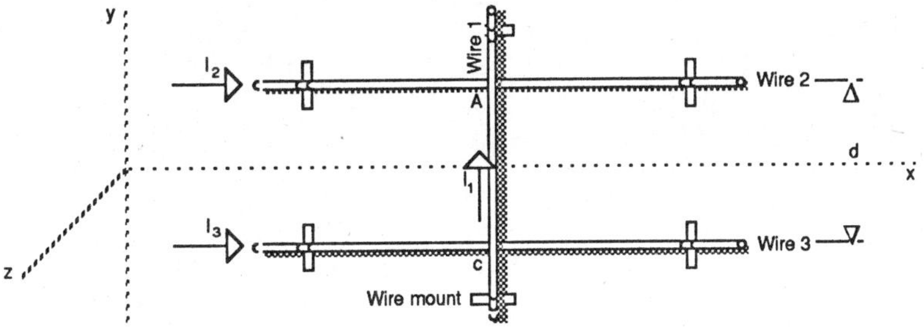


Determine the following:

| | |
|---|---|
| 1. Magnitude and direction of the $\vec{B}$-field experienced by wire 2 due to the current in wire 3 ($\vec{B}_3$) | $B_3 = \mu_0 I_3/2\pi d = 6.00 \times 10^{-6}$ T<br>$\vec{B}_3$ is in the -Z direction at wire 2. |
| 2. Magnetic force per unit length felt by wire 2 due to wire 3 ($\vec{F}_2/\ell$) | $F_2 = B_3 I_2 \ell \sin 90.0°$<br>$F_2/\ell = B_3 I_2 = 12.0 \times 10^{-6}$ N/m<br>The force experienced by wire 2 ($\vec{F}_2$) is in the +Y direction. |
| 3. Magnitude and direction of the $\vec{B}$-field experienced by wire 3 due to the current in wire 2 ($\vec{B}_2$) | $B_2 = \mu_0 I_2/2\pi d = 4.00 \times 10^{-6}$ T<br>$\vec{B}_2$ is in the -Z direction at wire 3. |
| 4. Magnetic force per unit length felt by wire 3 due to wire 2 ($\vec{F}_3/\ell$) | $F_3 = B_2 I_3 \ell \sin 90.0°$<br>$F_3/\ell = B_2 I_3 = 12.0 \times 10^{-6}$ N/m<br>The force experienced by wire 3 ($\vec{F}_3$) is in the -Y direction. |

| | |
|---|---|
| 5. If the current in wire 1 is cut off and the mounts holding wire 2 are released, what will happen to wire 2? | It will be repelled upward (+Y direction) from wire 3 |
| 6. If the current in wire 3 is cut off and the mounts holding wire 2 are released, what will happen to wire 2? | 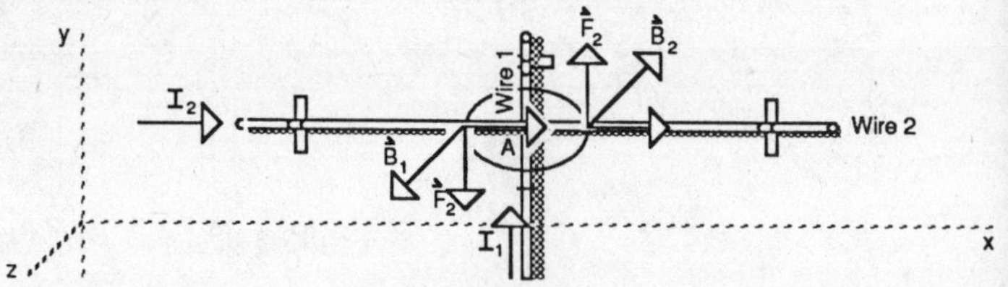 <br>Wire 2 will initially rotate counterclockwise in the plane of the paper about point A |
| 7. If the current in wire 2 is cut off and the mounts holding wire 1 are released, what will happen to wire 1? | 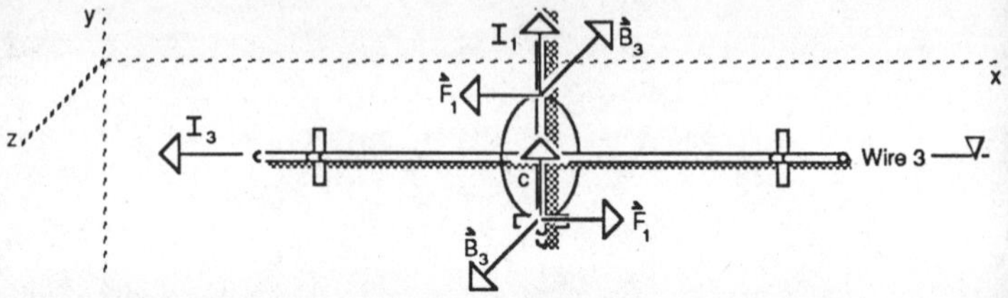 <br>Wire 1 will initially rotate counterclockwise in the plane of the paper about point C |
| 8. What must be done to the current in wire 2 in order for wires 2 and 3 to attract each other with a force per unit length of $2.40 \times 10^{-5}$ N? | $F/\ell = \mu_0 I_2 I_3 / 2\pi d$<br>$I_2 = (F/\ell) 2\pi d / \mu_0 I_3 = 4.00$ A<br>The current in wire 2 must double in magnitude and reverse in direction. |

<u>Example 21.4</u>: A long straight wire has a current of 10.0 A. A square coil 20.0 cm on a side, having a resistance of 5.00 Ω and containing a 10.0-V battery, is positioned 10.0 cm from the wire, as shown in Fig. 21.13. Calculate the net force experienced by the square coil.

Figure 21.13

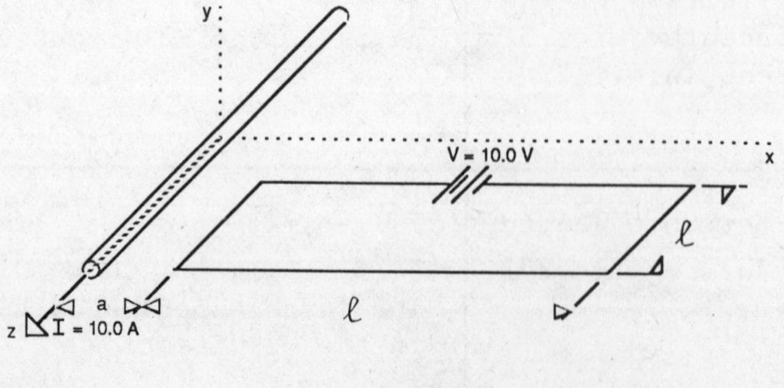

Given: $I = 10.0$ A -- current in the long straight wire
$V = 10.0$ V -- terminal potential difference of the battery
$R = 5.00\ \Omega$ -- resistance of the coil
$b = 2.00 \times 10^{-1}$ m -- length of the sides of the square coil
$a = 1.00 \times 10^{-1}$ m -- distance of the coil from the wire

Determine: The net force on the square coil

Strategy: We can first establish the fact that we only need to be concerned with those segments of the square coil that are parallel to the long straight wire. Next, we can determine the magnitude and direction of the magnetic field experienced by the parallel segments of the coil. Knowing V and R, we can determine the current in the coil. Finally, we can combine the information about the magnetic field and the current to find the force on each segment and hence the net force on the coil.

Solution: First let's investigate the effect of the segments of the coil that are perpendicular to the long straight wire.

Figure 21.14

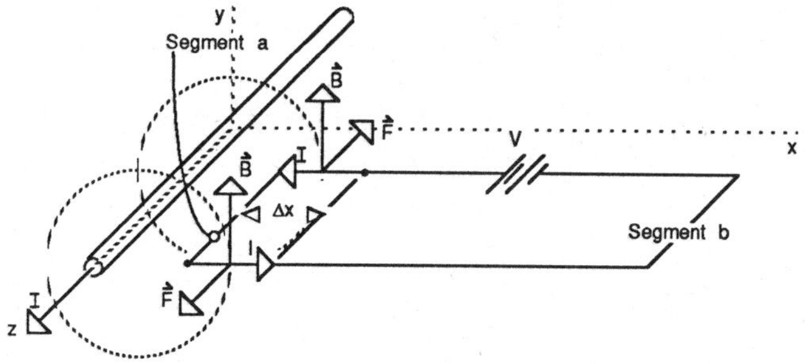


Let's look at the tiny segments of length $\Delta x$, shown in Fig. 21.14. These two segments are an equal distance from the long straight wire, hence they experience the same $\vec{B}$-field. Since these segments have the same current and experience the same $\vec{B}$-field, they will feel magnetic forces equal in magnitude and opposite in direction. If we continue in this manner along the two perpendicular segments, we see that their net effect is zero (i.e., they cancel each other). For convenience, let's refer to the parallel segments as segment a and segment b. The magnetic field experienced by these segments is given by

$$\vec{B}_a = \mu_0 I/2\pi a = 2.00 \times 10^{-5}\ \text{T} \qquad +Y \text{ direction}$$

$$\vec{B}_b = \mu_0 I/2\pi(a+b) = 6.67 \times 10^{-6} \qquad +Y \text{ direction}$$

The current in the coil is $I_c = V/R = 2.00$ A

The magnetic force experienced by each of these segments is

$$\vec{F}_a = -B_a I_c b \sin 90.0° = -8.00 \times 10^{-6}\ \text{N} \qquad (-X \text{ direction})$$

$$\vec{F}_b = +B_b I_c b \sin 90.0° = +2.67 \times 10^{-6}\ \text{N} \qquad (+X \text{ direction})$$

$$\vec{F}_{Net} = \vec{F}_a + \vec{F}_b = -5.33 \times 10^{-6}\ \text{N} \qquad (-X \text{ direction})$$

Related Text Problems: 21-23 through 21-29.

## 4. Induced Voltages (Section 21.6)

Review: Figure 21.15 shows a conductor of length $\ell$ moving at a constant speed v through a magnetic field $\vec{B}$.

Figure 21.15

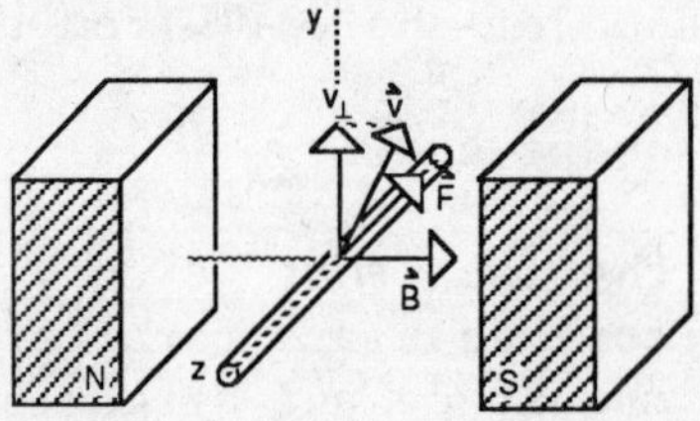


- v -- represents the velocity of the conductor
- $\ell$ -- represents the length of the conductor in the $\vec{B}$-field
- B -- represents the magnetic field
- $\theta$ -- represents the angle between the direction of $\vec{v}$ and $\vec{B}$
- F -- represents the force on positive charges in the conductor
- $v_\perp$ -- represents the component of $\vec{v}$ perpendicular to $\vec{B}$ ($v_\perp = v \sin\theta$)

The induced motional emf is given by

$$\varepsilon = B\ell v_\perp$$

As shown in Fig. 21.16, the magnetic flux over an area in space is the product of area times the component of the $\vec{B}$-field perpendicular to that area.

Figure 21.16

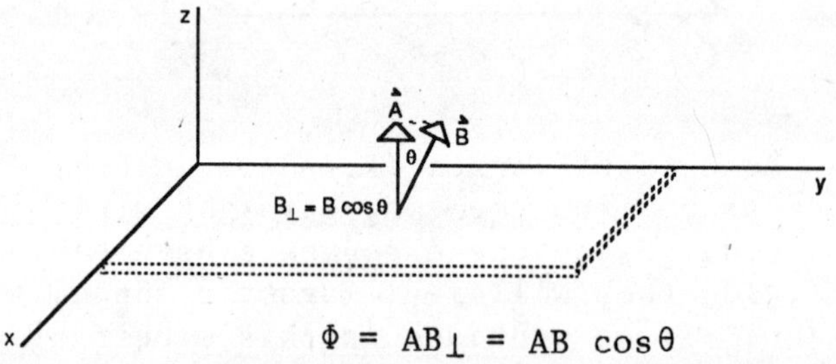


$$\Phi = AB_\perp = AB\cos\theta$$

If a circuit experiences a time rate of change of magnetic flux, an emf will be induced. Faraday's law expresses this mathematically as

$$\varepsilon = -\Delta\Phi/\Delta t$$

The minus sign is due to Lenz and reminds us that the induced current in the circuit must be in such a direction as to oppose the change that caused it.

Practice: Suppose that the moving conductor slides along the U-shaped conductor in a uniform $\vec{B}$-field, as in Fig. 21.17.

Figure 21.17

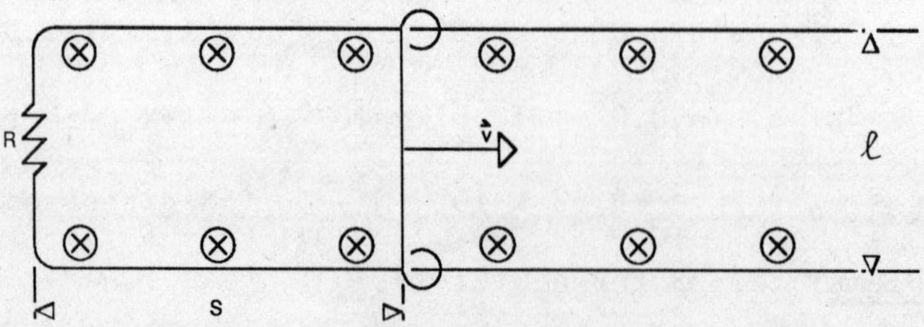

Determine expressions for the following:

| | |
|---|---|
| 1. Motional emf developed across the length of the sliding wire | $\varepsilon = B\ell v$ |
| 2. Magnetic flux through the circuit at any instant | $\Phi = BA = B\ell s$ |
| 3. Rate of change of magnetic flux in the circuit at any instant | $\frac{\Delta\Phi}{\Delta t} = B\,\frac{\Delta A}{\Delta t} = B\ell\,\frac{\Delta s}{\Delta t} = B\ell v$ |

Note: We have just demonstrated Faraday's law, which states that the induced emf is equal to the time rate of change of magnetic flux.

| | |
|---|---|
| 4. The magnitude of the induced current | $I = \varepsilon/R = B\ell v/R$ |
| 5. The direction of the induced current | The induced current must oppose what caused it. It is caused by an increasing flux in the circuit. If the current is counterclockwise, it produces a magnetic flux that opposes this increase. |
| 6. The magnitude of the force required to keep the sliding wire traveling at a constant speed | The magnetic force on the sliding wire is $F = I\ell B\sin 90.0° = (\frac{B\ell v}{R})\ell B = B^2\ell^2 v/R$ A force of this magnitude parallel to $\vec{v}$ is required. |

Figure 21.18 shows a rectangular loop (L long and w wide) of wire with a resistance R traveling at a constant speed as it (a) enters, (b) travels across, and (c) exits the uniform $\vec{B}$-field direction to the page.

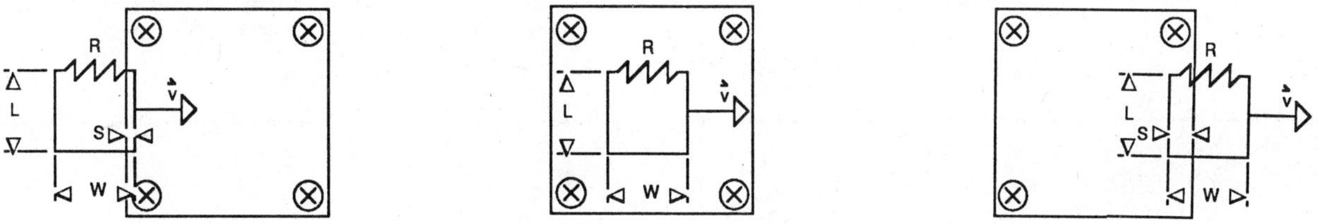


(a) entering (b) traveling across (c) exiting

Figure 21.18

Determine expressions for the following:

| | |
|---|---|
| 7. Magnetic flux through the coil at any instant while the coil is entering the $\vec{B}$-field | $\Phi = BLs$<br>Note that the value of s is changing |
| 8. Time rate of change of magnetic flux through the coil as it enters the $\vec{B}$-field | $\frac{\Delta\Phi}{\Delta t} = BL\frac{\Delta s}{\Delta t} = BLv$ |
| 9. Emf induced as the coil enters the $\vec{B}$-field | $\varepsilon = \frac{-\Delta\Phi}{\Delta t} = B\ell v$ |

Note: The minus sign on the emf has nothing to do with the magnitude of the emf. The minus sign is just Lenz's way of reminding us that the induced emf is in such a direction as to oppose the change that caused it.

| | |
|---|---|
| 10. Direction of the induced current as the coil enters the $\vec{B}$-field | Method A: The induced emf is the result of an increase in the downward magnetic flux in the coil. A counterclockwise current in the coil creates an upward magnetic flux, hence opposes the increase in downward magnetic flux.<br>Method B: A counterclockwise current establishes a magnetic force ($F = BIL \sin 90.0°$) on the leading segment of the coils and this force opposes the motion of the coil. |
| 11. Induced current as the coil enters the $\vec{B}$-field | $I = \varepsilon/R$ |
| 12. Force required to move the coil into the $\vec{B}$-field at a constant speed | The induced current causes a retarding force<br>$F = BIL = B(\varepsilon/R)L = B(BLv/R)L = B^2L^2v/R$<br>A force of this magnitude is required to move the coil into the $\vec{B}$-field at a constant speed. |
| 13. The magnetic flux through the coil as it travels across the $\vec{B}$-field | $\Phi = BLw$ |

| | |
|---|---|
| 14. The time rate of change of magnetic flux through the coil as it travels across the $\vec{B}$-field | $\frac{\Delta\Phi}{\Delta t} = 0$<br>Consequently, the induced emf is zero. |

Note: The magnetic flux through the coil is a constant as it travels across the magnetic field. Consequently, no emf is induced.

| | |
|---|---|
| 15. Induced emf as the coil leaves the $\vec{B}$-field | $\Phi = B\ell s$<br>$\Delta\Phi/\Delta t = B\ell\Delta s/\Delta t = B\ell v$<br>$\varepsilon = -\Delta\Phi/\Delta t = -B\ell v$ |
| 16. Direction of the induced current as the coil leaves the $\vec{B}$-field | Method A: The induced emf is the result of a decrease in the downward magnetic flux in the coil. A clockwise current in the coil creates a downward magnetic flux, hence opposes the decrease in downward magnetic flux.<br>Method B: A clockwise current establishes a magnetic force ($F = BI\ell \sin 90°$) on the trailing segment of the coil, and this force opposes the motion of the coil. |

Consider the rectangular coil (length a and width b) rotating with a constant angular speed $\omega$ in a uniform magnetic field B, as shown in Fig. 21.19. The coil is in the XY plane at t = 0.

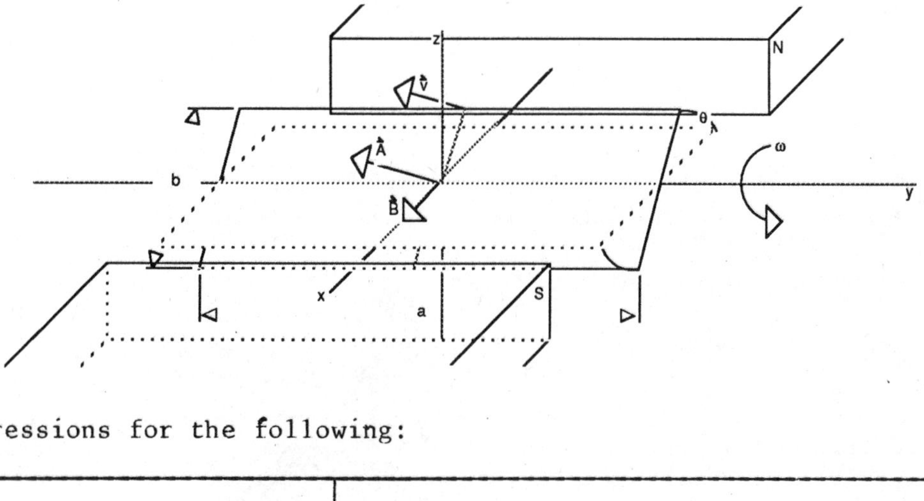


Figure 21.19

Determine expressions for the following:

| | |
|---|---|
| 17. Motional emf for the top wire | $\varepsilon = Bav_{\perp} = Bav \cos\theta$<br>$v = \omega b/2$, $\theta = \omega t$, $A = ab$<br>$\varepsilon = (BA\omega \cos\omega t)/2$ |

| | | |
|---|---|---|
| 18. | Motional emf for the bottom wire | $\varepsilon = (BA\omega \cos\omega t)/2$ |
| 19. | Motional emf for the end wires | Use the right-hand rule to determine that the end wires do not contribute to the motion of charged particles around the coil. |
| 20. | Motional emf for the entire coil | By summing the emf's from steps 17 and 18, we obtain $\varepsilon = BA\omega \cos\omega t$ |
| 21. | Maximum emf generated | $\varepsilon_{max} = BA\omega$ |
| 22. | Orientation of the coil for $\varepsilon=0$ and $\varepsilon=\varepsilon_{max}$ | $\varepsilon=0$ occurs when the coil is in the YZ plane. At the instant $v_\perp$ is zero and the flux through the coil is not changing.<br>$\varepsilon=\varepsilon_{max}$ occurs when the coil is in the XY plane. At this instant $v_\perp$ and $\Delta\Phi_t/\Delta t$ have maximum values. |
| 23. | A plot of the emf as a function of time | 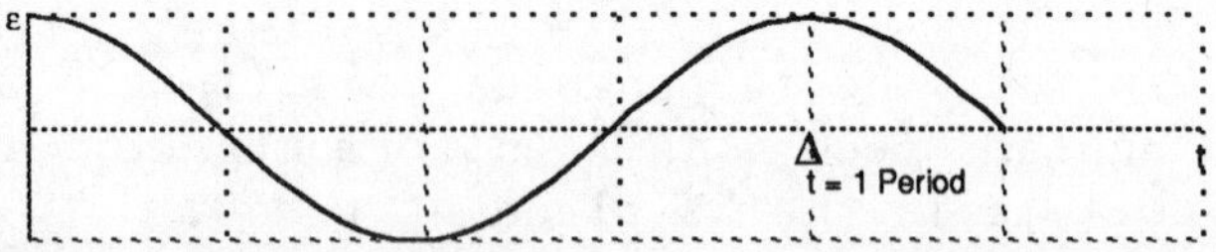 |

Example 21.5: The solenoid shown in Fig. 21.20 is 20.0 cm long, has a 2.00-cm radius, and 1000 turns of wire. A flat, 10-turn circular coil with a 0.500-cm radius is mounted inside the solenoid in a concentric manner and attached to a 10.0-Ω resistor. If the current in the solenoid is increasing at the rate of 10.0 A/s, determine the magnitude and direction of the induced current through the resistor.

Figure 21.20

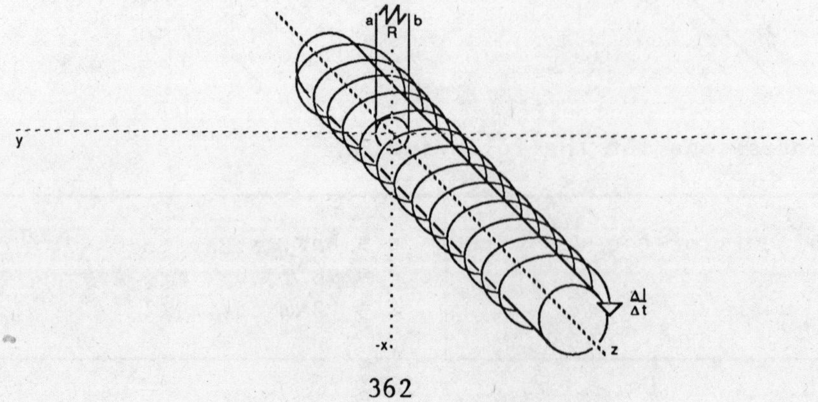

Given: $\ell$ = 20.0 cm -- length of the solenoid
$r_s$ = 2.00 cm -- radius of the solenoid
$N_s$ = 1000 -- turns of wire on the solenoid
$N_c$ = 10 -- turns of wire on the coil
$r_c$ = 0.500 cm -- radius of the coil
$\Delta I/\Delta t$ = 10.0 A/s -- rate of change of current in the solenoid
R = 10.0 Ω -- resistance of the resistor

Determine: Magnitude and direction of the induced current through the resistor.

Strategy: We can use our knowledge about the magnetic field of a solenoid and magnetic flux to develop an expression for the flux through the small coil as a function of the current in the solenoid. Using this expression and $\Delta I/\Delta t$, we can determine the induced emf in the coil. Knowing the induced emf and R, we can determine the current. We can use Lenz's law to determine the direction of the induced current.

Solution: The magnetic field inside the solenoid is given by

$$B = \mu_0 N_s I/\ell$$

The magnetic flux through the coil is

$$\Phi = BA_{coil} = (\mu_0 N_s I/\ell)\pi r_c^2$$

The time rate of change of the magnetic flux through the coil, and hence the induced emf, is

$$\varepsilon = N_c \Delta\Phi/\Delta t = (\mu_0 N_s N_c \pi r_c^2/\ell)(\Delta I/\Delta t) = 4.93 \times 10^{-6}\ V$$

The current is

$$I_c = \varepsilon/R = 4.93 \times 10^{-7}\ A$$

As the current in the solenoid increases the magnetic flux increases in the -Z direction. The induced current can oppose this by rotating counterclockwise in the coil (b to a in the resistor).

Related Text Problems: 21-29 through 21-38.

---

## PRACTICE TEST

Take and grade the practice test. Doing so will allow you to determine any weak spots in your understanding of the concepts taught in this chapter. The following section prescribes what you should study further to strengthen your understanding.

Figure 21.21 shows a region where a magnetic field is directed down into the page, several moving particles, and some information about the particles.

Figure 21.21

$m_2 = m_3 = 2.00 \times 10^{-26}$ kg
$q_3 = q_4 = 3.00 \times 10^{-16}$ C
$v_2 = v_3 = 1.00 \times 10^{6}$ m/s
$r_2 = r_4 = 10.0$ cm
$B = 0.500$ T

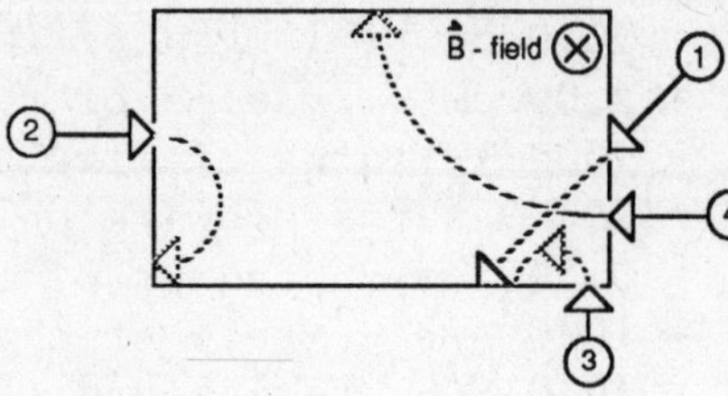


Determine the following:

______ 1. The particle(s) with no charge
______ 2. The particle(s) with positive charge
______ 3. The particle(s) with negative charge
______ 4. The central force on particle 2
______ 5. The charge of particle 2
______ 6. The radius of curvature for particle 3
______ 7. The momentum of particle 4

Figure 21.22 shows two parallel current-carrying wires held rigidly in place by wire mounts.

Figure 21.22

$I_1 = 1.00$ A
$I_2 = 2.00$ A
$I_3 = 1.00$ A
$d = 2.00$ cm
$+$ = out of page
$-$ = into page

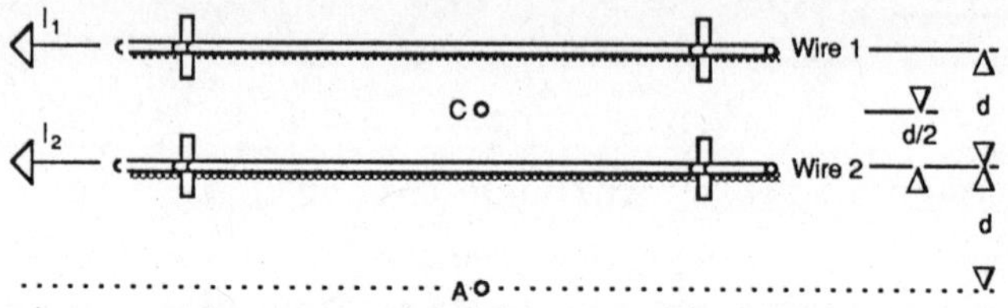


Determine the following:

______ 8. Magnitude and direction of the magnetic field at position A
______ 9. Magnitude and direction of the magnetic field midway between the wires
______ 10. Distance from wire 1 to where B = 0
______ 11. Magnitude and direction of the force per unit length on wire 1
______ 12. What would need to be done to the current in wire 1 in order to get B = 0 at position A

Figure 21.23 shows a flat circular coil inside a solenoid and concentric to the axis of the solenoid.

Figure 21.23

$r_s = 4.00$ cm
$r_c = 1.00$ cm
$N_s = 1000$
$N_c = 10$
$L_s = 40.0$ cm
$R_s = 100\ \Omega$
$V_b = 25.0$ V

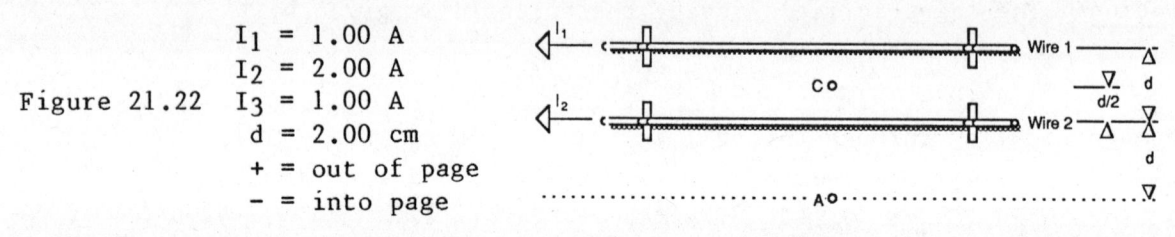

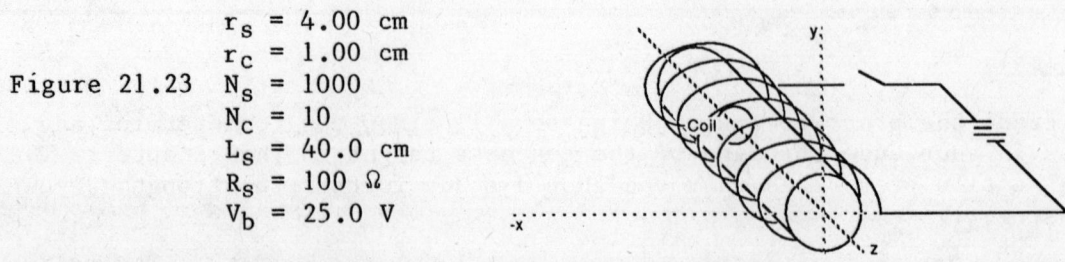

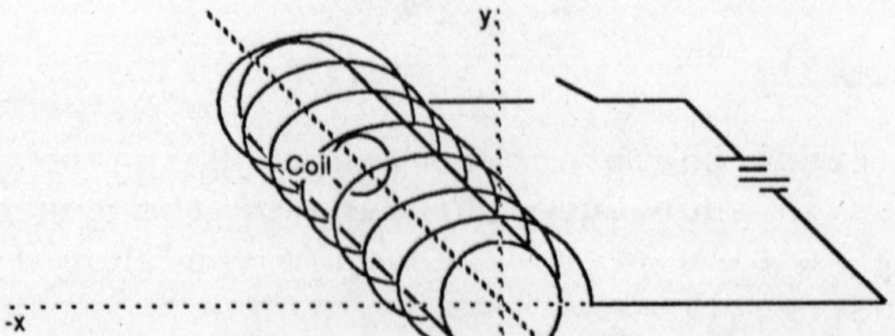


Determine the following:

______ 13. Magnitude and direction of the final magnetic field established in the solenoid

After the switch is closed, the current through the solenoid goes from zero to its maximum value. At the instant when the current is changing at the rate of 1.00 A/s, determine the following:

_______ 14. The rate at which the magnetic field in the solenoid is changing

_______ 15. The rate at which the magnetic flux in the solenoid is changing

_______ 16. The rate at which the magnetic flux in the coil is changing

_______ 17. The emf induced in each turn of wire in the coil

_______ 18. The total emf induced in the coil

(See Appendix I for answers.)

---

## PRINCIPAL CONCEPTS AND EQUATIONS PRESCRIPTION

Your score on the practice is an excellent measure of your understanding of this chapter. You should now use the following chart to write your own prescription for curing any of your physics ills. Look down the leftmost column to the number of the question(s) you answered incorrectly, read across that row to see which section(s) of the study guide you should return to for further study, and then do the suggested text problems to gain additional experience in working with the particular concept.

| Practice Test Number | Concepts and Equations | Prescription | |
|---|---|---|---|
| | | Principal Concept | Text Problems |
| 1 | Right-hand rule | 1 | 21-1,6 |
| 2 | Right-hand rule | 1 | 21-6,7 |
| 3 | Right-hand rule | 1 | 21-7,12 |
| 4 | Uniform circular motion: $F_c = mv^2/r$ | 4 of Ch. 5 | 21-1,8 |
| 5 | Magnetic force: $F = qvB \sin\theta$ | 1 | 21-2,3 |
| 6 | Magnetic force | 1 | 21-4,5 |
| 7 | Magnetic force | 1 | 21-9,10 |
| 8 | B-field of straight wire: $B = \mu_0 I/2\pi a$ | 2 | 21-12,15 |
| 9 | B-field of straight wire | 2 | 21-12,15 |
| 10 | B-field of straight wire | 2 | 21-12,15 |
| 11 | Force between currents: $F = \mu_0 I_1 I_2 \ell/2\pi a$ | 3 | 21-24,25 |
| 12 | B-field of straight wire | 2 | 21-12,15 |
| 13 | B-field of a solenoid: $B = \mu_0 NI/\ell$ | 2 | 21-16,20 |
| 14 | B-field of a solenoid | 2 | 21-16,20 |
| 15 | Magnetic flux: $\Phi = BA \cos\theta$ | 4 | 21-29,30 |
| 16 | Magnetic flux | 4 | 21-32,33 |
| 17 | Induced emf: $\varepsilon = -\Delta\Phi/\Delta t$ | 4 | 21-34,35 |
| 18 | Induced emf | 4 | 21-36,38 |

---

# 22 Inductance, Motors, and Generators

RECALL FROM PREVIOUS CHAPTERS

| Previously learned concepts and equations frequently used in this chapter | Text Section | Study Guide Page |
|---|---|---|
| Torque: $\tau = rF \sin\theta$ | 6.2 | 87 |
| Magnetic field of a coil: $B = \mu_0 NI/2r$ | 21.3 | 350 |
| Magnetic field of a solenoid: $B = \mu_0 NI/L$ | 21.3 | 350 |
| Magnetic forces on currents: $F = I\ell B \sin\theta$ | 21.4 | 353 |
| Motional emf: $\varepsilon = B\ell v_\perp$ | 21.6 | 358 |
| Magnetic flux: $\Phi = BA \cos\theta$ | 21.6 | 358 |
| Induced emf: $\varepsilon = -\Delta\Phi/\Delta t$ | 21.6 | 358 |

NEW IDEAS IN THIS CHAPTER

| Concepts and equations introduced | Text Section | Study Guide Page |
|---|---|---|
| Mutual inductance: $M_{21} = -\varepsilon_2/(\Delta I_1/\Delta t)$ | 22.2 | 366 |
| Self inductance: $L = -\varepsilon_1/(\Delta I_1/\Delta t)$ | 22.2 | 366 |
| Energy in an inductor: $U = LI^2/2$ | 22.2 | 371 |
| Inductor energy density: $\mu = U/V = B^2/2\mu_0$ | 22.2 | 371 |
| Current in a RL circuit: $I = (V/R)(1-e^{-[t/(L/R)]})$ and $I = I_0 e^{-[t/(L/R)]}$ | 22.3 | 373 |
| Time constant for a RL circuit: $T = L/R$ | 22.3 | 373 |
| Torque on a current loop in a $\vec{B}$ field: $\tau = NIAB \cos\theta$ and $\tau_{avg} = 0.637NIAB$ | 22.4,22.5 | 376 |
| Emf of an ac generator: $\varepsilon = NBA\omega \cos 2\pi ft$ | 22.6 | 382 |
| Transformers: $\varepsilon_s/\varepsilon_p = N_s/N_p = I_p/I_s$ | 22.8 | 385 |

PRINCIPAL CONCEPTS AND EQUATIONS

1. Mutual and Self Inductance (Section 22.2)

Review: Figure 22.1 shows a coil of wire wrapped around a solenoid.

Figure 22.1

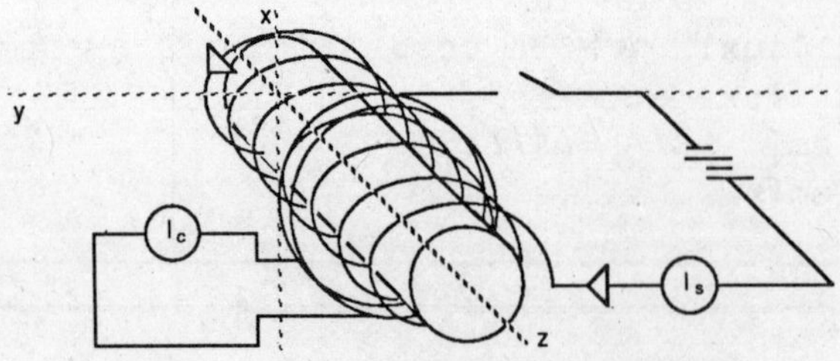

A changing current in the solenoid induces an emf in the coil and a changing current in the coil induces an emf in the solenoid. This property is called mutual inductance. The mutual inductance of the coil with respect to the solenoid is the emf $\varepsilon_c$ induced in the coil per unit rate of change of current $\Delta I_s/\Delta t$ in the solenoid. This can be expressed mathematically as

$$M_{cs} = -\varepsilon_c/(\Delta I_s/\Delta t)$$

The mutual inductance of the solenoid with respect to the coil is the emf $\varepsilon_s$ induced in the solenoid per unit rate of change of current $\Delta I_c/\Delta t$ in the coil. This can be expressed mathematically as

$$M_{sc} = -\ \varepsilon_s/(\Delta I_c/\Delta t)$$

The minus sign in the above expressions reminds us that the induced emf opposes the change that caused it.

The mutual inductance of the coil with respect to the solenoid is equal to the mutual inductance of the solenoid with respect to the coil. That is

$$M_{sc} = M_{cs}$$

When the current changes in the solenoid in Fig. 22.1, the flux through the solenoid also changes, causing a self-induced emf in the solenoid. The self inductance of a coil in a circuit is equal to the self-induced emf $\varepsilon_c$ in the coil per unit rate of change of current $\Delta I_c/\Delta t$ in the coil. This is expressed mathematically as

$$L = -\varepsilon_c/(\Delta I_c/\Delta t)$$

The minus sign in the above expression reminds us that the induced emf opposes the change that caused it.

Practice: Consider Fig. 22.1 and the following data

$\ell_s$ = 20.0 cm -- length of the solenoid
$N_s$ = 1000 -- turns of wire on the solenoid
$r_s$ = 2.00 cm -- radius of the solenoid
$\Delta I_s/\Delta t$ = 10.0 A/s -- rate of change of current in the solenoid
$N_c$ = 20 -- turns of wire on the coil
$r_c$ = 4.00 cm -- radius of the coil
$R_c$ = 20.0 Ω -- resistance of the coil

Determine the following:

| | |
|---|---|
| 1. An expression for the magnetic field inside the solenoid at any time | $B_s = \mu_0 N_s I_s/\ell_s$ where $I_s$ represents the current at any time |

| | |
|---|---|
| 2. An expression for the magnetic flux inside the solenoid at any time | $\Phi_s = B_s A_s = \mu_0 N_s I_s \pi r_s^2 / \ell_s$ |
| 3. An expression for the magnetic flux through the coil at any time | $\Phi_c = \Phi_s = B_s A_s = \mu_0 N_s I_s \pi r_s^2 / \ell_s$ |
| 4. An expression for the time rate of change of flux through the solenoid and coil | $\Delta\Phi_c/\Delta t = \Delta\Phi_s/\Delta t = (\mu_0 N_s \pi r_s^2/\ell_s)(\Delta I_s/\Delta t)$ |
| 5. An expression for the emf induced in each turn of the wire of the coil and each turn of wire of the solenoid | $\varepsilon_{c1turn} = \varepsilon_{s1turn} = -\Delta\Phi_c/\Delta t$<br>$= -(\mu_0 N_s \pi r_s^2/\ell_s)(\Delta I_s/\Delta t)$ |
| 6. An expression for the total emf induced in the coil (i.e., in all $N_c$ turns) | $\varepsilon_c = -N_c(\Delta\Phi_c/\Delta t) = N_c \varepsilon_{c1turn}$<br>$\varepsilon_c = -N_c(\mu_0 N_s \pi r_s^2/\ell_s)(\Delta I_s/\Delta t)$ |
| 7. An expression for the total emf induced in the solenoid (i.e., in all $N_s$ turns) | $\varepsilon_s = -N_s(\Delta\Phi_s/\Delta t) = N_s \varepsilon_{s1turn}$<br>$\varepsilon_s = -(N_s \mu_0 N_s \pi r_s^2/\ell_s)(\Delta I_s/\Delta t)$ |
| 8. An expression for the mutual inductance of the coil with respect to the solenoid | $M_{cs} = -\varepsilon_c/(\Delta I_s/\Delta t) = N_c \mu_0 N_s \pi r_s^2/\ell_s$<br><br>Note: This expression shows that the mutual inductance can be expressed as a function of the parameters for the two circuits. |
| 9. An expression for the self inductance of the solenoid | $L_s = -\varepsilon_s/(\Delta I_s/\Delta t) = N_s^2 \mu_0 \pi r_s^2/\ell_s$<br><br>Note: This expressions shows that the self inductance can be expressed as a function of the parameters of the circuit. |
| 10. Time rate of change of flux through the solenoid and coil | From step 4:<br>$\Delta\Phi_s/\Delta t = \Delta\Phi_c/\Delta t = (\mu_0 N_s \pi r_s^2/\ell_s)(\Delta I_s/\Delta t)$<br>$= 7.90 \times 10^{-5}$ V |

| | | |
|---|---|---|
| 11. | Emf induced in each turn of wire and in the solenoid and coil | $\varepsilon_{s1turn} = -\Delta\Phi_s/\Delta t = -7.90 \times 10^{-5}$ V<br>$\varepsilon_{c1turn} = -\Delta\Phi_c/\Delta t = -7.90 \times 10^{-5}$ V |
| 12. | Emf induced in the entire coil (all $N_c$ turns) | $\varepsilon_c = N_c\varepsilon_{c1turn} = -1.58 \times 10^{-3}$ V |
| 13. | The direction of the induced current in the coil | When the switch is closed, the flux through the solenoid and coil increases in the +Z direction. An induced current flowing clockwise through the coil will oppose this change. |
| 14. | Current through the coil | $I_c = \varepsilon_c/R_c = 7.90 \times 10^{-5}$ A |
| 15. | Mutual inductance of the coil with respect to the solenoid | From step 8:<br>$M_{cs} = -\varepsilon_c/(\Delta I_s/\Delta t) = 1.58 \times 10^{-4}$ V·s/A<br>or<br>$M_{cs} = N_c\mu_0 N_s\pi r_s{}^2/\ell_s = 1.58 \times 10^{-4}$ V·s/A |
| 16. | Rate at which the current in the solenoid must change in order to induce a $1.00 \times 10^{-3}$-V emf in the coil | $\Delta I_s/\Delta t = -\varepsilon_c/M_{cs} = 6.33$ A/s |
| 17. | Mutual inductance of the solenoid with respect to the coil | $M_{sc} = M_{cs} = 1.58 \times 10^{-4}$ V·s/A |
| 18. | Emf induced in the solenoid when the current in the coil is changing at the rate of 5.00 A/s | $M_{sc} = -\varepsilon_s/(\Delta I_c/\Delta t)$ or<br>$\varepsilon_s = -M_{sc}\Delta I_c/\Delta t = -7.90 \times 10^{-4}$ V |
| 19. | Self induced emf in the solenoid | From steps 7 and 9:<br>$L_s = -\varepsilon_s/(\Delta I_s/\Delta t) = N_s\varepsilon_{1turn}/(\Delta I_s/\Delta t)$<br>$= 7.90 \times 10^{-3}$ V·s/A<br>or<br>$L_s = N_s{}^2\mu_0\pi r_s{}^2/\ell_s = 7.90 \times 10^{-3}$ V·s/A |

| | | |
|---|---|---|
| 20. | Rate at which the current in the solenoid must change in order to induce a 1.00 x $10^{-3}$-V emf in the coil | $\Delta I_s/\Delta t = -\varepsilon_s/L_s = 0.127$ A/s |
| 21. | Rate at which the current in the coil must change in order to induce 1.00 x $10^{-3}$-V emf in the solenoid | $\Delta I_c/\Delta t = -\varepsilon_s/M_{sc} = 6.33$ A/s |

Example 22.1. A 1.00-cm radius coil of 20 turns is placed inside a solenoid in a concentric manner. The solenoid has a 4.00-cm radius, 50.0-cm length, and 1000 turns of wire. Determine the emf induced in the coil when the current in the solenoid is changing at the rate of 10.0 A/s.

Figure 22.2

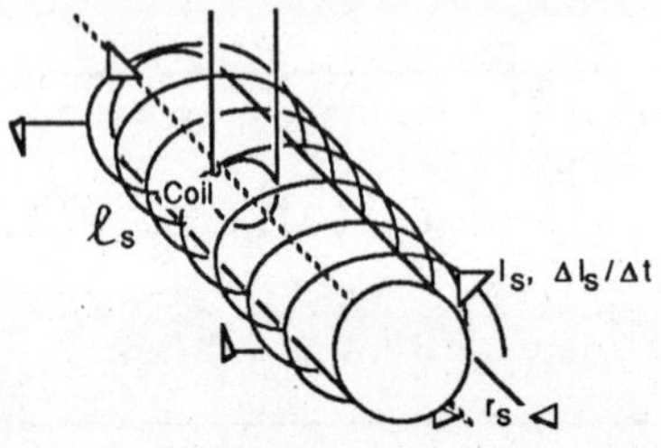


Given:
$r_c$ = 1.00 cm -- radius of the coil
$r_s$ = 4.00 cm -- radius of the solenoid
$\ell_s$ = 50.0 cm -- length of the solenoid
$N_c$ = 20 -- turns of wire on the coil
$N_s$ = 1000 -- turns of wire on the solenoid
$\Delta I_s/\Delta t$ = 10.0 A -- rate of change of current in the solenoid

Determine: The emf induced in the coil.

Strategy: We can write an expression for the magnetic field in the solenoid, and subsequently the flux through the coil, as a function of the current in the solenoid. From this expression, we can obtain an expression for the rate of change of flux in the coil as a function of the rate of change of current in the solenoid. We can now proceed by either of the following two methods.

(A) Knowing the rate of change of the flux through the coil and $N_c$, we can determine the emf induced in the coil.

(B) Knowing the rate of change of the flux through the coil and the definition of mutual inductance, we can determine the emf induced in the coil.

Solution: An expression for the magnetic field inside the solenoid is

$$B_s = \mu_0 N_s I_s/\ell_s$$

An expression for the magnetic flux through the coil is

$$\Phi_c = A_c B_c = A_c B_s = \pi r_c^2 \mu_o N_s I_s/\ell_s$$

An expression for the time rate of change of magnetic flux through the coil is

$$\Delta\Phi_c/\Delta t = (\pi r_c^2 \mu_o N_s/\ell_s)(\Delta I_s/\Delta t)$$

(A) The emf induced in the coil is

$$\varepsilon_c = -N_c(\Delta\Phi_c/\Delta t) = -(N_c \pi r_c^2 \mu_o N_s/\ell_s)(\Delta I_s/\Delta t) = -1.58 \times 10^{-4}\ V$$

(B) The mutual inductance of the coil with respect to the solenoid (see step 15 of the practice on page 369) is

$$M_{cs} = -\varepsilon_c/(\Delta I_s/\Delta t) = N_c \pi r_c^2 \mu_o N_s/\ell_s \quad \text{hence}$$

$$\varepsilon_c = -M_{cs}(\Delta I_s/\Delta t) = -(N_c \pi r_c^2 \mu_o N_s/\ell_s)(\Delta I_s/\Delta t) = -1.58 \times 10^{-4}\ V$$

Related Text Problems: 22-1 through 22-6.

## 2. Energy Stored in an Inductor (Section 22.2)

Review: Figure 22.3 shows an inductor of inductance L in a simple dc circuit. When the switch is closed, the current in the inductor increases from 0 to its final value I, and the magnetic field increases from 0 to its final value B in a time $\Delta t$.

Figure 22.3

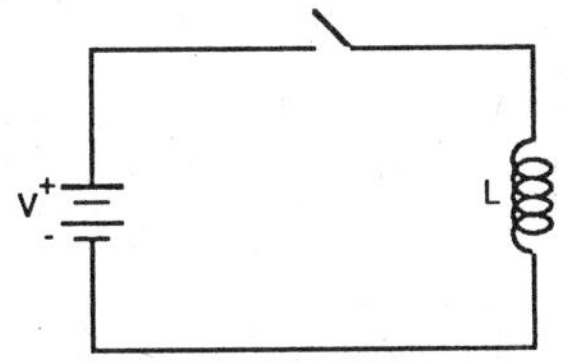


As the current is changing, an induced emf ($\varepsilon = -L\Delta I/\Delta t$) opposite the direction of the applied voltage is produced across the inductor. The voltage supplied to the inductor by the power source is

$$V_L = -\varepsilon = L\Delta I/\Delta t$$

The average current and the change in current are

$$I_{av} = (0 + I)/2 = I/2 \text{ and } \Delta I = (I-0) = I$$

The average power delivered to the inductor is

$$P_{av} = I_{av}V_L = I_{av}L\Delta I/\Delta t = LI^2/2\Delta t$$

The energy stored in the inductor is the amount of work done on the inductor by the source to establish the B-field.

$$U = W = P_{av}\Delta t = I_{av}L\Delta I = LI^2/2$$

The energy density (i.e., the energy per unit volume) in the inductor is

$$\mu = U/V = B^2/2\mu_o$$

Practice: The following is known for the situation shown in Fig. 22.3.

The inductor is a solenoid
$I = 3.00$ A -- final current in the inductor
$\Delta t = 1.00 \times 10^{-5}$ s -- time to establish the final $\vec{B}$-field
$\ell = 0.100$ m -- length of the solenoid
$r = 0.500$ cm -- radius of the solenoid
$n = 2.00 \times 10^{3}$/m -- turns of wire on the solenoid per m of length

Determine the following:

| | |
|---|---|
| 1. Magnitude of final magnetic field | $B = \mu_o nI = 7.54 \times 10^{-3}$ T |
| 2. Inductance of the solenoid | $N = n\ell = 200$<br>$L = N\mu_o nA = N\mu_o n\pi r^2$<br>$L = 3.95 \times 10^{-5}$ V·s/A |
| 3. The potential difference across the solenoid | $V_L = L(\Delta I/\Delta t) = 11.9$ V |
| 4. Average power delivered to the solenoid | $P_{av} = I_{av}V_L = IV_L/2 = 17.8$ W |
| 5. Work done by the power source to establish the final B-field | $W = P_{av}\Delta t = 1.78 \times 10^{-4}$ J<br>or<br>$W = LI^2/2 = 1.78 \times 10^{-4}$ J |
| 6. Energy stored in the magnetic field of the solenoid | $U = W = 1.78 \times 10^{-4}$ J |
| 7. Energy density of the energy stored in the solenoid | $V = A\ell = \pi r^2 \ell = 7.85 \times 10^{-6}$ m$^3$<br>$\mu = U/V = 22.7$ J/m$^3$ or<br>$\mu = B^2/2\mu_o = 22.6$ J/m$^3$ |

Example 22.2. We have a 20.0-cm-long solenoid with a 2.00-cm radius and $10^3$ turns of wire. Determine the current in the solenoid to produce a magnetic energy density equal to that of a $10^4$-V/m $\vec{E}$ field.

Given: $\ell = 20.0$ cm, $r = 2.00$ cm, $N = 10^3$ turns, $E = 10^4$ V/m, $\mu_E = \mu_B$

Determine: The current in the solenoid so that $\mu_B = \mu_E$ if $E = 10^4$ V/m.

Strategy: We can equate expressions for the energy density of the $\vec{E}$ and $\vec{B}$ field and solve for the strength of the $\vec{B}$ field. Knowing B and the dimensions of the solenoid, we can determine the value for the current.

Solution: Expressions for the energy density of an $\vec{E}$ and $\vec{B}$ field are

$$\mu_E = \varepsilon_o E^2/2 \quad \text{and} \quad \mu_B = B^2/2\mu_o$$

We want the value of B when $\mu_B = \mu_E$, hence

$$B^2/2\mu_o = \varepsilon_o E^2/2 \quad \text{or} \quad B = (\mu_o \varepsilon_o)^{1/2}\, E$$

The current required to give this value of B in the solenoid is obtained as follows:

$$B = \mu_o NI/\ell \quad \text{or} \quad I = B\ell/\mu_o N = (\varepsilon_o/\mu_o)^{1/2}\, E\ell/N = 5.30 \times 10^{-3}\ \text{A}$$

Related Text Problems: 22-5, 22-6, 22-8.

## 3. Inductors In Electric Circuits (Section 22.3)

Review: Figure 22.4 shows an inductor in a circuit with a resistor and a battery

Figure 22.4 $R = 20.0\ \Omega$, $L = 5.00$ H, $V = 10.0$ V

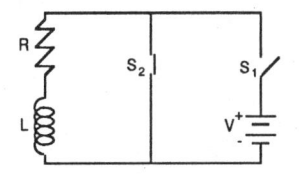

(a) $S_1$ closed and $S_2$ open (b) $S_1$ open and $S_2$ closed

If $S_2$ is left open and $S_1$ closed at $t = 0$ (Fig. 22.4a), applying Kirchhoff's loop rule to the circuit we obtain

$$V - V_R - V_L = 0 \quad \text{where} \quad V_R = IR \text{ and } V_L = L\Delta I/\Delta t$$

The current in the circuit at any time is

$$I = (V/R)(1 - e^{-[t/(L/R)]})$$

The time constant for the circuit is $T = L/R$.

If after the current has been established $S_1$ is opened and $S_2$ closed (Fig. 22.4b) applying Kirchhoff's loop rule we obtain

$$-V_R - V_L = 0$$

The current in the circuit at any time is

$$I = I_o e^{-[t/(L/R)]}$$

where $I_o$ is the current in the circuit the instant $S_2$ is closed.

Practice: For the situation shown in Fig. 22.4, determine the following:

| | |
|---|---|
| 1. An expression for the potential drop across the resistor at any time for Fig. 22.4a | $V_R = IR$<br>$V_R = V(1 - e^{-[t/(L/R)]})$ |
| 2. An expression for the potential drop across the inductor at any time for Fig. 22.4a | $V_L = V - V_R$<br>$V_L = Ve^{-[t/(L/R)]}$ |
| 3. Current at t = 0 for Fig. 22.4a | $I = (V/R)(1 - e^{-[t/(L/R)]})$<br>$I = (V/R)(1 - e^0) = 0$ |
| 4. Current as $t \to \infty$ for Fig. 22.4a | $I = V/R(1 - e^{-[t/(L/R)]})$<br>As $t \to \infty$, $e^{-t/(L/R)} \to 0$ and<br>$I \to V/R = 0.500$ A |
| 5. The current at t = L/R for Fig. 22.4a | $I = (V/R)(1 - e^{-[t/(L/R)]})$<br>At t = L/R, we obtain<br>$I = (V/R)(1 - e^{-1}) = 0.316$ A |
| 6. Potential drop across the resistor and the inductor at t = L/R for Fig. 22.4a | $V_R = V(1 - e^{-1}) = 6.32$ V<br>$V_L = Ve^{-1} = 3.68$ V<br>Note that $V_R + V_L = V$ |
| 7. Time constant for Figs. 22.4a and 22.4b | $T = L/R = 0.250$ s |
| 8. Current in the circuit after ten time constants for Fig. 22.4a | $I = V/R(1 - e^{-10}) = 0.500$ A |
| 9. Potential drop across the resistor and the inductor after 10 time constants for Fig. 22.4a | $V_R = V(1 - e^{-10}) = 10.0$ V<br>$V_L = Ve^{-10} = 0.000$ |
| 10. Time t' for the current to reach one-half its maximum value for Fig. 22.4a | $I_{max} = V/R$ (step 4)<br>$I = I_{max}/2 = (V/R)/2$ when $t = t'$<br>$I = V/R(1 - e-[t/(L/R])$<br>$(V/R)/2 = (V/R)(1 - e^{-[t'/(L/R)]})$<br>$e^{-[t'/(L/R)]} = 1/2$<br>$t' = (L/R)\ln 2 = 0.173$ s |

| | | |
|---|---|---|
| 11. | Rate at which the current is changing at t = L/R for Fig. 22.4a | After $t = L/R$, $V_L = Ve^{-1} = 3.68$ V<br>$\Delta I/\Delta t = V_L/L = 0.736$ A/s |
| 12. | Current at t = 0 s for Fig. 22.4b | $I = I_o e^{-[t/(L/R)]}$<br>At $t = 0$, $I = I_o = 0.500$ A |
| 13. | Current after 0.100 s for Fig. 22.4b | $I_o = 0.500$ A<br>$I = I_o e^{-[t/(L/R)]} = 0.335$ A |
| 14. | Potential drop across the resistor and the inductor at t = 0.100 s for Fig. 22.4b | $V_R = IR = 6.70$ V<br>$V_L = V - V_R = 0 - V_R = -6.70$ V |
| 15. | Rate at which the current is changing after t = 0.100 s for Fig. 22.4b | $\Delta I/\Delta t = V_L/L = -1.34$ A/s<br>The minus sign tells us that the current is decreasing. |

Example 22.3. A 2.00-H inductor and a 4.00-Ω resistor are connected in series with a 1.50-V battery. At a time equal to one time constant, determine (a) the current in the circuit, (b) the rate of change of the current, (c) the rate at which energy is being stored in the magnetic field, (d) the rate at which energy is being lost to joule heating, (e) the rate at which energy is being delivered by the battery.

Given: $L = 2.00$ H, $R = 4.00$ Ω, $V = 1.50$ V, and $t = L/R$

Determine:

(a) I -- current in the circuit
(b) $\Delta I/\Delta t$ -- rate of change of current
(c) $\Delta U_B/\Delta t$ -- rate at which energy is being stored in the magnetic field
(d) $\Delta U_{joule}/\Delta t$ -- rate at which energy is being lost to joule heating
(e) $\Delta U_{battery}/\Delta t$ -- rate at which energy is being supplied by the battery

Strategy: Knowing V, R, L, and t, we can determine the current I. Once I is known, we can use Kirchhoff's loop rule to determine $\Delta I/\Delta t$. Knowing the current, we can also use our knowledge of power to determine the rate at which energy is stored in the magnetic field, the rate at which energy is lost to joule heating, and the rate at which energy is supplied by the battery. As a quick check on our work, we can use the fact that the rate at which energy is supplied by the battery equals the sum of the rate at which it is stored in the B field and lost to joule heating.

Solution: (a) The current in the circuit at a time equal to one time constant ($t = L/R$) is obtained by

$$I = (V/R)(1 - e^{-[t/(L/R)]}) = (V/R)(1 - e^{-1}) = 0.237\ A$$

(b) The rate of change of current can be obtained by applying Kirchhoff's loop rule to the circuit.

$$V - V_R - V_L = 0 \text{ or } V - IR - L(\Delta I/\Delta t) = 0$$

$$\Delta I/\Delta t = (V - IR)/L = 0.276\ A/s$$

(c) The rate at which energy is being stored in the magnetic field is obtained as follows:

$$P = IV_L = IL(\Delta I/\Delta t) = 0.131\ W$$

(d) The rate at which energy is being lost to joule heating is

$$P = I^2R = 0.225\ W$$

(e) The rate at which energy is being supplied by the battery is

$$P = IV = 0.356\ W$$

As a check, note that the power supplied by the battery is the sum of the power stored in the magnetic field and the power lost due to joule heating.

Related Text Problems: 22-11 through 22-15.

## 4. Torque On a Current Loop In a $\vec{B}$ Field (Sections 22.4, 22.5)

Review: Figure 22.5 shows a rectangular current loop of length $\ell$ and width w in a uniform $\vec{B}$ field. At this particular instant, the plane of the loop is parallel to the $\vec{B}$ field.

Figure 22.5

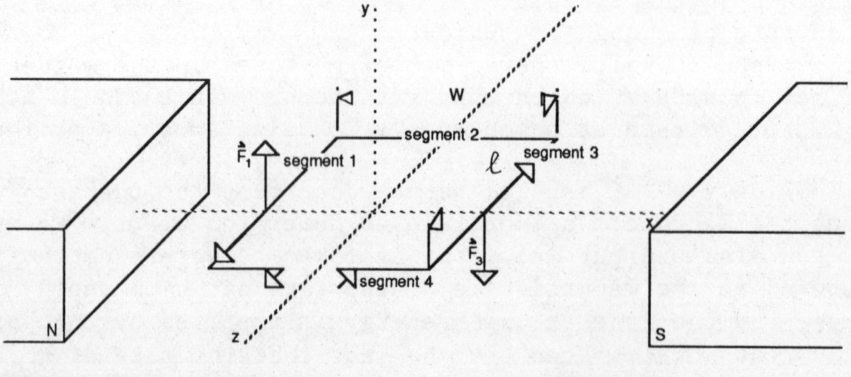


For convenience, the segments of the wire are labeled as shown in Fig. 22.5. Fig. 22.6 shows how to obtain the force on each segment of the loop.

| | Segment of the loop | Magnitude of the Force | Direction of the Force |
|---|---|---|---|
| (a) | 1 | $F_1 = I\ell B \sin 90°$<br>$F_1 = I\ell B$ | 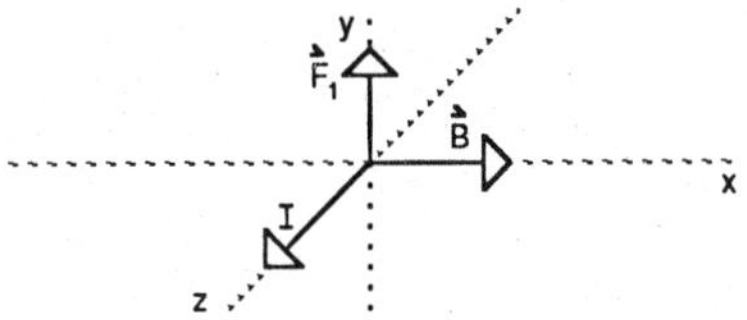 |
| (b) | 2 | $F_2 = IwB \sin 0$<br>$F2 = 0$ | 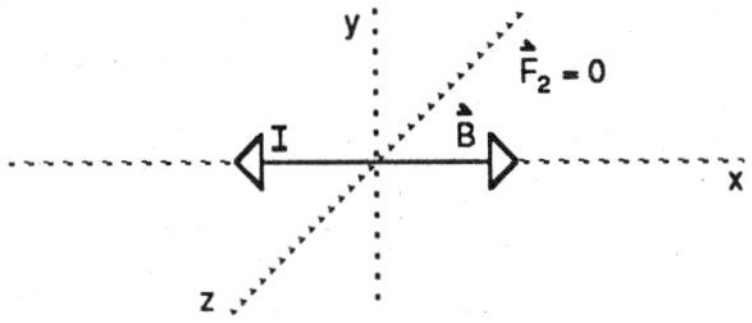 |
| (c) | 3 | $F_3 = I\ell B \sin 90°$<br>$F_3 = I\ell B$ | 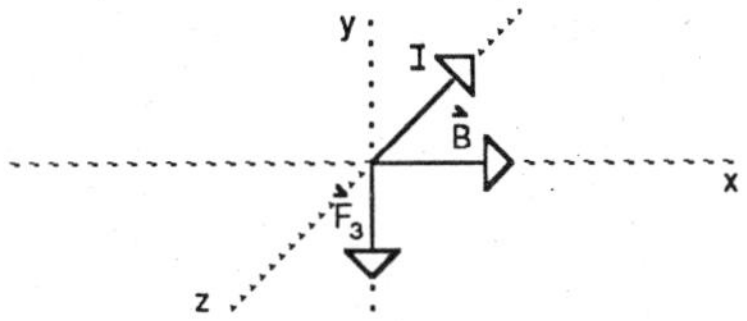 |
| (d) | 4 | $F_4 = IwB \sin 180°$<br>$F_4 = 0$ | 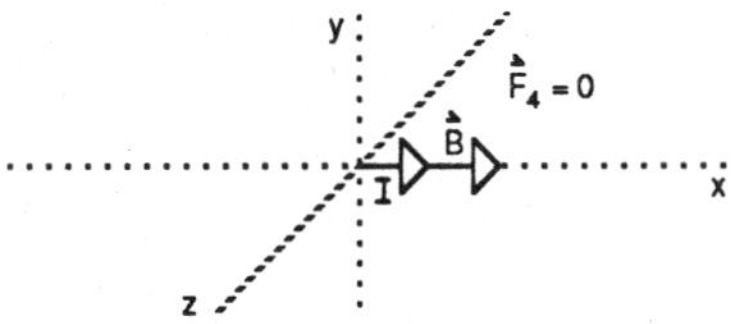 |



Figure 22.6.

The forces $\vec{F}_1$ and $\vec{F}_3$ exert an unbalanced clockwise torque on the loop. As a result, an instant later it appears as shown in Fig. 22.7.

Figure 22.7

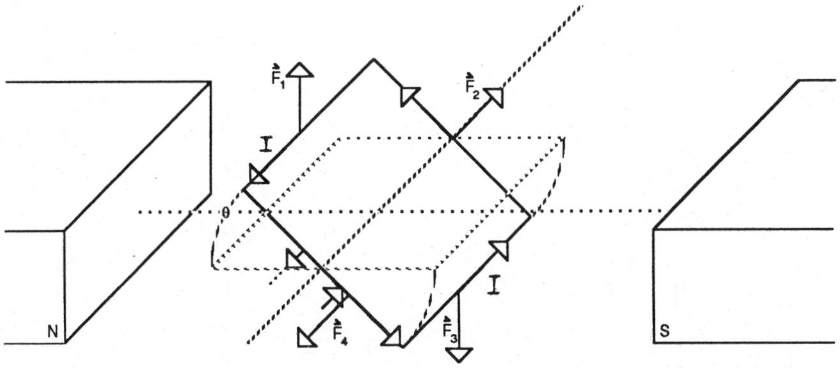


Figure 22.8 shows how to obtain the force on each segment of the loop

| | Segment of the loop | Magnitude of the Force | Direction of the Force |
|---|---|---|---|
| (a) | 1 | $F_1 = I\ell B \sin 90°$<br>$F_1 = I\ell B$ | 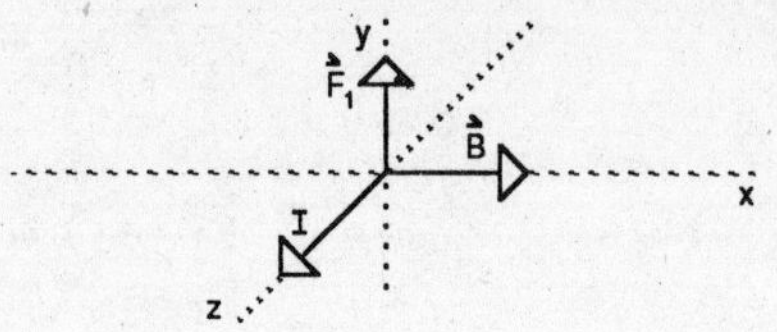 |
| (b) | 2 | $F_2 = IwB \sin(\pi - \theta)$<br>$F_2 = IwB \sin\theta$ | 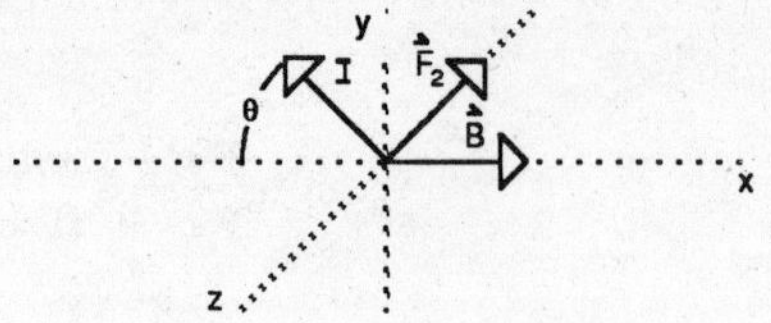 |
| (c) | 3 | $F_3 = I\ell B \sin 90°$<br>$F_3 = I\ell B$ | 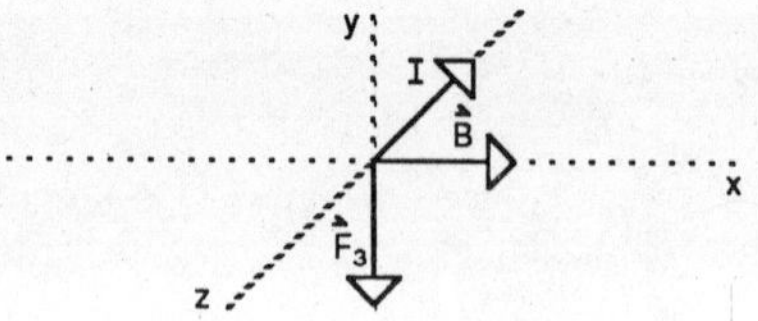 |
| (d) | 4 | $F_4 = IwB \sin\theta$ | 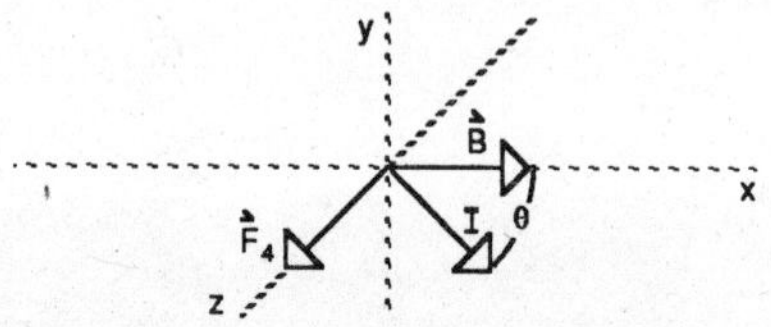 |

Figure 22.8

From Figs. 22.7 and 22.8, we see that for values of $\theta$ between 0° and 90°, the forces $\vec{F}_2$ and $\vec{F}_4$ cancel and the forces $\vec{F}_1$ and $\vec{F}_3$ exert an unbalanced clockwise torque on the loop about the z-axis. The torque due to $\vec{F}_1$ and $\vec{F}_3$ is obtained by multiplying the magnitude, the forces ($\vec{F}_1$ and $\vec{F}_3$) by the moment arm $[(w/2)\cos\theta]$.

$$\vec{\tau}_1 = -I\ell B(w/2)\cos\theta \qquad (- = \text{clockwise})$$

$$\vec{\tau}_2 = -I\ell B(w/2)\cos\theta$$

The total unbalanced torque on the rectangular loop is

$$\vec{\tau} = \vec{\tau}_1 + \vec{\tau}_2 = -I\ell wB\cos\theta$$

Noting that $\ell w$ is just the area A of the loop, we obtain

$$\vec{\tau} = -IAB\cos\theta$$

If the coil has N turns, we obtain

$$\vec{\tau}_N = -NIAB\cos\theta$$

Figure 22.9 shows that if the direction of the current is reversed at $\theta = 90°$ and again at $\theta = 270°$, the coil continues to experience a clockwise torque.

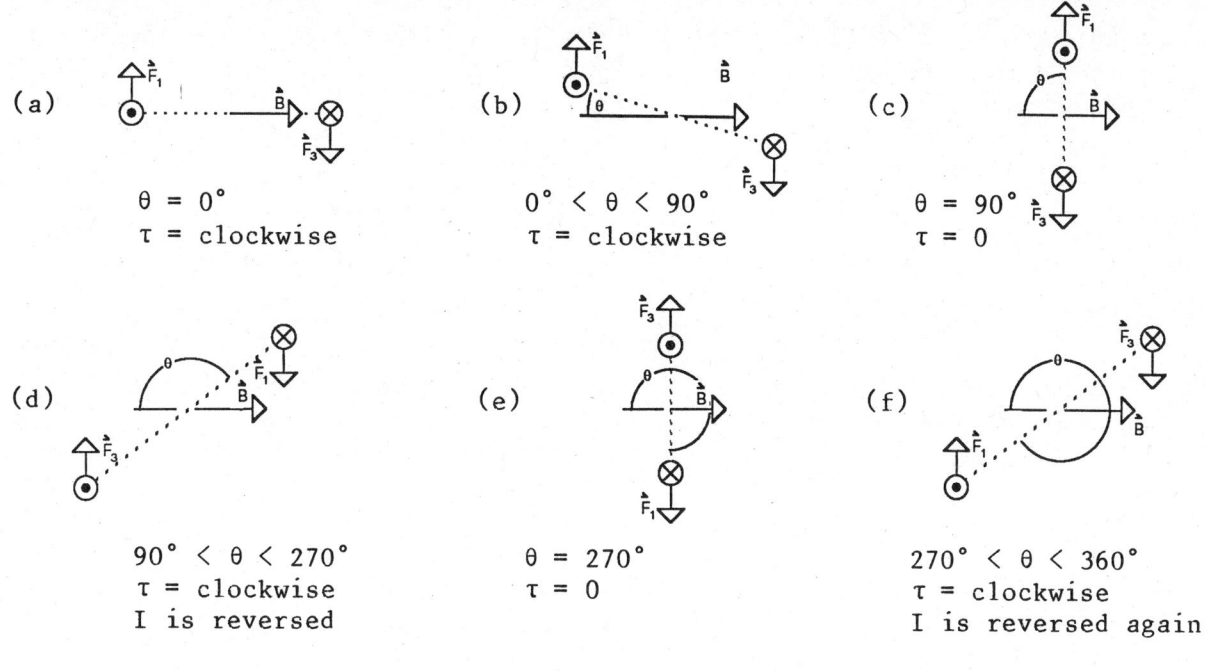

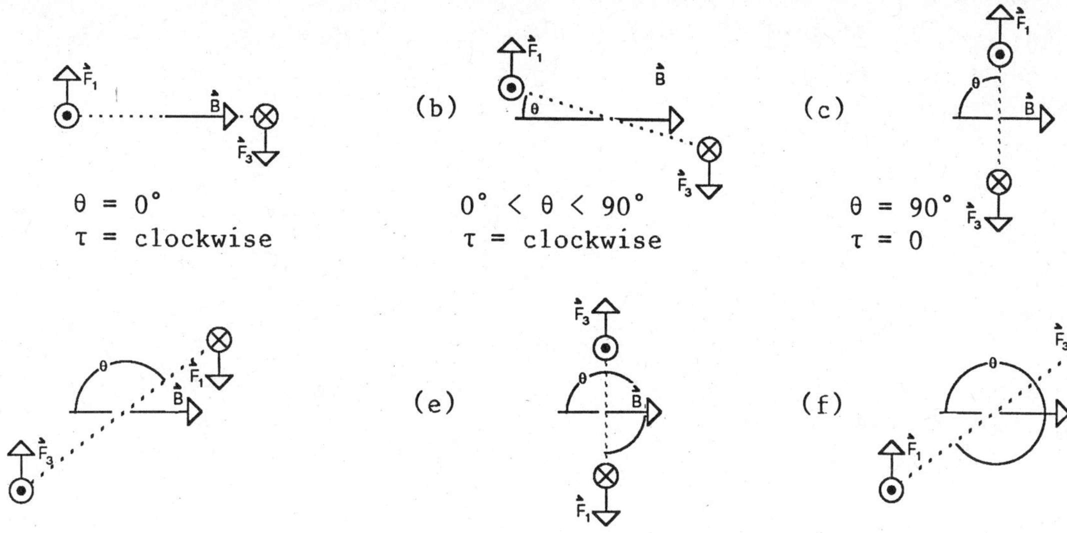


Figure 22.9

As the coil rotates, $\cos\theta$ has different values. If the coil rotates at a constant speed, then

$$(\cos\theta)_{av} = 0.637 \text{ and } \tau_N = 0.637NIAB$$

If the magnetic field is radial, $\theta = 0$, $\cos\theta = 1$, and the torque is constant for all orientations. That is

$$\tau_N = NIAB$$

<u>Practice:</u> The coil shown in Fig. 22.10 is released from $\theta = 10°$ at $t = 0$.

Figure 22.10

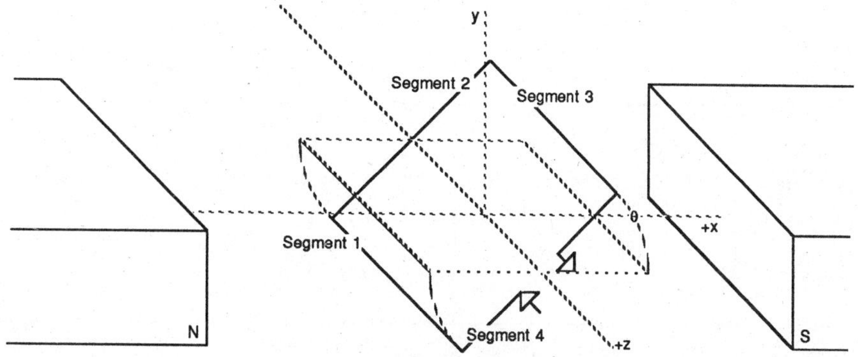


$N = 100$ turns  $w = 5.00$ cm  $B = 1.00$ T

$\ell = 10.0$ cm  $\theta = 10°$  $I = 10.0$ A

Determine the following:

| | |
|---|---|
| 1. Magnitude and direction of the force on each segment of one turn of the wire when $\theta = 10°$ | $\vec{F}_1 = I\ell B \sin 90° = 1.00$ N (-y dir.)<br>$\vec{F}_2 = IwB \sin 10° = 8.68 \times 10^{-2}$N (-z dir.)<br>$\vec{F}_3 = I\ell B \sin 90° = 1.00$ N (+y dir.)<br>$\vec{F}_4 = IwB \sin 170° = 8.68 \times 10^{-2}$N (+z dir.) |
| 2. Net force on the coil in the x, y, and z directions | $F_{xNet} = F_{yNet} = F_{zNet} = 0$ |
| 3. Torque due to each segment of one turn of the wire when $\theta = 10°$ | $\vec{\tau}_1 = +F_1(w/2)\cos 10° = +2.46 \times 10^{-2}$N·m<br>$\vec{\tau}_2 = 0$<br>$\vec{\tau}_3 = +F_3(w/2)\cos 10° = +2.46 \times 10^{-2}$N·m<br>$\vec{\tau}_4 = 0$<br>The + sign indicates a counterclockwise torque |
| 4. Torque due to one turn of wire when $\theta = 10°$ | $\vec{\tau} = \vec{\tau}_1 + \vec{\tau}_2 + \vec{\tau}_3 + \vec{\tau}_4 = +4.92 \times 10^{-2}$ N·m |
| 5. Torque due to all 100 turns of wire on the coil when $\theta = 10°$ | $\vec{\tau}_N = N\vec{\tau} = +4.92$ N·m or<br>$\tau_N = NIAB \cos\theta = +4.92$ N·m |
| 6. Values for $\theta$ when the torque is zero | $\tau = NIAB \cos\theta$<br>$\tau = 0$ when $\cos\theta = 0$<br>$\cos\theta = 0$ when $\theta = 90°$ and $270°$ |
| 7. Values for $\theta$ when the torque is a maximum | $\tau = NIAB \cos\theta$<br>$\tau$ = maximum when $\cos\theta = 1$<br>$\cos\theta = 1$ when $\theta = 0°$ and $180°$ |
| 8. Values for $\theta$ when the torque has a value one-half the maximum value | $\tau = NIAB \cos\theta = \tau_{max} \cos\theta$<br>$\tau_{max} = \pm NIAB$<br>$\theta = \theta'$ when $\tau = \tau max/2$<br>$\theta' = \cos^{-1}(\pm 0.500) = 60°, 120°, 240°,$ and $300°$ |
| 9. Maximum value of the torque | $\tau = NIAB \cos\theta$<br>$\tau_{max} = NIAB = 5.00$ N·m |

| | |
|---|---|
| 10. Average torque exerted on the coil | $\tau_{avg} = 0.637NIAB = 0.318\ N\cdot m$ |
| 11. A plot of torque vs. θ for values of θ from 0° to 360° | 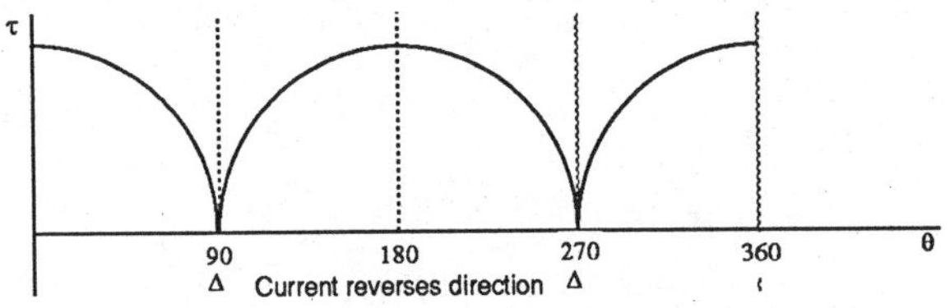  |

Example 22.4. The rectangular coil of wire shown in Fig. 22.11 has a linear mass density of $2.00 \times 10^{-2}$ kg/m and is pivoted about side a-d as a frictionless axis. The coil is 5.00 cm wide, 10.00 cm long, and has 20 turns. A uniform 0.200-T magnetic field is in the +y direction. Determine the angle between the plane of the coil and the $\vec{B}$ field when the current in the coil is 10.0 A.

Figure 22.11

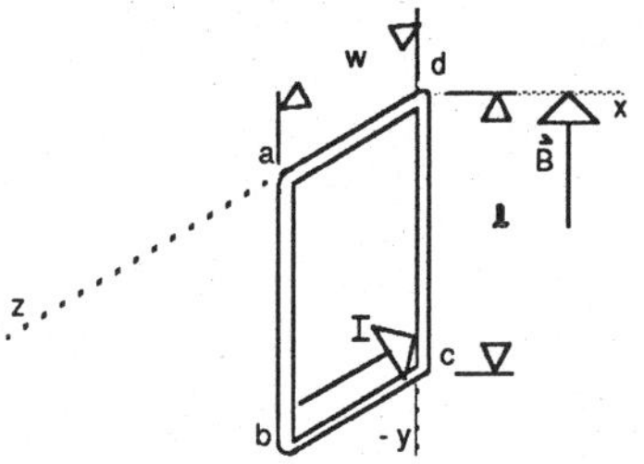


Given: $w = 5.00 \times 10^{-2}$ m, $\ell = 1.00 \times 10^{-1}$ m, $\lambda = M/\ell = 2.00 \times 10^{-2}$ kg/m, $B = 2.00 \times 10^{-1}$ T, $N = 20$, $I = 10.0$ A

Determine: The angle θ between the plane of the coil and the $\vec{B}$ field when the current is 10.0 A.

Strategy: The coil will rotate counterclockwise about the Z axis until the torque due to the $\vec{B}$ field is equal to the torque due the weight of the wire. We can write and equate expressions for these torques and then solve for θ.

Solution: As shown in Fig. 22.12, the coil rotates counterclockwise about the Z axis until the torque due to the $\vec{B}$ field is equal to the torque due to the weight of the coil.

Figure 22.12. $M = \lambda \ell N$
$M' = \lambda w N$

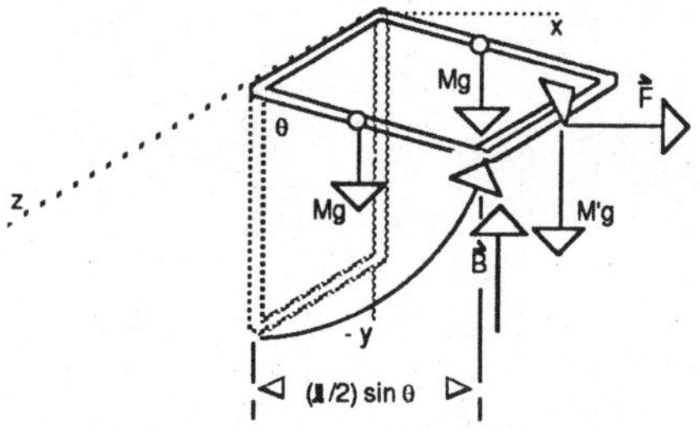

The clockwise torque due to the weight of the coil (two lengths each of mass M and one width of mass M') is

$$\tau_{cw} = 2(Mg)(\frac{\ell}{2}\sin\theta) + M'g\ell\sin\theta = (\ell + w)\lambda Ng\ell\sin\theta$$

The counterclockwise torque due to the $\vec{B}$ field (acting on the width w) is

$$\tau_{cw} = NF\ell\cos\theta = NBIw\ell\cos\theta$$

Since the coil is in rotational equilibrium, we can equate the clockwise and counterclockwise torques and solve for θ

$$(\ell + w)\lambda Ng\ell\sin\theta = NBIw\ell\cos\theta$$

$$\tan\theta = BIw/(\ell + w)\lambda g = 3.40$$

$$\theta = \tan^{-1}(3.40) = 73.6°$$

Related Text Problems: 22-19 through 22-22, 22-24.

## 5. Electric Generators (Section 22.6)

Review: The N-turn coil of wire in Fig. 22.13 is mechanically turned at a constant angular speed ω in a uniform magnetic field $\vec{B}$.

Figure 22.13

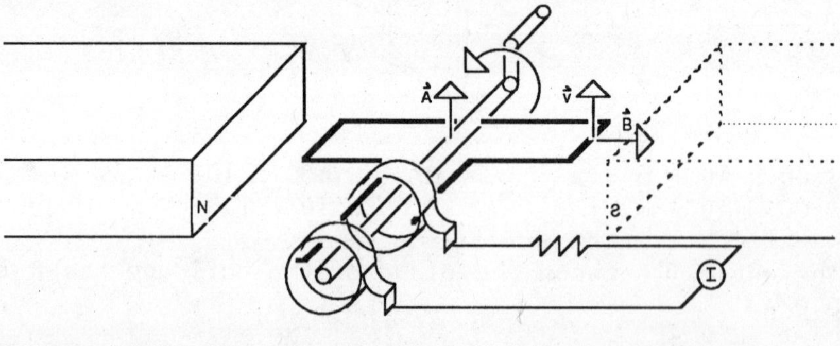


As the coil turns, the following quantities vary in time.

1. The magnetic flux through the coil (Φ)
2. The rate of change of magnetic flux through the coil (ΔΦ/Δt)
3. The induced emf (ε)
4. The induced current (I)
5. The angle between $\vec{v}$ and $\vec{B}$ (θ)
6. The angle between $\vec{A}$ and $\vec{B}$ (θ)
7. The angle between the plane of the coil and $\vec{B}$ (α)

If we agree to start our record of time when the angle between $\vec{v}$ and $\vec{B}$ is 0° (Fig. 22.14), the magnetic flux, rate of change of magnetic flux, induced emf, and induced current are obtained as follows:

Figure 22.14

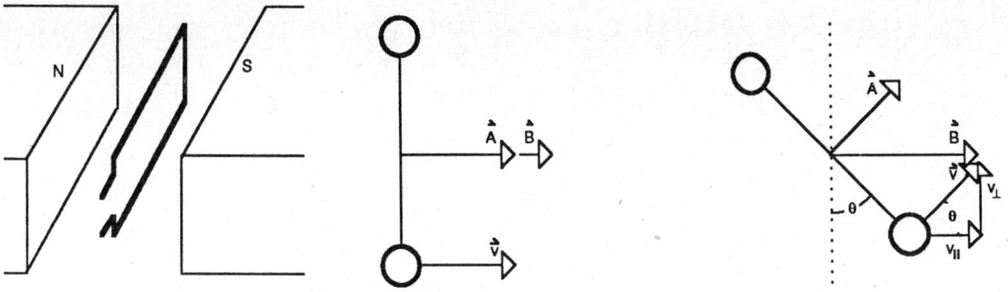


(a) t = 0 (b) t = 0 (c) t > 0

The magnetic flux at any time is $\Phi = BA \cos\theta = BA \cos\omega t$

The induced emf of one wire of side 1 of the coil is

$$\varepsilon_1 = B\ell v_\perp = B\ell v \sin\theta = B\ell\omega(w/2) \sin\theta = (BA\omega/2) \sin\theta$$

The induced emf due to N turns of both sides is $\varepsilon = 2N\varepsilon_1 = NBA\omega \sin\theta$

The induced current is $I = \varepsilon/R = (NBA\omega/R) \sin\theta$

Figure 22.15 shows the coil in four different orientations and a graphical representation of $\Phi$, $\Delta\Phi/\Delta t$, $\varepsilon$, and I.

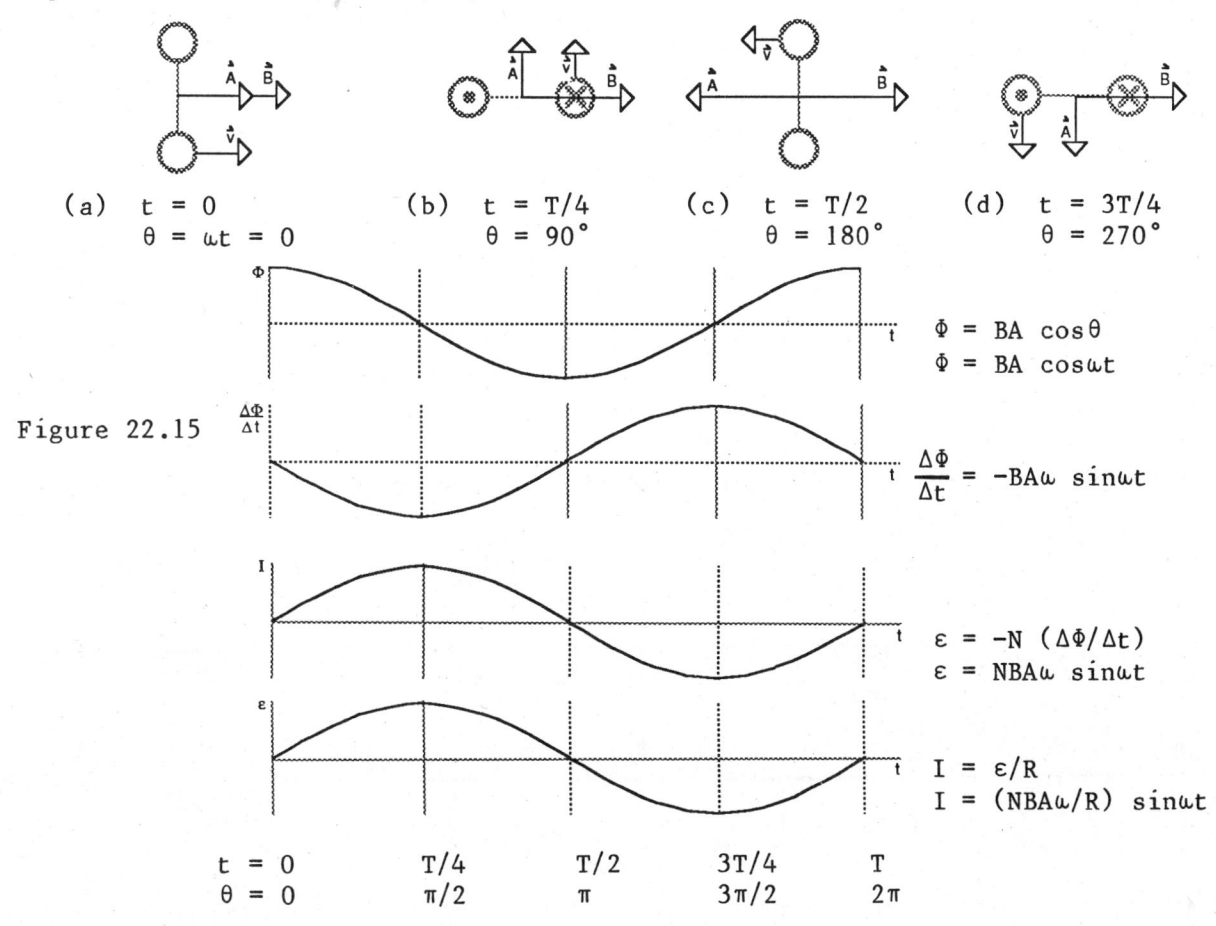

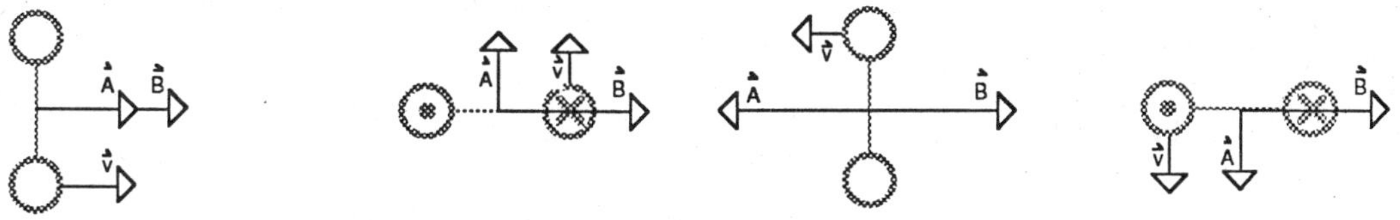


(a) $t = 0$, $\theta = \omega t = 0$ (b) $t = T/4$, $\theta = 90°$ (c) $t = T/2$, $\theta = 180°$ (d) $t = 3T/4$, $\theta = 270°$

Figure 22.15

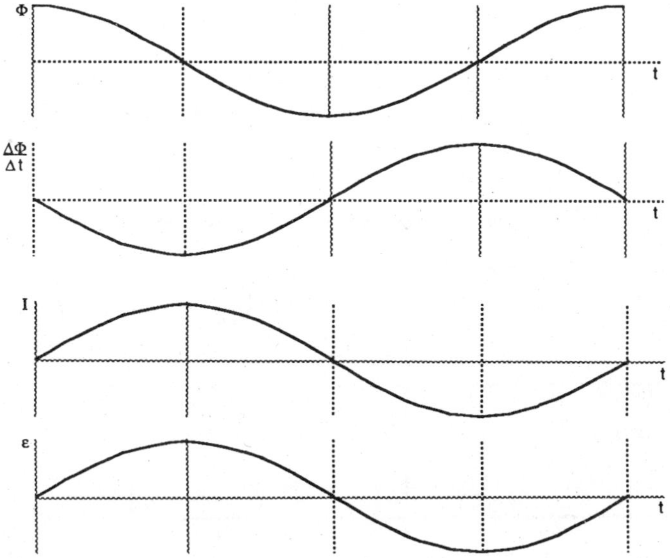


$\Phi = BA \cos\theta$
$\Phi = BA \cos\omega t$

$\frac{\Delta\Phi}{\Delta t} = -BA\omega \sin\omega t$

$\varepsilon = -N (\Delta\Phi/\Delta t)$
$\varepsilon = NBA\omega \sin\omega t$

$I = \varepsilon/R$
$I = (NBA\omega/R) \sin\omega t$

| $t = 0$ | $T/4$ | $T/2$ | $3T/4$ | $T$ |
|---|---|---|---|---|
| $\theta = 0$ | $\pi/2$ | $\pi$ | $3\pi/2$ | $2\pi$ |

Practice: The following information is known for the coil in Fig. 22.13.

$A = 50.0\ cm^2$, $B = 5.00\ T$, $N = 100$ turns, $\omega = 10.0$ rad/s, $t = 0$ when $\theta = 0$

Determine the following:

| | |
|---|---|
| 1. Maximum emf generated by the coil | $\varepsilon = NBA\omega \sin\theta$<br>$\varepsilon_{max} = NBA\omega = 25.0$ V |
| 2. Angular speed required to induce a maximum emf of 50.0 V | $\varepsilon_{max} = NBA\omega$<br>$\omega = \varepsilon_{max}/NBA = 20.0$ rad/s |
| 3. Angle between $\vec{v}$ and $\vec{B}$ when $\varepsilon = \pm\varepsilon_{max}$ | $\varepsilon = \varepsilon_{max} \sin\theta$<br>$\varepsilon = +\varepsilon_{max}$ when $\theta = \pi/2$<br>$\varepsilon = -\varepsilon_{max}$ when $\theta = 3\pi/2$ |
| 4. Angle between the plane of the coil and $\vec{B}$ when $\varepsilon = \pm\varepsilon_{max}$ | $\varepsilon = \varepsilon_{max}$ when $\Delta\Phi/\Delta t = (\Delta\Phi/\Delta t)_{max}$<br>$\Delta\Phi/\Delta t = (\Delta\Phi/\Delta t)_{max}$ when $\Phi = 0$<br>$\Phi = 0$ when the plane of the coil is parallel to $\vec{B}$, that is $\alpha = 0°$ |
| 5. Angle between $\vec{v}$ and $\vec{B}$ when $\varepsilon = 0$ | $\varepsilon = \varepsilon_{max} \sin\theta$<br>$\varepsilon = 0$ when $\theta = 0$ or $\pi$<br>The angle between $\vec{v}$ and $\vec{B}$ is also $\theta$, hence the angle between $\vec{v}$ and $\vec{B}$ is $\theta = 0$ or $\pi$ when $\varepsilon = 0$ |
| 6. Angle between the plane of the coil and $\vec{B}$ when $\varepsilon = 0$ | $\varepsilon = 0$ when $\Delta\Phi/\Delta t = 0$<br>$\Delta\Phi/\Delta t = 0$ when $\Phi = \pm\Phi_{max}$<br>$\Phi = \pm\Phi_{max}$ when the plane of the coil is perpendicular to $\vec{B}$, that is $\alpha = 90°$ |
| 7. Angle between $\vec{v}$ and $\vec{B}$ when $\varepsilon = 0.500\ \varepsilon_{max}$ | $\varepsilon = \varepsilon_{max} \sin\theta$<br>$0.500\ \varepsilon_{max} = \varepsilon_{max} \sin\theta$<br>$\theta = \sin^{-1}(0.500) = 30°$ and $150°$ |
| 8. Magnetic flux through the coil when $\theta = 60°$ | $\Phi = NBA \cos 60° = 1.25\ T\cdot m^2$ |
| 9. Emf induced in the coil when $\theta = 60°$ | $\varepsilon = NBA\omega \sin\theta = 21.6$ V |

| | |
|---|---|
| 10. Rate of change of magnetic flux when $\theta = 60°$ | $\varepsilon = -N(\Delta\Phi/\Delta t)$<br>$\Delta\Phi/\Delta t = -\varepsilon/N = -0.216$ V |

Example 22.5. In a model ac generator, a 100-turn rectangular coil of dimensions 10.0 cm by 20.0 cm rotates at 120 rev/min in a uniform magnetic field of 0.500 T. (a) What is the maximum emf induced in the coil? (b) What is the instantaneous value of the emf in the coil at $t = (1/16\pi)$s? (c) If we choose $t = 0$ when the emf is zero, what is the smallest value of t for which the emf will have its maximum value?

Given: $N = 100$ turns, $w = 10.0$ cm, $\ell = 20.0$ cm, $B = 0.500$ T, $\omega = 120$ rev/min

Determine: (a) $\varepsilon_{max}$, (b) $\varepsilon$ at $t = (1/16\pi)$ s, (c) time to go from $\varepsilon = 0$ to $\varepsilon = \varepsilon_{max}$

Strategy: Knowing the expression for the induced emf ($\varepsilon$) as a function of time, we can determine (a) $\varepsilon_{max}$, (b) $\varepsilon$ at $t = (1/16\pi)$s, and the time to go from $\varepsilon = 0$ to $\varepsilon = \varepsilon_{max}$.

Solution: $\omega = (120 \text{ rev/min})(\text{min}/60 \text{ s})(2\pi \text{ rad/rev}) = 4\pi$ rad/s

(a) $\varepsilon = NAB\,\omega\sin\omega t = \varepsilon_{max}\sin\omega t$ where $\varepsilon_{max} = NBA\omega = 12.6$ V
(b) $\varepsilon = \varepsilon_{max}\sin\omega t = \varepsilon_{max}\sin[(4\pi \text{ rad/s})(\text{s}/16\pi)] = 3.12$ V
(c) $\varepsilon = \varepsilon_{max}$ at $t = t'$
$\varepsilon_{max} = \varepsilon_{max}\sin\omega t'$
$\omega t' = \sin^{-1}(1) = 1.57$ rad or $t' = 1.57 \text{ rad}/\omega = 0.125$ s
Also, $\varepsilon$ will go from 0 to $\varepsilon_{max}$ in a time equal to one fourth period.

$$\omega = 2\pi f = 2\pi/T \quad \text{or } T = 2\pi/\omega = 0.500 \text{ s and } t' = T/4 = 0.125 \text{ s}$$

Related Text Problems: 22-26 through 22-31.

## 6. Transformers (Section 22.8)

Review: A schematic illustration of a transformer is shown in Fig. 22.16.

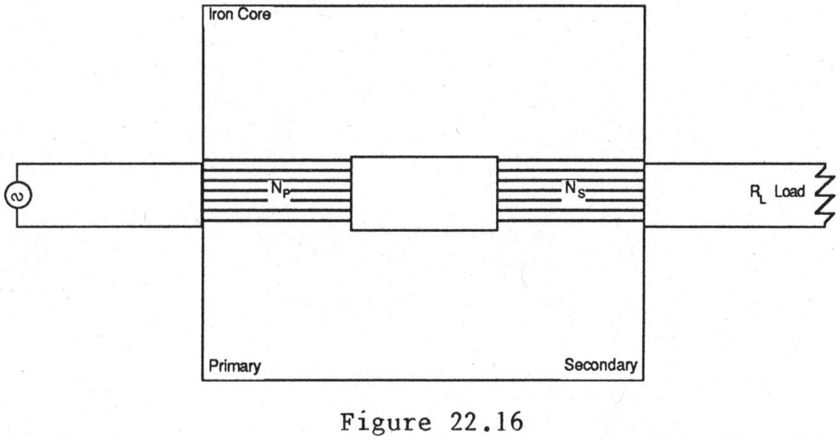


Figure 22.16

An alternating current source supplies a time-varying current to the primary coil. This current creates a time-varying magnetic flux in the iron core and hence through the secondary coil. The changing flux at the secondary induces a time-varying emf in the secondary, which is transmitted to the load. If we assume no flux losses and no energy losses, we obtain

$$\varepsilon_s/\varepsilon_p = N_s/N_p = I_p/I_s$$

Practice: The following data are given for the situation shown in Fig. 22.16.

$$N_p = 100;\; N_s = 10{,}000;\; R_L = 1000\ \Omega$$

Determine the following:

| | |
|---|---|
| 1. Time rate of change of magnetic flux in the primary coil when $\varepsilon_p = 10.0$ V | $(\Delta\Phi/\Delta t)p = -\varepsilon_p/N_p = 0.100\ T\cdot m^2/s = 0.100$ V |
| 2. Time rate of change of magnetic flux in the secondary coil when $\varepsilon_p = 10.0$ V | $(\Delta\Phi/\Delta t)_s = (\Delta\Phi/\Delta t)p = 0.100\ T\cdot m^2/s = 0.100$ V |
| 3. Emf induced in the secondary coil when $\varepsilon_p = 10.0$ V | $\varepsilon_s = -N_s(\Delta\Phi/\Delta t)_s = 1.00 \times 10^3$ V or<br>$\varepsilon_s = \varepsilon_p N_s/N_p = 1.00 \times 10^3$ V |
| 4. Current in the secondary coil when $\varepsilon_p = 10.0$ V | When $\varepsilon_p = 10.0$ V, $\varepsilon_s = 1{,}000$ V<br>$I_s = \varepsilon_s/R_L = 1.00$ A |
| 5. Current in the primary coil when $\varepsilon p = 10.0$ V | When $\varepsilon_p = 10.0$ V, $\varepsilon_s = 1{,}000$ V and<br>$I_s = 1.00$ A<br>$I_p = \varepsilon_s I_s/\varepsilon_p = 100$ A<br>$I_p = I_s N_s/N_p = 100$ A |

Figure 22.17 shows a typical power distribution system as you would see it driving along the road (Fig. 22.17a) and as the power plant engineer sees it on paper (Fig. 22.17b). Also shown is a house fuse box (Fig. 22.17c) and the circuit to the kitchen plug-ins.

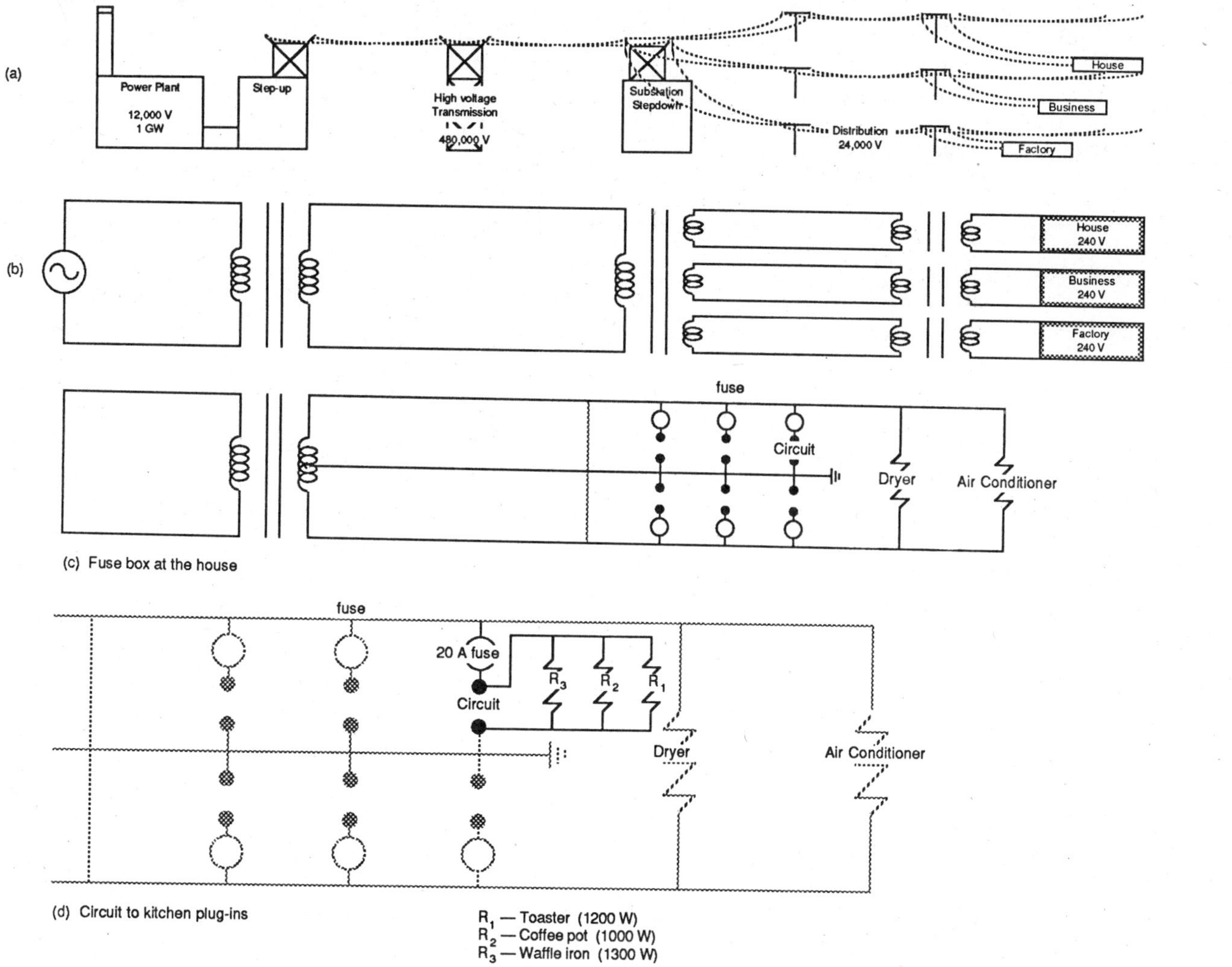


Figure 22.17

As shown in the Fig. 22.17, the plant capacity is 1 GW (1 gigawatt = $1 \times 10^9$ W) at 12,000 V ac, stepped up to 480,000 V for transmission to the substation, distributed via utility poles at 24,000 V, and delivered to customers at 200 A and 240 V. It is 20 miles from the power plant to the substation, and the high-voltage transmission lines have a resistance of 0.150 Ω/mile. The power company sells energy at the rate of 10¢/kW·h.

Determine the following:

| | |
|---|---|
| 6. $N_s/N_p$ for the step-up transformer at the plant | $N_s/N_p = \varepsilon_s/\varepsilon_p$<br>$= 4.80 \times 10^5$ V/$1.20 \times 10^4$ V = 40/1 |

| | | |
|---|---|---|
| 7. | $N_s/N_p$ for the step-down transformer at the substation | $N_s/N_p = \varepsilon_s/\varepsilon_p$<br>$= 2.40 \times 10^4$ V/4.80 x $10^5$ V = 1/20 |
| 8. | $N_s/N_p$ for the transformer on the utility pole | $N_s/N_p = \varepsilon_s/\varepsilon_p$<br>$= 2.40 \times 10^2$ V/2.40 x $10^4$ V = 1/100 |
| 9. | Maximum power available to the homeowner | $P_{max} = IV$<br>= (200 A)(240 V) = 4.80 x $10^4$ W |
| 10. | Current in the service line to the house when the power usage is one half its maximum value | $P = P_{max}/2 = 2.40 \times 10^4$ W<br>$I = P/V = 2.40 \times 10^4$ W/2.40 x $10^2$ V<br>= 100 A |
| 11. | Whether the fuse for the kitchen's plug-in circuit will blow | $P = P_1 + P_2 + P_3 = 3500$ W<br>$I = P/V = 3500$ W/120 V = 29.2 A<br>The 20.0-A fuse will blow |
| 12. | The current in the high-voltage transmission lines at a time when the plant output is 500 megawatt | $I_p = P_p/V_p$<br>$= 5.00 \times 10^8$ W/1.20 x $10^4$ V<br>$= 4.17 \times 10^4$ A<br>$I_s = I_pN_p/N_s$<br>$= (4.17 \times 10^4$ A)(1/40) = 1.04 x $10^3$ A<br>or<br>$P_s = P_p = 5.00 \times 10^8$ W and<br>$I_s = P_s/V_s$<br>$= 5.00 \times 10^8$ W/4.80 x $10^5$ V<br>$= 1.04 \times 10^3$ A |
| 13. | Rate of loss of energy due to joule heating in the high-voltage transmission lines. Assume the plant operates continually at half capacity | $R = (\frac{0.1500\ \Omega}{\text{mile}})(\frac{20\ \text{miles}}{\text{wire}})(2\ \text{wire}) = 6.00\ \Omega$<br>$P = I^2R = (1.04 \times 10^3\ \text{A})^2(6.00\ \Omega)$<br>$= 6.49 \times 10^6$ W |
| 14. | Financial loss per year due to joule heating. Assume the plant operates continually at half its capacity. | $E_{loss} = P_{loss}\Delta t$<br>$= (6.49 \times 10^6\ \text{W})(8.76 \times 10^3\ \text{h})$<br>$= 5.69 \times 10^{10}$ W·h<br><br>Loss = $(E_{loss})$(Cost)<br>$= (5.69 \times 10^{10}\ \text{W·h})(\frac{10¢}{\text{kw·h}})(\frac{\text{kW}}{10^3\ \text{W}})(\frac{\$}{10^2¢})$<br>$= \$5.69 \times 10^6$ |

Example 22.6. An electric generating plant produces electric energy at 12,000 V. We have a piece of electrical equipment that requires 75 A at 220 V. This equipment is housed a distance of 100 km from the generating plant. The transmission line used has a resistance of $1.50 \times 10^{-4}\ \Omega/m$. Electrical energy costs 10.0¢/kW·h. We have two options for getting power to the equipment.

Option A -- Transmit the power at 12,000 V and then use a step-down transformer at the equipment site.

Option B -- Use a step-up transformer to transmit the power at 60,000 V and then use a step-down transformer at the equipment site.

Determine the annual financial loss to the power company if they use Option A over Option B.

Given:

$V_p = 1.2 \times 10^4$ V -- voltage output at the plant
$I_e = 75.0$ A -- current requirement of the equipment
$V_e = 220$ V -- voltage requirement of the equipment
$d = 1.00 \times 10^5$ m -- distance between plant and equipment
$R/\ell = 1.50 \times 10^{-4}\ \Omega/m$ -- resistance per meter of transmission line
Cost = 10.0¢/kW·h -- cost of electrical energy
Option A -- transmit power at $1.20 \times 10^4$ V
Option B -- transmit power at $6.00 \times 10^4$ V

Determine: The annual financial loss to the power company if they use Option A over Option B.

Strategy: From the given information, we can determine the resistance of the transmission line and then the current in the transmission line for Option A and Option B. Knowing the resistance and the current for each option, we can establish the rate of loss of energy for each option. Knowing the rate of loss of energy, the time involved, and the price of energy to the customer, we can determine the financial loss to the power company if they choose Option A over Option B.

Solution: The resistance of the transmission wire is

$$R = (2\text{ wires})(R/\ell)(d) = (2)(1.50 \times 10^{-4}\ \Omega/m)(1.00 \times 10^5\ m) = 30.0\ \Omega$$

Option A -- The power is transmitted at $1.20 \times 10^4$ V and stepped down to $2.20 \times 10^2$ V at the pole transformer. The current in the transmission line ($I_A$) is the same as the current to the primary ($I_p$) of the transformer at the equipment site. Since we have been given $\varepsilon_p$, $\varepsilon_s$, and $I_s$, we can determine $I_p$.

$$I_A = I_p = I_s(\varepsilon_s/\varepsilon_p) = (75.0\ A)(2.20 \times 10^2\ V/1.20 \times 10^4\ V) = 1.38\ A$$

The rate of loss of energy due to joule heating is

$$P_{lossA} = I_A{}^2R = (1.38\ A)^2(30.0\ \Omega) = 57.1\ W$$

Option B -- The power is transmitted at $6.00 \times 10^4$ V and stepped down to $2.20 \times 10^2$ V at the pole transformer. The current in the transmission line ($I_B$) is the same as the current to the primary ($I_p$) of the transformer at the equipment site. Since we have been given $\varepsilon_p$, $\varepsilon_s$, and $I_s$, we can determine $I_p$.

$$I_B = I_p = I_s(\varepsilon_s/\varepsilon_p) = (75.0\ \text{A})(2.20 \times 10^2\ \text{V}/6.00 \times 10^4\ \text{V}) = 0.275\ \text{A}$$

The rate of loss of energy due to joule heating is

$$P_{lossB} = I_B{}^2R = (0.275\ \text{A})^2(30.0\ \Omega) = 2.27\ \text{W}$$

The difference in the rate of loss of energy for these two options is

$$P_{loss} = P_{lossA} - P_{lossB} = 54.8\ \text{W}$$

The difference in energy loss for these two options over a year is

$$E_{loss} = P_{loss}\Delta t = (54.8\ \text{W})(1\ \text{yr})(365\ \text{d/y})(24\ \text{h/d}) = 4.80 \times 10^5\ \text{W}\cdot\text{h}$$

If this energy can be sold for 10¢/kW·h, the financial loss to the power company over a period of a year is

$$\$lost = (E_{loss})(\text{Cost}) = (4.80 \times 10^5\ \text{W}\cdot\text{h})(10¢/\text{kW}\cdot\text{h})(\text{kW}/10^3\ \text{W})(\$1.00/100¢) = \$48.00$$

Related Text Problems: 22-32 through 22-38.

---

## PRACTICE TEST

Take and grade this practice test. Doing so will allow you to determine any weak spots in your understanding of the concepts taught in this chapter. The following section prescribes what you should study further to strengthen your understanding.

Figure 22.18 18 shows two different solenoids wrapped on the same coil form and some associated data.

Figure 22.18

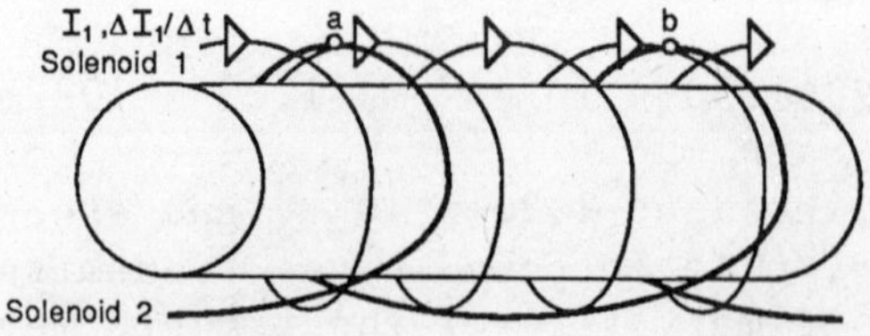


$\ell$ = 10.0 cm -- length of the solenoids
$A = 10.0 \times 10^{-4}\ \text{m}^2$ -- cross-sectional area of the solenoids
$n_1 = 10^3$/m -- turns per unit length for solenoid 1
$n_2 = 10^2$/m -- turns per unit length for solenoid 2
At the instant of interest $I_1 = 2.00$ A and $\Delta I_1/\Delta t = 5.00$ A/s

Determine the following for the instant of interest:

_______ 1. Magnitude of the magnetic field due to solenoid 1
_______ 2. Magnetic flux inside solenoid 1 and solenoid 2
_______ 3. Time rate of change of magnetic flux through each solenoid
_______ 4. Emf induced in each turn of wire of solenoid 1 and 2
_______ 5. Total emf induced in solenoid 1
_______ 6. Total emf induced in solenoid 2
_______ 7. Mutual inductance of the solenoids
_______ 8. Self inductance of solenoid 1
_______ 9. Direction of the current induced in solenoid 2
_______ 10. Energy stored in the magnetic field of solenoid 1

Figure 22.19 shows an inductor in a circuit with a resistor and battery.

Figure 22.19

R = 30.0 Ω
L = 10.0 H
V = 5.00 V

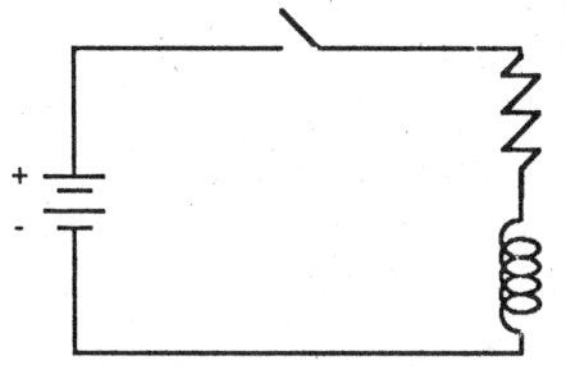


Determine the following:

_______ 11. Time constant for the circuit
_______ 12. Current in the circuit after two time constants
_______ 13. Potential difference across the resistor after two time constants
_______ 14. Potential difference across the inductor after two time constants
_______ 15. The rate of change of current in the inductor after two time constants.

Figure 22.20 shows a current-carrying rectangular coil in a uniform magnetic field at t = 0 s.

Figure 22.20

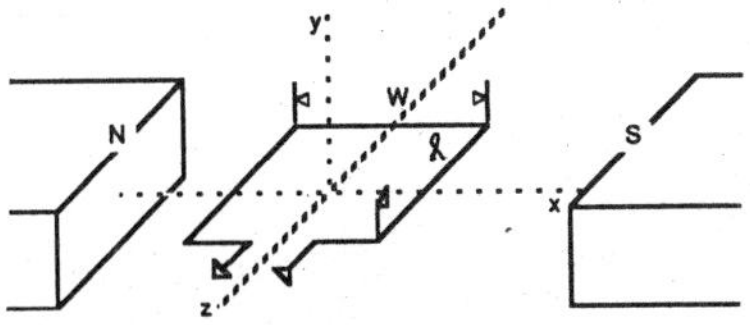


N = 100 turns    ℓ = 10.0 cm
I = 10.0 A    W = 5.00 cm
B = 1.00 T

Determine the following:

_______ 16. Magnitude of the force on each wire of length ℓ at t = 0
_______ 17. Magnitude of the torque due to one of the wires of length ℓ at t = 0
_______ 18. Total torque on the coil at t = 0
_______ 19. Total torque on the coil at a time equal to one-twelfth of the period
_______ 20. Maximum emf which could be generated by turning this coil at the rate of 10 rad/s

(See Appendix I for answers.)

## PRINCIPAL CONCEPTS AND EQUATIONS PRESCRIPTION

Your score on the practice is an excellent measure of your understanding of this chapter. You should now use the following chart to write your own prescription for curing any of your physics ills. Look down the leftmost column to the number of the question(s) you answered incorrectly, read across that row to see which section(s) of the study guide you should return to for further study, and then do the suggested text problems to gain additional experience in working with the particular concept.

| Question Practice Test | Concepts and Equations | Prescription: Principal Concept | Prescription: Text Problems |
|---|---|---|---|
| 1 | $\vec{B}$-field of a solenoid: $B = \mu_0 nI$ | 2 of Ch. 21 | 21-16,20 |
| 2 | Magnetic flux: $\Phi = BA\cos\theta$ | 4 of Ch. 21 | 21-29,30 |
| 3 | Rate of change of flux: $\Delta\Phi/\Delta t$ | 4 of Ch. 21 | 21-32,34 |
| 4 | Induced emf: $\varepsilon = -\Delta\Phi/\Delta t$ | 4 of Ch. 21 | 21-35,36 |
| 5 | Total emf: $\varepsilon = -N\Delta\Phi/\Delta t$ | 4 of Ch. 21 | 21-29,32 |
| 6 | Total emf: $\varepsilon = -N\Delta\Phi/\Delta t$ | 4 of Ch. 21 | 21-34,35 |
| 7 | Mutual inductance: $M_{21} = -\varepsilon_2/(\Delta I_1/\Delta t)$ | 1 | 22-1,2 |
| 8 | Self-inductance: $L = -\varepsilon_1/(\Delta I_1/\Delta t)$ | 1 | 22-3,5 |
| 9 | Induced current | 4 of Ch. 21 | 22-30,32 |
| 10 | Energy in an inductor: $U = LI^2/2$ | 2 | 22-5,6 |
| 11 | RL circuit time constant: $T = L/R$ | 3 | 22-12,15 |
| 12 | RL circuit current: $I = (V/R)(1 - e^{-[t/(L/R]})$ | 3 | 22-13,14 |
| 13 | Potential difference: $V_R = IR$ | 2 of Ch. 19 | 22-11,12 |
| 14 | Potential difference: $V_L = V - V_R$ | 3 | 22-13,14 |
| 15 | Potential difference: $V_L = L\Delta I/\Delta t$ | 3 | 22-11,15 |
| 16 | Magnetic force on a current: $F = I\ell B\sin\theta$ | 3 of Ch. 21 | 21-22,23 |
| 17 | Torque: $\tau = rF\sin\theta$ | 2 of Ch. 6 | 6-12,16 |
| 18 | Total torque: $\vec{\tau} = \vec{\tau}_1 + \vec{\tau}_2$ | 2 of Ch. 6 | 22-19,20 |
| 19 | Torque: $\tau = NIAB\cos\omega t$ | 4 | 22-21,22 |
| 20 | Emf of an ac generator: $\varepsilon = NBA\omega\cos\omega t$ | 5 | 22-28,29 |

# 23 Alternating Current and Electrical Safety

## RECALL FROM PREVIOUS CHAPTERS

| Previously learned concepts and equations frequently used in this chapter | Text Section | Study Guide Page |
|---|---|---|
| Ohm's law: $R = V/I$ | 19.3 | 308 |
| Power: $P = IV = I^2R = V^2/R$ | 19.7 | 315 |
| emf of an ac generator: $\varepsilon = NBA\omega \cos 2\pi ft$ | 22.6 | 382 |

## NEW IDEAS IN THIS CHAPTER

| Concepts and equations introduced | Text Section | Study Guide Page |
|---|---|---|
| Phasors | 23.2 | 393 |
| Initial phase angle | 23.2 | 393 |
| Phase angle difference | 23.2 | 393 |
| RLC series circuits: | 23.3,4 | 398 |
| Circuit current: $I = I_o \sin 2\pi ft$ | | |
| Circuit voltage: | | |
| $V = V_o \sin(2\pi ft + \phi_o)$, $V_o = I_oZ$ | | |
| $Z = [R^2 + (X_L - X_C)^2]^{1/2}$ | | |
| Resistor voltage: | | |
| $V_R = V_{Ro} \sin 2\pi ft$, $V_{Ro} = I_oR$ | | |
| Capacitor voltage: | | |
| $V_C = -V_{Co} \cos 2\pi ft$, $V_{Co} = I_oX_C$, $X_C = 1/2\pi fC$ | | |
| Inductor voltage: | | |
| $V_L = V_{Lo} \cos 2\pi ft$, $V_{Lo} = I_oX_L$, $X_L = 2\pi fL$ | | |
| Phasor Relationship: $\vec{V}_o = \vec{V}_{Ro} + \vec{V}_{Co} + \vec{V}_{Lo}$ | | |
| AC meter readings: | 23.5 | 406 |
| $I_{meter} = I_{rms} = [(I^2)_{av}]^{1/2} = I_o/\sqrt{2} = 0.707I_o$ | | |
| $V_{meter} = V_{rms} = [(V^2)_{av}]^{1/2} = V_o/\sqrt{2} = 0.707V_o$ | | |
| Power: $P = (I_oV_o/2)\cos\phi_o = IV\cos\phi_o = I_o^2R/2$ | 23.5 | 406 |

## PRINCIPAL CONCEPTS AND EQUATIONS

### 1. Phasor Diagrams (Section 23.2)

Review: Phasors are rotating vectors representing voltages or currents. Figure 23.1 shows the phasor representing the voltage $V = V_o \sin(2\pi ft + \phi_i)$.

where $V_o$ = magnitude of the phasor
$2\pi f$ = angular speed of the phasor
$t$ = any time
$\phi_i$ = initial phase angle, tells us the orientation of the phasor at $t = 0$
$V = V_o \sin((2\pi ft + \phi_i)$ = projection of the phasor onto the y axis

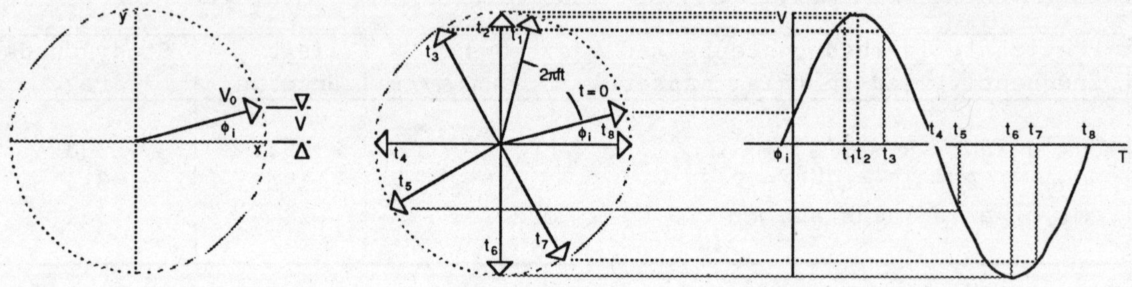


(a) $t > 0$ (b) various t (c) V vs t

Figure 23.1

Figure 23.2 shows two phasors representing the voltage $V = V_o \sin(2\pi ft + \phi_i)$ and the current $I = I_o \sin(2\pi ft + \phi_i + \phi_o) = I_o \sin(2\pi ft + \phi_i')$

where $V_o$ and $I_o$ = magnitude of the phasors
$2\pi f$ = angular speed of the phasors
$t$ = any time
$\phi_i$ = initial phase angle of $V_o$, tells the orientation of $V_o$ at $t = 0$
$\phi_i'$ = initial phase angle of $I_o$, tells the orientation of $I_o$ at $t = 0$ ($\phi_i' = \phi_i + \phi_o$)
$\phi_o$ = phase angle difference, tells the angular seperation of the two phasors at all times
$V = V_o \sin(2\pi ft + \phi_i)$ = projection of the phasor $V_o$ onto the y axis
$I = I_o \sin(2\pi ft + \phi_i')$ = projection of the phasor $I_o$ onto the y axis

Notice that in this example, the current leads the voltage by $\pi/2$ rad; hence $\phi_o = \pi/2$.

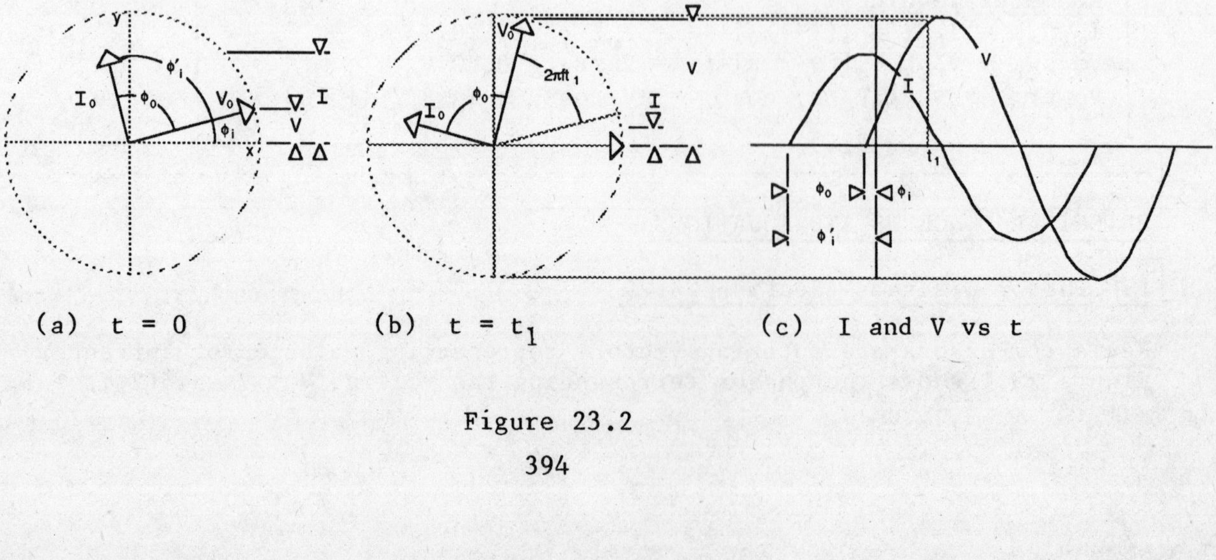


(a) $t = 0$ (b) $t = t_1$ (c) I and V vs t

Figure 23.2

As a matter of convenience, we choose the value of the initial phase angle for the phasor $I_o$ such that the current has the form $I = I_o \sin 2\pi ft$. This may be accomplished for this case by choosing the initial phase angle $\phi_i'$ to be zero.

If $\phi_i'$ is zero, the phasors $I_o$ and $V_o$ appear as shown in figure 23.3. Recall that

$$\phi_i' = \phi_i + \phi_o$$

If we choose $\phi_i' = 0$, then

$$\phi_o = -\phi_i$$

For this case $\phi_i = -90°$, hence $\phi_o = -\phi_i = 90°$, because we chose $V_o$ as a rference. However, if we use $I_o$ as the reference, then $\phi_o = 90°$ and we have

$$\phi_i = \phi_o = -90°$$

That is, the initial phase angle $\phi_i$ for the phasor $V_o$ is now equal to the phase angle difference $\phi_o$.

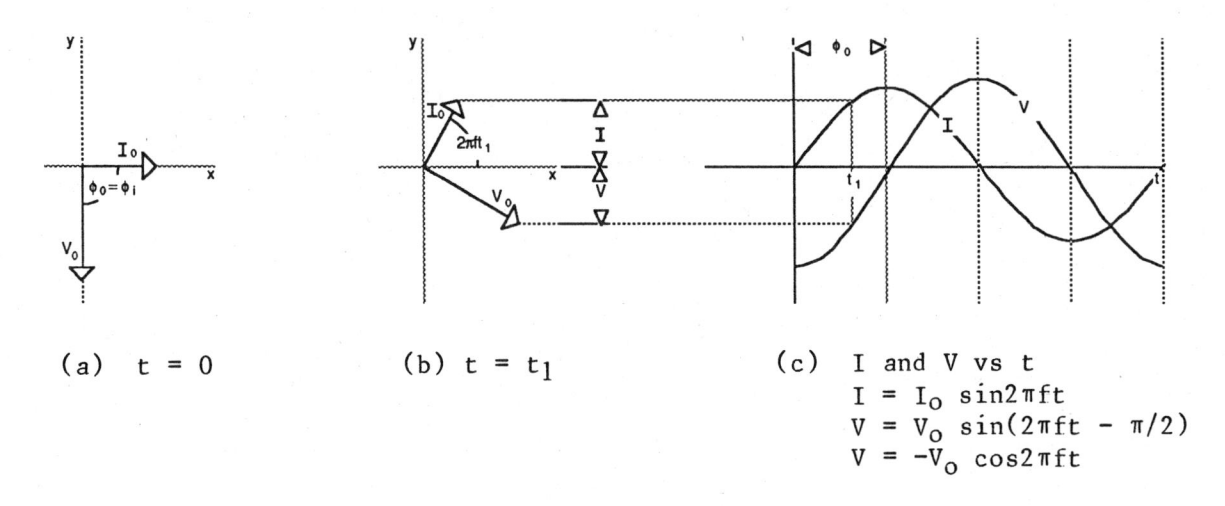

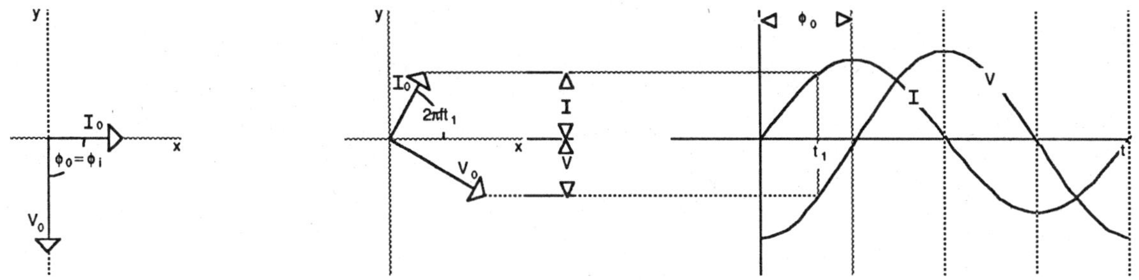


(a) $t = 0$ (b) $t = t_1$ (c) I and V vs t

$I = I_o \sin 2\pi ft$

$V = V_o \sin(2\pi ft - \pi/2)$

$V = -V_o \cos 2\pi ft$

Figure 23.3

Practice: For the situation shown in Fig. 23.1, we have that $V_o = 10.0$ V, $f = 60.0$ Hz, and $V = 8.66$ V at $t = T/4$ (i.e., at one-fourth the period).

Determine the following:

| | |
|---|---|
| 1. Period for the phasor | $T = 1/f = 0.0167$ s |
| 2. Initial phase angle | $V = V_o \sin(2\pi ft + \phi_i)$<br>$V = 8.66$ V at $t = T/4$<br>$8.66\text{ V} = (10.0\text{ V}) \sin(\frac{2\pi}{T}\frac{T}{4} + \phi_i)$ |

| | |
|---|---|
| | $0.866 = \sin(\frac{\pi}{2} + \phi_i)$<br>$\frac{\pi}{2} + \phi_i = \sin^{-1}(0.866) = \frac{\pi}{3}, \frac{2\pi}{3}$<br>$\phi_i = -\pi/6,\ \pi/6$<br>Since Fig. 23.1 shows a positive value for $\phi_i$, use $\phi_i = \pi/6$. |
| 3. The time when V = 5.00 V | $V = V_o \sin(2\pi ft + \phi_i)$<br>$5.00\ V = (10.0\ V) \sin(\frac{2\pi}{T} t + \frac{\pi}{6})$<br>$0.500 = \sin(\frac{2\pi}{T} t + \frac{\pi}{6})$<br>$\frac{2\pi}{T} t + \frac{\pi}{6} = \sin^{-1}(0.500) = \frac{\pi}{6}, \frac{5\pi}{6}$<br>$\frac{2\pi}{T} t = 0, \frac{2\pi}{3}$<br>$t = 0,\ T/3 = 5.57 \times 10^{-3}$ s |
| 4. The voltage at t = 3T/4 | $V = V_o \sin(2\pi ft + \phi_i)$<br>$V = (10.0\ V) \sin(\frac{2\pi}{T}\frac{3T}{4} + \frac{\pi}{6})$<br>$V = (10.0\ V) \sin(\frac{3\pi}{2} + \frac{\pi}{6}) = -8.66\ V$ |

The voltage phasor in Fig. 23.2 is the same as in Fig. 23.1. In addition, we know that $I_o = 5.00$ A and $\phi_o = \pi/2$ radian in Fig. 23.2.

| | |
|---|---|
| 5. Whether the currents leads the voltage or lags it | I leads V by 90° = $\pi/2$ rad |
| 6. An expression for I as a function of t but without $\phi_o$ | $I = I_o \sin(2\pi ft + \phi_o + \phi_i)$<br>$= I_o \sin[(2\pi ft + \phi_i) + \frac{\pi}{2}]$<br>$= I_o [\sin(2\pi ft + \phi_i) \cos\frac{\pi}{2} + \cos(2\pi ft + \phi_i) \sin\frac{\pi}{2}]$<br>$= I_o \cos(2\pi ft + \phi_i)$ |
| 7. The current at t = T/2 | $I = I_o \cos(2\pi ft + \phi_i)$<br>$= (5.00\ A) \cos(\frac{2\pi}{T}\frac{T}{2} + \frac{\pi}{6}) = -2.50\ A$ |

Example 23.1. The alternating current and voltage in a circuit are

$$I = 10 \cos(60\pi t + \pi/12), \quad V = 120 \sin(60\pi t + \pi/3)$$

Determine the (a) amplitude of the current and voltage, (b) frequency of the current and voltage, (c) period of the oscillation, (d) current and voltage at $t = 0$ and $t = T/2$, and (e) phase angle between the current and voltage.

Given: $I = 10 \cos(60\pi t + \pi/12)$, $V = 120 \sin(60\pi t + \pi/3)$

Determine:

(a) $I_o$ and $V_o$ = amplitude of current and voltage
(b) $f$ = frequency of the current and voltage
(c) $T$ = period of oscillation
(d) $I$ and $V$ at $t = 0$ and $t = T/2$
(e) $\phi_o$ = phase angle difference between $I$ and $V$

Strategy: Knowing that $I = I_o \cos(2\pi ft + \phi_i)$ and $V = V_o \sin(2\pi ft + \phi_i)$, we can determine $I_o$, $V_o$, and $f$. Knowing the frequency, we can determine the period. From the given expressions for I and V, we can determine I and V at any time and also the phase angle between them.

Solution:

(a) $I = I_o \cos(2\pi ft + \phi_i) = 10 \cos(60\pi t + \pi/12)$
$V = V_o \sin(2\pi ft + \phi_i) = 120 \sin(60\pi t + \pi/3)$
Comparing these, we see that $I_o = 10.0$ A and $V_o = 120$ V.

(b) Comparing the preceding expressions for either I or V, we have

$$2\pi ft = 60\pi t \qquad f = 30 \text{ Hz}$$

(c) The period of oscillation is

$$T = 1/f = 3.33 \times 10^{-2} \text{ s}$$

(d) At $t = 0$, $I = 10 \cos(\pi/12) = 9.66$ A
$V = 120 \sin(\pi/3) = 104$ V

At $t = T/2$, $I = 10 \cos(60\pi t + \frac{\pi}{12}) = 10 \cos(\frac{2\pi}{T} t + \frac{\pi}{12})$

$$= 10 \cos(\frac{2\pi}{T}\frac{T}{2} + \frac{\pi}{12}) = 10 \cos(13\pi/12) = -9.66 \text{ A}$$

$$V = 120 \sin(60\pi t + \frac{\pi}{3}) = 120 \sin(\frac{2\pi}{T} t + \frac{\pi}{3})$$

$$= 120 \sin(\frac{2\pi}{T}\frac{T}{2} + \frac{\pi}{3}) = 120 \sin \frac{4\pi}{3} = -104 \text{ V}$$

(e) The voltage is given by $V = 120 \sin(60\pi t + \pi/3)$. If we can express the current as $I = 10 \sin(60\pi t + \pi/3 + \phi_o)$, we can determine the phase angle $\phi_o$ between I and V.

$$\sin(60\pi t + \pi/3 + \phi_o) = \cos(60\pi t + \pi/12)$$
$$= \cos(60\pi t + \pi/12 + 3\pi/12 - 3\pi/12)$$
$$= \cos(60\pi t + \pi/3 - 3\pi/12)$$
$$= \sin(60\pi t + \pi/3 - 3\pi/12 + \pi/2)$$
$$= \sin(60\pi t + \pi/3 + \pi/4)$$
$$\phi_o = \pi/4$$

Related Text Problems: 23-1, 23-2.

## 2. Resistors, Capacitors, and Inductors in ac Circuits (Sections 23.3, 23.4).

Review: As a matter of convenience, we choose the value of the initial phase angle for the current such that it is of the form

$$I = I_o \sin 2\pi ft$$

As we have seen in the previous section, this choice for the initial phase angle for the current forces the initial phase angle for the voltage to be the phase angle difference between the phasors representing the current and voltage.

Resistance:

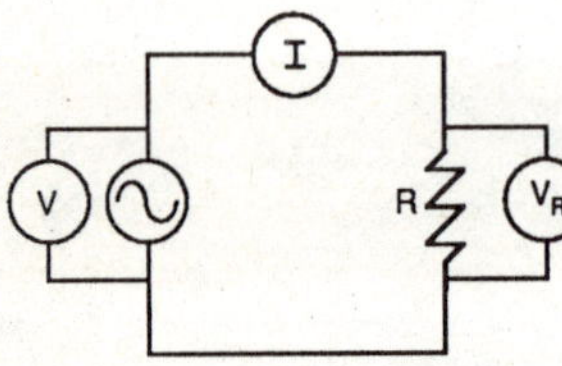


Figure 23.4

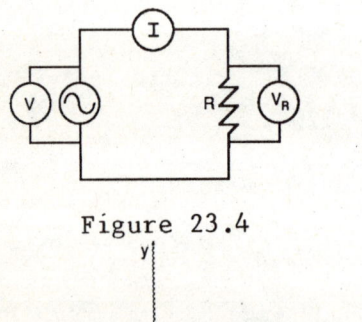

$I = I_o \sin 2\pi ft$ = circuit current

$V = V_R = V_o \sin(2\pi ft + \phi_o)$ = circuit and resistor voltage. Since $\phi_o = 0$, we have

$V = V_R = V_o \sin 2\pi ft$ = circuit and resistor voltage

$V_o = I_oR$

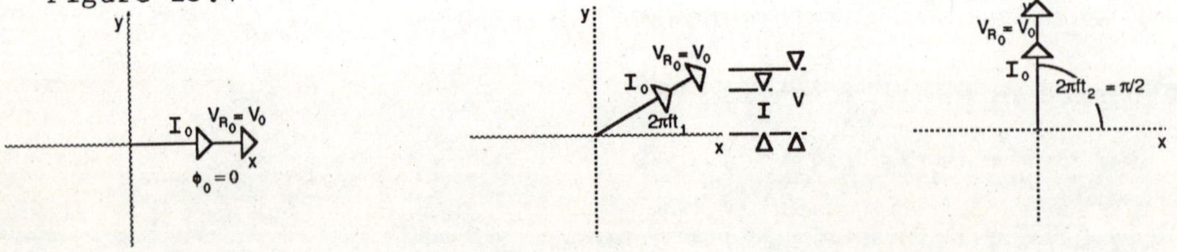


(a) $t = 0$ (b) $t = t_1$ (c) $t = t_2 = T/4$

Figure 23.5

Figure 23.6

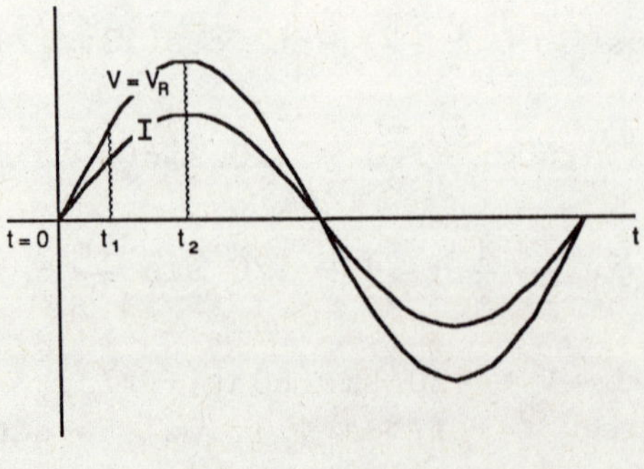

Capacitance:

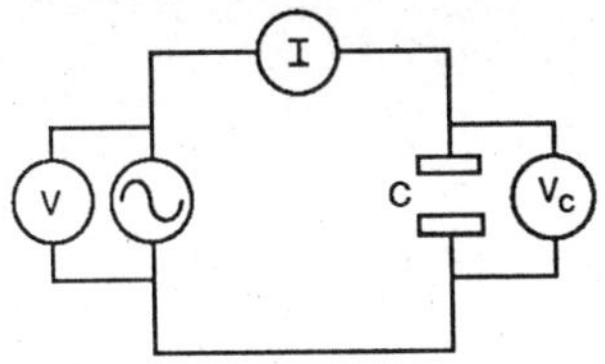


Figure 23.7

$I = I_o \sin 2\pi ft$ = circuit current

$V = V_C = V_{Co} \sin(2\pi ft + \phi_o)$ = circuit and capacitor voltage. Since $\phi_o = -90°$,

$V = V_C = -V_{Co} \cos 2\pi ft$ = circuit and capacitor voltage

$V_{Co} = I_oX_C$ = maximum capacitor voltage

$X_C = 1/2\pi fC$ = capacitive reactance ·

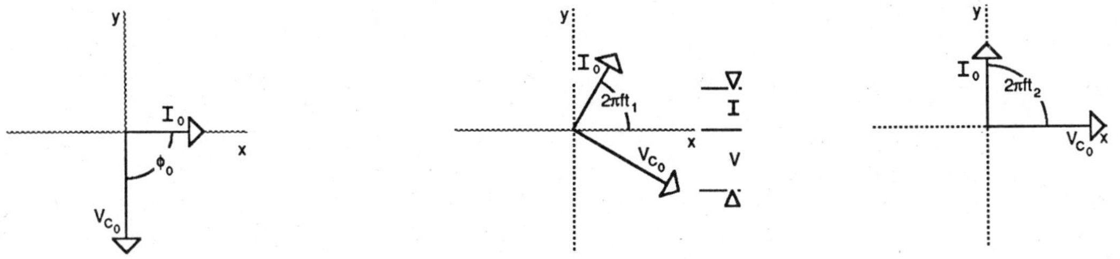


(a) $t = 0$ (b) $t = t_1$ (c) $t = t_2 = T/4$

Figure 23.8

Figure 23.9

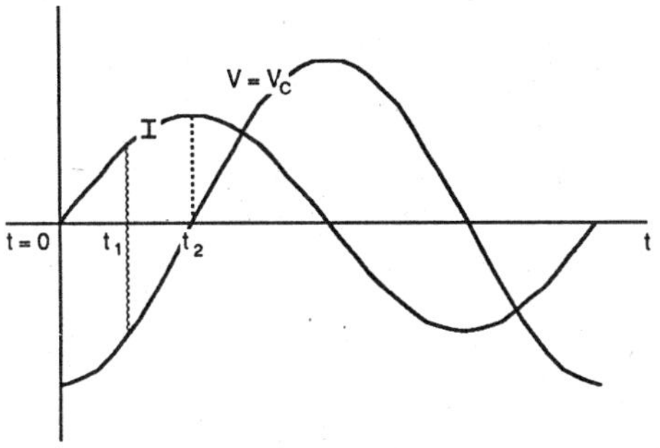


Inductance:

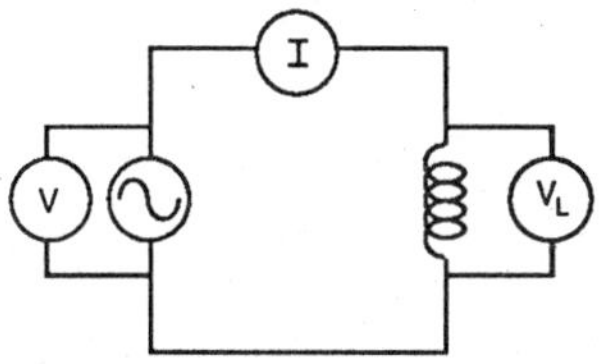


Figure 23.10

$I = I_o \sin 2\pi ft$ = circuit current

$V = V_L = V_{Lo} \sin(2\pi ft + \phi_o)$ = circuit and inductor voltage. Since $\phi = +90°$,

$V = V_L = V_{Lo} \cos 2\pi ft$ = circuit and inductor voltage

$V_{Lo} = I_oX_L$ = maximum inductor voltage

$X_L = 2\pi fL$ = inductive reactance

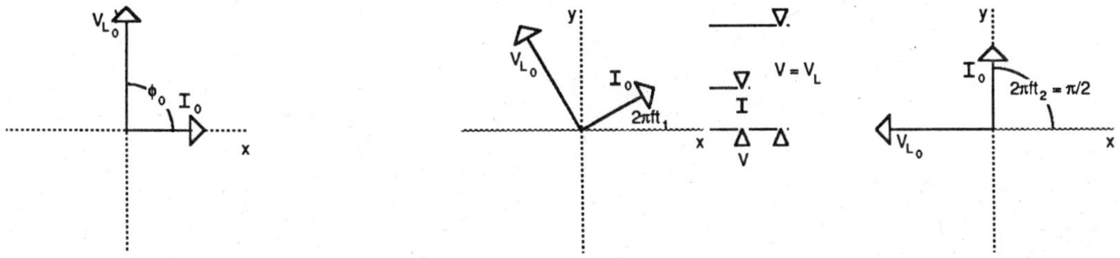


(a) $t = 0$ (b) $t = t_1$ (c) $t = t_2 = T/4$

Figure 23.11

Figure 23.12

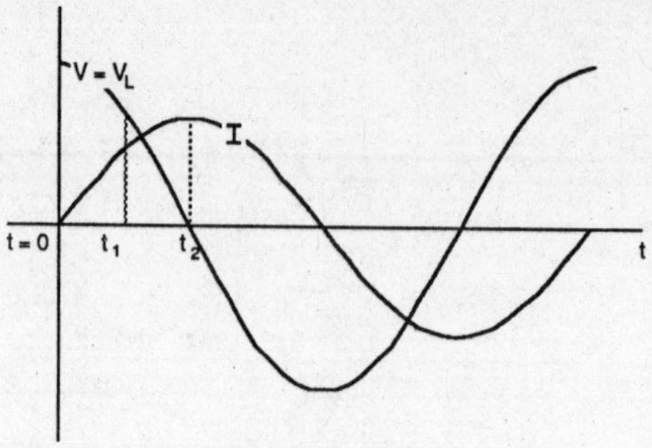


Resistance and Capacitance in Series:

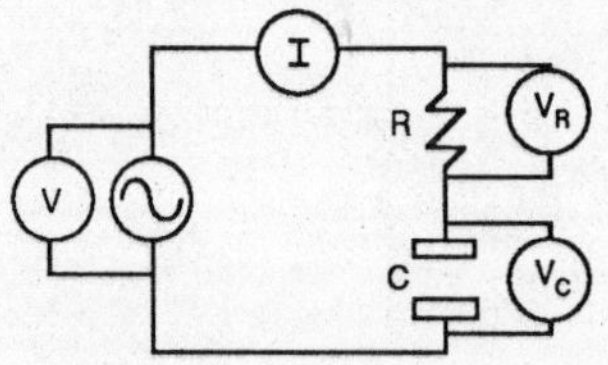


Figure 23.13

$I = I_0 \sin 2\pi ft$ = circuit current

$V = V_0 \sin(2\pi ft + \phi_0)$ = circuit voltage

$V_R = V_{Ro} \sin 2\pi ft$ = resistor voltage

$V_C = -V_{Co} \cos 2\pi ft$ = capacitor voltage

$V_{Ro} = I_0 R$

$V_{Co} = I_0 X_C$, where $X_C = 1/2\pi fC$

$V_0 = I_0 Z$, where $Z = [R^2 + X_C^2]^{1/2}$

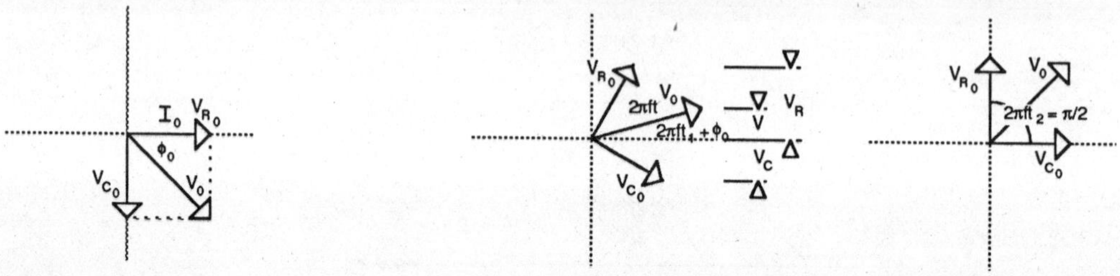


(a) $t = 0$ (b) $t = t_1$ (c) $t = t_2 = T/4$

Figure 23.14

From Fig. 23.14 it is easy to see the following:

$\vec{V}_0 = \vec{V}_{Ro} + \vec{V}_{Co}$

$V = V_R + V_C$, where

$V = V_0 \sin(2\pi ft + \phi_0)$;  $V_R = V_{Ro} \sin 2\pi ft$;  $V_C = -V_{Co} \cos 2\pi ft$

Figure 23.15

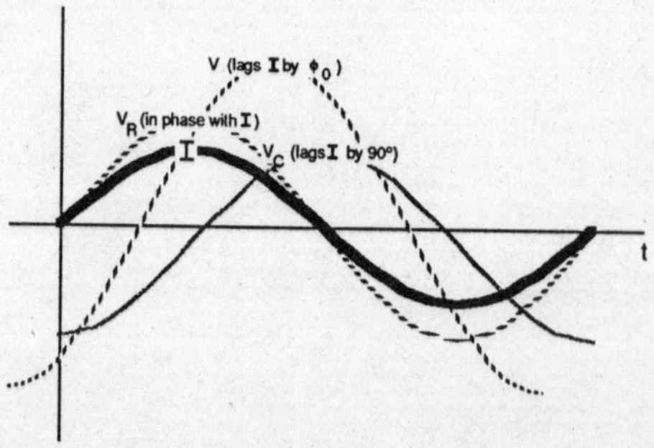

Resistance and Inductance in Series:

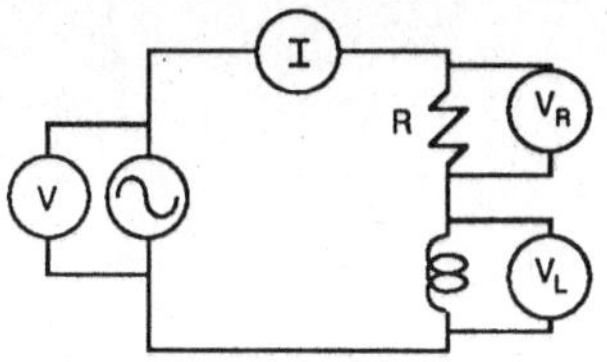


Figure 23.16

$I = I_o \sin 2\pi ft$ = circuit current
$V = V_o \sin(2\pi ft + \phi_o)$ = circuit voltage
$V_R = V_{Ro} \sin 2\pi ft$ = resistor voltage
$V_L = V_{Lo} \cos 2\pi ft$ = inductor voltage
$V_{Ro} = I_o R$
$V_{Lo} = I_o X_L$, where $X_L = 2\pi fL$
$V_o = I_o Z$, where $Z = [R^2 + X_L{}^2]^{1/2}$

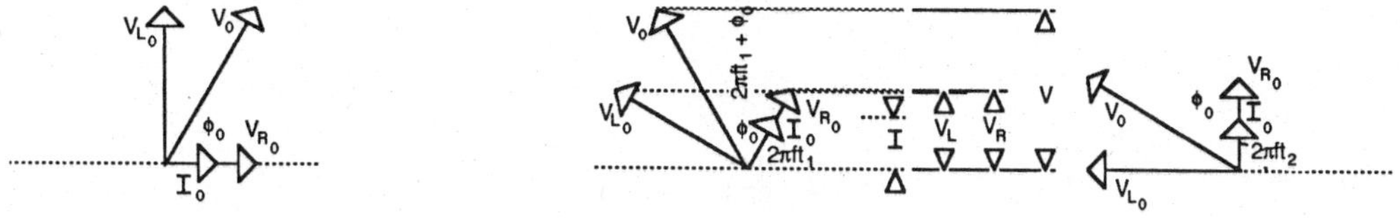

(a) $t = 0$ (b) $t = t_1$ (c) $t = t_2/T/4$

Figure 23.17

From Fig. 23.17, it is easy to see the following:

$\vec{V}_o = \vec{V}_{Ro} + \vec{V}_{Lo}$
$V = V_L + V_R$, where
$V = V_o \sin(2\pi ft + \phi_o)$; $V_R = V_{Ro}\sin 2\pi ft$; $V_L = V_{Lo}\cos 2\pi ft$

Figure 23.18

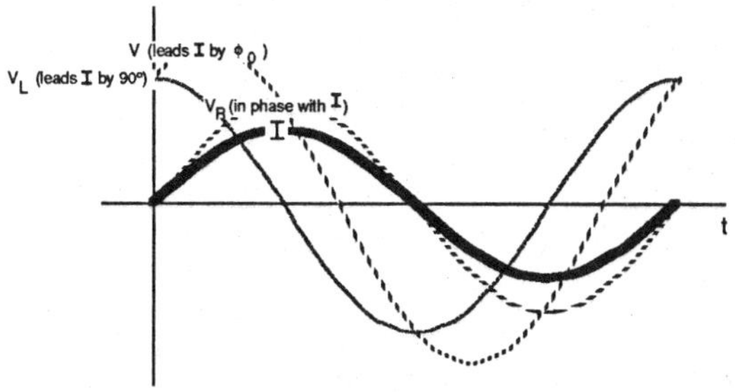


Resistance, Capacitance, and Inductance in Series:

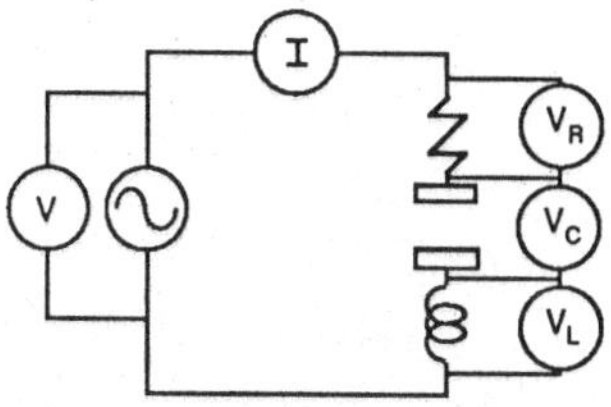


Figure 23.19

$I = I_o \sin 2\pi ft$ = circuit current
$V = V_o \sin(2\pi ft + \phi_o)$ = circuit voltage
$V_R = V_{Ro} \sin 2\pi ft$ = resistor voltage
$V_C = -V_{Co} \cos 2\pi ft$ = capacitor voltage
$V_L = V_{Lo} \cos 2\pi ft$ = inductor voltage
$V_{Ro} = I_o R$
$V_{Co} = I_o X_C$, where $X_C = 1/2\pi fC$
$V_{Lo} = I_o X_L$, where $X_L = 2\pi fL$
$V_o = I_o Z$, where $Z = [R^2 + (X_L - X_C)^2]^{1/2}$

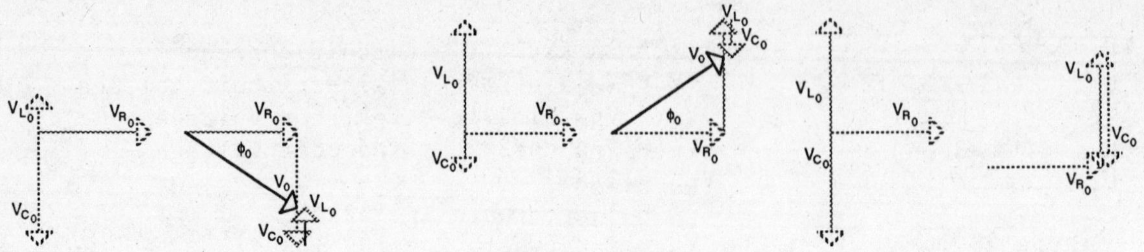


(a) $V_{Co} > V_{Lo}$
$\phi_o$ is -

(b) $V_{Lo} > V_{Co}$
$\phi_o$ is +

(c) $V_{Lo} = V_{Co}$
$\phi$ is 0

Figure 23.20

From Fig. 23.20 it is easy to see the following:

$$\vec{V}_o = \vec{V}_{Ro} + \vec{V}_{Co} + \vec{V}_{Lo}$$
$$V = V_R + V_C + V_L$$
$$V = V_o \sin(2\pi ft + \phi_o); \; V_R = V_{Ro} \sin 2\pi ft; \; V_C = V_{Co} \cos 2\pi ft$$
$$V_L = V_{Lo} \cos 2\pi ft$$

Figure 23.21 shows the curves for only Fig. 23.20b.

Figure 23.21

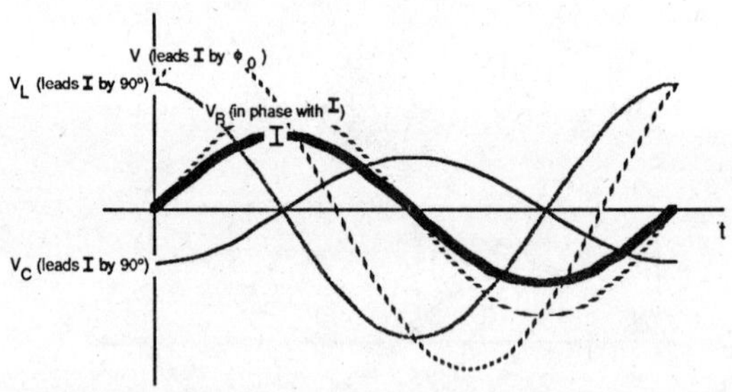


Practice: Consider the series RC circuit of Fig. 23.20. Values for the maximum resistor voltage, maximum capacitor voltage, current amplitude, and frequency are given.

Figure 23.22

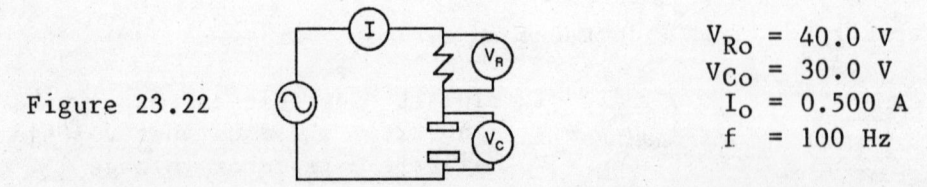

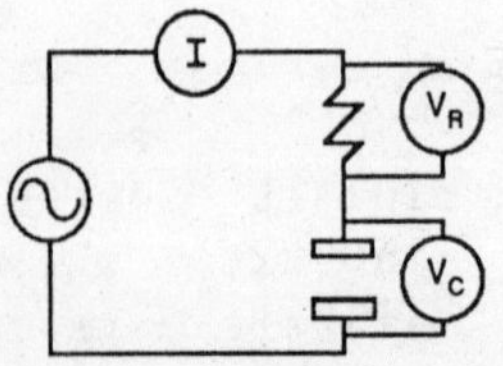


$V_{Ro}$ = 40.0 V
$V_{Co}$ = 30.0 V
$I_o$ = 0.500 A
f = 100 Hz

Determine the following:

| | |
|---|---|
| 1. Resistance of the resistor | $R = V_{Ro}/I_o = 80.0\ \Omega$ |
| 2. Capacitive reactance | $X_C = V_{Co}/I_o = 60.0\ \Omega$ |
| 3. Capacitance of the capacitor | $X_C = 1/2\pi fC$ or $C = 2.65 \times 10^{-5}$ F |

| | |
|---|---|
| 4. Amplitude of the ac voltage source | $V_o = [V_{Ro}^2 + V_{Co}^2]^{1/2} = 50.0$ V |
| 5. Phase angle between the circuit current and the voltage across the capacitor | The voltage across the capacitor lags the circuit current by 90°. |
| 6. Phase angle between the circuit current and the circuit voltage | 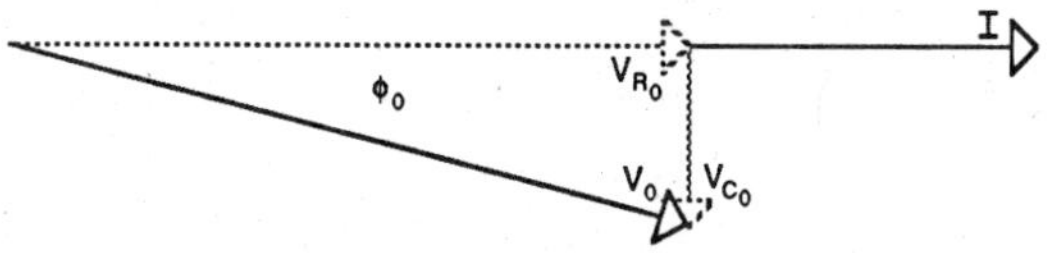<br>$\phi_o = \tan^{-1}(V_{Co}/V_{Ro}) = 36.9°$<br>The voltage lags the current |

Consider the series RL circuit shown in Fig. 23.23.

Figure 23.23

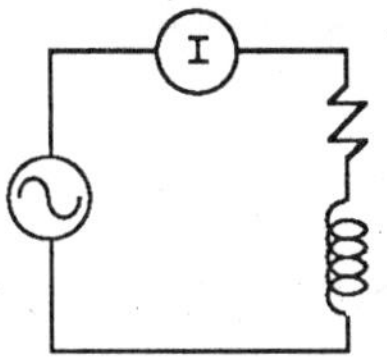


$R = 200\ \Omega$
$L = 1.00$ H
$f = 100$ Hz
$I_o = 0.200$ A

Determine the following:

| | |
|---|---|
| 7. Maximum resistor voltage | $V_{Ro} = I_oR = 40.0$ V |
| 8. Maximum inductor voltage | $V_{Lo} = I_oX_L = I_o2\pi fL = 126$ V |
| 9. Impedance of the circuit | $R = 200\ \Omega$, $X_L = 2\pi fL = 628\ \Omega$<br>$Z = [R^2 + X_L^2]^{1/2} = 659\ \Omega$ |
| 10. Amplitude of the ac voltage source | $V_o = [V_{Ro}^2 + V_{Lo}^2]^{1/2} = 132$ V<br>$V_o = I_oZ = 132$ V |
| 11. Phase angle between the circuit current and the voltage across the inductor | The voltage across the inductor leads the circuit current by 90°. |

12. Phase angle between the circuit current and the circuit voltage

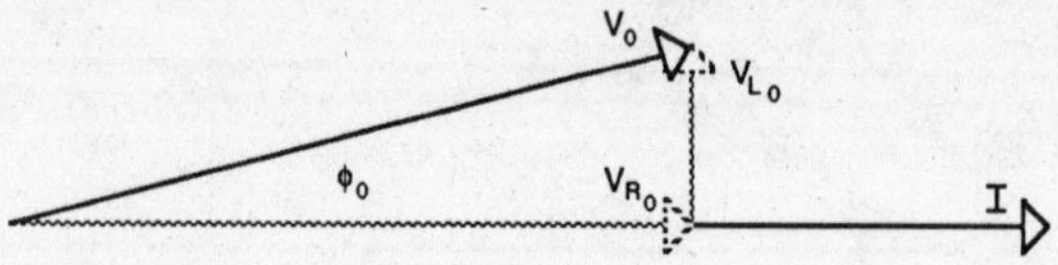


$\phi_o = \tan^{-1}(V_{Lo}/V_{Ro}) = 72.4°$
The voltage leads the current.

---

Consider the series RLC circuit and values shown in Fig. 23.24.

Figure 23.24

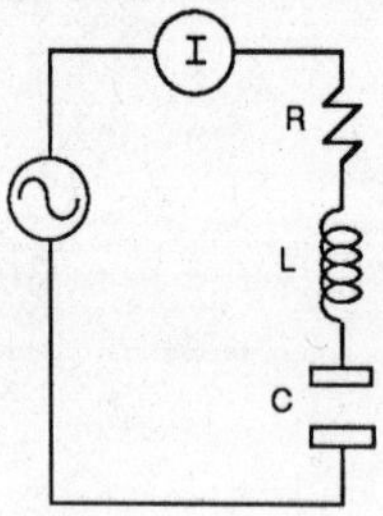


$R = 200\ \Omega$
$V_{Ro} = 100$ V
$V_{Lo} = 150$ V

$C = 1.00 \times 10^{-5}$ f
f = 100 Hz
Resonance exists

Determine the following:

| | |
|---|---|
| 13. Amplitude of the current in the circuit | $I_o = V_{Ro}/R = 0.500$ A |
| 14. Capacitive reactance | $X_C = 1/2\pi fC = 159\ \Omega$ |
| 15. Maximum capacitor voltage | $V_{Co} = I_oX_C = 79.5$ V |
| 16. Inductive reactance | $X_L = V_{Lo}/I_o = 300\ \Omega$ |
| 17. Inductance | $L = X_L/2\pi f = 0.478$ H |

18. Amplitude of the ac voltage source

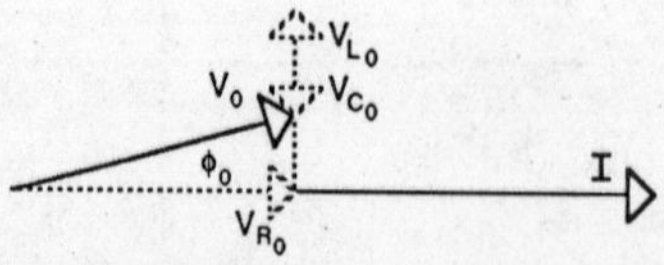


$V_o = [V_{Ro}{}^2 + (V_{Lo} - V_{Co}{}^2]^{1/2} = 122$ V

| | |
|---|---|
| 19. Total impedance | $Z = V_o/I_o = 244\ \Omega$<br>$Z = [R^2 + (X_L - X_C)^2]^{1/2} = 245\ \Omega$ |

| | |
|---|---|
| 20. Phase angle between the circuit current and voltage | $\phi_o = \tan^{-1}\dfrac{(V_{Lo} - V_{Co})}{V_{Ro}} = \tan^{-1}(0.705)$ <br> $= 35.2°$ |

Example 23.2. An RLC series circuit has elements with the following values: $R = 100.0\ \Omega$, $L = 0.200$ H, $C = 2.50 \times 10^{-5}$ F. The alternating voltage has an amplitude of 50.0 V and a frequency of 100 Hz. Calculate the (a) capacitive and inductive reactance, (b) impedance of the circuit, (c) current amplitude, (d) maximum voltage across the resistor, capacitor, and inductor, (e) phase angle difference between the circuit current and voltage, and (f) amount we must change the inductor to create resonance.

Given: $R = 100\ \Omega$, $L = 0.200$ H, $C = 2.50 \times 10^{-5}$ F, $V_o = 50.0$ V, $f = 100$ Hz

Determine: (a) $X_C$ , $X_L$, (b) $Z$, (c) $I_o$, (d) $V_{Ro}$, $V_{Co}$, $V_{Lo}$, (e) $\phi_o$, (f) the value of L to cause resonance

Strategy: (a) Knowing C, L, and f, we can determine $X_C$ and $X_L$. (b) Knowing R, $X_C$, and $X_L$, we can determine Z. (c) Knowing Z and $V_o$, we can determine $I_o$. (d) Knowing $I_o$, R, $X_C$, and $X_L$, we can determine $V_{Ro}$, $V_{Co}$, and $V_{Lo}$. (e) Knowing $V_{Ro}$, $V_{Co}$, and $V_{Lo}$, we can determine $\phi_o$. (f) Knowing $X_C$, we can determine the value of $X_L$ and hence L for resonance.

Solution:

(a) $X_C = 1/2\pi fC = 63.7\ \Omega$ $\quad X_L = 2\pi fL = 126\ \Omega$

(b) $Z = [R^2 + (X_L - X_C)^2]^{1/2} = 118\ \Omega$

(c) $I_o = V_o/Z = 0.424$ A

(d) $V_{Ro} = I_oR = 424$ V; $\quad V_{Co} = I_oX_C = 27.0$ V; $\quad V_{Lo} = I_oX_L = 53.4$ V

(e) $\phi_o = \tan^{-1}[(V_{Lo} - V_{Co})/V_{Ro}] = 31.9°$

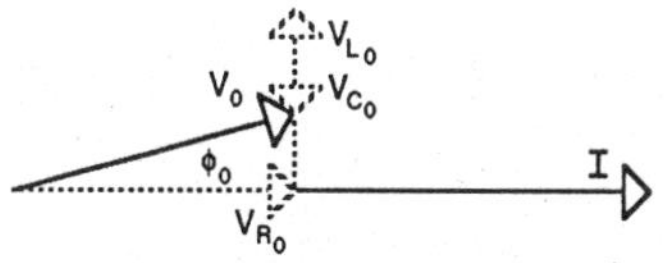


(f) For resonance, $X_C = X_L = 2\pi fL$ $\quad L = X_C/2\pi f = 0.101$ H

The inductance must be reduced by 0.0990 H (0.200 H - 0.101 H).

Related Text Problems: R only -- 23-3
C only -- 23-4, 27-7
L only -- 23-12, 23-13, 23-14
RC -- 23-8 through 23-11, 23-26
RL -- 23-15, 23-16, 23-18, 23-19, 23-20, 23-24, 23-25
RLC -- 23-21, 23-22, 23-23, 23-27, 23-28

## 3. Power and RMS Values (Section 23.5)

Review: So far, we have considered maxmimum current and voltage values:

$I_o$ = maximum circuit current
$V_o$ = maximum circuit voltage
$V_{Ro}$, $V_{Co}$, $V_{Lo}$ = maximum voltage across the resistor, capacitor and inductor

We have also considered instantaneous current and voltage values, for instance

$I = I_o \sin 2\pi ft$ = instantaneous circuit current
$V = V_o \sin(2\pi ft + \phi_o)$ = instantaneous circuit voltage
$V_R = V_{Ro} \sin 2\pi ft$ = instantaneous voltage across the resistor
$V_C = -V_{Co} \cos 2\pi ft$ = instantaneous voltage across the capacitor
$V_L = V_{Lo} \cos 2\pi ft$ = instantaneous voltage across the inductor

Now we are interested in average values for these quantities. The average values are root-mean-square (rms) values and are the values read on ac meters. They are given by

$$I = I_{rms} = [(I^2)av]^{1/2} = I_o/\sqrt{2} = 0.707 I_o$$
$$V = V_{rms} = [(V^2)av]^{1/2} = V_o/\sqrt{2} = 0.707 I_o$$
$$V_R = V_{Rrms} = [V_R{}^2)av]^{1/2} = V_{Ro}/\sqrt{2} = 0.707 V_{Ro}$$
$$V_C = V_{Crms} = [V_C{}^2)av]^{1/2} = V_{Co}/\sqrt{2} = 0.707 I_o$$
$$V_L = V_{Lrms} = [V_L{}^2)av]^{1/2} = V_{Lo}/\sqrt{2} = 0.707 I_o$$

As a matter of convenience, we drop the rms subscript and understand that any quantity we refer to is the rms value unless specifically stated otherwise.

Using this notation, we can write the power as

$$P_{av} = (I_o V_o \cos\phi_o)/2 = (I_o \sqrt{2})(V_o/\sqrt{2}) \cos\phi_o = IV \cos\phi_o$$

If $\phi_o = 90°$ (as with a pure inductor) or $\phi_o = -90°$ (as with a pure capacitor), the average power delivered is zero. Energy flows into and out of a capacitor and inductor but is not lost. If $\phi_o = 0$ (as with either a pure resistor or a resonate situation), the average power delivered is

$$P_{av} = I_o V_o/2 = (I_o/\sqrt{2})(V_o/\sqrt{2}) = IV$$
$$P_{av} = I_o{}^2 R/2 = (I_o/\sqrt{2})^2 R = I^2 V$$
$$P_{av} = V_o{}^2/2R^2 = (V_o/\sqrt{2})^2/R = V^2/R$$

The term $\cos\phi_o$ is called the power factor for the circuit and may vary from 0 to 1. A lower power factor means voltage and current are very much out of phase. When this happens, a large current must be applied for a given voltage to deliver a large power load. This large current causes unwanted joule heating, and the situation can be corrected by adding capacitors to the circuit.

Practice: A series RL circuit has a resistance of 200 Ω and an inductance of 0.500 H. The circuit is driven by a 120-V, 60-Hz source.

Determine the following:

| | |
|---|---|
| 1. Inductive reactance of the inductor | $X_L = 2\pi fL = 188\ \Omega$ |
| 2. Impendance of the circuit | $Z = [R^2 + X_L{}^2]^{1/2} = 274\ \Omega$ |
| 3. Circuit current | $I = V/Z = 0.438$ A |
| 4. Phase angle between the circuit voltage and current | $\phi_o = \tan^{-1}(X_L/R) = 43.2°$<br>$\phi_o = \sin^{-1}(X_L/Z) = 43.3°$<br>$\phi_o = \cos^{-1}(R/Z) = 43.1°$<br>The difference is due to rounding off |
| 5. Power factor for the circuit | $\cos\phi_o = 0.729$ |
| 6. Potential difference across the resistor read by an ac voltmeter | $V_R = V\cos\phi_o = 87.5$ V<br>$V_R = IR = 87.6$ V |
| 7. Power available to the circuit | $P = P_R = I^2R = V_R{}^2/R = IV_R = 38.3$ W |
| 8. Size of capacitor needed to obtain a power factor of 1 | $\cos\phi_o = 1$ when $\phi_o = 0$<br>$\phi_o = 0$ when $X_C = X_L = 188\ \Omega$<br>$X_C = 188\ \Omega$ when $C = 1/2\pi fX_C$<br>$= 1.41 \times 10^{-5}$ F |
| 9. Circuit current when the power factor is 1 | $I = V/R = 0.600$ A |
| 10. Power available to the circuit when the power factor is 1 | $P = IV = I^2R = V_R{}^2/R = 72.0$ W |

Example 23.3. A circuit draws 300 W from a 110-V, 60-Hz ac line. The power factor is 0.800, and the voltage leads the current. Find (a) the capacitance of a series capacitor that results in a power factor of 1 and (b) the power that is then drawn from the supply line.

Given: $P = 300$ W, $V = 110$ V, $f = 60$ Hz, $\cos\phi_o = 0.800$, voltage leads current

Determine: (a) The capacitance of a series capacitor that results in a power factor of 1 and (b) the power drawn from the supply line with this capacitor in series with the existing circuit.

Strategy: We can use the given information to obtain the circuit current and the potential difference across the inductor. From this information, we can determine $X_L$, $X_C$, and finally C for a power factor of 1. We can determine the resistance for the first circuit and then the power drawn for the second circuit.

Solution: Knowing $\cos\phi_o$, we can determine $\sin\phi_o$:

$$\cos^2\phi_o + \sin^2\phi_o = 1 \qquad \sin^2\phi_o = 1 - \cos^2\phi_o = 0.360 \qquad \sin\phi_o = 0.600$$

$$V_L = V \sin\phi_o = (110\ \text{V})(0.600) = 66.0\ \text{V}$$
$$P = IV \cos\phi_o \qquad I = P/V\cos\phi_o = 3.41\ \text{A}$$
$$X_C = X_L = V_L/I = (66.0\ \text{V})/(3.41\ \text{A}) = 19.4\ \Omega$$
$$X_C = 1/2\pi fC \qquad C = 1/2\pi fX_C = 1.37 \times 10^{-4}\ \text{f}$$
$$V_R = V \cos\phi_o = IR \qquad R = V \cos\phi/I = (110\ \text{V})(0.800)/3.41\ \text{A} = 25.8\ \Omega$$
$$P = IV \cos\phi_o = (V^2/R) \cos\phi = (110\ \text{V})^2(1.00)/25.8\ \Omega = 558\ \text{W}$$

Related Text Problems: 23-29 through 23-36.

---

PRACTICE TEST

Take and grade the practice test. Doing so will allow you to determine any weak spots in your understanding of the concepts taught in this chapter. The following section prescribes what you should study further to strengthen your understanding.

The alternating current and voltage in a circuit are $I = 10 \sin(120\pi t + \pi/6)$ and $V = 120 \sin(120\pi t + \pi/2)$

Determine the following:

_____ 1. Amplitude of the circuit current
_____ 2. Period of the time-varying current and voltage
_____ 3. Initial phase angle that would allow us to write $I = I_o \sin 2\pi ft$
_____ 4. Time when V = 60.0 V
_____ 5. Phase angle difference between I and V

The components of an RLC series circuit have the values R = 100 Ω, L = 0.300 H, C = $4.00 \times 10^{-5}$ F, and the circuit is powered by a 120-V, 60 Hz source.

Determine the following:

_____ 6. Inductive reactance
_____ 7. Capacitance reactance
_____ 8. Total impendance
_____ 9. Circuit current
_____ 10. Phase angle between circuit current and voltage

____ 11. Potential drop across the resistor read by an ac voltmeter
____ 12. Potential drop across the capacitor read by an ac voltmeter
____ 13. Potential drop across the inductor
____ 14. Power factor
____ 15. Value of the capacitance that brings this circuit into resonance
____ 16. Circuit current at resonance
____ 17. Power factor at resonance
____ 18. Power delivered to circuit at resonance
____ 19. Potential drop across the resistor at resonance
____ 20. Power loss to the capacitor at resonance

(See Appendix I for answers.)

PRINCIPAL CONCEPTS AND EQUATIONS PRESCRIPTION

Your score on the practice is an excellent measure of your understanding of this chapter. You should now use the following chart to write your own prescription for curing any of your physics ills. Look down the leftmost column to the number of the question(s) you answered incorrectly, read across that row to see which section(s) of the study guide you should return to for further study, and then do the suggested text problems to gain additional experience in working with the particular concept.

| Question Number | Concepts and Equations | Prescription: Principal Concept | Prescription: Text Problems |
|---|---|---|---|
| 1 | Amplitude of the current $I_o$ | 1 | 23-1 |
| 2 | Period: $T = 1/f$ | 1 | 23-1 |
| 3 | Initial phase $\phi_i$ | 1 | 23-1,2 |
| 4 | Instantaneous voltage | 1 | 23-1,2 |
| 5 | Phase angle difference $\phi_o$ | 1 | 23-2 |
| 6 | Inductive reactance: $X_L = 2\pi fL$ | 2 | 23-12,21 |
| 7 | Capacitative reactance: $X_C = 1/2\pi fC$ | 2 | 23-3,7 |
| 8 | Impedance: $Z = [R^2 + (X_L - X_C)^2]^{1/2}$ | 2 | 23-22,27 |
| 9 | Current: $I = V/Z$ | 2 | 23-21,28 |
| 10 | Phase angle: $\phi_o = \tan^{-1}[(X_L - X_C)/R]$ | 2 | 23-23,27 |
| 11 | $V_R = IR$ | 2 | 23-18,19 |
| 12 | $V_C = IX_C$ | 2 | 23-21,23 |
| 13 | $V_L = IX_L$ | 2 | 23-16,18 |
| 14 | Power factor: $\cos\phi_o$ | 3 | 23-17,29 |
| 15 | Series resonance: $X_L = X_C$, $\phi_o = 0$ | 2 | 23-25,33 |
| 16 | Current at resonance: $I = V/R$ | 2 | 23-21,23 |
| 17 | Power factor at resonance: $\cos\phi_o = 1$ | 3 | 23-33 |
| 18 | Power at resonance: $P = IV = I^2R = V^2/R$ | 3 | 27-17,22 |
| 19 | $V_R = IR$ | 2 | 23-18,19 |
| 20 | AC power to a capacitor | 3 | 23-22 |

# 24 Light and Geometric Optics

## RECALL FROM PREVIOUS CHAPTERS

You should be able to proceed with this chapter without reviewing any previous concepts and equations.

## NEW IDEAS IN THIS CHAPTER

| Concepts and equations introduced | Text Section | Study Guide Page |
|---|---|---|
| Index of refraction: $n = c/v$ | 24.5 | 410 |
| Snell's law: $\sin\theta_1/\sin\theta_2 = v_1/v_2 = n_2/n_1$ | 24.5 | 410 |
| Apparent depth: $h' = h/n$ | 24.5 | 410 |
| Critical angle: $\sin\theta_c = n_2/n_1$ | 24.5 | 410 |
| Mirror equation: $1/s + 1/s' = 1/f$ | 25.7 | 416 |
| Magnification: $m = h'/h = -s'/s$ | 24.7 | 416 |

## PRINCIPAL CONCEPTS AND EQUATIONS

### 1. Refraction (Section 24.5)

Review: Figure 24.1 shows a ray of light refracted first as it travels from air to glass and again as it travels from glass to air.

Figure 24.1

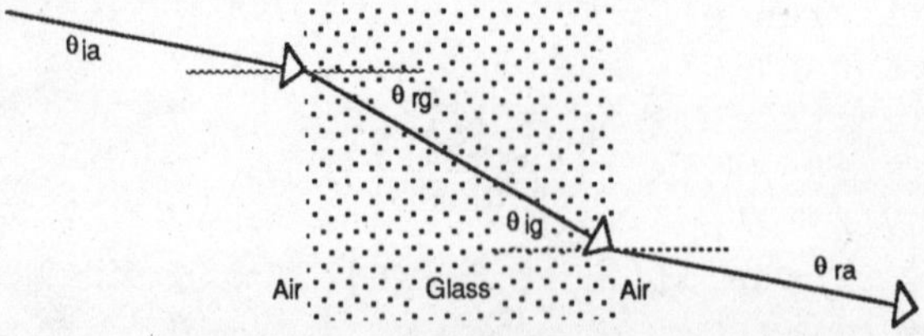


The angles of incidence in air and glass are $\theta_{ia}$ and $\theta_{ig}$. The angles of refraction in glass and air are $\theta_{rg}$ and $\theta_{ra}$.

Note: The authors of your text represent the index of refraction and the angle of incidence in medium 1 by $n_1$ and $\theta_1$. In like manner they represent the index of refraction and the angle of refraction in medium 2 by $n_2$ and $\theta_2$. This is fine for a generic discussion of two media, however when the media are for example air and water you will find it convenient to use subscripts a and w rather than 1 and 2. Also in order to avoid any chance of confusion, you might find it convenient to add another subscript to explicitly indicate incidence and refraction. For example, we use the following notation:

$\theta_{ia}$ = angle of incidence in air
$\theta_{rg}$ = angle of refraction in glass
$\theta_{ig}$ = angle of incidence in glass
$\theta_{ra}$ = angle of refraction in air

The subscripts make it clear which medium we are considering and whether we are interested in an angle of incidence or refraction.

Refraction is the bending of light as it passes from one transparent medium to another. It is caused by the fact that light travels at different speeds in the two media. The index of refraction of a transparent medium is equal to the ratio of the speed of light in a vacuum to the speed of light in that medium:

$$n = c/v$$

The greater the index of refraction of the second medium, the slower the light travels and hence the more its path is bent. Snell's law,

$$\sin\theta_i/\sin\theta_r = v_i/v_r = (c/n_i)/(c/n_r) = n_r/n_i$$

shows how the angle of refraction is related to

(1) the angle of incidence: $\sin\theta_r \propto \sin\theta_i$
large $\theta_i \rightarrow$ large $\theta_r$, small $\theta_i \rightarrow$ small $\theta_r$

(2) the speed of light in each medium: $\sin\theta_r \propto (v_r/v_i)$
small $(v_r/v_i) \rightarrow$ small $\theta_r \rightarrow$ large refraction

(3) the index of refraction of each medium: $\sin\theta_r \propto (n_i/n_r)$
small $(n_i/n_r) \rightarrow$ small $\theta_r \rightarrow$ large refraction

Because of refraction, objects in water have an apparent depth that is different from their real depth. This is summarized in Fig. 24.2.

Figure 24.2

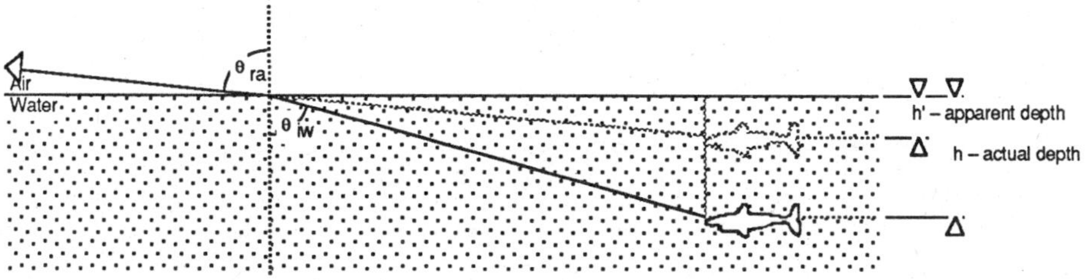


The apparent depth is related to the actual depth and the respective indices of refraction by

$$h' = h(n_a/n_w)$$

Substituting $n_a = 1.00$, we have

$$h' = h/n_w$$

The subscript on the n will of course change to be descriptive of the medium.

When light travels from one medium to another of lower index of refraction, we may observe total internal reflection, (Figure 24.3)

Figure 24.3.

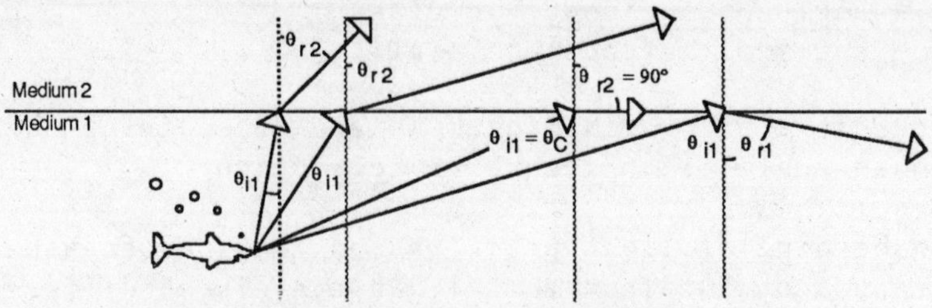


$\theta_{i1}$ = angle of incidence in medium 1
$\theta_{r2}$ = angle of refraction in medium 2
$\theta_{r1}$ = angle of reflection in medium 1
$\theta_c$ = critical angle

As the angle of incidence in medium 1 increases, the angle of refraction in medium 2 increases. The angle of incidence in medium 1 is equal to the critical angle when the angle of refraction in medium 2 is 90°. We may determine $\theta_c$ by applying Snell's Law to this situation.

$$n_1 \sin\theta_c = n_2 \sin 90°$$

$$\sin\theta_c = n_2/n_1$$

Notice that, in order for this to make sense, we much have $\theta_c < 90°$, which means $\sin\theta_c < 1$. This is possible only if $n_2 < n_1$. Subsequently, we expect a critical angle and total internal reflection only when light goes from one medium to another of lower index of refraction.

<u>Practice:</u> Figure 24.4 shows a ray of light traveling through air, flint glass, and acrylic plastic.

Figure 24.4

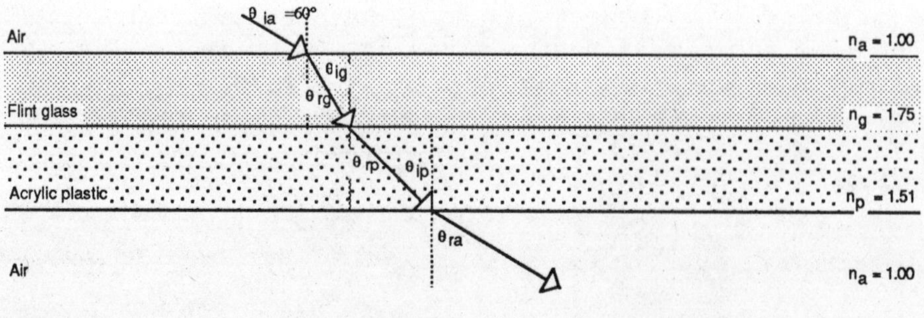


Determine the following:

| | |
|---|---|
| 1. Speed of the light in each medium | $v_a = c = 3.00 \times 10^8$ m/s<br>$v_g = c/n_g = 1.71 \times 10^8$ m/s<br>$v_p = c/n_p = 1.99 \times 108$ m/s |
| 2. Frequency of the light in each medium | $f_a = f_g = f_p = c/\lambda_a = 4.62 \times 10^{14}$ Hz |

| | | |
|---|---|---|
| 3. | Wavelength of the light in each medium | $\lambda_a = 6.50 \times 10^{-7}$ m<br>$\lambda_g = v_g/f_g = c/n_g f_g = \lambda_a/n_g$<br>$= 3.71 \times 10^{-7}$ m<br>$\lambda_p = v_p/f_p = c/n_p f_p = \lambda_a/n_p$<br>$= 4.30 \times 10^{-7}$ m |
| 4. | Angle of refraction in glass | $n_a \sin\theta_{ia} = n_g \sin\theta_{rg}$<br>$\sin\theta_{rg} = n_a \sin\theta_{ia}/n_g = 0.495$<br><br>$\theta_{rg} = \sin^{-1}(0.495) = 29.7°$ |
| 5. | Angle of refraction in plastic | $\theta_{ig} = \theta_{rg}$ (geometry)<br>$n_g \sin\theta_{ig} = n_p \sin\theta_{rp}$<br>$\sin\theta_{rp} = n_g \sin\theta_{ig}/n_p = 0.574$<br><br>$\theta_{rp} = \sin^{-1}(0.574) = 35.0°$ |
| 6. | Angle of refraction in air | $\theta_{ip} = \theta_{rp}$ (geometry)<br>$n_p \sin\theta_{ip} = n_a \sin\theta_{ra}$<br>$\sin\theta_{ra} = n_p \sin\theta_{ip}/n_a = 0.866$<br><br>$\theta_{ra} = \sin^{-1}(0.866) = 60.0°$ |
| 7. | Critical angle for the glass relative to the plastic | $n_g \sin\theta_c = n_p \sin 90°$<br>$\sin\theta_c = n_p/n_g = 0.863$<br><br>$\theta_c = \sin^{-1}(0.863) = 59.6°$ |
| 8. | Critical angle for the plastic relative to air | $n_p \sin\theta_c = n_a \sin 90°$<br>$\sin\theta_c = n_a/n_p = 0.662$<br><br>$\theta_c = \sin^{-1}(0.662) = 41.5°$ |
| 9. | Minimum value for $\theta_{ia}$ in order to have total internal reflec- at the glass-plastic boundary | We will have total internal reflection at this boundary if $\theta_{rg} = \theta_{ig} = \theta_c = 59.6°$.<br>$n_a \sin\theta_{ia} = n_g \sin\theta_{rg}$<br>$\sin\theta_{ia} = n_g \sin\theta_c/n_a = 1.51$<br>This is impossible because $\sin\theta_{ia}$ cannot be greater than 1. We cannot get a value of $\theta_{ia}$ great enough to cause total internal reflection at the glass-plastic boundary. |

| | |
|---|---|
| 10. Apparent depth of the object embedded in the glass | $h' = h(n_a/n_g) = 0.571$ cm |

Example 24.1. A beam of light composed of red and blue is incident on a 30°-60°-90° prism in such a manner that the red light goes through the prism parallel to the base (Fig. 24.5).

Figure 24.5

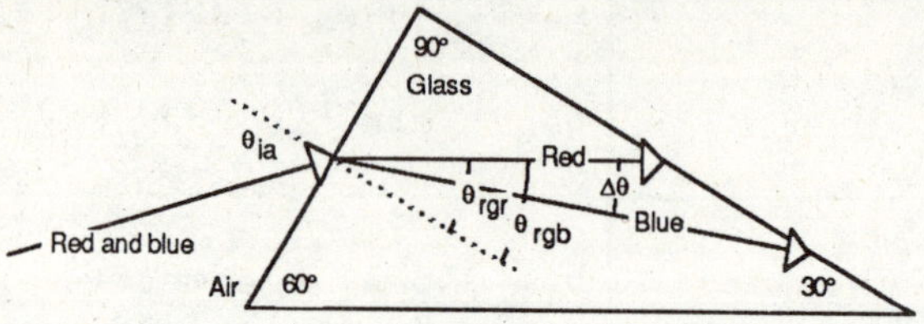


The index of refraction in the glass is 1.46 for red light and 1.47 for blue light. Determine the angle of incidence for the incident beam and the angular separation ($\Delta\theta$) for the red and blue light in the prism and establish whether refraction or internal reflection takes place at the second glass-air boundary.

The notation used is

$\theta_{iar}$, $\theta_{iab}$, $\theta_{ia}$ = angle of incidence in air for red light, blue light, and the combined beam
$\theta_{rgr}$, $\theta_{rgb}$ = angle of refraction in glass for red and blue light
$\Delta\theta$ = angular separation between the refracted red and blue light
$n_{gr}$, $n_{rb}$ = index of refraction in glass for red and blue light
$\theta_{cr}$, $\theta_{cb}$ = critical angle at the glass-air interface for red and blue light

Given: The red light goes through the prism parallel to the base.
The prism is a 30°-60°-90°
$n_{gr}$ = 1.46 = index of refraction in glass for red
$n_{gb}$ = 1.47 = index of refraction in glass for blue

Determine: The angle of incidence for the incident beam, the angular separation for the red and blue light in the prism, whether refraction or internal reflection takes place at the second glass-air boundary.

Strategy: From the geometry, we can determine the angle of refraction in glass for red light ($\theta_{rgr}$). Knowing this angle and the indices of refraction, we can determine the angle of incidence in air for the red light ($\theta_{iar}$) and hence for the beam ($\theta_{ia}$). Since the blue light is part of the beam, it has the same angle of incidence as the beam. Knowing the angle of incidence in air for the blue light and the indices of refraction, we can determine the angle of refraction in glass for blue light ($\theta_{rgb}$). We can determine the angular separation of the red and blue light from their respective angles of refraction in glass. We can determine the angles of incidence at the second boundary by geometry. Finally, we can determine the critical angles for the red and blue light and compare them to the respective angles of incidence to see if refraction or internal reflection takes place at the second glass-air boundary.

Solution: Looking at Fig. 24.5, we see that the angle of refraction in glass for red light ($\theta_{rgr}$) is 30.0°. Knowing this, we can determine the angle of incidence in air for red light ($\theta_{iar}$) and hence the angle of incidence in air ($\theta_{ia}$) for the beam.

$$n_a \sin\theta_{iar} = n_{gr} \sin\theta_{rgr}$$

$$\theta_{iar} = \sin^{-1}(n_{gr} \sin\theta_{rgr}/n_a) = \sin^{-1}(0.730) = 46.9°$$

$$\theta_{ia} = \theta_{iab} = \theta_{iar} = 46.9°$$

The angle of refraction in glass for blue light may now be determined.

$$n_a \sin\theta_{iab} = n_{gb} \sin\theta_{rgb}$$

$$\theta_{rgb} = \sin^{-1}(n_a \sin\theta_{iab}/n_{gb}) = \sin^{-1}(0.497) = 29.8°$$

The angular separation of the red and blue light in the glass is

$$\Delta\theta = \theta_{rgr} - \theta_{rgb} = 0.200°$$

From Fig. 24.6, we can determine the angles of incidence in glass at the second glass-air boundary.

$$\theta_{igr} = 90.0° - 30.0° = 60.0° \qquad \theta_{igb} = 90.0° = 29.8° = 60.2°$$

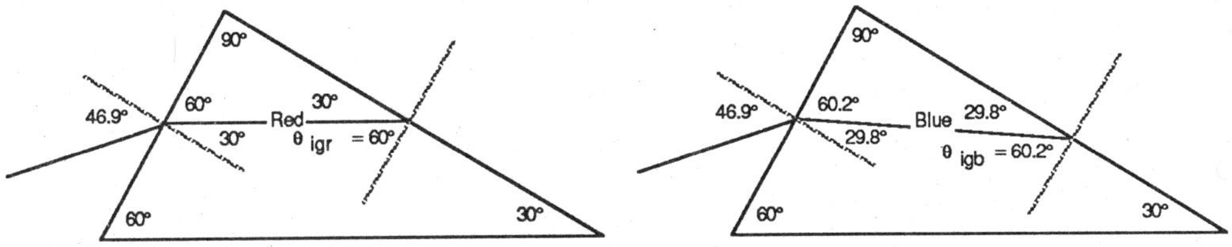


Figure 24.6

The critical angles can now be determined.

$$\begin{aligned} n_{gr} \sin\theta_{cr} &= n_a \sin 90° \\ \theta_{cr} &= \sin^{-1}(1/n_{gr}) \\ \theta_{cr} &= 43.2° \end{aligned} \qquad \text{and} \qquad \begin{aligned} n_{gb} \sin\theta_{cb} &= n_a \sin 90° \\ \theta_{cb} &= \sin^{-1}(1/n_{gb}) \\ \theta_{cb} &= 42.9° \end{aligned}$$

Since $\theta_{igr} > \theta_{cr}$ and $\theta_{igb} > \theta_{cb}$, the red and blue light rays experience total internal reflection at the second glass-air boundary.

Related Problems: 24-11 through 24-28.

## 2. Mirrors and Reflected Images (Secs. 24.4 and 24.7)

Review: Figure 24.7 reviews reflection and image formation for a planar mirror.

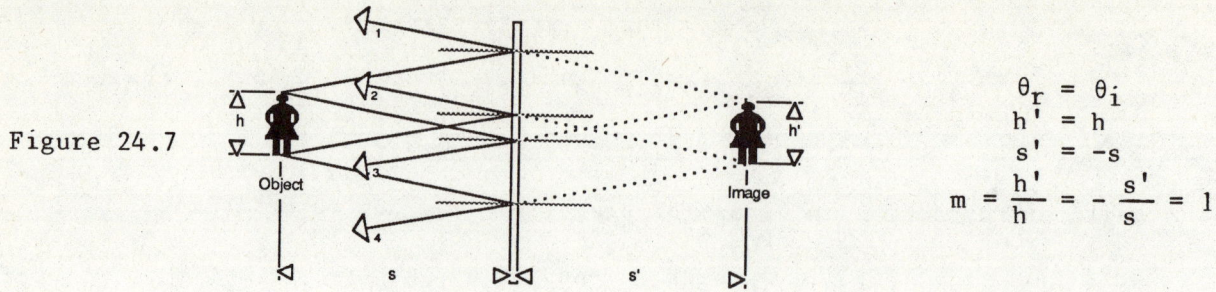

Figure 24.7

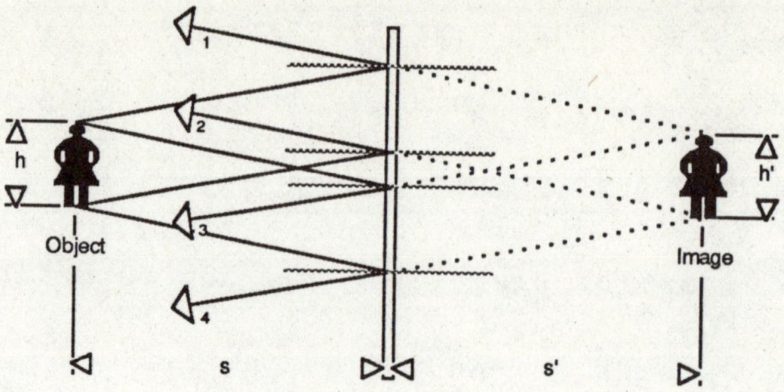


$$\theta_r = \theta_i$$
$$h' = h$$
$$s' = -s$$
$$m = \frac{h'}{h} = -\frac{s'}{s} = 1$$

We see that, for all rays (only four are shown), the angle of reflection is equal to the angle of incidence. The image size is equal to the object size. The image distance is equal to the object distance. The image is erect (has the same orientation as the object). The image is virtual (the reflected rays do not go through the image).

Figure 24.8 reviews reflection and image formation for a concave mirror.

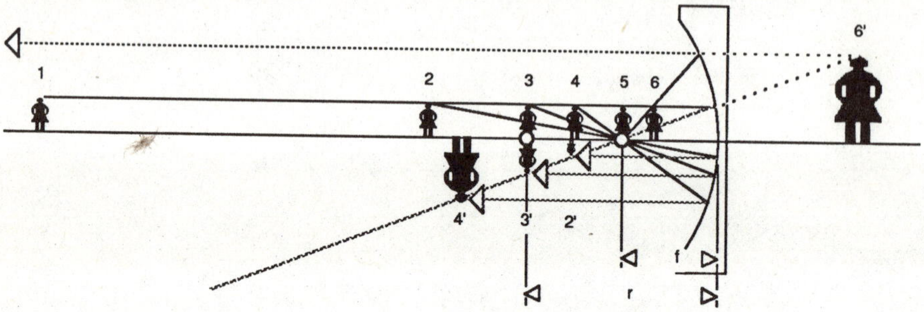


$$f = r/2$$
$$\frac{1}{s} + \frac{1}{s'} = \frac{1}{f}$$
$$m = \frac{h'}{h} = -\frac{s'}{s}$$

Figure 24.8

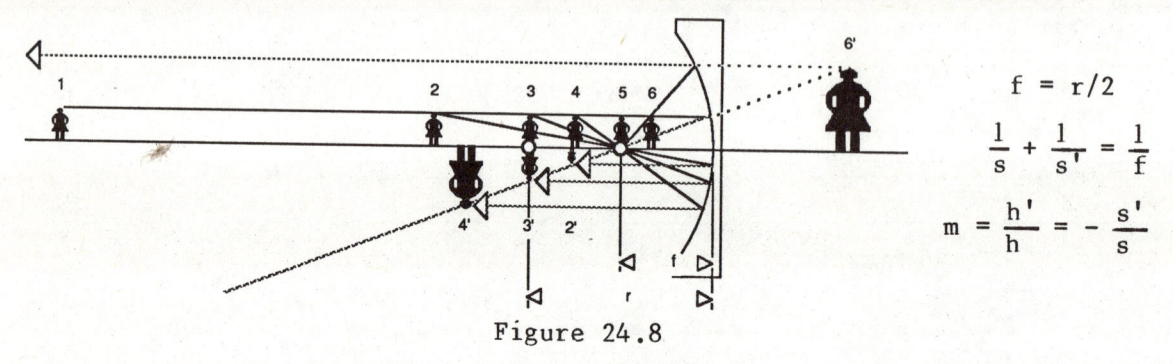

Figure 24.9 reviews reflection and image formation for a convex mirror.

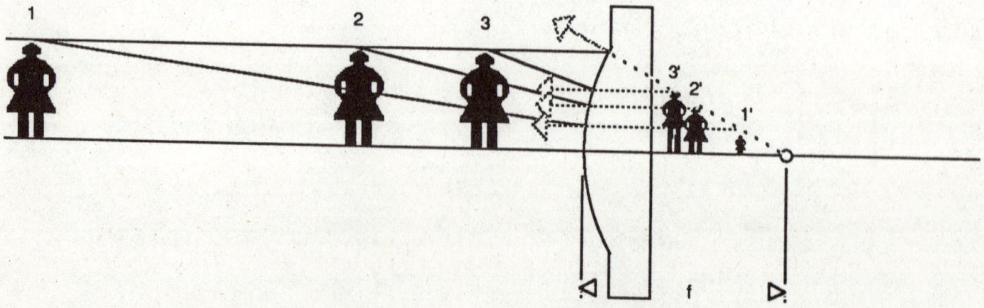

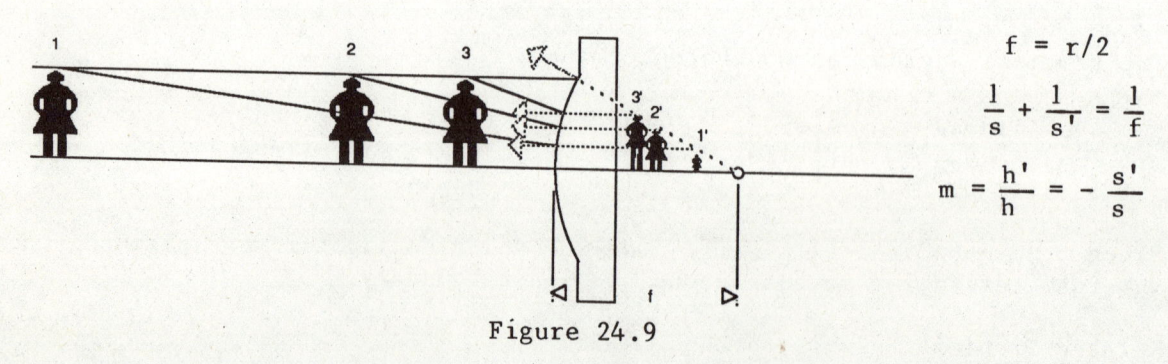

$$f = r/2$$
$$\frac{1}{s} + \frac{1}{s'} = \frac{1}{f}$$
$$m = \frac{h'}{h} = -\frac{s'}{s}$$

Figure 24.9

Note: The ray diagrams in Figs. 24.8 and 24.9 are easy to construct and contain a large quantity of information. In fact, the table, the sign conventions, and the summary statements that follow are all obtained from these two figures. Obviously, you can't hope to memorize all of this information. You have no choice but to learn how to do ray diagrams like those shown in Fig. 24.8 and 24.9 and to study them until you can write down all of this information from the diagrams.

A summary of the information obtained in Figs. 24.8 and 24.9 is shown below.

| | Concave Mirror | | | | | | Convex Mirror | | |
|---|---|---|---|---|---|---|---|---|---|
| Object number | 1 | 2 | 3 | 4 | 5 | 6 | 1 | 2 | 3 |
| Image number | 1' | 2' | 3' | 4' | NI* | 6' | 1' | 2' | 3' |
| Object location | $s \gg r$ | $s > r$ | $s = r$ | $r > s > f$ | $s = f$ | $s < f$ | $s \gg r$ | $s \simeq r$ | $s < r$ |
| Image location | $s' \geqslant f$ | $r > s' > f$ | $s' = r$ | $s' > r$ | NI | $s' > -s$ | $s' < -f$ | $s' < -f$ | $s' < -f$ |
| Sign of $s$ | + | + | + | + | + | + | + | + | + |
| Sign of $s'$ | + | + | + | + | NI | − | − | − | − |
| Sign of $-s'/s$ | − | − | − | − | NI | + | + | + | + |
| $\lvert s'/s \rvert$ | $\ll 1$ | $< 1$ | $= 1$ | $> 1$ | NI | $> 1$ | $< 1$ | $< 1$ | $< 1$ |
| Sign of $h$ | + | + | + | + | + | + | + | + | + |
| Sign of $h'$ | − | − | − | − | NI | + | + | + | + |
| Sign of $h'/h$ | − | − | − | − | NI | + | + | + | + |
| $\lvert h'/h \rvert$ | $< 1$ | $< 1$ | $= 1$ | $> 1$ | NI | $> 1$ | $< 1$ | $< 1$ | $< 1$ |
| Sign of $m$ | − | − | − | − | NI | + | + | + | + |
| $\lvert m \rvert$ | $\ll 1$ | $< 1$ | $= 1$ | $> 1$ | NI | $> 1$ | $< 1$ | $< 1$ | $< 1$ |

*NI = no image is found.

$s$ is +; the object is always in front of the mirror, and so is always positive
$f$ is + for concave mirrors.
$f$ is − for convex mirrors.
If $s'$ is +, the image is in front of mirror and is real.
If $s'$ is −, the image is behind the mirror and is virtual.
If $-s'/s$ is +, the image is virtual (the reflected rays do not go through it).
If $-s'/s$ is −, the image is real (the reflected rays go through it).
If $h$ is +, the object height is above the principal axis.
If $h$ is −, the object height is below the principal axis.
If $h'$ is +, the image height is above the principal axis.
If $h'$ is −, the image height is below the principal axis.
If $h'/h$ is +, the image is erect (it has the same orientation as the object).
If $h'/h$ is −, the image is inverted (its orientation is opposite that of the object).
If $m = h'/h = -s'/s$ is +, the image is virtual and erect.
If $m = h'/h = -s'/s$ is −, the image is real and inverted.

For concave mirrors:
- All real images are inverted.
- All virtual images are erect and have $|m| > 1$.
- If $s > r$, then $r > s' > f$, $|m| < 1$, and the image is real and inverted.
- If $s = r$, then $s' = r$, $|m| = 1$, and the image is real and inverted.
- If $r > s > f$, then $s' > r$, $|m| > 1$, and the image is real and inverted.
- If $s = f$, no image is found.
- If $f > s > 0$, then $s' > -s$, $|m| > 1$, and the image is virtual and erect.

For convex mirrors:
- All images are virtual.
- All images are erect.
- All images are formed inside the focal length.
- All images have $|m| < 1$.

Practice: Figure 24.10 shows objects in front of planar, concave, and convex mirrors. Information is also given about focal lengths, object size, and object distance.

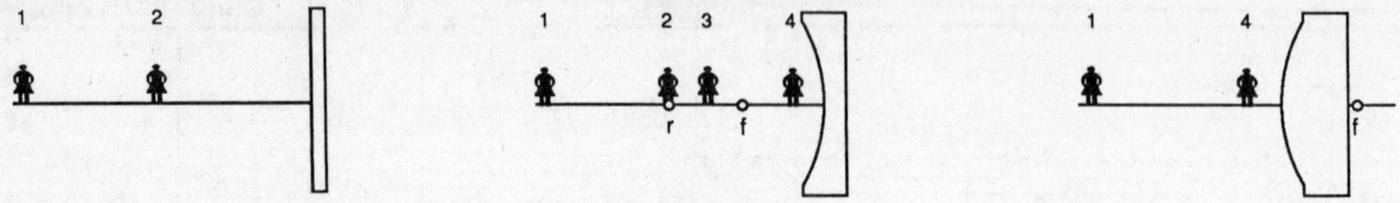


(a) Planar mirror (b) Concave mirror (c) Convex mirror

$|f| = 10.0$ cm
$h_1 = h_2 = h_3 = h_4 = 2.00$ cm
$s_1 = 40.0$ cm, $s_2 = 20.0$ cm, $s_3 = 15.0$ cm, $s_4 = 5.0$ cm

Figure 24.10

Determine the following for the planar mirror:

| | |
|---|---|
| 1. Image distance for each object | $s_1' = -s_1 = -40.0$ cm<br>$s_2' = -s_2 = -20.0$ cm |
| 2. Magnification of each object | $m_1 = -s_1'/s_1 = +1.00$<br>$m_2 = -s_2'/s_2 = +1.00$ |
| 3. Size and orientation of each image | $h_1' = m_1 h_1 = +2.00$ cm<br>$h_2' = m_2 h_2 = +2.00$ cm<br>+ means erect (same orientation as object) |
| 4. Whether the image is real or virtual for each object | Both images are virtual (the reflected rays do not pass through them). |

Determine the following for the concave mirror:

| | |
|---|---|
| 1. Image distance for each object | $1/s_1' = 1/f - 1/s_1 = 1/10 - 1/40 = 3/40$<br>$s_1' = 40/3 = 13.3$ cm<br>$s_2' = 20.0$ cm, $s_3' = 30.0$ cm,<br>$s_4' = -10.0$ cm |
| 2. Magnification of each object | $m_1 = -s_1'/s_1 = (-13.3 \text{ cm})/(40.0 \text{ cm})$<br>$= -0.333$<br>$m_2 = -1.00$, $m_3 = -2.00$, $m_4 = +2.00$ |

| | |
|---|---|
| 3. Size and orientation of each image | $h_1' = m_1h_1 = (-0.333)(2.00 \text{ cm}) = -0.666$ cm<br>$h_2' = -2.00$ cm, $h_3' = -4.00$ cm,<br>$h_4' = +4.00$ cm<br>+ means erect and - means inverted. |
| 4. Whether the image is real or virtual for each object | The image is real for objects 1, 2, and 3 and virtual for object 4. |

Determine the following for the convex mirror:

| | |
|---|---|
| 1. Image distance for each object | $1/s_1' = 1/f - 1/s_1 = -1/10 - 1/40 = -1/8$<br>$s_1' = -8.00$ cm<br>$s_4' = -3.33$ cm |
| 2. Magnification of each object | $m_1 = -s_1'/s_1 = +0.200$<br>$m_4 = -s_4'/s_4 = +0.667$ |
| 3. Size and orientation of each image | $h_1' = m_1h_1 = +0.400$ cm<br>$h_2' = m_2h_2 = +1.33$ cm<br>+ means erect |
| 4. Whether the image is real or virtual for each object | The image is virutal for both objects. |

Example 24.2. Two students have identical concave mirrors with 30.0-cm focal lengths. Each student places an object in front of her mirror and reports that the image is three times the size of the object. However, when they compare data, they find that their object distances are not the same. What are the object distances?

Given: $f = 30.0$ cm = focal length for each mirror
Image size is three times object size for each case

Determine: Object distance for each case

Strategy: Figure 24.11 shows the circumstances that allow this situation

Figure 24.11

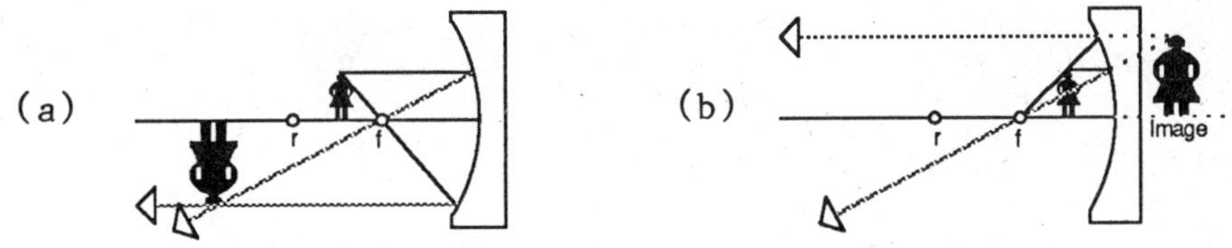

We can use the image size and orientation to obtain the magnification. We can use our knowledge of magnification to eliminate the image distance in the mirror equation. Finally, we can solve for the object distance in terms of the focal length.

Solution:

Case a

$h' = -3h$ (image is inverted)

$m = h'/h = -3$

$s' = -ms = +3s$

$$\frac{1}{s} + \frac{1}{s'} = \frac{1}{s} + \frac{1}{3s} = \frac{1}{f}$$

$s = 4f/3 = 40.0$ cm

Case b

$h' = +3h$ (image is erect)

$m = h'/h = +3$

$s' = -ms = -3s$

$$\frac{1}{s} + \frac{1}{s'} = \frac{1}{s} - \frac{1}{3s} = \frac{1}{f}$$

$s = 2f/3 = 20.0$ cm

Related Text Problems:

Planar mirror -- 24-6 through 24-10.
Concave mirror -- 24-30, 24-31, 24-33, 24-35, 24-37, 24-39, 24-40, 24-42.
Convex mirror -- 24-29, 24-32, 24-36, 24-38, 24-41.

---

## PRACTICE TEST

Take and grade the practice test. Doing so will allow you to determine any weak spots in your understanding of the concepts taught in this chapter. The following section prescribes what you should study further to strengthen your understanding.

A block of clear glass with opposite faces parallel is placed successively in various transparent, colorless liquids referred to as liquids 1, 2, and 3. The refractive indices are: air = 1.00, glass = 1.50, liquid 1 = 1.30, liquid 2 = 1.50, liquid 3 = 1.70.

Figure 24.12 shows five situations involving the glass and liquids.

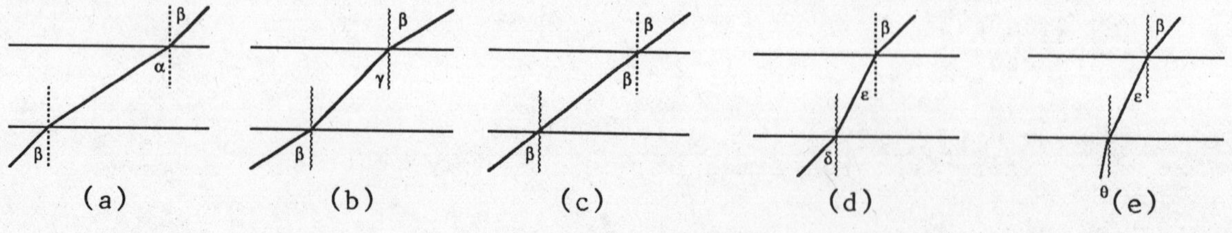


$\alpha > \beta > \gamma > \delta > \varepsilon > \theta$

Figure 24.12

For each of the situations given below, select the diagram that best represents the path of a ray passing through the glass under the conditions described.

__________ 1. The block is submerged in liquid 1.

______ 2. The block is submerged in liquid 2.
______ 3. The block is submerged in liquid 3.
______ 4. The block half submerged in liquid 1 (the top half is in air).
______ 5. The block half submerged in liquid 3 (the top half is in air).

Figure 24.13 shows a ball bearing embedded in a block of ice. The ray incident on the block undergoes total internal reflection at the second ice-air boundary.

Figure 24.13.

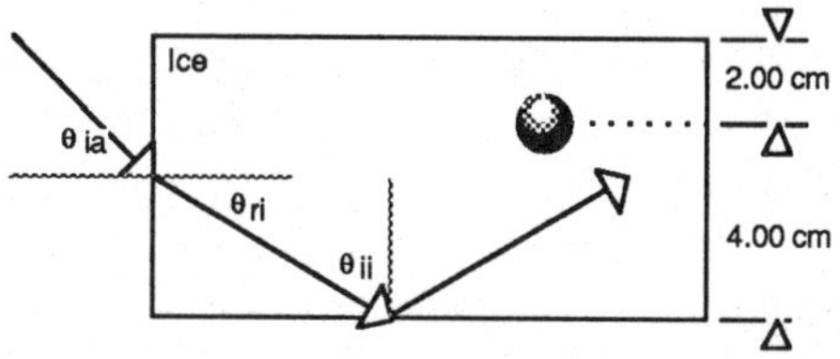


$n_i$ = 1.31 = index of refraction of ice
$\theta_{ri}$ = angle of refraction in ice
$\theta_{ia}$ = angle of incidence in air
$\theta_c$ = critical angle for ice relative to air
$\theta_{ii} = \theta_c + 2.20°$ = angle of incidence in ice

Determine the following:

______ 6. $\theta_c$
______ 7. $\theta_{ri}$
______ 8. $\theta_{ia}$
______ 9. Apparent depth of the ball bearing when viewed from the top
______ 10. Speed of light in the ice

Figure 24.14 shows objects in front of concave and convex mirrors.

Figure 24.14.

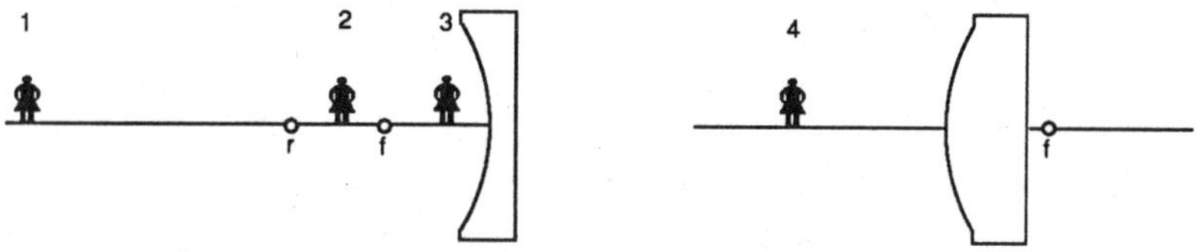


$|f|$ = 20.0 cm for both mirrors
$s_1$ = 100 cm, $s_2$ = 30.0 cm, $s_3$ = 10.0 cm, $s_4$ = 40.0 cm
$h_1 = h_2 = h_3 = h_4$ = 20.0 cm

Determine the following:

______ 11. Location of the image of object 1
______ 12. Location of the image of object 2
______ 13. Location of the image of object 3
______ 14. Location of the image of object 4
______ 15. Magnification of object 1
______ 16. Magnification of object 3
______ 17. Size and orientation of the image of object 2
______ 18. Size and orientation of the image of object 4

_______ 19. Whether the image of object 3 is real or virtual
_______ 20. Whether the image of object 4 is real or virtual

(See Appendix I for answers.)

---

PRINCIPAL CONCEPTS AND EQUATIONS PRESCRIPTION

Your score on the practice is an excellent measure of your understanding of this chapter. You should now use the following chart to write your own prescription for curing any of your physics ills. Look down the leftmost column to the number of the question(s) you answered incorrectly, read across that row to see which section(s) of the study guide you should return to for further study, and then do the suggested text problems to gain additional experience in working with the particular concept.

| Question Number | Concepts and Equations | Prescription: Principal Concept | Prescription: Text Problems |
|---|---|---|---|
| 1 | Snell's law: $\sin\theta_i/\sin\theta_r = n_r/n_i$ | 1 | 24-15,16 |
| 2 | Snell's law | 1 | 24-18,19 |
| 3 | Snell's law | 1 | 24-22,23 |
| 4 | Snell's law | 1 | 24-25,26 |
| 5 | Snell's law | 1 | 24-19,27 |
| 6 | Critical angle: $\sin\theta_c = n_2/n_1$ | 1 | 24-16,21 |
| 7 | Geometry | 1 | 24-19,24 |
| 8 | Snell's law | 1 | 24-15,26 |
| 9 | Apparent depth: $h' = h(n_1/n_2)$ | 1 | 24-11,22 |
| 10 | Index of refraction: $n = c/v$ | 1 | 24-15,18 |
| 11 | Mirror equation: $1/s + 1/s' = 1/f$ | 2 | 24-33,37 |
| 12 | Mirror equation | 2 | 24-34,39 |
| 13 | Mirror equation | 2 | 24-35,40 |
| 14 | Mirror equation | 2 | 24-32,36 |
| 15 | Magnification: $m = -s'/s$ | 2 | 24-37,39 |
| 16 | Magnification | 2 | 24-39,40 |
| 17 | Image size: $m = h'/h$ | 2 | 24-37,40 |
| 18 | Image size | 2 | 24-36,38 |
| 19 | Real vs. virtual | 2 | 24-39,40 |
| 20 | Real vs. virtual | 2 | 24-38,41 |

# 25 Lenses and Optical Instruments

RECALL FROM PREVIOUS CHAPTERS

| Previously learned concepts and equations frequently used in this chapter | Text Section | Study Guide Page |
|---|---|---|
| Mirror equation: $1/s + 1/s' = 1/f$ | 24.7 | 416 |
| Magnification: $m = -s'/s = h'/h$ | 24.7 | 416 |
| Sign conventions for s, s', f, h, h', and m | 24.7 | 416 |

New Ideas In This Chapter

| Concepts and equations introduced | Text Section | Study Guide Page |
|---|---|---|
| Thin lens equation: $1/s + 1/s' = 1/f$ | 25.3 | 423 |
| Magnification: $m = -s'/s = h'/h$ | 25.3 | 423 |
| Combination lens magnification: $m = m_1 m_2$ | 25.6 | 427 |
| Refractive power: $P = 1/f$ | 25.6 | 427 |
| Two lenses placed together ($d \approx 0$): $1/s_1 + 1/s_2 = 1/f_1 + 1/f_2 = 1/f$; $P = P_1 + P_2$ | 25.6 | 427 |

PRINCIPAL CONCEPTS AND EQUATIONS

## 1. Thin Lenses (Section 25.3)

Review: Figure 25.1 reviews image formation for a convex (converging) lens. Note that the image is located by tracing two rays from the tip of the object. One ray is incident parallel to the optical axis and exits the lens toward the focal point. The other ray is incident through the focal point and exits the lens parallel to the optical axis.

Figure 25.1

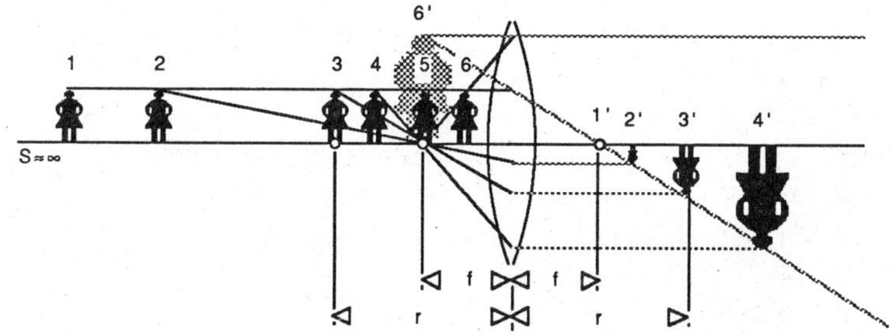


$$f = r/2$$

$$\frac{1}{s} + \frac{1}{s'} = \frac{1}{f}$$

$$m = \frac{h'}{h} = -\frac{s'}{S}$$

s = object distance
s' = image distance
f = focal length
r = lens radius of curvature

Figure 25.2 reviews image formation for a concave (diverging) lens. One ray is incident parallel to the optical axis and exits as though it came from the focal point on the incident side. The other ray is incident toward the focal point on the exiting side but exits parallel to the optical axis.

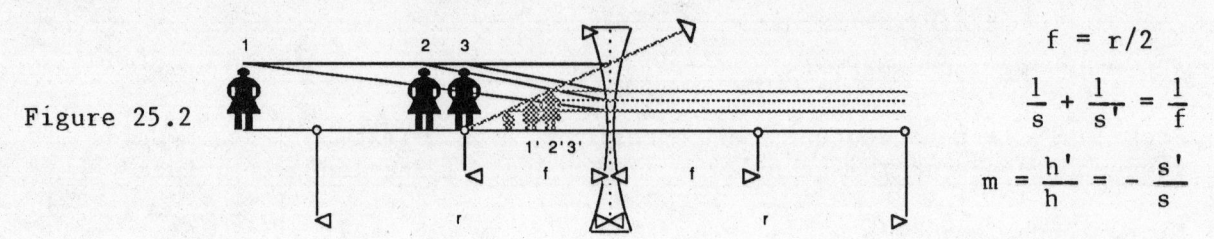

Figure 25.2

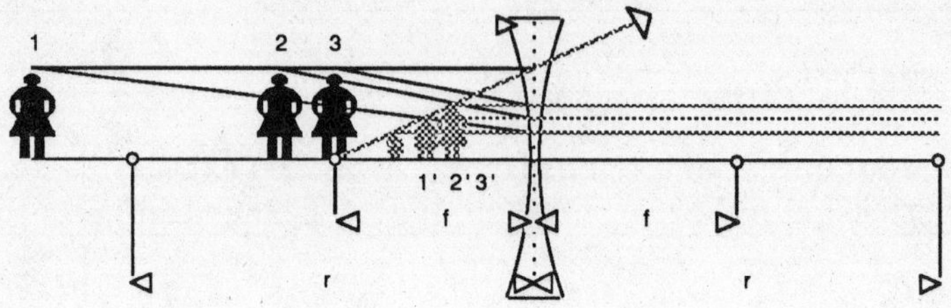


$$f = r/2$$

$$\frac{1}{s} + \frac{1}{s'} = \frac{1}{f}$$

$$m = \frac{h'}{h} = -\frac{s'}{s}$$

<u>Note</u>: The ray diagrams in Figs. 25.1 and 25.2 are easy to construct and contain much information. In fact, the table, sign convention, and summary statements that follow are all obtained from these two figures. Obviously, you can't hope to memorize all this information. You have no choice but to learn how to do ray diagrams like those shown in Figs. 25.1 and 25.2 and to study them until you can write down all of this information from the diagrams.

A summary of the information obtained in Figs. 25.1 and 25.2 is shown below.

| | Convex Lens | | | | | | Concave Lens | | |
|---|---|---|---|---|---|---|---|---|---|
| Object number | 1 | 2 | 3 | 4 | 5 | 6 | 1 | 2 | 3 |
| Image number | 1' | 2' | 3' | 4' | NI* | 6' | 1' | 2' | 3' |
| Object location | s>>r | s>r | s=r | r>s>f | s=f | s<f | s>r | r>s>f | s=f |
| Image location | s'⩾f | r>s'>f | s'=r | s'>r | NI | -s'>s | -s'<f | -s'<f | -s'<f |
| Sign of s | + | + | + | + | + | + | + | + | + |
| Sign of s' | + | + | + | + | NI | − | − | − | − |
| Sign of -s'/s | − | − | − | − | NI | + | + | + | + |
| $\lvert s'/s \rvert$ | <1 | <1 | 1 | >1 | NI | >1 | <1 | <1 | <1 |
| Sign of h | + | + | + | + | + | + | + | + | + |
| Sign of h' | − | − | − | − | NI | + | + | + | + |
| Sign of h'/h | − | − | − | − | NI | + | + | + | + |
| $\lvert h'/h \rvert$ | <1 | <1 | =1 | >1 | NI | >1 | <1 | <1 | <1 |
| Sign of m | − | − | − | − | NI | + | + | + | + |
| $\lvert m \rvert$ | <1 | <1 | =1 | >1 | NI | >1 | <1 | <1 | <1 |

*NI = no image found.

s is positive because the object is in front of the lens.
f is + for converging (convex) lenses.
f is - for diverging (concave lenses.
If s' is +, the image is real (the refracted rays go through it).
If s' is -, the image is virtual (the refracted rays do not go through it).
If -s'/s is +, the image is virtual.
If -s'/s is -, the image is real.
If h is +, the object height is above the optical axis.
If h is -, the object height is below the optical axis.
If h' is +, the image height is above the optical axis.

If h' is −, the image height is below the optical axis.
If h'/h is +, the image is erect.
If h'/h is −, the image is inverted.
If $m = h'/h = -s'/s$ is +, the image is virtual and erect.
If $m = h'/h = -s'/s$ is −, image is real and inverted.

For a converging (convex) lens:

All real images are inverted.
All virtual images are erect and have $|m| > 1$.
If $s > r$, then $r > s' > f$, $|m| < 1$, and the image is real and inverted.
If $s = r$, then $s' = r$, $|m| = 1$, and the image is real and inverted.
If $r > s > f$, then $s' > r$, $|m| > 1$, and the image is real and inverted.
If $s = f$, no image is found.
If $f > s > 0$, then $s' > -s$, $|m| > 1$, and the image is virtual and erect.

For a diverging (concave) lens:

All images are virtual.
All images are erect.
All images are formed inside the focal length.
All images have $|m| < 1$.

Practice: Figure 25.3 shows objects in front of converging and diverging lenses. Information is also given about focal lengths, object size, and object distance.

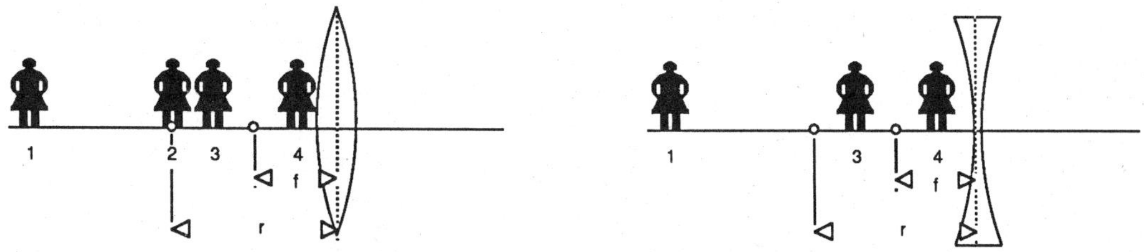

Converging lens    Diverging lens

$|f| = 10.0$ cm
$h_1 = h_2 = h_3 = h_4 = 2.00$ cm
$s_1 = 40.0$ cm, $s_2 = 20.0$ cm, $s_3 = 15.0$ cm, $s_4 = 5.0$ cm

Figure 25.3

Determine the following for the converging lens:

| | |
|---|---|
| 1. Image distance for each object | $1/s_1' = 1/f - 1/s = 1/10 - 1/40 = 3/40$<br>$s_1' = 40/3 = 13.3$ cm<br>$s_2' = 20.0$ cm, $s_3' = 30.0$ cm,<br>$s_4' = -10.0$ cm |
| 2. Magnification for each object | $m_1 = -s_1'/s_1 = -13.3$ cm/40.0 cm $= -0.333$<br>$m_2 = -1.00$, $m_3 = -2.00$, $m_4 = +2.00$ |

| | |
|---|---|
| 3. Size and orientation of each image | $h_1' = m_1h_1 = -0.666$ cm<br>$h_2' = -2.00$ cm, $h_3' = -4.00$ cm,<br>$h_4' = +4.00$ cm<br>+ means erect and - means inverted |
| 4. Whether the image is real or virtual for each object | The image is real for objects 1, 2, and 3 and virtual for object 4. |

Determine the following for the diverging lens:

| | |
|---|---|
| 5. Image distance for each object | $1/s_1' = 1/f - 1/s_1 = -1/10 - 1/40 = -1/8$,<br>$s_1' = -8.00$ cm<br>$s_3' = -6.00$ cm, $s_4' = -3.33$ cm |
| 6. Magnification for each object | $m_1 = -s_1'/s_1 = +0.200$<br>$m_3 = +0.400$, $m_4 = +0.666$ |
| 7. Size and orientation of each image | $h_1' = m_1h_1 = = +0.400$<br>$h_3' = +0.800$ cm, and $h_4' = +1.33$ cm<br>+ means erect. |
| 8. Whether the image is real or virtual | All images for diverging (concave) lenses are virtual. |

Example 25.1. A small lightbulb is placed 60.0 cm from a screen and serves as an object for a lens having a +20.0-cm focal length. (a) At what two distances from the screen should the lens be placed in order that a real image can be focused on the screen? (b) Calculate the magnification for each case. (c) Calculate the image size for each case if the lightbulb is 2.00 cm high.

Given: $s + s' = 60.0$ cm, $f = +10.0$ cm, $h = 2.00$ cm, image for each case is real

Determine: (a) The two locations of the lens relative to the screen that allow a real image. (b) The magnification for each case ($m_1$ and $m_2$). (c) The image size for each case ($h_1'$ and $h_2'$).

Strategy: (a) Since the images are formed on the screen, the two distances of interest (i.e. the distance from the lens to the screen) are the image distances $s_1'$ and $s_2'$. For this problem, the lens equation amounts to one equation and two unknowns ($s$ and $s'$). The relationship $s + s' = 60.0$ cm also amounts to one equation and the same two unknowns. We can solve these two equations simultaneously to obtain $s$ and $s'$. The solution involves a quadratic, the two roots of which are the two possible distances of the lens from the screen. (b) Knowing the object and image distances, we can determine the magnification. (c) Knowing the magnification and the object size, we can determine the image size.

Solution:

(a) We can write the following two equations involving the unknowns s and s'.

$$s + s' = 60.0 \text{ cm} \qquad 1/s + 1/s' = 1/f$$

When these equations are combined, we obtain

$$\frac{1}{60 - s'} + \frac{1}{s'} = \frac{1}{10} \qquad s'^2 - 60s' + 600 = 0$$

Solving this quadratic in s', we obtain

$$s' = \frac{-(-60) \pm [(-60)^2 - 4(600)]^{1/2}}{2} = \frac{60 \pm (1200)^{1/2}}{2} = 47.3 \text{ cm and } 12.7 \text{ cm}$$

The two possible cases are as follows:

Case I. $s_1' = 47.3$ cm, $s_1 = 12.7$ cm  Case II. $s_2' = 12.7$ cm, $s_2 = 47.3$ cm

(b) The magnification for each case is

Case I. $m_1 = -s_1'/s_1 = -3.72$  Case II. $m_2 = -s_2'/s_2 = -0.268$

(c) The size of the image for each case is

Case I. $h_1' = m_1h_1 = -7.44$ cm  Case II. $h_2' = m_2h_2 = -0.536$

Related Text Problems:

Converging lens -- 25-1 through 25-4, 25-6 through 25-10, 25-15.
Diverging lens -- 25-5, 25-12.

## 2. Lens Combinations (Section 25.6)

Review: When two lenses are used, light from an object strikes the first lens and an image is formed. The image of the first lens becomes the object for the second lens, which forms a final image. Figure 25.4 reviews image formation for two convex lenses with focal lengths $f_1$ and $f_2$ and a separation d.

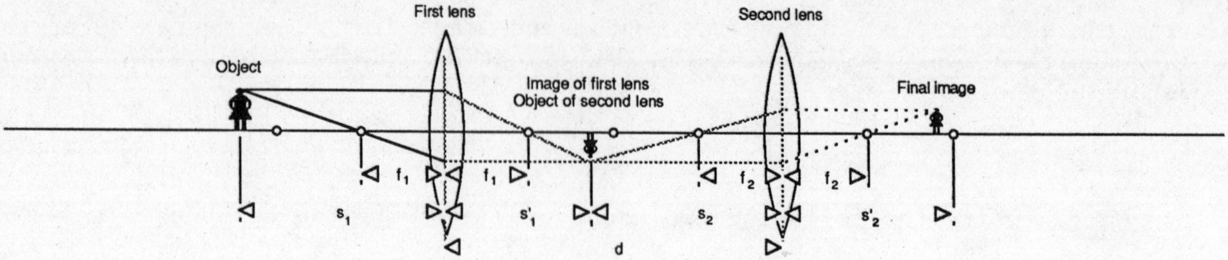


Figure 25.4

Note: The image for the first lens is found exactly as we found the image for a single converging lens (i.e., without regard to the second lens). The image for the first lens becomes the object for the second lens. The final image is found by tracing rays through the second lens without regard to the first lens.

Note: All sign conventions and relationships developed for a single lens are still applicable. It should be noted for the lens combination that the lens separation is related to $s_1'$ and $s_2$ by

$$d = s_1' + s_2$$

Also, the magnification of a combination of lenses is equal to the product of the individual magnifications. For two lenses with magnifications $m_1$ and $m_2$,

$$m = m_1 m_2$$

If two lenses are placed together ($d \approx 0$; hence $s_2 \approx -s_1'$), we obtain

$$1/s_1 + 1/s_2' = 1/f_1 + 1/f_2$$

Since the lenses are together, $s_1$ is the object distance and $s'_2$ is the image distance for the combination. We may then write

$$1/s_1 + 1/s_2' = 1/f$$

where f is the focal length for the combination. Combining the above two equations, we have

$$1/f = 1/f_1 + 1/f_2$$

where f is the focal length for the combination of two thin lenses in contact ($d \approx 0$) and $f_1$ and $f_2$ are the individual focal lengths.

The focal length of a lens is an indication of its refractive power. However, a more convenient unit for describe the refractive power of a lens is the diopter, defined as

$$P = 1/f \text{ (in meters)}$$

The refractive power of a lens combination is then

$$P = P_1 + P_2$$

<u>Practice:</u> The following information is given about the situation shown in Fig. 25.4:

$d = 40.0$ cm, $s_1 = 30.0$ cm, $h_1 = 2.00$ cm, $f_1 = 10.0$ cm, $f_2 = 8.00$ cm

Determine the following:

| | |
|---|---|
| 1. Location of the image from the first lens | $1/s_1' = 1/f_1 - 1/s_1 = 1/15$<br>$s_1' = 15.0$ cm |
| 2. Magnification of the image from the first lens | $m_1 = -s_1'/s_1 = -0.500$ |
| 3. Size of the image from the first lens | $m_1 = h_1'/h_1$<br>$h_1' = m_1h_1 = -1.00$ cm |
| 4. Nature of the image from the first lens | Real, inverted, and smaller than the object |
| 5. Object distance for the second lens | $d = s_1' + s_2$<br>$s_2 = d - S_1' = 25.0$ cm |
| 6. Image distance for the second lens | $1/s_2' = 1/f_2 - 1/s_2 = 17/200$<br>$s_2' = 11.8$ cm |
| 7. Magnification of the image from the second lens | $m_2 = -s_2'/s_2 = -0.472$ |
| 8. Nature of the image from the second lens | Real, inverted, and smaller than the object |
| 9. Magnification of the combination | $m = m_1m_2 = 0.236$ |

| | |
|---|---|
| 10. Size and orientation of the final image | First, let's use the information for the second lens:<br>$h_2' = m_2h_2$ and $h_2 = h_1' = -1.00$ cm<br>$h_2' = (-0.472)(-1.00 \text{ cm}) = 0.472$ cm<br>The final image is 0.472 cm tall, inverted with respect to the object for lens 2, and erect with respect to the initial object.<br>Second, let's use the information for the lens combination:<br>$h_2' = mh_1$ and $m = m_1m_2$<br>$h_2' = (0.236)(2.00 \text{ cm}) = 0.472$ cm<br>The final image is 0.472 cm tall and erect with respect to the initial object. |

Consider the lens combination shown in Fig. 25.5.

Figure 25.5

$h_1 = 2.00$ cm
$s_1 = 15.0$ cm
$f_1 = +10.0$ cm
$f_2 = -10.0$ cm
$d = 25.0$ cm

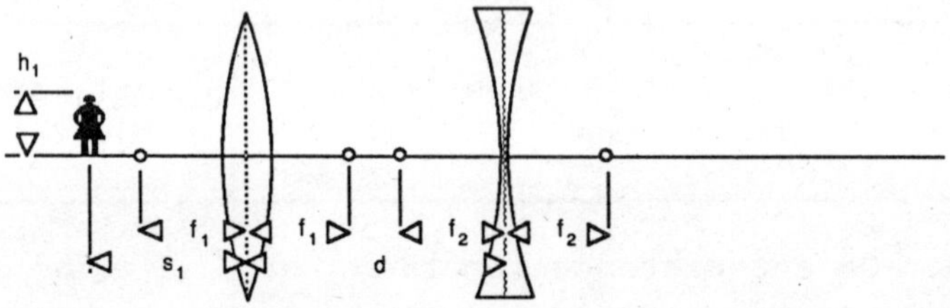


Determine the following:

| | |
|---|---|
| 11. Ray diagram to determine the image from the first lens | 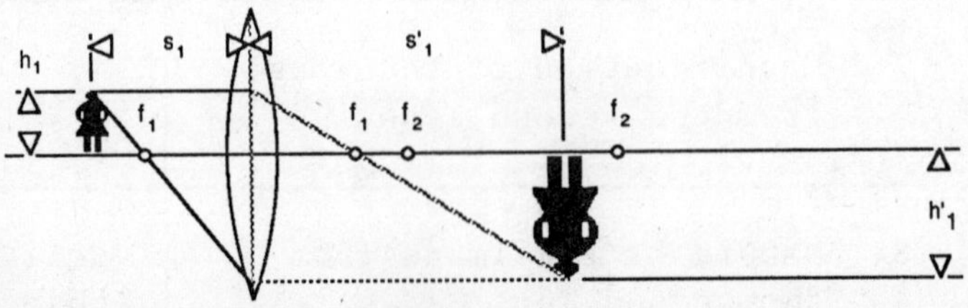  |
| 12. Location of the image from the first lens | $1/s_1' = 1/f_1 - 1/s_1 = 1/30$<br>$s_1' = 30.0$ cm |
| 13. Magnification of the image from the first lens | $m_1 = -s_1'/s_1 = -2.00$ |

| | |
|---|---|
| 14. Size and orientation of the image from the first lens | $h_1' = m_1 h_1 = -4.00$ cm<br>The minus sign tells us the image is inverted. |
| 15. Nature of the image from the first lens | Real, inverted, and larger than the object |
| 16. Object distance for the second lens | $d = s_1' + s_2$<br>$s_2 = d - s_1' = -5.00$ cm |
| 17. Location of the image from the second lens | $1/s_2' = 1/f_2 - 1/s_2 = 1/(-10) - 1/(-5)$<br>$= 1/10$<br>$s_2' = 10.0$ cm |
| 18. Magnification of the image from the second lens | $m_2 = -s'_2/s_2 = -10.0/-5.00 = +2.00$ |
| 19. Magnification of the combination | $m = m_1 m_2 = (-2.00)(+2.00) = -4.00$ |
| 20. Size of the image from the second lens | $h_2' = m_2 h_2$ and $h_2 = \|h_1'\|$<br>$h_2' = (2.00)(4.00 \text{ cm}) = 8.00$ cm<br>The final image is 8.00 cm tall and erect with respect to the object for the second lens.<br>$h_2' = mh_1 = m_1 m_2 h_1 = -8.00$ cm<br>The final image is 8.00 cm tall and inverted with respect to the initial image. |
| 21. Ray diagram to see how the final image is formed | 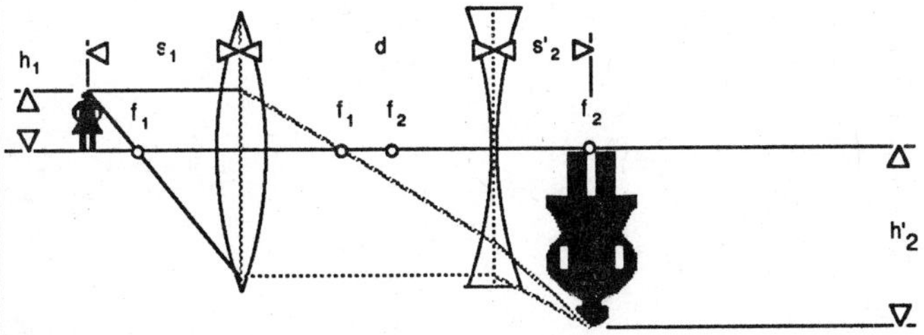<br>The two traced rays diverge at the second lens and intersect as shown. |

Example 25.5. An object 2 cm high is placed 120 cm in front of two thin lenses in contact, and the image is focused on a screen 60.0 cm from the combination. The optical power of one of the lenses is -3.5 diopters. Calculate (a) the focal length of the combination, (b) the optical power of the combination, (c) the focal length of the unknown lens, and (d) the size of the image.

Given: $h = 2.00$ cm, $s_1 = 120$ cm, $s_2' = 60.0$ cm, $d = 0$ cm, $P_1 = -3.50$ diopters

Determine: (a) f for combination, (b) P for combination, (c) $f_2$, (d) $h_2'$

Strategy: (a) Knowing $s_1$ and $s_2'$, we can determine f for the combination. (b) Knowing f, we can determine P for the combination. (c) Knowing P and $P_1$, we can determine $P_2$ and hence $f_2$. (d) Knowing $s_1$ and $s_2'$, we can determine the magnification for the combination and hence $h_2'$.

Solution:

(a) $1/f = 1/s_1 + 1/s_2' = 1/120 \text{ cm} + 1/60.0 \text{ cm} = 1/40.0 \text{ cm}$ $\quad f = 40.0$ cm
(b) $P = 1/f$ (in meters) $= 1/0.400 \text{ m} = 2.50/\text{m}$
(c) $P_2 = P - P_1 = 2.50/\text{m} - (-3.50/\text{m}) = 6.00/\text{m}$ and $f_2 = 1/P_2 = 0.167$ m
(d) $m = -s_2'/s_1 = -60.0 \text{ cm}/120 \text{ cm} = -0.500$ and $h'_2 = mh_1 = -1.00$ cm

Related Text Problems: 25-23 to 25-27.

---

PRACTICE TEST

Take and grade the practice test. Doing so will allow you to determine any weak spots in your understanding of the concepts taught in this chapter. The following section prescribes what you should study further to strengthen your understanding.

Figure 25.6 shows objects in front of converging and diverging lenes.

$|f| = 20.0$ cm
$h_1 = h_2 = h_3 = h_4 = h_5 = 2.00$ cm
$s_1 = 60.0$ cm, $s_2 = 30.0$ cm, $s_3 = 10.0$ cm, $s_4 = 50.0$ cm, and $s_5 = 20.0$ cm

Figure 25.6

Determine the following:

__________ 1. Image distance for object 1
__________ 2. Image distance for object 4

3. Magnification of the image of object 2
4. Magnification of the image of object 5
5. Size and orientation of the image of object 3
6. Size and orientation of the image of object 5
7. Objects with virtual images

Figure 25.7 shows two converging lenses being used to view a small object.

Figure 25.7

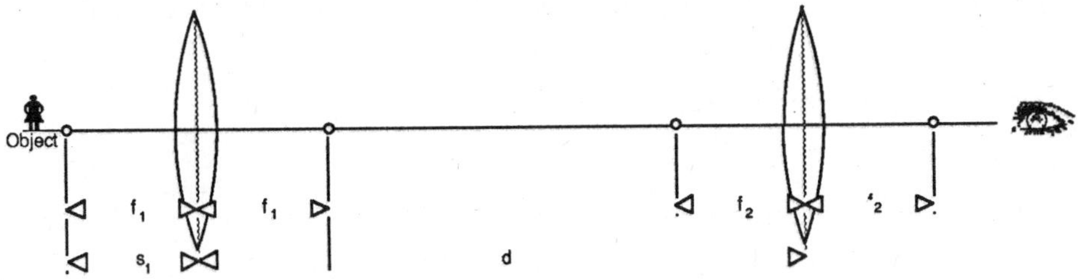


$f_1$ = +2.00 cm, $f_2$ = +2.50 cm, d = 2.4 cm, $s_1$ = 2.20 cm, $h_1$ = 1.00 mm

Determine the following:

8. Image distance for the first lens
9. Magnification due to the first lens
10. Size of the image created by the first lens
11. Object distance for the second lens
12. Image distance for the second lens
13. Magnification due to the second lens
14. Magnification due to the combination of lenses
15. Size and orientation of the final image

Figure 25.8 shows diverging and converging lenses being used to view an object.

Figure 25.8

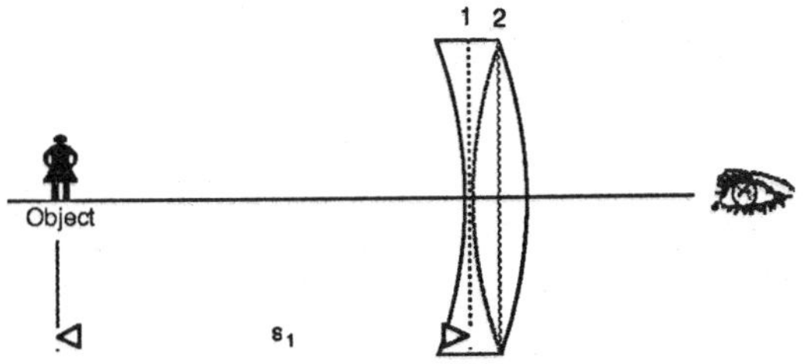


$P_i$ = -10.0 diopters, $f_2$ = +20.0 cm, $s_1$ = +20.0 cm, $h_1$ = +2.00 cm

Determine the following:

16. Focal length of the lens combination
17. Refracting power of the lens combination
18. Magnification due to the second lens
19. Magnification of the combination
20. Size and orientation of the final image

(See Appendix I for answers.)

## PRINCIPAL CONCEPTS AND EQUATIONS PRESCRIPTION

Your score on the practice is an excellent measure of your understanding of this chapter. You should now use the following chart to write your own prescription for curing any of your physics ills. Look down the leftmost column to the number of the question(s) you answered incorrectly, read across that row to see which section(s) of the study guide you should return to for further study, and then do the suggested text problems to gain additional experience in working with the particular concept.

| Practice Test Question | Concepts and Equations | Prescription: Principal Concept | Prescription: Text Problems |
|---|---|---|---|
| 1 | Thin lens equation: $1/s + 1/s' = 1/f$ | 1 | 25-1,2 |
| 2 | Thin lens equation | 1 | 25-5,12 |
| 3 | Magnification: $m = -s'/s$ | 1 | 25-3,4 |
| 4 | Magnification | 1 | 25-6,12 |
| 5 | Image size: $m = h'/h$ | 1 | 25-7,8 |
| 6 | Image size | 1 | 25-5,12 |
| 7 | Virtual images | 1 | 25-6 |
| 8 | Thin lens equation | 1 | 25-24,26 |
| 9 | Magnification | 1 | 25-26,28 |
| 10 | Image size | 1 | 25-28,29 |
| 11 | Lens separation: $d = s_1' + s_2$ | 2 | 25-29,30 |
| 12 | Thin lens equation | 1 | 25-30,31 |
| 13 | Magnification | 1 | 25-31,32 |
| 14 | Combination magnification: $m = m_1 m_2$ | 2 | 25-24,30 |
| 15 | Combination size: $m = h_2'/h_1$ | 2 | 25-26,29 |
| 16 | Combination focal length: $1/f = 1/f_1 + 1/f_2 = 1/s_1 + 1/s_2'$ | 2 | 25-23,25 |
| 17 | Refracting power: $P = 1/f$ | 2 | 25-25,27 |
| 18 | Magnification: $m_2 = -s_2'/s_2$ | 1 | 25-23,27 |
| 19 | Combination magnification: $m = m_1 m_2$ | 2 | 25-25,27 |
| 20 | Combination image size: $m = h_2'/h_1$ | 2 | 25-23,25 |

# 26 Physical Optics

## RECALL FROM PREVIOUS CHAPTERS

You should be able to proceed with this chapter without reviewing any previous concepts and equations.

## NEW IDEAS IN THIS CHAPTER

| Concepts and equations introduced | Text Section | Study Guide Page |
|---|---|---|
| Double-slit interference pattern: | 26.3 | 435 |
| Constructive when $\Delta s = m\lambda = d \sin\theta_m$ | | |
| at $y_m = R \tan\theta_m \simeq R \sin\theta_m = Rm\lambda/d$ | | |
| Destructive when $\Delta s = (m - 1/2)\lambda = d \sin\theta_m$ | | |
| at $y_m = R \tan\theta_m \simeq R \sin\theta_m = R(m - 1/2)\lambda/d$ | | |
| Thin film interference: | 26.4 | 438 |
| Case I: neither or both rays change phase | | |
| constructive $2t = m\lambda/n$ | | |
| destructive $2t = (m + 1/2)\lambda/n$ | | |
| Case II: one or the other changes phase | | |
| constructive $2t = (m + 1/2)\lambda/n$ | | |
| destructive $2t = m\lambda/n$ | | |
| Single-slit diffraction: | 26.5 | 441 |
| Destructive when $w \sin\theta_m = m\lambda$ | | |
| at $y_m = R \tan\theta_m \simeq R \sin\theta_m = Rm\lambda/w$ | | |
| Planar diffraction grating: | 26.6 | 444 |
| Constructive when $m\lambda = d \sin\theta_m$, $d = 1/N$ | | |
| at $y_m = R \tan\theta_m$ | | |
| Brewster's law: $\tan\theta_p = n_2/n_1$ | 26.7 | 447 |
| Law of Malus: $I = I_o \cos^2\theta$ | 26.7 | 447 |

## PRINCIPAL CONCEPTS AND EQUATIONS

### 1. Double Slit (Section 26.3)

Review: Figure 26.1 reviews what happens to a double-slit interference pattern as the width of the slit goes from $w << \lambda$ to $w > \lambda$.

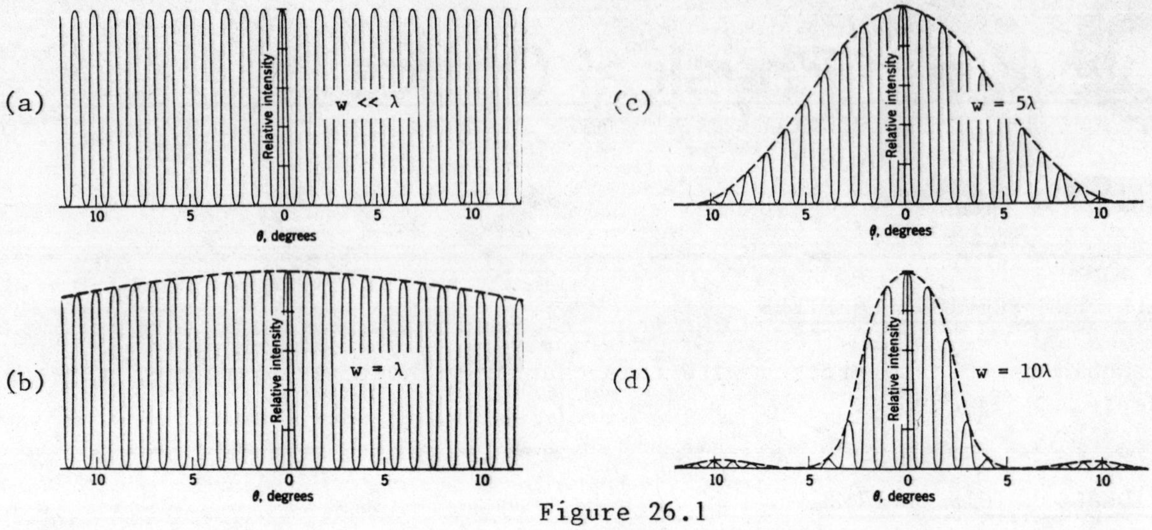


Figure 26.1

Figure 26.2 reviews constructive and destructive interference from a double slit for the case $a > \lambda$.

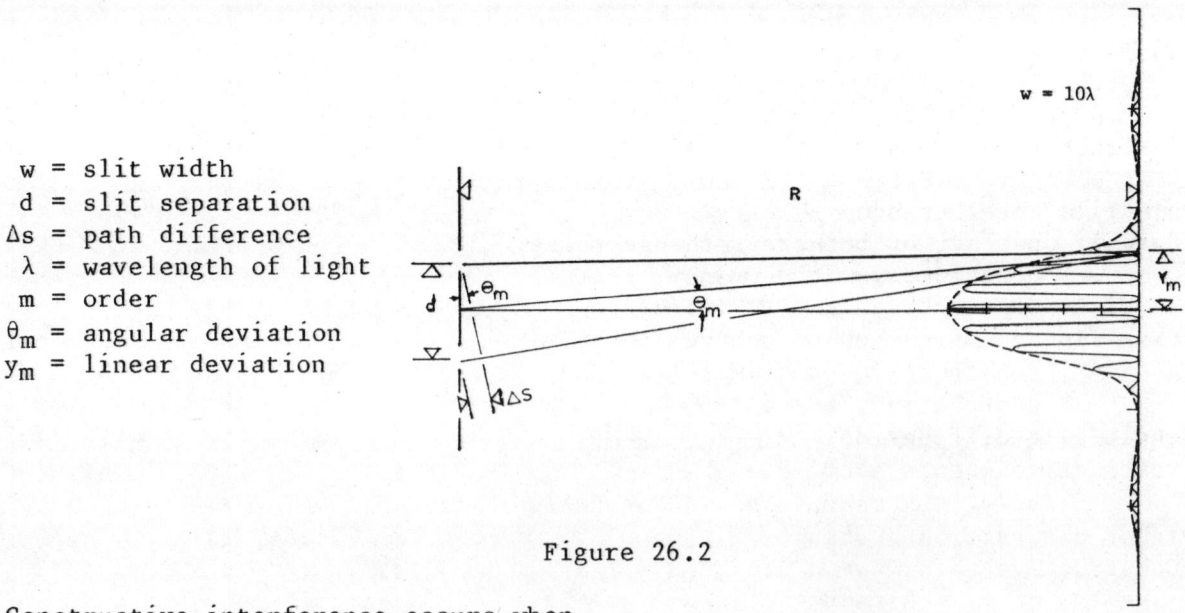

w = slit width
d = slit separation
$\Delta s$ = path difference
$\lambda$ = wavelength of light
m = order
$\theta_m$ = angular deviation
$y_m$ = linear deviation

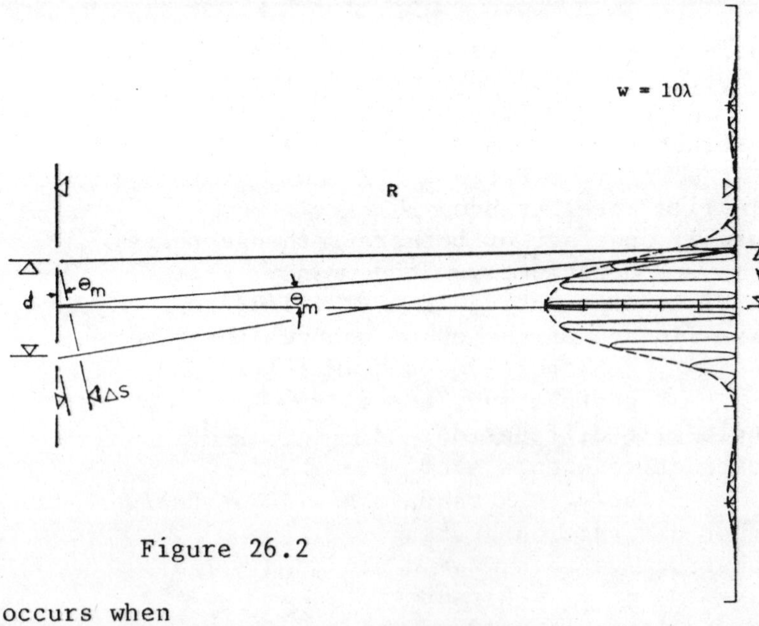


Figure 26.2

Constructive interference occurs when

$$\Delta s = m\lambda = d \sin\theta_m \qquad m = 0,1,2,3$$

Since the angular deviation is small, constructive interference occurs at

$$y_m = R \tan\theta_m \approx R \sin\theta_m = Rm\lambda/d \qquad m = 1,2,3$$

Destructive interference occurs when

$$\Delta s = (m - 1/2)\lambda = d \sin\theta_m \qquad m = 1,2,3$$

Since the angular deviation is small, destructive interference occurs at

$$y_m = R \tan\theta_m \approx R \sin\theta_m = R(M - 1/2)\lambda/d \qquad m = 1,2,3$$

Practice: The following information is given for the situation shown in Fig. 26.2.

$$d = 2.00 \times 10^{-5}\ \text{m},\ \lambda = 6.00 \times 10^{-7}\ \text{m},\ R = 2.00\ \text{m}$$

Determine the following:

| | |
|---|---|
| 1. Angular deviation $\theta_2$ of the second order constructive interference band | $m\lambda = d \sin\theta_m$, $m = 2$<br>$\theta_2 = \sin^{-1}(m\lambda/d) = 3.44°$ |
| 2. Linear distance between central and second-order maxima | $\tan\theta_m = y_m/R$, $m = 2$<br>$y_2 = R \tan\theta_2 = 0.120$ m<br>$y_2 = Rm\lambda/d = 0.120$ m |
| 3. Linear distance between any two adjacent constructive maxima | $y_m = Rm\lambda/d$<br>$y_{m+1} = R(m + 1)\lambda/d$<br>$\Delta y = y_{m+1} - y_m = R\lambda/d = 6.00 \times 10^{-2}$ m |
| 4. Where we should reposition the screen so that the linear separation of the two second-order bright bands is 10.0 cm | $y_2 = Rm\lambda/d$; $2y_2 = 0.100$ m<br>$R = y_2 d/m\lambda = 0.833$ m |
| 5. Wavelength of light that would place the third-order maxima 12.5 cm from the central maximum | $\lambda = y_m d/Rm = 4.17 \times 10^{-7}$ m |
| 6. The wavelength of light that would have fifth-order minima at the site of the present third-order maxima | $y_m = Rm\lambda/d$ constructive<br>$y_m = R(m - 1/2)\lambda/d$ destructive<br>We want<br>$y_3$(constructive) = $y_5$(destructive)<br>$3R\lambda/d = (5 - 1/2)R\lambda'/d$<br>$\lambda' = 2\lambda/3 = 4.00 \times 10^{-7}$ m |

Example 26.1. Light is incident on a double slit, and the interference pattern is formed on a distant screen. The separation of the slits changes, and the second-order bright fringe occurs where the third-order bright fringe originally occurred. Determine the new slit separation if the old separation is 0.150 mm.

Given: $d_i = 1.50 \times 10^{-4}$ m, $y_{3initial} = y_{2final}$

Determine: The final slit width.

Strategy: We can write an expression for the location of the third-order bright fringe for the initial slit separation $d_i$. We can write an expression for the location of the second-order bright fringe for the final slit separation $d_f$. We can equate these two expressions and determine the new slit separation.

Solution: The initial location of the third-order bright fringe is

$$y_{3initial} = 3R\lambda/d_i$$

The final location of the second order bright fringe is

$$y_{2final} = 2R\lambda/d_f$$

Equating these two expressions and solving for $d_f$, we have

$$3R\lambda/d_i = 2R\lambda/d_f$$

$$d_f = (2/3)d_i = (2/3)(1.50 \times 10^{-4}\text{ m}) = 1.00 \times 10^{-4}\text{ m}$$

Related Text Problems: 26-1 through 26-7.

## 2. Thin-Film Interference (Section 26.4)

Review: Figure 26.3 shows light incident on a thin film and partially reflected from each surface.

Figure 26.3

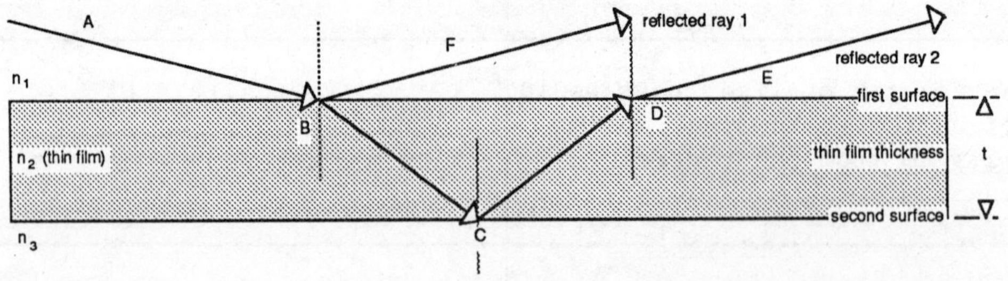


Light that reaches the eye is composed of light reflected from the top and bottom surfaces. These two parts, which we consider as separate rays or waves, are superimposed at the eye and produce an interference phenomenon. The interference may be constructive, destructive, or something in between, depending on such factors as

(a) the length of the path difference of the two rays
(b) the wavelength of the light in the thin film (this depends on the index of refraction of the thin film material)
(c) phase changes at the reflecting surface.

(a) If the length of the path difference is an integral number of wavelengths, constructive interference will occur if neither or both rays change phase on reflection, destructive interference will occur if one or the other of the rays changes phase on reflection. If the length of the path difference is an odd integral number of half wavelengths, constructive interference will will occur if one or the other of the rays changes phase on reflection, destructive interference will occur if neither or both rays change phase on reflection.

(b) The wavelength of the light in the film is not equal to its wavelength in a vacuum (or air). As light goes from one medium to another, the frequency remains constant but the speed and wavelength change. Light of wavelength $\lambda$ in a vacuum has a wavelength of $\lambda/n$ in a medium with an index of refraction n. This may be written as

$$\lambda_n = \lambda/n$$

(c) If the medium from which the light reflects has a larger index of refraction than the medium in which it is traveling, there will be a 180° phase change. If the light is reflected from a medium that has a smaller index of refraction than that medium in which the light is traveling, there will be no phase change. The preceding is reviewed for two cases.

Case I - neither or both rays change phase on reflection

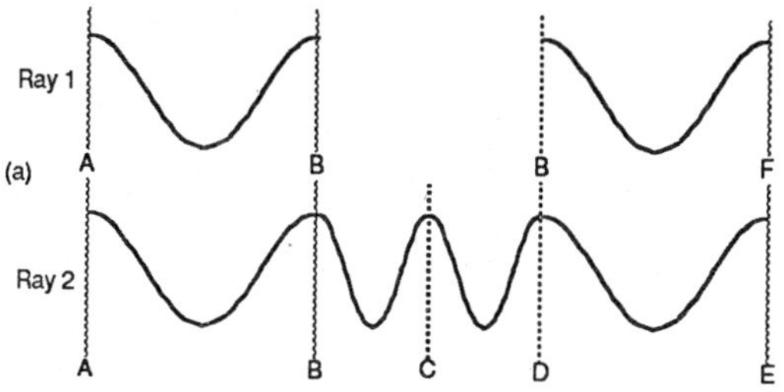

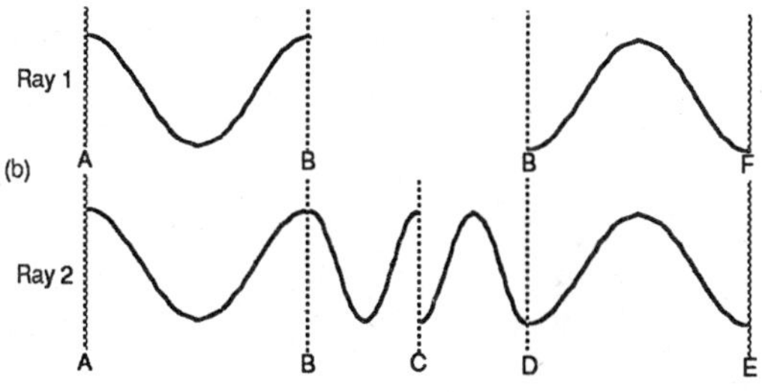

Case II - one or the other of the rays changes phase on reflection

Neither ray changes phase on reflection: $n_1 > n_2 > n_3$

Phase change on reflection from the first boundary $n_1 < n_2 > n_3$

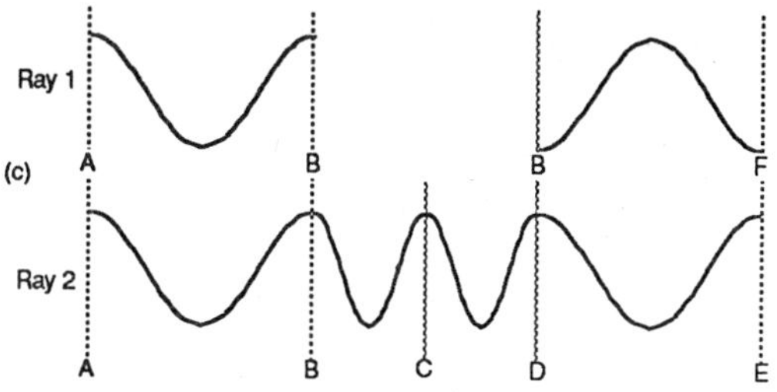

Both rays change phase on reflection: $n_1 < n_2 < n_3$

Phase change on reflection from the secondary boundary: $n_1 > n_2 < n_3$

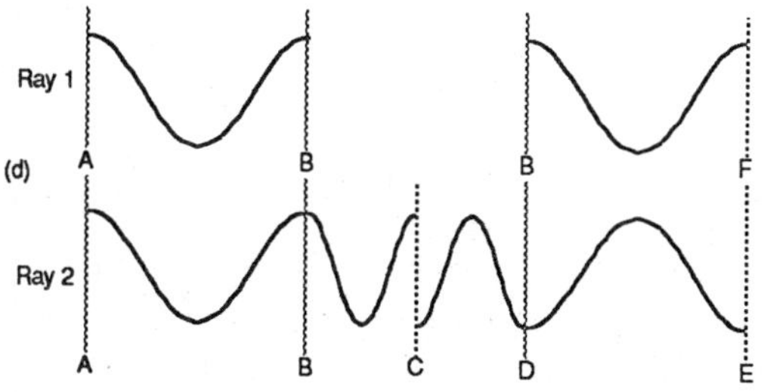

Constructive Interference

$2t = m\lambda/n \quad m = 0,1,2,3,\ldots$  |  $2t = (m + 1/2)\lambda/n \quad m = 0,1,2,3,\ldots$

Destructive Interference

$2t = (m + 1/2)\lambda/n \quad m = 0,1,2,3,\ldots$  |  $2t = m\lambda/n \quad m = 0,1,2,3,\ldots$

Figure 26.4

Practice: The following information is given for Fig. 26.3.

$$n_1 = 1.00, \quad n_2 = 1.36, \quad n_3 = 1.65, \quad t = 4.00 \times 10^{-7}\ \text{m}$$

Determine the following:

| | |
|---|---|
| 1. Phase change for ray 1 and ray 2 at first surface | The reflected ray (1) undergoes a 180° phase change. The transmitted ray (2) does not undergo a phase change. |
| 2. Phase change for ray 2 at the second surface | Experiences a 180° phase change. |
| 3. Wavelengths in the visible range that experience destructive interference | This is a case-I situation (both rays change phase on reflection). Destructive interference occurs for $2t = (m + 1/2)\lambda/n_2$, or $\lambda = 2n_2t/(m + 1/2)$<br>$\lambda_0 = 21.8 \times 10^{-7}$ m  m = 0<br>$\lambda_1 = 7.25 \times 10^{-7}$ m  m = 1<br>$\lambda_2 = 4.35 \times 10^{-7}$ m  m = 2<br>$\lambda_3 = 3.11 \times 10^{-7}$ m  m = 3<br>Only $\lambda_1$ and $\lambda_2$ are in the visible range. |
| 4. Wavelengths in the visible range that experience constructive interference | For case I, constructive interference occurs for<br>$2t = m\lambda/n_2$, or $\lambda = 2tn_2/m$<br>$\lambda_0 = \infty$  m = 0<br>$\lambda_1 = 10.9 \times 10^{-7}$  m = 1<br>$\lambda_2 = 5.44 \times 10^{-7}$  m = 2<br>$\lambda_3 = 3.63 \times 10^{-7}$  m = 3<br>Only $\lambda_2$ is in the visible range. |
| 5. The thinnest the film can be and destructive interference occur for reflected light of $5.00 \times 10^{-7}$ m wavelength | For case I, destructive interference occurs for $2t = (m + 1/2)\lambda/n_2$; hence $t = (m + 1/2)\lambda/2n_2$<br>The thinnest film occurs for m = 0, or $t = \lambda/4n_2 = 9.19 \times 10^{-8}$ m |

Example 26.2. A spacer separates one end of two 12.0-cm-long plates of flat glass, and the other ends of the two pieces are touching each other. When light of wavelength $5.46 \times 10^{-7}$ m is incident on the plates, the distance between consecutive dark fringes is 0.500 cm. What is the thickness of the spacer?

Given: $\ell$ = 12.0 cm = length of the two plates of flat glass
$\lambda = 5.46 \times 10^{-7}$ = wavelength of the incident light
s = 0.500 cm = separation of consecutive dark fringes

Determine: T = thickness of the spacer

Strategy: Figure 26.5 shows the air wedge and reflected rays 1 and 2.

Figure 26.5

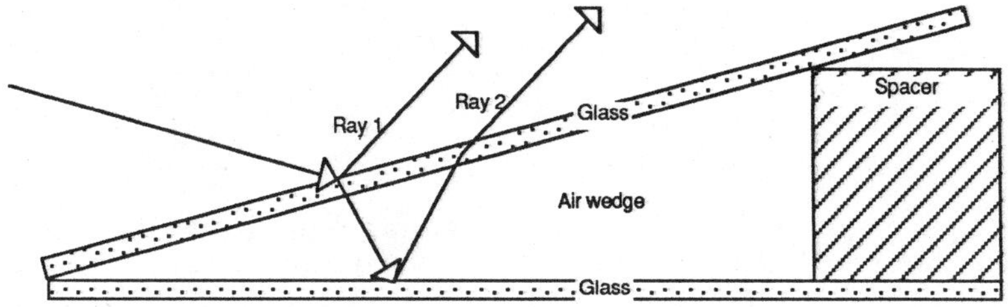


First we need to establish that a dark fringe exists at the contact end of the glass plates and that this is a case-II situation. Then we can determine the separation of the plates at the next dark fringe and hence the angular separation of the plates. Knowing the angular separation and the length of the plates, we can determine the thickness of the spacer.

Solution: At the contact end of the glass plates, ray 1 (reflects off the top layer of the air wedge) experiences no phase change and ray 2 (reflects off the top of the second glass plate) experiences a 180° phase change. Because of this phase change, a dark fringe occurs at the contact end. Also, since only one of the rays (2) experiences a phase change, this is a case-II situation, for which the thin film thickness (i.e., separation of the glass plates due to the spacer) at which destructive interference occurs is obtained as follows:

$$2t = M\lambda/n \qquad m = 0,1,2,3,\ldots$$
$$t = m\lambda/2n$$
$$t_0 = 0 \text{ for } m = 0 \text{ (contact end)}$$
$$t_1 = \lambda/2n = 2.73 \times 10^{-7} \text{ m for } m = 1 \text{ (second dark fringe)}$$

The angular separation of the plates is

$$\alpha = \tan^{-1}(t_1/s) = (3.13 \times 10^{-3})^\circ$$

The thickness of the spacer is

$$T = \ell \tan\alpha = 6.56 \times 10^{-6} \text{ m}$$

Related Text Problems: 26-8 through 26-12.

## 3. Single-Slit Diffraction (Section 26.5)

Review: Figure 26.6 reviews the diffraction pattern of a single slit.

Figure 26.6

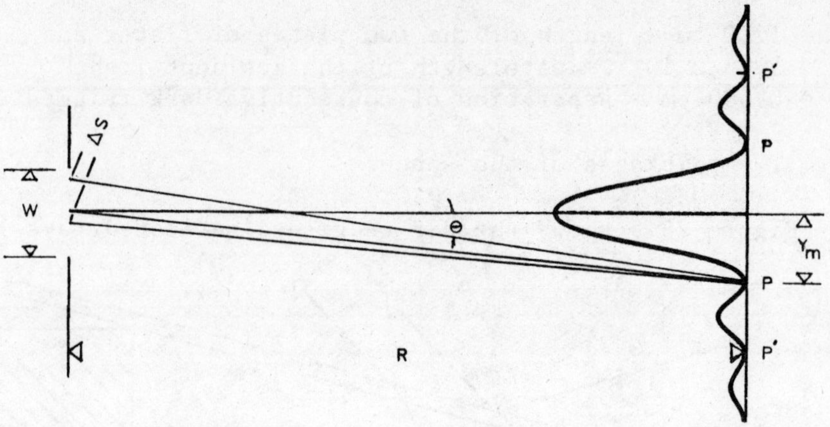


For every point along the top half of the aperture, there is a corresponding point along the bottom half for which the path difference to point P is equal to a $\lambda/2$. Consequently, we can write

$$\Delta s = (w/2)\sin\theta = \lambda/2 \quad \text{or} \quad w \sin\theta = \lambda$$

For point P', this expression is

$$w \sin\theta = 2\lambda$$

In general, for the dark spaces

$$w \sin\theta = m\lambda \qquad m = 1,2,3,\ldots \text{ (order of the dark band)}$$

If the distance R is large relative to $y_m$, we can write

$$\sin\theta_m \approx \tan\theta_m = y_m/R$$

The dark spaces are located at

$$y_m = R \sin\theta_m = R(m\lambda/w) \qquad m = 1,2,3,\ldots$$

Practice:

Determine the following:

| | |
|---|---|
| 1. Width of the single slit if the first minimum occurs at 10° for light of wavelength $\lambda = 5.00 \times 10^{-7}$ m | $m\lambda = w \sin\theta_m$<br>$w = m\lambda/\sin\theta m = 2.88 \times 10^{-6}$ m |
| 2. Angular separation between first and third minima for a slit width of $5.00 \times 10^{-6}$ m and light of wavelength $\lambda = 5.00 \times 10^{-7}$ m | $m\lambda = w \sin\theta_m$<br>$\theta_1 = \sin^{-1}(\lambda/w) = 5.74°$<br>$\theta_3 = \sin^{-1}(3\lambda/w) = 17.5°$<br>$\Delta\theta = \theta_3 - \theta_1 = 11.8°$ |

| | |
|---|---|
| 3. Wavelength of light that has its second minimum coincide with the third minimum of light of wavelength $\lambda = 4.00 \times 10^{-7}$ m | $y_m = Rm\lambda/w$<br>$y_2 = y_3$<br>$2R\lambda'/w = 3R\lambda/w$<br>$\lambda' = 3\lambda/2 = 6.00 \times 10^{-7}$ m |
| 4. Width of central maximum for light of wavelength $\lambda = 4.00 \times 10^{-7}$ m if slit is $5.00 \times 10^{-6}$ m wide and screen is 2.00 m from slit | $m\lambda = w \sin\theta_m$<br>$\sin\theta_m \approx \tan\theta_m = y_m/R$<br>Combining these, obtain $y_m = m\lambda R/w$<br>Width of central maxima is<br>$2y_1 = 2(\lambda R/w) = 0.320$ m |
| 5. Distance between successive dark fringes for the situation given in step 4 | $y_m = m\lambda R/w$ (step 4)<br>$y_{m+1} = (m + 1)\lambda R/w$<br>$\Delta y = y_{m+1} - y_m = \lambda R/w = 0.160$ m |
| 6. Maximum number of fringes expected on either side of central maximum if slit slit width is 20 $\lambda$ | $m\lambda = w \sin\theta_m$<br>$\theta_m = 90°$ (this value lets us look to the left and right the maximum amount)<br>$m = w \sin 90°/\lambda = 20\lambda/\lambda = 20$ |

Example 26.3. Light of wavelength 600 nm is incident on a slit 0.100 mm wide and the resulting diffraction pattern observed on a screen 2.00 m away. (a) What is the distance between consecutive dark fringes? (b) If the entire apparatus is immersed in water, what is the distance between consecutive dark fringes?

Given: $\lambda = 6.00 \times 10^{-7}$ m, $w = 1.00 \times 10^{-4}$ m, $R = 2.00$ m

Determine: (a) The dark-fringe spacing when the apparatus is in air. (b) The dark fringe spacing when the apparatus is immersed in water.

Strategy: Using our knowledge of single-slit theory, we can develop an expression for the location of two successive minima and hence the fringe spacing. We can use this expression and the given information to determine the fringe spacing in air. Next we can calculate the wavelength in water and then use the previously developed expression to determine the fringe spacing in water.

Solution: For a single slit, the $m^{th}$ dark space occurs at

$$y_m = mR\lambda/w$$

and the next dark space occurs at

$$y_{m+1} = (m + 1)R\lambda/w$$

Hence the fringe spacing is given by

$$\Delta y = y_{m+1} - y_m = R\lambda/w = 1.20 \times 10^{-2}\ \text{m}$$

If the apparatus is immersed in water, R and w do not change but the wavelength becomes

$$\lambda_n = \lambda/n = 4.51 \times 10^{-7}\ \text{m}$$

The fringe spacing in water is

$$\Delta y_n = R\lambda_n/w = R\lambda/nw = \Delta y/n = 9.02 \times 10^{-3}\ \text{m}$$

Related Text Problems: 26-13 through 26-16.

## 4. Planar Diffraction Grating (Section 26.6)

Review: Figure 26.7 reviews the interference pattern for two slits and six slits for the case where the slit width is very small relative to the wavelength. The six slits have the same separation as the two slits.

Figure 26.7.

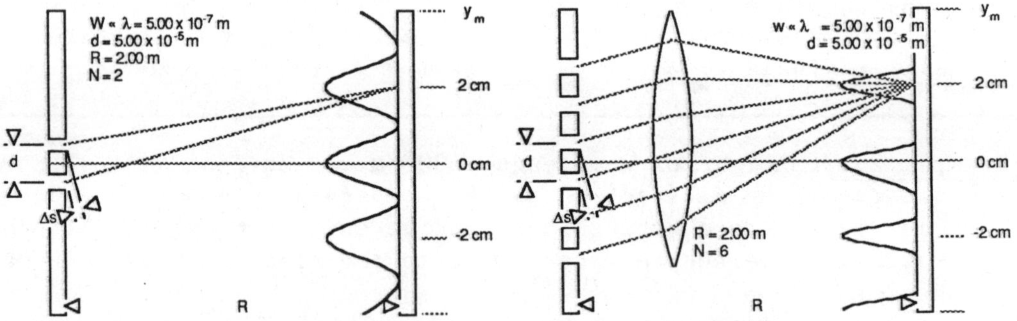


Constructive interference occurs when

$$\Delta s = m\lambda = d\sin\theta_m \qquad m = 0,1,2,3,...$$

Since the angular deviation is small, constructive interference occurs at

$$y_m = R\tan\theta_m \approx R\sin\theta_m = Rm\lambda/d \qquad m = 0,1,2,3,...$$

Destructive interference occurs when

$$\Delta s = (m - 1/2)\lambda = d\sin\theta_m \qquad m = 1,2,3,...$$

Since the angular deviation is small, destructive interference occurs at

$$y_m = R\tan\theta_m \approx R\sin\theta_m = R(m - 1/2)\lambda/d \qquad m = 1,2,3,...$$

Notice that this is identical to the double slit information. In general, gratings do not have a spacing much less than the wavelength of light, and let's see what happens to the interfernce pattern when we use more typical values.

Typical values for a diffraction grating are (let's agree to use light of wavelength $\lambda = 5.00 \times 10^{-7}$ m)

$N = 5000/\text{cm}$ = lines or scratches per cm on the grating
$d = 1/N = 2.00 \times 10^{-6}$ m = distance between scratches
$w = d = 2.00 \times 10^{-6}$ m = slit width ($w = 4\lambda$ for this case)
$R = 2.00$ m = distance from grating to screen

The angular deviation for successive maxima is

$$\theta_m = \sin^{-1}(m\lambda/d) = 14.5° \text{ for } m = 1$$
$$= 30.0° \text{ for } m = 2$$
$$= 48.6° \text{ for } m = 3$$

The diffraction pattern will look like that shown in Fig. 26.8.

Figure 26.8

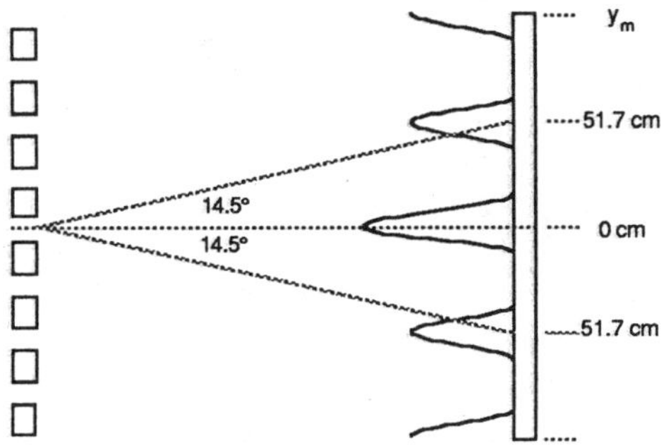


For this case, where the slit width is not much smaller than the wavelength (here $w = d = 4\lambda$), the angular deviation is large and the maxima are far apart. Consequently, the location of the site of destructive interference is of little interest to us (actually, it locates the center of the large dark space between the maxima) and it is inappropriate to use the approximation $\tan\theta_m \approx \sin_m$.

For such a diffraction grating, constructive interference occurs when

$$m\lambda = d \sin\theta_m \qquad m = 0,1,2,3,\ldots$$

and the sites of constructive interference are located at

$$y_m = R \tan\theta_m \qquad m = 0,1,2,3,\ldots$$

<u>Practice:</u> A diffraction grating with 5000 lines/cm is located 2.00 m from a viewing screen.

Determine the following:

| | |
|---|---|
| 1. Grating spacing (i.e., distance between the scratches) | $d = 1/N = 2.00 \times 10^{-4}$ cm |

| | | |
|---|---|---|
| 2. | Number of orders you may view using light of wavelength $5 \times 10^{-7}$ m | $m = d \sin\theta m/\lambda$<br>If we look to the left or right as far as possible ($\theta_m = 90°$), we see all orders.<br>$m = d \sin 90°/\lambda = 4.00$ |
| 3. | Longest wavelength for which the third order can be observed | $m\lambda = d \sin\theta_m$<br>$\lambda = d \sin 90°/m = 6.67 \times 10^{-7}$ m |
| 4. | Distance on the screen between second-order images for light of wavelength $4.00 \times 10^{-7}$ m and $6.00 \times 10^{-7}$ m | $m\lambda = d \sin\theta m$<br>$\theta_m = \sin^{-1}(m\lambda/d) = \sin^{-1}(m\lambda N)$<br>$\theta_2 = \sin^{-1}[(2)(4.00 \times 10^{-7}\text{ m})(5 \times 10^5/\text{m})]$<br>$= 23.6°$<br>$y_2 = R \tan\theta_2 = 0.874$ m<br>$\theta_2 = \sin^{-1}[(2)(6.00 \times 10^{-7}\text{ m})(5 \times 10^5/\text{m}]$<br>$= 36.9°$<br>$y_2 = R\tan\theta_2 = 1.50$ m<br>$\Delta y = y_2 - y_2 = 0.626$ m |
| 5. | Spacing for a grating whose third-order image has the same angular deviation as the second-order image of this grating under similar conditions (i.e., same $\lambda$ and R) | $m\lambda = d \sin\theta_m$<br>$d = 2\lambda/\sin\theta_2$<br>$\theta_3' = \theta_2$; hence $\sin\theta_3' = \sin\theta_2 = 2\lambda/d$<br>$d' = 3\lambda/\sin\theta_3'$<br>$d' = 3\lambda/(2\lambda/d) = 3d/2 = 3.00 \times 10^{-4}$ cm |

Example 26.4. White light containing wavelengths from $4.00 \times 10^{-7}$ m to $7.00 \times 10^{-7}$ m is directed at a planar diffraction grating of 6000 lines/cm. (a) Do the first-and second-order spectra overlap? (b) If they do, what is the angular overlap; if they don't, what is the angular separation?

Given: $\lambda_{min} = 4.00 \times 10^{-7}$ m = minimum wavelength
$\lambda_{max} = 7.00 \times 10^{-7}$ m = maximum wavelength
N = 6000 lines/cm = lines/cm on the grating

Determine: (a) Whether or not the first-order (m = 1) and second-order (m = 2) spectra overlap. (b) The angular overlap or separation.

Strategy: We can determine the angular deviation for the longest wavelength in the first-order and that for the shortest wavelength in the second-order. All desired information may be determined from these angles.

Solution: The angular deviation for any wavelength and order may be determined by

$$m\lambda = d \sin\theta_m \quad \text{or} \quad \theta_m = \sin^{-1}(m\lambda/d) = \sin^{-1}(m\lambda N)$$

The angular deviation for the longest wavelength in the first order is

$$\theta_{1max} = \sin^{-1}[(1)(7.00 \times 10^{-7}\ m)(6 \times 10^5/m)] = \sin^{-1}(0.420) = 24.8°$$

The angular deviation for the shortest wavelength in the second order is

$$\theta_{2min} = \sin^{-1}[(2)(4.00 \times 10^{-7}\ m)(6 \times 10^5/m)] = \sin^{-1}(0.480) = 28.7°$$

The first-order and second-order spectra do not overlap, and the angular separation is

$$\Delta\theta = \theta_{2min} - \theta_{1max} = 3.90°$$

Related Text Problems: 26-23 through 26-28

## 5. Polarized Light (Section 26.7)

Review: When unpolarized light is incident upon a nonmetallic material, the beam becomes partially polarized by reflection. For one particular angle of incidence, called the polarizing angle $\theta_p$ (or Brewsters angle), the reflected beam is at right angles to the refracted beam and complete polarization by reflection is achieved. The direction of the polarization of the reflected beam is parallel to the reflecting surface. Figure 26.9 shows this situation. The polarizing angle is given by

$$\tan\theta_p = n_2/n_1$$

This expression is known as Brewster's law.

Figure 26.9

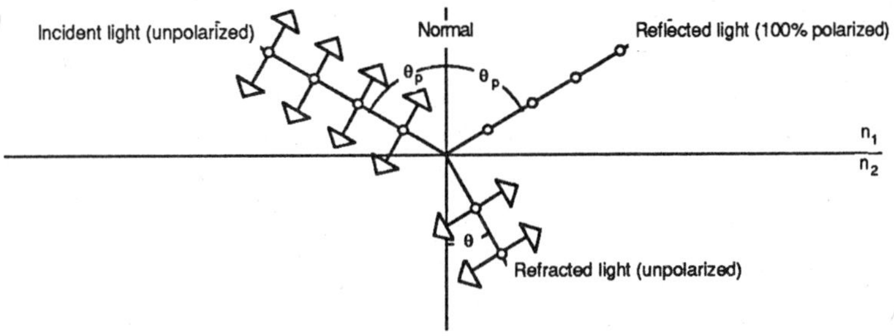


Figure 26.10 shows a beam of unpolarized light incident on a polarizer. The polarized light is then incident on an analyzer oriented at an angle θ with respect to the direction of polarization of the light.

Figure 26.10

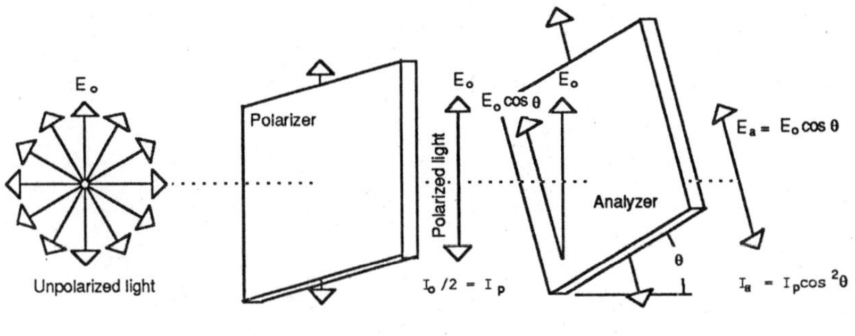

If the amplitude of the beam after the polarizer is $E_o$, the amplitude after the analyzer is $E_o \cos\theta$. The intensity of the wave is proportional to the square of the amplitude. If the intensity of the beam after the polarizer is $I_p$, according to the law of Malus the intensity after the analyzer is

$$I_a = I_p \cos^2\theta$$

Practice: Refer to Fig. 26.9 and determine the following:

| | |
|---|---|
| 1. Polarizing angle for light reflected off crown glass | $\tan\theta_p = n_{glass}/n_{air} = (1.58/1.33) = 1.19$<br>$\theta_p = \tan^{-1}(1.19) = 49.9°$ |
| 2. For light traveling in water, angle of incidence that causes complete polarization when reflected off flint glass? | $\tan\theta_p = n_{glass}/n_{water}$<br>$\theta_{incidence} = \theta_p = \tan^{-1}(1.75/1.33) = 52.8°$ |
| 3. Index of refraction of a medium for which the Brewster angle for light incident in air is 60° | $\tan\theta_p = n_{medium}/n_{air}$<br>$n_{medium} = n_{air} \tan\theta_p = 1.73$ |
| 4. Polarizing angle of a medium for which the reflected beam is completely polarized when the refraction angle is 35° | If the refraction angle is 35°, the reflection angle is 55° for complete polarization.<br>$\theta_p = \theta_{incidence} = \theta_{reflection} = 55°$ |
| 5. Index of refraction of a medium for which the reflected beam is completely polarized when the refraction angle is 35° | For this case, $\theta_p = 55°$ (step 4)<br>$n_{medium} = n_{air}\tan\theta_p = 1.43$ |

Refer to Fig. 26.10 and determine the following:

| | |
|---|---|
| 6. Intensity of light after the polarizer | Incident light can be resolved into components perpendicular and parallel to the direction of polarization. Because the incident light is a random mixture of all directions of polarization, these two components are equal, and hence half the light gets through: $I_p = I_o/2$ |

| | |
|---|---|
| 7. Angle of orientation of analyzer in order for $I_a$ to equal $I_p/8$ | $I_a = I_p \cos^2\theta$, $I_a = I_p/8$<br>Combine these to obtain<br>$\theta = [\cos^{-1}(1/8)]^{1/2} = 9.10°$ |
| 8. Final intensity ($I_a$) if the polarizer is vertical as shown andd the angle $\theta$ for the analyzer is 30° | $I_o$ = intensity of source<br>$I_p = I_o/2$ = intensity after polarizer (step 6)<br>$I_a = I_p \cos^2\theta = (I_o/2) \cos^2 30° = 0.375 I_o$ |

Example 26.5. A beam of light is incident at an angle of 53° on a reflecting non metallic surface. The reflected beam is completely linearly polarized. Determine the angle of refraction of the transmitted beam and the refractive index of the reflecting material.

Given: $\theta_p = 53°$ and the incident medium is air.

Determine: The angle of refraction ($\theta_r$) and the index of refraction (n) of the reflecting material.

Stratety: Knowing $\theta_p$ and that the angle between the reflected and refracted beams is 90°, we can determine $\theta_r$. Knowing the polarization angle and the index of refraction of the medium for the incident beam, we can determine the refractive index of the reflecting medium.

Solution:

$$\theta_p = \theta_{reflection} = 53°,\ \theta_r = 90° - 53° = 37°$$
$$\tan\theta_p = n/n_{air};\ n = n_{air} \tan\theta_p = \tan 53° = 1.33$$

Alternatively, using Snell's law, $n_{air} \sin\theta_p = n \sin\theta_r$,

$$n = n_{air} \sin\theta_p/\sin\theta_r = \sin 53°/\sin 37° = 1.33$$

Example 26.6. Three polarizing filters are stacked with the polarizing axes of the second and third at 30° and 90°, respectively, with that of the first. If unpolarized light of intensity $I_o$ is incident on the stack, find the intensity of the emerging light.

Given: $\theta_2 = 30°$, $\theta_3 = 90°$ (Fig. 26.11), intensity of incident light is $I_o$.

Figure 26.11

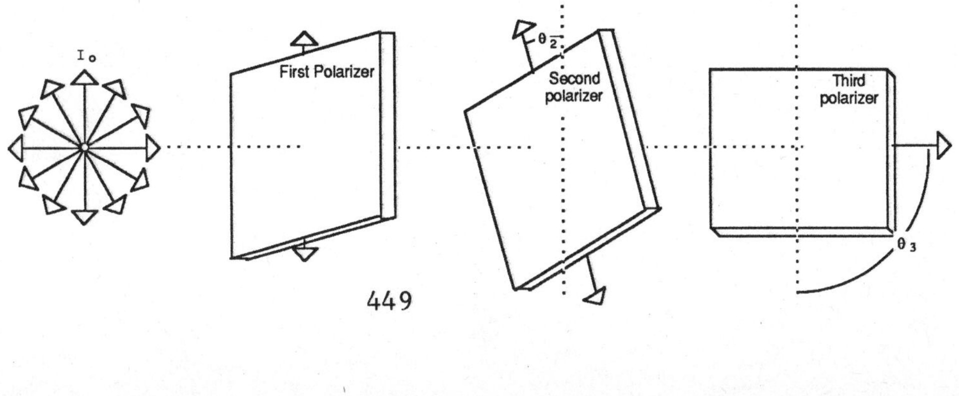

Determine: Intensity of the emerging light.

Strategy: We can establish the intensity of light ($I_1$) after the first polarizer. Knowing this and $\theta_2$, we can establish the intensity of light ($I_2$) after the second polarizer. Knowing this and $\theta_3$, we can establish the intensity of light ($I_3$) after the third polarizer.

Solution: The light incident on the first polarizer can be resolved into components perpendicular and parallel to the direction of polarization. Because the incident light is a random mixture of all directions of polarization, these two components are equal and hence half the light gets through. We can write

$$I_1 = I_o/2$$

Applying the law of Malus, we obtain

$$I_2 = I_1 \cos^2 30° = I_1(0.750)$$

$$I_3 = I_2 \cos^2 60° = [I_1(0.750)](0.250) = 0.188 I_1 = 0.188(I_o/2) = 0.094\ I_o$$

Related Text Problems: 26-29 through 26.35.

---

## PRACTICE TEST

Take and grade this practice test. Doing so will allow you to determine any weak spots in your understanding of the concepts taught in this chapter. The following section prescribes what you should study further to strengthen your understanding.

Light of wavelength $6.00 \times 10^{-7}$ m is incident on two slits separated by 0.100 mm, and the resulting interference pattern is viewed on a screen 1.00 m away. Determine the following:

_____ 1. Angular deviation for the second-order maximum
_____ 2. Distance of the third dark space from the central maximum
_____ 3. Interference fringe spacing on the screen
_____ 4. Where screen should be relocated relative to slits in order to have third-order maxima separated by 5.00 cm

A thin film of oil on water is shown in Fig. 26.12.

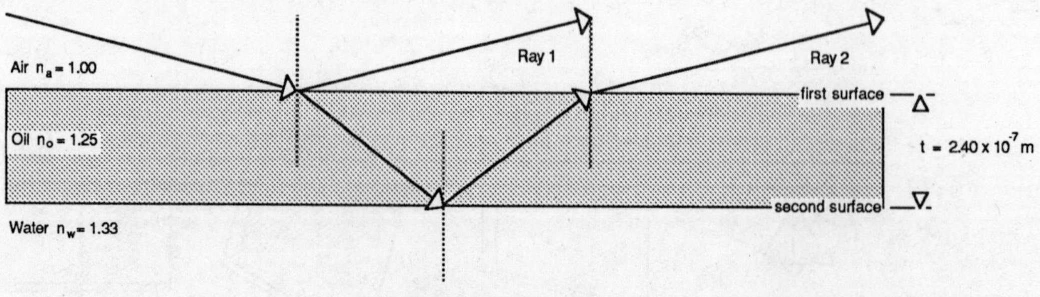


Figure 26.12

Determine the following:

______ 5. Phase change of ray 1 at first surface
______ 6. Phase change of ray 2 at second surface
______ 7. Wavelengths in visible range that experience constructive interference
______ 8. Thinnest film of this oil that can cause destructive interference for light of wavelength $\lambda = 5.00 \times 10^{-7}$ m

Light of wavelength $6.00 \times 10^{-7}$ m is incident on a single slit 0.100 mm wide, and the resulting diffraction pattern is viewed on a screen 2.00 m away. Determine the following:

______ 9. Width of central maximum
______ 10. Angular deviation between first and second dark spaces
______ 11. Approximate distance of first secondary maximum from central maximum
______ 12. Where screen should be relocated relative to slit in order to double width of central maximum

Light of wavelength $6.66 \times 10^{-7}$ m is incident on a planar diffraction grating that has 5000 lines/cm. The resulting diffraction pattern is viewed on a screen 0.500 m away. Determine the following:

______ 13. Grating spacing
______ 14. Number of diffraction images that can be observed
______ 15. Angular deviation of first-order diffraction image
______ 16. Distance on screen between first- and second-order images

Unpolarized light traveling in air is incident upon a liquid with a refractive index of 1.60. Determine the following:

______ 17. Polarizing angle for the liquid
______ 18. Angle of refraction of the transmitted beam with the reflected beam is completely polarized

Unpolarized light of intensity $I_o$ is incident on a stack of two polarizers that have an angle of 45° between the directions of polarization. Determine the following:

______ 19. Intensity of light after first polarizer
______ 20. Intensity of light after second polarizer

(See Appendix I for answers.)

## PRINCIPAL CONCEPTS AND EQUATIONS PRESCRIPTION

Your score on the practice is an excellent measure of your understanding of this chapter. You should now use the following chart to write your own prescription for curing any of your physics ills. Look down the leftmost column to the number of the question(s) you answered incorrectly, read across that row to see which section(s) of the study guide you should return to for further study, and then do the suggested text problems to gain additional experience in working with the particular concept.

| Practice Test Question | Concepts and Equations | Prescription: Principal Concept | Prescription: Text Problems |
|---|---|---|---|
| 1 | Double-slit interference: $\lambda = d \sin\theta_m$ | 1 | 26-1,4 |
| 2 | Double-slit interference: $y_m = R(m - 1/2)\lambda/d$ | 1 | 26-6,7 |
| 3 | Double-slit interference: $\Delta y = R\lambda/d$ | 1 | 26-2,3 |
| 4 | Double-slit interference: $y_m = Rm\lambda/d$ | 1 | 26-2,4 |
| 5 | Thin-film phase change | 2 | 26-8,9 |
| 6 | Thin-film phase change | 2 | 26-9,10 |
| 7 | Thin-film case-I constructive interference: $2t = m\lambda/n$ | 2 | 26-10,11 |
| 8 | Thin-film case-1 destructive interference: $2t = (m + 1/2)\lambda/n$ | 2 | 26-9,12 |
| 9 | Single-slit diffraction: $m\lambda = w \sin\theta_m$ | 3 | 26-25,16 |
| 10 | Single-slit diffracttion: $\theta_m = \sin^{-1}(m\lambda/w)$ | 3 | 26-13,14 |
| 11 | Single-slit diffraction: $Y_m = Rm\lambda/w$ | 3 | 26-15,16 |
| 12 | Single-slit diffraction: $R = Y_m w/m\lambda$ | 3 | 26-13,15 |
| 13 | Diffraction grating: $d = 1/N$ | 4 | 26-26,28 |
| 14 | Diffraction grating: $m = d \sin\theta m/\lambda$ | 4 | 26-23,28 |
| 15 | Diffraction grating: $\theta_m = \sin^{-1}(m\lambda/d)$ | 4 | 26-27,28 |
| 16 | Diffraction grating: $R = Y_m/\tan\theta_m$ | 4 | 26-24,25 |
| 17 | Brewster's law: $\tan\theta_p = n_2/n_1$ | 5 | 26-29,30 |
| 18 | Polarization by reflection or Snell's law | 5 | 26-30,34 |
| 19 | Polarized light: $I = I_o/2$ | 5 | 26-31,32 |
| 20 | Law of Malus: $I = I_o \cos^2\theta$ | 5 | 26-32,35 |

# 27 Theory of Relativity

RECALL FROM PREVIOUS CHAPTERS

| Previously learned concepts and equations frequently used in this chapter | Text Section | Study Guide Page |
|---|---|---|
| Relationship between velocity, displacement, and time: $v = \Delta s/\Delta t$ | 2.3 | 15 |
| Kinetic energy: $KE = mv^2/2$ | 7.5 | 108 |
| Work-energy theorem: $W_{net} = \Delta KE$ | 7.5 | 108 |
| Linear momentum: $p = mv$ | 8.2 | 123 |

NEW IDEAS IN THIS CHAPTER

| Concepts and equations introduced | Text Section | Study Guide Page |
|---|---|---|
| Time dilation: $\Delta t' = \gamma\Delta t$ | 27.5 | 453 |
| Length contraction: $\ell' = \ell/\gamma$ | 27.5 | 457 |
| Relativistic mass: $m = \gamma m_o$ | 27.6 | 459 |
| Relativistic velocities: $V_x' = (V_x - v)/(1 - V_x v/c^2)$ | 27.6 | 460 |
| Rest energy: $E_o = m_o c^2$ | 27.7 | 462 |
| Relativistic energy: $E = mc^2 = \gamma m_o c^2 = \gamma E_o$ | 27.7 | 462 |
| Kinetic energy: $KE = E - E_o = (\gamma - 1)E_o$ | 27.7 | 462 |
| Momentum: $p = mv = \gamma m_o v$ | 27.7 | 462 |
| Relationship between energy and momentum: $E^2 = p^2c^2 + E_o{}^2$ | 27.7 | 462 |
| Relationship between kinetic energy and momentum: $p^2c^2 = (KE)^2 - 2(KE)E_o$ | 27.7 | 462 |

PRINCIPAL CONCEPTS AND EQUATIONS

## 1. Time Dilation (Section 27.5)

Review: Consider observer O in a frame of reference that is earthbound. Another frame of reference, attached to a spaceship, moves past the earth frame at a speed of 0.900c. Observer O' is in the spaceship. Two events (A and B) take place on the spaceship, and both O and O' record when and where these events take place. For event A, a passenger on the spaceship raises her hand. For event B, she lowers her hand.

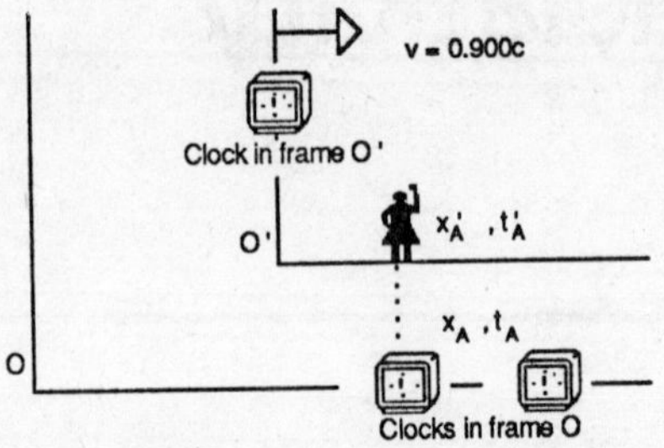


(a) When the girl raises her hand, O says she is at $x_A$ and the time is $t_A$, but O' says she is at $x_A'$ and the time is $t_A'$.

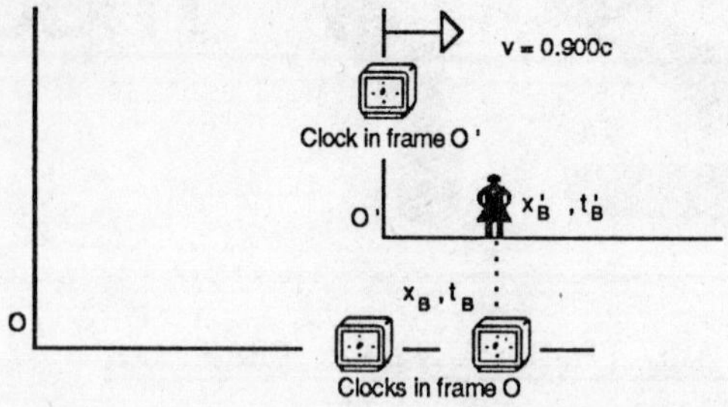


(b) When the girl lowers her hand, O says she is at $x_B$ and the time is $t_B$, but O' says she is at $x_B'$ (note that $x_B' = x_A'$) and the time is $t_B'$.

Figure 27.1

Since the two events occur at the same location in the frame O' is in, the time interval can be determined with a single clock in this frame. This time is proper time; hence

$$\Delta t = t_B' - t_A'$$

Since the relationship between $\Delta t$ and $\Delta t'$ is

$$\Delta t' = \gamma \Delta t$$

and the time between events A and B as measured O (who is in motion relative to the ship) is

$$\Delta t' = t_B - t_A = 5.00 \text{ min}$$

then the time between the two events according to O' is

$$\Delta t = \Delta t'/\gamma = \Delta t'[1 - (v/c)^2]^{1/2} = (5.00 \text{ min})(0.436) = 2.18 \text{ min}$$

If O' (who measures proper time) claims that

$$\Delta t = 5.00 \text{ min}$$

then O would claim that the time between the events is

$$\Delta t' = \gamma \Delta t = \Delta t/[1 - (v/c)^2]^{1/2} = 5.00 \text{ min}/0.436 = 11.5 \text{ min}$$

Note: The time dilation formula ($\Delta t' = \gamma \Delta t$) is simple to use if you are able to establish the values of $\Delta t'$ and $\Delta t$. Recall that $\Delta t$ is proper time. An observer measuring an interval of time between events occurring in his or her own reference frame measures the proper time $\Delta t$, and this time interval is shorter than that measured by an observer in motion with respect to her or him.

Practice: A student and professor are moving at 0.800c with respect to each other. The student is given an examination to be completed in 1.00 h by the professor's clock.

Determine the following:

| | |
|---|---|
| 1. The value of $\gamma$ | $\gamma = 1/[1 - (v/c)^2]^{1/2} = 1.67$ |
| 2. The two events of interest | Event A - the professor's clock reads hour zero<br>Event B - the professor's clock reads hour one |
| 3. The frame in which events A and B occur at the same location | Events A and B occur at the same location in the professor's frame. |
| 4. The proper frame | The professor's frame is the proper frame. |
| 5. The proper time | $\Delta t = 1.00$ h |
| 6. The time the student has to take the exam according to the student's clock | $\Delta t' = (1.67)(1.00 \text{ h}) = 1.67$ h |

Let's redo the above steps, only this time the examination is to be completed in 1.00 h according to the student's clock.

| | |
|---|---|
| 7. The two events of interest | Event C - the student's clock reads hour zero<br>Event D - the student's clock reads hour one |
| 8. The frame in which events C and D occur at the same location | Events C and D occur at the same location in the student's frame. |
| 9. The proper frame | The student's frame is the proper frame. |
| 10. The proper time | $\Delta t = 1.00$ h |

| | |
|---|---|
| 11. The time the student has to take the exam according to the professor's clock | $\Delta t' = \gamma\Delta t = (1.67)(1.00\text{ h}) = 1.67\text{ h}$ |

Example 27.1. A light on your spaceship flashes every 12.0 h according to a clock on the spaceship. The light flashes every 24.0 h according to a clock in the earth frame. How fast is the ship moving relative to the earth?

Given: $\Delta t_{spaceship} = 12.0\text{ h}$ and $\Delta t_{earth} = 24.0\text{ h}$

Determine: How fast the ship is moving relative to the earth

Strategy: We must first establish which observer measures proper time. Once this is established, we can obtain $\gamma$ and then v.

Solution: The flashing light is on the spaceship with you. You see both events occurring in your frame at the same place. You measure proper time, and hence we may write

$$\Delta t = \Delta t_{spaceship} = 12.0\text{ h}$$
$$\Delta t' = \Delta t_{earth} = 24.0\text{ h}$$

The value for $\gamma$ can be determined by

$$\Delta t' = \gamma\Delta t$$
$$\gamma = \Delta t'/\Delta t = 2.00$$

Finally, we can obtain v as follows:

$$\gamma = 1/[1 - (v/c)^2]^{1/2}$$

$$v = c[1 - 1/\gamma^2] = 0.750c = 2.25 \times 10^8\text{ m/s}$$

Example 27.2. A woman has a pulse rate of 70.0 beats/min. What would her pulse rate be according to an observer moving with a velocity of 0.800c relative to her?

Given: Pulse rate = 70.0 beats/min, v = 0.800c

Determine: The woman's pulse rate according to an observer moving with a velocity of 0.800c relative to her

Strategy: We must first establish which observer measures proper time. Knowing that her pulse rate is 70.0 beats/min, we can establish the time between two heart beats in the woman's frame and in the frame moving relative to her. Knowing the time between two heart beats in the frame moving relative to her, we can establish her pulse rate in that frame.

Solution: The two events of interest are two successive heart beats. These two events occur at the same place in the woman's frame, and hence she and any observer at rest relative to her measure proper time.

The time between two heart beats in the woman's frame (the proper frame) is

$$\Delta t = 1/(70.0\ \text{min}) = 1.43 \times 10^{-2}\ \text{min}$$

The time between two heart beats in the frame moving relative to the woman is

$$\gamma = 1/[1 - (v/c)^2]^{1/2} = 1.67$$
$$\Delta t' = \gamma \Delta t = (1.67)(1.43 \times 10^{-2}\ \text{min}) = 2.39 \times 10^{-2}\ \text{min}$$

The woman's pulse rate in the frame moving relative to her is

$$\text{Pulse rate} = 1/\Delta t' = 1/2.39 \times 10^{-2}\ \text{min} = 41.8\ \text{beats/min}$$

Related Text Problems: 27-3 through 27-7.

## 2. Length Contraction (Section 27.5)

Review: The length of an object as measured in a frame of reference in which it is at rest is called its proper length ($\ell$). If an object is at rest with respect to you and you measure its length to be $\ell$, then $\ell$ is its proper length. If that object travels past you at a speed v, the apparent length (as seen by you) in the direction of the motion is decreased (length contraction) and you claim that it is

$$\ell' = \ell/\gamma \quad \text{where} \quad \gamma = 1/[1 - (v/c)^2]^{1/2}$$

Note: The length contraction formula ($\ell' = \ell/\gamma$) is simple to use if you can establish the value of $\ell$ and $\ell'$. Recall that $\ell$ is proper length. An observer at rest with respect to an object measures its proper length $\ell$, and this length is greater than that measured by any other observer in motion with respect to the object.

Practice: Observers O and O' have metersticks mounted in their reference frames, as shown in Fig. 27-2, and O' moves past O with a speed of $v = 0.800c$.

Figure 27.2

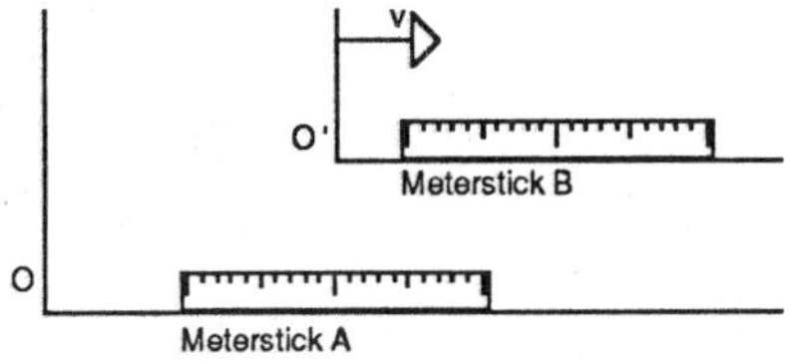


Determine the following:

| | |
|---|---|
| 1. $1/\gamma$ | $1/\gamma = [1 - (v/c)^2]^{1/2} = 0.600$ |
| 2. The proper frame for meterstick A | A is at rest with respect to O, and so O is its proper frame. |

| | | |
|---|---|---|
| 3. | Length of A as measured by O | O measures A to be 1.00 m, $\ell_A = 1.00$ m |
| 4. | Length of A as measured by O' | $\ell'_A = \ell_A/\gamma = 0.600$ m |
| 5. | Proper length of B | B is at rest with respect to O', and so O' measures the proper length of B to be 1.00 m. $\ell_B = 1.00$ m |
| 6. | Length of B as measured by O | $\ell'_B = \ell_B/\gamma = 0.600$ m |

A spaceship has a length of 100 m in its rest frame and appears to be 80.0 m to an observer in another frame.

Determine the following:

| | | |
|---|---|---|
| 7. | Proper length of the ship | $\ell = 100$ m |
| 8. | $1/\gamma$ | $1/\gamma = \ell'/\ell = 0.800$ |
| 9. | Relative velocity of the reference frames | $1/\gamma = [1 - (v/c)^2]^{1/2} = 0.800$, hence<br>$v = c[1 - 1/\gamma^2)]^{1/2} = 0.600c$ |
| 10. | Speed spaceship must have in order for its length to appear to be $0.5\ell$ | $\ell = 100$ m<br>$\ell' = \ell/2 = 50$ m<br>$\ell' = \ell/\gamma;\ 1/\gamma = [1 - (v/c)^2]^{1/2}$<br>$[1 - (v/c)^2]^{1/2} = \ell'/\ell$<br>$v = [1 - (\ell'/\ell)^2]^{1/2} = 0.866c$ |

Example 27.3. An observer O' holds a meterstick at an angle of 30° with respect to the positive x' axis (Fig. 27-3). If O' is moving to the right with a velocity of 0.800c with respect to O, what are the length and angle of the meterstick as measured by O?

Figure 27.3

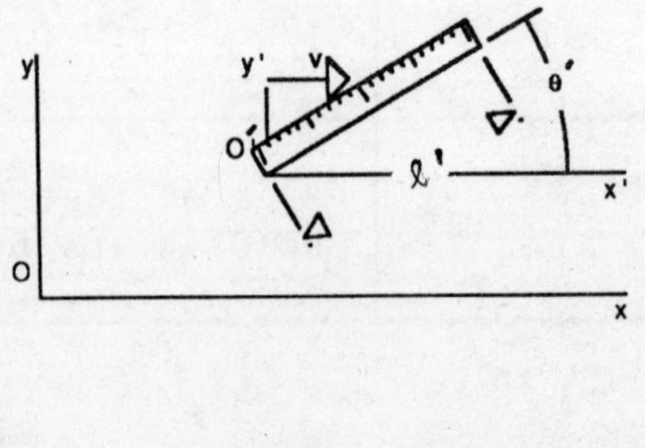

Given: $\ell' = 1.00$ m, $\theta' = 30.0°$, $v = 0.800c$

Determine: $\ell$ = length of meterstick according to O
$\theta$ = angle according to O

Strategy: Knowing $\ell'$ and $\theta'$, we can determine the ($x'$ and $y'$) components of $\ell'$ according to O'. Knowing $x'$ and $y'$, we can determine the ($x$ and $y$) components of the meterstick according to O. Knowing $x$ and $y$, we can determine the length of the stick and its orientation according to O.

Solution: The $x'$ and $y'$ components of the meterstick according to O' are

$$x' = \ell' \cos\theta' = (1.00 \text{ m}) \cos 30° = 0.866 \text{ m}$$
$$y' = \ell' \sin\theta' = (1.00 \text{ m}) \sin 30° = 0.500 \text{ m}$$

The $x$ and $y$ components of the meterstick according to O are

$$x = x'/\gamma = x'[1 - (v/c)^2]^{1/2} = 0.600x' = 0.520 \text{ m}$$
$$y = y' = 0.500 \text{ m}$$ (contraction occurs only along the direction of motion)

The length of the stick and its orientation according to O are

$$\ell = [x^2 + y^2]^{1/2} = 0.721 \text{ m}$$

$$\theta = \tan^{-1}(y/x) = \tan^{-1}(0.962) = 43.9°$$

Related Text Problems: 27-6a, 27-8, 27-13b.

## 3. Relativistic Mass (Section 27.6)

Review: The mass-transformation equation is

$$m = m_o/[1 - (v/c)^2]^{1/2}$$

where $m_o$ is the rest mass of an object (i.e., its mass as measured in its own frame) and $m$ is the mass of the object as measured by an observer moving at a speed $v$ with respect to the object.

Practice:

Determine the following:

| | |
|---|---|
| 1. Rest mass of an electron | $m_o = 9.11 \times 10^{-31}$ kg |
| 2. Mass of an electron traveling at a speed of 0.800c past us | $m = m_o/[1 - (v/c)^2]^{1/2}$<br>$m = m_o/0.600 = 15.2 \times 10^{-31}$ kg |

| | |
|---|---|
| 3. Speed of an electron that has an apparent mass increase of $0.100m_o$ | $m = 1.10m_o$<br>$m = m_o/[1 - (v/c)^2]^{1/2}$<br>$v = c[1 - (m_o/m)^2]^{1/2} = 0.417c$ |

Example 27.4. Determine the percentage increase in the mass of a rocket moving at a speed of 0.800c.

Given: $v = 0.800c$

Determine: The percentage increase in the mass of a rocket traveling at a speed of 0.800c.

Strategy: Knowing v and c, we can determine the ratio $m/m_o$ and then the percentage increase in the mass of the rocket.

Solution: $m = m_o/[1 - (v/c)^2]^{1/2}$
$m/m_o = 1/[1 - (v/c)^2]^{1/2} = 1.67$
$m = (m_o + \Delta m) = 1.67m_o$; $\Delta m = .67m_o$
$\Delta m/m_o = .67 = 67\%$

Related Text Problems: 27-13a, 27-14, 27-15, 27-21a.

## 4. Relativistic Velocities (Section 27.6)

Review: Figure 27-4 shows two different reference frames and a moving object whose velocity can be measured by observers in either frame.

Figure 27.4.

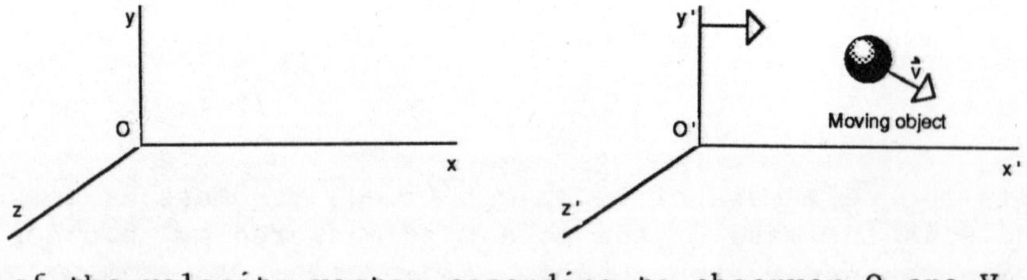


The components of the velocity vector according to observer O are $V_x, V_y, V_z$.

The components of the velocity vector according to observer O' are $V'_x, V'_y, V'_z$.

The transformation equations relating $V_x$, $V_y$, and $V_z$ to $V'_x$, $V'_y$, and $V'_z$ are

$$V'_x = (V_x - v)/(1 - V_x v/c^2)$$

$$V'_y = V_y [1 - (v/c)^2]^{1/2}/(1 - V_x v/c^2)$$

$$V'_z = V_z [1 - (v/c)^2]^{1/2}/(1 - V_x v/c^2)$$

Practice: Observer O' is moving relative to observer O with a speed of $v = 0.900c$. Both observers see spaceships A, B, and D. The speed of A according O is $v_A = 0.700c$, the speed of B according to O' is $v_B' = -0.800c$, and the speed of D according to O is c.

Determine the following:

| | |
|---|---|
| 1. Speed of A according to O' ($V_A'$) | $V_A' = (V_A - v)/(1 - V_A v/c^2) = -0.540c$<br>O' sees A approaching at a speed of 0.540c |
| 2. Speed of B according to O ($V_B$) | $V_B' = (V_B - v)/(1 - V_B v/c^2)$<br>This may be solved for $V_B$ to obtain<br>$V_B = (V_B' + v)/(1 + V_B' v/c^2) = 0.357c$ |
| 3. Speed of O according to O' ($V_O'$) | $V_O' = (V_O - v)/(1 - V_O v/c^2)$<br>$V_O = 0$ (speed of O according to O)<br>$V_O' = (0 - 0.900c)/(1 - 0) = -0.900c$ |
| 4. Speed of D according to O' ($V_D'$) | $V_D' = (V_D - v)/(1 - V_D v/c^2) = c$ |

Example 27.5. A particle moves with a speed 0.900c at an angle of 30° with respect to the x axis, according to O. What is the particle's velocity as determined by an observer O', who is moving with a speed 0.500c along the common x axis?

Figure 27.5

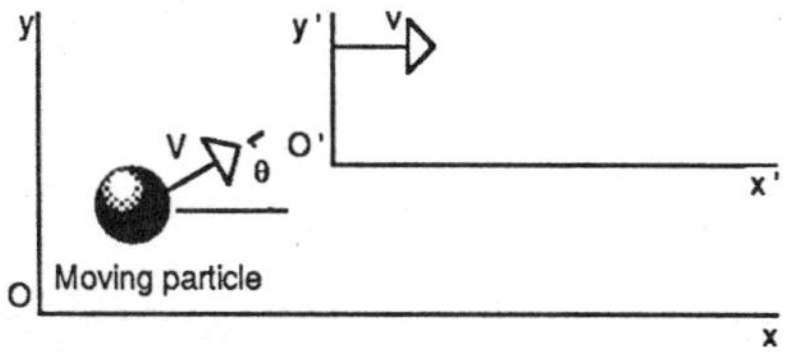


Given: $V = 0.900c$ = speed of particle according to O
$\theta = 30°$ = angle according to O
$v = 0.500c$ = speed of O' relative to O

Determine: The velocity (magnitude and direction) of the particle according to O'.

Strategy: Knowing V and $\theta$, we can determine $V_x$ and $V_y$. Knowing $V_x$ and $V_y$, we can use the x and y velocity transformation equations to obtain $V_x'$ and $V_y'$. Knowing $V_x'$ and $V_y'$, we can determine the magnitude and direction of the particle's velocity according to O'.

Solution: $V_x = V\cos\theta = 0.779c$, $V_y = V\sin\theta = 0.450c$

$$V'_x = \frac{V_x - v}{1 - V_x v/c^2} = \frac{0.779c - 0.500c}{1 - (0.779c)(0.500c)/c^2} = 0.457c$$

$$V'_y = \frac{V_y\,[(1 - (v/c)^2]^{1/2}}{1 - V_x v/c^2} = \frac{(0.450c)(0.866)}{1 - (0.779c)(0.500c)/c^2} = 0.638c$$

$$V' = [V'^2_x + V'^2_y]^{1/2} = 0.785c \qquad \theta_2 = \tan^{-1}(V'_y/V'_x) = 54.4°$$

Related Text Problems: 27-12, 27-16, 27-18, 27-19, 27-20.

## 5. Work, Energy, and Momentum (Section 27.7)

Review: An object with a rest mass $m_o$ has a rest energy

$$E_o = m_o c^2$$

If work is done on that object, causing it to move past an observer at a speed v, the observer says the mass, energy, kinetic energy, and momentum of the object are

$$m = \gamma m_o$$

$$E = mc^2 = \gamma m_o c^2 = \gamma E_o$$
$$KE = E - E_o = (\gamma - 1)E_o = (\gamma - 1)m_o c^2$$
$$p = mv = \gamma m_o v$$

These quantities are related as follows (see text problem 27-22):

$$E^2 = p^2c^2 + E_o^2$$
$$p^2c^2 = (KE)^2 - 2(KE)E_o$$

If the mass of an object decreases by an amount $\Delta m$, the energy decreases by an amount

$$\Delta E = \Delta m c^2$$

This energy may be lost by radiation or by a decrease in speed.

Practice: An electron is accelerated from rest through an electric potential difference of $5.00 \times 10^6$ V. Determine the following:

| | |
|---|---|
| 1. Rest mass of the electron | $m_o = 9.11 \times 10^{-31}$ kg |
| 2. Rest energy of the electron | $E_o = m_o c^2 = 8.20 \times 10^{-14}$J = 0.513 MeV |

| | |
|---|---|
| 3. Kinetic energy of the electron | $KE = eV = 5.00\ MeV = 8.00 \times 10^{-13}\ J$ |
| 4. Total energy of the electron | $E = KE + E_o = 5.51\ MeV = 8.82 \times 10^{-13}\ J$ |
| 5. Relativistic mass of the electron | $m = E/c^2 = 9.80 \times 10^{-30}\ kg$ |
| 6. Speed of the electron | $E = \gamma E_o,\ \gamma = 1/[1 - (v/c)^2]^{1/2}$<br>Combining these equations, obtain<br>$v = c[1 - (E_o/E)^2]^{1/2} = 0.996c$ |
| 7. Momentum of the electron | $p = mv = 2.93 \times 10^{-21}\ kg \cdot m/s$<br>$E^2 = p^2c^2 + E_o^2$, which gives<br>$p = (E^2 - E_o^2)^{1/2}/c = 2.93 \times 10^{-21}\ kg \cdot m/s$ |
| 8. Increase in mass of the electron if the kinetic energy increases by 10.0 keV | $\Delta E = \Delta KE = 10.0\ keV$<br>$\Delta m = \Delta E/c^2 = 1.78 \times 10^{-33}\ kg$ |
| 9. Magnitude of the magnetic field required to keep the electron in a circle of radius 1.00 m | $F_C = mv^2/r \quad F_B = qvB$<br>$F_B = F_C$<br>$qvB = mv^2/r$<br>$B = mv/qr = p/qr = 1.83 \times 10^{-2}\ T$ |

Each second, the sun gives off about $4.00 \times 10^{26}$ J of radiant energy through the conversion of mass to energy by nuclear reactions.

Determine the following:

| | |
|---|---|
| 10. Mass lost by the sun each second | $\Delta E = \Delta mc^2, \quad \Delta E/\Delta t = (\Delta m/\Delta t)c^2$<br>$\Delta m/\Delta t = (\Delta E/\Delta t)/c^2 = (4.00 \times 10^{26}\ J/s)/c^2$<br>$= 4.44 \times 10^9\ kg/s$ |
| 11. Mass lost by the sun in the last $1.00 \times 10^{12}$ years, assuming the same rate of radiation | $\Delta M = (\Delta m/\Delta t)\Delta t$<br>$= (4.44 \times 10^9\ kg/s)(1.00 \times 10^{12}\ y)$<br>$= (4.44 \times 10^{21}\ kg \cdot y/s)(3.15 \times 10^7\ s/y)$<br>$= 1.40 \times 10^{29}\ kg$ |

| | |
|---|---|
| 12. Percentage of the sun's mass lost due to radiation in the last $1.00 \times 10^{10}$ years. Assume the present mass to be $2.00 \times 10^{30}$ kg. | $\Delta M = 1.40 \times 10^{29}$ kg for the last $1.00 \times 10^{10}$ years. The mass $1.00 \times 10^{10}$ years ago was $M_{then} = M_{now} + \Delta M = 2.14 \times 10^{30}$ The percentage of the sun's mass lost in the last $1.00 \times 10^{10}$ years is $\frac{\Delta M}{M_{then}} \times 100 = \frac{1.40 \times 10^{29}}{2.14 \times 10^{30}} \times 100 = 6.54\%$ |

Example 27.6. An electron moves in the laboratory with a speed 0.800c. An observer moves with a speed 0.600c in a direction opposite the direction of electron's motion. What is the kinetic energy of the electron according to the observer?

Given: $m_o = 9.11 \times 10^{-31}$ kg = rest mass of electron
$V_{EL} = 0.800c$ = speed of electron in laboratory frame
$v = -0.600c$ = speed of observer in laboratory frame

Strategy: We can use the velocity transformation to determine the speed of the electron in the observer's frame. Knowing the speed with respect to the observer, we can establish the total energy and then the kinetic energy according to the observer.

Solution: The velocity of the electron with respect to the observer is

$$V_{EO} = (V_{EL} - v)/(1 - V_{EL}v/c^2) = 0.946c$$

We may now determine $\gamma$

$$\gamma = 1/[1 - (V_{EO}/c)^2]^{1/2} = 3.08$$

Finally, the kinetic energy is

$$KE = E - E_o = (\gamma - 1)E_o = (2.08)(0.511 \text{ MeV}) = 1.06 \text{ MeV}$$

Related Text Problems: 27-21 through 27-36.

## PRACTICE TEST

Take and grade this practice test. Doing so will allow you to determine any weak spots in your understanding of the concepts taught in this chapter. The following section prescribes what you should study further to strengthen your understanding.

Figure 27.6 shows observer O' moving past observer O with a relative speed 0.800c. A woman in frame O' throws a ball of rest mass $5.00 \times 10^{-1}$ kg against a wall which O claims is 4.00 m away, and it returns to her in 5.00 s according to O'.

Figure 27.6

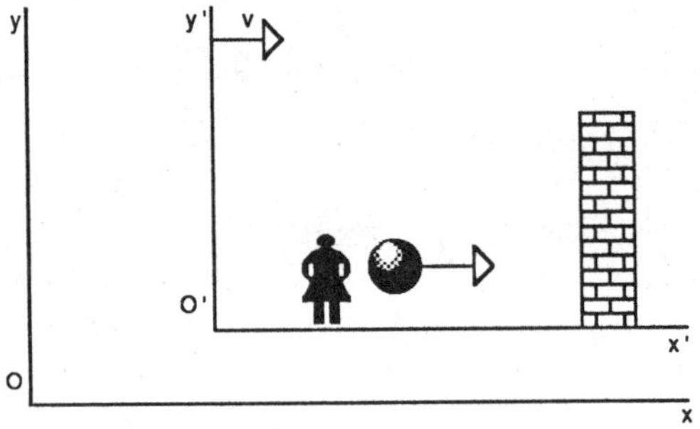


Determine the following:

__________ 1. The time it takes the ball to make a round trip according to O
__________ 2. The distance between the woman and the wall according to O'
__________ 3. The mass of the ball according to the woman
__________ 4. The mass of the ball according to O
__________ 5. The speed of O as measured by O'

Observer O' detects a spaceship traveling 0.900 x in the x' direction

__________ 6. The speed of the spaceship according to O

An electron is accelerated from rest through an electrical potential difference of $1.00 \times 10^6$ V.

Determine the following:

__________ 7. Final kinetic energy of the electron
__________ 8. Total relativistic energy of the electron
__________ 9. Relativistic mass of the electron
__________ 10. Speed of the electron
__________ 11. Momentum of the electron

A $2.00 \times 10^3$ MW nuclear power plant operating at full power delivers $2.00 \times 10^9$ J of electrical energy and $4.00 \times 10^9$ J of waste heat every second.

Determine the following:

__________ 12. The rate at which mass is converted to energy
__________ 13. The amount of mass "burned" in one year

A particle of $4.00 \times 10^{-19}$-C charge and $5.00 \times 10^{-28}$-kg rest mass moving perpendicular to a 0.500-T magnetic field traces out a 1.00-m radius circle.

Determine the following:

_______ 14. Magnitude of the velocity of the particle
_______ 15. Magnitude of the linear momentum of the particle
_______ 16. Kinetic energy of the particle
_______ 17. Total energy of the particle
_______ 18. Mass of the particle

(See Appendix I for answers.)

## PRINCIPAL CONCEPTS AND EQUATIONS PRESCRIPTION

Your score on the practice test is an excellent measure of your understanding of this chapter. You should now use the following chart to write your own prescription for curing any of your physics ills. Look down the leftmost column to the number of the question(s) you answered incorrectly, read across that row to see which section(s) of the study guide you should return to for further study, and then do the suggested text problems to gain additional experience in working with the particular concept.

| Question Number | Concepts and Equations | Prescription: Principal Concept | Prescription: Text Problems |
|---|---|---|---|
| 1 | Time dilation: $\Delta t' = \gamma \Delta t$ | 1 | 27-4,6 |
| 2 | Length contraction $\ell' = \ell/\gamma$ | 2 | 27-3,13b |
| 3 | Rest mass: $m_o$ | 3 | 27-13a,14 |
| 4 | Relativistic mass: $m = \gamma m_o$ | 3 | 27-15,21a |
| 5 | Relativistic velocities: | 4 | 27-16,18 |
| 6 | $V_x' = (V_x - v)/(1 - V_x v/c^2)$ | | |
| | Relativistic velocities | 4 | 27-19,20 |
| 7 | Kinetic energy: $KE = eV$ | 2 of Ch. 18 | 27-23a |
| 8 | Total energy: $E = KE + E_o$ | 5 | 27-23b |
| 9 | Total energy: $E = mc^2$ | 5 | 27-21b |
| 10 | Total energy: $E = \gamma E_o$ | 5 | 27-36d |
| 11 | Momentum: $p = mv$, $E^2 = p^2c^2 + E_o^2$, or $p^2c^2 = (KE)^2 + 2(KE)E_o$ | 5 | 27-22,25 |
| 12 | Mass-energy conversion: $p = \Delta m/t = \Delta mc^2/t$ | 5 | 27-27a,28a |
| 13 | $M = (\Delta m/t)t$ | 5 | 27-27b,28b |
| 14 | Magnetic force: $mv^2/r = qvB$ | 1 of Ch. 21 | 27-25,36 |
| 15 | Momentum: $p = mv = \gamma m_o v$, $p = qrN$ | 5 | 27-30,36b |
| 16 | Kinetic energy: $KE = E - E_o = (\gamma - 1)E_o$ | 5 | 27-21c,36c |
| 17 | Total energy: $E = KE + E_o$, $E = \gamma E_o$ | 5 | 27-36d |
| 18 | Relativistic mass: $m = \gamma m_o$ | 3 | 27-13a,21a |

# 28 Birth of Quantum Physics

RECALL FROM PREVIOUS CHAPTERS

| Previously learned concepts and equations frequently used in this chapter | Text Section | Study Guide Page |
|---|---|---|
| Power: $P = W/\Delta t$ | 7.4 | 106 |
| Elastic collisions | 8.4 | 129 |
| Intensity: $I = P/A$ | 11.8 | 187 |
| Stefan's Law: $\Delta Q/\Delta t = \varepsilon\sigma AT$ | 15.6 | 240 |
| Electric potential difference: $V = \Delta PE/q_o$ | 18.2 | 289 |
| Constructive interference: $\Delta s = m\lambda$ | 26.3 | 435 |

NEW IDEAS IN THIS CHAPTER

| Concepts and equations introduced | Text Section | Study Guide Page |
|---|---|---|
| Relationship between $\lambda_{max}$ and T for a blackbody: $\lambda_{max}T = 2.898 \times 10^{-3}$ m·K | 28.2 | 467 |
| Planck's radiation law: $I = 2hf^3/c^2[\exp(hf/kT) - 1]$ | 28.3 | 467 |
| Quantization of energy: $E_{emitted} = n\varepsilon = nhf$ | 28.3 | 467 |
| Photons: $\varepsilon = hf = hc/\lambda$ | 28.3 | 467 |
| Threshold frequency and work function: $W = hf_o$ | 28.4 | 470 |
| Photoelectric Equation: $eV_o = hf - W$ | 28.4 | 470 |
| Bragg diffraction: $m\lambda = 2d \sin\theta$ | 28.5 | 473 |
| Compton effect: $\lambda' - \lambda = h(1 - \cos\theta)/m_oc$ | 28.6 | 475 |
| Atomic Spectra: $\frac{1}{\lambda} = R[(1/n_1^2) - (1/n_2^2)]$ | 28.7 | 478 |

PRINCIPAL CONCEPTS AND EQUATIONS

1. Blackbody Radiation (Sections 28.2, 28.3)

Review: A blackbody is any object that can absorb and emit radiation of all wavelengths. A plot of radiated energy intensity versus wavelength is shown in Fig. 28-1 for a blackbody at various temperatures.

Figure 28-1

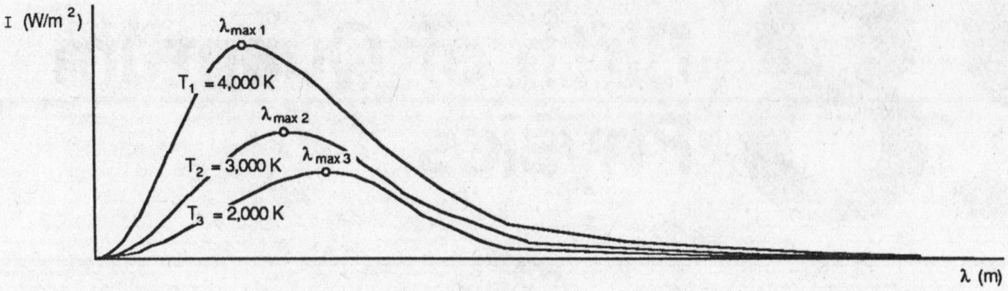


Wien discovered a simple empirical relationship between the wavelength at maximum intensity ($\lambda_{max}$) and the temperature of the blackbody:

$$\lambda_{max}T = 2.898 \times 10^{-3} \text{ m}\cdot\text{K}$$

Planck discovered that the intensity of radiation (I) at a given wavelength for a blackbody at a temperature T is given by

$$I = 2hf^3/c^2[\exp(hf/kT) - 1]$$

Planck derived this expression from theory. He treated the vibrating atoms which make up the interior walls of a blackbody as tiny oscillators that could have only certain discrete energies $E_n$ given by

$$E_n = nhf$$

where n is a positive integer (referred to as a quantum number in later chapters), f is the frequency of vibration of the oscillating atoms, and h is Planck's constant ($h = 6.63 \times 10^{-34}$ J·s). According to Planck, these oscillating atoms can emit or absorb energy in discrete units of light energy called quanta (or photons). If the oscillating atom goes from one quantum state (n) to the next highest quantum state (n + 1), it absorbs a quantum of energy given by

$$E_{n+1} - E_n = (n + 1)hf - nhf = hf$$

The energy of a single quantum, or photon, is

$$\varepsilon = hf = hc/\lambda$$

Practice: The filament of an incandescent light bulb has a surface area of $4.00 \times 10^{-1}$ cm² and radiates 150 W of power. Assume an average wavelength of $5.00 \times 10^{-7}$ m.

Determine the following:

| | |
|---|---|
| 1. Average frequency of the emitted photons | $f = c/\lambda = 6.00 \times 10^{14}$ Hz |
| 2. Average energy of the emitted photons | $\varepsilon = hf = hc/\lambda = 3.98 \times 10^{-19}$ J |

| | |
|---|---|
| 3. Intensity of the light emitted by the filament | $I = P/A = 3.75 \times 10^6\ W/m^2$ |
| 4. Temperature of the light bulb filament, assuming it radiates as a blackbody | $I = \varepsilon\sigma T^4$<br>$\varepsilon = 1.00,\ \sigma = 5.67 \times 10^{-8}\ W/m^2\cdot K^4$<br>$T = (I/\varepsilon\sigma)^{1/4} = 2.85 \times 10^3\ K$ |
| 5. The most intensively radiated wavelength | $\lambda_{max}T = 2.898 \times 10^{-3}\ m\cdot K$<br>$\lambda_{max} = (2.898 \times 10^{-3}\ m\cdot K)/T$<br>$\lambda_{max} = 1.02 \times 10^{-6}\ m$ |
| 6. Rate at which filament emits photons | $P = \frac{E}{t} = \frac{n\varepsilon}{t} = \frac{n}{t}\varepsilon$<br>$\frac{n}{t} = \frac{P}{\varepsilon} = (3.77 \times 10^{20})/s$ |
| 7. Power arriving at a square meter of surface 10 m from the filament | $A = 4\pi R^2 = 4\pi(10\ m)^2 = 1.26 \times 10^3\ m^2$<br>$\frac{P}{A} = \frac{150\ W}{1.26 \times 10^3\ m^2} = 1.19 \times 10^{-1}\ W/m^2$ |
| 8. Intensity of the light at a distance of 10 m | $I = \frac{P}{A} = 1.19 \times 10^{-1}\ W/m^2$ |
| 9. Rate of arrival of photons per square centimeter of surface at a distance of 10 m | $\frac{P}{A} = \frac{E/t}{A} = \frac{n\varepsilon/t}{A} = \frac{n}{t}\frac{\varepsilon}{A}$<br>$\frac{n/t}{A} = \frac{P/A}{\varepsilon} = 2.99 \times 10^{13}/s\cdot cm^2$ |

Example 28.1. A 1.00-kg mass is attached to a spring of force constant k = 10.0 N/m. The spring is stretched 0.200 m from its equilibrium position and released. (a) Find the frequency of oscillation and the total energy of oscillation according to classical mechanics. (b) Assuming that the energy is quantized, find the quantum number n for the system. (c) Find the amount of energy carried away if the oscillator undergoes a one-quantum change in energy (i.e., it jumps to the next lower quantum state). (d) The change in energy of the oscillator if n increases by 1000.

Given: $m = 1.00$ kg = mass attached to spring
$k = 10.0$ N/m = spring constant
$A = 0.200$ m = amplitude of oscillation
$\Delta n = 1000$ = change in quantum number

**Determine:** $f$ = frequency of oscillation of spring
$E$ = total energy of oscillator
$n$ = quantum number for oscillator
$\varepsilon$ = energy carried away by one quantum
$\Delta E$ = energy change of oscillator where $\Delta n = 1000$

**Strategy:** Knowing m, k, and A, we can determine f and E. Knowing f and E, we can determine n and $\varepsilon$. Knowing $\Delta n$ and f, we can determine $\Delta E$.

**Solution:**
(a) $f = (k/m)^{1/2}/2\pi = 5.03 \times 10^{-1}$ Hz
$E = kA^2/2 = 2.00 \times 10^{-1}$ J
(b) $E = nhf$; $n = E/hf = 6.00 \times 10^{32}$
(c) $\varepsilon = hf = 3.33 \times 10^{-34}$ J
(d) $\Delta E = \Delta nhf = 3.33 \times 10^{-31}$ J

**Related Problems:** 28-1 through 28-13.

## 2. Photoelectric Effect (Section 28.4)

**Review:** Figure 28-2 shows photons of frequency f incident on a photocell connected to a retarding potential.

Figure 28-2

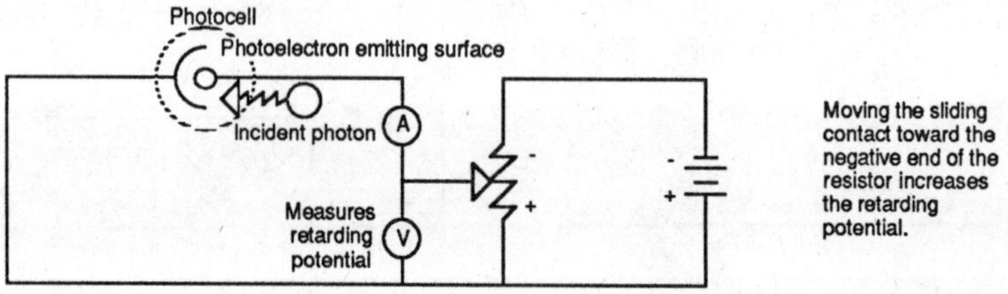


Given either the frequency or wavelength of the incident photons, we can determine their energy.

$$\varepsilon = hf = hc/\lambda$$

Electrons will not be emitted by the surface unless the incident photons can supply them with a minimum amount of energy called the work function W of the surface.

$$\varepsilon_{min} = hf_o = hc/\lambda_o = W$$

In the preceding expression, $f_o$ and $\lambda_o$ are called the threshold frequency and threshold wavelength, respectively. If the incident photon has more energy than the work function (i.e., if $\varepsilon > W$), the excess energy shows up as kinetic energy of the emitted electron. In fact, the maximum kinetic energy of a photoelectron is

$$KE_{max} = \varepsilon - W = hf - W$$

As the retarding potential is increased, electron flow (monitored by the ammeter) from the emitter to the collector decreases. When the electron flow ceases (i.e., the current drops to zero), the retarding potential is called the stopping potential ($V_o$). At this point, the kinetic energy of the most energetic electrons ($KE_{max}$) is changed to electrical potential energy ($eV_o$) and we can write

$$KE_{max} = eV_o$$

Combining the above expressions for $KE_{max}$, we obtain the photoelectric equation:

$$eV_o = hf - W$$

Comparing this expression with the equation for a straight line ($y = mx + b$), we see that a plot of $eV_o$ versus f should be a straight line (Fig. 28-3).

Figure 28-3

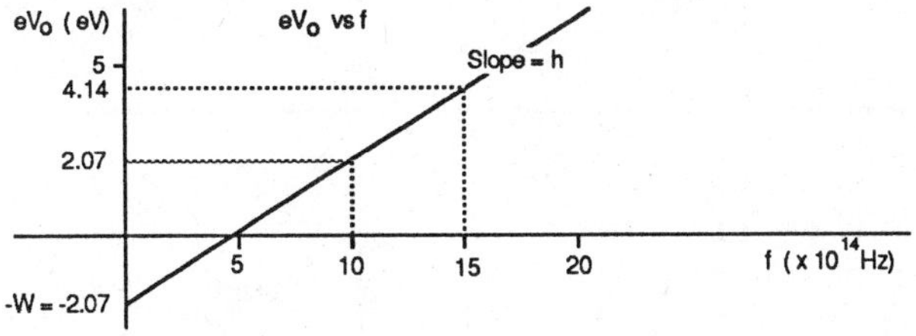


The slope of this straight line is Planck's constant, the ordinate intercept is the negative of work function, and the abscissa intercept is the threshold frequency.

<u>Practice:</u> Use Fig. 28-3 to determine the following.

| | |
|---|---|
| 1. Work function of the photoemitting surface | $eV_o = hf - W$ and $y = mx + b$<br>Comparing these equations, we see that the intercept of an $eV_o$ vs f plot is $-W$. Hence $W = 2.07$ eV. |
| 2. Planck's constant | $eV_o = hf - W$ and $y = mx + b$<br>Comparing these equations, we see that the slope of an $eV_o$ vs f plot is h.<br>Hence $h = (4.14\ \text{eV})/(10 \times 10^{14}\ \text{s}^{-1})$<br>$= 4.14 \times 10^{-15}$ eV·s<br>$= 6.62 \times 10^{-34}$ J·s<br>This value differs slightly from the accepted value due to the fact that the graph cannot be read accurately. |
| 3. Threshold wavelength | $f_o = 5.00 \times 10^{14}$ Hz<br>$\lambda_o = c/f_o = 6.00 \times 10^{-7}$ m |
| 4. Energy of a photon of frequency $10.0 \times 10^{14}$ Hz | $\varepsilon = hf = 6.63 \times 10^{-19}$ J $= 4.14$ eV<br>This may also be obtained from the $eV_o$ vs f plot by noting that photons with $f = 10.0 \times 10^{14}$ Hz cause electrons with $KE_{max} = 2.07$ eV to be emitted. Since these photoelectrons had to overcome the work function, the energy of the photons must be<br>$\varepsilon = KE_{max} + W = 4.14$ eV $= 6.63 \times 10^{-19}$ J |

| | |
|---|---|
| 5. The amount of energy photons of frequency $f = 3.00 \times 10^{14}$ must gain in order to cause photoelectrons to be emitted | The energy of these photons is $\varepsilon = hf = 1.99 \times 10^{-19}$ J = 1.24 eV<br>The energy needed to cause photoelectrons to be emitted is<br>$W = 2.07$ eV $= 3.31 \times 10^{-19}$ J<br>The additional energy needed by the photon is<br>$W - \varepsilon = 0.830$ eV $= 1.33 \times 10^{-19}$ J |
| 6. Maximum kinetic energy of the photoelecteons emitted when the incident radiation has a wavelength of $2.00 \times 10^{-7}$ m | The energy of the incident photons is<br>$\varepsilon = hf = hc/\lambda = 9.95 \times 10^{-19}$ J<br>$KE_{max} = \varepsilon - W = 6.64 \times 10^{-19}$ J |
| 7. Stopping potential for the photoelectrons emitted when the incident radiation has a wavelength of $2.00 \times 10^{-7}$ m | $KE_{max} = 6.64 \times 10^{-19}$ J = 4.15 eV<br>$eV_o = KE_{max} = 4.15$ eV<br>$V_o = 4.15$ V |

Example 28.2. The threshold wavelength for electron emission from a certain metallic surface is $5.50 \times 10^{-7}$ m. Calculate the binding energy, or work function, of an electron on this surface. Find the stopping potential for electrons emitted from this surface when light of wavelength $2.00 \times 10^{-7}$ m strikes it.

Given: $\lambda_o = 5.50 \times 10^{-7}$ m, $\lambda = 2.00 \times 10^{-7}$ m

Determine: $W$ = work function
$V_o$ = stopping potential for electrons emitted when the incident light has $\lambda = 2.00 \times 10^{-7}$ m

Strategy: Knowing the threshold wavelength, we can determine the work function for the surface. Knowing the wavelength of the incident light and the work function for the surface, we can determine the stopping potential of the emitted electrons.

Solution: The work function is

$$W = hf_o = hc/\lambda_o = 3.62 \times 10^{-19} \text{ J}$$

The stopping potential for electrons emitted when light of wavelength $\lambda = 2.00 \times 10^{-7}$ m is incident on the surface may be obtained as follows. The energy of the incident ($\lambda = 2.00 \times 10^{-7}$ m) photons is

$$\varepsilon = hf = hc/\lambda = 9.95 \times 10^{-19} \text{ J}$$

The maximum kinetic energy of the emitted electrons is

$$KE_{max} = \varepsilon - W = 6.33 \times 10^{-19} \text{ J}$$

The stopping potential is

$$V_o = KE_{max}/e = 3.96 \text{ V}$$

Related Problems: 28-14 through 28-20.

## 3. Production and Diffraction of X-Rays (Section 28.5)

Review: Figure 28-4 reviews the production of x-rays.

Figure 28-4

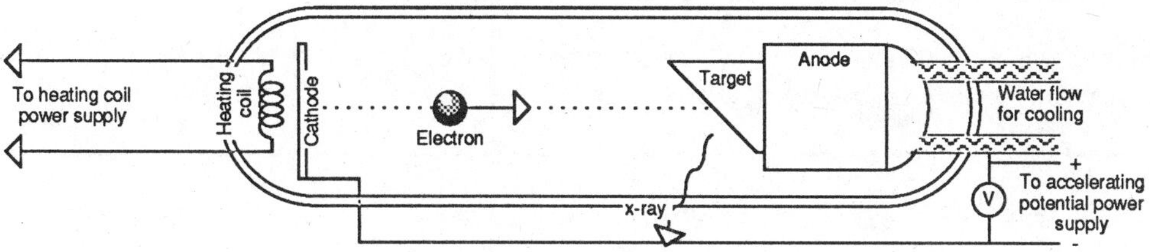


The heating coil causes the cathode to emit electrons. If the accelerating potential is V, the electrons have an electric potential energy eV at the cathode. As the electrons traverse the tube, this electric potential energy is converted to translational kinetic energy. At the anode, the kinetic energy of the electrons is

$$KE = eV$$

If all this energy is given to the x-ray (we are ignoring the relatively small work function) that is created when the electron slams into the target, the maximum energy of the x-ray is

$$E_{x\text{-ray max}} = KE_{electron} = eV$$

The maximum frequency and minimum wavelength of the x-rays are

$$E_{x\text{-ray max}} = eV = hf_{max} \qquad f_{max} = eV/h$$

$$E_{x\text{-ray max}} = eV = hc/\lambda_{min} \qquad \lambda_{min} = hc/eV$$

Figure 28-5

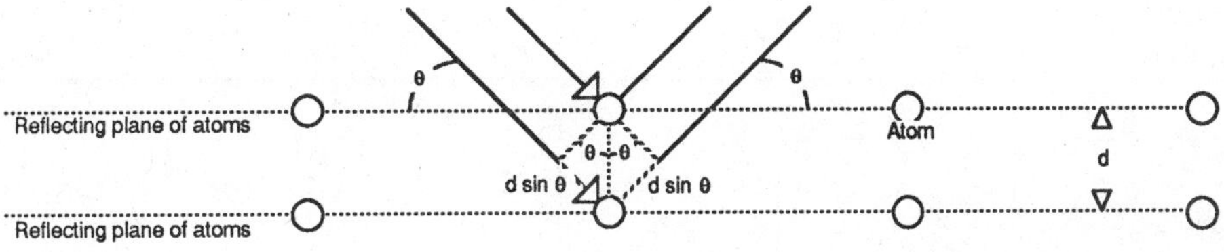


When x-rays are incident on a crystal, as shown in Fig. 28-5, we observe constructive interference in the reflected beam. The path difference between the two reflected beams is $2d \sin\theta$, when $\theta$ is the angle shown and d is the spacing between the planes of reflecting atoms. Constructive interference occurs whenever this path difference is an integral multiple of whole wavelengths, that is, whenever

$$m\lambda = 2d \sin\theta_m \qquad m = 1,2,3,\ldots$$

This equation is referred to as the Bragg diffraction equation.

Practice: If the accelerating potential for an x-ray tube is $6.00 \times 10^4$ V, determine the following:

| | |
|---|---|
| 1. Electric potential energy of the electrons at the cathode | $PE = eV = 6.00 \times 10^4 \text{ eV} = 9.60 \times 10^{-15}$ J |
| 2. Maximum kinetic energy of electrons as they slam into target | $KE_{max} = PE = eV = 9.60 \times 10^{-15}$ J |
| 3. Maximum energy of emitted x-rays | $E_{x\text{-ray max}} = KE_{max} = 9.60 \times 10^{-15}$ J |
| 4. Maximum frequency of emitted x-rays | $f_{max} = E_{max}/h = 1.45 \times 10^{19}$ Hz |
| 5. Minimum wavelength of emitted x-rays | $\lambda_{min} = c/f_{max} = 2.07 \times 10^{-11}$ m |

These minimum-wavelength x-rays are incident on a crystal, and the first-order diffraction maximum for the reflected x-rays occurs for an incident angle of $\theta = 20.0°$. For this situation, determine the following:

| | |
|---|---|
| 6. Spacing between reflecting planes of crystal | $m\lambda = 2d \sin\theta$<br>$m = 1.00$, $\lambda = 2.07 \times 10^{-11}$ m, $\theta = 20.0°$<br>$d = m\lambda/(2 \sin\theta) = 3.03 \times 10^{-11}$ m |
| 7. Angle of incidence needed to obtain second-order diffraction maximum | $m\lambda = 2d \sin\theta$, $m = 2$<br>$\theta = \sin^{-1}(m\lambda/2d) = 43.1°$ |
| 8. X-ray wavelength that gives third-order diffraction maximum when rays are incident at an angle of $\theta = 40°$ | $m\lambda = 2d \sin\theta$<br>$\lambda = 2d \sin\theta/m = 1.30 \times 10^{-11}$ m |

Example 28.3. A beam of x-rays is reflected from the lattice planes of a sodium chloride crystal. If x-rays of wavelength $\lambda = 1.41 \times 10^{-10}$ m give a first-order diffraction maximum at 30.0°, find the spacing between the reflection planes.

Given: $\lambda = 1.41 \times 10^{-10}$ m, m = 1.00, $\theta = 30.0°$

Determine: d = spacing between reflection planes

Strategy: Knowing $\lambda$, m, and $\theta$, we can use the Bragg diffraction equation to determine d.

Solution: $m\lambda = 2d \sin\theta$

$d = m\lambda/2 \sin\theta = 1.41 \times 10^{-10}$ m

Related Problems: 28-21 through 28-26.

## 4. Compton Effect (Section 28.6)

Review: Figure 28-6 shows the experimental arrangement for observing the Compton effect.

Figure 28-6

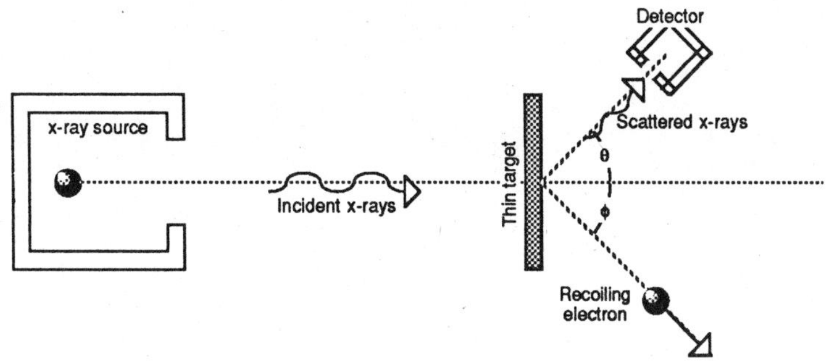


The incident x-rays may be classified by their energy ($\varepsilon$), frequency (f), or wavelength ($\lambda$):

$$\varepsilon = hf = hc/\lambda$$

The scattered x-rays may be classified by their energy ($\varepsilon'$), frequency (f'), or wavelength ($\lambda'$):

$$\varepsilon' = hf' = hc/\lambda'$$

The recoil electron may be classified by its kinetic energy (KE):

$$KE = \varepsilon - \varepsilon'$$

Compton observed that the wavelength of the scattered x-rays was greater than the wavelength of the incident x-rays and that the wavelength difference depended on the scattering angle $\theta$. He theoretically explained this phenomenon by treating both the x-rays and the electrons as particles. Using "billard-ball physics" (conservation of energy and momentum for elastic collisions), he derived the relationship

$$\lambda' - \lambda = \frac{h}{m_o c}(1 - \cos\theta)$$

where $\lambda'$ and $\lambda$ represent the wavelengths of the shifted and original x-rays, respectively, h is Planck's constant, $m_o$ is the electron rest mass, c is the speed of light, and $\theta$ is the angle of scatter.

Practice: The energy of the incident x-rays in Fig. 28-6 is 10.0 keV. Determine the following:

| | |
|---|---|
| 1. Wavelength of incident x-rays | $E = hc/\lambda$<br>$\lambda = hc/E = 1.24 \times 10^{-10}$ m |
| 2. Maximum wavelength shift $\Delta\lambda$ | Maximum $\Delta\lambda$ occurs when $\theta = 180°$<br>$\Delta\lambda = \lambda' - \lambda = h(1 - \cos\theta)/m_oc$<br>$\Delta\lambda = 2h/m_oc = 4.85 \times 10^{-12}$ m |
| 3. Wavelength of x-rays scattered at $\theta = 60.0°$ | $\lambda' = \lambda + h(1 - \cos\theta)/m_oc = 1.25 \times 10^{-10}$ m |
| 4. Kinetic energy of recoil electrons when x-rays are scattered at 60.0° | The energy of the incident x-rays is $\varepsilon = 10.0$ keV.<br>The energy of the scattered x-rays is $\varepsilon' = hc/\lambda' = 9.94$ keV.<br>The kinetic energy of the recoil electrons is $KE = \varepsilon - \varepsilon' = 60.0$ eV. |
| 5. Energy of incident x-rays that causes scattered x-rays to exit target at 135° with an energy of 20.0 keV | $\varepsilon' = 20.0$ keV<br>$\lambda' = hc/\varepsilon' = 6.22 \times 10^{-11}$ m<br>$\lambda = \lambda' - h(1 - \cos\theta)/m_oc = 5.81 \times 10^{-11}$ m<br><br>$\varepsilon = hc/\lambda = 3.42 \times 10^{-15}$ J |
| 6. Wavelength of scattered x-rays when electron recoils with 5.00-keV of kinetic energy when 10.0-keV x-rays are incident on the target | $\varepsilon' = \varepsilon - KE = 5.00 \text{ keV} = 8.00 \times 10^{-16}$ J<br>$\lambda' = hc/\varepsilon' = 2.49 \times 10^{-10}$ m |

Example 28.4. X-ray photons of wavelength $5.00 \times 10^{-11}$ m are incident on a copper target, and Compton-scattered photons are observed at an angle of 135° with respect to the direction of the incident beam. Find the (a) wavelength of the scattered photons, (b) magnitude of the linear momentum of the incident and scattered photons, (c) kinetic energy of the recoil electrons, and (d) magnitude and direction of the momentum of the recoil electrons.

Given: $\lambda = 5.00 \times 10^{-11}$ m = wavelength of incident photons
$\theta = 135°$ = scattering angle for photons

**Determine:** $\lambda'$ = wavelength of scattered photons
$p$ and $p'$ = momentum of incident and scattered photons
KE = kinetic energy of recoil electrons
$p_e$ = linear momentum of recoil electrons

**Strategy:** We can use the Compton equation and the known information to determine the wavelength of the scattered photons. We can determine the momentum of the incident and scattered photons from their wavelength. We can conserve energy to determine the kinetic energy of the recoil electrons. We can conserve momentum to determine the magnitude and direction of the linear momentum of the recoil electrons.

**Solution:** The wavelength of the scattered photons is obtained by using the Compton equation:

$$\lambda' = \lambda + h(1 - \cos\theta)/m_oc = 5.41 \times 10^{-11} \text{ m}$$

The momentum of the incident and scattered photons is

$$p = h/\lambda = 1.33 \times 10^{-23} \text{ kg·m/s} \qquad p' = h/\lambda' = 1.22 \times 10^{-23} \text{ kg·m/s}$$

The kinetic energy of the recoil electron is determined by conserving energy:

$$E + E_o = E' + KE + E_o$$

$$KE = E - E' = \frac{hc}{\lambda} - \frac{hc}{\lambda'} = hc\left(\frac{1}{\lambda} - \frac{1}{\lambda'}\right) = 3.01 \times 10^{-16} \text{ J}$$

The magnitude and direction of the linear momentum of the recoil electron are determined by conserving momentum, as shown in Fig. 28-7.

Figure 28.7

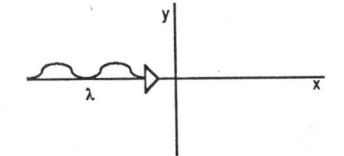

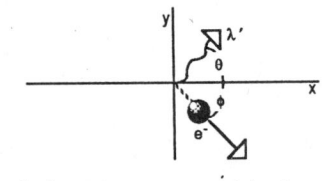

(a) Before collision (b) After collision

Conserving momentum in the x direction, we obtain

(i) $$\frac{h}{\lambda} = \frac{h}{\lambda'}\cos\theta + p\cos\phi \qquad p\cos\phi = \frac{h}{\lambda} - \frac{h\cos\theta}{\lambda'} = 2.19 \times 10^{-23}$$

Conserving momentum in the y direction, we obtain

(ii) $$0 = -\frac{h}{\lambda'}\sin\theta + p\sin\phi \qquad p\sin\phi = \frac{h}{\lambda'}\sin\theta = 0.866 \times 10^{-23}$$

Dividing (ii) by (i) we obtain

$$\frac{p\sin\phi}{p\cos\phi} = \tan\phi = 0.395 \qquad \phi = \tan^{-1}(0.395) = 21.6°$$

This value for $\phi$ can be inserted into either (i) or (ii) to obtain p.

From (i), $p = (2.19 \times 10^{-23}\ kg\cdot m/s)/\cos\phi = 2.35 \times 10^{-23}\ kg\cdot m/s$

From (ii), $p = (0.866 \times 10^{-23}\ kg\cdot m/s)/\sin\phi = 2.35 \times 10^{-23}\ kg\cdot m/s$

The final momentum vector for the recoil electron has a magnitude of $8.93 \times 10^{-23}$ kg·m/s and is directed at an angle $\phi = 75.8°$ from the direction of the incident photon.

Related Problems: 28-27 through 28-31.

## 5. Atomic Spectra (Section) 28.7)

Review: As a result of the work of J.J. Balmer and J.R. Rydberg, it was discovered that the wavelengths of the spectral lines of hydrogen can be determined by

$$\frac{1}{\lambda} = R\left(\frac{1}{n_1^2} - \frac{1}{n_2^2}\right) \qquad n_2 > n_1$$

The various series observed and reported are

| | |
|---|---|
| Lyman | $n_1 = 1$ and $n_2 = 2,3,4,\ldots$ |
| Balmer | $n_1 = 2$ and $n_2 = 3,4,5,\ldots$ |
| Paschen | $n_1 = 3$ and $n_2 = 4,5,6,\ldots$ |
| Brackett | $n_1 = 4$ and $n_2 = 5,6,7,\ldots$ |
| Pfund | $n_1 = 5$ and $n_2 = 6,7,8,\ldots$ |

Practice:

Determine the following:

| | |
|---|---|
| 1. Longest wavelength in Lyman series | For the Lyman series, $n_1 = 1$ and $n_2 = 2,3,4,\ldots$ The longest wavelength occurs for $n_2 = 2$. $R = 1.10 \times 10^7/m$ <br> $\frac{1}{\lambda} = R(\frac{1}{1^2} - \frac{1}{2^2}) = \frac{3R}{4} = 8.25 \times 10^6\ \frac{1}{m}$ <br> $\lambda = 1.21 \times 10^{-7}$ m |
| 2. Shortest wavelength in Lyman series | The shortest wavelength occurs for $n_2 = \infty$. <br> $\frac{1}{\lambda} = R(\frac{1}{1^2} - \frac{1}{\infty^2}) = R = 1.10 \times 10^7\ \frac{1}{m}$ <br> $\lambda = 9.09 \times 10^{-8}$ m |

| | |
|---|---|
| 3. Frequency of light when $n_1 = 2$ and $n_2 = 4$ | $\frac{1}{\lambda} = R(\frac{1}{2^2} - \frac{1}{4^2}) = \frac{3R}{16}$<br>$f = \frac{c}{\lambda} = \frac{3Rc}{16} = 6.19 \times 10^{14}$ Hz |
| 4. Energy of photons given off when $n_1 = 3$ and $n_2 = 4$ | $\frac{1}{\lambda} = R(\frac{1}{3^2} - \frac{1}{4^2}) = \frac{7R}{144}$<br>$E = \frac{hc}{\lambda} = \frac{7Rhc}{144} = 1.06 \times 10^{-19}$ J |

Example 28.5. Calculate the wavelength, frequency, and energy of the longest-wavelength photon for the Lyman series of hydrogen.

Given: Lymon series: $n_1 = 1$ and $n_2 = 2,3,4,\ldots$

Determine: The wavelength, frequency, and energy of the longest-wavelength photon for the Lyman series of hydrogen.

Strategy: We must first determine the value of $n_2$ that gives the longest-wavelength photon. Knowing $n_1$ and $n_2$, we can calculate the wavelength. We can then use the wavelength to determine the frequency and the energy of the photon.

Solution: The wavelength of the various photons of the Lyman series is obtained by

$$\frac{1}{\lambda} = R(\frac{1}{1^2} - \frac{1}{n_2^2}) = R(1 - \frac{1}{n_2^2})$$

The largest value of $\lambda$ occurs for the smallest value of $n_2$. For the Lyman series, the smallest value of $n_2$ is 2. Using $n_2 = 2$, we obtain $\lambda$, f and $\varepsilon$:

$$\frac{1}{\lambda} = R(1 - \frac{1}{n_2^2}) = R(1 - \frac{1}{2^2}) = \frac{3R}{4} \qquad \lambda = 4/3R = 1.21 \times 10^{-7} \text{ m}$$

$$f = c/\lambda = 2.48 \times 10^{15} \text{ Hz} \qquad \varepsilon = hf = 1.64 \times 10^{-18} \text{ J}$$

Related Problems: 28-32 through 28-36.

## PRACTICE TEST

Take and grade this practice test. Doing so will allow you to determine any weak spots in your understanding of the concepts taught in this chapter. The following section prescribes what you should study further to strengthen your understanding.

The filament of an incandescent light bulb has a surface area of $2.00 \times 10^{-1}$ $cm^2$ and radiates 100 W of power. Assume an average wavelength of $5.00 \times 10^{-7}$ m and determine the following:

__________ 1. Intensity of the light emitted by the filament
__________ 2. Temperature of the filament, assuming it radiates as a blackbody
__________ 3. Most intensively radiated wavelength
__________ 4. Rate of emission of photons by the filament
__________ 5. Rate of arrival of photons per square centimeter of surface at a distance of 10 m

Light of wavelength $\lambda = 3.00 \times 10^{-7}$ m is incident on a photocell that has a work function of 3.00 eV. Determine the following:

__________ 6. Energy of the incident photons
__________ 7. Threshold frequency
__________ 8. Maximum kinetic energy of the emitted electrons
__________ 9. Stopping potential for the emitted electrons

A potential difference of 10.0 keV exists between the cathode and anode of an x-ray tube. Determine the following:

__________ 10. Maximum kinetic energy of the electrons as they slam into the anode target
__________ 11. Maximum energy of the emitted x-rays
__________ 12. Minimum wavelength of the emitted x-rays

X-rays with an energy of $8.00 \times 10^{-15}$ J are incident on a crystal, and first-order diffraction for the reflected x-rays occurs for an incident angle of 10.0°. Determine the following:

__________ 13. Spacing between the reflecting planes of the crystal
__________ 14. Angle of incidence needed to obtain the second-order diffraction maximum

When 10.0-keV x-rays are incident on a thin target, recoil electrons leave the target with a kinetic energy of 60.0 eV. Determine the following:

__________ 15. Wavelength of the scattered x-rays
__________ 16. Angle of scatter for the x-rays
__________ 17. Angle of departure for the recoil electrons
__________ 18. Momentum of the recoil electrons

For the hydrogen atom, determine the following:

________ 19. Shortest-wavelength photon in the Brackett series
________ 20. Longest-wavelength photon in the Paschen series

(See Appendix I for answers.)

## PRINCIPAL CONCEPTS AND EQUATIONS PRESCRIPTION

Your score on the practice test is an excellent measure of your understanding of this chapter. You should now use the following chart to write your own prescription for curing any of your physics ills. Look down the leftmost column to the number of the question(s) you answered incorrectly, read across that row to see which section(s) of the study guide you should return to for further study, and then do the suggested text problems to gain additional experience in working with the particular concept.

| Question Number | Concepts and Equations | Prescription: Principal Concept | Prescription: Text Problems |
|---|---|---|---|
| 1 | Intensity: $I = P/A$ | 5 of Ch. 11 | 28-6a,9a |
| 2 | Intensity of a blackbody: $I = \varepsilon\sigma T^4$ | 5 of CH. 15 | 28-11a |
| 3 | $\lambda_{max}T = 2.898 \times 10^{-3}$ m·K | 1 | 28-11b |
| 4 | Rate of emission of photons: $n/t = P/\varepsilon$ | 1 | 28-10 |
| 5 | $(n/t)/A = (P/A)/\varepsilon$ | 1 | 28-6,12 |
| 6 | Energy of a photon: $\varepsilon = hc/\lambda$ | 2 | 28-14a,17c |
| 7 | Threshold frequency: $f_o = W/h$ | 2 | 28-14c,16a |
| 8 | Kinetic energy of photoelectrons: $KE = \varepsilon - W$ | 2 | 28-20,17b |
| 9 | Stopping potential: $V = KE/e$ | 2 | 28-15a,18b |
| 10 | Electric potential difference: $\Delta KE = -\Delta PE$ and $V = \Delta PE/q_o$ | 4 of Ch. 18 | 28-21a |
| 11 | Production of x-rays: $E_{x\text{-rays max}} = KE_{electrons}$ | 3 | 28-21b |
| 12 | Production of x-rays: $\lambda_{min} = hc/eV$ | 3 | 28-22 |
| 13 | Bragg diffraction: $m\lambda = d \sin\theta$ | 3 | 28-23,24 |
| 14 | Bragg diffraction: $m\lambda = d \sin\theta$ | 3 | 28-25,26 |
| 15 | Compton effect: $\varepsilon' = \varepsilon - KE$, $\lambda' = hc/\varepsilon'$ | 4 | 28-28 |
| 16 | Compton effect: $\lambda' - \lambda = (h/m_oc)(1 - \cos\theta)$ | 4 | 28-21 |
| 17 | Compton effect-consv. of momentum | 4 | 28-30 |
| 18 | Compton effect-consv. of momentum | 4 | 28-30 |
| 19 | Atomic spectra: $1/\lambda = R[(1/n_1^2) - (1/n_2^2)]$ | 5 | 28-32,35 |
| 20 | Atomic spectra: | 5 | 28-33,34 |

# 29 Atomic Physics

RECALL FROM PREVIOUS CHAPTERS

| Previously learned concepts and equations frequently used in this chapter | Text Section | Study Guide Page |
|---|---|---|
| Centripetal force: $F_c = mv^2/r$ | 5.5 | 73 |
| Kinetic energy: $KE = mv^2/2$ | 7.5 | 108 |
| Angular momentum: $L = I\omega$ | 9.5 | 152 |
| Coulomb's law: $F = k_e q_1 q_2/r^2$ | 17.4 | 271 |
| Electric potential energy: $PE = k_e q_1 q_2/r$ | 18.2 | 285 |
| Energy, frequency, and wavelength of a photon: $\varepsilon = hf = hc/\lambda$ | 28.3 | 467 |

NEW IDEAS IN THIS CHAPTER

| Concepts and equations introduced | Text Section | Study Guide Page |
|---|---|---|
| The postulates of Bohr | 29.3 | 482 |
| Radius of the allowed orbits: $r_n = n^2 r_1$ | 29.3 | 482 |
| Speed of electrons in allowed orbits: $v_n = v_1/n$ | 29.3 | 482 |
| Energy levels of the hydrogen atom: $E_n = E_1/n^2$ | 29.3 | 482 |
| Energy and wavelength of emitted photons: $\varepsilon = E_f - E_i = -E_1\left(\frac{1}{n_f^2} - \frac{1}{n_i^2}\right) \quad n_f < n_i$ ; $\frac{1}{\lambda} = R\left(\frac{1}{n_f^2} - \frac{1}{n_i^2}\right) \quad n_f < n_i$ | 29.3 | 482 |
| Extension of Bohr theory to other one-electron atoms: replace $e^2$ with $Ze^2$ | 29.3 | 482 |
| Energy level diagrams | 29.4 | 488 |

PRINCIPAL CONCEPTS AND EQUATIONS

## 1. The Bohr Atom (Section 29.3)

Review: Shown below are the postulates of Bohr and their consequences.

Postulate 1. The electron revolves about the nucleus in certain circular orbits because of the attraction of the Coulomb force.

An electron traveling in a circular orbit experiences a centripetal force given by

(a) $$F_c = mv^2/r$$

Since the centripetal force is supplied by the Coulomb attraction, we can write

(b) $$F_c = F_{coulb} = ke^2/r^2$$

Combining (a) and (b), we have

(c) $$mv^2/r = ke^2/r^2 \qquad mv^2 = ke^2/r$$

Postulate 2. The only orbits allowed are those in which the angular momentum is an integral multiple of Planck's constant divided by $2\pi$:

(d) $$L = nh/2\pi$$

Recall that the angular momentum of a particle traveling in a circle is

(e) $$L = mvr$$

Combining (d) and (e), we have

(f) $$mvr = nh/2\pi$$

Postulate 3. The atom does not radiate energy while the electron is in one of the allowed orbits.

We can obtain the speed of an electron in an allowed orbit by combining (c) and (f) and eliminating r:

(g) $$v_n = \frac{1}{n}\left(\frac{2\pi ke^2}{h}\right) = \frac{1}{n}v_1, \text{ where } v_1 = \frac{2\pi ke^2}{h} = 2.18 \times 10^6 \text{ m/s}$$

We can obtain the radius of the allowed orbits by combining (c) and (f) and eliminating v:

(h) $$r_n = n^2\left(\frac{h^2}{4\pi^2 mke^2}\right) = n^2 r_1, \text{ where } r_1 = \frac{h^2}{4\pi^2 mke^2} = 5.30 \times 10^{-11} \text{ m}$$

We can obtain the kinetic, potential, and total energy of an electron in an allowed orbit as follows:

(i) $$KE_n = \frac{1}{2}mv_n^2 = \frac{1}{2}\left(\frac{ke^2}{r_n}\right)$$ [using equation (c)]

(j) $$PE_n = -\frac{ke^2}{r_n}$$

(k) $$E_n = KE_n + PE_n = -\frac{ke^2}{2r_n} = \frac{E_1}{n^2}, \quad \text{where } E_1 = \frac{-2\pi^2 mk^2 e^4}{h^2} = -2.18 \times 10^{-18}\ \text{J}$$

Postulate 4. The atom emits a photon of energy hf when the electron drops from an orbit of energy $E_i$ to an orbit of lower energy $E_f$. Stated algebraically,

(l) $$\varepsilon = hf = E_i - E_f \qquad E_f < E_i$$

Further, if the atom absorbs a photon of energy hf, the electron jumps from an orbit of energy $E_i$ to an orbit of higher energy $E_f$. Stated algebraically,

(m) $$\varepsilon = hf = E_f - E_i \qquad E_f > E_i$$

We can combine (k) and (l) to express the energy of the emitted photon in terms of $n_i$ and $n_f$:

(n) $$\varepsilon_{\text{emitted}} = hf = E_i - E_f = -E_1\left(\frac{1}{n_f^2} - \frac{1}{n_i^2}\right) \qquad E_f < E_i;\ n_f < n_i$$

We can combine (k) and (m) to express the energy of the absorbed photon in terms of $n_i$ and $n_f$:

(o) $$\varepsilon_{\text{absorbed}} = hf = E_f - E_i = -E_1\left(\frac{1}{n_i^2} - \frac{1}{n_f^2}\right) \qquad E_f > E_i;\ n_f > n_i$$

The wavelength of the emitted and absorbed photons can be determined from the following:

(p) $$\frac{1}{\lambda_{\text{emitted}}} = \frac{f}{c} = \frac{1}{c}\left(\frac{E_i - E_f}{h}\right) = R\left(\frac{1}{n_f^2} - \frac{1}{n_i^2}\right) \qquad n_f < n_i$$

(q) $$\frac{1}{\lambda_{\text{absorbed}}} = \frac{f}{c} = \frac{1}{c}\left(\frac{E_f - E_i}{h}\right) = R\left(\frac{1}{n_i^2} - \frac{1}{n_f^2}\right) \qquad n_f > n_i$$

As a result of the preceding theory, we have expressions that allow us to determine the following for the one-electron hydrogen atom:

radius of the orbits available to the electron -------------- (h)
speed of the electron in the available orbits --------------- (g)
centripetal force on an electron in an available orbit ------ (a or b)
kinetic energy of an electron in an available orbit --------- (i)
electric potential energy of the atom when the electron is in one of the available orbits -------------------------- (j)
total energy of the atom when the electron is in one of the available orbits ------------------------------------------ (k)
angular momentum of the electron in any orbit --------------- (d or e)
energy, frequency, and wavelength of the photon emitted when an electron jumps to a lower orbit ------------------ (n and p)
energy, frequency, and wavelength of the photon absorbed when an electron jumps to a higher orbit ----------------- (o and q)

The Bohr theory can be extended to any one-electron (hydrogen-like) atom simply by replacing $e^2$ with $Ze^2$ in all of the above expressions. Singly ionized helium atoms and doubly ionized lithium atoms are one-electron atoms.

<u>Practice</u>: Given the following information:

$m = 9.11 \times 10^{-31}$ kg = mass of an electron
$e = 1.60 \times 10^{-19}$ C = charge on an electron
$h = 6.63 \times 10^{-34}$ J·s = Planck's constant
$k = 9.00 \times 10^{9}$ N·m²/C² = Coulomb's law constant
$c = 3.00 \times 10^{8}$ m/s = speed of light
$r_1 = 5.30 \times 10^{-11}$ m = radius of the first Bohr orbit
$v_1 = 2.18 \times 10^{6}$ m/s = speed of an electron in the first Bohr orbit
$E_1 = -2.18 \times 10^{-18}$ J = energy of a hydrogen atom when the electron is in the first Bohr orbit
$R = 1.10 \times 10^{7}$/m = Rydberg's constant

Determine the following for a hydrogen atom:

| | |
|---|---|
| 1. Radius of n = 2 orbit | $r_n = n^2 r_1 = 4r_1 = 2.12 \times 10^{-10}$ m |
| 2. Speed of electron in n = 2 orbit | $v_n = v_1/n = v_1/2 = 1.09 \times 10^{6}$ m/s |
| 3. Kinetic energy of electron in the n = 2 orbit | $KE_n = mv_n^2/2 = 5.41 \times 10^{-19}$ J<br>$KE_n = ke^2/2r_n = 5.43 \times 10^{-19}$ J<br>The difference is due to rounding off. |
| 4. Electric potential energy of an electron in n = 2 orbit | $PE_n = -ke^2/r_n = -1.09 \times 10^{-18}$ J |
| 5. Total energy of electron in n = 2 orbit | $E_2 = KE_2 + PE_2 = -5.47 \times 10^{-19}$ J<br>$E_2 = E_1/n^2 = E_1/4 = -5.45 \times 10^{-19}$ J<br>The difference is due to rounding off. |
| 6. Centripetal force on electron in n = 2 orbit | $F_c = mv_2^2/r_2 = 5.11 \times 10^{-9}$ N<br>$F_c = ke^2/r_2^2 = 5.13 \times 10^{-9}$ N<br>The difference is due to rounding off. |

| | |
|---|---|
| 7. Energy of photon in n = 2 to n = 1 transition | $\varepsilon_{2\to1} = E_2 - E_1 = 1.63 \times 10^{-18}$ J<br><br>$\varepsilon_{2\to1} = -E_1(\frac{1}{n_f^2} - \frac{1}{n_i^2}) = 1.63 \times 10^{-18}$ J |
| 8. Frequency of photon emitted in n = 2 to n = 1 transition | $f_{2\to1} = \varepsilon_{2\to1}/h = 2.46 \times 10^{15}$ Hz<br><br>$f_{2\to1} = \frac{-E_1}{h}(\frac{1}{n_f^2} - \frac{1}{n_i^2}) = 2.46 \times 10^{15}$ Hz |
| 9. Wavelength of photon emitted in n = 2 to n = 1 transition | $\lambda_{2\to1} = c/f_{2\to1} = 1.22 \times 10^{-7}$ m<br><br>$\frac{1}{\lambda_{2\to1}} = R(\frac{1}{n_f^2} - \frac{1}{n_i^2}) = \frac{3}{4}R = 8.25 \times 10^{-6}$/m<br><br>$\lambda_{2\to1} = 1.21 \times 10^{-7}$ m<br><br>The difference is due to rounding off |
| 10. Photon energy that causes atom to undergo n = 2 to n = 3 transition | $E_3 = -E_1/n_f^2 = -E_1/9 = -2.42 \times 10^{-19}$ J<br>$E_2 = -E_1/n_i^2 = -E_1/4 = -5.45 \times 10^{-19}$ J<br>$\varepsilon_{2\to3} = E_3 - E_2 = 3.03 \times 10^{-19}$ J<br>$\varepsilon_{2\to3} = -E_1(\frac{1}{n_i^2} - \frac{1}{n_f^2}) = 3.03 \times 10^{-19}$ J |
| 11. Photon frequency and wavelength that cause atoms to undergo n = 2 to n = 3 transition | $f_{2\to3} = \varepsilon_{2\to3}/h = 4.57 \times 10^{14}$ Hz<br>$\lambda_{2\to3} = c/f_{2\to3} = 6.56 \times 10^{-7}$ m<br><br>$\frac{1}{\lambda_{2\to3}} = R[\frac{1}{n_i^2} - \frac{1}{n_f^2}] = 1.53 \times 10^{6}$/m<br><br>$\lambda_{2\to3} = 6.54 \times 10^{-7}$ m<br><br>The difference is due to rounding off. |

Example 29.1. A hydrogen atom in the n = 4 state decays to the n = 2 state. (a) Determine what happens to the radius ($r_n$) of the electron's orbit, the linear speed ($v_n$) of the electron in its orbit, its angular speed ($\omega_n$), the centripetal force ($F_{cn}$) on it, its kinetic energy ($KE_n$), the electric

potential energy ($PE_n$) and total energy ($E_n$) of the atom, and the angular momentum ($L_n$) of the electron. (b) Determine the energy ($\varepsilon$), frequency (f), wavelength ($\lambda$), and angular momentum (L) of the emitted photon.

Given: The atom is a hydrogen atom.
$n_i = 4$ = initial state
$n_f = 2$ = final state

Determine: (a) The change in $r_n$, $v_n$, $\omega_n$, $F_{cn}$, $KE_n$, $PE_n$, $E_n$, and $L_n$. (b) $\varepsilon$, f, $\lambda$, and L of the emitted photon.

Strategy: (a) As a result of our development of the Bohr theory, we know expressions for $r_n$, $v_n$, $F_{cn}$, $KE_n$, $PE_n$, $E_n$, and $L_n$. We can develop an expression for $\omega_n$. We can use these expressions to determine the desired quantities for the initial and final states. We can then take the ratio of the desired quantities for each state in order to determine what happens as the atom decays from n = 4 to n = 2. (b) We can use the Bohr theory to determine $\varepsilon$. Knowing $\varepsilon$, we can use the information about photons to obtain f and $\lambda$. Finally, we can use conservation of angular momentum to obtain L.

Solution:

(a) the radius of an acceptable orbit is $r_n = n^2 r_1$. Hence $r_4 = 16r_1$, $r_2 = 4r_1 = r_4/4$.

The speed of an electron in an acceptable orbit is $v_n = v_1/n$. Hence $v_4 = v_1/4$, $v_2 = v_1/2 = 2v_4$.

From $v_n$ and $r_n$, we can obtain $\omega_n = v_n/r_n = (1/n^3)(v_1/r_1)$. Hence $\omega_4 = (1/64)(v_1/r_1)$, $\omega_2 = (1/8)(v_1/r_1) = 8\omega_4$.

The centripetal force is $F_{cn} = mv_n{}^2/r_n = (1/n^4)(mv_1{}^2/r_1)$. Hence $F_{c4} = (1/256)(mv_1{}^2/r_1)$, $F_{c2} = (1/16)(mv_1{}^2/r_1) = 16F_{c4}$.

Alternately the centripetal force is $F_{cn} = ke^2/r_n{}^2 = (1/n^4)(ke^2/r_1{}^2)$. Hence $F_{c4} = (1/256)(ke^2/r_1{}^2)$, $F_{c2} = (1/16)(ke^2/r_1{}^2) = 16F_{c4}$

The kinetic energy is $KE_n = mv_n{}^2/2 = (1/n^2)(mv_1{}^2/2)$. Hence $KE_4 = (1/16)(-ke^2/r_1)$, $KE_2 = (1/4)(-ke^2/r_1) = 4KE_4$.

The electric potential energy is $PE_n = -ke^2/r_n = (1/n^2)(-ke^2/r_1)$. Hence $PE_4 = (1/16)(-ke^2/r_1)$, $PE_2 = (1/4)(-ke^2/r_1) = 4PE_4$.

The total energy is $E_n = E_1/n^2$. Hence $E_4 = E_1/16$, $E_2 = E_1/4$, $= 4E_4$.

The angular momentum is $L_n = nh/2\pi$. Hence $L_4 = 4h/2\pi$, $L_2 = 2h/2\pi = L_4/2$.

$$\varepsilon_{4\to2} = E_4 - E_2 = -E_1 \left( \frac{\quad}{n_2^{\ 2}} - \frac{\quad}{n_4^{\ 2}} \right) = -(3/16)E_1 = 4.09 \times 10 \quad J$$

The frequency and wavelength of the photon are

$f_{4\to2} = \varepsilon_{4\to2}/h = 6.17 \times 10^{14}/s$ $\qquad$ $\lambda_{4\to2} = c/f_{4\to2} = 4.86 \times 10^{-7}$ m

The angular momentum of the photon is determined by assuming that the angular momentum lost by the atom is carried off by the photon:

$$L_{photon} = L_4 - L_2 = h/\pi = 2.11 \times 10^{-34}\ J\cdot s$$

Related Problems: 29-6 through 24-24.

## 2. Energy Level Diagrams (Section 29.4)

Review: The energy of the hydrogen atom in its nth state is

$$E_n = E_1/n^2 \quad \text{where} \quad E_1 = -2.18 \times 10^{-18}\ J = -13.6\ eV$$

Therefore $E_1 = -13.6$ eV, $E_2 = -3.40$ eV, $E_3 = -1.51$ eV, $E_4 = -0.850$ eV, $E_5 = -0.544$ eV, $E_6 = -0.378$ eV, and so on. It is useful to arrange these energy values in an energy level diagram as shown in Fig. 29-1.

Figure 29-1

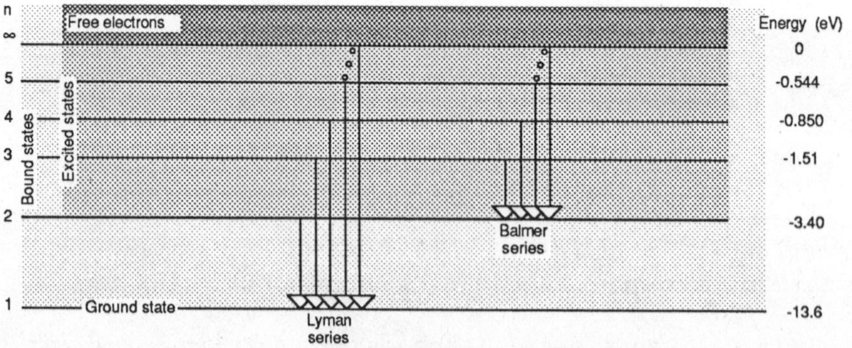


Practice: Refer to Fig. 29-1 and determine the following:

| | |
|---|---|
| 1. Energy of atom in ground state | For the ground state, n = 1, $E_1 = -13.6$ eV |
| 2. Energy of atom in second excited state | n = 3 for second excited state<br>$E_3 = -1.51$ eV |
| 3. Binding and excitation energy for atom when electron is in third excited state | n = 4 for third excited state<br>$E_{binding} = E_4 = -0.850$ eV<br>$E_{excited} = E_4 - E_1 = +12.75$ eV |

| | |
|---|---|
| 4. Energy, frequency, and wavelength of lowest-energy photon emitted in Lyman series | $\varepsilon = E_2 - E_1 = 10.2$ eV<br>$f = \varepsilon/h = 2.46 \times 10^{15}$/s<br>$\lambda = c/f = 1.22 \times 10^{-7}$ m |
| 5. Maximum number of photons omitted by atom in n = 4 state | The atom can decay to the ground state by any of the following schemes:<br>$n = 4 \rightarrow n = 1$<br>$n = 4 \rightarrow n = 2$; $n = 2 \rightarrow n = 1$<br>$n = 4 \rightarrow n = 3$; $n = 3 \rightarrow n = 1$<br>$n = 4 \rightarrow n = 3$; $n = 3 \rightarrow n = 2$; $n = 2 \rightarrow n = 1$<br>This amounts to eight possibilities but only six different photons. |
| 6. Wavelength of incident photon that causes atom to go from n = 2 to n = 4 state | $E_{2\rightarrow 4} = E_4 - E_2 = 2.55$ eV<br>$\lambda_{2\rightarrow 4} = hc/\varepsilon_{2\rightarrow 4} = 4.87 \times 10^{-7}$ m |
| 7. Whether or not a photon 2.00-eV can excite a hydrogen atom from n = 2 state to the n = 3 state | $E_2 = -3.40$ eV, $E_3 = -1.51$ eV<br>Excitation can occur only with a photon of energy $E_{2\rightarrow 3} = E_3 - E_2 = 1.89$ eV. Since this photon has an energy of $\varepsilon = 2.00$ eV, the excitation cannot occur. |

Example 29.2. Draw an energy level diagram for singly ionized helium and use it to determine the smallest-wavelength photon this atom can emit when it is in the n = 3 state.

Given: A singly ionized helium atom in the n = 3 state

Determine: Energy level diagram for singly ionized helium and smallest-wavelength photon such an atom can emit when it is in the n = 3 state.

Strategy: From the Bohr theory, we have an expression for the energy levels of an hydrogen atom. We can extend all expressions developed in the Bohr theory to any one-electron atom (e.g., singly ionized helium) by replacing $e^2$ with $Ze^2$. Once an expression for the energy levels of singly ionized helium is obtained, we can draw an energy level diagram, that can be used to determine the most energetic photon (hence the smallest wavelength) a singly ionized helium atom in the n = 3 state can emit.

Solution: The energy levels of hydrogen are

$$E_n = E_1/n^2 \quad \text{where} \quad E_1 = -2\pi mk^2e^4/h^2$$

This expression may be extended to any one-electron atom by replacing $e^2$ with $Ze^2$. Hence, the energy levels for singly ionized helium are

$$E_n = E_1'/n^2 \quad \text{where} \quad E_1' = -2\pi^2 mk^2 Z^2 e^4/h^2 = E_1 Z^2 = -54.5 \text{ eV}$$

The energy levels are then $E_1 = -54.5 \text{ eV}/(1)^2 = -54.5$ eV, $E_2 = -54.5 \text{ eV}/(2)^2 = -13.6$ eV, $E_3 = -6.06$ eV, $E_4 = -3.41$ eV, $E_5 = -2.18$ eV, and so on. The energy level diagram is shown in Fig. 29-2.

Figure 29-2

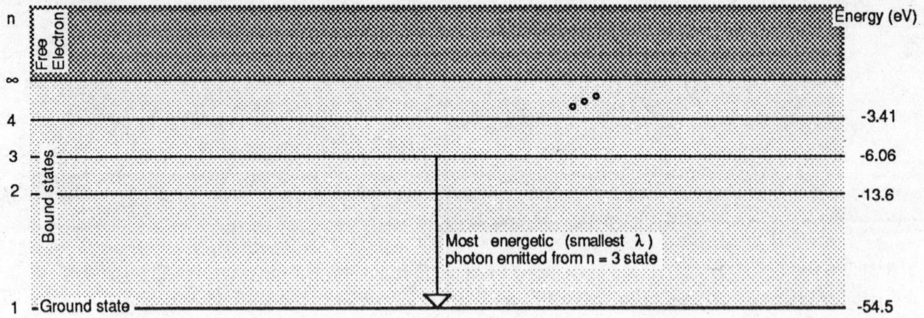


If the atom is in the n = 3 state, the most energetic (smallest-wavelength) photon is emitted in the n = 3 to n = 1 transition. The energy of this photon is

$$\varepsilon_{3\to1} = E_1 - E_3 = 48.4 \text{ eV}$$

The wavelength of this photon is

$$\lambda_{3\to1} = hc/\varepsilon_{3\to1} = 2.57 \times 10^{-8} \text{ m}$$

Related Problems: 29-7, 29-8, 29-17.

---

PRACTICE TEST

Take and grade this practice test. Doing so will allow you to determine any weak spots in your understanding of the concepts taught in this chapter. The following section prescribes what you should study further to strengthen your understanding.

Determine the following for a hydrogen atom:

_______ 1. Radius of the n = 3 orbit
_______ 2. Speed of an electron in the n = 3 orbit
_______ 3. Angular speed of an electron in the n = 3 orbit
_______ 4. Kinetic energy of an electron in the n = 3 orbit
_______ 5. Centripetal force experienced by an electron in the n = 3 orbit
_______ 6. Electric potential energy of an atom in the n = 3 state
_______ 7. Angular momentum of an electron in the n = 3 orbit
_______ 8. Energy of the photon emitted when the atom goes from the n = 3 to the n = 2 state

___________ 9. Angular momentum of the photon emitted when the atom goes from the n = 3 to the n = 2 state

___________ 10. Wavelength of the photon emitted when the atom goes from the n = 3 to the n = 2 state

Use the energy level diagram For doubly ionized lithium (Z=3) shown in Fig. 29-3 to determine the following:

Figure 29-3

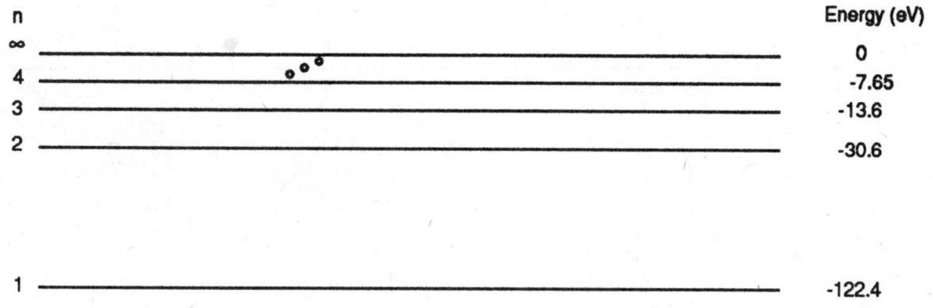


___________ 11. Ionization potential

___________ 12. Energy of the atom in the ground state

___________ 13. Binding energy of the atom in the second excited state

___________ 14. Excitation energy of the atom in the second excited state

___________ 15. Energy of the photon emitted when the electron jumps from the n = 3 to the n = 2 orbit

___________ 16. Wavelength of the photon emitted when the electron jumps from the n = 3 to the n = 2 orbit

___________ 17. Maximum number of different energy photons that could be emitted by numerous atoms going from the n = 4 to the n = 1 state

___________ 18. Frequency of the photon that must be absorbed in order to cause the electron to jump from the n = 3 to the n = 4 orbit

(See Appendix I for answers.)

## PRINCIPAL CONCEPTS AND EQUATIONS PRESCRIPTION

Your score on the practice test is an excellent measure of your understanding of this chapter. You should now use the following chart to write your own prescription for curing any of your physics ills. Look down the leftmost column to the number of the question(s) you answered incorrectly, read across that row to see which section(s) of the study guide you should return to for further study, and then do the suggested text problems to gain additional experience in working with the particular concept.

| Question Number | Concepts and Equations | Prescription: Principal Concept | Prescription: Text Problems |
|---|---|---|---|
| 1 | Radius of an orbit: $r_n = n^2 r_1$ | 1 | 29-8 |
| 2 | Speed of electron in orbit: $v_n = v_1/n$ | 1 | 29-12 |
| 3 | Angular speed of electron: $\omega_n = v_n/r_n$ | 1 | 5-3,5 |
| 4 | Kinetic energy of electron: $KE_n = mv_n^2/2$ | 1 | 7-36a,38a |
| 5 | Centripetal force on electron $F_{cn} = mn_n^2/r_n$ | 1 | 29-9 |
| 6 | Electric potential energy of atom: $PE_n = -k_e^2/r_n$ | 1 | 18-9, 19 |
| 7 | Angular momentum of electron: $L_n = nh/2\pi$ | 1 | 29-6,9 |
| 8 | Energy of emitted photon: $\varepsilon = E_i - E_f$ | 1 | 29-13,21a |
| 9 | Angular momentum of photon: $L = L_i - L_f$ | 1 | 29-6b |
| 10 | Wavelength of emitted photon: $\lambda = hc/\varepsilon$ | 2 | 29-11,18a |
| 11 | Ionization potential: $IP = -E_1$ | 2 | 29-7,17 |
| 12 | Ground state energy: $E_1$ | 2 | 29-7,17 |
| 13 | Binding energy: $E_n = E_1/n^2$ | 2 | 29-7,17 |
| 14 | Excitation energy: $E_{excit} = E_n - E_1$ | 2 | 29-7,17 |
| 15 | Energy of emitted photon: $\varepsilon = E_i - E_f$ | 2 | 29-7,17 |
| 16 | Wavelength of emitted photon: $\lambda = hc/\varepsilon$ | 2 | 29-20,21a |
| 17 | Energy level diagram | 2 | 29-21,22 |
| 18 | Frequency of absorbed photon: $f = E/h$ | 2 | 29-22,23 |

# 30 Quantum Mechanics

## RECALL FROM PREVIOUS CHAPTERS

| Previously learned concepts and equations frequently used in this chapter | Text Section | Study Guide Page |
|---|---|---|
| Kinetic energy: $KE = mv^2/2$ | 7.5 | 108 |
| Momentum: $p = mv$ | 8.2 | 123 |
| Electric potential difference: $V = \Delta PE/q_o$ | 18.2 | 289 |
| Single-slit diffraction: $m\lambda = W \sin\theta$ | 26.5 | 441 |
| Energy and momentum of a photon: $\varepsilon = hf$ and $p = \varepsilon/c$ | 27.7,28.3 | 462,467 |
| Bragg equation: $m\lambda = 2d \sin\theta$ | 28.5 | 473 |

## NEW IDEAS IN THIS CHAPTER

| Concepts and equations introduced | Text Section | Study Guide Page |
|---|---|---|
| The deBroglie equation: $\lambda = h/p$ | 30.2 | 493 |
| Quantum numbers: $n$, $\ell$, $m$, and $m_s$ | 30.3 | 498 |
| Orbital angular momentum and its z component: $L = \sqrt{\ell(\ell + 1)}(h/2\pi)$  $L_z = mh/2\pi$ | 30.3 | 498 |
| Spin angular momentum and its z component: $S = \sqrt{s(s + 1)}(h/2\pi)$  $S_z = m_s h/2\pi$ | 30.3 | 498 |
| Heisenberg uncertainty principal: $\Delta x \Delta p_x \geq (h/2\pi)/2$  $\Delta E \Delta t \geq (h/2\pi)/2$ | 30.4 | 501 |

## PRINCIPAL CONCEPTS AND EQUATIONS

### 1. Matter Waves (Section 30.2)

Review: Matter has a particle nature and is characterized by its energy and momentum. Light has a wave nature and is characterized by its wavelength and frequency. This is summarized in Fig. 30-1a.

After Einstein explained the photoelectric effect, physicists realized that light had a particle nature (photons). We can write expressions for the energy and momentum of the photons (Fig. 30-1b).

Louis de Broglie looked at all this information, considered the symmetry of nature, and hypothesized that, if light has a particle nature, then perhaps matter has a wave nature. We can rewrite the expressions for the energy and momentum of a photon to obtain the expressions for the wavelength and frequency of a particle (Fig. 30-1c). The de Broglie equation gives the wavelength of a particle in terms of its momentum:

$$\lambda = h/p = h/mv$$

| | Particle Nature | Wave Nature |
|---|---|---|
| Matter | E<br>p | |
| Light | | $\lambda$<br>f |

(a) Matter has a particle nature. Light has a wave nature.

| | Particle Nature | Wave Nature |
|---|---|---|
| Matter | E<br>p | |
| Light | $\varepsilon = hf$<br>$P = \varepsilon/c$<br>$= h/\lambda$ | $\lambda$<br>f |

(b) Matter has a particle nature. Light has a wave and a particle nature.

| | Particle Nature | Wave Nature |
|---|---|---|
| Matter | E<br>p | $f = E/h$<br>$\lambda = h/p$ |
| Light | $\varepsilon = hf$<br>$p = \varepsilon/c$<br>$= h/\lambda$ | $\lambda$<br>f |

(c) Matter has a particle and a wave nature. Light has a wave and a particle nature.

Figure 30-1

Experimental evidence for the existence of matter waves came from the Davisson and Germer experiment (Fig. 30-2), 54.0-eV electrons incident upon crystalline nickel were reflected in such a manner as to have the associated matter waves constructively interfer at an angle of 50.0° with respect to the incident beam.

Figure 30-2

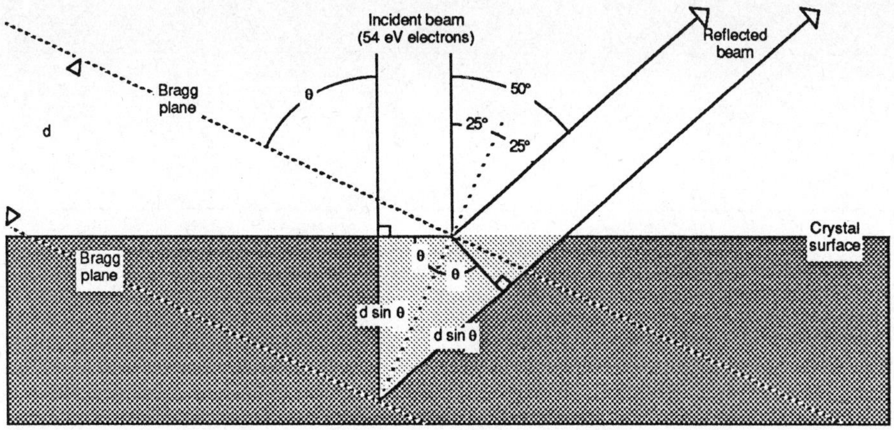


The path difference for the two waves is shown in figure 30-2 to be 2d sinθ. Constructive interference of the associated matter waves occurs when the path difference is an integral multiple of whole wavelengths; hence the Bragg equation gives

$$m\lambda = 2d \sin\theta \qquad m = 1,2,3,...$$

For crystalline nickel, $d = 9.09 \times 10^{-11}$ m and first-order interference ($m = 1$) occurs for $\theta = 65.0°$, hence

$$\lambda = 2d \sin\theta = 2(9.09 \times 10^{-11} \text{ m}) \sin 65.0° = 1.65 \times 10^{-10} \text{ m}$$

A particle with kinetic energy KE has a momentum p, and its associated matter wave has a wavelength λ:

$$p = (2mKE)^{1/2} \qquad \lambda = h/p = h/(2mKE)^{1/2}$$

The matter wave associated with 54.0-eV electrons has a wavelength

$$\lambda = h/p = h/(2mKE)^{1/2} = 1.67 \times 10^{-10} \text{ m}$$

Notice that the experimental value of λ for 54.0-eV electrons from the Davisson-Germer experiment is in excellent agreement with the theoretical value obtained using the de Broglie hypothesis.

Practice:

Determine the de Broglie wavelength of the following:

| | |
|---|---|
| 1. A 4000-kg truck moving at 5 m/s | $\lambda = h/p = h/mv = 3.32 \times 10^{-38}$ m |
| 2. A 10.0-keV neutron | $KE = 10.0 \text{ keV} = 1.00 \times 10^{4}$ eV<br>$p = (2mKE)^{1/2}$<br>$\lambda = h/p = h/(2mKE)^{1/2} = 2.87 \times 10^{-13}$ m |

| | | |
|---|---|---|
| 3. | A 10.0-keV x-ray | $E = 10.0\ \text{keV} = 1.00 \times 10^4\ \text{eV}$<br>$P = E/c$<br>$\lambda = h/p = hc/E = 1.24 \times 10^{-10}\ \text{m}$ |
| 4. | An electron accelerated through a $1.00 \times 10^3$-V potential difference | $p = (2mKE)^{1/2} = (2mqV)^{1/2}$<br>$\lambda = h/p = h/(2mqV)^{1/2} = 3.88 \times 10^{-11}\ \text{m}$ |

Electrons that have been accelerated from rest through a 50.0-V potential difference are incident normal to a crystaline surface. First-order constructive interference of the reflected matter waves occurs at an angle of 40.0° with respect to the incident beam.

Determine the following:

| | | |
|---|---|---|
| 5. | Final kinetic energy of electrons in eV and joules | $\Delta PE = -qV = -50.0\ \text{eV}$<br>$\Delta KE = -\ \Delta PE = 50.0\ \text{eV} = 8.00 \times 10^{-18}\ \text{J}$ |
| 6. | Final momentum of electrons | $KE = mv^2/2 = m^2v^2/2m = p^2/2m$<br>$p = (2mKE)^{1/2} = 3.82 \times 10^{-24}\ \text{kg·m/s}$ |
| 7. | Wavelength of the associated matter waves | $\lambda = h/p = 1.74 \times 10^{-10}\ \text{m}$ |
| 8. | Angle between incident beam and Bragg planes | 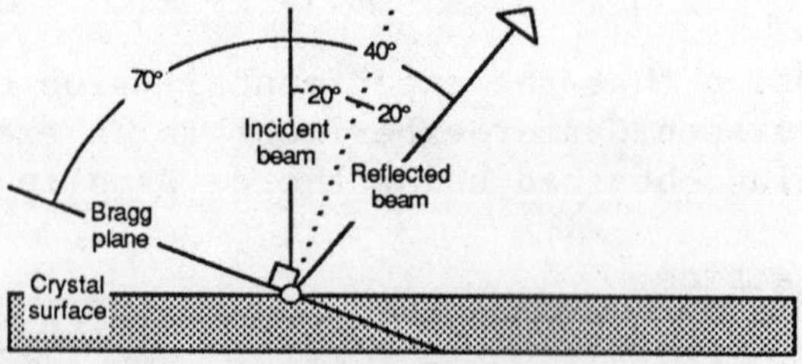 <br>40° = angle between incident and reflected beams<br>20° = angle between incident beam and normal to Bragg planes<br>70° = angle between incident beam and Bragg planes |
| 9. | Spacing between Bragg planes | $m = 1,\quad \theta = 70.0°$<br>$m\lambda = 2d\ \sin\theta$<br>$d = m\lambda/(2\ \sin\theta) = 9.26 \times 10^{-11}\ \text{m}$ |

| | |
|---|---|
| 10. Kinetic energy of a neutron with the same wavelength | $\lambda = h/p = h/(2mKE)^{1/2}$<br>$\lambda_n = \lambda_e$<br>$h/(2m_nKE_n)^{1/2} = h/(2m_eKE_e)^{1/2}$<br>$KE_n = m_eKE_e/m_n = 4.36 \times 10^{-21}$ J |
| 11. Voltage through which you would have to accelerate a proton in order for it to have the same wavelength | $\lambda = h/p = h/(2mqV)^{1/2}$<br>$\lambda_p = \lambda_e$<br>$h/(2m_peV_p)^{1/2} = h/(2m_eeV_e)^{1/2}$<br>$V_p = m_eV_e/m_p = 2.73 \times 10^{-2}$ V |

Example 30.1. What is the wavelength of the matter wave associated with an electron in the second Bohr orbit of a hydrogen atom? How does this compare with the circumference of the orbit?

Given: An electron in the n = 2 state of a hydrogen atom.

Determine: The wavelength associated with the electron and compare it with the circumference of the orbit.

Strategy: We can use the Bohr theory to determine the speed of the electron in its orbit and the circumference of the orbit. Knowing the speed and mass of the electron, we can determine the wavelength of the associated wave.

Solution: From Bohr theory, we know that the speed of an electron in the second Bohr orbit is

$$v_2 = v_1/2 = 1.09 \times 10^6 \text{ m/s, where } v_1 = 2.18 \times 10^6 \text{ m/s}$$

From Bohr theory, we know that the radius of the second Bohr orbit is

$$r_2 = (2)^2 r_1 = 2.12 \times 10^{-10} \text{ m, where } r_1 = 5.30 \times 10^{-11} \text{ m}$$

The wavelength of the matter wave associated with an electron in the second Bohr orbit is

$$\lambda = h/p = h/mv_2 = 6.68 \times 10^{-10} \text{ m}$$

The circumference of the second Bohr orbit is

$$c_2 = 2\pi r_2 = 1.33 \times 10^{-9} \text{ m}$$

Notice that $c_2 = 2\lambda$, and hence two de Broglie waves can fit into the circumference of the Bohr orbit.

Related Text Problems: 30-1 through 30-9.

## 2. Wave Mechanics (Section 30.3)

Review: Schrodinger's development of the quantum theory greatly expanded our knowledge of the atom. To summarize,

- $n$ = principal quantum number; gives information about the energy of the atom and the size of the electron orbit
  $n = 1,2,3,\ldots$
- $\ell$ = angular momentum quantum number; gives information about the shape of the electron orbit
  $\ell = 0,1,2,\ldots,n-1$ (n possible values)
  The angular momentum is given by $L = [\ell(\ell+1)]^{1/2}(h/2\pi)$
- $m$ = magnetic quantum number; gives information about the orientation of the electron orbit
  $m = 0, \pm 1, \pm 2, \ldots, \pm \ell$ ($2\ell + 1$ possible values)
  The z component of the angular momentum is given by $L_z = mh/2\pi$
- $s$ = spin quantum number = 1/2
  The spin angular momentum is given by $S = [s(s+1)]^{1/2}(h/2\pi)$
- $m_s$ = spin orientation quantum number; gives information about the orientation of the spin of the electron
  $m_s = \pm 1/2$
  The z component of the spin angular momentum is given by
  $S_z = m_s h/2\pi = \pm(1/2)h/2\pi$

$L$ and $L_z$ are shown in Fig. 30-3 for $\ell = 1$

Figure 30-3

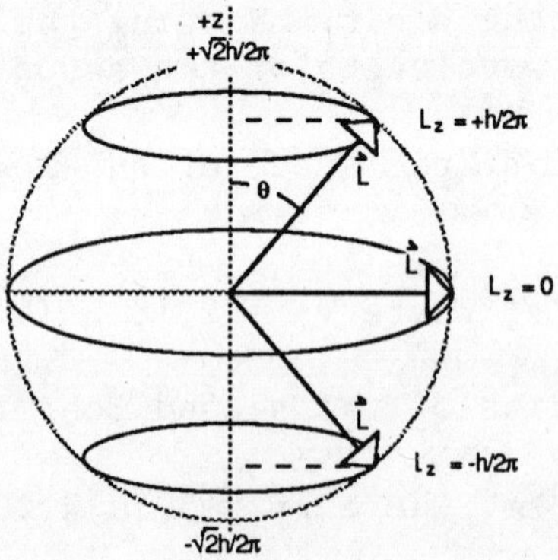


$\ell = 1$

$L = \sqrt{\ell(\ell+1)}\,(h/2\pi) = \sqrt{2}(h/2\pi)$

$L_z = 0, +h/2\pi, -h/2\pi$

$\cos\theta = L_z/L = m/\sqrt{\ell(\ell+1)}$

$S$ and $S_z$ are shown in Fig. 30-4

Figure 30-4

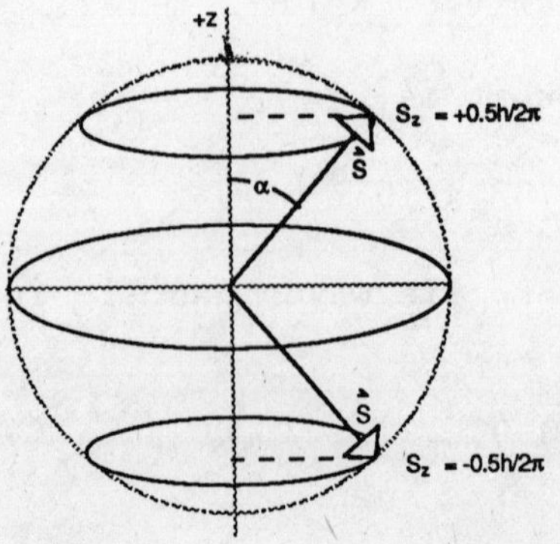


$s = 1/2$

$S = \sqrt{s(s+1)}(h/2\pi) = \sqrt{3/4}(h/2\pi)$

$S_z = m_s h/2\pi = \pm(1/2)h/2\pi$

$\cos\alpha = S_z/S = m_s/\sqrt{s(s+1)}$

Figure 30-5 gives the possible quantum numbers for a 28 electron atom in the ground state.

Figure 30-5

| n | $\ell$ | m | $m_s$ | Comments | |
|---|---|---|---|---|---|
| 1 | 0 | 0 | ±1/2 | $\ell$=0 subshell | n=1 shell 2 electrons |
| 2 | 0 | 0 | ±1/2 | $\ell$=0 subshell | n=2 shell 8 electrons |
| 2 | 1 | +1 | ±1/2 | $\ell$=1 subshell | |
| 2 | 1 | 0 | ±1/2 | | |
| 2 | 1 | -1 | ±1/2 | | |
| 3 | 0 | 0 | ±1/2 | $\ell$=0 subshell | n=3 shell 18 electrons |
| 3 | 1 | +1 | ±1/2 | $\ell$=1 subshell | |
| 3 | 1 | 0 | ±1/2 | | |
| 3 | 1 | -1 | ±1/2 | | |
| 3 | 2 | +2 | ±1/2 | $\ell$=2 subshell | |
| 3 | 2 | +1 | ±1/2 | | |
| 3 | 2 | 0 | ±1/2 | | |
| 3 | 2 | -1 | ±1/2 | | |
| 3 | 2 | -2 | ±1/2 | | |

Practice: Determine the following for a neon atom with 10 electrons:

| | |
|---|---|
| 1. Table showing ground-state quantum number | n ℓ m $m_s$<br>1 0 0 ±1/2<br>2 0 0 ±1/2<br>2 1 +1 ±1/2<br>2 1 0 ±1/2<br>2 1 -1 ±1/2 |
| 2. Number of shells occupied by the electrons | Since n = 2, the electrons occupy two shells. |
| 3. Number of possible spin orientations for each electron | Since $m_s$ = ±1/2, each electron has two possible spin orientations. |
| 4. Number of possible orientations for an orbit with $\ell$ = 1 | If $\ell$ = 1, then m = +1,0,-1 and the orbit has three possible orientations. |

| | |
|---|---|
| 5. Which electrons are, on average, nearer the nucleus | Electrons with $n = 1$ are, on average, nearer the nucleus |
| 6. Which electrons are in less eccentric orbits | Electrons with $\ell = 1$ have greater angular momentum and hence less eccentric orbits about the nucleus. |
| 7. Angle between z axis and orbital angular momentum vector if $\ell = 1$ and $m = \pm 1$ | $\cos\theta = \frac{L_z}{L} = \frac{m(h/2\pi)}{\sqrt{\ell(\ell + 1)}(h/2\pi)} = 0.707$ <br> $\theta = \cos^{-1}(0.707) = 45.0°$ |
| 8. Angle between z axis and spin angular momentum vector is $m_s = 1/2$ | $\cos\alpha = \frac{S_z}{S} = \frac{m_s(h/2\pi)}{\sqrt{s(s + 1)}(h/2\pi)} = 0.577$ <br> $\alpha = \cos^{-1}(0.577) = 54.7°$ |

Example 30.2. (a) Show that, for a particle moving in a one-dimensional box between $x = 0$ and $x = w$, the wave function is $\psi = A \sin(n\pi x/w)$, $n = 1,2,3,\ldots$. (b) Obtain an expression for the wavelength, linear momentum, and kinetic energy of the particle.

Given: A particle moving in a one-dimensional box between $x = 0$ and $x = w$ and the wave function $\psi = A \sin(n\pi x/w)$.

Determine: (a) That the wave function $\psi = A \sin(n\pi x/w)$ is appropriate. (b) An expression for the wavelength, linear momentum, and kinetic energy of the particle.

Strategy: (a) Since the particle is confined to the box, the wave function that describes it must vanish at $x = 0$ and $x = w$. We can insert these values for x into the wave function to see if it vanishes at $x = 0$ and $x = w$. If it does, it is a suitable wave function. (b) Assuming that only those wavelengths associated with standing waves in the box are allowed, we can determine the possible wavelengths of the associated matter waves. Knowing the wavelength, we can determine the linear momentum and kinetic energy.

Solution: (a) Since the particle is confined to the box, the wavefunction that describes it must vanish at the limits of the box (i.e., at $x = 0$ and $x = w$).

$$\psi = A \sin(n\pi x/w)$$
$$\text{At } x = 0 \quad \psi = A \sin 0 = 0$$
$$\text{At } x = w \quad \psi = A \sin(n\pi x/w) = A \sin(n\pi) = 0$$

Since this wave function vanishes at $x = 0$ and $x = w$, it can be used to describe a particle in the box.

(b) The matter wave associated with the particle can fit into the box as shown in Fig. 30-6.

Figure 30-6

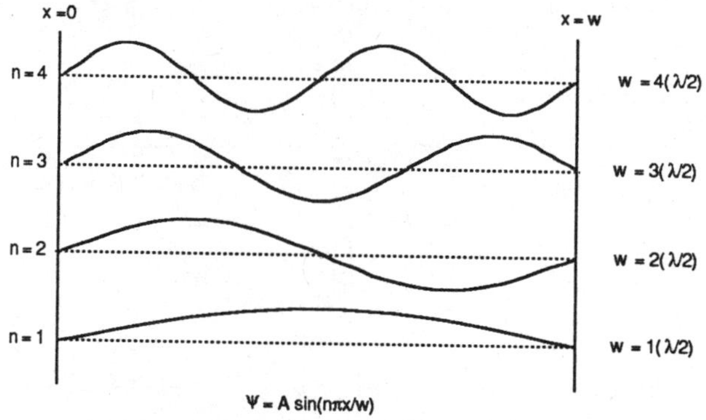


The only wavelengths allowed are those for which an integral number of half wavelengths will fit into the box. This can be expressed algebraically as

$$w = n(\lambda/2) \qquad n = 1,2,3,...$$

Hence an expression for the excepted wavelengths is

$$\lambda = 2w/n \qquad n = 1,2,3,...$$

The de Broglie equation is used to determine an expression for the momentum of the particle:

$$p = h/\lambda = nh/2w \qquad n = 1,2,3,...$$

The kinetic energy of the particle is

$$KE = mv^2/2 = m^2v^2/2m = p^2/2m = n^2h^2/8mw^2 \qquad n = 1,2,3,...$$

Related Text Problems: 30-10 through 30-15.

## 3. The Heisenberg Uncertainty Principle (Section 30.4)

Review: Heisenberg determined that there are limits on the precision that can be obtained when we simultaneously measure momentum and position (both linear and angular) and energy and time. These statements of uncertainty are stated algebraically as follows:

Linear position and momentum

$$\Delta x \Delta p_x \geq (h/2\pi)/2 \qquad \Delta y \Delta p_y \geq (h/2\pi)/2 \qquad \Delta z \Delta p_z \geq (h/2\pi)/2$$

Angular position and momentum

$$\Delta\theta \Delta L \geq (h/2\pi)/2$$

Energy and time

$$\Delta E \Delta t \geq (h/2\pi)/2$$

Practice: Electrons are incident upon a single slit as shown in Fig. 30-7.

Figure 30-7

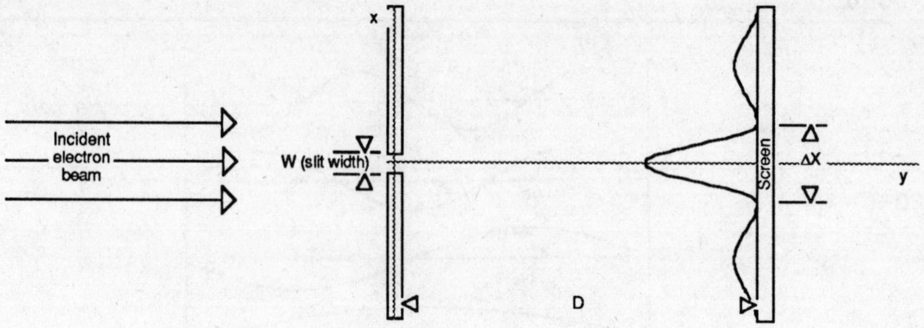


$v_y = 1.00 \times 10^6$ m/s = speed of electrons in y direction; electrons have no initial speed in x direction

$w = 1.00 \times 10^{-6}$ m = slit width

d = 2.00 m = distance between slit and screen

Determine the following:

| | |
|---|---|
| 1. Momentum of electrons in y direction | $p_y = mv_y = 9.11 \times 10^{-25}$ kg·m/s |
| 2. Wavelength of matter waves associated with electrons | $\lambda = h/p_y = 7.28 \times 10^{-10}$ m |
| 3. Location of first minimum in single-slit pattern for this wavelength | $m\lambda = W \sin\theta$<br>$\sin\theta \approx \tan\theta = x/d \quad m = 1$<br>$X_1 = d \sin\theta_1 = d\lambda/w = 1.46 \times 10^{-3}$ m |
| 4. Width of central maximum using single-slit theory | $\Delta X = 2X_1 = 2.92 \times 10^{-3}$ m |
| 5. Uncertainty in x position for electrons going through slit | $\Delta x \leq w = 1.00 \times 10^{-6}$ m |
| 6. Uncertainty in x component of momentum of electrons going through slit | $\Delta p_x \geq [(h/2\pi)/2]/\Delta x$<br>$\Delta p_x \geq 5.28 \times 10^{-29}$ kg·m/s |
| 7. Momentum of electrons in x direction after they go through slit | $p_x = \Delta p_x \geq 5.28 \times 10^{-29}$ kg·m/s |

| | | |
|---|---|---|
| 8. | Speed of electrons in x direction after they go through the slit | $v_x = p_x/m \geq 58.0$ m/s |
| 9. | Time it takes electrons to go from slit to screen | $t = d/v_y = 2.00 \times 10^{-6}$ s |
| 10. | Maximum displacement of electrons in x direction | $X = v_x t \geq 1.16 \times 10^{-4}$ m |
| 11. | Width of central maximum according to uncertainty principle | $\Delta X = 2X \geq 2.32 \times 10^{-4}$ m |

Note: As a consequence of the limit on the simultaneous determination of position and momentum of the electrons, when the position is uncertain by an amount less than or equal to $1.00 \times 10^{-6}$ m (step 5), the momentum is uncertain by an amount greater than or equal to $5.28 \times 10^{-28}$ kg·m/s (step 7), and the central maximum on a screen 2.00 m away must be at least $2.32 \times 10^{-4}$ m wide. According to single-slit diffraction theory, the width of the central maximum is $2.92 \times 10^{-3}$ m (step 11) which is consistent with the uncertainty principle. No ingenious subtlely in experiment design can remove this basic uncertainty.

Example 30.3. An electron in an excited state has a lifetime of $1.00 \times 10^{-6}$ s. What is the minimum uncertainty in measuring the energy of this excited state? What is the minimum uncertainty in the frequency of the observed spectral line?

Given: The minimum value for $\Delta E$ and $\Delta f$

Determine: $\Delta E$ = minimum uncertainty in energy of this excited state; $\Delta f$ = minimum uncertainty in frequency of the observed spectral line

Strategy: Knowing the lifetime, we can determine the maximum value for $\Delta t$. Knowing the maximum value for $\Delta t$, we can determine the minimum value for $\Delta E$. Knowing the minimum value for $\Delta E$, we can determine the minimum value for $\Delta f$.

Solution: Since the lifetime of the excited state is $1.00 \times 10^{-6}$ s, the uncertainty of this value can be no greater than the value itself. Hence, $\Delta t$ can be no larger than $1.00 \times 10^{-6}$ s. This allows us to write $\Delta t \leq 1.00 \times 10^{-6}$ s.

The minimum uncertainty in the energy is then

$$\Delta E \geq [(h/2\pi)/2]/\Delta t = 5.28 \times 10^{-29} \text{ J}$$

The minimum value for $\Delta f$ is determined from $E = hf$ or $\Delta E = h\Delta f$; hence

$$\Delta f = \Delta E/h = 7.96 \times 10^4/m$$

Related Text Problems: 30-16 through 30-26.

---

PRACTICE TEST

Take and grade this practice test. Doing so will allow you to determine any weak spots in your understanding of the concepts taught in this chapter. The following section prescribes what you should study further to strengthen your understanding.

Determine the de Broglie wavelength of the following:

______ 1. A $2.00 \times 10^3$-kg vehicle moving at 10 m/s
______ 2. A 5.00-keV proton
______ 3. A 5.00-keV photon
______ 4. An electron accelerated through a $1.00 \times 10^4$ V potential difference

Electrons that have been accelerated from rest through a 80.0-V potential difference are incident normal to a crystaline surface. First-order constructive interference of the reflected matter waves occurs at an angle of 70.0° with respect to the incident beam. Determine the following:

______ 5. Final kinetic energy of the electrons in joules
______ 6. Final momentum of the electrons
______ 7. Wavelength of the associated matter waves
______ 8. Angle between the incident beam and the reflecting Bragg planes
______ 9. Spacing between the reflecting Bragg planes
______ 10. Kinetic energy of a photon with the same wavelength
______ 11. Kinetic energy of a neutron with the same wavelength
______ 12. Voltage through which you would have to accelerate a proton in order for it to have the same wavelength

Determine the following for a neutral sulfur atom (16 electrons) in the ground state:

______ 13. The number of shells occupied by the electrons
______ 14. The number of subshells occupied by electrons with $n = 3$
______ 15. The number of possible orientations for the orbit of the 16th electron
______ 16. The number of possible spin orientations for the 16th electron
______ 17. The angle between the z axis and the orbital angular momentum vector of $\ell = 1$ and $m = 0$

The position of a photon is known to within $1.00 \times 10^{-10}$ m. Determine the following:

______ 18. Minimum uncertainty in its momentum
______ 19. Minimum uncertainty in its energy

__________ 20. The minimum amount of time that must be taken to make the position measurement

(See Appendix I for answers.)

---

PRINCIPAL CONCEPTS AND EQUATIONS PRESCRIPTION

Your score on the practice test is an excellent measure of your understanding of this chapter. You should now use the following chart to write your own prescription for curing any of your physics ills. Look down the leftmost column to the number of the question(s) you answered incorrectly, read across that row to see which section(s) of the study guide you should return to for further study, and then do the suggested text problems to gain additional experience in working with the particular concept.

| Question Number | Concepts and Equations | Prescription: Principal Concept | Prescription: Text Problems |
|---|---|---|---|
| 1 | de Broglie wavelength: $\lambda = h/mv$ | 1 | 30-1,5 |
| 2 | de Broglie wavelength: $\lambda = hc/E$ | 1 | 30-6,7 |
| 3 | de Broglie wavelength: $\lambda = h/\sqrt{2mKE}$ | 1 | 30-3 |
| 4 | de Broglie wavelength: $\lambda = h/\sqrt{2meV}$ | 1 | 30-2a,4 |
| 5 | Electric potential difference: $\Delta KE = -\Delta PE = qV$ | 4 of Ch. 18 | 30-2a,4 |
| 6 | Momentum and kinetic energy: $KE = mv^2/2 = p^2/2m$ | 3 of Ch. 7 | 30-2b |
| 7 | de Broglie wavelength: $\lambda = h/p$ | 1 | 30-2c,3b |
| 8 | Reflection: $\theta_i = \theta_r$ | 2 of Ch. 24 | 30-8 |
| 9 | Bragg equation: $m\lambda = 2d \sin\theta$ | 3 of Ch. 28 | 30-9 |
| 10 | de Broglie wavelength: $\lambda = hc/E$ | 1 | 30-3a |
| 11 | de Broglie wavelength: $\lambda = h/\sqrt{2mKE}$ | 1 | 30-6,7 |
| 12 | de Broglie wavelength $\lambda = h/\sqrt{2meV}$ | 1 | 30-2,4 |
| 13 | Quantum number: $n$ | 2 | 30-10,11 |
| 14 | Quantum number: $\ell$ | 2 | 30-11,12 |
| 15 | Quantum number: $m$ | 2 | 30-10,12 |
| 16 | Quantum number: $m_s$ | 2 | 30-10,11 |
| 17 | Orbital angular momentum: $L = \sqrt{\ell(\ell + 1)}\, h/2\pi$ | 2 | -- |
| 18 | Uncertainty principle: $\Delta x \Delta p_x \geq (h/2\pi)/2$ | 2 | 39-17,18a |
| 19 | Energy of a photon: $\Delta E = c\Delta p$ | 3 | 30-21,23 |
| 20 | Uncertainty principle: $\Delta E \Delta t \geq (h/2\pi)/2$ | 3 | 30-21,23 |

# 31 The Nucleus

## RECALL FROM PREVIOUS CHAPTERS

Your success with this chapter will be affected by your understanding of previous chapters. However, you should be able to proceed with this chapter without reviewing any previous concepts and equations.

## NEW IDEAS IN THIS CHAPTER

| Concepts and equations introduced | Text Section | Study Guide Page |
|---|---|---|
| Notation: ${}^{A}_{Z}X$ | 31.2 | 506 |
| Alpha decay: ${}^{A}_{Z}P \rightarrow {}^{A-4}_{Z-2}D + {}^{4}_{2}He$ | 31.2 | 506 |
| Beta decay: ${}^{A}_{Z}P \rightarrow {}^{A}_{Z+1}D + {}^{0}_{-1}e$ | 31.2 | 506 |
| Gamma decay: ${}^{A}_{Z}P \rightarrow {}^{A}_{Z}P + \gamma$ | 31.2 | 506 |
| Mass difference: $\Delta M = Zm_H + (A - Z)m_n - M$ | 31.3 | 509 |
| Binding energy: $BE = \Delta Mc^2$ | 31.3 | 509 |
| Radioactive decay equation: $N = N_o e^{-\lambda t} = N_o e^{-t/\tau}$ | 31.5 | 514 |
| Half-life: $T = \ell n2/\lambda = \tau\ \ell n2$ | 31.5 | 514 |
| Activity: $A = \Delta N/\Delta t = \lambda N = \lambda N_o e^{-\lambda t} = A_o e^{-\lambda t}$ | 31.5 | 514 |

### 1. Nuclear Events (Section 31.2)

Review: The notation used in nuclear physics is:

$${}^{A}_{Z}X$$

where X represents the symbol for an element, Z is the charge on the nucleus, and A (the atomic mass number) is the number of mass units in the nucleus. Since each proton and each neutron contribute one mass unit apiece, A is equal to the number of protons plus the number of neutrons in the nucleus.

The radioactive particles we will study are the following:

1. Alpha particle. An alpha particle is just like the nucleus of a He atom. It consists of two protons and two neutrons, and hence Z = 2 and A = 4. The symbol for an alpha particle is ${}^{4}_{2}He$.

2. Beta particle. A beta particle is just like an electron. Because the charge on an electron is minus one electronic charge, Z = -1. Because the mass of an electron is essentially zero, A = 0. The symbol for a beta particle is ${}^{0}_{-1}e$.

3. Gamma ray. A gamma ray is just like any other photon or any other electromagnetic wave. Because it has no charge or mass, the symbol for a gamma ray is $\gamma$.

Note: If an alpha particle is just like the nucleus of a helium atom, a beta particle just like an electron, and a gamma ray just like any other photon or electromagnetic wave, why do we give them special names? You cannot, for example, tell a beta particle from an electron. However, the label "beta particle" tells us that a given electron originated from a nuclear event rather than being an atomic (outside the nucleus) electron. In like manner, the label "alpha particle" allows us to distinguish between a helium atom stripped of its electrons and something that looks and acts just like it but originated from a nuclear event. The same is true of a gamma ray, as opposed to any x-ray, photon, or other electromagnetic wave.

In the process of looking at nuclear events and predicting results, we must conserve nuclear charge and mass.

When a parent nucleus emits an alpha particle, the daughter nucleus has two fewer protons and a mass number that is four less:

$${}^{A}_{Z}P \rightarrow {}^{A-4}_{Z-2}D + {}^{4}_{2}He$$

When a parent nucleus emits a beta particle, the daughter nucleus has one more proton and the same mass number:

$${}^{A}_{Z}P \rightarrow {}^{A}_{Z+1}D + {}^{0}_{-1}e$$

During beta emission, we believe that the following takes place in the nucleus

$${}^{1}_{0}n \rightarrow {}^{1}_{1}p + {}^{0}_{-1}e$$

and the beta particle is emitted.

If the parent nucleus is in an excited state, it can dissipate its excitation energy by emitting a gamma ray:

$$ {}^{A}_{Z}P \rightarrow {}^{A}_{Z}P + \gamma $$

Practice: Determine the following:

| | |
|---|---|
| 1. ${}^{14}_{6}C \rightarrow \square + {}^{0}_{-1}e$ | Conserving charge and mass, we see that the unknown nucleus must have Z = 7 and A = 14. The nucleus is ${}^{14}_{7}N$. |
| 2. ${}^{60}_{27}Co \rightarrow \square + \gamma$ | ${}^{60}_{27}Co$ |
| 3. ${}^{210}_{83}Bi \rightarrow \square + {}^{4}_{2}He$ | ${}^{206}_{81}T\ell$ |
| 4. ${}^{206}_{81}T\ell \rightarrow {}^{206}_{82}Pb + \square$ | ${}^{0}_{-1}e$ |
| 5. ${}^{4}_{2}He + {}^{14}_{7}N \rightarrow {}^{17}_{8}O + \square$ | ${}^{1}_{1}H$ |
| 6. ${}^{214}_{84}Po \rightarrow {}^{210}_{82}Pb + \square$ | ${}^{4}_{2}He$ |
| 7. ${}^{226}_{88}Ra \rightarrow {}^{222}_{86}Rm + \square$ | ${}^{4}_{2}He$ |
| 8. ${}^{29}_{12}Mg \rightarrow {}^{0}_{-1}e + \square$ | ${}^{29}_{13}A\ell$ |
| 9. ${}^{47}_{21}Sc \rightarrow {}^{47}_{21}Sc + \square$ | $\gamma$ |
| 10. ${}^{236}_{92}U \rightarrow {}^{131}_{53}I + 3{}^{1}_{0}n + \square$ | ${}^{102}_{39}Y$ |
| 11. ${}^{4}_{2}He + {}^{9}_{4}Be \rightarrow {}^{12}_{6}C + \square$ | ${}^{1}_{0}n$ |

| | |
|---|---|
| 12. $^{10}_{5}B$ + [ ] $\rightarrow$ $^{7}_{3}Li$ + $^{4}_{2}He$ | $^{1}_{0}n$ |

Example 31.1. Thorium-229 undergoes alpha decay, and its daughter nucleus undergoes beta decay. What are the daughter and grandaughter nuclei?

Given: $^{229}_{90}Th$ - the parent nucleus
The parent decays by alpha emission.
The daughter decays by beta emission.

Determine: The daughter and the granddaughter nuclei

Strategy: By conserving nuclear charge and mass, we can determine the daughter and granddaughter nuclei.

Solution:

$$^{229}_{90}Th \rightarrow {}^{4}_{2}He + {}^{225}_{88}Ra$$

$$^{225}_{88}Ra \rightarrow {}^{0}_{-1}e + {}^{225}_{89}Ac$$

The daughter nucleus is $^{225}_{88}Ra$, and the granddaughter nucleus is $^{225}_{89}Ac$.

Related Problems: 31-1, 31-4, 31-5, 31-7, 31-8.

## 2. Binding Energy (Section 31.3)

Review: If a particular nucleus consists of Z protons and N neutrons, we find that its composite mass is less than the sum of the mass of its components. This mass difference is given by

$$\Delta M = Zm_p + Nm_n - M$$

where

$Zm_p$ is the mass of all the protons
$Nm_n$ is the mass of all the neutrons
M is the mass of the composite nucleus
$\Delta M$ is the difference between the mass of all the protons plus all the neutrons and the mass of the composite nucleus

Since the mass number A represents the total number of mass units (protons plus neutrons) in the nucleus, we can write

$$N + Z = A \quad \text{or} \quad N = A - Z$$

This allows us to rewrite the mass difference as

$$\Delta M = Zm_p + (A - Z)m_n - M$$

We can determine the amount of energy that would be released when these Z protons and A - Z neutrons come together by using Einstein's statement of mass-energy equivalence:

$$E = \Delta mc^2$$

Or, looking at it another way, we have determined the amount of energy required to break the composite nucleus up into protons and neutrons. That is, we have determined that the binding energy of the nucleus is

$$BE = (\Delta M)c^2 = [Zm_p + (A - Z)m_n - M]c^2$$

If we take the expression for ΔM and add and then subtract the mass of Z electrons (so that we retain the equality), we have

$$\Delta M = Zm_p + (A - Z)m_n - M + Zm_e - Zm_e$$

Rearranging, we obtain

$$\Delta M = Z(m_p + m_e) + (A - Z)m_n - (M + Zm_e)$$

Now note that

$m_p + m_e$ = atomic mass of a hydrogen atom = $m_H$

All hydrogen atoms have one proton; however, they may have zero, one, or two neutrons ($^1_1H$, $^2_1H$, $^3_1H$). In this case, we are interested in the mass of the $^1_1H$ atom and we will represent it by $m_H$.

$M + Zm_e$ = mass of the atom containing the nucleus under consideration

As a result of this, we can write ΔM as

$$\Delta M = Zm_H + (A - Z)m_n - M$$

where

$m_H$ = mass of a hydrogen atom
$m_n$ = mass of a neutron
$M$ = mass of the entire atom (not just the nucleus)

The expression for the binding energy then becomes

$$BE = \Delta Mc^2 = [Zm_H + (A - Z)m_n - M]c^2$$

and the binding energy per nucleon is

$$BE/A = [Zm_H + (A-Z)m_n - M]c^2/A$$

In working problems, it will prove convenient to realize that one atomic mass unit (1 u) has an energy equivalence of 931.5 MeV.

$$E = mc^2 = (1\ u)c^2$$
$$= (1.660566 \times 10^{-27}\ kg)(299{,}792{,}458\ m/s)^2$$
$$= 1.4924 \times 10^{-19}\ \frac{kg \cdot m^2}{s^2}\left(\frac{N}{\frac{kg \cdot m}{s^2}}\right)\left(\frac{J}{N \cdot m}\right)\left(\frac{eV}{1.6019 \times 10^{-19}\ J}\right)\left(\frac{MeV}{10^6\ eV}\right)$$
$$= 931.5\ MeV$$

<u>Note</u>: When doing nuclear calculations, it is frequently necessary to use more than three significant figures. For example, if we use only three significant figures, a proton and a neutron have the same mass. When available, we will record all information to the sixth decimal place, do the calculations, and then round off to three significant figures, unless the rounding off will obscure some of the nuclear information.

<u>Practice</u>: Given the following masses:

$m_e$ = 0.000548 u = mass of electron
$m_n$ = 1.008665 u = mass of neutron
$m_p$ = 1.007277 u = mass of proton
$m_H$ = 1.007825 u = mass of a neutral $^1_1H$ atom
$m_{He}$ = 4.002603 u = mass of neutral $^4_2He$ atom
$m_O$ = 15.994915 u = mass of a neutral $^{16}_8O$ atom

Determine the following:

| | |
|---|---|
| 1. Expression for mass difference ($\Delta M$) between individual nucleons and the composite $^4_2He$ nucleus | $\Delta M = Zm_H + (A - Z)m_n - M_{He}$ |

| | |
|---|---|
| 2. Mass difference for the above case | $Zm_H = 2(1.007825\ u) = 2.015650\ u$<br>$(A - Z)m_n = 2(1.008665\ u) = 2.017330\ u$<br>$M_{He} = 4.002603\ u$<br>$\Delta M = Zm_H + (A - Z)m_n - M_{He}$<br>$= 0.030377\ u$ |
| 3. Energy given off when two neutrons and two protons are brought together to form a $^4_2He$ nucleus | $E = \Delta mc^2$<br>$= (0.030377\ u)(931.5\ MeV/u)$<br>$= 28.30\ MeV$ |
| 4. Energy required to break the $^4_2He$ nucleus up into individual nucleons. That is, the binding energy for $^4_2He$. | $BE = \Delta Mc^2 = 28.30\ MeV$ |
| 5. Binding energy per nucleon for $^4_2He$ | $BE = 28.30\ MeV$<br>$A = 4$ nucleons<br>$BE/A = 7.08\ MeV$ |
| 6. Expression for mass difference between the individual nucleons and composite $^{16}_8O$ nucleus | $\Delta M = Zm_H + (A - Z)m_n - M_O$<br>where $Z = 8$ and $A = 16$ |
| 7. Mass difference for the above case | $Zm_H = 8m_H = 8.062600\ u$<br>$(A - Z)m_n = 8m_n = 8.069320\ u$<br>$M_O = 15.994915\ u$<br>$\Delta M = 0.137005\ u$ |
| 8. Binding energy for $^{16}_8O$ nucleus | $BE = \Delta Mc^2$<br>$= (0.137005\ u)(931.5\ MeV/u)$<br>$= 127.6\ MeV$ |

| | |
|---|---|
| 9. Binding energy per nucleon for $^{16}_{8}O$ | BE = 127.6 MeV, A = 16<br>BE/A = 7.98 MeV |

Example 31.2. The binding energy per nucleon for $^{238}_{92}U$ is 7.570 MeV. Determine the total binding energy for this nucleus and the mass of the nucleus in atomic mass units.

Given: BE/A = 7.570 MeV for $^{238}_{92}U$

Determine: The total binding energy for the $^{238}_{92}U$ nucleus and the mass of this nucleus.

Strategy: Knowing that the nucleus of interest is $^{238}_{92}U$, we can determine the number of nucleons involved. Knowing the binding energy per nucleon and the number of nucleons, we can determine the total binding energy. Knowing the total binding energy and the mass-energy equivalence, we can determine $\Delta M$ for this atom. Knowing $\Delta M$, Z, A, $m_H$, $m_n$ and $m_e$, we can determine the mass of the nucleus.

Solution: The nucleus has 238 nucleons (i.e., A = 238). Hence the total binding energy is

$$BE = (BE/A)A = \left(\frac{7.570 \text{ MeV}}{\text{nucleon}}\right)(238 \text{ nucleons}) = 1.802 \times 10^3 \text{ MeV}$$

The mass difference is determined by using the mass-energy equivalence:

$$\Delta M = (1802 \text{ MeV})(u/931.5 \text{ MeV}) = 1.935 \text{ u}$$

The mass of a neutral $^{238}_{92}U$ atom is determined by

$$\Delta M = Zm_H + (A - Z)m_n - M_U$$

or

$$M_U = (92)(1.007825 \text{ u}) + (238 - 92)(1.008665 \text{ u}) - 1.935 \text{ u} = 238.050 \text{ u}$$

The mass of the nucleus may be determined by subtracting the mass of 92 electrons

$$(M_U)_{nucleus} = (M_U)_{atom} - 92\ m_e = 238.050 \text{ u} - 92(0.000548 \text{ u}) = 238.000 \text{ u}$$

Example 31.3. Which nucleus is more stable, $^{14}_{6}C$ (14.003242 u) or $^{14}_{7}N$ (14.003074 u)?

Given: $^{14}_{6}C$ (14.003242 u) and $^{14}_{7}N$ (14.003074 u)

Determine: Which of these two nuclei is more stable?

Strategy: We can determine the binding energy per nucleon for each nucleus. The nucleus with the larger binding energy per nucleon is more stable.

Solution: First let's obtain $\Delta M$, BE, and BE/A for each nucleus.

$^{14}_{6}C$ $\quad \Delta M = 6m_H + 8m_n - M_C$

$$= 6.046950 \text{ u} + 8.069320 \text{ u} - 14.003242 \text{ u} = 0.113028 \text{ u}$$

$$BE = (0.113028 \text{ u})(931.5 \text{ MeV/u}) = 105.3 \text{ MeV}$$
$$BE/A = 105.3 \text{ MeV}/14 = 7.521 \text{ MeV}$$

$^{14}_{7}N$ $\quad \Delta M = 7m_H + 7m_n - M_N$

$$\Delta M = 7.054775 \text{ u} + 7.060655 \text{ u} - 14.003074 \text{ u} = 0.112356 \text{ u}$$

$$BE = (0.112356 \text{ u})(931.5 \text{ MeV/u}) = 104.7 \text{ MeV}$$
$$BE/A = 104.7 \text{ MeV}/14 = 7.479 \text{ MeV}$$

Since the binding energy per nucleon is greater for $^{14}_{6}C$, it is the more stable of the two nuclei.

Related Text Problems: 31-9 through 31-12 and 31-14 through 31-16.

## 3. Radioactive Decay (Section 31.5)

Review: The equation that describes radioactive decay is

$$N = N_o e^{-\lambda t} = N_o e^{-t/\tau}$$

where $N$ = number of radioactive nuclei present at any time $t$
$N_o$ = number of radioactive nuclei present at time $t = 0$
$\lambda$ = the decay constant = probability per unit time that a particular nucleus will decay
$t$ = time
$\tau$ = time constant $\tau = 1/\lambda$

The half-life of a radioactive sample is the time it takes for one half of the nuclei to decay.

If at $t = 0$ we have $N = N_o$, then after one half-life we have

$$t = T \qquad N = N_o/2$$

That is, the number of radioactive nuclei left after a time equal to one half-life is just one half of the original number. As shown below when this information is inserted into the radioactive decay equation, we obtain a relationship between T and $\lambda$.

$$N = N_o e^{-\lambda t}$$

Insert $N = N_o/2$ at $t = T$ to obtain

$$N_o/2 = N_o e^{-\lambda T}$$

which may be rearranged to obtain

$$2 = e^{\lambda T}$$

Taking the natural log of this equation, obtain

$$\ell n2 = \lambda T$$

$$T = \ell n2/\lambda = \tau \ell n2$$

Using this expression for T in terms of $\lambda$, we see that after one half-life (i.e., $t = T$), we have

$$N = N_o e^{-\lambda T} = N_o e^{-\lambda \ell n2/\lambda} = N_o e^{-\ell n2} = N_o/2$$

After two half-lives (i.e., $t = 2T$), we have

$$N = N_o e^{-\lambda(2T)} = N_o e^{-\lambda 2\ell n2/\lambda} = N_o e^{-2\ell n2} = N_o(e^{-\ell n2})^2 = N_o/2^2 = N_o/4$$

After n half-lives (i.e., $t = nT$), we have

$$N = N_o e^{-\lambda(nT)} = N_o e^{-\lambda n\ell n2/\lambda} = N_o e^{-n\ell n2} = N_o(e^{-\ell n2})^n = N_o/2^n$$

Notice that after seven half-lives,

$$N = N_o/2^7 = N_o/128 = 0.78\%\ N_o$$

we have less than 1% of the original number of radioactive nuclei left. Consequently, we say that the exponential decay process is essentially complete after seven half-lives.

The activity of a radioactive sample is the rate at which the nuclei are decaying:

$$A = \Delta N/\Delta t$$

The activity may also be expressed as the probability that any nucleus will decay in the next time interval ($\lambda$) times the number of radioactive nuclei present (N):

$$A = \lambda N = \lambda N_o e^{-\lambda t} = A_o e^{-\lambda t}$$

where $A_o = \lambda N_o$ is just the activity at $t = 0$.

Practice: A sample of radioactive material shows a measured activity of $5.00 \times 10^3$ disintegrations per minute when first measured and $3.00 \times 10^3$ disintegrations per minute when measured 1 h later.

Determine the following:

| | |
|---|---|
| 1. Activity at t = 0 | $A_o = 5.00 \times 10^3$ disintegrations/min |
| 2. Activity after t = 3600 s | $A = 3.00 \times 10^3$ disintegrations/min |
| 3. Decay constant | $A = A_o e^{-\lambda t}$<br>$\frac{3.00 \times 10^3}{\text{min}} = \frac{5.00 \times 10^3}{\text{min}} e^{-\lambda(3.60 \times 10^3\ s)}$<br>$0.600 = e^{-\lambda(3.60 \times 103\ s)}$<br>$1.67 = e^{\lambda(3.60 \times 103\ s)}$<br>$\ell n(1.67) = \lambda(3.60 \times 10^3\ s)$<br>$\lambda = \ell n(1.67)/3.60 \times 10^3\ s$<br>$= 1.42 \times 10^{-4}\ s$ |
| 4. Half-life | $T = \ell n2/\lambda = 4.88 \times 10^3\ s$ |
| 5. Time constant | $\tau = 1/\lambda = 7.04 \times 10^3\ s$ |
| 6. Number of radioactive nuclei present at t = 0 | $A_o = \lambda N_o$<br>$N_o = A_o/\lambda$<br>$= (5.00 \times 10^3\ \text{dis/min})/(1.42 \times 10^{-4}\ s)$<br>$= (3.52 \times 10^7\ \text{s/min})(\text{min}/60.0\ s)$<br>$= 5.87 \times 10^5$ |
| 7. Activity after three half-lives | $A = A_o e^{-\lambda t}$; when $t = 3T = 3\ell n2/\lambda$<br>$A = A_o e^{-\lambda(3\ell n2/\lambda)} = A_o e^{-3\ell n2}$<br>$= A_o/(2)^3 = A_o/8 = 625$ dis/min |
| 8. Activity after 2 h | $A = A_o e^{-\lambda t}$<br>$\lambda t = (1.42 \times 10^{-4}/s)(7.20 \times 10^3\ s) = 1.02$<br>$A = (5.00 \times 10^3\ \text{dis/min})e^{-1.02}$<br>$= (5.00 \times 10^3\ \text{dis/min})(0.361)$<br>$= 1.81 \times 10^3$ dis/min |

| | |
|---|---|
| 9. Number of radioactive nuclei left after 2 h | $A = \lambda N$<br>$N = A/\lambda = (1.81 \times 10^3/\text{min})/(1.42 \times 10^{-4}/\text{s})$<br>$= (1.27 \times 10^7\ \text{s/min})(\text{min}/60.0\ \text{s})$<br>$= 2.12 \times 10^5$<br>Also<br>$N = N_o e^{-\lambda t}$<br>$\lambda t = 1.02$ (step 8)<br>$N_o = 5.87 \times 10^5$ (step 6)<br>$N = 5.87 \times 10^5\ e^{-1.02} = 2.12 \times 10^5$ |

Example 31.4. A sample of strontium 90 ($T = 28.0$ years) has an activity of $2.50 \times 10^3$ disintegrations per second. (a) What is the initial mass ($m_o$) of the sample? (b) How long will it be until only 10% of the sample is left?

Given: $T = 28.0$ years = half-life for strontium 90
$A_o = 2.50 \times 10^3$ dis/s = initial activity of sample
$m_f = 0.100\ m_o$ = final mass of sample

Determine: (a) Initial mass of the sample and time when the mass of the sample is 10% of the initial mass.

Strategy: Knowing the half-life, we can determine the decay constant. Knowing the decay constant and the initial activity, we can determine the initial number of nuclei and hence the initial mass of the sample. Knowing the initial mass and the final mass, we can determine the time.

Solution:

$$\lambda = \frac{\ell n2}{T} = \frac{2.48 \times 10^{-2}}{1\ \text{year}}\left(\frac{1\ \text{year}}{3.65 \times 10^2\ \text{days}}\right)\left(\frac{1\ \text{day}}{8.64 \times 10^4\ \text{s}}\right) = \frac{7.86 \times 10^{-10}}{\text{s}}$$

$$N_o = \frac{A_o}{\lambda} = \frac{2.50 \times 10^3/\text{s}}{7.86 \times 10^{-10}/\text{s}} = 3.18 \times 10^{12}\ \text{nuclei}$$

If we initially have $N_o$ nuclei present, we have $N_o$ atoms present. The initial number of moles of the sample and the initial mass are

$$n_o = \frac{N_o}{N_A} = \frac{m_o}{M} \qquad m_o = N_o M/N_A$$

where $n_o$ = initial number of moles of sample
$N_o$ = initial number of nuclei and atoms
$N_A$ = Avagadro's number
$m_o$ = initial mass of sample
$M$ = molecular mass of sample

Inserting values, we obtain

$$m_o = \frac{N_o M}{N_A} = \frac{(3.18 \times 10^{12}\ \text{atoms})(87.6\ \text{g/mole})}{(6.02 \times 10^{23}\ \text{atoms/mol})} = 4.63 \times 10^{-10}\ \text{g}$$

Since the mass of the sample is directly proportional to the number of nuclei present, we can write

$$m_f = m_o e^{-\lambda t}$$

We are interested in the value of t when $m_f = 0.100 m_o$

$$0.100 m_o = m_o e^{-\lambda t}$$

$$10.0 = e^{\lambda t}$$

$$\ell n 10.0 = \lambda t$$

$$t = \ell n 10.0/\lambda = 2.30/(7.86 \times 10^{-10}/\text{s}) = 2.93 \times 10^{9}\ \text{s}$$

Related Text Problems: 31-18 through 31-27, 31-29 through 31-32.

---

PRACTICE TEST

Take and grade this practice test. Doing so will allow you to determine any weak spots in your understanding of the concepts taught in this chapter. The following section prescribes what you should study further to strengthen your understanding:

Determine the missing nucleus or particle for each of the following nuclear reactions.

______ 1. $^{40}_{19}K \rightarrow {}^{40}_{20}Ca + \square$

______ 2. $^{214}_{84}Po \rightarrow \square + {}^{4}_{2}He$

______ 3. $^{61}_{28}Ni \rightarrow {}^{61}_{28}Ni + \square$

______ 4. $^{12}_{5}B \rightarrow \square + {}^{0}_{-1}e$

______ 5. $^{10}_{5}B + {}^{1}_{0}n \rightarrow {}^{7}_{3}Li + \square$

______ 6. $^{210}_{84}Po \rightarrow \square + \gamma$

Given the following masses:

$m_e$ = 0.000548 u = mass of electron
$m_n$ = 1.008665 u = mass of neutron
$m_p$ = 1.007277 u = mass of proton
$m_H$ = 1.007825 u -- mass of neutral ${}^1_1H$ atom
$m_{He}$ = 4.002603 u = mass of a nuetral ${}^4_2He$ atom
$m_C$ = 13.003354 u = mass of neutral ${}^{13}_6C$ atom

Determine the following:

__________ 7. The mass difference between four nucleons (two protons and two neutrons) and the ${}^4_2He$ nucleus

__________ 8. The amount of energy given off if you brought two neutrons and two protons together to form a ${}^4_2He$ nucleus

__________ 9. The binding energy per nucleon for ${}^4_2He$

__________ 10. The mass difference for ${}^{13}_6C$

__________ 11. The binding energy for ${}^{13}_6C$

A sample of radioactive material is monitored by a counting system that has an efficiency of 50%. The system records $6.00 \times 10^5$ counts/min when the sample is first placed in the counting chamber and $4.00 \times 10^5$ counts/min after 30.0 min. Determine the following:

__________ 12. Count rate (in $s^{-1}$) at $t = 0$
__________ 13. Activity ($A_o$) at $t = 0$
__________ 14. Activity (A) after 30.0 min
__________ 15. Decay constant for these radioactive nuclei
__________ 16. Half-life for these radioactive nuclei
__________ 17. Number of radioactive nuclei present at $t = 0$
__________ 18. Number of moles of the sample at $t = 0$
__________ 19. Count rate after 4 h
__________ 20. Time at which 10% of the sample is left

(See Appendix I for answers.)

## PRINCIPAL CONCEPTS AND EQUATIONS PRESCRIPTION

Your score on the practice test is an excellent measure of your understanding of this chapter. You should now use the following chart to write your own prescription for curing any of your physics ills. Look down the leftmost column to the number of the question(s) you answered incorrectly, read across that row to see which section(s) of the study guide you should return to for further study, and then do the suggested text problems to gain additional experience in working with the particular concept.

| Question Number | Concepts and Equations | Prescription | |
|---|---|---|---|
| | | Principal Concept | Text Problems |
| 1 | Beta decay: ${}^{A}_{Z}P \rightarrow {}^{A}_{Z+1}D + {}^{0}_{-1}e$ | 1 | 31-1,4 |
| 2 | Alpha decay: ${}^{A}_{Z}P \rightarrow {}^{A-4}_{Z-2}D + {}^{4}_{2}He$ | 1 | 31-5,7 |
| 3 | Gamma decay: ${}^{A}_{Z}P \rightarrow {}^{A}_{Z}D + \gamma$ | 1 | 31-8,9 |
| 4 | Beta decay: ${}^{A}_{Z}P \rightarrow {}^{A}_{Z+1}D + {}^{0}_{-1}e$ | 1 | 31-1,7 |
| 5 | Alpha decay: ${}^{A}_{Z}P \rightarrow {}^{A-4}_{Z-2}D + {}^{4}_{2}He$ | 1 | 31-5,9 |
| 6 | Gamma decay: ${}^{A}_{Z}P \rightarrow {}^{A}_{Z}D + \gamma$ | 1 | 31-8,4 |
| 7 | Mass difference: $\Delta M = Zm_H + (A - Z)m_n - M$ | 2 | 31-9,10 |
| 8 | Binding energy: $BE = \Delta Mc^2$ | 2 | 31-14,15 |
| 9 | Binding energy per nucleon: BE/A | 2 | 31-11,12 |
| 10 | Mass difference $\Delta M = Zm_H + (A-Z)m_n - M$ | 2 | 31-10,15 |
| 11 | Binding energy: $BE = \Delta Mc^2$ | 2 | 31-10,15 |
| 12 | Count rate | 3 | 31-24 |
| 13 | Activity: $A_o = CR_o/\varepsilon$ | 3 | 31-24 |
| 14 | Activity: $A = CR/\varepsilon$ | 3 | 31-24 |
| 15 | Decay constant: $A = A_o e^{-\lambda t}$ | 3 | 31-19a,24b |
| 16 | Half life: $T = \ell n2/\lambda$ | 3 | 31-19b,20a |
| 17 | Activity: $A_o = N_o\lambda$ | 3 | 31-23,24c |
| 18 | Moles of sample: $n = N/N_A$ | 3 | 31-24c,27 |
| 19 | Radioactive decay: $CR = CR_o e^{-\lambda t}$ | 3 | 31-20c,22 |
| 20 | Radioactive decay: $N = N_o e^{-\lambda t}$ | 3 | 31-18,26 |

# 32 Ionizing Radiation, Safety, and Nuclear Medicine

## RECALL FROM PREVIOUS CHAPTERS

| Previously learned concepts and equations frequently used in this chapter | Text Section | Study Guide Page |
|---|---|---|
| Half-life: $T = \ln 2/\lambda$ | 31.5 | 514 |
| Activity: $A = \lambda N$ | 31.5 | 514 |

## NEW IDEAS IN THIS CHAPTER

| Concepts and equations introduced | Text Section | Study Guide Page |
|---|---|---|
| Source activity: $1\ \text{Ci} = 3.70 \times 10^{10}/\text{s} = 3.70 \times 10^{10}\ \text{Bq}$ | 32.4 | 521 |
| Exposure: $1\ \text{R} = 2.082 \times 10^{9}$ ion pairs/cm$^3$ $= 1.61 \times 10^{12}$ ion pairs/g $= 2.58 \times 10^{-4}$ C/kg | 32.4 | 524 |
| Absorbed dose: $1\ \text{rad} = 10^{-2}\ \text{J/kg} = 10^{-2}\ \text{Gy}$ | 32.4 | 525 |
| Effective dose: 1 rem = 1 rad x QF; 1 Sv = 1 Gy x QF, 1 rem = $10^{-2}$ Sv | 32.4 | 527 |
| Equivalent dose rate: dose rate = $(\text{Rhm})(A/r^2)$ | 32.4 | 529 |

## 1. Source Activity (Section 32.4)

Review: The activity of a sample tells us the rate at which its nuclei are decaying. The units of activity are the curie (Ci) and the becquered (Bq).

$$1\ \text{Ci} = 3.70 \times 10^{10}/\text{s}$$

$$1\ \text{Bq} = 1.00/\text{s}$$

The curie and becquerel are related by

$$1\ \text{Ci} = 3.70 \times 10^{10}\ \text{Bq}$$

Practice:

A sample of radium 226 ($T = 1.60 \times 10^3$ years) being counted by a detector that is 30% efficient gives $6.75 \times 10^5$ counts/min.

Determine the following:

| | |
|---|---|
| 1. Activity of sample in counts per second | Count rate $= \frac{6.75 \times 10^5}{\text{min}} \left(\frac{\text{min}}{60.0\ \text{s}}\right)$<br>Count rate $= 1.13 \times 10^4/\text{s}$<br>$A = \frac{\text{count rate}}{\text{detector eff.}} = \frac{1.13 \times 10^4/\text{s}}{0.300}$<br>$= 3.77 \times 10^4/\text{s}$ |
| 2. Activity in becquerel | $A = 3.77 \times 10^4/\text{s}(\text{Bq}/\text{s}^{-1})$<br>$= 3.77 \times 10^4\ \text{Bq}$ |
| 3. Activity in curies and microcuries | $A = 3.77 \times 10^4/\text{s}(\text{Ci}/3.70 \times 10^{10}/\text{s})$<br>$= 1.02 \times 10^{-6}\ \text{Ci}$<br>$A = (1.02 \times 10^{-6}\ \text{Ci})(10^6\ \mu\text{Ci}/\text{Ci})$<br>$= 1.02\ \mu\text{Ci}$ |
| 4. Decay constant | $\lambda = \frac{\ell n2}{T} = \frac{\ell n2}{1.60 \times 10^3\ \text{years}} \left(\frac{1\ \text{year}}{3.15 \times 10^7\ \text{s}}\right)$<br>$= 1.38 \times 10^{-11}/\text{s}$ |
| 5. Number of radioactive nuclei present | $N = \frac{A}{\lambda} = \frac{3.77 \times 10^4/\text{s}}{1.38 \times 10^{-11}/\text{s}} = 2.73 \times 10^{15}$ |
| 6. Number of moles of sample | $n = N/N_A$, $N_A$ = Avagadro's Number<br>$= \frac{2.73 \times 10^{15}\ \text{nuclei}}{6.023 \times 10^{23}\ \text{nuclei/mol}}$<br>$= 4.53 \times 10^{-9}\ \text{mol}$ |
| 7. Atomic mass of radium 226 | $AM = 2.24 \times 10^2\ \text{g/mol}$ |
| 8. Mass of sample | $m = n(AM)$<br>$= (4.53 \times 10^{-9}\ \text{mol})(2.24 \times 10^2\ \text{g/mol})$<br>$= 1.01 \times 10^{-6}\ \text{g} = 1.01\ \mu\text{g}$ |

Example 32.1. Determine the activity in Ci and Bq of $1.00 \times 10^{-3}$ g of tritium (T = 12.3 years).

Given: $m = 1.00 \times 10^{-3}$ g, T = 12.3 years, sample is tritium ($^3_1H$).

Determine: The activity in Ci and Bq.

Strategy: Knowing the mass and identity of the sample, we can determine the number of nuclei. Knowing the half-life of the sample, we can determine the decay constant. Knowing the number of nuclei and the decay constant, we can determine the activity.

Solution: The atomic mass of tritium is

$$AM = 3.02 \text{ g/mol}$$

The number of moles of sample is

$$n = m/AM = (1.00 \times 10^{-3} \text{ g})/(3.02 \text{ g/mol}) = 3.31 \times 10^{-4} \text{ mol}$$

The number of nuclei in the sample is

$$N = nN_A = (3.31 \times 10^{-4} \text{ mol})(6.023 \times 10^{23} \text{ nuclei/mol}) = 1.99 \times 10^{20} \text{ nuclei}$$

The decay constant is

$$\lambda = \frac{\ell n2}{T} = \frac{\ell n2}{12.3 \text{ years}} \left(\frac{1 \text{ year}}{3.15 \times 10^{7} \text{ s}}\right) = 1.79 \times 10^{-9}/\text{s}$$

The activity of the sample is

$$A = \lambda N = (1.79 \times 10^{-9}/\text{s})(1.99 \times 10^{20}) = 3.56 \times 10^{11}/\text{s}$$

The activity in Bq is

$$A = (3.56 \times 10^{11}/\text{s})\ (\text{Bq}/\text{s}^{-1}) = 3.56 \times 10^{11} \text{ Bq}$$

The activity in Ci is

$$A = (3.56 \times 10^{11}/\text{s})\ (\text{Ci}/3.70 \times 10^{10} \text{ s}^{-1}) = 9.62 \text{ Ci}$$

Related Text Problems: 32-1 through 32-5.

## 2. Exposure (Section 32.4)

Review: Two different radioactive samples can have the same activity and still have their respective radiations differ widely in the ability to produce ions. Ion-producing ability gives a measure of exposure. The roentgen (R) is the unit of exposure and is defined as the amount of x-radiation or gamma radiation that produces $2.082 \times 10^9$ ion pairs in one cubic centimeter of dry air. This is equivalent to producing $1.61 \times 10^{12}$ ion pairs per gram of air or an ion charge per unit mass of $2.58 \times 10^{-4}$ C/kg.

$R = 2.082 \times 10^9$ ion pairs/cm$^3$

$$R = 2.082 \times 10^9 \frac{\text{ion pairs}}{\text{cm}^3} \left(\frac{\text{m}^3}{1.29\ \text{kg}}\right)\left(\frac{10^{-3}\ \text{kg}}{\text{g}}\right)\left(\frac{10^6\ \text{cm}^3}{\text{m}^3}\right) = 1.61 \times 10^{12} \frac{\text{ion pairs}}{\text{g}}$$

$$R = 1.61 \times 10^{12} \frac{\text{ion pairs}}{\text{g}} \left(\frac{10^3\ \text{g}}{\text{kg}}\right)\left(\frac{1.60 \times 10^{-19}\ \text{C}}{\text{ion pair}}\right) = 2.58 \times 10^{-4} \frac{\text{C}}{\text{kg}}$$

Practice: An x-ray beam produces $5.00 \times 10^{10}$ ion pairs in 10.0 s in $5.00 \times 10^{-4}$ kg of dry air. ($\rho_{air} = 1.29$ kg/m$^3$)

Determine the following:

| | |
|---|---|
| 1. Exposure in ion pairs per gram | Exposure $= \dfrac{5.00 \times 10^{10}\ \text{ion pairs}}{5.00 \times 10^{-4}\ \text{kg}}$ <br> $= 1.00 \times 10^{14} \dfrac{\text{ion pairs}}{\text{kg}} \left(\dfrac{10^{-3}\ \text{kg}}{\text{g}}\right)$ <br> $= 1.00 \times 10^{11}$ ion pairs/g |
| 2. Exposure in ion pairs per cubic centimeter | Exposure $= \dfrac{1.00 \times 10^{14}\ \text{ion pairs}}{\text{kg}} \left(\dfrac{1.29\ \text{kg}}{\text{m}^3}\right)$ <br> $= \dfrac{1.29 \times 10^{14}\ \text{ion pairs}}{\text{m}^3} \left(\dfrac{\text{m}^3}{10^6\ \text{cm}^3}\right)$ <br> $= 1.29 \times 10^8$ ion pairs/cm$^3$ |
| 3. Exposure in coulombs per kilogram | Exposure $= \dfrac{1.00 \times 10^{14}\ \text{ion pairs}}{\text{kg}} \times$ <br> $\left(\dfrac{1.60 \times 10^{-19}\ \text{C}}{\text{ion pair}}\right) = 1.60 \times 10^{-5}$ C/kg |
| 4. Exposure in roentgen | Exposure $= 1.00 \times 10^{11} \dfrac{\text{ion pair}}{\text{g}} \times$ <br> $\dfrac{\text{R}}{1.61 \times 10^2\ \text{ion pair/g}}$ <br> $= 6.21 \times 10^8$ R |

| | |
|---|---|
| 5. Exposure rate in C/kg·s | $\text{Exposure rate} = \frac{1.60 \times 10^{-5}\ \text{C/kg}}{10.0\ \text{s}}$ <br> $= 1.60 \times 10^{-6}\ \text{C/kg·s}$ |

Example 32.2. If it requires 34.0 eV to produce one ion pair, how much energy per gram is imparted to air when it completely absorbs 1 R of radiation?

Given: It requires 34.0 eV to produce one ion pair.

Determine: The energy per gram imparted to air when it completely absorbs 1 R of radiation.

Strategy: Knowing the energy required per ion pair and the number of ion pairs created in 1 g of air when it absorbs 1 R, we can determine the energy per gram imparted to air when it completely absorbs 1 R of radiation.

Solution: The energy imparted per gram when 1 R of radiation is absorbed is

$$\frac{\text{Energy}}{\text{g}} = \left(\frac{\text{Energy}}{\text{ion pair}}\right)\left(\frac{\text{ion pairs}}{\text{g}}\right) = \left(\frac{34.0\ \text{eV}}{\text{ion pair}}\right)\text{R}$$

$$= \left(\frac{34.0\ \text{eV}}{\text{ion pair}}\right)(\text{R})\left(\frac{1.61 \times 10^{12}\ \text{ion pair/g}}{\text{R}}\right)\left(\frac{1.60 \times 10^{-19}\ \text{J}}{\text{eV}}\right)$$

$$= 8.76 \times 10^{-6}\ \text{J/g}$$

Related Text Problems: 32-13 through 32-16.

## 3. Absorbed Dose (Section 32.4)

Review: The absorbed dose tells us the amount of energy released by the radiation in the absorbing tissue. The units of absorbed energy are rad (radiation absorbed dose) and gray (Gy).

The rad is defined as the amount of radiation necessary to deposit 100 ergs of energy ($1.00 \times 10^{-5}$ J) in 1 g of material:

$$1\ \text{rad} = 100\ \text{erg/g} = 1.00 \times 10^{-2}\ \text{J/kg}$$

The gray is the amount of radiation that deposits 1 J of energy in 1 kg of material

$$1\ \text{Gy} = 1.00\ \text{J/kg}$$

The rad and Gray are related by

$$1\ \text{rad} = 1.00 \times 10^{-2}\ \text{Gy}$$

Practice: In a medical diagnostic procedure, a patient absorbs 8.00 rad in a 3.00-kg body section. The radiation used is 80.0-keV photons.

Determine the following:

| | |
|---|---|
| 1. Radiation absorbed in gray | $(8.00\ \text{rad})\left(\frac{1.00 \times 10^{-2}\ \text{Gy}}{\text{rad}}\right) = 8.00 \times 10^{-2}\ \text{Gy}$ |
| 2. Total energy absorbed | $E_{abs} = (8.00 \times 10^{-2}\ \text{Gy})\left(\frac{1.00\ \text{J/kg}}{\text{Gy}}\right)(3.00\ \text{kg})$ <br> $= 0.240\ \text{J}$ |
| 3. Number of photons absorbed per kilogram of tissue | Let N represent the number of photons. <br> $E_{abs} = E_{photon}\ N$ <br> $N = \frac{E_{abs}}{E_{photon}} = \frac{0.240\ \text{J}}{80.0\ \text{keV}}\left(\frac{\text{keV}}{10^3\ \text{eV}}\right)\left(\frac{\text{eV}}{1.60 \times 10^{-19}\ \text{J}}\right)$ <br> $N = 1.88 \times 10^{13}$ |
| 4. Change in temperature of body section due to radiation absorbed (Assume the specific heat of the body section is the same as that of water.) | $\Delta Q = E_{abs} = (0.240\ \text{J})$ <br> $\Delta Q = mc\Delta T$ ; $c = 4.18 \times 10^3\ \text{J/kg·K}$ <br> $\Delta T = \frac{\Delta Q}{mc} = \frac{0.240\ \text{J}}{(3.00\ \text{kg})(4.18 \times 10^3\ \text{J/kg·K})}$ <br> $\Delta T = 1.91 \times 10^{-5}\ \text{K}$ |

Example 32.3. A technician working at a nuclear reactor facility is exposed to slow neutron radiation and receives a dose of 2.00 rad. How much energy is absorbed by 100 g of the worker's tissue?

Given: The source of radiation is slow neutrons and the dose is 2.00 rad.

Determine: The energy absorbed per 100 g of tissue.

Strategy: Knowing the dose in rad, we can establish the amount of energy deposited per kilogram of tissue. Knowing this energy, we can determine the energy absorbed per 100 g of tissue.

Solution: The amount of energy deposited per kilogram of tissue by this dose is

$$(2.00\ \text{rad})\left(\frac{1.00 \times 10^{-2}\ \text{J/kg}}{\text{rad}}\right) = 2.00 \times 10^{-2}\ \text{J/kg}$$

The energy absorbed per 100 g of tissue is

$$E_{abs} = (2.00 \times 10^{-2}\ J/kg)(1.00 \times 10^{2}\ g)(kg/10^{3}\ g) = 2.00 \times 10^{-3}\ J$$

Related Text Problems: 32-6 through 32-9.

## 4. Effective Dose (Section 32.4)

Review: The effective dose tells us the biological effect of a given absorbed dose. Since different types of radiation lose energy to ionization at different rates along their paths, a quality factor (QF) relates the relative biological effectiveness of various types of radiation. For example, a low-energy gamma ray ($E < 4$ MeV) has a QF of 1 rem/rad, a beta particle with $E < 30$ keV has a QF of 1.7 rem/rad, and an alpha particle has a QF of 10 rem/rad. This says that the alpha particle will do ten times the biological damage of the low-energy gamma ray per centimeter of path length in tissue.

The rem (roentgen equivalent man) tells us the effective dose or dose equivalent of radiation exposure. The rem is equal to the product of the absorbed dose in rads and the quality factor of the radiation involved in the exposure:

$$1\ rem = 1\ rad \times QF$$

The sievert (Sv) is the SI unit of effective dose. It is equal to the effective dose in grays times the quality factor:

$$Sv = Gy \times QF$$

The rem and the Sv are related by

$$1\ rem = 10^{-2}\ Sv$$

Practice: In a medical diagnostic procedure, a patient absorbs 6.00 rad in a 3.00-kg body section. The radiation used is 100-keV photons.

Determine the following:

| | |
|---|---|
| 1. Quality factor for the photons | QF = 1.00 rem/rad = 1.00 Sv/Gy (Table 32-2 of text) |
| 2. Effective dose in rem | $Dose_{eff} = rad \times QF$ <br> $= (6.00\ rad)(1.00\ rem/rad)$ <br> $= 6.00\ rem$ |
| 3. Absorbed dose in gray | $Dose_{abs} = (6.00\ rad)(10^{-2}\ Gy/rad)$ <br> $= 6.00 \times 10^{-2}\ Gy$ |

| | |
|---|---|
| 4. Effective dose in sieverts | $Dose_{eff} = Gy \times QF$<br>$= (6.00 \times 10^{-2}\ Gy)(1.00\ Sv/Gy)$<br>$= 6.00 \times 10^{-2}\ Sv$<br>$Dose_{eff} = (6.00\ rem)(10^{-2}\ Sv/rem)$<br>$= 6.00 \times 10^{-2}\ Sv$ |
| 5. Energy absorbed per gram of tissue by the body section | $(E_{abs}/g) = dose_{abs}$<br>$(E_{abs}/g) = (6.00\ rad)(\frac{10^{-2}\ J/kg}{rad})(\frac{kg}{10^3\ g})$<br>$= 6.00 \times 10^{-5}\ J/g$ |
| 6. Total energy absorbed by body section | $E_{abs\ total} = (E_{abs}/g)(M)$<br>$= (6.00 \times 10^{-5}\ \frac{J}{g})(3.00\ kg)(\frac{10^3\ g}{kg})$<br>$= 1.80 \times 10^{-1}\ J$ |
| 7. Number of photons absorbed by tissue | Let N = number of photons<br>$E_{abs\ total} = N(E/photons)$<br>$N = (\frac{1.8 \times 10^{-1}\ J}{10^2\ keV})(\frac{keV}{10^3\ eV})(\frac{eV}{1.60 \times 10^{-19}\ J})$<br>$= 1.13 \times 10^{13}$ |
| 8. Change in temperature of body section ($c = 4.18 \times 10^3$ J/kg·K) | $\Delta T = \frac{\Delta Q}{mc} = \frac{\Delta Q/m}{c} = \frac{6.00 \times 10^{-2}\ J/kg}{4.18 \times 10^3\ J/kg \cdot K}$<br>$\Delta T = 1.44 \times 10^{-5}\ K$ |

Example 32.4. A man exposed to radiation receives a dose of 50 mrad of fast neutrons. (a) What is his effective dose in rem? (b) Did he exceed the recommended weekly dose?

Given: Absorbed dose = 50.0 mrad and source is fast neutrons.

Determine: The effective dose in rem and whether or not the man exceeded the recommended weekly dose.

Strategy: Knowing that the source is fast neutrons, we can determine the quality factor for the radiation. Knowing the absorbed dose and the QF, we can determine the effective dose. The effective dose may be compared with the above-background radiation dose limits in Table 32-4 of the text to see whether the weekly dose has been exceeded.

Solution: From Table 32-2 of the text, the quality factor for fast neutrons is

$$QF = 10.0 \text{ rem/rad} = 10.0 \text{ Sv/Gy}$$

$$\text{Dose}_{eff} = \text{dose}_{abs} \times QF = (50.0 \text{ mrad})(10.0 \text{ rem/rad})(10^{-3} \text{ rad/mrad}) = 0.500 \text{ rem}$$

According to Table 32-4, the maximum above-background radiation dose limit for the whole body is

$$(5 \text{ rem/year})(1 \text{ year}/52 \text{ wk}) = 0.0962 \text{ rem/wk}$$

According to this, the man was exposed to five times (0.500/0.0962 = 5) the maximum.

Related Text Problems: 32-6, 32-7, 32-10, 32-14, 32-18.

## 5. Equivalent Dose Rate (Section 32.7)

Review: For determining dose rates from exposures to different isotopes, a specific equivalent dose rate, labeled Rhm, has been defined and measured for a number of commonly used isotopes. The Rhm for a particular isotope is the absorbed dose rate in rads per hour at a distance of 1 m from a 1-ci source. The SI unit of Rhm is $(\text{rad/h})(\text{m}^2/\text{Ci})$. Thus, the dose rate in rad/h received by a person at a distance r from an isotope of activity A is

$$\text{dose rate} = (\text{Rhm})(A/r^2)$$

Table 32-5 of your text gives Rhm values for several commonly used isotopes.

Practice: A cancer patient is placed 0.700 m from a 300-Ci cesium 137 source.

Determine the following:

| | |
|---|---|
| 1. Rhm value of source | $\text{Rhm} = 0.32 \text{ rad}\cdot\text{m}^2/\text{h}\cdot\text{Ci}$ |
| 2. Dose rate received by patient in rad/h | $\text{Dose rate} = (\text{Rhm})(A/r^2)$ <br> $= (0.32 \frac{\text{rad}\cdot\text{m}^2}{\text{h}\cdot\text{Ci}}) \frac{(3.00 \times 10^2 \text{ Ci})}{(0.700 \text{ m})^2}$ <br> $= 1.96 \times 10^2 \text{ rad/h}$ |
| 3. Time necessary for 60-rad exposure | $\text{Amount} = (\text{rate})(\text{time})$ <br> $\text{Time} = \frac{60.0 \text{ rad}}{1.96 \times 10^2 \text{ rad/h}} = 0.306 \text{ h}$ |

| | |
|---|---|
| 4. Distance source should be from patient to provide a dose rate of 6.00 Gy/h | Dose rate $= (6.00\ \frac{Gy}{h})(\frac{1.00\ rad}{10^{-2}\ Gy})$<br>$= 6.00 \times 10^{2}$ rad/h<br><br>Dose rate = $(Rhm)(A/r^2)$<br><br>$r^2 = (Rhm)A/(\text{dose rate})$<br>$= \dfrac{(\frac{0.32\ rad \cdot m^2}{h \cdot Ci})(3.00 \times 10^2\ Ci)}{6.00 \times 10^2\ \frac{rad}{h}}$<br>$= 0.160\ m^2$<br>$r = 0.400$ m |
| 5. Absorbed dose after 10.0 min | $Dose_{abs}$ = (dose rate)(time)<br>$= (600\ \frac{rad}{h})(10.0\ min)(\frac{h}{60.0\ min})$<br>= 100 rad |
| 6. The effective dose after 10.0 min, given that dominate radiation is 0.825 meV gamma rays | $Dose_{eff} = dose_{abs} \times QF$<br>= (100 rad)(1.00 rem/rad)<br>= 100 rem |

Example 32.5. It is desired to deliver a dose of 80.0 rad to a radiation therapy patient in a 15.0-min interval. What must the activity of the sample be if it is cobalt 60 and is to be positioned 1.00 m from the patient?

Given: $dose_{abs}$ = 80.0 rad = desired absorbed dose
t = 15.0 min = exposure time
cobalt 60 = source
r = 1.00 m = patient distance from source

Determine: The activity of the sample that gives a dose of 80.0 rad in 15.0 min at a distance of 1.00 m.

Strategy: Knowing that the source is cobalt 60, we can determine the Rhm value. Knowing the dose and the time interval, we can determine the dose rate. Knowing the Rhm value, dose rate, and distance, we can determine the needed activity of the source.

Solution: The Rhm value for cobalt 60 is

$$\text{Rhm} = 1.3 \frac{\text{rad}\cdot\text{m}^2}{\text{h}\cdot\text{Ci}}$$

The dose rate is

$$\text{Dose rate} = \frac{80.0\ \text{rad}}{15.0\ \text{min}}\left(\frac{60.0\ \text{min}}{\text{h}}\right) = \frac{320\ \text{rad}}{\text{h}}$$

The activity is determined by

$$\text{Dose rate} = (\text{Rhm})\left(\frac{A}{r^2}\right)$$

$$A = (\text{dose rate})(r^2/\text{Rhm})$$

$$= \left(320\ \frac{\text{rad}}{\text{h}}\right)\left(\frac{1.00\ \text{m}^2}{1.3\ \frac{\text{rad}\cdot\text{m}^2}{\text{h}\cdot\text{Ci}}}\right)$$

$$= 2.5 \times 10^2\ \text{Ci}$$

Related Problems: 32-19, 32-20, 32-22, 32-23.

---

## PRACTICE TEST

Take and grade this practice test. Doing so will allow you to determine any weak spots in your understanding of the concepts taught in this chapter. The following section prescribes what you should study further to strengthen your understanding.

A sample of plutonium 239 ($T = 2.41 \times 10^4$ years) is counted by a detector that is 50% efficient and registers $8.00 \times 10^6$ counts in 2.00 min. Determine the following:

__________ 1. The activity of the sample in disintegrations per second
__________ 2. The activity of the sample in curies
__________ 3. The activity of the sample in becquerels
__________ 4. The decay constant for the isotope
__________ 5. The number of nuclei present in the sample
__________ 6. The number of moles of the sample
__________ 7. The mass of the sample

Determine the following for a 5-R x-ray beam:

__________ 8. The number of ion pairs created per cubic meter of air by this beam
__________ 9. The number of ion pairs created per kilogram of air
__________ 10. The coulombs of ion charge created per kilogram of air

In a medical diagnostic procedure, a patient absorbs 6.00 rad in a 3.00-kg body section. The radiation used is 60.0-keV photons.
Determine the following:

__________ 11. The absorbed dose in grays
__________ 12. The total energy absorbed by the body section
__________ 13. The number of incident photons required to supply this absorbed energy
__________ 14. The increase in temperature of the body section
__________ 15. The effective dose in rem
__________ 16. The effective dose in sieverts

A 200-Ci cesium 137 source is placed 1.50 m from a patient.
Determine the following:

__________ 17. The Rhm value for the isotope
__________ 18. The dose rate received by the patient in rad/h
__________ 19. The time necessary for a 50.0-rad exposure
__________ 20. The distance the source should be from the patient to provide a dose rate of 6.00 Gy/h

(See Appendix I for answers.)

PRINCIPAL CONCEPTS AND EQUATIONS PRESCRIPTION

Your score on the practice test is an excellent measure of your understanding of this chapter. You should now use the following chart to write your own prescription for curing any of your physics ills. Look down the leftmost column to the number of the question(s) you answered incorrectly, read across that row to see which section(s) of the study guide you should return to for further study, and then do the suggested text problems to gain additional experience in working with the particular concept.

| Question Number | Concepts and Equations | Prescription: Principal Concept | Prescription: Text Problems |
|---|---|---|---|
| 1 | Activity: $A$ = count rate/$\varepsilon$ | 1 | 32-1,2 |
| 2 | Activity: 1 Ci = $3.70 \times 10^{10}$/s | 1 | 32-2,3a |
| 3 | Activity: 1 Bq = 1.00/s | 1 | 32-1,5 |
| 4 | Decay constant: $\lambda = \ell n2/T$ | 3 of Ch. 31 | 32-3b,4 |
| 5 | Activity: $A = N\lambda$ | 1 | 32-3b,4 |
| 6 | Moles of sample: $n = N/N_A$ | 3 of Ch. 14 | 32-3b,4 |
| 7 | Mass of sample: $m = n(AM)$ | 3 of Ch. 14 | 32-3b,4 |
| 8 | Exposure: 1 R = $2.082 \times 10^9$ ion pairs/$cm^3$ | 2 | 32-11,14 |
| 9 | Exposure: 1 R = $1.61 \times 10^{12}$ ion pairs/g | 2 | 32-13,15 |
| 10 | Exposure: 1 R = $2.58 \times 10^{-4}$ C/kg | 2 | 32-16 |
| 11 | Absorbed dose: 1 rad = $1.00 \times 10^{-2}$ Gy | 3 | 32-6a,7b |
| 12 | Absorbed energy: $E_{abs} = dose_{abs}m$ | 3 | 32-7a |
| 13 | $E_{abs} = NE_{photon}$ | 3 | 32-6c |
| 14 | $E_{abs} = \Delta Q = mc\Delta T$ | 3 | 32-8,9 |
| 15 | Effective dose: $Dose_{eff} = dose_{abs} \times QF$; rem = rad x QF | 4 | 32-6a,7c |
| 16 | Effective dose: $Dose_{eff} = dose_{abs} \times QF$; Sv = Gy x QF | 4 | 32-6c,7c |
| 17 | Rhm value (Table 32-5) | 5 | 32-20,22 |
| 18 | Dose rate: dose rate = (Rhm)($A/r^2$) | 5 | 32-20,23 |
| 19 | Amount = (rate)(time) | - | 32-19a,22 |
| 20 | Dose rate: dose rate = (Rhm)($A/r^2$) | 5 | 32-19b,20 |

# 33 Nuclear Fission and Fusion

RECALL FROM PREVIOUS CHAPTERS

| Previously learned concepts and equations used in this chapter | Text Section | Study Guide Page |
|---|---|---|
| Kinetic theory of gases: $KE = 3kT/2$ | 14.8 | 227 |
| Mass difference: $\Delta M = Zm_H + (A - Z)m_n - M$ | 31.3 | 509 |
| Binding energy: $BE = \Delta mc^2$ | 31.3 | 509 |

NEW IDEAS IN THIS CHAPTER

| Concepts and equations introduced | Text Section | Study Guide Page |
|---|---|---|
| Nuclear fission | 33.3 | 534 |
| Nuclear fusion | 33.6 | 538 |
| Energy release | 33.6 | 538 |
| Method I - mass-energy equivalence | | |
| Method II - binding energy per nucleon | | |

PRINCIPAL CONCEPTS AND EQUATIONS

1. Nuclear Fission (Section 33.3)

Review: Fission refers to the process of a heavy nucleus splitting into two lighter nuclei and emitting two or more neutrons. It occurs spontaneously in only a few of the very heavy elements, such as U-238. Fission can be induced by a neutron. The isotopes that fission with either a fast or a slow neutron are U-233, U-235, Pu-239, and Pu-241. An example of such a fission process is

$$^{235}_{92}U + ^{1}_{0}n \rightarrow ^{236}_{92}U \rightarrow ^{139}_{56}Ba + ^{95}_{36}Kr + 2\ ^{1}_{0}n$$

We can determine the amount of energy released in this fission by finding the difference in mass ($\Delta m$) of the reactants and products and then using mass-energy equivalence ($E = \Delta mc^2$):

| Mass of reactants | Mass of products |
|---|---|
| ${}^{235}_{92}U$ = 235.043915 u | ${}^{139}_{56}Ba$ = 138.908830 u |
| ${}^{1}_{0}n$ = 1.008665 u | ${}^{95}Kr$ = 94.897331 u |
| sum = 236.052580 u | $2\,{}^{1}_{0}n$ = 2.017330 u |
| | sum = 235.823491 u |

The mass difference is

$$\Delta m = m_{reactants} - m_{products} = 0.229089 \text{ u}$$

The energy released is

$$\Delta E = \Delta mc^2 = (0.229089 \text{ u})(931.5 \text{ MeV/u}) = 213 \text{ MeV}$$

Practice: Consider the following neutron induced fission of a U-235 nucleus in which two neutrons are ejected:

$${}^{235}_{92}U + {}^{1}_{0}n \rightarrow {}^{96}_{40}Zr + \boxed{\phantom{X}} + 2\,{}^{1}_{0}n + 4\,{}^{0}_{-1}e + \text{energy}$$

Determine the following:

| | |
|---|---|
| 1. Atomic number of the unknown isotope | Let $Z_x$ represent the atomic number of the unknown isotope. Conserving Z number, we have<br>$92 + 0 = 40 + Z_x + 0 + (-4)$<br>$Z_x = 92 - 36 = 56$ |
| 2. Atomic mass number of the unknown isotope | Let $A_x$ represent the atomic mass number of the unknown isotope. Conserving A number, we have<br>$235 + 1 = 96 + A_x + 2$<br>$A_x = 236 - 98 = 138$ |
| 3. The unknown isotope | For the unknown isotope, we have Z = 56 and A = 138; hence it is barium-138 |
| 4. Mass of the reactants | ${}^{235}_{92}U$ = 235.043915 u<br>${}^{1}_{0}n$ = 1.008665 u<br>236.052580 u |

| | |
|---|---|
| 5. Mass of the products | ${}^{96}_{40}Zr$ = 95.908286 u<br>${}^{138}_{56}Ba$ = 137.905000 u<br>$2\,{}^{1}_{0}n$ = 2.017330 u<br>235.830616 u |
| 6. $\Delta m$ for the reaction | $\Delta m = m_{reactants} - m_{products}$<br>= 236.052580 u - 235.830616 u<br>= 0.221964 u |
| 7. Energy released | $E = \Delta m(931.5\ MeV/u) = 207\ MeV$ |
| 8. Number of moles in 1 kg of U=-235 | $n = m/AM$<br>$= \frac{(1.00\ kg)(10^3\ g/kg)}{235\ g/mol} = 4.26\ mol$ |
| 9. Number of nuclei in 1 kg of U-235 | $N = nN_A$<br>$= (4.26\ mol)(6.023 \times 10^{23}\ \frac{nuclei}{mol})$<br>$= 2.57 \times 10^{24}$ nuclei |
| 10. Total energy released when 1.00 kg of U-235 fissions | $E_{total} = N(E/fission)$<br>$= (2.57 \times 10^{24})(2.07 \times 10^2\ MeV)$<br>$= 5.32 \times 10^{26}\ MeV$ |
| 11. Mass converted to energy when 1.00 kg of U-235 fissions | $E = 5.32 \times 10^{26}\ MeV = 8.51 \times 10^{13}\ J$<br>The mass converted to energy is<br>$\Delta m/kg = E/c^2$<br>$= 8.51 \times 10^{13}\ J/(9.00 \times 10^{16}\ m^2/s^2)$<br>$= 9.46 \times 10^{-4}\ kg$ |
| 12. Mass of U-235 that must fission to meet annual energy consumption on the earth ($4.00 \times 10^{20}$ J) | $\frac{E}{kg} = 5.32 \times 10^{26}\ \frac{MeV}{kg} = 8.51 \times 10^{13}\ \frac{J}{kg}$<br>$m = E_{total}/(E/kg)$<br>$= 4.00 \times 10^{20}\ J/(8.51 \times 10^{13}\ J/kg)$<br>$= 4.70 \times 10^6\ kg$ |

| | |
|---|---|
| 13. Mass converted to energy in meeting the earth's annual energy need by this fission process | $\Delta m/kg = 9.46 \times 10^{-4}$ kg (step 11)<br><br>$\Delta m_{total} = (\Delta m/kg)(\text{number of kg})$<br>$= (9.46 \times 10^{-4}\ kg)(4.70 \times 10^6)$<br>$= 4.45 \times 10^3$ kg<br>or<br>$\Delta m = E_{total}/c^2$<br>$= 4.00 \times 10^{20}\ J/(9 \times 10^{16}\ m^2/s^2)$<br>$= 4.44 \times 10^3$ kg |

Example 33.1. Determine the mass of U-235 fuel needed per year by a 50% efficient 200-MW power plant. Assume an energy release of 200 MeV/fission.

Given: E/fission = 200 MeV/fission
$P = 200 \times 10^6\ W$ = power output of plant
${}^{235}_{92}U$ = fissionable material used as fuel
$\varepsilon = 0.500$ = efficiency of plant

Determine: Mass of U-235 needed per year by a 50% efficient 200-MW power plant.

Strategy: Knowing the power, time, and efficiency of the power plant, we can determine the total amount of energy released. Knowing the total amount of energy released and the energy released per fission, we can determine the number of fissions that must occur. Knowing the number of fissions, we can determine the mass of U-235 needed.

Solution: The total amount of energy to be released in 1 year is

$$E_{released} = (\text{power})(\text{time})/\varepsilon$$

$$= (2.00 \times 10^8\ W)(1\ y)(3.15 \times 10^7\ s/y)/0.500 = 1.26 \times 10^{16}\ J$$

The energy released per fission is

$$E/\text{fission} = 200\ MeV = 3.20 \times 10^{-11}\ J$$

The number of fissions that must occur is

$$E_{released} = N(E/\text{fission})$$

$$N = E_{released}/(E/\text{fission}) = 1.26 \times 10^{16}\ J/3.20 \times 10^{-11}\ J = 3.94 \times 10^{26}$$

The mass of U-235 needed is

$$m = N(AM)/N_A$$

$$= (3.94 \times 10^{26})(0.235\ kg/mol)/(6.02 \times 10^{23}/mol) = 154\ kg$$

Related Text Problems: 33-1 through 33-15.

## 2. Nuclear Fusion (Section 33.6)

Review: Fusion refers to the process of merging (fusing together) several light nuclei to form one larger nucleus and release energy. There are two methods for determining the amount of energy available.

Method I - mass-energy equivalence. When using this method, we find the mass of the reacting nuclei that fuse, the mass of the product nucleus, the mass of any other products, and the mass difference ($\Delta m$) between the reactants and the product(s) and then use mass-energy equivalence ($E = \Delta mc^2$) to determine the energy released.

Example: $${}^{2}_{1}H + {}^{2}_{1}H \rightarrow {}^{4}_{2}He + \gamma$$

Note: We will use $m_H$, $m_D$, and $m_T$ to represent the mass of an atom of ${}^{1}_{1}H$, ${}^{2}_{1}H$, and ${}^{3}_{1}H$ respectively.

Mass of the reactants is

$$2m_H = 2(2.014102 \text{ u}) = 4.028204 \text{ u}$$

Mass of the product nucleus is

$$m_{He} = 4.002603 \text{ u}$$

The mass difference is

$$\Delta M = 2m_H - m_{He} = 0.025601 \text{ u}$$

The binding energy is

$$BE = \Delta Mc^2 = (0.025601 \text{ u})(931.5 \frac{\text{MeV}}{\text{u}}) = 23.8 \text{ MeV}$$

Method II -binding energy per nucleon. When using this method, we find the binding energy per nucleon before fusion and the binding energy per nucleon after fusion. We then multiply the difference in binding energies by the number of nucleons involved to obtain the energy released.

Example: ${}^{2}_{1}H + {}^{2}_{1}H \rightarrow {}^{4}_{2}He + \gamma$

The binding energy for ${}^{2}_{1}H$ is

$$BE_D = (1m_H + 1m_n - M_D)c^2$$

$$= (1.007825 \text{ u} + 1.008665 \text{ u} - 2.014102 \text{ u})(931.5 \text{ MeV/u})$$
$$= (0.002388 \text{ u})(931.5 \text{ MeV/u})$$
$$= 2.22 \text{ MeV}$$

The binding energy per nucleon for $^{2}_{1}H$ is

$$(BE/A)_D = 2.22 \text{ MeV/2 nucleons} = 1.11 \text{ MeV/nucleon}$$

The binding energy for $^{4}_{2}He$ is

$$BE_{He} = (2m_H + 2m_n - M_{He})\ c^2 = 28.3 \text{ MeV}$$

The binding energy per nucleon for $^{4}_{2}He$ is

$$(BE/A)_{He} = 28.3 \text{ MeV/4 nucleons} = 7.07 \text{ MeV/nucleon}$$

The total energy released is

$$\text{Energy} = (7.07 \text{ MeV/nucleon} - 1.11 \text{ MeV/nucleon})(4 \text{ nucleons}) = 23.8 \text{ MeV}$$

Note: In general, you will find method I more useful except for very simple cases of fusion (such as this example) where we can easily determine the binding per nucleon before and after fusion. If the binding energies are not already known, it is much quicker to use the mass-energy equivalence method.

Practice: Consider the fusion reaction

$$^{2}_{1}H + {}^{2}_{1}H \rightarrow {}^{4}_{2}He$$

Determine the following:

| | |
|---|---|
| 1. Initial kinetic energy of each deuteron that allows them sufficient energy to get close enough to fuse together.<br>(A nuclear radius is of the order of $1.00 \times 10^{-14}$ m.) | Assume they will fuse if they get within $2.00 \times 10^{-14}$ m of each other. Also assume the initial potential energy of the deuteron pair is zero and they come to rest just as they fuse. Using conservation of energy, we have<br>$\Delta KE + \Delta PE = 0$<br>$(KE_f - KE_i) + (PE_f - PE_i) = 0$<br>$KE_i = PE_f = ke^2/r$<br>$KE_i = \dfrac{(9.00 \times 10^9 \text{ N}\cdot\text{m}^2/\text{c}^2)(1.60 \times 10^{-19} \text{ c})^2}{2.00 \times 10^{-14} \text{ m}}$<br>$KE_i = 1.15 \times 10^{-14} \text{ J} = 0.0720 \text{ MeV}$<br>$(KE/\text{deuteron})_i = 5.75 \times 10^{-15} \text{ J}$<br>$= 0.0360 \text{ MeV}$ |

| | |
|---|---|
| 2. Absolute temperature required to give deuterons this amount or energy | $KE = 3kT/2$; $T = 2(KE)/3k$<br>$T = \frac{2(5.75 \times 10^{-15}\ J)}{3(1.38 \times 10^{-23}\ J/K)} = 2.78 \times 10^{8}\ K$ |
| 3. Energy released | $m_D = 2.014102\ u$ ; $m_{He} = 4.002603\ u$<br>$\Delta m = 2m_D - m_{He} = 0.025601\ u$<br>$E_{released} = \Delta mc^2 = 23.8\ MeV$ |
| 4. Ratio of energy released to energy invested | $E_{released} = 23.8\ MeV$ (Step 3)<br>$E_{invested} = 0.0720\ MeV$ (Step 1)<br>$\frac{E_{released}}{E_{invested}} = \frac{23.8\ MeV}{0.0720\ MeV} = 331$ |
| 5. The number of such fusion reactions required to supply the annual energy consumption of the earth ($4.00 \times 10^{20}$ J) | $E_T = 4.00 \times 10^{20}\ J = 2.50 \times 10^{33}\ MeV$<br>Let N represent the number of fusion reactions.<br>$E_T = N(E/reaction)$<br>$N = E_T/(E/reaction)$<br>$N = 2.50 \times 10^{33}\ MeV/(23.8\ MeV/reaction)$<br>$N = 1.05 \times 10^{32}$ reactions |
| 6. Number of deuterium nuclei required to cause this number of reactions | Since each reaction requires 2 deuterium nuclei, we need<br>$(1.05 \times 10^{32}\ \text{reactions})\left(\frac{2\ {}^2_1H\ \text{nuclei}}{\text{reaction}}\right) =$<br>$2.10 \times 10^{32}$ nuclei |
| 7. Number of water molecules in 1.00 kg of water | The number of moles of water in 1.00 kg is<br>$n = \frac{1.00\ kg\ (10^3\ g/kg)}{(18.0\ g/mol)} = 55.6\ mol$<br>The number of molecules in 55.6 mol of water is<br>$N = nN_A = 3.35 \times 10^{25}$ molecules |
| 8. Number of hydrogen nuclei in 1.00 kg of water | Since water molecules each contain two hydrogen nuclei, we have $2(3.35 \times 10^{25})$ $= 6.70 \times 10^{25}$ nuclei |

| | |
|---|---|
| 9. Number of deuterium nuclei in 1.00 kg of water | One out of every 6700 hydrogen nuclei in water is deuterium. Therefore $(6.70 \times 10^{25}\ \text{hydrogen})(\frac{1\ \text{deuterium}}{6.70 \times 10^{3}\ \text{hydrogen}})$ $1.00 \times 10^{32}$ deuterium nuclei |
| 10. Mass of water needed to supply the deuterium for a sufficient number of fusion reactions to provide the earth's annual energy consumption ($4.00 \times 10^{20}$ J) by this fusion process | This energy may be provided by $N = 1.05 \times 10^{32}$ reactions of the type being studied. This requires $2.10 \times 10^{32}$ deuterium nuclei. Since 1.00 kg of water contains $1.00 \times 10^{22}$ deuterium nuclei, the mass of water needed is $M = 2.10 \times 10^{10}$ kg |

Example 33.2. Two deuterium nuclei fuse to produce tritium and a proton. Calculate the rate at which deuterium is consumed to produce 10.0 MW of power. (Assume all energy from the reaction is available.)

Given: ${}^{2}_{1}H + {}^{2}_{1}H \rightarrow {}^{3}_{1}H + {}^{1}_{1}p$ , $P = 10.0$ MW

Determine: The rate of consumption of deuterium to produce 10.0 MW of power.

Strategy: We can look up the mass of the reactants and products and then determine the energy of released per fusion. Knowing the energy released per fusion and the rate at which energy must be produced ($P = 10.0$ MW $= 1.00 \times 10^{6}$ J/s $= E/t$), we can determine the rate at which the fusion reactions must occur and subsequently the rate at which deuterium is consumed.

Solution: The mass of the reactants and products is

$$m_p = 1.007277\ \text{u}\ ,\quad m_D = 2.104102\ \text{u}\ ,\quad m_T = 3.016050\ \text{u}$$

The energy released per fusion is

$$E_{release}/\ \text{fusion} = [2m_D - (m_T + m_p)]c^2$$

$$= [2(2.014102\ \text{u})-(3.016050\ \text{u} + 1.007277\ \text{u})]\ (931.5\ \frac{\text{MeV}}{\text{u}}) = 4.54\ \text{MeV/fusion}$$

The rate at which energy must be supplied is

$$E/t = P = 10.0\ \text{MW}\ (\frac{10^6\ \text{W}}{\text{MW}})(\frac{\text{J/s}}{\text{W}})(\frac{\text{eV}}{1.60 \times 10^{-19}\ \text{J}})(\frac{\text{MeV}}{10^6\ \text{eV}}) = 6.25 \times 10^{19}\ \text{MeV/s}$$

The rate at which the fusion reactions must occur is

$$\text{Fusion/s} = (6.25 \times 10^{19}\ \frac{\text{MeV}}{\text{s}})(\frac{\text{Fusion}}{4.54\ \text{MeV}}) = 1.38 \times 10^{19}\ \text{fusions/s}$$

The rate at which deuterium nuclei are consumed is

$$\text{Deuterium nuclei/s} = (1.38 \times 10^{19}\ \frac{\text{fusions}}{\text{s}})(\frac{2\ \text{nuclei}}{\text{fusion}}) = 2.76 \times 10^{19}\ \text{nuclei/s}$$

The rate at which deuterium is consumed is

$$(\frac{\Delta m}{\Delta t})_D = (\frac{2.76 \times 10^{19}\ \text{nuclei}}{\text{s}})(\frac{1\ \text{mol}}{6.02 \times 10^{26}\ \text{nuclei}})(\frac{2.01\ \text{g}}{\text{mol}})(\frac{\text{kg}}{10^3\ \text{g}})$$

$$= 9.22 \times 10^{-11}\ \text{kg/s}$$

Related Text Problems: 32-14 through 33-26.

PRACTICE TEST

Take and grade this practice test. Doing so will allow you to determine any weak spots in your understanding of the concepts taught in this chapter. The following section prescribes what you should study further to strengthen your understanding.

Consider the following fission reaction

$$^{235}_{92}U + ^{1}_{0}n \rightarrow \boxed{\phantom{XX}} + ^{101}_{44}Ru + 3\,^{1}_{0}n + 6\,^{0}_{-1}e + \text{Energy}$$

__________ 1. Atomic number of the unknown isotope

__________ 2. Atomic mass of the unknown isotope

__________ 3. Mass difference between reactants and products

__________ 4. Energy release for this fission reaction

__________ 5. Number of moles in 1 kg of U-235

__________ 6. Number of nuclei in 1 kg of U-235

__________ 7. Total amount of energy (in MeV) released when 1 kg of U-235 fissions in this manner

__________ 8. Mass (in kg) converted to energy when 1 kg of U-235 fissions in this manner

__________ 9. Rate of consumption of U-235 if this fission reaction is to supply 1000 MW power reactor that is 40% efficient

Consider the reaction $^{4}_{2}He + ^{4}_{2}He + ^{4}_{2}He \rightarrow ^{12}_{6}C + \text{energy}$

Determine the following:

__________ 10. Final electric potential energy (in joules) of the three helium nuclei just before they fuse

__________ 11. Initial kinetic energy of each helium nucleus so that they get close enough to fuse

__________ 12. Absolute temperature required to give the helium nuclei the needed kinetic energy

__________ 13. Mass difference between the reactants and products

__________ 14. Energy released in this fusion reaction

__________ 15. Binding energy per nucleon before fusion

__________ 16. Binding energy per nculeon after fusion

__________ 17. Energy release for this fusion reaction calculated from binding energy per nucleon information

__________ 18. Number of such fusion reactions required to supply the annual energy consumption of the earth ($4.00 \times 10^{20}$ J)

__________ 19. Number of He-4 nuclei needed to supply the annual energy consumption of the earth by the this reaction

__________ 20. Mass of He-4 needed to supply the annual energy consumption of the earth by this reaction

(See Appendix I for answers.)

PRINCIPAL CONCEPTS AND EQUATIONS PRESCRIPTION

Your score on the practice test is an excellent measure of your understanding of this chapter. You should now use the following chart to write your own prescription for curing any of your physics ills. Look down the leftmost column to the number of the question(s) you answered incorrectly, read across that row to see which section(s) of the study guide you should return to for further study, and then do the suggested text problems to gain additional experience in working with the particular concept.

| Question Number | Concepts and Equations | Prescription: Principal Concept | Prescription: Text Problems |
|---|---|---|---|
| 1 | Conservation of atomic number (Z) | 1 | 31-1,3 |
| 2 | Conservation of atomic mass number (A) | 1 | 31-4,5 |
| 3 | Mass difference: $\Delta m = m_{reactants} - m_{products}$ | 2 of Ch. 31 | 33-1,2 |
| 4 | Mass energy equivalence: $E = \Delta mc^2$ | 2 of Ch. 31 | 33-3,10 |
| 5 | Number of moles: $n = M/AM$ | 3 of Ch. 14 | 33-5a,6a |
| 6 | Number of nuclei: $N = nN_A$ | 3 of Ch. 14 | 33-5a,6a |
| 7 | $E_{total} = (E/\text{fission})(N)$ | 1 | 33-5a,9 |
| 8 | Mass energy equivalence: $\Delta m = E/c^2$ | 2 of Ch. 31 | 33-5b,6b |
| 9 | $\Delta M/\Delta t = (E_{total}/t)/(E/M)$ | 1 | 33-6,13 |
| 10 | Electric potential energy: $PE = ke^2/r$ | 1 of Ch. 18 | 31-3 |
| 11 | Conservation of energy: $\Delta KE + \Delta PE = 0$ | 5 of Ch. 7 | 31-6 |
| 12 | Kinetic theory of gases: $KE = 3kT/2$ | 4 of CH. 14 | 33-18 |
| 13 | Mass difference: $\Delta m = m_{reactants} - m_{products}$ | 2 | 3-17,19 |
| 14 | Mass energy equivalence: $E = \Delta mc^2$ | 2 of Ch. 31 | 33-20,22 |
| 15 | Binding energy per nucleon: $(BE/A) = [Zm_H + (A - Z)M_n - M)c^2/A$ | 2 of Ch. 31 | 31-12a,13 |
| 16 | Binding energy per nucleon | 2 of Ch. 31 | 31-12b,17a |
| 17 | $E_{release} = [(BE/A)_{after} - (BE/A)_{before}]\text{Nucleons}$ | 2 | -- |
| 18 | $N = E_{Total}/(E/\text{fusion})$ | 2 | 33-23,24 |
| 19 | Nuclei = (number of fusions)(nuclei/fusion) | | 33-23,24 |
| 20 | $\text{Mass} = \dfrac{(\text{number of nuclei})(AM)}{N_A}$ | 3 of CH. 14 | 33-23,24 |

# Appendix I

Chapter 1

1. kg·m/s
2. $kg \cdot m^2/s^2$
3. m
4. $kg \cdot m/s^2$
5. 60.0 km
6. $53.33
7. 5.00 $\ell$/h
8. 16.7 m/s
9. 3.14 $m^2$
10. 0.524 $m^3$
11. 4.02 $kg/m^3$
12. 18 m/s
13. $1.29 \times 10^3$ $cm^2$
14. 22 mi/h
15. $3.94 \times 10^6$ in
16. $3.0 \times 10^6$
17. 2.8
18. 102.60

Chapter 2

1. +24.0 m
2. 72.0 m
3. -16.0 m
4. 80.0 m
5. -1.33 m/s
6. 0
7. 6.67 m/s
8. 6.00 m/s
9. -12.0 m/s
10. 12.0 m/s
11. 0
12. 2.00 $m/s^2$
13. +4.47 m/s
14. 1.02 m
15. 0.456 s
16. -9.80 $m/s^2$
17. 5.00 s
18. -44.5 m/s

Chapter 3

1. -2.57 m
2. -6.93 m
3. -2.24 m
4. -1.37 m
5. 2.63 m
6. 31.5° S of W
7. 15.0 m/s, 26.0 m/s
8. 0, -9.80 $m/s^2$
9. 2.65 s
10. 84.5 m
11. 15 m/s horizontal
12. -9.80 $m/s^2$
13. 5.30 s
14. 15.0 m/s, -26.0 m/s
15. 6.80 s
16. 43.3 m/s
17. 69.7° below horiz.
18. 112 m

Chapter 4

1. 147 N
2. 146 N
3. 14.6 N
4. 14.7 N
5. 255 N
6. +214.3 N
7. 3.57 $m/s^2$
8. 133.7 N
9. 0.600 N
10. 71.4 N
11. 107 N
12. 35.7 N
13. 3.57 m/s
14. 1.79 m
15. 2.67 m/s

Chapter 5

1. $4.00 \times 10^{-2}$ m/s$^2$
2. 4.00 rad/s$^2$
3. 0.800 m/s$^2$
4. 0.200 m/s
5. 20.0 rad/s
6. 4.00 m/s
7. 0.500 m
8. 7.96 rev
9. 10.0 m
10. 22.9 N
11. 19.4 N
12. 12.5 N
13. 7.68 N
14. 3.10 N
15. 15.7 m/s$^2$
16. $1.10 \times 10^3$ N
17. $1.74 \times 10^7$ m
18. 9.21 N

Chapter 6

1. -866 N
2. -433 N
3. +1299 N
4. -500 N
5. +250 N
6. +250 N
7. 0
8. Yes
9. -866 N·m
10. -346 N·m
11. +1212 N·m
12. 1212 N
13. +606 N
14. -350 N
15. 0.100 m, 0.150 m
16. 0.250 m, 0.0500 m
17. 0.121 m
18. 0.136 m

Chapter 7

1. $1.00 \times 10^4$ J
2. $2.00 \times 10^3$ J
3. 0
4. 0
5. -500 J
6. $1.50 \times 10^3$ J
7. $1.50 \times 10^3$ J
8. 5.42 m/s
9. $1.15 \times 10^4$ J
10. $1.00 \times 10^4$ J
11. $-1.00 \times 10^4$ J
12. $-1.73 \times 10^3$ J
13. $8.27 \times 10^3$ J
14. $8.27 \times 10^3$ J
15. $9.77 \times 10^3$ J
16. 100 N
17. 0.980 m/s$^2$
18. 97.7 m

Chapter 8

1. 4.00 kg·m/s
2. -20.0 m/s
3. +12.0 m/s
4. 19.1°
5. 3.66 kg·m/s
6. +3.20 kg·m/s
7. +32.0 N
8. 1.62 m/s
9. 0
10. 1.62 m/s
11. 4.54 m/s
12. 1.51 m/s
13. 6.05 m/s
14. -1.40 m/s
15. 1.80 m/s
16. 9.09 m/s
17. -8.26 m/s$^2$
18. -1.82 N
19. 0.844

Chapter 9

1. $mr^2/2$
2. $mr^2\omega_i/2$
3. $mr^2\omega_i^2/4$
4. $mr^2/4$
5. $3mr^2/4$
6. $mr^2\omega_i/2$
7. $2\omega_i/3$
8. $mr^2\omega_i^2/6$
9. $-mr^2\omega_i/12$
10. Distortion of Putty and friction
11. $2h/t$
12. $2h/rt$
13. $2mh^2/t^2$
14. $-mgh$
15. $mgh - 2mh^2/t^2$
16. $mr^2[(gt^2/2h) -1]$
17. $2h/t^2$
18. $2h/rt^2$
19. $m[g - (2h/t^2)]$
20. $mr[g - (2h/t^2)]$
21. $mr^2](gt^2/2h) - 1]$

## Chapter 10

1. 200 N/m
2. 0.200 m
3. 10.0 rad/s
4. 1.59 Hz
5. 0.629 s
6. 2.00 m/s
7. 20.0 $m/s^2$
8. $3\pi/2$
9. 0
10. -2.00 m/s
11. 0
12. 0
13. 4.00 J
14. 0
15. 4.00 J
16. 0.0282 m
17. -1.98 m/s
18. -2.82 $m/s^2$
19. 0.0795 J
20. 3.92 J

## Chapter 11

1. 1.00 m
2. 0.200 Hz
3. 10.0 m
4. 2.00 m/s
5. 2.00 m/s
6. 2.00 Hz
7. 0
8. 0.125 s
9. 0.200 kg/m
10. 200 m/s
11. $8.00 \times 10^3$ N
12. 1.00 m
13. 300 m/s
14. $1.8 \times 10^3$ N
15. $7.20 \times 10^3$ N
16. 0.667 m
17. 170 Hz
18. 0.40 m
19. 0.167 m

## Chapter 12

1. $7.78 \times 10^6$ $N/m^2$
2. $7.50 \times 10^{-4}$
3. $1.04 \times 10^{10}$ $N/m^2$
4. $3.78 \times 10^{-15}$ $m^2$
5. $1.89 \times 10^{-6}$ N
6. 100 $N/m^2$
7. $5.29 \times 10^{-2}$
8. $1.89 \times 10^3$
9. $1.94 \times 10^4$ $N/m^2$
10. $1.21 \times 10^{-6}$
11. $2.42 \times 10^{-8}$ m
12. $7.78 \times 10^{-16}$ $m^3$
13. $2.48 \times 10^{-14}$ m

## Chapter 13

1. $4.90 \times 10^4$ $N/m^2$
2. $9.80 \times 10^4$ $N/m^2$
3. $2.48 \times 10^5$ $N/m^2$
4. 5.00 m
5. 613 $N/m^2$
6. 613 $N/m^2$
7. 245 N
8. 25.0 kg
9. 700 $kg/m^3$
10. 0.862 N
11. 0.862 N
12. 0.862 N
13. $1.05 \times 10^3$ $kg/m^3$
14. 0.617 N
15. 0.617 N
16. 500 $kg/m^3$
17. $7.85 \times 10^{-3}$ $m^3/s$
18. 1.00 m/s
19. 16.0 m/s
20. $1.53 \times 10^5$ $N/m^2$
21. 1.15 m

## Chapter 14

1. 50.0°C
2. 60.0°S
3. 343 K
4. $4.80 \times 10^{-4}$
5. $1.01 \times 10^8$ $N/m^2$
6. -27.8 K
7. 22.1 $cm^3$
8. $10.0 \times 10^5$ $N/m^2$
9. 3.61 mol
10. $1.80 \times 10^6$ $N/m^2$
11. $5.39 \times 10^6$ $N/m^2$
12. $3.20 \times 10^{-2}$ kg/mole
13. $5.32 \times 10^{-26}$ kg
14. 5.00 mol
15. $3.01 \times 10^{24}$
16. 321 K
17. $1.33 \times 10^7$ $N/m^2$
18. $2.00 \times 10^4$ J

## Chapter 15

1. 49.0 J
2. 11.7 cal
3. $2.34 \times 10^{-2}$°C
4. 450 cal
5. $7.20 \times 10^{3}$ cal
6. 2318 cal
7. 0°C
8. 66.6 g
9. 60.4°C
10. 1.89 cal/s
11. 35.2°C
12. 11.0 cal/s
13. 8.07 cal/s
14. 9.59¢
15. 25.6 $g/m^3$
16. 15.4 $g/m^3$
17. $2.00 \times 10^{4}$ g

## Chapter 16

1. 4.00 mol
2. $4.00 \times 10^{-3}$ $m^3$
3. $4.00 \times 10^{3}$ J
4. $9.995 \times 10^{3}$ J
5. $5.996 \times 10^{3}$ J
6. $8.00 \times 10^{-3}$ $m^3$
7. $1.109 \times 10^{4}$ J
8. 230 J/K
9. $1.211 \times 10^{-3}$ $m^3$
10. 364.6 K
11. 0
12. $5.816 \times 10^{3}$ J
13. $-9.12 \times 10^{3}$ J
14. $-1.367 \times 10^{4}$ J
15. $-2.278 \times 10^{4}$ J
16. 0
17. $1.179 \times 10^{4}$ J
18. $1.179 \times 10^{4}$ J

## Chapter 17

1. -135 N
2. +135 N
3. -180 N
4. +30.0 N
5. -150 N
6. $1.50 \times 10^{6}$ N/C
7. 90.0 N repulsive
8. 135 N attractive
9. 162 N 11.3° N of E
10. $1.62 \times 10^{6}$ N/C 11.3° N of E
11. 9.09 m (to the left of $q_1$)
12. 9.09 m (to the left of $q_1$)
13. $9.95 \times 10^{-2}$ N
14. $1.73 \times 10^{2}$ N/C
15. $8.00 \times 10^{-17}$ N
16. $-8.78 \times 10^{13}$ $m/s^2$
17. $5.69 \times 10^{-11}$ m
18. $1.14 \times 10^{-12}$ s

## Chapter 18

1. $8.18 \times 10^{-1}$ J
2. $8.18 \times 10^{-2}$ J
3. $9.00 \times 10^{-1}$ J
4. $5.11 \times 10^{17}$ V
5. $5.63 \times 10^{18}$ V
6. +12.0 eV
7. 0
8. 0
9. +24.0 eV
10. +36.0 eV
11. 600 V/m
12. $1.92 \times 10^{-16}$ N
13. 36.0 eV
14. 3.87 m/s
15. $4.43 \times 10^{-12}$ F
16. $1.06 \times 10^{-10}$ C
17. $2.22 \times 10^{-11}$ F
18. $5.30 \times 10^{-10}$ C
19. $5.30 \times 10^{-10}$ C
20. 120 V

## Chapter 19

1. 3.00 V
2. 2.00 V
3. 2.00 V
4. 2.00 Ω
5. 1.00 Ω
6. 10.0 C
7. 3.00 W
8. 2.00 W
9. 1.00 W
10. 0.667
11. 30.0 J
12. 120 Ω
13. $2.40 \times 10^{4}$ m
14. 121 Ω
15. 0.983 Ω·m
16. 133 Ω
17. 248°C
18. 108 W

## Chapter 20

1. 1.55 Ω
2. 0.500 A
3. 0.278 A
4. 2.50 V
5. 1.72 A
6. 2.57 A
7. 0.86 A
8. -0.88 V
9. +12.0 V
10. $3.00 \times 10^{-2}$ Ω
11. $4.97 \times 10^{3}$ Ω
12. 0.251 A
13. 10.0 μF
14. 5.00 V
15. $4.00 \times 10^{-5}$ C
16. $5.00 \times 10^{-5}$ C
17. 10.0 V
18. $2.00 \times 10^{-3}$ s
19. $1.84 \times 10^{-2}$ A
20. $8.65 \times 10^{-5}$ C

## Chapter 21

1. 1
2. 3
3. 2, 4
4. $2.00 \times 10^{-13}$ N
5. $4.00 \times 10^{-19}$ C
6. $1.33 \times 10^{-4}$ m
7. $1.50 \times 10^{-17}$ kg·m/s
8. $2.50 \times 10^{-5}$ T
9. $-2.00 \times 10^{-5}$ T
10. 0.667 cm
11. $-2.00 \times 10^{-5}$ T (toward bottom of page)
12. Multiply by 4 and reverse the direction
13. $7.85 \times 10^{-4}$ T (+Z dir)
14. $3.15 \times 10^{-3}$ T/S
15. $1.58 \times 10^{-5}$ T·m²/s
16. $9.86 \times 10^{-7}$ T·m²/s
17. $9.86 \times 10^{-7}$ V
18. $9.86 \times 10^{-6}$ V

## Chapter 22

1. $2.52 \times 10^{-3}$ T
2. $2.52 \times 10^{-6}$ T·m²
3. $6.30 \times 10^{-6}$ T·m²/s
4. $6.30 \times 10^{-6}$ V
5. $6.30 \times 10^{-4}$ V
6. $6.30 \times 10^{-5}$ V
7. $1.26 \times 10^{-5}$ V·s/A
8. $1.26 \times 10^{-4}$ V·s/A
9. b → a
10. $2.52 \times 10^{-4}$ J
11. 0.333 s
12. 0.144 A
13. 4.32 V
14. 0.680 V
15. $6.80 \times 10^{-2}$ A/s
16. 1.00 N
17. $5.00 \times 10^{-2}$ N·m
18. 5.00 N·m
19. 4.33 N·m
20. 5.00 V

## Chapter 23

1. 10.0 A
2. $1.67 \times 10^{-2}$ s
3. 0°
4. $2.78 \times 10^{-3}$ s
   $1.39 \times 10^{-2}$ s
5. 60.0°
6. 113 Ω
7. 66.3 Ω
8. 110 Ω
9. 1.09 A
10. 25.0°
11. 109 V
12. 72.3 V
13. 123 V
14. 0.906
15. $2.35 \times 10^{-5}$ F
16. 1.20 A
17. 1.00
18. 144 W
19. 120 V
20. 0

Chapter 24

1. 30.0 cm
2. -14.3 cm
3. -3.00
4. 0.500
5. 4.00 cm, erect
6. 1.00 cm, erect
7. 3, 4, 5
8. 2.20 cm
9. -10.0
10. -10.0 mm
11. 2.00 cm
12. -10.0 cm
13. 5.00
14. -50.0
15. -50.0 mm
16. -0.200 m
17. -5.00
18. 1.50
19. 0.500
20. 0.500 mm, erect

Chapter 25

1. a
2. c
3. b
4. d
5. e
6. 47.8°
7. 40.0°
8. 57.4°
9. 1.53 cm
10. $2.29 \times 10^8$ m/s
11. 25.0 cm
12. 60.0 cm
13. -20.0 cm
14. -13.3 cm
15. -0.250
16. 2.00
17. -40.0 cm
18. 6.66 cm
19. virtual
20. virtual

Chapter 26

1. 0.688°
2. $1.50 \times 10^{-2}$ m
3. $6.00 \times 10^{-3}$ m
4. 1.39 m
5. 180°
6. 180°
7. $6.00 \times 10^{-7}$ m
8. $1.00 \times 10^{-7}$ m
9. $2.42 \times 10^{-2}$ m
10. 0.344°
11. $1.80 \times 10^{-2}$ m
12. 4.00 m
13. $2.00 \times 10^{-6}$ m
14. 3.00
15. 19.4°
16. 0.271 m
17. 60.0°
18. 30.0°
19. $I_o/2$
20. $I_o/4$

Chapter 27

1. 8.35 s
2. 6.67 m
3. $5.00 \times 10^{-1}$ kg
4. $8.33 \times 10^{-1}$ kg
5. $-2.40 \times 10^3$ m/s
6. 0.988 c
7. 1.00 MeV
8. 1.51 MeV
9. $2.69 \times 10^{-30}$ kg
10. $2.82 \times 10^8$ m/s
11. $7.59 \times 10^{-22}$ kg·m/s
12. $6.67 \times 10^{-8}$ kg/s
13. 2.10 kg
14. $2.40 \times 10^8$ m/s
15. $2.00 \times 10^{-19}$ kg·m/s
16. $3.02 \times 10^{-11}$ J
17. $7.52 \times 10^{-11}$ J
18. $8.35 \times 10^{-28}$ kg

## Chapter 28

1. $5.00 \times 10^{6}$ W/m$^{2}$
2. $3.06 \times 10^{3}$ K
3. $9.47 \times 10^{-7}$ m
4. $2.51 \times 10^{20}$/s
5. $1.99 \times 10^{13}$/s·cm$^{2}$
6. $6.63 \times 10^{-19}$ J
7. $7.24 \times 10^{14}$/s
8. $1.83 \times 10^{-19}$ J
9. 1.14 V
10. $1.00 \times 10^{4}$ eV
11. $1.00 \times 10^{4}$ eV
12. $1.24 \times 10^{-10}$ m
13. $7.17 \times 10^{-11}$ m
14. 20.3°
15. $1.25 \times 10^{-10}$ m
16. 54.0°
17. 62.5°
18. $4.83 \times 10^{-24}$ kg·m/s
19. $1.45 \times 10^{-6}$ m
20. $1.87 \times 10^{-6}$ m

## Chapter 29

1. $4.77 \times 10^{-10}$ m
2. $7.27 \times 10^{5}$ m/s
3. $1.52 \times 10^{5}$ rad/s
4. $2.41 \times 10^{-19}$ J
5. $1.01 \times 10^{-9}$ N
6. $-4.83 \times 10^{-19}$ J
7. $3.16 \times 10^{-34}$ J·s
8. $3.03 \times 10^{-19}$ J
9. $1.06 \times 10^{-34}$ J·s
10. $6.57 \times 10^{-7}$ m
11. 122.4 ev
12. -122.4 ev
13. 13.6 ev
14. 108.8 ev
15. 17.0 ev
16. $7.32 \times 10^{-8}$ m
17. 6
18. $1.44 \times 10^{15}$ s$^{-1}$

## Chapter 30

1. $3.32 \times 10^{-38}$ m
2. $4.07 \times 10^{-13}$ m
3. $2.49 \times 10^{-13}$ m
4. $1.23 \times 10^{-11}$ m
5. $1.28 \times 10^{-17}$ J
6. $4.83 \times 10^{-25}$ kg·m/s
7. $1.37 \times 10^{-9}$ m
8. 55.0°
9. $8.36 \times 10^{-10}$ m
10. $1.45 \times 10^{-16}$ J
11. $7.01 \times 10^{-23}$ J
12. $4.38 \times 10^{-4}$ V
13. 3
14. 2
15. 3
16. 2
17. 90.0°
18. $5.28 \times 10^{-25}$ kg·m/s
19. $1.58 \times 10^{-19}$ J
20. $3.36 \times 10^{-16}$ s

## Chapter 31

1. ${}^{0}_{-1}e$
2. ${}^{210}_{82}Pb$
3. $\gamma$
4. ${}^{12}_{6}C$
5. ${}^{4}_{2}He$
6. ${}^{210}_{84}Po$
7. 0.030377 u
8. 28.3 Mev
9. 7.08 Mev
10. 0.104251 u
11. 97.1 Mev
12. $1.00 \times 10^{4}$/s
13. $2.00 \times 10^{4}$/s
14. $1.33 \times 10^{4}$/s
15. $2.27 \times 10^{-4}$/s
16. $3.07 \times 10^{3}$ s
17. $8.85 \times 10^{7}$
18. $1.47 \times 10^{-16}$
19. 386/s
20. $1.02 \times 10^{4}$ s

## Chapter 32

1. $1.33 \times 10^5$/s
2. $3.59 \times 10^{-6}$ Ci
3. $1.33 \times 10^5$ Bq
4. $9.13 \times 10^{-13}$/s
5. $1.46 \times 10^7$
6. $2.43 \times 10^{-7}$ mol
7. $5.81 \times 10^{-8}$ kg
8. $1.04 \times 10^{16}$ ion pair/$m^3$
9. $8.06 \times 10^{15}$ ion pair/kg
10. $1.29 \times 10^{-3}$ C/kg
11. $6.00 \times 10^{-2}$ Gy
12. 0.180 J
13. $1.88 \times 10^{13}$
14. $1.44 \times 10^{-6}$ K
15. 6.00 rem
16. $6.00 \times 10^{-2}$ Sv
17. 0.32 rad·$m^2$/h·Ci
18. 28.4 rad/h
19. 1.76 h
20. 0.327 m

## Chapter 33

1. 54.0
2. 132
3. 0.216866 u
4. 202 MeV
5. 4.26 mol
6. $2.56 \times 10^{24}$ nuclei
7. $5.17 \times 10^{26}$ MeV
8. $9.19 \times 10^{-4}$ kg
9. $3.02 \times 10^{-5}$ kg/s
10. $3.46 \times 10^{-14}$ J
11. $1.15 \times 10^{-14}$ J
12. $5.56 \times 10^8$ K
13. 0.007809 u
14. 7.27 MeV
15. 7.07 MeV
16. 7.68 MeV
17. 7.32 MeV
18. $3.42 \times 10^{32}$
19. $1.03 \times 10^{33}$
20. $6.85 \times 10^6$ kg

# Appendix II

Factors for Converting to and from SI Units

| One Non-SI Unit | Equals in SI |
|---|---|
| acre | $4.047 \times 10^{3}$ m$^2$ |
| angstrom | $1.000 \times 10^{-10}$ m |
| astronomical unit (AU) | $1.496 \times 10^{11}$ m |
| atmosphere (standard) | $1.013 \times 10^{5}$ N/m$^2$ |
| atomic mass unit (u) | $1.661 \times 10^{-27}$ kg |
| British thermal unit | $1.054 \times 10^{3}$ J |
| calorie | 4.184 J |
| day | $8.640 \times 10^{4}$ s |
| dyne | $1.000 \times 10^{-5}$ N |
| electronvolt | $1.602 \times 10^{-19}$ J |
| erg | $1.000 \times 10^{-7}$ J |
| foot | $3.048 \times 10^{-1}$ m |
| gallon | $3.785 \times 10^{-3}$ m$^3$ |
| gauss | $1.000 \times 10^{-4}$ T |
| horsepower | $7.457 \times 10^{2}$ W |
| hour | $3.600 \times 10^{3}$ s |
| inch | $2.540 \times 10^{-2}$ m |
| light-year | $9.461 \times 10^{15}$ m |
| liter | $1.000 \times 10^{-3}$ m$^3$ |
| mile (statute) | $1.609 \times 10^{3}$ m |
| mile (nautical) | $1.852 \times 10^{3}$ m |
| ounce (avoirdupois) | $2.780 \times 10^{-1}$ N |
| ounce (troy) | $3.050 \times 10^{-1}$ N |
| ounce (U.S. fluid) | $2.957 \times 10^{-5}$ m$^3$ |
| pound | 4.448 N |
| quart | $9.464 \times 10^{-4}$ m$^3$ |
| slug | $1.459 \times 10^{1}$ kg |
| ton (metric), tonne | $1.000 \times 10^{3}$ kg |
| torr (mmHg 0°C) | $1.333 \times 10^{2}$ N/m$^2$ |
| yard | $9.144 \times 10^{-1}$ m |

Often-Used Physical Constants

| Quantity | Symbol | Magnitude | Unit |
|---|---|---|---|
| Avogadro number | $N_A$ | $6.022 \times 10^{23}$ | $mol^{-1}$ |
| Bohr radius | $r_1$ | $5.292 \times 10^{-11}$ | m |
| Boltzmann constant | k | $1.381 \times 10^{-23}$ | J/K |
| Coulomb constant | $k_e$ | $8.988 \times 10^{9}$ | $N \cdot m^2/C^2$ |
| Electric permittivity (vacuum) | $\varepsilon_o$ | $8.854 \times 10^{-12}$ | F/m |
| Electron rest mass | $m_e$ | $9.110 \times 10^{-31}$ | kg |
| Elementary charge | e | $1.602 \times 10^{-19}$ | C |
| Gas constant | R | 8314 | J/(mol·K) |
| Gravitational constant | G | $6.672 \times 10^{-11}$ | $N \cdot m^2/kg^2$ |
| Mass of earth | $M_e$ | $5.979 \times 10^{24}$ | kg |
| Neutron rest mass | $m_n$ | $1.675 \times 10^{-27}$ | kg |
| Planck's constant | h | $6.626 \times 10^{-34}$ | J·s |
| Proton rest mass | $m_p$ | $1.673 \times 10^{-27}$ | kg |
| Radius of earth (av) | $R_e$ | $6.376 \times 10^{6}$ | m |
| Rydberg constant | R | $1.097 \times 10^{7}$ | $m^{-1}$ |
| Speed of light (vacuum) | c | $2.998 \times 10^{8}$ | m/s |
| Standard gravity | g | 9.807 | $m/s^2$ |
| Stefan-Boltzmann constant | σ | $5.670 \times 10^{-8}$ | $W/(m^2 \cdot k^4)$ |

Derived SI Units (Common)

| Quantity | Unit Name | Symbol | Expressed in Fundamental Units | Expressed in Other SI Units |
|---|---|---|---|---|
| Capacitance | farad | F | $A^2 \cdot S^4/(kg \cdot m^2)$ | |
| Electric charge | coulomb | C | A·s | |
| Electric potential | volt | V | $kg \cdot m^2/A \cdot s^3$ | J/C |
| Electric resistance | ohm | Ω | $kg \cdot m^2/A^2 \cdot s^3$ | V/A |
| Force | newton | N | $kg \cdot m/s^2$ | |
| Frequency | hertz | Hz | $s^{-1}$ | |
| Inductance | henry | H | $kg \cdot m^2/(A^2 \cdot s^2)$ | V·s/A |
| Magnetic field intensity | tesla | T | $kg/(A \cdot s^2)$ | N·s/(C·m) |
| Power | watt | W | $kg \cdot m^2/s^3$ | J/s |
| Pressure | pascal | Pa | $kg/(m \cdot s^2)$ | $N/m^2$ |
| Viscosity | poiseuille | Pl | kg/(m·s) | $N \cdot s/m^2$ |
| Work and energy | joule | J | $kg \cdot m^2/s^2$ | N·m |